P9-BXY-646

MEDICAL INSTRUMENTATION

Application and Design

MEDICAL INSTRUMENTATION
Application and Design

THIRD EDITION

John G. Webster, Editor

Contributing Authors

John W. Clark, Jr.
Rice University

Michael R. Neuman
Case Western Reserve University

Walter H. Olson
Medtronic, Inc.

Robert A. Peura
Worcester Polytechnic Institute

Frank P. Primiano, Jr.
Amethyst Research, Inc.

Melvin P. Siedband
University of Wisconsin–Madison

John G. Webster
University of Wisconsin–Madison

Lawrence A. Wheeler
Nutritional Computing Concepts

JOHN WILEY & SONS, INC.
New York • Chichester • Weinheim • Brisbane • Singapore • Toronto

COVER PHOTO Leonardo da Vinci. Drawing of ideal proportions of the human figure according to Vitruvius' 1st. century A.D. treatise "De Architect ura."

ACQUISITIONS EDITOR Charity Robey
MARKETING MANAGER Harper Mooy
PRODUCTION MANAGER Charlotte Hyland
PRODUCTION EDITOR Tony VenGraitis
SENIOR DESIGNER Kevin Murphy
ILLUSTRATION Jaime Perea
OUTSIDE PRODUCTION Ingrao Associates

This book was set in Times Ten by Bi-Comp, Inc. and printed and bound by Malloy Lithography, Inc. The cover was printed by Phoenix Color.

Recognizing the importance of preserving what has been written, it is a policy of John Wiley & Sons, Inc. to have books of enduring value published in the United States printed on acid-free paper, and we exert our best efforts to that end.

The paper in this book was manufactured by a mill whose forest management programs include sustained yield harvesting of its timberlands. Sustained yield harvesting principles ensure that the number of trees cut each year does not exceed the amount of new growth.

ISBN 0-471-15368-0

Printed in the United States of America

10 9 8 7 6 5 4

PREFACE

Medical Instrumentation: Application and Design, Third Edition, is written for a senior to graduate-level course in biomedical engineering. It describes the principles, applications, and design of the medical instruments most commonly used in hospitals. Because equipment changes with time, we have stressed fundamental principles of operation and general types of equipment, avoiding detailed descriptions and photographs of specific models. And because biomedical engineering is an interdisciplinary field, requiring good communication with health-care personnel, we have provided some applications for each type of instrument. However, to keep the book to a reasonable length, we have omitted much of the physiology.

Most of those who use this text have had an introductory course in chemistry, are familiar with mathematics through differential equations, have a strong background in physics, and have taken courses in electric circuits and electronics. However, readers without this background will gain much from the descriptive material and should find this text a valuable reference. In addition, we recommend reading background material from an inexpensive physiology text, such as W. F. Ganong's *Review of Medical Physiology,* 17th edition (Los Altos, CA: Lange Medical Publications, 1995).

EMPHASIS ON DESIGN

Throughout the book, we emphasize *design.* A scientist or engineer who has some background in electronics and instrumentation will glean enough information, in many of the areas we address, to be able to design medical instruments. This ability should be especially valuable in those situations—so frequently encountered—where special instruments that are not commercially available are required.

PEDAGOGY

The book provides both in-text worked examples and homework problems at the end of each chapter (more than 300). Problems are designed to cover a wide variety of applications ranging from analysis of the waves of the electrocardiogram to circuit design of biopotential amplifiers and identification of electric safety hazards.

REFERENCES

Rather than giving an exhaustive list of references, we have provided a list of review articles and books that can serve as a point of departure for further study on any given topic.

ORGANIZATION

Each chapter has been carefully reviewed and updated for the third edition, and many new problems and references are included.

Chapter 1 covers general concepts that are applicable to all instrumentation systems, including the commercial development of medical instruments, on biostatistics, and on the regulation of medical devices. Chapter 2 describes basic sensors, and Chapter 3 presents the design of amplifiers for them. Chapters 4–6 deal with biopotentials, tracing the topic from the origin of biopotentials, through electrodes, to the special amplifier design required.

Chapters 7 and 8 cover the measurement of cardiovascular dynamics—pressure, sound, flow, and volume of blood. Chapter 9 presents the measurement of respiratory dynamics—pressure, flow, and concentration of gases.

Chapter 10 describes the developing field of biosensors: sensors that measure chemical concentrations within the body via catheters or implants. Chapter 11 describes that area in the hospital where the greatest number of measurements are made, the clinical laboratory. Chapter 12 starts with general concepts of medical imaging and shows their applications to x-ray techniques, magnetic resonance imaging, positron emission tomography, and Doppler ultrasonic imagers.

Chapter 13 deals with devices used in therapy, such as the pacemaker, defibrillator, cochlear prosthesis, transcutaneous electrical nerve stimulation, implantable automatic defibrillators, the total artificial heart, lithotripsy, high-frequency ventilators, infant radiant warmers, drug infusion pumps, and anesthesia machines. Chapter 14 presents a guide both to electric safety in the hospital and to minimization of hazards.

We have used the internationally recommended SI units throughout this book. In the case of units of pressure, we have presented both the commonly used millimeters of mercury and the SI unit, the pascal. To help the reader follow the trend toward employing SI units, the Appendix provides the most common conversion factors. The Appendix also provides a number of physical constants used in the book and a list of abbreviations.

A *Solutions Manual* containing complete solutions to all problems is available free to adopters of this text.

ACKNOWLEDGMENTS

We would like to thank the reviewers of both the first and the second editions.

First Edition Reviewers
David Arnett, *Pennsylvania State University*
Robert B. Northrup, *University of Connecticut (Storrs)*
Kenneth C. Mylrea, *University of Arizona*
Curran S. Swift, *Iowa State University*

Second Edition Reviewers
Jonathan Newell, *Rensselaer Polytechnic Institute*
Robert B. Northrup, *University of Connecticut (Storrs)*

Third Edition Reviewers
Noel Thompson, *Stanford University*
W. Ed Hammond, *Duke University*
Robert B. Northrup, *University of Connecticut*
Richard Jendrucko, *University of Tennessee, Knoxville*

The authors welcome your suggestions for improvement of subsequent printings and editions.

John G. Webster

LIST OF SYMBOLS

This list gives single-letter symbols for quantities, without subscripts or modifiers. Symbols for physical constants are given in Appendix A.1, multiletter symbols in Appendix A.4, and chemical symbols in Appendix A.5.

Symbol	Quantity	Introduced in Section
a	absorptivity	10.3
a	activity	5.2
a	coefficient	1.10
$\mathbf{a}$	lead vector	6.2
A	absorbance	10.3
A	area	2.2
A	coefficient	1.9
A	gain	3.1
A	percent	1.7
b	coefficient	1.9
b	intercept	1.9
B	coefficient	1.10
B	percent	1.9
B	viscous friction	1.10
$\mathbf{B}$	magnetic flux density	8.3
c	coefficient	7.13
c	specific heat	8.1
c	velocity of sound	8.4
C	capacitance	1.10
C	compilance	7.3
C	concentration	10.3
C	contrast	12.1
d	derivative	1.10
d	diameter	1.10
d	distance	4.1
D	density	12.4
D	detector responsivity	2.17
D	d/dt	1.10
D	diameter	5.8
D	diffusing capacity	9.8
D	distance	4.4
E	emf	2.7

Table *(Continued)*

Symbol	Quantity	Introduced in Section
E	energy	2.13
E	exposure	12.4
E	irradiance	2.17
E	modulus of elasticity	7.3
f	force	2.6
f	frequency	1.10
f	function	4.2
F	filter transmission	2.17
F	flow	7.3
F	force	2.2
F	fraction	12.1
F	molar fraction	9.3
g	conductance/area	4.1
G	conductance	2.9
G	form factor	2.4
G	gage factor	2.2
G	gain	1.7
h	height	7.13
H	feedback gain	1.7
i	current	2.6
I	current	3.7
I	intensity	10.3
j	$+(-1)^{1/2}$	1.10
J	number of standard deviations	12.1
k	constant	6.7
k	piezoelectric constant	2.6
K	constant	1.10
K	number	12.1
K	sensitivity	1.10
K	solubility product	5.3
K	spring constant	1.10
L	inductance	2.4
L	inertance	7.3
L	length	2.2
L	line-source response	12.10
m	average number	12.1
m	mass	7.3
m	slope	1.9
M	mass	1.10
$\underline{M}$	measured values	12.2
$\overline{M}$	modulation	12.1
M	cardiac vector	6.2
n	number	1.8
n	refractive index	2.14
N	noise equivalent bandwidth	12.3
N	number	5.3
N	turns ratio	3.13

Table *(Continued)*

Symbol	Quantity	Introduced in Section
p	change in pressure	9.1
p	probability	12.1
P	power	1.9
P	pressure	7.3
P	projection	12.8
q	charge	2.6
q	rate of heat	8.1
q	change in volume flow	9.1
Q	heat content	8.2
Q	volume flow	9.1
r	correlation coefficient	1.8
r	radius	7.3
r	resistance/length	4.2
R	range	8.4
R	ratio	10.3
R	resistance	1.10
S	standard deviation	1.8
S	modulation transfer function	12.2
S	saturation	10.1
S	slew rate	3.11
S	source output	2.17
t	thickness	5.8
t	time	1.10
T	interval	1.10
T	temperature	2.8
T	transmittance	11.1
u	velocity	4.2
u	work function	12.6
U	molar uptake	9.1
v	voltage	1.10
v	change in volume	9.1
V	voltage	1.10
V	volume	2.2
W	power	2.10
W	weight	10.3
W	weighting factor	12.8
x	constant	10.3
x	distance	2.4
x	input	1.7
X	chemical species	9.1
X	effort variable	1.9
X	value	1.8
y	constant	10.3
y	output	1.7
Y	admittance	1.9
Y	flow variable	1.9
Y	value	1.8

Table *(Continued)*

Symbol	Quantity	Introduced in Section
z	distance	4.1
Z	atomic number	12.6
Z	impedance	1.9

Greek Letters

Symbol	Quantity	Introduced in Section
α	polytropic constant	9.5
α	thermistor coefficient	2.9
α	thermoelectric sensitivity	2.8
β	thermistor constant	2.9
Δ	deviation	10.3
ϵ	emissivity	2.10
ϵ	dielectric constant	2.5
ζ	damping ratio	1.10
η	viscosity	7.3
θ	angle	2.14
Λ	logarithmic decrement	1.10
λ	wavelength	2.10
μ	attenuation coefficient	12.8
μ	mobility	5.2
μ	permeability	2.4
μ	Poisson's ratio	2.2
ν	frequency	2.13
ρ	density	7.3
ρ	mole density	9.1
ρ	resistivity	2.2
σ	conductance	13.4
σ	conductivity/distance	4.7
σ^2	variance	12.1
τ	time constant	1.10
ϕ	number of photons	12.6
ϕ	phase shift	1.10
ϕ	divergence	8.4
Φ	potential	4.6
ω	frequency	1.10

CONTENTS

12 MEDICAL IMAGING SYSTEMS 518

Melvin P. Siedband

13 THERAPEUTIC AND PROSTHETIC DEVICES 577

Michael R. Neuman

14 ELECTRICAL SAFETY 623

Walter H. Olson

1

BASIC CONCEPTS OF MEDICAL INSTRUMENTATION

Walter H. Olson

The invention, prototype design, product development, clinical testing, regulatory approval, manufacturing, marketing, and sale of a new medical instrument add up to a complex, expensive, and lengthy process. Very few new ideas survive the practical requirements, human barriers, and inevitable setbacks of this arduous process. Usually there is one person who is the "champion" of a truly new medical instrument or device. This person—who is not necessarily the inventor—must have a clear vision of the final new product and exactly how it will be used. And most important, this person must have the commitment and persistence to overcome unexpected technical problems, convince the nay-sayers, and cope with the bureaucratic apparatus that is genuinely needed to protect patients.

One of five inventors' stories from *New Medical Devices: Invention, Development and Use* (Eckelman, 1988) is reprinted here to illustrate this process. The automated biochemical analyzer uses spectrophotometric methods in a continuous-flow system to measure the amount of many clinically important substances in blood or urine samples (Section 11.2).

Development of Technicon's Auto Analyzer

Edwin C. Whitehead

In 1950 Alan Moritz, chairman of the department of pathology at Case Western Reserve University and an old friend of mine, wrote to tell me about Leonard Skeggs, a young man in his department who had developed an instrument that Technicon might be interested in. I was out of my New York office on a prolonged trip, and my father, cofounder with me of Technicon Corporation, opened the letter. He wrote to Dr. Moritz saying that Technicon was always interested in new developments and enclosed a four-page confidential disclosure form. Not surprisingly, Dr. Moritz thought that Technicon was not really interested in Skeggs's instrument, and my father dismissed the matter as routine.

Three years later Ray Roesch, Technicon's only salesman at the time, was visiting Joseph Kahn at the Cleveland Veterans Administration Hospital. Dr. Kahn asked Ray why Technicon had turned down Skeggs's invention. Ray responded that he had never heard of it and asked, "What invention?" Kahn replied, "A machine to automate chemical analysis." When Ray called me and asked why I had turned Skeggs's idea down, I said I had not heard of it either. When he told me that Skeggs's idea was to automate clinical chemistry, my reaction was, "Wow! Let's look at it and make sure Skeggs doesn't get away."

That weekend, Ray Roesch loaded some laboratory equipment in his station wagon and drove Leonard Skeggs and his wife Jean to New York. At Technicon, Skeggs set up a simple device consisting of a peristaltic pump to draw the specimen sample and reagent streams through the system, a continuous dialyzer to remove protein molecules that might interfere with the specimen-reagent reaction, and a spectrophotometer equipped with a flow cell to monitor the reaction. This device demonstrated the validity of the idea, and we promptly entered negotiations with Skeggs for a license to patent the Auto Analyzer. We agreed on an initial payment of $6,000 and royalties of 3 percent after a certain number of units had been sold.

After Technicon "turned-down" the project in 1950, Skeggs had made arrangements first with the Heinecke Instrument Co. and then the Harshaw Chemical Co. to sell his device. Both companies erroneously assumed that the instrument was a finished product. However, neither company had been able to sell a single instrument from 1950 until 1953. This was not surprising, because Skeggs's original instrument required an expensive development process to make it rugged and reliable, and to modify the original, manual chemical assays. Technicon spent 3 years refining the simple model developed by Skeggs into a commercially viable continuous-flow analyzer.

A number of problems unique to the Auto Analyzer had to be overcome. Because the analyzer pumps a continuous-flow stream of reagents interrupted by specimen samples, one basic problem was the interaction between specimen samples. This problem was alleviated by introducing air bubbles as physical barriers between samples. However, specimen carryover in continuous-flow analyzers remains sensitive to the formation and size of bubbles, the inside diameter of the tubing through which fluids flow, the pattern of peristaltic pumping action, and other factors.

Development of the Auto Analyzer was financed internally at Technicon. In 1953 Technicon had ongoing business of less than $10 million per year: automatic tissue processors and slide filing cabinets for histology laboratories, automatic fraction collectors for chromatography, and portable respirators for polio patients. Until it went public in 1969, Technicon had neither borrowed money nor sold equity. Thus, Technicon's patent on Skegg's original invention was central to the development of the Auto Analyzer. Without patent protection, Technicon could never have afforded to pursue the expensive development of this device.

Early in the instrument's development, I recognized that traditional marketing techniques suitable for most laboratory instruments would not work for something as revolutionary as the Auto Analyzer. At that time, laboratory instruments were usually sold by catalog salesmen or by mail from specification sheets listing instrument specifications, price, and perhaps product benefits. In contrast, we decided that Technicon had to market the Auto Analyzer as a complete system—instrument, reagents, and instruction.

Technicon's marketing strategy has been to promote the Auto Analyzer at professional meetings and through scientific papers and journal articles. Technicon employs only direct salesmen. The company has never used agents or distributors, except in countries where the market is too small to support direct sales.

To introduce technology as radical as the Auto Analyzer into conservative clinical laboratories, Technicon decided to perform clinical evaluations. Although unusual at that time, such evaluations have since become commonplace. An important condition of the clinical evaluations was Technicon's insistence that the laboratory conducting the evaluation call a meeting of its local professional society to announce the results. Such meetings generally resulted in an enthusiastic endorsement of the Auto Analyzer by the laboratory director. I believe this technique had much to do with the rapid market acceptance of the Auto Analyzer.

Other unusual marketing strategies employed by Technicon to promote the Auto Analyzer included symposia and training courses. Technicon sponsored about 25 symposia on techniques in automated analytical chemistry. The symposia were generally 3-day affairs, attracting between 1,000 and 4,500 scientists, and were held in most of the major countries of the world including the United States.

Because we realized that market acceptance of the Auto Analyzer could be irreparably damaged by incompetent users, Technicon set up a broad-scale training program. We insisted that purchasers of Auto Analyzers come to our training centers located around the world for a 1-week course of instruction. I estimate that we have trained about 50,000 people to use Auto Analyzers.

Introduction of Technicon's continuous-flow Auto Analyzer in 1957 profoundly changed the character of the clinical laboratory, allowing a hundredfold increase in the number of laboratory tests performed over a 10-year period. When we began to develop the Auto Analyzer in 1953, I estimated a potential market of 250 units. Currently, more than 50,000 Auto Analyzer Channels are estimated to be in use around the world.

In reviewing the 35-year history of the Auto Analyzer, I have come to the conclusion that several factors significantly influenced our success. First, the Auto Analyzer allowed both an enormous improvement in the quality of laboratory test results and an enormous reduction in the cost of doing chemical analysis. Second, physicians began to realize that accurate laboratory data are useful in diagnosis. Last, reimbursement policies increased the availability of health care.

This success story demonstrates that important new ideas rarely flow smoothly to widespread clinical use. There are probably one hundred untold failure stories for each success story! New inventions usually are made by the wrong person with the wrong contacts and experience, in the wrong place at the wrong time. It is important to understand the difference between a crude feasibility prototype and a well-developed, reliable, manufacturable product. Patents are important to protect ideas during the development process and to provide incentives for making the financial investments needed. Many devices have failed because they were too hard to use, reliability and ruggedness were inadequate, marketing was misdirected, user education was lacking, or service was poor and/or slow.

An evolutionary product is a new model of an existing product that adds new features, improves the technology, and reduces the cost of production. A revolutionary new product either solves a totally new problem or uses a new principle or concept to solve an old problem in a better way that displaces old methods. A medical instrument that improves screening, diagnosis, or monitoring may not add value by improving patient outcome unless improvements in the application of therapy occur as a result of using the medical instrument.

1.1 TERMINOLOGY OF MEDICINE AND MEDICAL DEVICES

Most biomedical engineers learn the physical sciences first in the context of traditional engineering, physics, or chemistry. When they become interested in medicine, they usually take at least a basic course in physiology, which does not describe disease or pathologic terminology. The book *Medical Terminology: An Illustrated Guide* (Cohen, 1994) is recommended. It emphasizes the Latin and Greek roots in a workbook format (with answers) that includes clinical case studies, flash cards, and simple, clear illustrations. An unabridged medical dictionary such as *Dorland's Illustrated Medical Dictionary*, 28th ed. (Dorland, 1994), is often useful. Physicians frequently use abbreviations and acronyms that are difficult to look up, and ambiguity or errors result. Six references on medical abbreviations are given (Cohen, 1989; Davis, 1988; Firkin and Whitworth, 1987; Haber, 1988; Hamilton and Guidos, 1988; Heister, 1989). Medical eponyms are widely used to describe diseases and syndromes by the name of the person who first identified them. Refer to *Dictionary of Medical Eponyms* (Firkin and Whitworth, 1987).

The name used to describe a medical instrument or device should be informative, consistent, and brief. The annual *Health Devices Sourcebook* (Anonymous, 1995) is a directory of U.S. and Canadian medical device prod-

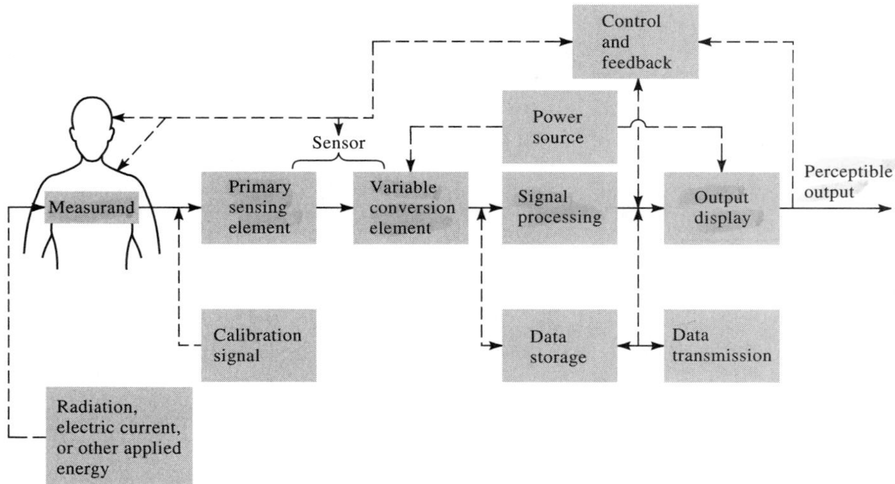

Figure 1.1 Generalized instrumentation system The sensor converts energy or information from the measurand to another form (usually electric). This signal is then processed and displayed so that humans can perceive the information. Elements and connections shown by dashed lines are optional for some applications.

ucts, tradenames, manufacturers, and related services. This book uses internationally accepted nomenclature and a numerical coding system for over 5000 product categories. The *Product Development Directory* (1996) lists all specific medical products by the FDA standard product category name since enactment of the Medical Devices Amendments in April, 1976. The *Encyclopedia of Medical Devices and Instrumentation* (Webster, 1988), vols. 1–4 has many detailed descriptions. But beware of borrowing medical terminology to describe technical aspects of devices or instruments. Confounding ambiguities can result.

Recent information on medical instrumentation can be found by searching World Wide Web servers such as www.altavista.digital.com or www.uspto.gov, Library Online Catalogs, and journal electronic databases such as INSPEC and Science Citation Index.

1.2 GENERALIZED MEDICAL INSTRUMENTATION SYSTEM

Every instrumentation system has at least some of the functional components shown in Figure 1.1. The primary flow of information is from left to right. Elements and relationships depicted by dashed lines are not essential. The major difference between this system of medical instrumentation and conventional instrumentation systems is that the source of the signals is living tissue or energy applied to living tissue.

MEASURAND

The physical quantity, property, or condition that the system measures is called the *measurand.* The accessibility of the measurand is important because it may be internal (blood pressure), it may be on the body surface (electrocardiogram potential), it may emanate from the body (infrared radiation), or it may be derived from a tissue sample (such as blood or a biopsy) that is removed from the body. Most medically important measurands can be grouped in the following categories: biopotential, pressure, flow, dimensions (imaging), displacement (velocity, acceleration, and force), impedance, temperature, and chemical concentrations. The measurand may be localized to a specific organ or anatomical structure.

SENSOR

Generally, the term *transducer* is defined as a device that converts one form of energy to another. A *sensor* converts a physical measurand to an electric output. The sensor should respond only to the form of energy present in the measurand, to the exclusion of all others. The sensor should interface with the living system in a way that minimizes the energy extracted, while being minimally invasive. Many sensors have a primary sensing element such as a diaphragm, which converts pressure to displacement. A variable-conversion element, such as a strain gage, then converts displacement to an electric voltage. Sometimes the sensitivity of the sensor can be adjusted over a wide range by altering the primary sensing element. Many variable-conversion elements need external electric power to obtain a sensor output.

SIGNAL CONDITIONING

Usually the sensor output cannot be directly coupled to the display device. Simple signal conditioners may only amplify and filter the signal or merely match the impedance of the sensor to the display. Often sensor outputs are converted to digital form and then processed by specialized digital circuits or a microcomputer (Tompkins and Webster, 1981). For example, signal filtering may reduce undesirable sensor signals. It may also average repetitive signals to reduce noise, or it may convert information from the time domain to the frequency domain.

OUTPUT DISPLAY

The results of the measurement process must be displayed in a form that the human operator can perceive. The best form for the display may be numerical or graphical, discrete or continuous, permanent or temporary—depending on the particular measurand and how the operator will use the information. Although most displays rely on our visual sense, some information (Doppler ultrasonic signals, for example) is best perceived by other senses (here, the auditory sense). User controls and the output display should conform to the *Human factors engineering guidelines and preferred practices for the design of medical devices* (AAMI, 1993).

AUXILIARY ELEMENTS

A calibration signal with the properties of the measurand should be applied to the sensor input or as early in the signal-processing chain as possible. Many forms of control and feedback may be required to elicit the measurand, to adjust the sensor and signal conditioner, and to direct the flow of output for display, storage, or transmission. Control and feedback may be automatic or manual. Data may be stored briefly to meet the requirements of signal conditioning or to enable the operator to examine data that precede alarm conditions. Or data may be stored before signal conditioning, so that different processing schemes can be utilized. Conventional principles of communications can often be used to transmit data to remote displays at nurses' stations, medical centers, or medical data-processing facilities.

1.3 ALTERNATIVE OPERATIONAL MODES

DIRECT-INDIRECT MODES

Often the desired measurand can be interfaced directly to a sensor because the measurand is readily accessible or because acceptable invasive procedures are available. When the desired measurand is not accessible, we can use either another measurand that bears a known relation to the desired one or some form of energy or material that interacts with the desired measurand to generate a new measurand that *is* accessible. Examples include cardiac output (volume of blood pumped per minute by the heart), determined from measurements of respiration and blood gas concentration or from dye dilution; morphology of internal organs, determined from x-ray shadows; and pulmonary volumes, determined from variations in thoracic impedance plethysmography.

SAMPLING AND CONTINUOUS MODES

Some measurands—such as body temperature and ion concentrations—change so slowly that they may be sampled infrequently. Other quantities—such as the electrocardiogram and respiratory gas flow— may require continuous monitoring. The frequency content of the measurand, the objective of the measurement, the condition of the patient, and the potential liability of the physician all influence how often medical data are acquired. Many data that are collected may go unused.

GENERATING AND MODULATING SENSORS

Generating sensors produce their signal output from energy taken directly from the measurand, whereas modulating sensors use the measurand to alter the flow of energy from an external source in a way that affects the output of the sensor. For example, a photovoltaic cell is a generating sensor because it provides an output voltage related to its irradiation, without any additional external energy source. However, a photoconductive cell is a modulating sensor; to measure its change in resistance with irradiation, we must apply external energy to the sensor.

ANALOG AND DIGITAL MODES

Signals that carry measurement information are either *analog*, meaning continuous and able to take on any value within the dynamic range, or *digital*, meaning discrete and able to take on only a finite number of different values. Most currently available sensors operate in the analog mode, although some inherently digital measuring devices have been developed. Increased use of digital signal processing has required concurrent use of analog-to-digital and digital-to-analog converters to interface computers with analog sensors and analog display devices. Researchers have developed indirect digital sensors that use analog primary sensing elements and digital variable-conversion elements (optical shaft encoders). Also quasi-digital sensors, such as quartz-crystal thermometers, give outputs with variable frequency, pulse rate, or pulse duration that are easily converted to digital signals.

The advantages of the digital mode of operation include greater accuracy, repeatability, reliability, and immunity to noise. Furthermore, periodic calibration is usually not required. Digital numerical displays are replacing many analog meter movements because of their greater accuracy and readability. Many clinicians, however, prefer analog displays when they are determining whether a physiological variable is within certain limits and when they are looking at a parameter that can change quickly, such as beat-to-beat heart rate. In the latter case, digital displays often change numbers so quickly that they are very difficult and annoying to observe.

REAL-TIME AND DELAYED-TIME MODES

Of course sensors must acquire signals in real time as the signals actually occur. The output of the measurement system may not display the result immediately, however, because some types of signal processing, such as averaging and transformations, need considerable input before any results can be produced. Often such short delays are acceptable unless urgent feedback and control tasks depend on the output. In the case of some measurements, such as cell cultures, several days may be required before an output is obtained.

1.4 MEDICAL MEASUREMENT CONSTRAINTS

The medical instrumentation described throughout this book is designed to measure various medical and physiological parameters. The principal measurement and frequency ranges for each parameter are major factors that affect the design of all the instrument components shown in Figure 1.1. To get a brief overview of typical medical parameter magnitude and frequency ranges, refer to Table 1.1. Shown here are approximate ranges that are intended to include normal and abnormal values. Most of the parameter measurement ranges are quite low compared with nonmedical parameters. Note, for example, that most voltages are in the microvolt range and that pressures are low (about 100 mm Hg = 1.93 psi = 13.3 kPa). Also note that all the signals listed are in the audiofrequency range or below and that many signals contain dc and very low frequencies. These general properties of medical parameters limit the practical choices available to designers for all aspects of instrument design.

Many crucial variables in living systems are inaccessible because the proper measurand–sensor interface cannot be obtained without damaging the system. Unlike many complex physical systems, a biological system is of such a nature that it is not possible to turn it off and remove parts of it during the measurement procedure. Even if interference from other physiological systems can be avoided, the physical size of many sensors prohibits the formation of a proper interface. Either such inaccessible variables must be measured indirectly, or corrections must be applied to data that are affected by the measurement process. The cardiac output is an important measurement that is obviously quite inaccessible.

Variables measured from the human body or from animals are seldom deterministic. Most measured quantities vary with time, even when all controllable factors are fixed. Many medical measurements vary widely among normal patients, even when conditions are similar. This inherent *variability* has been documented at the molecular and organ levels, and even for the whole body. Many internal anatomical variations accompany the obvious external differences among patients. Large tolerances on physiological measurements are partly the result of interactions among many physiological systems. Many feedback loops exist among physiological systems, and many of the interrelationships are poorly understood. It is seldom feasible to control or neutralize the effects of these other systems on the measured variable. The most common method of coping with this variability is to assume empirical statistical and probabilistic distribution functions. Single measurements are then compared with these *norms* (see Section 1.8).

Nearly all biomedical measurements depend either on some form of energy being applied to the living tissue or on some energy being applied as an incidental consequence of sensor operation. X-ray and ultrasonic imaging techniques and electromagnetic or Doppler ultrasonic blood flow-meters depend on externally applied energy interacting with living tissue. Safe levels

Table 1.1 Medical and Physiological Parameters

Parameter or Measuring Technique	Principal Measurement Range of Parameter	Signal Frequency Range, Hz	Standard Sensor or Method
Ballistocardiography (BCG)	0–7 mg	dc–40	Accelerometer, strain gage
	0–100 μm	dc–40	Displacement (LVDT)
Bladder pressure	1–100 cm H_2O	dc–10	Strain-gage manometer
Blood flow	1–300 ml/s	dc–20	Flowmeter (electromagnetic or ultrasonic)
Blood pressure, arterial			
Direct	10–400 mm Hg	dc–50	Strain-gage manometer
Indirect	25–400 mm Hg	dc–60	Cuff, auscultation
Blood pressure, venous	0–50 mm Hg	dc–50	Strain gage
Blood gases			
P_{O_2}	30–100 mm Hg	dc–2	Specific electrode, volumetric or manometric
P_{CO_2}	40–100 mm Hg	dc–2	Specific electrode, volumetric or manometric
P_{N_2}	1–3 mm Hg	dc–2	Specific electrode, volumetric or manometric
P_{CO}	0.1–0.4 mm Hg	dc–2	Specific electrode, volumetric or manometric
Blood pH	6.8–7.8 pH units	dc–2	Specific electrode
Cardiac output	4–25 liter/min	dc–20	Dye dilution, Fick
Electrocardiography (ECG)	0.5–4 mV	0.01–250	Skin electrodes
Electroencephalography (EEG)	5–300 μV	dc–150	Scalp electrodes
(Electrocorticography and brain depth)	10–5000 μV	dc–150	Brain-surface or depth electrodes
Electrogastrography (EGG)	10–1000 μV	dc–1	Skin-surface electrodes
	0.5–80 mV	dc–1	Stomach-surface electrodes
Electromyography (EMG)	0.1–5 mV	dc–10,000	Needle electrodes
Eye potentials			
EOG	50–3500 μV	dc–50	Contact electrodes
ERG	0–900 μV	dc–50	Contact electrodes
Galvanic skin response (GSR)	1–500 kΩ	0.01–1	Skin electrodes
Gastric pH	3–13 pH units	dc–1	pH electrode; antimony electrode

Table 1.1 *(Continued)*

Parameter or Measuring Technique	Principal Measurement Range of Parameter	Signal Frequency Range, Hz	Standard Sensor or Method
Gastrointestinal pressure	0–100 cm H_2O	dc–10	Strain-gage manometer
Gastrointestinal forces	1–50 g	dc–1	Displacement system, LVDT
Nerve potentials	0.01–3 mV	dc–10,000	Surface or needle electrodes
Phonocardiography (PCG)	Dynamic range 80 dB, threshold about 100 μPa	5–2000	Microphone
Plethysmography (volume change)	Varies with organ measured	dc–30	Displacement chamber or impedance change
Circulatory	0–30 ml	dc–30	Displacement chamber or impedance change
Respiratory functions			
Pneumotachography (flow rate)	0–600 liter/min	dc–40	Pneumotachograph head and differential pressure
Respiratory rate	2–50 breaths/min	0.1–10	Strain gage on chest, impedance, nasal thermistor
Tidal volume	50–1000 ml/breath	0.1–10	Above methods
Temperature of body	32–40 °C 90–104 °F	dc–0.1	Thermistor, thermocouple

SOURCE: Revised from *Medical Engineering.* C. D. Ray (ed.). Copyright © 1974 by Year Book Medical Publishers, Inc., Chicago. Used by permission.

of these various types of energy are difficult to establish, because many mechanisms of tissue damage are not well understood. A fetus is particularly vulnerable during the early stages of development. The heating of tissue is one effect that must be limited, because even reversible physiological changes can affect measurements. Damage to tissue at the molecular level has been demonstrated in some instances at surprisingly low energy levels.

Operation of instruments in the medical environment imposes important additional constraints. Equipment must be reliable, easy to operate, and capable of withstanding physical abuse and exposure to corrosive chemicals. Electronic equipment must be designed to minimize electric-shock hazards (Chapter 14). The safety of patients and medical personnel must be considered in all phases of the design and testing of instruments. The Medical Device Amendments of 1976 and the Safe Medical Devices Act of 1990 amend the Federal Food, Drug, and Cosmetics Act to provide for the safety and effectiveness of medical devices intended for human use (Section 1.13).

1.5 CLASSIFICATIONS OF BIOMEDICAL INSTRUMENTS

The study of biomedical instruments can be approached from at least four viewpoints. Techniques of biomedical measurement can be grouped according to the *quantity that is sensed*, such as pressure, flow, or temperature. One advantage of this classification is that it makes different methods for measuring any quantity easy to compare.

A second classification scheme uses the *principle of transduction*, such as resistive, inductive, capacitive, ultrasonic, or electrochemical. Different applications of each principle can be used to strengthen understanding of each concept; also, new applications may be readily apparent.

Measurement techniques can be studied separately for each *organ system*, such as the cardiovascular, pulmonary, nervous, and endocrine systems. This approach isolates all important measurements for specialists who need to know only about a specific area, but it results in considerable overlap of quantities sensed and principles of transduction.

Finally, biomedical instruments can be classified according to the *clinical medicine specialties,* such as pediatrics, obstetrics, cardiology, or radiology. This approach is valuable for medical personnel who are interested in specialized instruments. Of course, certain measurements—such as blood pressure—are important to many different medical specialties.

1.6 INTERFERING AND MODIFYING INPUTS

Desired inputs are the measurands that the instrument is designed to isolate. *Interfering inputs* are quantities that indavertently affect the instrument as a consequence of the principles used to acquire and process the desired inputs. If spatial or temporal isolation of the measurand is incomplete, the interfering input can even be the same quantity as the desired input. *Modifying inputs* are undesired quantities that indirectly affect the output by altering the performance of the instrument itself. Modifying inputs can affect processing of either desired or interfering inputs. Some undesirable quantities can act as both a modifying input and an interfering input.

A typical electrocardiographic recording system, shown in Figure 1.2, will serve to illustrate these concepts. The desired input is the electrocardiographic voltage v_{ecg} that appears between the two electrodes on the body surface.

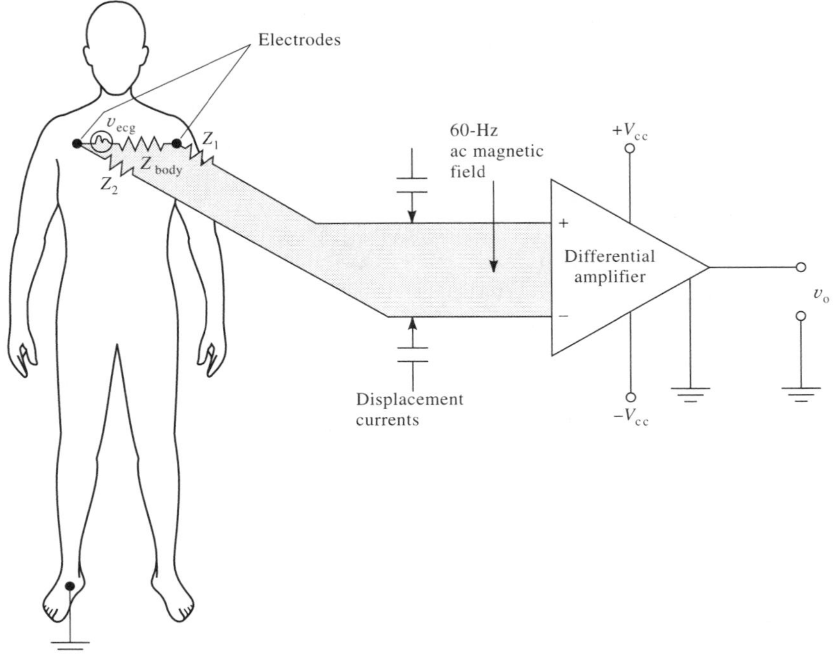

Figure 1.2 Simplified electrocardiographic recording system Two possible interfering inputs are stray magnetic fields and capacitively coupled noise. Orientation of patient cables and changes in electrode–skin impedance are two possible modifying inputs. Z_1 and Z_2 represent the electrode–skin interface impedances.

One interfering input is 60-Hz noise voltage induced in the shaded loop by environmental ac magnetic fields. The desired and the interfering voltages are in series, so both components appear at the input to the differential amplifier. Also, the difference between the capacitively coupled displacement currents flowing through each electrode and the body to ground causes an interfering voltage to appear across Z_{body} between the two electrodes and two interfering voltages across Z_1 and Z_2, the electrode impedances.

An example of a modifying input is the orientation of the patient cables. If the plane of the cables is parallel to the ac magnetic field, magnetically introduced interference is zero. If the plane of the cables is perpendicular to the ac magnetic field, magnetically introduced interference is maximal.

1.7 COMPENSATION TECHNIQUES

The effects of most interfering and modifying inputs can be reduced or eliminated either by altering the design of essential instrument components or by adding new components designed to offset the undesired inputs. The former alternative is preferred when it is feasible, because the result is usually simpler. Unfortunately, designers of instruments can only rarely eliminate the actual source of the undesired inputs. We shall discuss several compensation methods for eliminating the effects of interfering and modifying inputs.

INHERENT INSENSITIVITY

If all instrument components are inherently sensitive only to desired inputs, then interfering and modifying inputs obviously have no effect. For the electrocardiograph an example is the twisting of the electrode wires to reduce the number of magnetic flux lines that cut the shaded loop in Figure 1.2. The voltage of the induced noise is proportional to the area of that loop. The effects of electrode motion can be reduced by techniques described in Section 5.5.

NEGATIVE FEEDBACK

When a modifying input cannot be avoided, then improved instrument performance requires a strategy that makes the output less dependent on the transfer function G_d. The negative feedback method takes a portion of the output, H_f, at any instant of time and feeds it back to the input of the instrument. This output-dependent signal is subtracted from the input, and the difference becomes the effective system input. For input, x_d, and output y, we can write

$$(x_d - H_f y)G_d = y \tag{1.1}$$

$$x_d G_d = y(1 + H_f G_d) \tag{1.2}$$

$$y = \frac{G_d}{1 + H_f G_d} x_d \tag{1.3}$$

G_d usually includes amplification, so $H_f G_d \gg 1$ and $y \cong (1/H_f)(x_d)$. This well-known relationship shows that only the feedback element, H_f, determines the output for a given input. Of course, this strategy fails if H_f is also affected by modifying inputs. Usually the feedback device carries less power, so it is more accurate and linear. Less input-signal power is also needed for this feedback scheme, so less loading occurs. The major disadvantage of using this feedback principle is that dynamic instability leading to oscillations can occur, particularly if G_d contains time delays. The study of feedback systems is a well-developed discipline that cannot be pursued further here (Kuo, 1994).

SIGNAL FILTERING

A filter separates signals according to their frequencies. Most filters accomplish this by attenuating the part of the signal that is in one or more frequency bands. A more general definition for a filter is "a device or program that separates data, signals, or material in accordance with specified criteria" (Jay, 1988).

Filters may be inserted at the instrument input, at some point within the instrument, or at the output of the instrument. In fact, the limitations of people's senses may be used to filter unwanted signal components coming from display devices. An example is the utilization of flicker fusion for rapidly changing images from a real-time ultrasonic scanner.

Input filtering blocks interfering and modifying inputs but does not alter the desired input. Filter elements may be distinct devices that block or pass all inputs, or they may be embodied in a single device that selectively blocks only undesired inputs. Many designers do not use input filters that are electric circuits but use instead mechanical, pneumatic, thermal, or electromagnetic principles to block out undesired environmental inputs. For example, instruments are often shock-mounted to filter vibrations that affect sensitive instrument components. Electromagnetic shielding is often used to block interfering electric and magnetic fields, such as those indicated in Figure 1.2.

Electronic filters are often incorporated at some intermediate stage within the instrument. To facilitate filtering based on differences in frequency, mixers and modulators are used to shift desired and/or undesired signals to another frequency range where filtering is more effective. Digital computers are used to filter signals on the basis of template-matching techniques and various time-domain signal properties. These filters may even have time- or signal-dependent criteria for isolating the desired signal.

Output filtering is possible, though it is usually more difficult because desired and undesired output signals are superimposed. The selectivity needed may be easier to achieve with higher-level output signals.

OPPOSING INPUTS

When interfering and/or modifying inputs cannot be filtered, additional interfering inputs can be used to cancel undesired output components. These extra intentional inputs may be the same as those to be canceled. In general, the unavoidable and the added opposing inputs can be quite different, as long as the two output components are equal so that cancelation results. The two outputs must cancel despite variations in all the unavoidable interfering inputs and variations in the desired inputs. The actual cancelation of undesired output components can be implemented either before or after the desired and undesired outputs are combined. Indeed, either the intentional or the unavoidable interfering input signal might be processed by G_d. The method of opposing inputs can also be used to cancel the effects of modifying inputs.

Automatic real-time corrections are implied for the method of opposing inputs just described. Actually, output corrections are often calculated manually or by computer methods and applied after data are collected. This requires quantitative knowledge of the interfering and/or modifying input at the time of the measurement and also of how these inputs affect the output. This method is usually cumbersome, loses real-time information, and is often used only for rather static interfering inputs, such as temperature and atmospheric pressure.

An example of using the opposing-input method is to intentionally induce a voltage from the same 60-Hz magnetic field present in Figure 1.2 to be amplified and inverted until cancelation of the 60-Hz noise in the output is achieved. An obvious disadvantage of this method is that the amplifier gain has to be adjusted whenever the geometry of the shaded loop in Figure 1.2 changes. In electronic circuits that must operate over a wide temperature range, *thermistors* (temperature-dependent resistors) are often used to counteract unavoidable temperature-dependent changes in characteristics of active circuit elements, such as transistors and integrated circuits.

1.8 BIOSTATISTICS

The application of statistics to medical data is used to design experiments and clinical studies; to summarize, explore, analyze, and present data; to draw inferences from data by estimation or hypothesis testing; to evaluate diagnostic procedures; and to assist clinical decision making (Dawson-Saunders and Trapp, 1990).

Medical research studies can be *observational studies*, wherein characteristics of one or more groups of patients are observed and recorded, or *experimental intervention studies*, wherein the effect of a medical procedure or treatment is investigated. The simplest observational studies are *case-series* studies that describe some characteristics of a group. These studies are without control subjects, in order only to identify questions for further research. *Case-control* observational studies use individuals selected because they have (or do not have) some outcome or disease and then look backward to find possible causes or risk factors. *Cross-sectional* observational studies analyze characteristics of patients at one particular time to determine the status of a disease or condition. *Cohort* observational studies prospectively ask whether a particular characteristic is a precursor or risk factor for an outcome or disease. Experimental clinical trials are *controlled* if the outcome for patients administered a drug, device, or procedure is compared to the outcome for patients given a placebo or another accepted treatment. The trials are *uncontrolled* if there is no such comparison. *Concurrent controls* are best, because patients are selected in the same way and for the same time period. A *double-blind* study with *randomized* selection of patients to treatment options is preferred, because this design minimizes investigator and patient bias. Medical outcome studies show cost effective improvements in patient health are increasingly required prior to adoption and reimbursement for new medical technologies (Anonymous, 1994).

Quantitative data are measured on a continuous or discrete *numerical scale* with some precision. Qualitative data values that fit into categories are measured on *nominal scales* that show the names of the categories. An *ordinal scale* is used when the categories exhibit an inherent order. Descriptive statistics are useful to summarize data and their attributes. *Distributions* of empirical or theoretical data reflect the values of a variable or characteristic and the frequency of occurrence of those values.

Measures of the middle, or central tendency, include the well-known *mean*, which is the sum of observed values divided by the number of observations. The mean, found as follows,

$$\overline{X} = \frac{\sum X_i}{n} \tag{1.4}$$

works best as the measure of central tendency for symmetric distributions. The *median* is the value for which half the observations are smaller and half are larger; it is used for skewed numerical data or ordinal data. The *mode* is the observation that occurs most frequently; it is used for bimodal distributions. The *geometric mean* is the *n*th root of the product of the observations:

$$GM = \sqrt[n]{X_1 X_2 X_3 \cdot \cdot \cdot X_n} \tag{1.5}$$

It is used with data on a logarithmic scale.

Measures of spread or dispersion of data describe the variation in the observations. The *range*, which is the difference between the largest and smallest observations, is used to emphasize extreme values. The *standard deviation* is a measure of the spread of data about the mean. It is computed as follows:

$$s = \sqrt{\frac{\sum (X_i - \overline{X})^2}{n - 1}} \tag{1.6}$$

It is used with the mean for symmetric distributions of numerical data. Regardless of the type of symmetric distribution, at least 75% of the values always lie between $\overline{X} - 2s$ and $\overline{X} + 2s$. The *coefficient of variation* is calculated as follows:

$$CV = \left(\frac{s}{\overline{X}}\right)(100\%) \tag{1.7}$$

It standardizes the variation, making it possible to compare two numerical distributions that are measured on different scales. A *percentile* gives the percentage of a distribution that is less than or equal to the percentile number; it may be used with the median for ordinal data or skewed numerical data. The *interquartile range* is the difference between the 25th and 75th percentiles, so it describes the central 50% of a distribution with any shape. The *standard error of the mean*, SEM, expresses the variability to be expected among the *means* in future samples, whereas the *standard deviation* describes the variability to be expected among *individuals* in future samples.

Often we need to study relationships between two numerical characteristics. The *correlation coefficient* r is a measure of the relationship between numerical variables X and Y for paired observations.

$$r = \frac{\sum (X_i - \overline{X})(Y_i - \overline{Y})}{\sqrt{\sum (X_i - \overline{X})^2}\sqrt{\sum (Y_i - \overline{Y})^2}} \tag{1.8}$$

The correlation coefficient ranges from -1 for a negative linear relationship to $+1$ for a positive linear relationship; 0 indicates that there is no linear relationship between X and Y. The correlation coefficient is independent of the units employed to measure the variables, which can be different. Like the standard deviation, the correlation coefficient is strongly influenced by outlying values. Because the correlation coefficient measures only a straight-line relationship, it may be small for a strong curvilinear relationship. Of course, a high correlation does *not* imply a cause-and-effect relationship between the variables.

Estimation and hypothesis testing are two ways to make an inference about a value in a population of subjects from a set of observations drawn from a sample of such subjects. In estimation, *confidence intervals* are calculated for a statistic such as the mean. The confidence intervals indicate that a percentage—say 95%—of such confidence intervals contain the true value of the population mean. The confidence intervals indicate the degree of confidence we can have that they contain the true mean. Hypothesis testing reveals whether the sample gives enough evidence for us to reject the *null hypothesis*, which is usually cast as a statement that expresses the opposite of what we think is true. A *P-value* is the probability of obtaining, if the null hypothesis is true, a result that is at least as extreme as the one observed. The P-value indicates how often the observed difference would occur by chance alone if, indeed, nothing but chance were affecting the outcome. Recent trends favor using estimation and confidence intervals rather than hypothesis testing.

Methods for measuring the accuracy of a diagnostic procedure use three pieces of information. The *sensitivity* of a test is the probability of its yielding positive results in patients who actually have the disease. A test with high sensitivity has a low *false-negative* rate. The *specificity* of a test is the probability of its yielding negative results in patients who do not have the disease. A test with high specificity has a low *false-positive* rate; it does not give a false positive result in many patients who do not have the disease. The third piece of information is the *prior probability*, or prevalence of the condition prior to the test. There are several methods for revising the probability that a patient has a condition on the basis of the results of a diagnostic test. Taking into consideration the results of a diagnostic procedure is only one part of the complex clinical decision-making process. Decision tree analysis and other forms of decision analysis that include economic implications are also used in an effort to make optimal decisions.

1.9 GENERALIZED STATIC CHARACTERISTICS

To enable purchasers to compare commercially available instruments and evaluate new instrument designs, quantitative criteria for the performance of instruments are needed. These criteria must clearly specify how well an instrument measures the desired input and how much the output depends on interfering and modifying inputs. Characteristics of instrument performance are usually subdivided into two classes on the basis of the frequency of the input signals.

Static characteristics describe the performance of instruments for dc or very low frequency inputs. The properties of the output for a wide range of constant inputs demonstrate the quality of the measurement, including nonlinear and statistical effects. Some sensors and instruments, such as piezoelectric devices, respond only to time-varying inputs and have no static characteristics.

Dynamic characteristics require the use of differential and/or integral equations to describe the quality of the measurements. Although dynamic characteristics usually depend on static characteristics, the nonlinearities and statistical variability are usually ignored for dynamic inputs, because the differential equations become difficult to solve. Complete characteristics are approximated by the sum of static and dynamic characteristics. This necessary oversimplification is frequently responsible for differences between real and ideal instrument performance.

ACCURACY

The *accuracy* of a single measured quantity is the difference between the true value and the measured value divided by the true value. This ratio is usually expressed as a percent. Because the true value is seldom available, the accepted true value or reference value should be traceable to the National Institute of Standards and Technology.

The accuracy usually varies over the normal range of the quantity measured, usually decreases as the full-scale value of the quantity decreases on a multirange instrument, and also often varies with the frequency of desired, interfering, and modifying inputs. Accuracy is a measure of the total error without regard to the type or source of the error. The possibility that the measurement is low and that it is high are assumed to be equal. The accuracy can be expressed as percent of reading, percent of full scale, $\pm$ number of digits for digital readouts, or $\pm\frac{1}{2}$ the smallest division on an analog scale. Often the accuracy is expressed as a sum of these. For example, on a digital device: $\pm 0.01\%$ of reading $\pm 0.015\%$ of full scale $\pm$ 1 digit. If accuracy is expressed simply as a percentage, full scale is usually assumed. Some instrument manufacturers specify accuracy only for a limited period of time.

PRECISION

The *precision* of a measurement expresses the number of distinguishable alternatives from which a given result is selected. For example, a meter that displays a reading of 2.434 V is more precise than one that displays a reading of 2.43 V. High-precision measurements do not imply high accuracy, however, because precision makes no comparison to the true value.

RESOLUTION

The smallest incremental quantity that can be measured with certainty is the *resolution.* If the measured quantity starts from zero, the term *threshold* is synonymous with *resolution.* Resolution expresses the degree to which nearly equal values of a quantity can be discriminated.

REPRODUCIBILITY

The ability of an instrument to give the same output for equal inputs applied over some period of time is called *reproducibility* or *repeatability.* Reproducibility does not imply accuracy. For example, a broken digital clock with an AM or PM indicator gives very reproducible values that are accurate only once a day.

STATISTICAL CONTROL

The accuracy of an instrument is not meaningful unless all factors, such as the environment and the method of use, are considered. Statistical control ensures that random variations in measured quantities that result from all factors that influence the measurement process are tolerable. Any systematic errors or bias can be removed by calibration and correction factors, but random variations pose a more difficult problem. The measurand and/or the instrument may introduce statistical variations that make outputs unreproducible. If the cause of this variability cannot be eliminated, then statistical analysis must be used to determine the error variation. The estimate of the true value can be improved by making multiple measurements and averaging the results.

STATIC SENSITIVITY

A static calibration is performed by holding all inputs (desired, interfering, and modifying) constant except one. This one input is varied incrementally over the normal operating range, resulting in a range of incremental outputs. The static sensitivity of an instrument or system is the ratio of the incremental output quantity to the incremental input quantity. This ratio is the static component of G_d for desired inputs within the range of the incremental inputs. The incremental slope can be obtained from either the secant between two

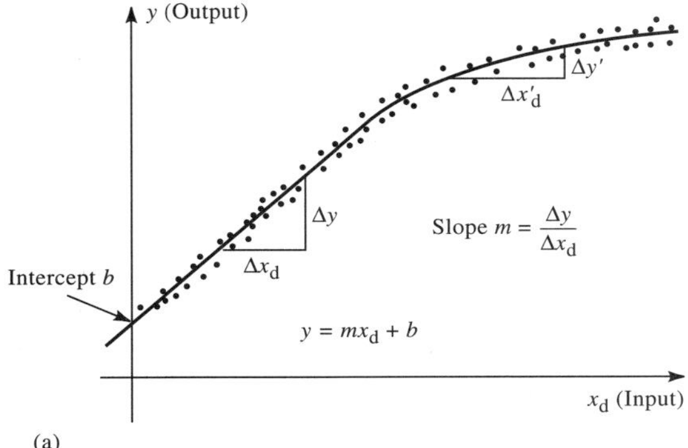

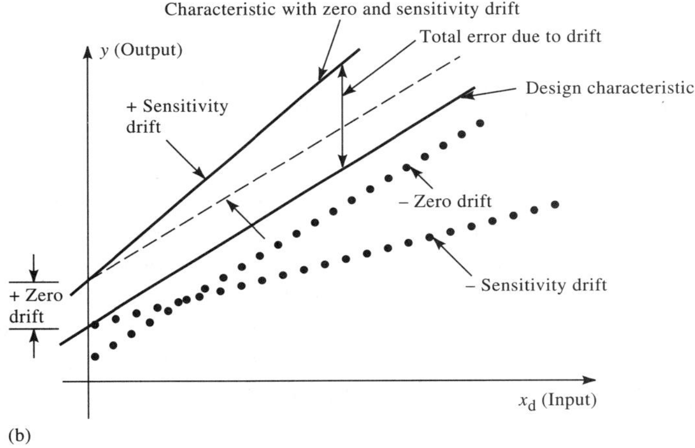

Figure 1.3 (a) Static-sensitivity curve that relates desired input x_d to output y. Static sensitivity may be constant for only a limited range of inputs. (b) Static sensitivity: zero drift and sensitivity drift. Dotted lines indicate that zero drift and sensitivity drift can be negative. [Part (b) modified from *Measurement Systems: Application and Design,* by E. O. Doebelin. Copyright © 1990 by McGraw-Hill, Inc. Used with permission of McGraw-Hill Book Co.]

adjacent points or the tangent to one point on the calibration curve. The static sensitivity may be constant for only part of the normal operating range of the instrument, as shown in Figure 1.3(a). For input–output data that indicate a straight-line calibration curve, the slope m and intercept b for the line with the minimal sum of the squared differences between data points and the line are given by the following equations:

$$m = \frac{n\sum x_d y - \left(\sum x_d\right)\left(\sum y\right)}{n\sum x_d^2 - \left(\sum x_d\right)^2} \tag{1.9}$$

$$b = \frac{\left(\sum y\right)\left(\sum x_d^2\right) - \left(\sum x_d y\right)\left(\sum x_d\right)}{n\sum x_d^2 - \left(\sum x_d\right)^2} \tag{1.10}$$

$$y = mx_d + b \tag{1.11}$$

where n is the total number of points and each sum is for all n points. The static sensitivity for modulating sensors is usually given per volt of excitation, because the output voltage is proportional to the excitation voltage. For example, the static sensitivity for a blood-pressure sensor containing a strain-gage bridge might be $50\ \mu V \cdot V^{-1} \cdot mm\ Hg^{-1}$.

ZERO DRIFT

Interfering and/or modifying inputs can affect the static calibration curve shown in Figure 1.3(a) in several ways. Zero drift has occurred when all output values increase or decrease by the same absolute amount. The slope of the sensitivity curve is unchanged, but the output-axis intercept increases or decreases as shown in Figure 1.3(b). The following factors can cause zero drift: manufacturing misalignment, variations in ambient temperature, hysteresis, vibration, shock, and sensitivity to forces from undesired directions. A change in the dc-offset voltage at the electrodes in the electrocardiograph example in Figure 1.2 is an example of zero drift. Slow changes in the dc-offset voltage do not cause a problem, because the ECG amplifier is ac-coupled. Fast changes due to motion of the subject do cause low-frequency artifact to appear at the output.

SENSITIVITY DRIFT

When the slope of the calibration curve changes as a result of an interfering and/or modifying input, a drift in sensitivity results. Sensitivity drift causes error that is proportional to the magnitude of the input. The slope of the calibration curve can either increase or decrease, as indicated in Figure 1.3(b). Sensitivity drift can result from manufacturing tolerances, variations in power supply, nonlinearities, and changes in ambient temperature and pressure. Variations in the electrocardiograph-amplifier voltage gain as a result of fluctuations in dc power-supply voltage or change in temperature are examples of sensitivity drift.

LINEARITY

A system or element is linear if it has properties such that, if y_1 is the response to x_1, and y_2 is the response to x_2, then $y_1 + y_2$ is the response to $x_1 + x_2$, and Ky_1 is the response to Kx_1 (Jay, 1988). These two requirements for system

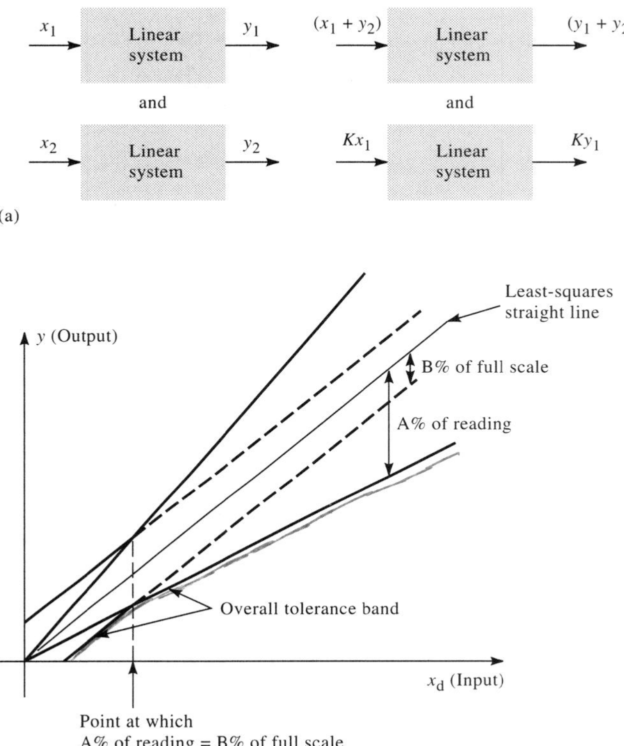

Figure 1.4 (a) Basic definition of linearity for a system or element. The same linear system or element is shown four times for different inputs. (b) A graphical illustrataion of independent nonlinearity equals ±A% of the reading, or ±B% of full scale, whichever is greater (that is, whichever permits the larger error). [Part (b) modified from *Measurement Systems: Application and Design,* by E. O. Doebelin. Copyright © 1990 by McGraw-Hill, Inc. Used with permission of McGraw-Hill Book Co.]

linearity are restated in Figure 1.4(a). They are clearly satisfied for an instrument with a calibration curve that is a straight line.

Keep in mind, however, that high accuracy does not necessarily imply linearity. In practice, no instrument has a perfect linear response, so a measure of deviation from linearity is needed. *Independent nonlinearity* expresses the maximal deviation of points from the least-squares fitted line as either $\pm A\%$ of the reading or $\pm B\%$ of full scale, whichever is greater (that is, whichever permits the larger error). This linearity specification is shown in Figure 1.4(b). For up-scale readings, the percent-of-reading figure is desirable because most errors are proportional to the reading. For small readings near zero, however, percentage of full scale is more realistic because it is not feasible to test for small percent-of-reading deviations near zero. All data points must fall inside the "funnel" shown in Figure 1.4(b). For most instruments that are essentially linear, if other sources of error are minimal, the accuracy is equal to the nonlinearity.

INPUT RANGES

Several maximal ranges of allowed input quantities are applicable for various conditions. Minimal resolvable inputs impose a lower bound on the quantity to be measured. The normal linear operating range specifies the maximal or near-maximal inputs that give linear outputs.

The static linear range and the dynamic linear range may be different. The maximal operating range is the largest input that does not damage the instrument. Operation in the upper part of this range is more likely to be nonlinear. Finally, storage conditions specify environmental and interfering input limits that should not be exceeded when the instrument is not being used. These ranges are not always symmetric with respect to zero input, particularly for storage conditions. Typical operating ranges for blood-pressure sensors have a positive bias, such as +200 mm Hg to −60 mm Hg (+26.6 to −8.0 kPa).

INPUT IMPEDANCE

Because biomedical sensors and instruments usually convert nonelectric quantities into voltage or current, we introduce a generalized concept of input impedance. This is necessary so that we can properly evaluate the degree to which instruments disturb the quantity being measured. For every desired input X_{d1} that we seek to measure, there is another implicit input quantity X_{d2} such that the product $X_{d1} \cdot X_{d2}$ has the dimensions of power. This product represents the instantaneous rate at which energy is transferred across the tissue–sensor interface. The generalized input impedance Z_x is the ratio of the phasor equivalent of a steady-state sinusoidal *effort* input variable (voltage, force, pressure) to the phasor equivalent of a steady-state sinusoidal *flow* input variable (current, velocity, flow).

$$Z_x = \frac{X_{d1}}{X_{d2}} = \frac{\text{effort variable}}{\text{flow variable}} \tag{1.12}$$

The power P is the time rate of energy transfer from the measurement medium.

$$P = X_{d1} \cdot X_{d2} = \frac{X_{d1}^2}{Z_x} = Z_x X_{d2}^2 \tag{1.13}$$

To minimize P, when measuring effort variables X_{d1}, we should make the generalized input impedance as large as possible. This is usually achieved by minimizing the flow variable. However, most instruments function by measuring minute values of the flow variable, so the flow variable cannot be reduced to zero. On the other hand, when we are measuring flow variables X_{d2}, small input impedance is needed to minimize P. The loading caused by measuring devices depends on the magnitude of the input impedance $|Z_x|$ compared with the magnitude of the source impedance $|Z_s|$ for the desired input. Unfortunately, biological source impedances are usually unknown, variable, and difficult to measure and control. Thus the instrument designer must usually focus on maximizing the input impedance Z_x for effort-variable measurement. When the measurand is a flow variable instead of an effort variable, it is more convenient to use the admittance $Y_x = 1/Z_x$ than the impedance.

1.10 GENERALIZED DYNAMIC CHARACTERISTICS

Only a few medical measurements, such as body temperature, are constant or slowly varying quantities. Most medical instruments must process signals that are functions of time. It is this time-varying property of medical signals that requires us to consider dynamic instrument characteristics. Differential or integral equations are required to relate dynamic inputs to dynamic outputs for continuous systems. Fortunately, many engineering instruments can be described by ordinary linear differential equations with constant coefficients. The input $x(t)$ is related to the output $y(t)$ according to the following equation:

$$a_n \frac{d^n y}{dt^n} + \cdots + a_1 \frac{dy}{dt} + a_0 y(t) = b_m \frac{d^m x}{dt^m} + \cdots + b_1 \frac{dx}{dt} + b_0 x(t) \tag{1.14}$$

where the constants a_i $(i = 0, 1, \ldots, n)$ and b_j $(j = 0, 1, \ldots, m)$ depend on the physical and electric parameters of the system. By introducing the differential operator $D^k \equiv d^k(\ )/dt^k$, we can write this equation as

$$(a_n D^n + \cdots + a_1 D + a_0) y(t) = (b_m D^m + \cdots + b_1 D + b_0) x(t) \tag{1.15}$$

Readers familiar with Laplace transforms may recognize that D can be replaced by the Laplace parameter s to obtain the equation relating the transforms $Y(s)$ and $X(s)$. This is a *linear* differential equation, because the linear properties stated in Figure 1.4(a) are assumed and the coefficients a_i and b_j are not functions of time or the input $x(t)$. The equation is *ordinary,* because there is only one independent variable y. Essentially such properties mean that the instrument's methods of acquiring and analyzing the signals do not change as a function of time or the quantity of input. For example, an autoranging instrument may violate these conditions.

Most practical instruments are described by differential equations of zero, first, or second order; thus $n = 0, 1, 2$, and derivatives of the input are usually absent, so $m = 0$.

The input $x(t)$ can be classified as transient, periodic, or random. No general restrictions are placed on $x(t)$, although, for particular applications, bounds on amplitude and frequency content are usually assumed. Solutions for the differential equation depend on the input classifications. The step function is the most common transient input for instrumentation. Sinusoids are the most common periodic function to use because, through the Fourier-series expansion, any periodic function can be approximated by a sum of sinusoids. Band-limited white noise (uniform-power spectral content) is a common random input because one can test instrument performance for all frequencies in a particular bandwidth.

TRANSFER FUNCTIONS

The transfer function for a linear instrument or system expresses the relationship between the input signal and the output signal mathematically. If the transfer function is known, the output can be predicted for any input. The *operational transfer function* is the ratio $y(D)/x(D)$ as a function of the differential operator D.

$$\frac{y(D)}{x(D)} = \frac{b_m D^m + \cdots + b_1 D + b_0}{a_n D^n + \cdots + a_1 D + a_0} \tag{1.16}$$

This form of the transfer function is particularly useful for transient inputs. For linear systems, the output for transient inputs, which occur only once and do not repeat, is usually expressed directly as a function of time, $y(t)$, which is the solution to the differential equation.

The *frequency transfer function* for a linear system is obtained by substituting $j\omega$ for D in (1.16).

$$\frac{Y(j\omega)}{X(j\omega)} = \frac{b_m (j\omega)^m + \cdots + b_1 (j\omega) + b_0}{a_n (j\omega)^n + \cdots + a_1 (j\omega) + a_0} \tag{1.17}$$

where $j = +\sqrt{-1}$ and ω is the angular frequency in radians per second. The input is usually given as $x(t) = A_x \sin(\omega t)$, and all transients are assumed to have died out. The output $y(t)$ is a sinusoid with the same frequency, but the amplitude and phase depend on ω; that is, $y(t) = B(\omega) \sin[\omega t + \phi(\omega)]$. The frequency transfer function is a complex quantity having a magnitude that is the ratio of the magnitude of the output to the magnitude of the input and a phase angle ϕ that is the phase of the output $y(t)$ minus the phase of the input $x(t)$. The phase angle for most instruments is negative. We do not usually express the output of the system as $y(t)$ for each frequency, because we know that it is just a sinusoid with a particular magnitude and phase. Instead, the amplitude ratio and the phase angle are given separately as functions of frequency.

The dynamic characteristics of instruments are illustrated below by examples of zero-, first-, and second-order linear instruments for step and sinusoidal inputs.

ZERO-ORDER INSTRUMENT

The simplest nontrivial form of the differential equation results when all the a's and b's are zero except a_0 and b_0.

$$a_0 y(t) = b_0 x(t) \tag{1.18}$$

This is an algebraic equation, so

$$\frac{y(D)}{x(D)} = \frac{Y(j\omega)}{X(j\omega)} = \frac{b_0}{a_0} = K = \text{static sensitivity} \tag{1.19}$$

where the single constant K replaces the two constants a_0 and b_0. This zero-order instrument has ideal dynamic performance, because the output is proportional to the input for all frequencies and there is no amplitude or phase distortion.

A linear potentiometer is a good example of a zero-order instrument. Figure 1.5 shows that if the potentiometer has pure uniform resistance, then the output voltage $y(t)$ is directly proportional to the input displacement $x(t)$, with no time delay for any frequency of input. In practice, at high frequencies, some parasitic capacitance and inductance might cause slight distortion. Also, low-resistance circuits connected to the output can load this simple zero-order instrument.

FIRST-ORDER INSTRUMENT

If the instrument contains a single energy-storage element, then a first-order derivative of $y(t)$ is required in the differential equation.

$$a_1 \frac{dy(t)}{dt} + a_0 y(t) = b_0 x(t) \tag{1.20}$$

This equation can be written in terms of the differential operator D as

$$(\tau D + 1) y(t) = K x(t) \tag{1.21}$$

where $K = b_0/a_0$ = static sensitivity, and $\tau = a_1/a_0$ = time constant.

Exponential functions offer solutions to this equation when appropriate constants are chosen. The operational transfer function is

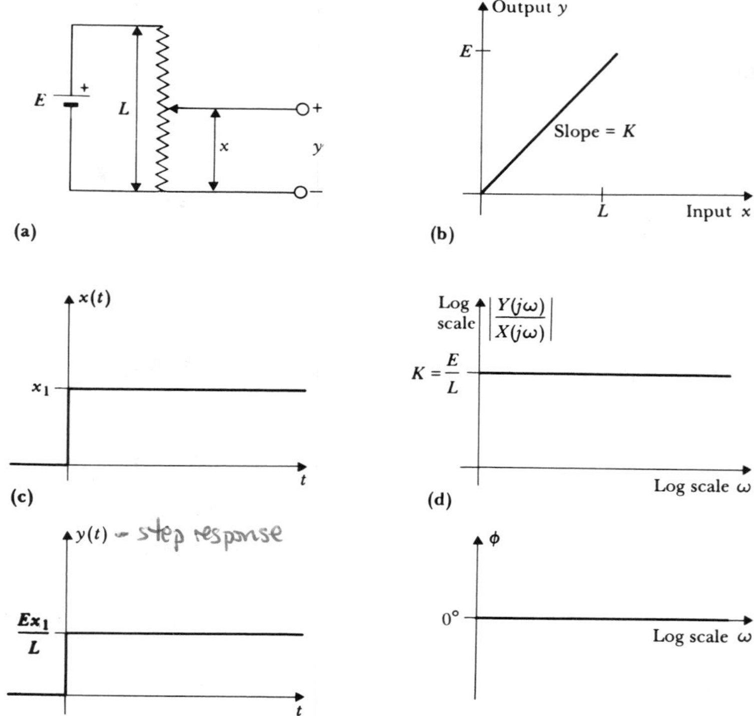

Figure 1.5 (a) A linear potentiometer, an example of a zero-order system. (b) Linear static characteristic for this system. (c) Step response is proportional to input. (d) Sinusoidal frequency response is constant with zero phase shift.

$$\frac{y(D)}{x(D)} = \frac{K}{1 + \tau D} \tag{1.22}$$

and the frequency transfer function is

$$\frac{Y(j\omega)}{X(j\omega)} = \frac{K}{1 + j\omega\tau} = \frac{K}{\sqrt{1 + \omega^2\tau^2}} \underline{/\phi = \arctan\,(-\omega\tau/1)} \tag{1.23}$$

The RC low-pass filter circuit shown in Figure 1.6(a) is an example of a first-order instrument. The input is the voltage $x(t)$, and the output is the voltage $y(t)$ across the capacitor. The first-order differential equation for this circuit is $RC[dy(t)/dt] + y(t) = Kx(t)$. The static-sensitivity curve given in Figure 1.6(b) shows that static outputs are equal to static inputs. This is verified by the differential equation because, for static conditions, $dy/dt = 0$. The step response in Figure 1.6(c) is exponential with a time constant $\tau = RC$.

$$y(t) = K(1 - e^{-t/\tau}) \tag{1.24}$$

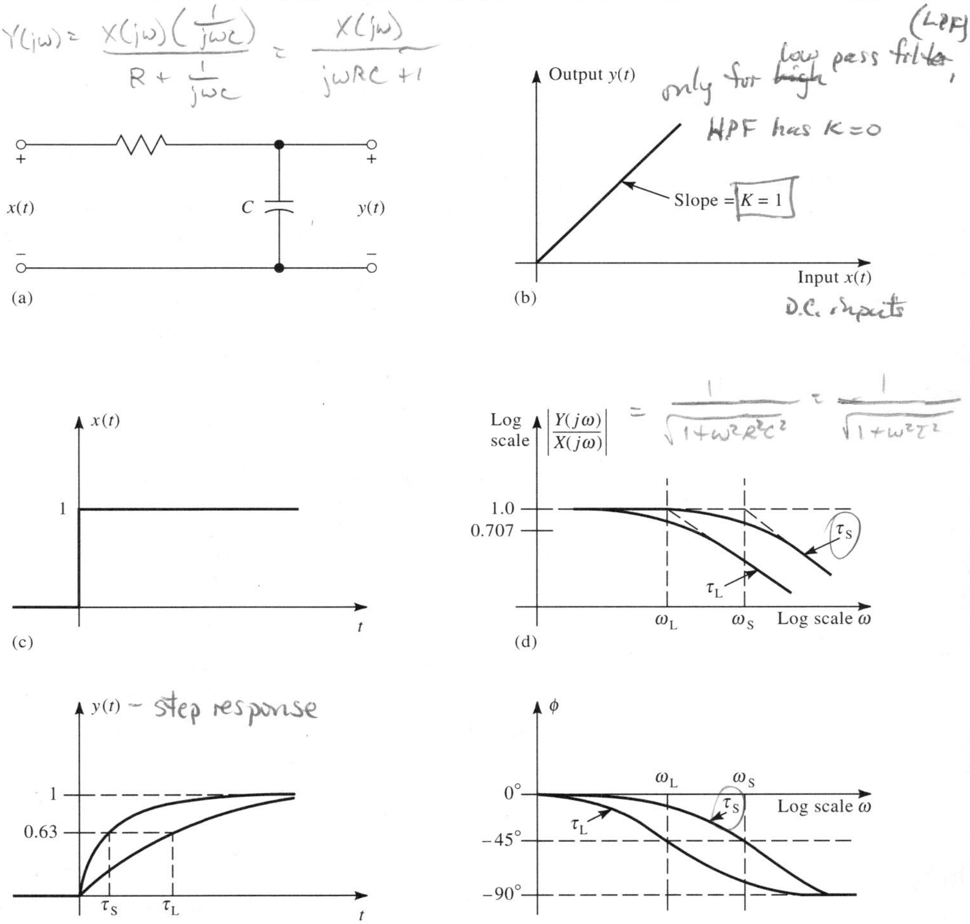

Figure 1.6 (a) A low-pass *RC* filter, an example of a first-order instrument. (b) Static sensitivity for constant inputs. (c) Step response for large time constants (τ_L) and small time constants (τ_S). (d) Sinusoidal frequency response for large and small time constants.

The smaller the time constant, the faster the output approaches the input. For sinusoids, (1.23) and Figure 1.6(d) show that the magnitude of the output decreases as frequency increases. For larger time constants, this decrease occurs at lower frequency.

When $\omega = 1/\tau$, the magnitude is $1/\sqrt{2} = 0.707$ times smaller, and the phase angle is $-45°$. This particular frequency ω is known as the *corner, cutoff,* or *break* frequency. Figure 1.6(d) verifies that this is a low-pass filter; low-frequency sinusoids are not severely attenuated, whereas high-frequency sinusoids produce very little output voltage. The ordinate of the frequency-response magnitude in Figure 1.6(d) is usually plotted on a log scale and may be given in units of decibels (dB), which are defined as dB $= 20 \log_{10}|Y(j\omega)/$

$X(j\omega)|$. A mercury-in-glass thermometer is another example of a low-pass first-order instrument.

EXAMPLE 1.1 A first-order low-pass instrument has a time constant of 20 ms. Find the maximal sinusoidal input frequency that will keep output error due to frequency response less than 5%. Find the phase angle at this frequency.

ANSWER

$$\frac{Y(j\omega)}{X(j\omega)} = \frac{K}{1 + j\omega\tau}$$

$$\left|\frac{K}{1 + j\omega\tau}\right| = \frac{K}{\sqrt{1 + \omega^2\tau^2}} = 0.95K$$

$$(\omega^2\tau^2 + 1)(0.95)^2 = 1$$

$$\omega^2 = \frac{1 - (0.95)^2}{(0.95)^2(20 \times 10^{-3})^2}$$

$$\omega = 16.4 \text{ rad/s}$$

$$f = \frac{\omega}{2\pi} = 2.62 \text{ Hz}$$

$$\phi = \tan^{-1}\left(\frac{-\omega\tau}{1}\right) = -18.2°$$

high pass filter

If R and C in Figure 1.6(a) are interchanged, the circuit becomes another first-order instrument known as a *high-pass filter*. The static characteristic is zero for all values of input, and the step response jumps immediately to the step voltage but decays exponentially toward zero as time increases. Thus $y(t) = Ke^{-t/\tau}$. Low-frequency sinusoids are severely attenuated, whereas high-frequency sinusoids are little attenuated. The sinusoidal transfer function is $Y(j\omega)/X(j\omega) = j\omega\tau/(1 + j\omega\tau)$.

SECOND-ORDER INSTRUMENT

An instrument is second order if a second-order differential equation is required to describe its dynamic response.

$$a_2 \frac{d^2y(t)}{dt^2} + a_1 \frac{dy(t)}{dt} + a_0 y(t) = b_0 x(t) \tag{1.25}$$

Many medical instruments are second order or higher, and low pass. Furthermore, many higher-order instruments can be approximated by second-order characteristics if some simplifying assumptions can be made. The four constants in (1.25) can be reduced to three new ones that have physical significance:

$$\left[\frac{D^2}{\omega_n^2} + \frac{2\zeta D}{\omega_n} + 1\right] y(t) = Kx(t) \tag{1.26}$$

where

$$K = \frac{b_0}{a_0} = \text{static sensitivity, output units divided by input units}$$

$$\omega_n = \sqrt{\frac{a_0}{a_2}} = \text{undamped natural frequency, rad/s}$$

$$\zeta = \frac{a_1}{2\sqrt{a_0 a_2}} = \text{damping ratio, dimensionless}$$

Again exponential functions offer solutions to this equation, although the exact form of the solution varies as the damping ratio becomes greater than, equal to, or less than unity. The operational transfer function is

$$\frac{y(D)}{x(D)} = \frac{K}{\dfrac{D^2}{\omega_n^2} + \dfrac{2\zeta D}{\omega_n} + 1} \tag{1.27}$$

and the frequency transfer function is

$$\frac{Y(j\omega)}{X(j\omega)} = \frac{K}{(j\omega/\omega_n)^2 + (2\zeta j\omega/\omega_n) + 1}$$

$$= \frac{K}{\sqrt{[1-(\omega/\omega_n)^2]^2 + 4\zeta^2\omega^2/\omega_n^2}} \angle \phi = \arctan\frac{2\zeta}{\omega/\omega_n - \omega_n/\omega} \tag{1.28}$$

A mechanical force-measuring instrument illustrates the properties of a second-order instrument (Doebelin, 1990). Mass, spring, and viscous-damping elements oppose the applied input force $x(t)$, and the output is the resulting displacement $y(t)$ of the movable mass attached to the spring [Figure 1.7(a)]. If the natural frequency of the spring is much greater than the frequency components in the input, the dynamic effect of the spring can be included by adding one-third of the spring's mass to the mass of the moving elements to obtain the equivalent total mass M.

Hooke's law for linear springs is assumed, so the spring constant is K_s. Dry friction is neglected and perfect viscous friction is assumed, with constant B.

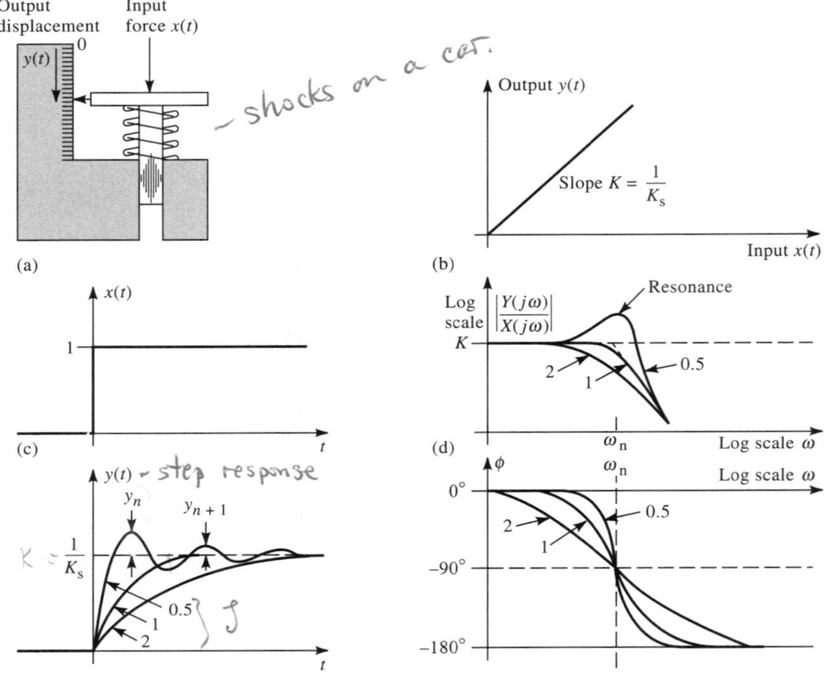

Figure 1.7 (a) Force-measuring spring scale, an example of a second-order instrument. (b) Static sensitivity. (c) Step response for overdamped case $\zeta = 2$, critically damped case $\zeta = 1$, underdamped case $\zeta = 0.5$. (d) Sinusoidal steady-state frequency response, $\zeta = 2$, $\zeta = 1$, $\zeta = 0.5$. [Part (a) modified from *Measurement Systems: Application and Design,* by E. O. Doebelin. Copyright © 1990 by McGraw-Hill, Inc. Used with permission of McGraw-Hill Book Co.]

To eliminate gravitational force from the equation, we adjust the scale until $y = 0$ when $x = 0$. Then the sum of the forces equals the product of mass and acceleration.

$$x(t) - B\frac{dy(t)}{dt} - K_s y(t) = M\frac{d^2y(t)}{dt^2} \tag{1.29}$$

This equation has the same form as (1.26) when the static sensitivity, undamped natural frequency, and damping ratio are defined in terms of K_s, B, and M, as follows:

$$K = 1/K_s \tag{1.30}$$

$$\omega_n = \sqrt{K_s/M} \tag{1.31}$$

$$\zeta = \frac{B}{2\sqrt{K_{\mathrm{m}} M}} \tag{1.32}$$

The static response is $y(t) = Kx(t)$, as shown in Figure 1.7(b). The step response can have three forms, depending on the damping ratio. For a unit-step input, these three forms are

Overdamped, $\zeta > 1$:

$$y(t) = -\frac{\zeta + \sqrt{\zeta^2 - 1}}{2\sqrt{\zeta^2 - 1}} K e^{(-\zeta + \sqrt{\zeta^2 - 1})\omega_{\mathrm{n}} t} + \frac{\zeta - \sqrt{\zeta^2 - 1}}{2\sqrt{\zeta^2 - 1}} K e^{(-\zeta - \sqrt{\zeta^2 - 1})\omega_{\mathrm{n}} t} + K \tag{1.33}$$

Critically damped, $\zeta = 1$:

$$y(t) = -(1 + \omega_{\mathrm{n}} t) K e^{-\omega_{\mathrm{n}} t} + K \tag{1.34}$$

Underdamped, $\zeta < 1$:

$$y(t) = -\frac{e^{-\zeta\omega_{\mathrm{n}} t}}{\sqrt{1 - \zeta^2}} K \sin\left(\sqrt{1 - \zeta^2}\, \omega_{\mathrm{n}} t + \phi\right) + K$$
$$\phi = \arcsin \sqrt{1 - \zeta^2} \tag{1.35}$$

Examples of these three step responses are represented in Figure 1.7(c). Only for damping ratios less than unity does the step response overshoot the final value. Equation (1.35) shows that the frequency of the oscillations in the underdamped response in Figure 1.7(c) is the damped natural frequency $\omega_{\mathrm{d}} = \omega_{\mathrm{n}}\sqrt{1 - \zeta^2}$. A practical compromise between rapid rise time and minimal overshoot is a damping ratio of about 0.7.

EXAMPLE 1.2 For underdamped second-order instruments, find the damping ratio ζ from the step response.

ANSWER To obtain the maximums for the underdamped response, we take the derivative of (1.35) and set it to zero. For $\zeta < 0.3$ we approximate positive maximums when the sine argument equals $3\pi/2$, $7\pi/2$, and so forth. This occurs at

$$t_n = \frac{3\pi/2 - \phi}{\omega_{\mathrm{n}}\sqrt{1 - \zeta^2}} \quad \text{and} \quad t_{n+1} = \frac{7\pi/2 - \phi}{\omega_{\mathrm{n}}\sqrt{1 - \zeta^2}} \tag{1.36}$$

The ratio of the first positive overshoot y_n to the second positive overshoot y_{n+1} [Figure 1.7(c)] is

$$\frac{y_n}{y_{n+1}} = \frac{\left(\dfrac{K}{\sqrt{1-\zeta^2}}\right)\left(\exp\left\{-\zeta\omega_n\left[\dfrac{(3\pi/2-\phi)}{\omega_n\sqrt{1-\zeta^2}}\right]\right\}\right)}{\left(\dfrac{K}{\sqrt{1-\zeta^2}}\right)\left(\exp\left\{-\zeta\omega_n\left[\dfrac{(7\pi/2-\phi)}{\omega_n\sqrt{1-\zeta^2}}\right]\right\}\right)}$$

$$= \exp\left(\frac{2\pi\zeta}{\sqrt{1-\zeta^2}}\right) \tag{1.37}$$

$$\ln\left(\frac{y_n}{y_{n+1}}\right) = \Lambda = \frac{2\pi\zeta}{\sqrt{1-\zeta^2}}$$

where Λ is defined as *logarithmic decrement.* Solving for ζ yields

$$\zeta = \frac{\Lambda}{\sqrt{4\pi^2+\Lambda^2}} \tag{1.38}$$

For sinusoidal steady-state responses, the frequency transfer function (1.28) and Figure 1.7(d) show that low-pass frequency responses result. The rate of decline in the amplitude frequency response is twice the rate of that decline for first-order instruments. Note that resonance phenomena can occur if the damping ratio is too small. Also note that the output phase lag can be as much as 180°, whereas for single-order instruments, the maximal phase lag is 90°.

TIME DELAY

Instrument elements that give an output that is exactly the same as the input, except that it is delayed in time by τ_d, are defined as *time-delay elements.* The mathematical expression for these elements is

$$y(t) = Kx(t-\tau_d), \qquad t > \tau_d \tag{1.39}$$

These elements may also be called analog delay lines, transport lags, or dead times. Although first-order and second-order instruments have negative phase angles that imply time delays, the phase angle varies with frequency, so the delay is not constant for all frequencies. For time delays, the static characteristic is the constant K, the step response is specified by (1.39), and the sinusoidal frequency response for magnitude and phase is

$$\frac{Y(j\omega)}{X(j\omega)} = Ke^{-j\omega\tau_d} \tag{1.40}$$

Time delays are present in transmission lines (electric, mechanical, hydraulic blood vessels, and pneumatic respiratory tubing), magnetic tape recorders, and some digital signal-processing schemes. Usually these time delays are to be avoided, especially in instruments or systems that involve feedback, because undesired oscillations may result.

If the instrument is used strictly for measurement and is not part of a feedback-control system, then some time delay is usually acceptable. The transfer function for undistorted signal reproduction with time delay becomes $Y(j\omega)/X(j\omega) = K \underline{/- \omega\tau_d}$. Our previous study of time-delay elements shows that the output magnitude is K times the input magnitude for all frequencies and that the phase lag increases linearly with frequency.

The transfer-function requirements concern the *overall* instrument transfer function. The overall transfer function of linear elements connected in series is the product of the transfer functions for the individual elements. Many combinations of nonlinear elements can produce the overall linear transfer function required. Various forms of modulation and demodulation are used, and unavoidable sensor nonlinearities can sometimes be compensated for by other instrument elements.

1.11 DESIGN CRITERIA

As shown, many factors affect the design of biomedical instruments. The factors that impose constraints on the design are of course different for each type of instrument. However, some of the general requirements can be categorized as signal, environmental, medical, and economic factors. Figure 1.8 shows how these factors are incorporated into the initial design and development of an instrument.

Note that the type of sensor selected usually determines the signal-processing equipment needed, so an instrument specification includes more than just what type of sensor to use. To obtain a final design, some compromises in specifications are usually required. Actual tests on a prototype are always needed before final design decisions can be made. Changes in performance and interaction of the elements in a complex instrument often dictate design modifications. Good designs are frequently the result of many compromises throughout the development of the instrument. In Chapter 2, we shall examine basic methods of sensing biomedical quantities to ensure that many sensor design alternatives are considered.

1.12 COMMERCIAL MEDICAL INSTRUMENTATION DEVELOPMENT PROCESS

A commercial medical instrument project has many phases, and the size of the team needed grows as the project progresses. Ideas often come from people working where health care is delivered, because clinical needs are most

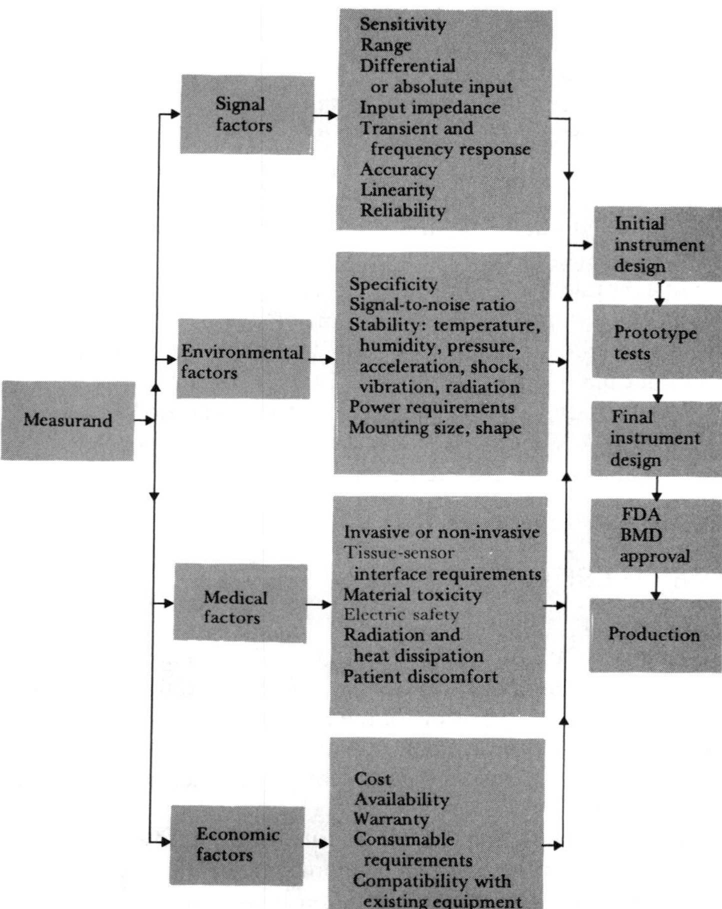

Figure 1.8 Design process for medical instruments Choice and design of instruments are affected by signal factors, and also by environmental, medical, and economic factors. (Revised from *Transducers for Biomedical Measurements: Application and Design,* by R. S. C. Cobbold. Copyright © 1974, John Wiley and Sons, Inc. Used by permission of John Wiley and Sons, Inc.)

evident there. Physicians, nurses, clinical engineers, and sales personnel are therefore good sources of ideas. Industry engineers and marketing personnel should spend time in hospitals observing how their products are actually being used (and misused). Unsolicited inventions are often presented to industry. Most companies have one or more people who are responsible for evaluating new ideas, so it is very important for those who want to propose a new idea to be in contact with a person who can understand the idea and who has the authority to proceed with a detailed evaluation. A simple, signed disclosure is adequate (but essential) to protect patent rights when detailed ideas are discussed.

At this early stage even the best ideas are crudely defined, and they are often not appreciated until at least a preliminary *feasibility analysis* is done. The feasibility analysis and written *product description* should consider the medical need for the product, its technical feasibility, and how well it fits with the company's current or planned product lines and sales methods. Determining medical feasibility entails examining the clinical need for the product, including patient indications, number of patients, clinical specialty, and exactly how, why, and by whom the device will be utilized. Marketing research and analysis can be useful for evolutionary products but are not reliable for revolutionary products, especially when a change in medical practice is required. Technical feasibility analysis should include a description of the core technologies needed, suitability of existing or readily modifiable components, breakthroughs or inventions required, systems analysis, and preliminary cost estimates. A functional prototype that includes the sensor, primary signal processing elements, and other new or unique components should be built and tested on animals and humans as early as possible. The manufacturing feasibility analysis should be done early, when major alternative technologies can still be selected. A brief business plan should assess financial aspects, personnel requirements, competition, patents, standards, manufacturing requirements, sales and service requirements, and the time schedule for the project. Experience has shown that abbreviating the feasibility phase or adding features after this phase usually causes disasters later. A detailed feasibility review should be required at the end of this phase. If the results are not clearly favorable, the project should be abandoned or the feasibility phase should be continued.

After a favorable feasibility review is complete, a detailed *product specification* must be written to describe all the product features. Included should be numerical values for performance, many internal testing requirements, user interface requirements, environmental testing requirements, and even the size, weight, and color of the instrument. The product specification should say everything about "what" is required but nothing about "how" it is to be achieved. This gives design engineers important flexibility. The product specification is used by design and development engineers, test equipment design engineers, quality assurance engineers, and manufacturing engineers throughout the product life cycle. The product specification remains a confidential document in most companies; only the main external performance specifications appear in user technical manuals and promotional materials.

Design and development uses the product specification to make a manufacturable prototype. For most instruments, functions must be partitioned into hardware and software before detailed design can begin. Circuit diagrams, detailed software requirements, and mechanical designs are made by means of computer-aided-design (CAD) tools that allow such detailed simulations of functional operation that there is little need for "breadboard" prototypes.

Design reviews focus on systems issues and problems that the company may encounter in meeting the product specification. True prototypes that are very similar to the final product are thus developed much earlier and can be used by test, design assurance, and manufacturing engineers. These prototypes are also tested on animals or human subjects to verify expected operation and performance at a few clinical testing centers. Data from these *clinical feasibility trials* (including repeated observations of users) should be studied closely, because design flaws that cause user frustration or errors must be discovered at this critical stage. A final *design review* should include test results for all specifications, clinical results including subjective user feedback, an assessment of manufacturability, and more detailed cost estimates.

Production engineers must be involved throughout the design and development process to avoid a redesign for manufacturability. Special production equipment usually has a long lead time. Many fixtures and automatic tests must be designed. All final documents that describe how the product is made must be released by the design and development engineers. Any changes made after the final design review must be authorized via an *engineering change order* (ECO). Small production runs are usually needed to discover problems and improve efficiency. Most medical instruments and devices are not produced in high volume, as consumer products are, so full automation of production is not cost-effective. Even after full production begins, some ECOs may be needed to correct problems that occur in the field. Any failures of medical devices must be documented, along with the corrective action taken.

Throughout the product life cycle, which includes the period after the product is no longer offered for sale, technical support for user questions and problems must be provided. Successful companies must be committed to repairing or replacing their products throughout the entire expected lifetime of the product.

1.13 REGULATION OF MEDICAL DEVICES

In 1976 the United States Congress passed what are known as the **Medical Device Amendments** (Public Law 94-295) to the Federal Food, Drug, and Cosmetics Act that dates back to the 1930s. In the Safe Medical Devices Act of 1990, further amendments were made (Pacela, 1991). The primary purpose was to ensure the safety and efficacy of new medical devices prior to marketing of the device. The law's definition of the term *medical device* is "any item promoted for a medical purpose that does not rely on chemical action to achieve its intended effect" (Kessler *et al.,* 1987). Medical devices were classified in two ways. First, the division of such devices into Class I, II, and III was based on the principle that devices that pose greater potential hazards should be subject to more regulatory requirements. Second, seven categories were established: preamendment, postamendment, substantially equivalent, implant, custom, investigational, and transitional.

The seven categories into which medical devices are divided are described in Table 1.2, which also includes classification rules and examples. Software used in medical devices has become an area of increasing concern; several serious accidents have been traced to software bugs (Murfitt, 1990). Increased requirements for maintaining traceability of devices to the ultimate customer, postmarketing surveillance for life-sustaining and life-supporting implants, and hospital reporting requirements for adverse incidents were added to the law in 1990.

Class I general controls. Manufacturers are required to perform registration, premarketing notification, record keeping, labeling, reporting of adverse experiences, and good manufacturing practices. These controls apply to all three classes.

Class II performance standards. These standards were to be defined by the federal government, but the complexity of procedures called for in the amendments and the enormity of the task have resulted in little progress having been made toward defining the 800 standards needed. The result has been overreliance on the postamendment "substantial equivalence" known as the 510(k) process.

Class III premarketing approval. Such approval is required for devices used in supporting or sustaining human life and preventing impairment of human health. The FDA has extensively regulated these devices by requiring manufacturers to prove their safety and effectiveness prior to market release.

PROBLEMS

1.1 Find the independent nonlinearity, as defined in Figure 1.4(b), for the following set of inputs and outputs from a nearly linear system. Instrument was designed for an output equal to twice the input. Full scale is 20.

Inputs	0.50	1.50	2.00	5.00	10.00
Outputs	0.90	3.05	4.00	9.90	20.50

1.2 Find the correlation coefficient r for the set of five inputs and outputs given in Problem 1.1.

1.3 Derive the operational transfer function and the sinusoidal transfer function for an RC high-pass filter. Plot the step response and the complete frequency response (magnitude and phase).

1.4 A first-order low-pass instrument must measure hummingbird wing displacements (assume sinusoidal) with frequency content up to 100 Hz with an amplitude inaccuracy of less than 5%. What is the maximal allowable time constant for the instrument? What is the phase angle at 50 Hz and at 100 Hz?

1.5 A mercury thermometer has a cylindrical capillary tube with an internal diameter of 0.2 mm. If the volume of the thermometer and that of the bulb

Table 1.2 FDA Medical Device Categories

Category	Description	Classification Rules	Examples
Preamendment devices (or old devices)	Devices on the market before May 28, 1976, when the Medical Device Amendments were enacted.	Devices are assigned to 1 of 3 classes. A presumption exists that preamendment devices should be placed in Class I unless their safety and effectiveness cannot be ensured without the greater regulation afforded by Classes II and III. A manufacturer may petition the FDA for reclassification.	Analog electrocardiography machine; electrohydraulic lithotriptor; contraceptive intrauterine device and accessories; infant radiant warmer; contraceptive tubal-occlusion device; automated heparin analyzer; automated differential cell counter; automated blood-cell separator; transabdominal aminoscope.
Postamendment devices (or new devices)	Devices put on the market after May 28, 1976.	Unless shown to be substantially equivalent to a device that was on the market before the amendments took effect, these devices are automatically placed in Class III. A manufacturer may petition the FDA for reclassification.	Magnetic resonance imager; extracorporeal shock-wave lithotriptor; absorbable sponge; YAG laser; AIDS-antibody test kit; hydrophilic contact lenses; percutaneous catheter for transluminal coronary angioplasty; implantable defibrillator; bone-growth stimulator; alpha-fetoprotein RIA kit; hepatitis B–antibody detection kit.
Substantially equivalent devices	Postamendment devices that are substantially equivalent to preamendment devices.	Devices are assigned to the same class as their preamendment counterparts and subject to the same requirements. If and when the FDA requires testing and approval of preamendment devices, their substantially equivalent counterparts will also be subject to testing and approval.	Digital electrocardiography machines; YAG lasers for certain uses; tampons; ELISA diagnostic kits; devices used to test for drug abuse.
Implanted devices	Devices that are inserted into a surgically formed or natural body cavity and intended to remain there for ≥30 days.	Devices are assumed to require placement in Class III unless a less-regulated class will ensure safety and effectiveness.	Phrenic-nerve stimulator; pacemaker pulse generator; intracardiac patch; vena cava clamp.

Table 1.2 *(Continued)*

Category	Description	Classification Rules	Examples
Custom devices	Devices not generally available to other licensed practitioners and not available in finished form. Product must be specifically designed for a particular patient and may not be offered for general commercial distribution.	Devices are exempt from premarketing testing and performance standards but are subject to general controls.	Dentures; orthopedic shoes.
Investigational devices	Unapproved devices undergoing clinical investigation under the authority of an Investigational Device Exemption.	Devices are exempt if an Investigational Device Exemption has been granted.	Artificial heart; ultrasonic hyperthermia equipment; DNA probes; laser angioplasty devices; positron emission tomography machines.
Transitional devices	Devices that were regulated as drugs before enactment of the statute but are now defined as medical devices.	Devices are automatically assigned to Class III but may be reclassified in Class I or II.	Antibiotic susceptibility disks; bone heterografts; gonorrhea diagnostic products; injectable silicone; intraocular lenses; surgical sutures; soft contact lenses.

SOURCE: Reprinted with permission from *The New England Journal of Medicine* 317(6), 357–366, 1987.

are not affected by temperature, what volume must the bulb have if a sensitivity of 2 mm/°C is to be obtained? Assume operation near 24 °C. Assume that the stem volume is negligible compared with the bulb internal volume. Differential expansion coefficient of Hg = 1.82×10^{-4} ml/(ml·°C).

1.6 For the spring scale shown in Figure 1.7(a), find the transfer function when the mass is negligible.

1.7 Find the time constant from the following differential equation, given that x is the input, y is the output, and a through h are constants.

$$a\frac{dy}{dt} + bx + c + hy = e\frac{dy}{dt} + fx + g$$

1.8 A low-pass first-order instrument has a time constant of 20 ms. Find the frequency, in hertz, of the input at which the output will be 93% of the dc output. Find the phase angle at this frequency.

1.9 A second-order instrument has a damping ratio of 0.4 and an undamped natural frequency of 85 Hz. Sketch the step response, and give numerical values for the amplitude and time of the first two positive maxima. Assume that the input goes from 0 to 1 and that the static sensitivity is 10.

1.10 Consider an underdamped second-order system with step response as shown in Figure 1.7(c). A different way to define logarithmic decrement is

$$\Gamma = \ln \frac{y_n}{y_{n+1}}$$

Here n refers not to the number of positive peaks shown in Figure 1.7(c) but to both positive and negative peaks. That is, n increases by 1 for each half-cycle. Derive an equation for the damping ratio ζ in terms of this different definition of logarithmic decrement.

REFERENCES

Anonymous, *Designers Handbook: Medical Electronics. A Resource and Buyers Guide for Medical Electronics Engineering and Design,* 3rd ed. Santa Monica, CA: Canon Communications, Inc. 1994.

Anonymous, *Health Devices Sourcebook.* Plymouth Meeting, PA: Emergency Care Research Institute, 1996.

Anonymous, *Human Factors Engineering Guidelines and Preferred Practices for the Design of Medical Devices.* Arlington, VA: Association for the Advancement of Medical Instrumentation, 1993.

Anonymous, *Introduction to Transducers for Instrumentation.* Oxnard, CA: Statham Instruments, 1966.

Anonymous, *Medical Outcomes and Guidelines Sourcebook.* New York, NY: Faulkner and Gray, Inc., 1994.

Anonymous, *Product Development Directory from Medical Device Register.* Montvale, NJ: Medical Economics Co., 1996.

Atherton, D. P., *Stability of Nonlinear Systems.* New York: Wiley, 1981.

Bronzino, J. D. (ed.), *The Biomedical Engineering Handbook.* Boca Raton, FL: CRC Press, 1995.

Brush, L. C. *et al., The Guide to Biomedical Standards.* 20th ed. Brea, CA: Quest Publishing Co., 1995.

Cobbold, R. S. C., *Transducers for Biomedical Measurements: Principles and Applications.* New York: Wiley, 1974.

Cohen, B. J., *Medical Terminology: An Illustrated Guide.* 2nd ed. Philadelphia: Lippincott, 1995.

Cromwell, L., F. J. Weibell, and E. A. Pfeiffer, *Biomedical Instrumentation and Measurements,* 2nd ed. Englewood Cliffs, NJ: Prentice-Hall, 1980.

Davis, N. M., *Medical Abbreviations: 12,000 Conveniences at the Expense of Communication and Safety,* 8th ed. Huntington Valley, PA: Neil M. Davis Associates, 1997.

Dawson-Saunders, B., and R. G. Trapp, *Basic and Clinical Biostatistics.* 2nd ed. Norwalk, CT: Appleton & Lange, 1994.

Doebelin, E. O., *Measurement Systems: Application and Design,* 4th ed. New York: McGraw-Hill, 1990.

Dorland, N. W. (ed.), *Dorland's Illustrated Medical Dictionary,* 28th ed. Philadelphia: Saunders, 1994.

Ekelman, K. B., *New Medical Devices: Invention, Development, and Use.* Washington, DC: National Academy Press, 1988.

Firkin, B. G., and J. A. Whitworth, *Dictionary of Medical Eponyms,* 2nd ed. Park Ridge, NJ: Parthenon, 1996.

Geddes, L. A., and L. E. Baker, *Principles of Applied Biomedical Instrumentation,* 3rd ed. New York: Wiley, 1989.

Haber, K., *Common Abbreviations in Clinical Medicine.* New York: Raven Press, 1988.

Hamilton, B., and B. Guidos, *Medical Acronyms, Symbols and Abbreviations.* 2nd ed. New York: Neal-Schuman, 1988.

Heister, R., *Dictionary of Abbreviations in Medical Sciences.* New York: Springer-Verlag, 1989.

Institute of Electrical and Electronics Engineers, *IEEE Standard Dictionary of Electrical and Electronic Terms,* 6th ed. Piscataway, NJ: Institute of Electrical and Electronics Engineers, 1997.

Jacobson, B., and J. G. Webster, *Medicine and Clinical Engineering.* Englewood Cliffs, NJ: Prentice-Hall, 1977.

Kessler, D. A., *et al.,* "The federal regulation of medical devices." *N. Engl. J. Med.,* 1987, 317(6), 357–366.

Kuo, B., *Automatic Control Systems,* 7th ed. Englewood Cliffs, NJ: Prentice-Hall, 1994.

Murfitt, R. R., "United States government regulation of medical device software: A review." *J. Med. Eng. Technol.,* 1990, 14(3), 111–113.

Pacela, A. F. (ed.), "Safe medical devices law: Sweeping changes for hospitals and the standards industry," *Biomed. Safety and Stand.,* Special Supplement No. 12, February 1991.

Rabbitt, J. T., and P. A. Bergh, *The ISO 9000 Book,* 2nd ed. White Plains, NY: Quality Resources, 1994.

Ray, C. D. (ed.), *Medical Engineering.* Chicago: Year Book, 1974.

Scott, D., and J. G. Webster, "Development phases for medical devices." *Med. Instrum.,* 1986, 20(1), 48–52.

Stanaszek, M. J. *et al., The Inverted Medical Dictionary,* 2nd ed. Lancaster, PA: Technomic Publishing Co., 1991.

Tompkins, W. J., and J. G. Webster (eds.), *Design of Microcomputer-Based Medical Instrumentation.* Englewood Cliffs, NJ: Prentice-Hall, 1981.

Webster, J. G. (ed.), *Encyclopedia of Medical Devices and Instrumentation,* Vols. 1–4. New York: Wiley, 1988.

Webster, J. G., and A. M. Cook (eds.), *Clinical Engineering: Principles and Practices.* Englewood Cliffs, NJ: Prentice-Hall, 1979.

2

BASIC SENSORS AND PRINCIPLES

Robert A. Peura and John G. Webster

This chapter deals with basic mechanisms and principles of the sensors used in a number of medical instruments. A *transducer* is a device that converts energy from one form to another. A *sensor* converts a physical parameter to an electric output. An *actuator* converts an electric signal to a physical output. An electric output from the sensor is normally desirable because of the advantages it gives in further signal processing (Pallás-Areny and Webster, 1991). As we shall see in this chapter, there are many methods used to convert physiological events to electric signals. Dimensional changes may be measured by variations in resistance, inductance, capacitance, and piezoelectric effect. Thermistors and thermocouples are employed to measure body temperatures. Electromagnetic-radiation sensors include thermal and photon detectors. In our discussion of the design of medical instruments in the following chapters, we shall use the principles described in this chapter.

2.1 DISPLACEMENT MEASUREMENTS

The physician and biomedical researcher are interested in measuring the size, shape, and position of the organs and tissues of the body. Variations in these parameters are important in discriminating normal from abnormal function. Displacement sensors can be used in both direct and indirect systems of measurement. Direct measurements of displacement are used to determine the change in diameter of blood vessels and the changes in volume and shape of cardiac chambers.

Indirect measurements of displacement are used to quantify movements of liquids through heart valves. An example is the movement of a microphone diaphragm that detects the movement of the heart indirectly and the resulting heart murmurs.

Here we will describe the following types of displacement-sensitive measurement methods: resistive, inductive, capacitive, and piezoelectric.

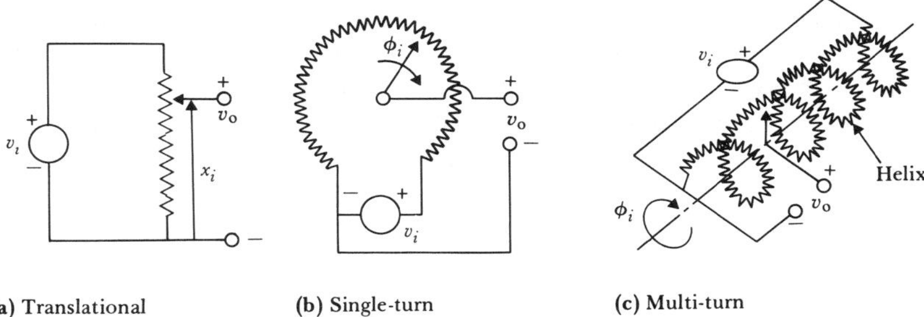

Figure 2.1 Three types of potentiometric devices for measuring displacements (a) Translational. (b) Single-turn. (c) Multi-turn. (From *Measurement Systems: Application and Design,* by E. O. Doebelin. Copyright © 1990 by McGraw-Hill, Inc. Used with permission of McGraw-Hill Book Co.)

2.2 RESISTIVE SENSORS

POTENTIOMETERS

Figure 2.1 shows three types of potentiometric devices for measuring displacement. The potentiometer shown in Figure 2.1(a) measures translational displacements from 2 to 500 mm. Rotational displacements ranging from 10° to more than 50° are detected as shown in Figure 2.1(b) and (c). The resistance elements (composed of wire-wound, carbon-film, metal-film, conducting-plastic, or ceramic material) may be excited by either dc or ac voltages. These potentiometers produce a linear output (within 0.01% of full scale) as a function of displacement, provided that the potentiometer is not electrically loaded.

The resolution of these potentiometers is a function of the construction. It is possible to achieve a continuous stepless conversion of resistance for low-resistance values up to 10 Ω by utilizing a straight piece of wire. For greater variations in resistance, from several ohms to several megohms, the resistance wire is wound on a mandrel or card. The variation in resistance is thereby not continuous, but rather stepwise, because the wiper moves from one turn of wire to the next. The fundamental limitation of the resolution is a function of the wire spacing, which may be as small as 20 μm. The frictional and inertial components of these potentiometers should be low in order to minimize dynamic distortion of the system.

STRAIN GAGES

When a fine wire (25 μm) is strained within its elastic limit, the wire's resistance changes because of changes in the diameter, length, and resistivity. The resulting strain gages may be used to measure extremely small displacements, on the order of nanometers. The following derivation shows how each of these parameters influences the resistance change. The basic equation for the resistance R of a wire with resistivity ρ (ohms · meter), length L (meters), and cross-sectional area A (meters squared) is given by

$$R = \frac{\rho L}{A} \tag{2.1}$$

The differential change in R is found by taking the differential

$$dR = \frac{\rho dL}{A} - \rho A^{-2} L \, dA + L \frac{d\rho}{A} \tag{2.2}$$

We shall modify this expression so that it represents finite changes in the parameters and is also a function of standard mechanical coefficients. Thus dividing members of (2.2) by corresponding members of (2.1) and introducing incremental values, we get

$$\frac{\Delta R}{R} = \frac{\Delta L}{L} - \frac{\Delta A}{A} + \frac{\Delta \rho}{\rho} \tag{2.3}$$

Poisson's ratio μ relates the change in diameter ΔD to the change in length, $\Delta D/D = -\mu \, \Delta L/L$. Substituting this into the center term of (2.3) yields

$$\frac{\Delta R}{R} = \underbrace{(1 + 2\mu)\frac{\Delta L}{L}}_{\text{Dimensional effect}} + \underbrace{\frac{\Delta \rho}{\rho}}_{\text{Piezoresistive effect}} \tag{2.4}$$

Note that the change in resistance is a function of changes in dimension—length ($\Delta L/L$) and area ($2\mu \, \Delta L/L$)—plus the change in resistivity due to strain-induced changes in the lattice structure of the material, $\Delta\rho/\rho$. The gage factor G, found by dividing (2.4) by $\Delta L/L$, is useful in comparing various strain-gage materials.

$$G = \frac{\Delta R/R}{\Delta L/L} = (1 + 2\mu) + \frac{\Delta \rho/\rho}{\Delta L/L} \tag{2.5}$$

Table 2.1 Properties of Strain-gage Materials

Material	Composition (%)	Gage Factor	Temperature Coefficient of Resistivity (°C^{-1} – 10^{-5})
Constantan (advance)	Ni_{45}, Cu_{55}	2.1	±2
Isoelastic	Ni_{36}, Cr_8 (Mn, Si, Mo)$_4$ Fe_{52}	3.52 to 3.6	+17
Karma	Ni_{74}, Cr_{20}, Fe_3 Cu_3	2.1	+2
Manganin	Cu_{84}, Mn_{12}, Ni_4	0.3 to 0.47	±2
Alloy 479	Pt_{92}, W_8	3.6 to 4.4	+24
Nickel	Pure	−12 to −20	670
Nichrome V	Ni_{80}, Cr_{20}	2.1 to 2.63	10
Silicon	(*p* type)	100 to 170	70 to 700
Silicon	(*n* type)	−100 to −140	70 to 700
Germanium	(*p* type)	102	
Germanium	(*n* type)	−150	

SOURCE: From R. S. C. Cobbold, *Transducers for Biomedical Measurements,* 1974, Wiley; used with permission of John Wiley and Sons, Inc., New York.

Table 2.1 gives the gage factors and temperature coefficient of resistivity of various strain-gage materials. Note that the gage factor for semiconductor materials is approximately 50 to 70 times that of the metals. Also note that the gage factor for metals is primarily a function of dimensional effects. For most metals, $\mu = 0.3$ and thus G is at least 1.6, whereas for semiconductors, the piezoresistive effect is dominant. The desirable feature of higher gage factors for semiconductor devices is offset by their greater resistivity-temperature coefficient.

Designs for instruments that use semiconductor materials must incorporate temperature compensation.

Strain gages can be classified as either unbonded or bonded. An unbonded strain-gage unit is shown in Figure 2.2(a). The four sets of strain-sensitive wires are connected to form a Wheatstone bridge, as shown in Figure 2.2(b). These wires are mounted under stress between the frame and the movable armature such that preload is greater than any expected external compressive load. This is necessary to avoid putting the wires in compression. This type of sensor may be used for converting blood pressure to diaphragm movement, to resistance change, then to an electric signal.

A bonded strain-gage element, consisting of a metallic wire, etched foil, vacuum-deposited film, or semiconductor bar, is cemented to the strained surface. Figure 2.3 shows typical bonded strain gages. The deviation from linearity is approximately 1%. One method of temperature compensation for

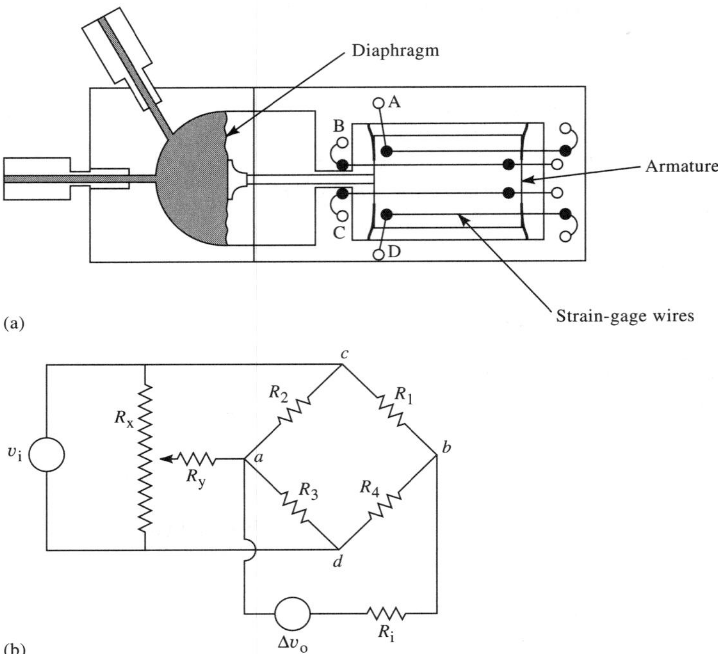

Figure 2.2 (a) Unbonded strain-gage pressure sensor. The diaphragm is directly coupled by an armature to an unbonded strain-gage system. With increasing pressure, the strain on gage pair B and C is increased, while that on gage pair A and D is decreased. (b) Wheatstone bridge with four active elements. $R_1 = B$, $R_2 = A$, $R_3 = D$, and $R_4 = C$ when the unbonded strain gage is connected for translational motion. Resistor R_y and potentiometer R_x are used to initially balance the bridge. v_i is the applied voltage and Δv_o is the output voltage on a voltmeter or similar device with an internal resistance of R_i.

the natural temperature sensitivity of bonded strain gages involves using a second strain gage as a dummy element that is also exposed to the temperature variation, but not to strain. When possible, the four-arm bridge shown in Figure 2.2 should be used, because it not only provides temperature compensation but also yields four times greater output if all four arms contain active gages. Four bonded metal strain gages can be used on cantilever beams to measure bite force in dental research (Dechow, 1988).

Strain-gage technology advanced in the 1960s with the introduction of the semiconductor strain-gage element, which has the advantage of having a high gage factor, as shown in Table 2.1. However, it is more temperature-sensitive and inherently more nonlinear than metal strain gages because the piezoresistive effect varies with strain. Semiconductor elements can be used as bonded, unbonded, or integrated strain-gage units. These integrated devices

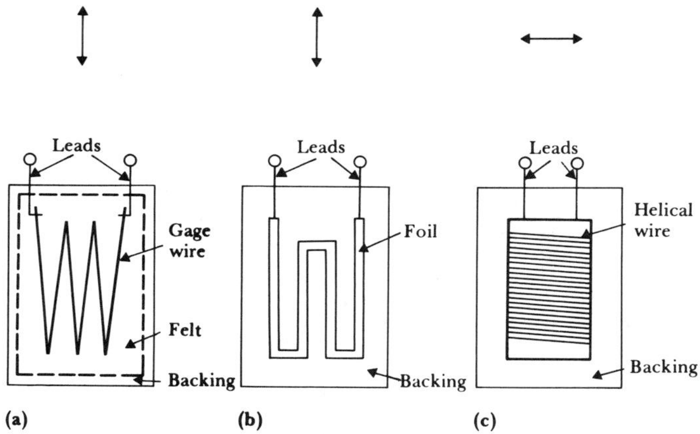

Figure 2.3 Typical bonded strain-gage units (a) Resistance-wire type. (b) Foil type. (c) Helical-wire type. Arrows above units show direction of maximal sensitivity to strain. [Parts (a) and (b) are modified from *Instrumentation in Scientific Research,* by K. S. Lion. Copyright © 1959 by McGraw-Hill, Inc. Used with permission of McGraw-Hill Book Co.]

can be constructed with either silicon or germanium *p* or *n* type as the substrate that forms the structural member. The opposite-type material is diffused into the substrate. Opposite signs for the gage factor result for *n*- and *p*-type substrate gages. A large gage factor can be attained with lightly doped material. Figure 2.4 shows typical semiconductor strain-gage units.

The integrated-type sensor has an advantage in that a pressure sensor can be fabricated by using a silicon substrate for the structural member of the diaphragm. The gages are diffused directly onto the diaphragm. When pressure is applied to the diaphragm, a radial stress component occurs at the edge. The sign of this component is opposite to that of the tangential stress component near the center. The placement of the eight diffused strain-gage units shown in Figure 2.4(c) gives high sensitivity and good temperature compensation (Cobbold, 1974). Figure 14.15 shows a disposable blood-pressure sensor that uses an integrated silicon chip. Silicon strain-gage pressure sensors can be placed on the tip of a catheter and inserted directly into the blood, resulting in more accurate measurements and faster response times (Korites, 1987).

Elastic-resistance strain gages are extensively used in biomedical applications, especially in cardiovascular and respiratory dimensional and plethysmographic (volume-measuring) determinations. These systems normally consist of a narrow silicone-rubber tube (0.5 mm ID, 2 mm OD) from 3 to 25 cm long and filled with mercury or with an electrolyte or conductive paste. The ends of the tube are sealed with electrodes (amalgamated copper, silver, or platinum). As the tube stretches, the diameter of the tube decreases and the length increases, causing the resistance to increase. The resistance per unit

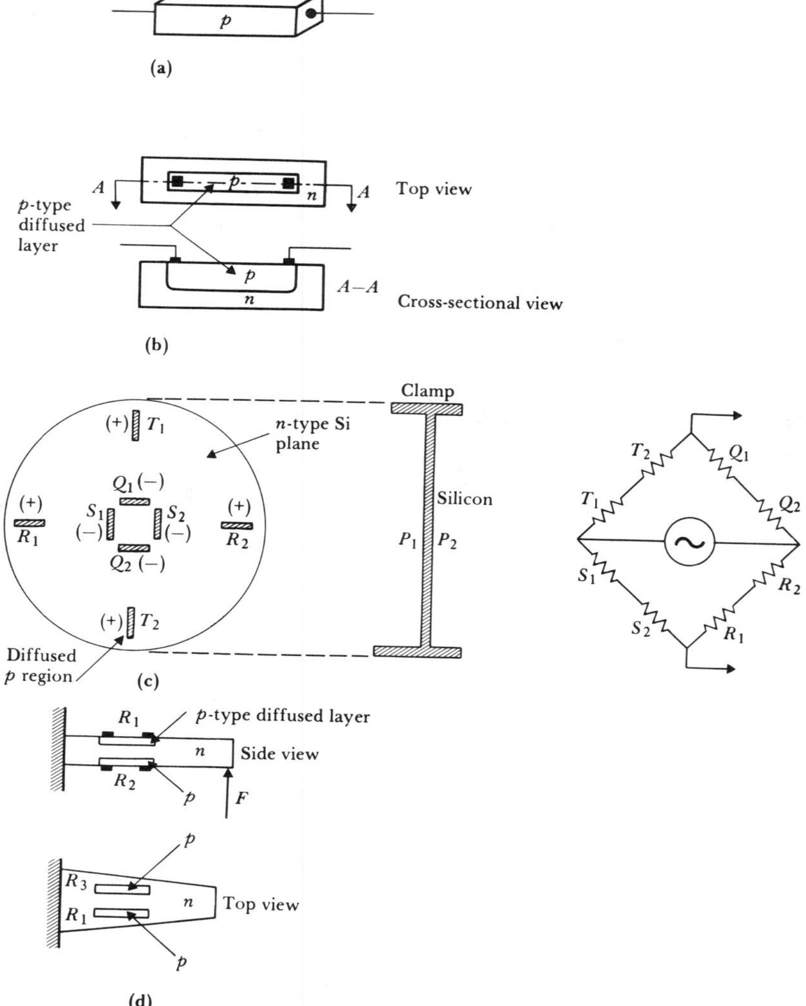

Figure 2.4 Typical semiconductor strain-gage units (a) Unbonded, uniformly doped. (b) Diffused p-type gage. (c) Integrated pressure sensor. (d) Integrated cantilever-beam force sensor. (From *Transducers for Medical Measurements: Application and Design,* by R. S. C. Cobbold. Copyright © 1974, John Wiley and Sons, Inc. Reprinted by permission of John Wiley and Sons, Inc.)

length of typical gages is approximately 0.02–2 Ω/cm. These units measure much higher displacements than other gages.

The elastic strain gage is linear within 1% for 10% of maximal extension. As the extension is increased to 30% of maximum, the nonlinearity reaches 4% of full scale. The initial nonlinearity (dead band) is ascribed to slackness of the unit. Long-term creep is a property of the rubber tubing. This is not a problem for dynamic measurements.

Operational problems include maintaining a good contact between the mercury and the electrodes, ensuring continuity of the mercury column, and controlling the drift in resistance due to a relatively large gage temperature coefficient. In addition, accurate calibration is difficult because of the mass-elasticity and stress-strain relations of the tissue–strain-gage complex. The low value of resistance means more power is required to operate these strain-gage units.

Lawton and Collins (1959) determined the static and dynamic response of elastic strain gages. They found that the amplitude and phase were constant up to 10 Hz. Significant distortion occurred for frequencies greater than 30 Hz. Cobbold (1974) indicated that a problem not fully appreciated is that the gage does not distend fully during pulsations when diameter of the vessel is being measured. The mass of the gage and its finite mechanical resistance can cause it to dig into the vessel wall as the vessel expands, so it can give a reading several times lower than that measured using ultrasonic or cineangiographic methods.

Hokanson *et al.* (1975) described an electrically calibrated mercury-in-rubber strain gage. Lead-wire errors are common with these devices because of the low resistance of the strain gage. In Hokanson's design, the problem was eliminated by effectively placing the strain gage at the corners of the measurement bridge. A constant-current source causes an output that is linear for large changes in gage resistance. Figure 2.5 shows the device and its output when applied to the human calf.

2.3 BRIDGE CIRCUITS

The Wheatstone-bridge circuit is ideal for measuring small changes in resistance. Figure 2.2(b) shows a Wheatstone bridge with an applied dc voltage of v_i and a readout meter Δv_o with internal resistance R_i. It can be shown by the voltage-divider approach that Δv_o is zero—that is, the bridge is balanced—when $R_1/R_2 = R_4/R_3$.

Resistance-type sensors may be connected in one or more arms of a bridge circuit. The variation in resistance can be detected by either null-balance or deflection-balance bridge circuits. The null-balance bridge results when the resistance change of the sensor is balanced (zero output) by a variable resistance in an adjacent arm of the bridge. The calibrated adjustment required for the null is an indication of the change in resistance of the sensor. The deflection-balance method, on the other hand, utilizes the amount of bridge unbalance to determine the change in sensor resistance.

Assume that all values of resistance of the bridge are initially equal to R_0 and that $R_0 \ll R_i$. An increase in resistance, ΔR, of all resistances still results

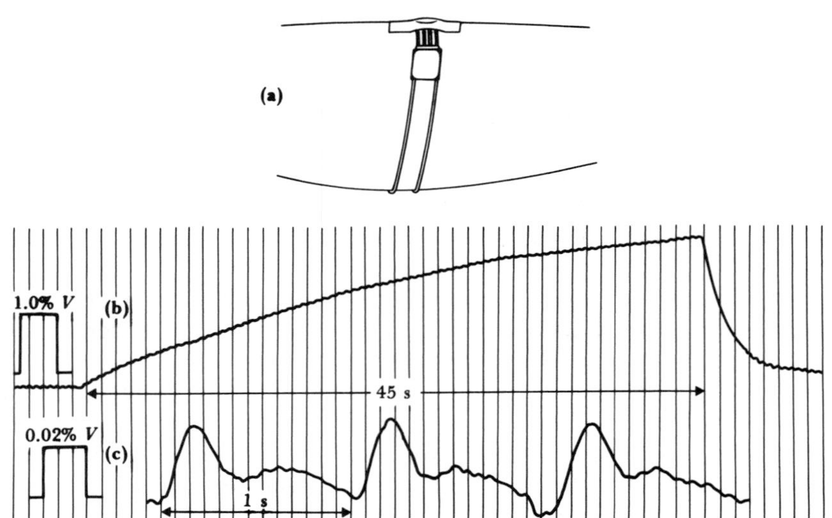

Figure 2.5 Mercury-in-rubber strain-gage plethysmography (a) Four-lead gage applied to human calf. (b) Bridge output for venous-occlusion plethysmography. (c) Bridge output for arterial-pulse plethysmography. [Part (*a*) is based on D. E. Hokanson, D. S. Sumner, and D. E. Strandness, Jr., "An electrically calibrated plethysmograph for direct measurement of limb blood flow." 1975, BME-22, 25–29; used with permission of *IEEE Trans. Biomed. Eng.*, 1975, New York.]

in a balanced bridge. However, if R_1 and R_3 increase by ΔR, and R_2 and R_4 decrease by ΔR, then

$$\Delta v_o = \frac{\Delta R}{R_0} v_i \tag{2.6}$$

Because of the symmetry a similar expression results if R_2 and R_4 increase by ΔR, and R_1 and R_3 decrease by ΔR. Note that (2.6), for the four-active-arm bridge, shows that Δv_o is linearly related to ΔR. A nonlinearity in $\Delta R/R_0$ is present even when $R_0/R_i = 0$.

It is common practice to incorporate a balancing scheme in the bridge circuit [see Fig. 2.2(b)]. Resistor R_y and potentiometer R_x are used to change the initial resistance of one or more arms. This arrangement brings the bridge into balance so that zero voltage output results from "zero" (or "base-level") input of the measured parameter.

To minimize loading effects, R_x is approximately 10 times the resistance of the bridge leg, and R_y limits the maximal adjustment. Strain-gage applications normally use a value of R_y = 25 times the resistance of the bridge leg. Ac

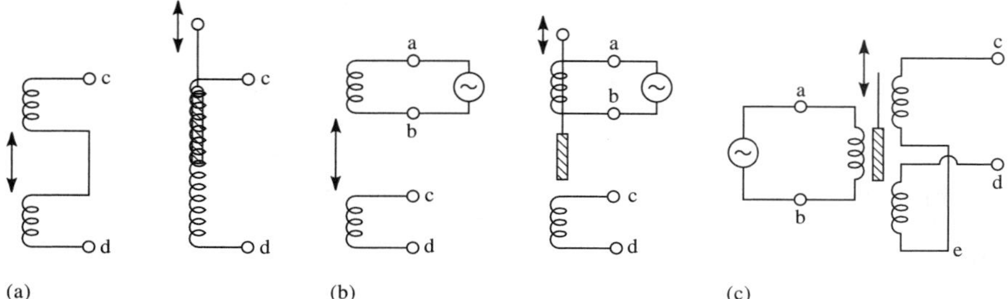

Figure 2.6 Inductive displacement sensors (a) Self-inductance. (b) Mutual inductance. (c) Differential transformer.

balancing circuits are more complicated because a reactive as well as a resistive imbalance must be compensated.

2.4 INDUCTIVE SENSORS

An inductance L can be used to measure displacement by varying any three of the coil parameters:

$$L = n^2 G \mu \tag{2.7}$$

where

n = number of turns of coil

G = geometric form factor

μ = effective permeability of the medium

Each of these parameters can be changed by mechanical means.

Figure 2.6(a) shows self-inductance, Figure 2.6(b) mutual-inductance, and Figure 2.6(c) differential transformer types of inductive displacement sensors. It is usually possible to convert a mutual-inductance system into a self-inductance system by series of parallel connections of the coils. Note in Figure 2.6 that the mutual-inductance device (b) becomes a self-inductance device (a) when terminals b–c are connected.

An inductive sensor has an advantage in not being affected by the dielectric properties of its environment. However, it may be affected by external magnetic fields due to the proximity of magnetic materials.

The variable-inductance method employing a single displaceable core is shown in Figure 2.6(a). This device works on the principle that alterations in the self-inductance of a coil may be produced by changing the geometric form factor or the movement of a magnetic core within the coil. The change in inductance for this device is not linearly related to displacement. The fact that these devices have low power requirements and produce large variations in inductance makes them attractive for radiotelemetry applications.

The mutual-inductance sensor employs two separate coils and uses the variation in their mutual magnetic coupling to measure displacement [Figure 2.6(b)]. Cobbold (1974) describes the application of these devices in measuring cardiac dimensions, monitoring infant respiration, and ascertaining arterial diameters.

Van Citters (1966) provides a good description of applications of mutual inductance transformers in measuring changes in dimension of internal organs (kidney, major blood vessels, and left ventricle). The induced voltage in the secondary coil is a function of the geometry of the coils (separation and axial alignment), the number of primary and secondary turns, and the frequency and amplitude of the excitation voltage. The induced voltage in the secondary coil is a nonlinear function of the separation of the coils. In order to maximize the output signal, a frequency is selected that causes the secondary coil (tuned circuit) to be in resonance. The output voltage is detected with standard demodulator and amplifier circuits.

The *linear variable differential transformer* (LVDT) is widely used in physiological research and clinical medicine to measure pressure, displacement, and force (Reddy and Kesavan, 1988). As shown in Figure 2.6(c), the LVDT is composed of a primary coil (terminals a–b) and two secondary coils (c–e and d–e) connected in series. The coupling between these two coils is changed by the motion of a high-permeability alloy slug between them. The two secondary coils are connected in opposition in order to achieve a wider region of linearity.

The primary coil is sinusoidally excited, with a frequency between 60 Hz and 20 kHz. The alternating magnetic field induces nearly equal voltages v_{ce} and v_{de} in the secondary coils. The output voltage $v_{cd} = v_{ce} - v_{de}$. When the slug is symmetrically placed, the two secondary voltages are equal and the output signal is zero.

LVDT characteristics include linearity over a large range, a change of phase by 180° when the core passes through the center position, and saturation on the ends. Specifications of commercially available LVDTs include sensitivities on the order of 0.5–2 mV for a displacement of 0.01 mm/V of primary voltage, full-scale displacement of 0.1–250 mm, and linearity of ±0.25%. Sensitivity for LVDTs is much higher than that for strain gages.

A disadvantage of the LVDT is that it requires more complex signal-processing instrumentation. Figure 2.7 shows that essentially the same magnitude of output voltage results from two very different input displacements. The direction of displacement may be determined by using the fact that there is a 180° phase shift when the core passes through the null position. A phase-sensitive demodulator is used to determine the direction of displacement. Figure 3.17 shows a ring-demodulator system that could be used with the LVDT.

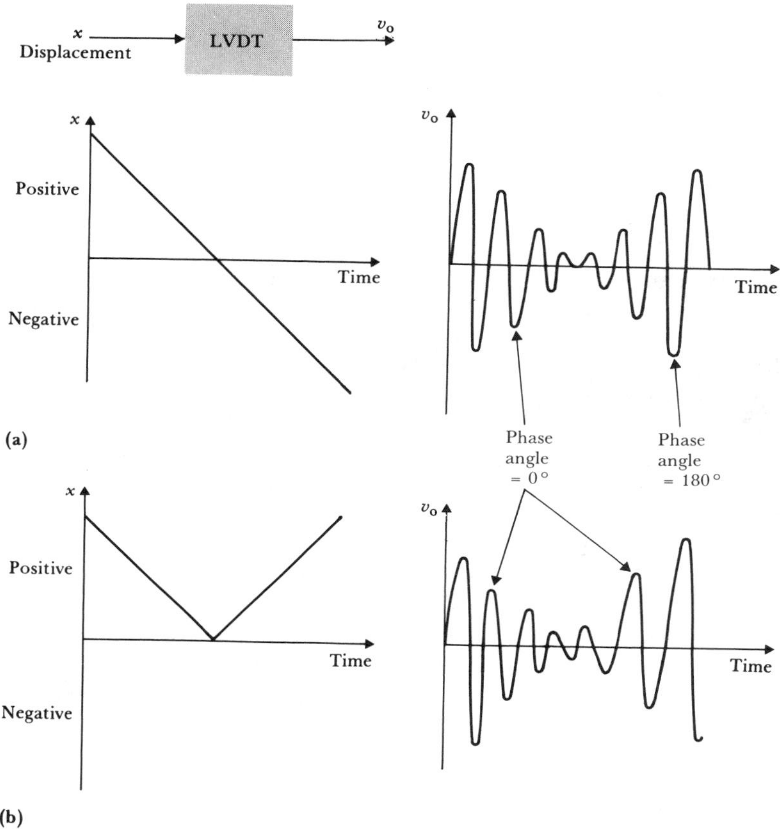

Figure 2.7 (a) As x moves through the null position, the phase changes 180°, while the magnitude of v_o is proportional to the magnitude of x. (b) An ordinary rectifier-demodulator cannot distinguish between (a) and (b), so a phase-sensitive demodulator is required.

2.5 CAPACITIVE SENSORS

The capacitance between two parallel plates of area A separated by distance x is

$$C = \epsilon_0 \epsilon_r \frac{A}{x} \tag{2.8}$$

where ϵ_0 is the dielectric constant of free space (Appendix A.1) and ϵ_r is the relative dielectric constant of the insulator (1.0 for air) (Bowman and Meindl, 1988). In principle it is possible to monitor displacement by changing any of

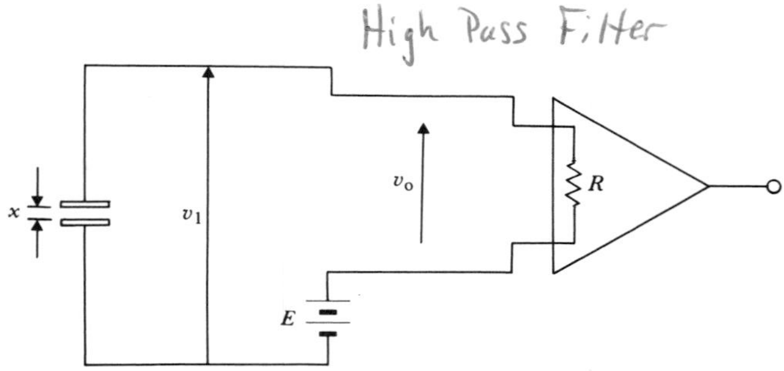

Figure 2.8 **Capacitance sensor for measuring dynamic displacement changes**

the three parameters: ϵ_r, A, or x. However, the method that is easiest to implement and that is most commonly used is to change the separation between the plates.

The sensitivity K of a capacitive sensor to changes in plate separation Δx is found by differentiating (2.8).

$$K = \frac{\Delta C}{\Delta x} = -\epsilon_0 \epsilon_r \frac{A}{x^2} \tag{2.9}$$

Note that the sensitivity increases as the plate separation decreases.

By substituting (2.8) into (2.9), we can develop an expression showing that the percent change in C about any neutral point is equal to the per-unit change in x for small displacements. Thus

$$\frac{dC}{dx} = \frac{-C}{x} \tag{2.10}$$

or

$$\frac{dC}{C} = \frac{-dx}{x} \tag{2.11}$$

The capacitance microphone shown in Figure 2.8 is an excellent example of a relatively simple method for detecting variation in capacitance (Doebelin, 1990, Cobbold, 1974). This is a dc-excited circuit, so no current flows when the capacitor is stationary (with separation x_0) and thus $v_1 = E$. A change in position $\Delta x = x_1 - x_0$ produces a voltage $v_o = v_1 - E$. The output voltage V_o is related to x_1 by

$$\frac{V_o(j\omega)}{X_1(j\omega)} = \frac{(E/x_0)\, j\omega\tau}{j\omega\tau + 1} \tag{2.12}$$

where $\tau = RC = R\epsilon_0\epsilon_r A/x_0$.

Typically, R is 1 MΩ or higher, and thus the readout device must have a high (10 MΩ or higher) input impedance.

For $\omega\tau \gg 1$, $V_o(j\omega)/X_1(j\omega) \cong E/x_0$, which is a constant. However, the response drops off for low frequencies, and it is zero when $\omega = 0$. Thus (2.12) describes a high-pass filter. This frequency response is quite adequate for a microphone that does not measure sound pressures at frequencies below 20 Hz. However, it is inadequate for measuring most physiological variables because of their low-frequency components.

Compliant plastics of different dielectric constants may be placed between foil layers to form a capacitive mat to be placed on a bed. Patient movement generates charge, which is amplified and filtered to display respiratory movements from the lungs and ballistographic movements from the heart (Alihanka *et al.,* 1982).

A capacitance sensor can be fabricated from layers of mica insulators sandwiched between corrugated metal layers. Applied pressure flattens the corrugations and moves the metallic plates closer to each other, thus increasing the capacitance. The sensor is not damaged by large overloads, because flattening of the corrugations does not cause the metal to yield. The sensor measures the pressure between the foot and the shoe (Patel *et al.,* 1989).

2.6 PIEZOELECTRIC SENSORS

Piezoelectric sensors are used to measure physiological displacements and record heart sounds. Piezoelectric materials generate an electric potential when mechanically strained, and conversely an electric potential can cause physical deformation of the material. The principle of operation is that, when an asymmetrical crystal lattice is distorted, a charge reorientation takes place, causing a relative displacement of negative and positive charges. The displaced internal charges induce surface charges of opposite polarity on opposite sides of the crystal. Surface charge can be determined by measuring the difference in voltage between electrodes attached to the surfaces.

Initially, we assume infinite leakage resistance. Then, the total induced charge q is directly proportional to the applied force f.

$$q = kf \tag{2.13}$$

where k is the piezoelectric constant, C/N. The change in voltage can be found by assuming that the system acts like a parallel-plate capacitor where the voltage v across the capacitor is charge q divided by capacitance C. Then, by substitution of (2.8), we get

$$v = \frac{kf}{C} = \frac{kfx}{\epsilon_0 \epsilon_r A} \tag{2.14}$$

Tables of piezoelectric constants are given in the literature (Lion, 1959; and Cobbold, 1974).

Typical values for k are 2.3 pC/N for quartz and 140 pC/N for barium titanate. For a piezoelectric sensor of 1-cm^2 area and 1-mm thickness with an applied force due to a 10-g weight, the output voltage v is 0.23 mV and 14 mV for the quartz and barium titanate crystals, respectively.

There are various modes of operation of piezoelectric sensors, depending on the material and the crystallographic orientation of the plate (Lion, 1959). These modes include the thickness or longitudinal compression, transversal compression, thickness-shear action, and face-shear action.

Also available are piezoelectric polymeric films, such as polyvinylidene fluoride (PVDF) (Hennig, 1988; Webster, 1988). These films are very thin, lightweight, and pliant, and they can be cut easily and adapted to uneven surfaces. The low mechanical quality factor does not permit resonance applications, but it permits acoustical broadband applications for microphones and loudspeakers.

Piezoelectric materials have a high but finite resistance. As a consequence, if a static deflection x is applied, the charge leaks through the leakage resistor (on the order of 100 GΩ). It is obviously quite important that the input impedance of the external voltage-measuring device be an order of magnitude higher than that of the piezoelectric sensor. It would be helpful to look at the equivalent circuit for the piezoelectric sensor [Figure 2.9(a)] in order to quantify its dynamic-response characteristics.

This circuit has a charge generator q defined by

$$q = Kx \tag{2.15}$$

where

K = proportionality constant, C/m
x = deflection

The circuit may be simplified by converting the charge generator to a current generator, i_s.

$$i_s = \frac{dq}{dt} = K\frac{dx}{dt} \tag{2.16}$$

The modified circuit is shown in Figure 2.9(b), where the resistances and capacitances have been combined. Assuming that the amplifier does not draw any current, we then have

$$i_s = i_C + i_R \tag{2.17}$$

$$v_o = v_C = \left(\frac{1}{C}\right)\int i_C\, dt \tag{2.18}$$

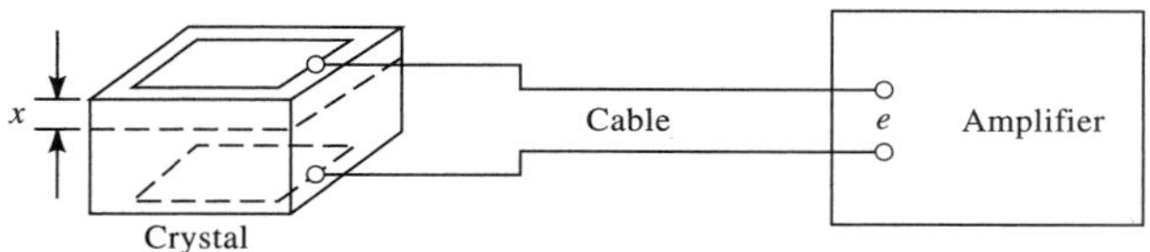

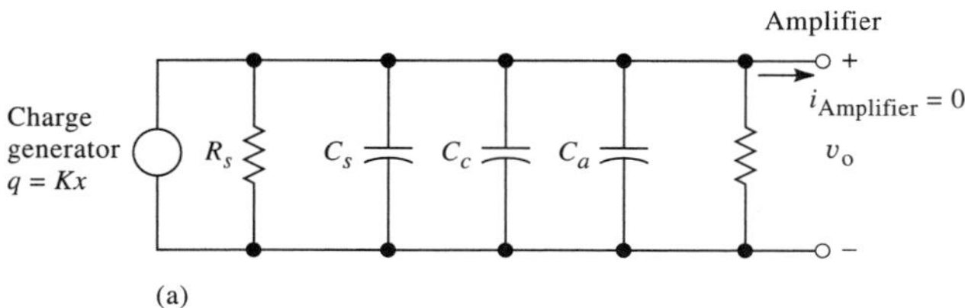

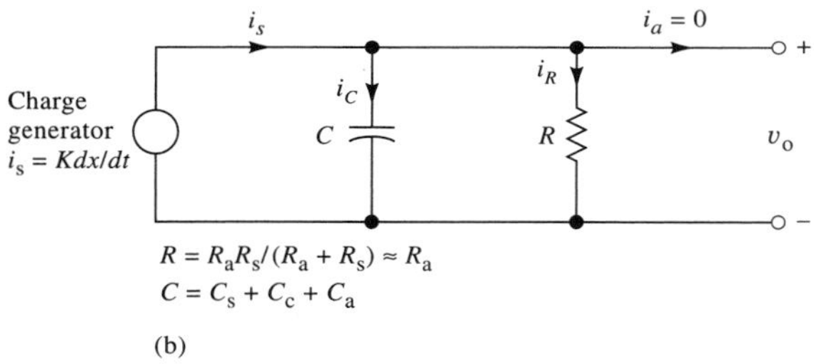

Figure 2.9 (a) Equivalent circuit of piezoelectric sensor, where R_s = sensor leakage resistance, C_s = sensor capacitance, C_c = cable capacitance, C_a = amplifier input capacitance, R_a = amplifier input resistance, and q = charge generator. (b) Modified equivalent circuit with current generator replacing charge generator. (From *Measurement Systems: Application and Design,* by E. O. Doebelin. Copyright © 1990 by McGraw-Hill, Inc. Used with permission of McGraw-Hill Book Co.)

$$i_s - i_R = C\left(\frac{dv_o}{dt}\right) = K\frac{dx}{dt} - \frac{v_o}{R} \tag{2.19}$$

or

$$\frac{V_o(j\omega)}{X(j\omega)} = \frac{K_s j\omega\tau}{j\omega\tau + 1} \tag{2.20}$$

where

$$K_s = K/C \quad \text{(sensitivity, V/m)}$$
$$\tau = RC \text{ (time constant)}$$

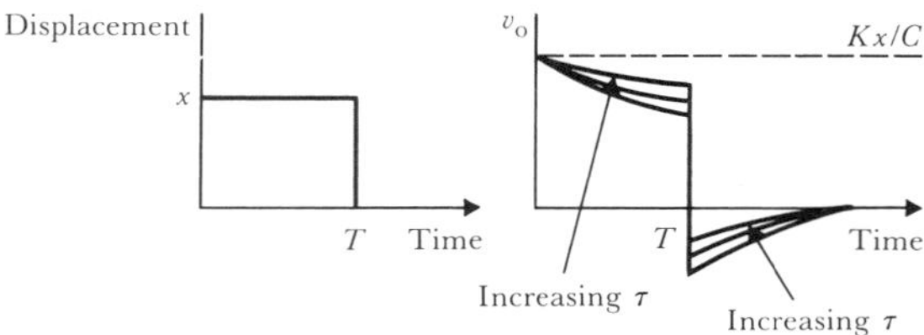

Figure 2.10 Sensor response to a step displacement (From Doebelin, E. O. 1990. *Measurement Systems: Application and Design,* New York: McGraw-Hill.)

EXAMPLE 2.1 A piezoelectric sensor has $C = 500$ pF. The sensor leakage resistance is 10 GΩ. The amplifier input impedance is 5 MΩ. What is the low-corner frequency?

ANSWER We may use the modified equivalent circuit of the piezoelectric sensor given in Figure 2.9(b) for this calculation.

$$f_c = 1/(2\pi RC) = 1/[2\pi(5 \times 10^6)(500 \times 10^{-12})] = 64 \text{ Hz}$$

Note that by increasing the input impedance of the amplifier by a factor of 100, we can lower the low-corner frequency to 0.64 Hz.

Figure 2.10 shows the voltage-output response of a piezoelectric sensor to a step displacement x. The output decays exponentially because of the finite internal resistance of the piezoelectric material. At time equal to T the force is released, and a displacement restoration results that is equal and opposite to the original displacement. This causes a sudden decrease in voltage of magnitude Kx/C, with a resulting undershoot equal to the decay prior to the release of the displacement. The decay and undershoot can be minimized by increasing the time constant, $\tau = RC$. The simplest approach to increasing τ is to add a parallel capacitor. However, doing so reduces the sensitivity in the midband frequencies according to (2.20).

Another approach to improving the low-frequency response is to use the charge amplifier described in Section 3.8.

Because of its mechanical resonance, the high-frequency equivalent circuit for a piezoelectric sensor is complex. This effect can be represented by adding a series *RLC* circuit in parallel with the sensor capacitance and leakage resistance. Figure 2.11 shows the high-frequency equivalent circuit and its frequency response. Note that in some applications—for example, in the case of crystal filters—the mechanical resonance is useful for accurate frequency control.

Piezoelectric sensors are used quite extensively in cardiology for external (body-surface), and internal (intracardiac) phonocardiography. They are also used in the detection of Korotkoff sounds in blood-pressure measurements (Chapter 7). Additional applications of piezoelectric sensors involve their use

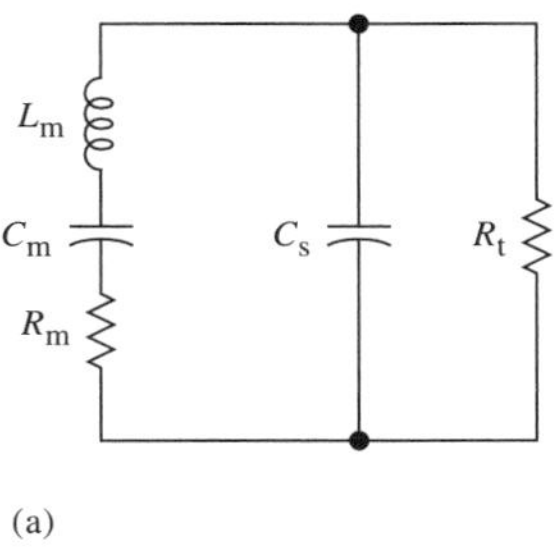

(a)

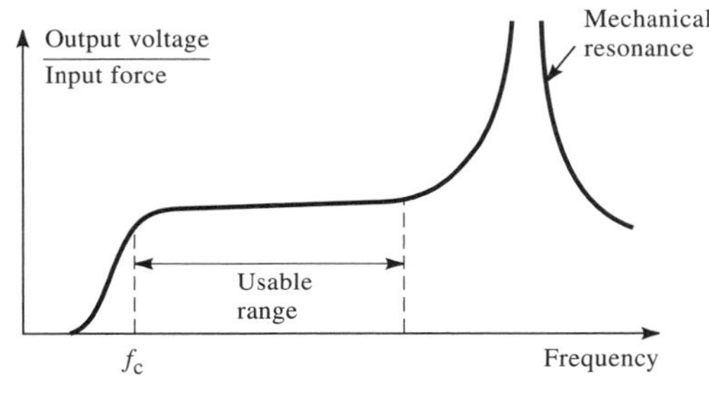

(b)

Figure 2.11 (a) High-frequency circuit model for piezoelectric sensor. R_s is the sensor leakage resistance and C_s the capacitance. L_m, C_m, and R_m represent the mechanical system. (b) Piezoelectric sensor frequency response. (From *Transducers for Biomedical Measurements: Principles and Applications,* by R. S. C. Cobbold. Copyright © 1974, John Wiley and Sons, Inc. Reprinted by permission of John Wiley and Sons, Inc.)

in measurements of physiological accelerations. A piezoelectric sensor and circuit can measure the acceleration due to human movements and provide an estimate of energy expenditure (Servais *et al.,* 1984). Section 8.4 describes ultrasonic blood-flow meters in which the piezoelectric element operating at mechanical resonance emits and senses high-frequency sounds.

2.7 TEMPERATURE MEASUREMENTS

A patient's body temperature gives the physician important information about the physiological state of the individual (Vaughan, 1988). External body temperature is one of many parameters used to evaluate patients in shock, because the reduced blood pressure of a person in circulatory shock results in low blood flow to the periphery. A drop in the big-toe temperature is a good early clinical warning of shock. Infections, on the other hand, are usually reflected by an increase in body temperature, with a hot, flushed skin and loss of fluids.

Increased ventilation, perspiration, and blood flow to the skin result when high fevers destroy temperature-sensitive enzymes and proteins. Anesthesia decreases body temperature by depressing the thermal regulatory center. In fact, physicians routinely induce hypothermia in surgical cases in which they wish to decrease a patient's metabolic processes and blood circulation.

In pediatrics, special heated incubators are used for stabilizing the body temperature of infants. Accurate monitoring of temperature and regulatory control systems are used to maintain a desirable ambient temperature for the infant.

In the study of arthritis, physicians have shown that temperatures of joints are closely correlated with the amount of local inflammation. The increased blood flow due to arthritis and chronic inflammation can be detected by thermal measurements.

The specific site of body-temperature recording must be selected carefully so that it truly reflects the patient's temperature. Also, environmental changes and artifacts can cause misleading readings. For example, the skin and oral-mucosa temperature of a patient seldom reflects true body-core temperature.

The following types of thermally sensitive methods of measurement will be described here: thermocouples, thermistors, and radiation and fiber-optic detectors (Christensen, 1988). The voltage across a p–n junction changes about 2 mV/°C so temperature sensors that use this principle are available (Togawa, 1988).

2.8 THERMOCOUPLES

Thermoelectric thermometry is based on the discovery of Seebeck in 1821. He observed that an *electromotive force* (emf) exists across a junction of two dissimilar metals. This phenomenon is due to the sum of two independent effects. The first effect, discovered by Peltier, is an emf due solely to the contact of two unlike metals and the junction temperature. The net Peltier emf is roughly proportional to the difference between the temperatures of the two junctions. The second effect, credited to Thomson (Lord Kelvin), is an emf due to the temperature gradients along each single conductor. The net Thomson emf is proportional to the difference between the squares of the absolute junction temperatures (T_1 and T_2). The magnitudes of the Peltier and Thomson emfs can be derived from thermodynamic principles (Anonymous, 1974) and either may predominate, depending on the metals chosen.

Knowledge of these two effects is not generally useful in practical applications, so empirical calibration data are usually curve-fitted with a power series expansion that yields the Seebeck voltage,

$$E = aT + \tfrac{1}{2}bT^2 + \cdots \tag{2.21}$$

where T is in degrees Celsius and the reference junction is maintained at 0 °C.

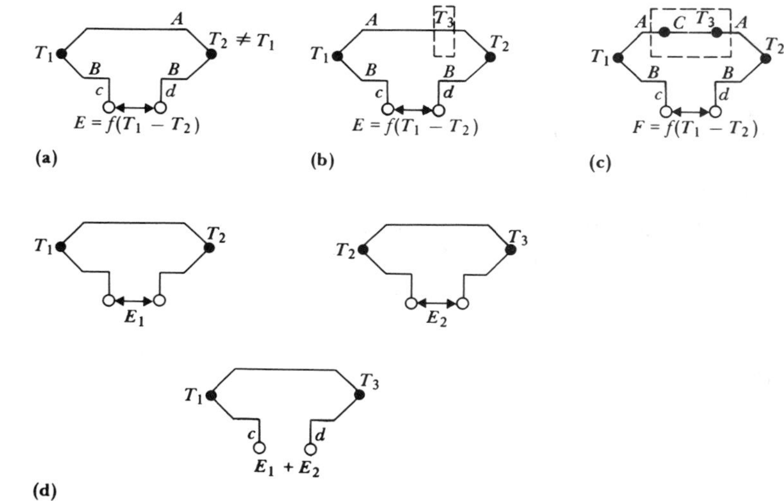

Figure 2.12 Thermocouple circuits (a) Peltier emf. (b) Law of homogeneous circuits. (c) Law of intermediate metals. (d) Law of intermediate temperatures.

Figure 2.12(a) is a thermocouple circuit with two dissimilar metals, A and B, at two different temperatures, T_1 and T_2. The net emf at terminals c–d is a function of the difference between the temperatures at the two junctions and the properties of the two metals. In the practical situation, one junction is held at a constant known temperature (by an ice bath or controlled oven) for a reference in order to determine the desired or unknown temperature.

An understanding of the three empirical thermocouple laws leads to using them properly. The first law, *homogeneous circuits,* states that in a circuit composed of a single homogeneous metal, one cannot maintain an electric current by the application of heat alone. In Figure 2.12(b), the net emf at c–d is the same as in Figure 2.12(a), regardless of the fact that a temperature distribution (T_3) exists along one of the wires (A).

The second law, *intermediate metals,* states that the net emf in a circuit consisting of an interconnection of a number of unlike metals, maintained at the same temperature, is zero. The practical implication of this principle is that lead wires may be attached to the thermocouple without affecting the accuracy of the measured emf, provided that the newly formed junctions are at the same temperature [Figure 2.12(c)].

The third law, *successive* or *intermediate temperatures,* is illustrated in Figure 2.12(d), where emf E_1 is generated when two dissimilar metals have junctions at temperatures T_1 and T_2 and emf E_2 results for temperatures T_2 and T_3. It follows that an emf $E_1 + E_2$ results at c–d when the junctions are at temperatures T_1 and T_3. This principle makes it possible for calibration curves derived for a given reference-junction temperature to be used to determine the calibration curves for another reference temperature.

The *thermoelectric sensitivity* α (also called the thermoelectric power or the Seebeck coefficient) is found by differentiating (2.21) with respect to T. Then

$$\alpha = dE/dT = a + bT + \cdots \tag{2.22}$$

Note that α is not a constant but varies (usually increases) with temperature. The sensitivities of common thermocouples range from 6.5 to 80 μV/°C at 20 °C, with accuracies from $\frac{1}{4}$% to 1%.

For accurate readings, the reference junction should be kept in a triple-point-of-water device the temperature of which is 0.01 ± 0.0005 °C (Doebelin, 1990). Normally the accuracy of a properly constructed ice bath, 0.05 °C with a reproducibility of 0.001 °C is all that is necessary. Temperature-controlled ovens can maintain a reference temperature to within ± 0.4 °C. Modern thermocouple signal conditioners contain an electronic cold junction (Tompkins and Webster, 1988; Sheingold, 1980).

Increased sensitivity may be achieved by connecting a number of thermocouples in series, all of them measuring the same temperature and using the same reference junction. An arrangement of multiple-junction thermocouples is referred to as a *thermopile*. Parallel combinations may be used to measure average temperature.

It is easy to obtain a direct readout of the thermocouple voltage using a digital voltmeter. Chart recordings may be secured by using a self-balancing potentiometer system. The linearity of this latter device is dependent only on the thermocouple and potentiometer; it is independent of the other circuitry.

Thermocouples have the following advantages: fast response time (time constant as small as 1 ms), small size (down to 12 μm diameter), ease of fabrication, and long-term stability. Their disadvantages are small output voltage, low sensitivity, and the need for a reference temperature.

Numerous examples of the use of thermocouples in biomedical research are given in the literature (Welch and Pearce, 1988). Thermocouples can be made small in size, so they can be inserted into catheters and hypodermic needles.

2.9 THERMISTORS

Thermistors are semiconductors made of ceramic materials that are thermal resistors with a high negative temperature coefficient. These materials react to temperature changes in a way that is opposite to the way metals react to such changes. The resistance of thermistors decreases as temperature increases and increases as temperature decreases (Coremans, 1988).

Sapoff (1971) reviewed the various types of thermistors that have been found to be most suitable for biomedical use. The resistivity of thermistor semiconductors used for biomedical applications is between 0.1 and 100 $\Omega \cdot$m.

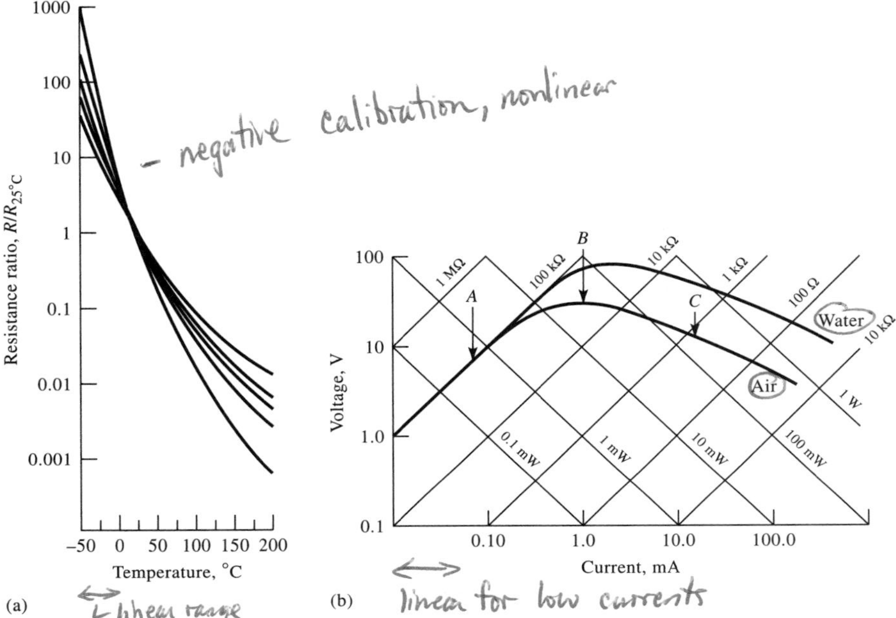

Figure 2.13 (a) Typical thermistor zero-power resistance ratio–temperature characteristics for various materials. (b) Thermistor voltage-versus-current characteristic for a thermistor in air and water. The diagonal lines with a positive slope give linear resistance values and show the degree of thermistor linearity at low currents. The intersection of the thermistor curves and the diagonal lines with negative slope give the device power dissipation. Point *A* is the maximal current value for no appreciable self-heat. Point *B* is the peak voltage. Point *C* is the maximal safe continuous current in air. [Part (b) is from *Thermistor Manual,* EMC-6, © 1974, Fenwal Electronics, Framingham, MA; used by permission.]

These devices are small in size (they can be made less than 0.5 mm in diameter), have a relatively large sensitivity to temperature changes (−3 to −5%/°C), and have excellent long-term stability characteristics (±0.2% of nominal resistance value per year).

Figure 2.13(a) shows a typical family of resistance-versus-temperature characteristics of thermistors. These properties are measured for the thermistor operated at a very small amount of power such that there is negligible self-heating. This resistance is commonly referred to as *zero-power resistance.* The empirical relationship between the thermistor resistance R_t and absolute temperature T in kelvins (K) (the SI unit *kelvin* does not use a degree sign) is

$$R_t = R_0 e^{[\beta(T_0 - T)/TT_0]} \tag{2.23}$$

where

β = material constant for thermistor, K
T_0 = standard reference temperature, K

The value of β increases slightly with temperature. However, this does not present a problem over the limited temperature spans for biomedical work (10–20 °C). β, also known as the characteristic temperature, is in the range of 2500–5000 K. It is usually about 4000 K.

The temperature coefficient α can be found by differentiating (2.23) with respect to T and dividing by R_t. Thus

$$\alpha = \frac{1}{R_t}\frac{dR_t}{dT} = -\frac{\beta}{T^2}\,(\%/\mathrm{K}) \tag{2.24}$$

Note from (2.24) that α is a nonlinear function of temperature. This nonlinearity is also reflected in Figure 2.13(a).

The voltage-versus-current characteristics of thermistors, as shown in Figure 2.13(b), are linear up to the point at which self-heating becomes a problem. When there is large self-heating, the thermistor voltage drop decreases as the current increases. This portion of the curve displays a negative-resistance characteristic.

In the linear portion, Ohm's law applies and the current is directly proportional to the applied voltage. The temperature of the thermistor is that of its surroundings. However, at higher currents a point is reached, because of increased current flow, at which the heat generated in the thermistor raises the temperature of the thermistor above ambient. At the peak of the v–i characteristics, the incremental resistance is zero, and for higher currents a negative-resistance relationship occurs. Operations in this region renders the device vulnerable to thermal destruction.

Figure 2.13(b) shows the difference in the self-heat regions for a thermistor in water and air due to the differences in thermal resistance of air and water. The principle of variation in thermal resistance can be used to measure blood velocity, as described in Section 8.5.

The current-time characteristics of a thermistor are important in any dynamic analysis of the system. When a step change in voltage is applied to a series circuit consisting of a resistor and a thermistor, a current flows. The time delay for the current to reach its maximal value is a function of the voltage applied, the mass of the thermistor, and the value of the series-circuit resistance. Time delays from milliseconds to several minutes are possible with thermistor circuits. Similar time delays occur when the temperature surrounding the thermistor is changed in a step fashion.

Various circuit schemes for linearizing the resistance-versus-temperature characteristics of thermistors have been proposed (Cobbold, 1974; and Doebelin, 1990). Modern instruments use microcomputers to correct for nonlinearities, rather than the former circuit schemes.

The circuitry used for thermistor readout is essentially the same as for conductive sensors, and many of the same techniques apply. Bridge circuits give high sensitivity and good accuracy. The bridge circuit shown in Figure 2.2(b) could be used with $R_3 = R_t$ and R_4 = the thermistor resistance at the midscale value.

Very small differences in temperature can be found using a differential-temperature bridge. It is often necessary to measure such minute differences in biological work. An example is the need to determine the temperature difference between two organs or between multiple sites in the same organ.

A dc differential bridge can achieve a linearity of better than 1% of full-scale output when bead thermistors matched to within ±1% of each other at 25 °C are used. The dc stability of this bridge is not normally a problem, because the output voltage of the bridge—even for temperature differences of 0.01%—is larger than the dc drift of good integrated-circuit operational amplifiers (Cobbold, 1974).

Operational-amplifier circuits may be used to measure the current in a thermistor as a function of temperature. In essence, this circuit applies a constant voltage to the thermistor and monitors its current with a current-to-voltage converter.

Various shapes of thermistors are available: beads, chips, rods, and washers (Sapoff, 1971). The glass-encapsulated bead thermistor is the one most commonly used in biomedical applications. The glass coating protects the sensing element from the hostile environment of the body without significantly affecting the thermal response time of the system. The small size of these thermistors makes possible their placement at the tip of catheters or hypodermic needles. The thermodilution-catheter system discussed in Section 8.2 employs a four-lumen catheter with a thermistor located near the catheter tip.

An additional application of thermistors is in the clinical measurement of oral temperature. Thermistor probes with disposable sheaths are presently used, but these exhibit a first-order step response as shown in Figure 1.6(c). To yield the oral temperature prior to stabilization, a fixed correction of about 1 °C is added to the probe temperature when the rate of change of probe temperature decreases below 0.1 °C/s.

A problem with thermistor neonatal skin surface temperature-monitoring instruments is that the probes fall off. Thermal contact with the skin can be monitored by applying a 14-s pulse every 4.5 min and monitoring the resultant temperature rise (Re and Neuman, 1991).

2.10 RADIATION THERMOMETRY

The basis of *radiation thermometry* is that there is a known relationship between the surface temperature of an object and its radiant power. This principle makes it possible to measure the temperature of a body without physical contact with it. Medical thermography is a technique whereby the temperature distribution of the body is mapped with a sensitivity of a few tenths of a kelvin. It is based on the recognition that skin temperature can vary from place to place depending on the cellular or circulatory processes occurring at each location in the body. Thermography has been used for the early detection of breast cancer, but the method is controversial. It has also been used for determining the location and extent of arthritic disturbances, for gaging the depth of tissue destruction from frostbite and burns, and for detecting various peripheral circulatory disorders (venous thrombosis, carotid artery occlusions, and so forth) (Carr, 1988). Here we shall deal with the basic principles of thermal radiation and detector systems.

Every body that is above absolute zero radiates electromagnetic power, the amount being dependent on the body's temperature and physical properties. For objects at room temperature, the spectrum is predominantly in the far- and extreme-far-infrared regions.

A blackbody is an ideal thermal radiator; as such, it absorbs all incident radiation and emits the maximal possible thermal radiation. The radiation emitted from a body is given by Planck's law multiplied by emissivity ϵ. This expression relates the radiant flux per unit area per unit wavelength W_λ at a wavelength λ (μm) and is stated as

$$W_\lambda = \frac{\epsilon C_1}{\lambda^5(e^{C_2/\lambda T} - 1)} \qquad (\mathrm{W/cm^2 \cdot \mu m}) \tag{2.25}$$

where

$$C_1 = 3.74 \times 10^4 \qquad (\mathrm{W \cdot \mu m^4/cm^2})$$
$$C_2 = 1.44 \times 10^4 \qquad (\mathrm{\mu m \cdot K})$$
T = blackbody temperature, K
ϵ = emissivity, the extent by which a surface deviates from a blackbody ($\epsilon = 1$)

Figure 2.14(a) shows a plot of (2.25), the spectral radiant emittance versus wavelength for a blackbody at 300 K.

Wien's displacement law gives the wavelength λ_m for which W_λ is a maximum. It can simply be found by differentiating (2.25) and setting this to zero.

$$\lambda_m = \frac{2898}{T} \qquad (\mu\mathrm{m}) \tag{2.26}$$

Figure 2.14(a) indicates $\lambda_m = 9.66\ \mu$m ($T = 300$ K). Note from (2.25) that the maximal level of spectral emittance increases with T, and from (2.26) that λ_m is inversely related to T.

The total radiant power W_t can be found by integrating the area under the curve. This expression is known as the *Stefan–Boltzmann law.*

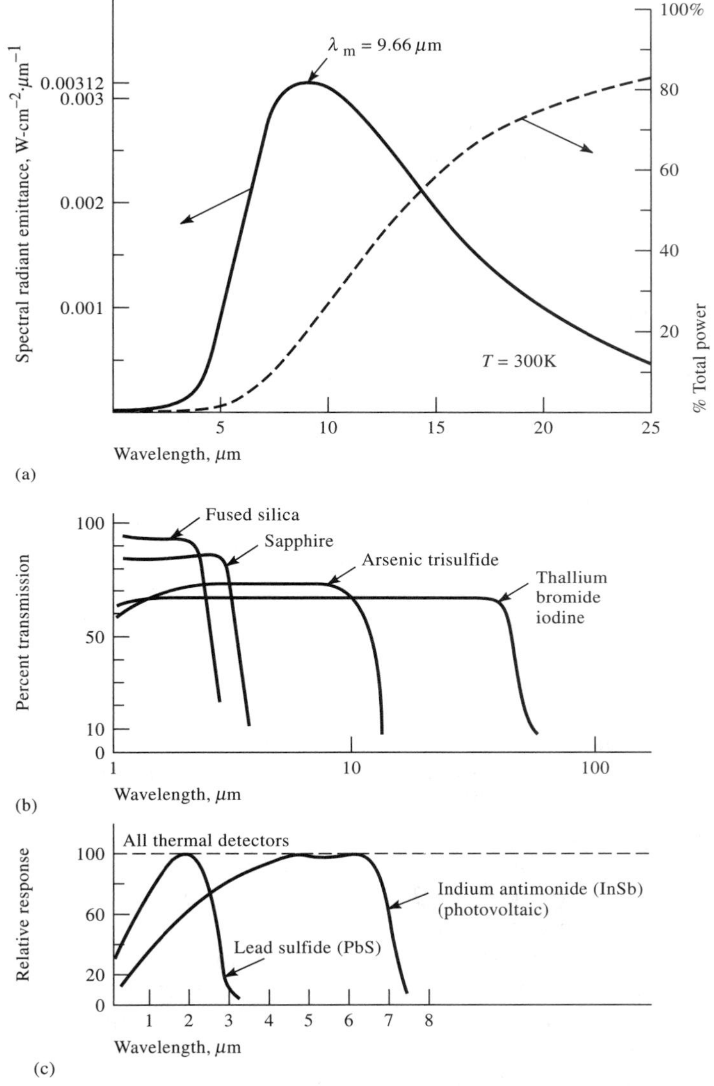

Figure 2.14 (a) Spectral radiant emittance versus wavelength for a blackbody at 300 K on the left vertical axis; percentage of total energy on the right vertical axis. (b) Spectral transmission for a number of optical materials. (c) Spectral sensitivity of photon and thermal detectors. [Part (a) is from *Transducers for Biomedical Measurements: Principles and Applications,* by R. S. C. Cobbold. Copyright © 1974, John Wiley and Sons, Inc. Reprinted by permission of John Wiley and Sons, Inc. Parts (b) and (c) are from *Measurement Systems: Application and Design,* by E. O. Doebelin. Copyright © 1990 by McGraw-Hill, Inc. Used with permission of McGraw-Hill Book Co.]

$$W_t = \epsilon\sigma T^4 \qquad (\text{W/cm}^2) \tag{2.27}$$

where σ is the Stefan-Boltzmann constant (see Appendix).

It is of interest to examine how the percentage of total radiant power varies with wavelength for room-temperature objects. This parameter, plotted in Figure 2.14(a), is found by dividing $\int_0^\lambda W_\lambda \, d\lambda$ by the total radiant power W_t (2.27). Note that approximately 80% of the total radiant power is found in the wavelength band from 4 to 25 μm.

Determining the effect of changes in surface emissivity with wavelength is important in order to accurately determine the temperature of a given source. It can be shown that for $T = 300$ K and $\lambda = 3$ μm, a 5% change in ϵ is equivalent to a temperature change of approximately 1 °C. Variations in ϵ with λ should be found in the case of absolute-temperature determinations, but they are less significant when relative temperature is desired, provided that ϵ remains constant over the surface area being measured. The data relating the variation of ϵ with λ for human skin are not consistent and show variations as large as 5% from unity over the λ span from 2 to 6 μm (Cobbold, 1974).

The lenses used in infrared instruments must be carefully selected for their infrared spectral properties. Special materials must be chosen because standard glass used for the visible spectrum does not pass wavelengths longer than 2 μm. On the other hand, some materials (such as arsenic trisulfide) readily pass infrared and not visible light. Figure 2.14(b) shows the spectral transmission for a number of optical materials.

Infrared detectors and instrument systems must be designed with a high sensitivity because of the weak signals. These devices must have a short response time and appropriate wavelength-bandwidth requirements that match the radiation source. Thermal and photon detectors are used as infrared detectors. The detectors are of two types, both of which are described in Section 2.16. The thermal detector has low sensitivity and responds to all wavelengths, as shown in Figure 2.14 (c), whereas quantum detectors respond only to a limited wavelength band.

Suitable instrumentation must be used to amplify, process, and display these weak signals from radiation detectors. Most radiometers make use of a beam-chopper system to interrupt the radiation at a fixed rate (several hundred hertz). This arrangement allows the use of high-gain ac amplifiers without the inherent problems of stability associated with dc amplifiers. In addition, comparison of reference sources and techniques of temperature compensation are more applicable to ac-instrumentation systems.

Figure 2.15 shows a typical chopped-beam radiation-thermometer system (Cobbold, 1974). A mirror focuses the radiation on the detector. However, a blackened chopper interrupts the radiation beam at a constant rate. The output of the detector circuit is a series of pulses with amplitude dependent on the strength of the radiation source. This ac signal is amplified, while the mean value, which is subject to drift, is blocked. A reference-phase signal, used to synchronize the phase-sensitive demodulator (Section 3.15), is generated in a special circuit consisting of a light source and detector. The signal is then

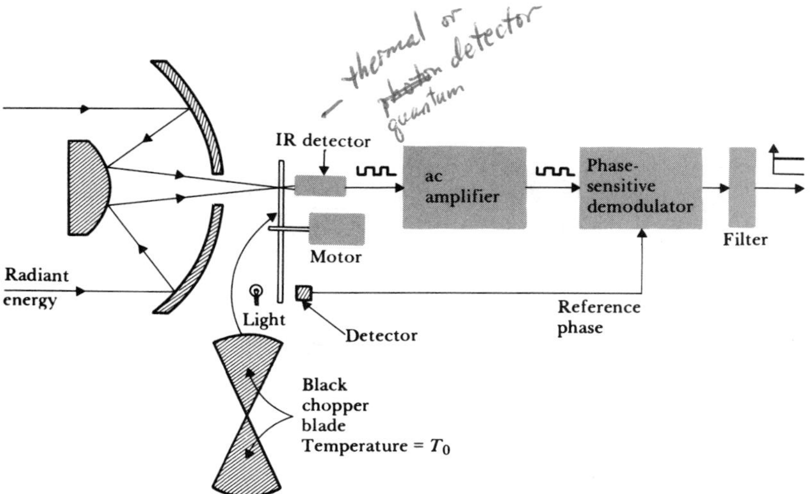

Figure 2.15 Stationary chopped-beam radiation thermometer (From *Transducers for Biomedical Measurements: Principles and Applications,* by R. S. C. Cobbold. Copyright © 1974, John Wiley and Sons, Inc. Reprinted by permission of John Wiley and Sons, Inc.)

filtered to provide a dc signal proportional to the target temperature. This signal can then be displayed or recorded. Infrared microscopes have also been designed using these techniques.

One application of radiation thermometry is an instrument that determines the internal or core body temperature of the human by measuring the magnitude of infrared radiation emitted from the tympanic membrane and surrounding ear canal. The tympanic membrane and hypothalamus are perfused by the same vasculature. The hypothalamus is the body's main thermostat, which regulates the core body temperature. This approach has advantages over using mercury thermometers, thermocouples, or thermistors. The standard temperature-measuring techniques measure the temperature of the sensor, not that of the subject. The sensor must be in contact with the patient long enough for its temperature to become the same as, or close to, that of the subject whose temperature is being measured. However, the infrared thermometry device detects emitted energy that is proportional to the actual temperature of the subject. There is negligible thermal time constant for the pyroelectric sensor (Fraden, 1997). The infrared tympanic temperature-monitoring system has a response time in the order of 0.1 s and an accuracy of approximately 0.1 °C. A disposable sanitary probe cover is used to prevent crosscontamination from patient to patient. Ear thermometry offers several clinical benefits over taking sublingual (oral) or rectal measurements. Response is rapid, and readings can be obtained independent of user technique and degree of patient activity or cooperation (Fraden, 1991).

Infrared tympanic temperature-monitoring systems require a calibration target in order to maintain their high accuracy.

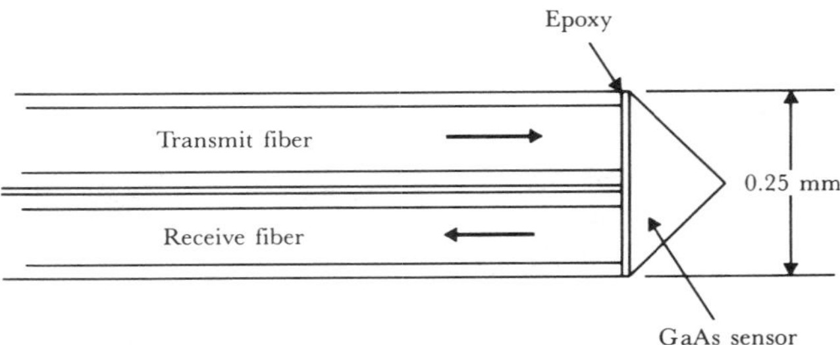

Figure 2.16 Details of the fiber/sensor arrangement for the GaAs semiconductor temperature probe.

2.11 FIBER-OPTIC TEMPERATURE SENSORS

Figure 2.16 shows the details of a GaAs semiconductor temperature probe (Christensen, 1988). A small prism-shaped sample of single-crystal undoped GaAs is epoxied at the ends of two side-by-side optical fibers. The sensors and fibers can be quite small, compatible with biological implantation after being sheathed. One fiber transmits light from a light-emitting diode source to the sensor, where it is passed through the GaAs and collected by the other fiber for detection in the readout instrument. Some of the optical power traveling through the semiconductor is absorbed, by the process of raising valence-band electrons, across the forbidden energy gap into the conduction band. Because the forbidden energy gap is a sensitive function of the material's temperature, the amount of power absorbed increases with temperature.

This nonmetallic probe is particularly suited for temperature measurement in the strong electromagnetic heating fields used in heating tissue for cancer therapy or in patient rewarming.

2.12 OPTICAL MEASUREMENTS

Optical systems are widely used in medical diagnosis. The most common use occurs in the clinical-chemistry lab, in which technicians analyze samples of blood and other tissues removed from the body. Optical instruments are also used during cardiac catheterization to measure the oxygen saturation of hemoglobin and to measure cardiac output.

Figure 2.17(a) shows that the usual optical instrument has a source, filter, and detector. Figure 2.17(b) shows a common arrangement of components. Figure 2.17(c) shows that in some cases, the function of source, filter, sample, and detector may be accomplished by solid-state components.

The remainder of this chapter is divided into sections that deal with sources, geometrical optics, filters, detectors, and combinations thereof.

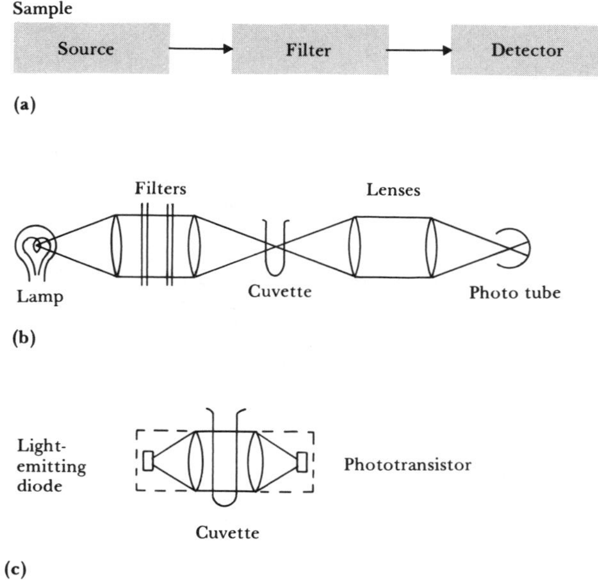

Figure 2.17 (a) General block diagram of an optical instrument. (b) Highest efficiency is obtained by using an intense lamp, lenses to gather and focus the light on the sample in the cuvette, and a sensitive detector. (c) Solid-state lamps and detectors may simplify the system.

2.13 RADIATION SOURCES

TUNGSTEN LAMPS

Incandescent tungsten-wire filament lamps are the most commonly used sources of radiation. Their radiant output varies with temperature and wavelength, as given by (2.25). For $\lambda < 1\ \mu m$, tungsten has an emissivity of about 0.4 and thus emits about 40% of what it would if the emissivity were 1.0. The relative-output spectrum shown in Figure 2.18(a) is only slightly altered. For higher temperatures, λ_m, the maximal wavelength of the radiant-output curves, shifts to a shorter wavelength, as given by (2.26) and as shown in Figure 2.18(a).

Low temperatures, then, yield a reddish color (infrared lamps), whereas high temperatures yield a bluish color (photoflood lamps). The total radiation is given by (2.27). Hence the radiant output increases rapidly with temperature, as do the efficiency, the evaporation of tungsten, and the blackening of the glass bulb. The life of the filament is thus drastically shortened by higher temperatures.

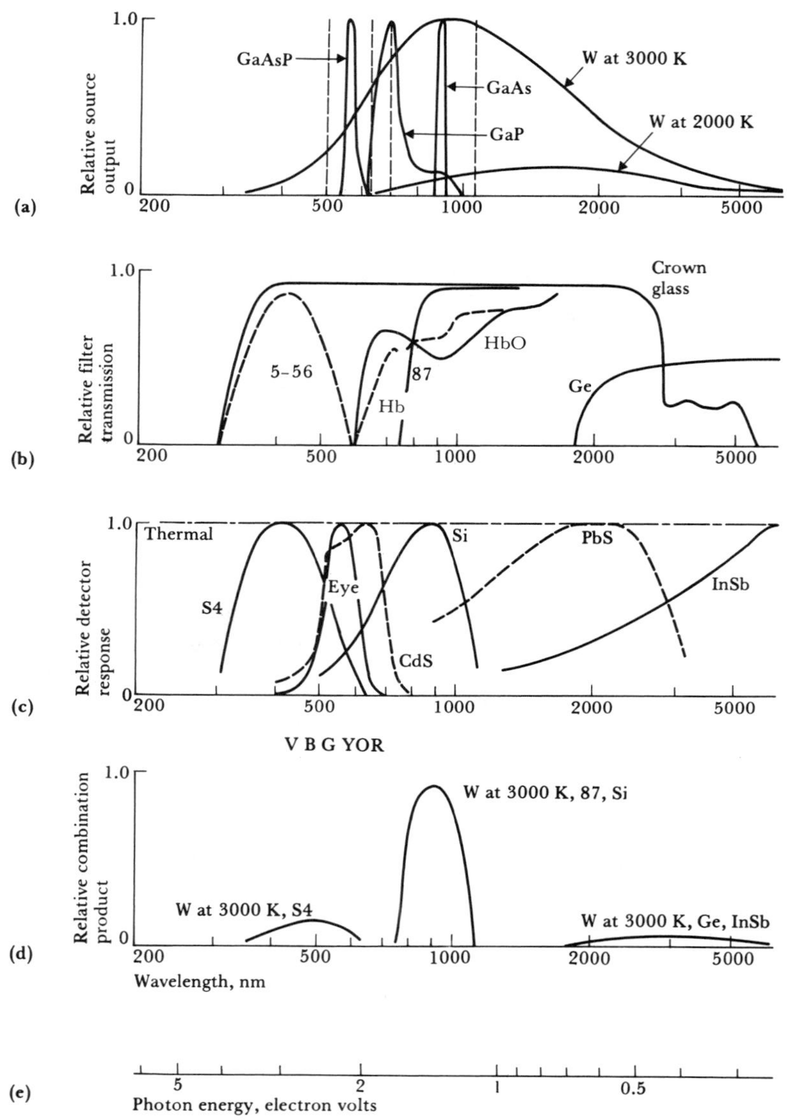

Figure 2.18 Spectral characteristics of sources, filters, detectors, and combinations thereof (a) Light sources. Tungsten (W) at 3000 K has a broad spectral output. At 2000 K, output is lower at all wavelengths and peak output shifts to longer wavelengths. Light-emitting diodes yield a narrow spectral output with GaAs in the infrared, GaP in the red, and GaAsP in the green. Monochromatic outputs from common lasers are shown by dashed lines: Ar, 515 nm; HeNe, 633 nm; ruby, 693 nm; Nd, 1064 nm; CO_2 (not shown), 10,600 nm. (b) Filters. A Corning 5-56 glass filter passes a blue wavelength band. A Kodak 87 gelatin filter passes infrared and blocks visible wavelengths. Germanium

Filaments are usually coiled to increase their emissivity and efficiency. For use in instruments, short linear coils may be arranged within a compact, nearly square area lying in a single plane. To produce a source of uniform radiant output over a substantial area, ribbon filaments may be used.

Tungsten-halogen lamps have iodine or bromine added to the gases normally used to fill the bulb. The small quartz bulbs operate at temperatures above 250 °C and usually require cooling by a blower. The halogen combines with tungsten at the wall. The resulting gas migrates back to the filament, where it decomposes and deposits tungsten on the filament. As a result, these lamps maintain more than 90% of their initial radiant output throughout their life. The radiant output of a conventional lamp, on the other hand, declines as much as 50% over its lifetime.

ARC DISCHARGES

The fluorescent lamp is filled with a low-pressure Ar–Hg mixture. Electrons are accelerated and collide with the gas atoms, which are raised to an excited level. As a given atom's electron undergoes a transition from a higher level to a lower level, the atom emits a quantum of energy.

Because the strongest transition of the mercury atom corresponds to about 5 eV, Figure 2.18(e) shows the resulting wavelength to be about 250 nm. A phosphor on the inside of the glass bulb absorbs this ultraviolet radiation and emits light of longer, visible wavelengths. The fluorescent lamp has low radiant output per unit area, so it is not used in optical instruments. However, it can be rapidly turned on and off in about 20 μs, so it is used in the *tachistoscope* (which presents brief stimuli to the eye) used in measurements of visual perception. Other low-pressure discharge lamps include the glowlamp (such as the neon lamp), the sodium-vapor lamp, and the laser.

High-pressure discharge lamps are more important for optical instruments because the arc is compact and the radiant output per unit area is high. The carbon arc has been in use for the longest time, but it has largely been replaced by the mercury lamp (bluish-green color), the sodium lamp (yellow color),

---

lenses pass long wavelengths that cannot be passed by glass. Hemoglobin Hb and oxyhemoglobin HbO pass equally at 805 nm and have maximal difference at 660 nm. (c) Detectors. The S4 response is a typical phototube response. The eye has a relatively narrow response, with colors indicated by VBGYOR. CdS plus a filter has a response that closely matches that of the eye. Si *p–n* junctions are widely used. PbS is a sensitive infrared detector. InSb is useful in far infrared. *Note:* These are only relative responses. Peak responses of different detectors differ by 10^7. (d) Combination. Indicated curves from (a), (b), and (c) are multiplied at each wavelength to yield (d), which shows how well source, filter, and detector are matched. (e) Photon energy: If it is less than 1 eV, it is too weak to cause current flow in Si *p–n* junctions.

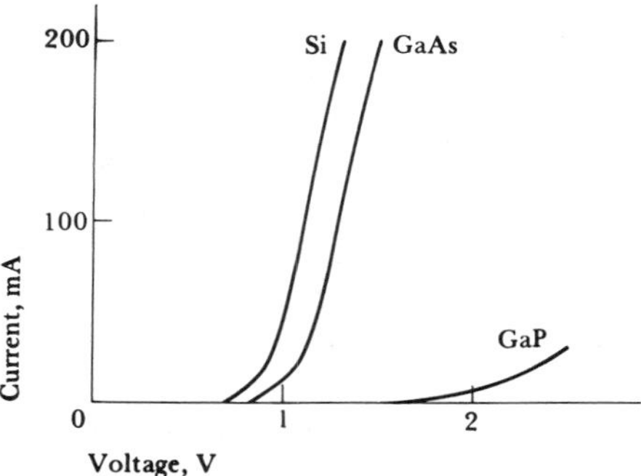

Figure 2.19 Forward characteristics for *p–n* junctions Ordinary silicon diodes have a band gap of 1.1 eV and are inefficient radiators in the near-infrared. GaAs has a band gap of 1.44 eV and radiates at 900 nm. GaP has a band gap of 2.26 eV and radiates at 700 nm.

and the xenon lamp (white color). These lamps usually have a clear quartz bulb with electrodes at both ends of the spherical bulb. The zirconium arc lamp provides an intense point source.

LIGHT-EMITTING DIODES (LEDs)

LEDs are *p–n* junction devices that are optimized for radiant output. The ordinary silicon *p–n* junction characteristic shown in Figure 2.19 emits radiant power when a current (typically 20 mA) passes in the forward direction. Spontaneous recombination of injected hole and electron pairs results in the emission of radiation. Because the silicon band gap is 1.1 eV, the wavelength is at about 1100 nm. The silicon device is not efficient. However, GaAs has a slightly higher band gap, as shown in Figure 2.19, and therefore radiates at 900 nm, as shown in Figure 2.18(a). Although the output is not visible, the efficiency is high and the GaAs device is widely used. It can be switched in less than 10 ns.

Figure 2.18(c) and Figure 2.18(e) show that, in order to produce visible light, the band gap of a *p–n* junction must exceed 1.9 eV. The GaP LED in Figure 2.19 has a band gap of 2.26 eV, requires a larger forward-bias voltage than silicon diodes, and is electroluminescent at 700 nm, as shown in Figure 2.18(a). It is an efficient visible LED and produces a bright red light. GaAsP LEDs make use of a special phosphor that absorbs two photons at one wavelength and emits a single photon at a shorter wavelength. GaAs is Si doped to emit radiation at 940 nm. Power at this wavelength is absorbed by the phosphor coating that emits green light at 540 nm, as shown in Figure 2.18(a). The decay time of the phosphor is about 1 ms.

LEDs are compact, rugged, economical, and nearly monochromatic. They are widely used in a variety of medical, transportation, and industrial circuits.

A variety of circuits are available for LEDs and photodetectors using either steady or modulated radiation.

LASERS

Laser (*L*ight *A*mplification by *S*timulated *E*mission of *R*adiation) action can occur in GaAs. The end faces that are perpendicular to the *p–n* junction are polished to serve as partial mirrors, thus forming a resonant optical cavity. The forward current pumps a large population of the molecules to an excited energy level. Radiation incident on the molecules causes the production of additional radiation that is identical in character. This phenomenon, known as *stimulated emisison,* is produced by the feedback from the mirrors. Laser output is highly monochromatic, collimated (parallel), and phase-coherent. However, *p-n* junction lasers are not widely used because they operate in the infrared and require current densities of 10^3 A/cm^2 or more, thus necessitating pulsed (10–100 ns) operation rather than CW (continuous wave).

The most common laser is the He–Ne laser that operates at 633 nm in the red region, as shown in Figure 2.18(a). The laser is operated by a low-pressure arc similar to a neon sign and provides up to 100 mW. Partially reflective mirrors at each end provide the resonant optical cavity and laser action.

Argon lasers provide the highest continuous-power levels (1–15 W) in the visible part of the spectrum at 515 nm [Figure 2.18(a)]. This high-power output permits photocoagulation of blood vessels in the eyes of patients suffering from diabetic retinopathy.

CO_2 lasers provide 50–500 W of CW output power and are used for cutting plastics, rubber, and metals up to 1 cm thick.

Two solid-state lasers—both usually operated in the pulsed mode—are widely used. The lasers are pumped by firing a flash tube that is wound around them. The ruby laser has a moderate (1-mJ) output in the red region of the spectum at 693 nm, as shown in Figure 2.18(a). The neodymium in yttrium aluminum garnet (Nd : YAG) laser has a high (2-W/mm) output in the infrared region at 1064 nm, as shown in Figure 2.18(a).

The most important medical use of the laser has been to mend tears in the retina. A typical photocoagulator uses a pulsed ruby laser with a controllable output. It is focused on a tear in the retina. The heat dissipated by the pulse forms a burn, which, on healing, develops scar tissue that mends the original tear. Section 13.10 provides further information on therapeutic applications of lasers.

Safety to the eye should be considered with respect to some light sources. It is safe to look at a 100-W frosted light bulb for long periods of time. However, looking at clear incandescent lamps, the sun, high-pressure arc sources, or lasers can cause burns on the retina. Protective eyewear worn by the physician to protect against lasers usually consists of a set of filters that attenuate at the specific wavelengths emitted by the laser but transmit as much visible radiation as possible.

2.14 GEOMETRICAL AND FIBER OPTICS

GEOMETRICAL OPTICS

There are a number of geometric factors that modify the power transmitted between the source and the detector. In Figure 2.17(b), the most obvious optical elements are the lenses. The lamp emits radiation in all directions. The first lens should have as small an *f number* (ratio of focal length to diameter) as practical. Thus it collects the largest practical solid angle of radiation from the lamp. The first lens is usually placed one focal length away from the lamp, so that the resulting radiation is *collimated* (that is, the rays are parallel). Thus, for a point source, the second lens can be placed at any distance without losing any radiation. Also, some interference filters operate best in collimated rays.

The second lens focuses the radiation on a small area of sample in the cuvette. Because the radiation now diverges, third and fourth lenses are used to collect all the radiation and focus it on a detector. Some spectrophotometers [Figure 2.17(c)] transmit collimated radiation through the sample section. The lenses can be coated with a coating that is a quarter-wavelength thick to prevent reflective losses at air–glass surfaces. Full mirrors may be used to fold the optical path to produce a compact instrument. Half-silvered mirrors enable users to split the beam into two beams for analysis or to combine two beams for analysis by a single detector. Curved mirrors may function as lenses for wavelengths that are absorbed by normal glass lenses.

Scattered radiation must be prevented from reaching the detector. Internal support structures and mechanical components of optical instruments are internally painted flat black to prevent scattered radiation. Stops (apertures that pass only the desired beam size) may be placed at several locations along the instrument's optical axis to trap scattered radiation.

FIBER OPTICS

Fiber optics are an efficient way of transmitting radiation from one point to another (Epstein, 1988). Transparent glass or plastic fiber with a refractive index n_1 is coated or surrounded by a second material of a lower refractive index n_2. By Snell's law,

$$n_2 \sin \theta_2 = n_1 \sin \theta_1 \tag{2.28}$$

where θ is the angle of incidence shown in Figure 2.20. Because $n_1 > n_2$, sin $\theta_2 >$ sin θ_1, so sin $\theta_2 = 1.0$ for a value of θ_1 that is less than 90°. For values of θ_1 greater than this, sin θ_2 is greater than unity, which is impossible, and the ray is internally reflected. The critical angle for reflection (θ_{ic}) is found by setting sin $\theta_2 = 1.0$, which gives

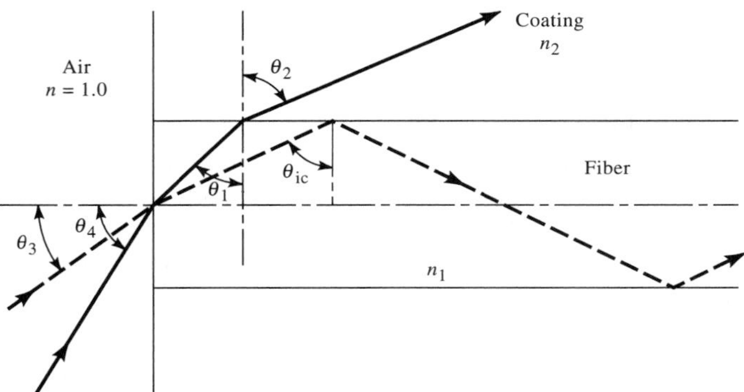

Figure 2.20 Fiber optics The solid line shows refraction of rays that escape through the wall of the fiber. The dashed line shows total internal reflection within a fiber.

$$\sin \theta_{ic} = \frac{n_2}{n_1} \tag{2.29}$$

A ray is internally reflected for all angles of incidence greater than θ_{ic}. Because rays entering the end of a fiber are usually refracted from air ($n = 1.0$) into glass ($n = 1.62$ for one type), a larger cone of radiation (θ_3) is accepted by a fiber than that indicated by calculations using $90° - \theta_{ic}$. Rays entering the end of the fiber at larger angles (θ_4) are not transmitted down the fiber; they escape through the walls.

Fiber-optic (FO) sensors are replacing some conventional sensors for measuring a variety of electrical, electronic, mechanical, pneumatic, and hydraulic variables (Sirohi and Kothiyal, 1991; Udd, 1991). They are chemically inert and have freedom from electromagnetic interference.

A 50-cm glass fiber exhibits a transmission exceeding 60% for wavelengths between 400 and 1200 nm. A 50-cm plastic fiber has a transmission exceeding 70% for wavelengths between 500 and 850 nm. Although a single fiber is useful for sampling incident radiation of a small area, most applications use flexible bundles of about 400 fibers. In *noncoherent bundles* (called *light guides*), the diameter of a fiber is typically 13 to 100 nm. There is no correlation between a fiber's spatial position at the input and at the output. These fibers are useful only for transmitting radiation. In one application, light is transmitted through the flexible bundle for viewing internal organs. In a second application, an instrument that measures blood oxygen saturation within the vessels alternately transmits radiation at two wavelengths down one bundle (Chapter 10). The radiation is backscattered by the red-blood cells and returned to the instrument for analysis through a second bundle.

In *coherent-fiber bundles,* the fibers occupy the same relative position at both end faces. An image at one end is faithfully transmitted to the other end. The most important medical application of these fibers is in the *endoscope* (a tube for examining body cavities through natural openings) (Sievert and Potter, 1988). A typical endoscope is 1 m long and 1 cm in diameter and may be used for viewing the lining of the stomach, intestines, and so forth. A noncoherent bundle transmits light for illumination. A small lens focuses the image of the lining onto the end of a coherent bundle, which transmits the image in such a way that it may be viewed or photographed. External levers make it possible to steer the internal end of the optical-fiber device over a 360° range so that the examining physician can look at cavity walls and around corners.

LIQUID CRYSTALS

Liquid crystals change their state in such a way that they modify passive scattering or absorption of light. As the crystals melt, the three-dimensional order becomes a two-dimensional or one-dimensional order. Layers or strands form that can be seen as a clarification of the previously turbid melt.

In one medical application, the patient's body is painted with a black water-soluble varnish to show up the color of the liquid crystals better. Liquid crystals are painted over the varnish, and any inflammation causes a rise in temperature that is indicated by a color pattern. Liquid crystals are also used, in disposable thermometers, in the measurement of oral temperatures. They are widely used in wristwatches, because a low-voltage (1–15 V), low-power (1 $\mu W/cm^2$) electric field causes observable changes in digital-display elements.

2.15 OPTICAL FILTERS

FILTERS

Filters are frequently inserted in the optical system to control the distribution of radiant power or wavelength. To reduce radiant power only, neutral-density filters are used. When glass is partially silvered, most of the power is reflected and the desired fraction of the power is transmitted. When carbon particles are suspended in plastic, most of the power is absorbed and the desired fraction of the power is transmitted. Two Polaroid filters may also be used to attenuate the light. Each filter transmits only that portion of the light that is in a particular state of polarization. As one is rotated with respect to the other, the optical transmission of the combination varies.

Color filters transmit certain wavelengths and reject others. Gelatin filters are the most common type of absorption filters. An organic dye is dissolved in an aqueous gelatin solution, and a thin film is dried on a glass substrate. An example shown in Figure 2.18(b) is the infrared Kodak 87 Wratten filter. Glass filters, made by combining additives with the glass itself in its molten state, are extensively used. They provide rather broad passbands, as illustrated by the blue Corning 5-56 filter shown in Figure 2.18(b).

Interference filters are formed by depositing a reflective stack of layers on both sides of a thicker spacer layer. This sandwich construction provides multiple reflection and interference effects that yield sharp-edge high, low, and bandpass filters with bandwidths from 0.5 to 200 nm. Interference filters are generally used with collimated radiation and cost more than those just mentioned. Interference coatings are used on dichroic mirrors (cold mirrors), which reflect visible radiation from projection lamps. The nonuseful infrared radiation is transmitted through the coating and mirror to the outside of the optical system. This reduces heat within the optical system without sacrificing the useful light.

Diffraction gratings are widely employed to produce a wavelength spectrum in the spectrometer. Plane gratings are formed by cutting thousands of closely spaced parallel grooves in a material. The grating is overcoated with aluminum, which reflects and disperses white light into a diffraction spectrum. A narrow slit selects a narrow band of wavelengths for use.

Although clear glass is not ordinarily thought of as a filter, Figure 2.18(b) shows that crown glass does not transmit below 300 nm. For instruments that operate in the ultraviolet, fused quartz (silica glass) is used. Most glasses do not transmit well above 2600 nm, so instrument makers use either curved mirrors for infrared instruments or lenses made of Ge, Si, AsS_3, CaF_2, or Al_2O_3.

Conway *et al.* (1984) describe an optical method for measuring the percentage of fat in the body. They found that fat has an absorption band at 930 nm and that water has an absorption band at 970 nm. From a single-beam rapid-scanning spectrophotometer, they conducted light to and reflected light from five sites on the body through a fiber-optic probe. The method successfully predicted percent body fat in 17 subjects ($r = 0.91$) when compared with the D_2O dilution technique.

2.16 RADIATION SENSORS

Radiation sensors may be classified into two general categories: thermal sensors and quantum sensors (Peterson, 1988).

THERMAL SENSORS

The *thermal sensor* absorbs radiation and transforms it into heat, thus causing a rise in temperature in the sensors. Typical thermal sensors are the thermistor and the thermocouple. The sensitivity of such a sensor does not change with (is flat with) wavelength, and the sensor has slow response [Figure 2.18(c)]. Changes in output due to changes in ambient temperature cannot be distinguished from changes in output due to the source, so a windmill-shaped mechanical chopper is frequently used to interrupt the radiation from the source periodically.

The *pyroelectric sensor* (Fraden, 1997) absorbs radiation and converts it into heat. The resulting rise in temperature changes the polarization of the crystals, which produces a current proportional to the rate of change of temperature. As it is for the piezoelectric sensor, dc response is zero, so a chopper is required for dc measurements.

QUANTUM SENSORS

Quantum sensors absorb energy from individual photons and use it to release electrons from the sensor material. Typical quantum sensors are the eye, the phototube, the photodiode, and photographic emulsion. Such sensors are sensitive over only a restricted band of wavelengths; most respond rapidly. Changes in ambient temperature cause only a second-order change in sensitivity of these sensors.

PHOTOEMISSIVE SENSORS

Photoemissive sensors—an example is the *phototube*—have photocathodes coated with alkali metals. If the energy of the photons of the incoming radiation is sufficient to overcome the work function of the photocathode, the forces that bind electrons to the photocathode are overcome, and it emits electrons. Electrons are attracted to a more positive anode and form a current that is measured by an external circuit. Photon energies below 1 eV are not large enough to overcome the work function, so wavelengths longer than 1200 nm cannot be detected. Figure 2.18(c) shows the spectral response of the most common photocathode, the S4, which has lower sensitivity in the ultraviolet region because of absorption of radiation in the glass envelope.

The *photomultiplier* shown in Figure 2.21 is a phototube combined with an electron multiplier (Lion, 1975). Each accelerated electron hits the first dynode with enough energy to liberate several electrons by secondary emission. These electrons are accelerated to the second dynode, where the process is repeated, and so on. Time response is less than 10 ns. Photomultipliers are the most sensitive photodetectors. When they are cooled (to prevent electrons from being thermally generated), they can count individual photons. The eye is almost as sensitive; under the most favorable conditions, it can detect six photons arriving in a small area within 100 ms.

PHOTOCONDUCTIVE CELLS

Photoresistors are the simplest solid-state photoelectric sensors. A photosensitive crystalline material such as CdS or PbS [Figure 2.18(c)] is deposited on a ceramic substrate, and ohmic electrodes are attached. If a photon of the incoming radiation has sufficient energy to jump the band gap, hole–electron pairs are produced because the electron is raised from the valence band to

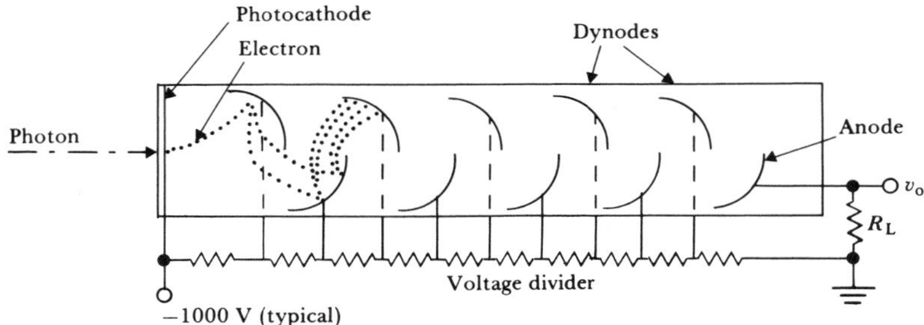

Figure 2.21 Photomultiplier An incoming photon strikes the photocathode and liberates an electron. This electron is accelerated toward the first dynode, which is 100 V more positive than the cathode. The impact liberates several electrons by secondary emission. They are accelerated toward the second dynode, which is 100 V more positive than the first dynode. This electron multiplication continues until it reaches the anode, where currents of about 1 μA flow through R_L.

the conduction band. The presence of the electrons in the conduction band and the presence of the holes in the valence band increases the conductivity of the crystalline material so that the resistance decreases with input radiation. Photocurrent is linear at low levels of radiation but nonlinear at levels that are normally used. It is independent of the polarity of applied voltage. After a step increase or decrease of radiation, the photocurrent response rises and decays with a time constant of from 10 to 0.01 s, depending on whether the radiation is low or high.

PHOTOJUNCTION SENSORS

Photojunction sensors are formed from *p–n* junctions and are usually made of silicon. If a photon has sufficient energy to jump the band gap, hole-electron pairs are produced that modify the junction characteristics, as shown in Figure 2.22. If the junction is reverse-biased, the reverse photocurrent flowing from the cathode to the anode increases linearly with an increase in radiation. The resulting photodiode responds in about 1 μs. In phototransistors, the base lead is not connected, and the resulting radiation-generated base current is multiplied by the current gain (beta) of the transistor to yield a large current from collector to emitter. The radiation-current characteristics have a nonlinearity of about 2% because beta varies with collector current. The response time is about 10 μs. Silicon *p–n* junctions are also manufactured as photo Darlington transistors, photo FETs, photo unijunction transistors, and photo SCRs. Photon couplers are LED–photodiode combinations that are used for isolating electric circuits. For example, they are used for breaking ground loops and for preventing dangerous levels of current from leaking out of equipment and entering the heart of a patient (Section 14.9).

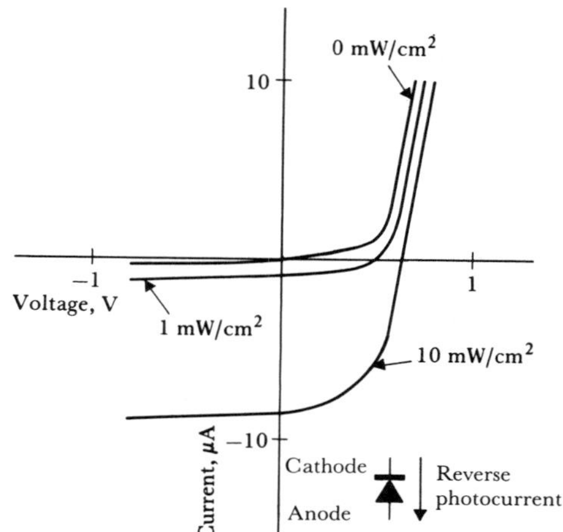

Figure 2.22 Voltage-current characteristics of irradiated silicon *p–n* junction For 0 irradiance, both foward and reverse characteristics are normal. For 1 mW/cm^2, open-circuit voltage is 500 mV and short-circuit current is 0.8 μA. For 10 mW/cm^2, open-circuit voltage is 600 mV and short-circuit current is 8 μA.

PHOTOVOLTAIC SENSORS

The same silicon *p–n* junction can be used in the photovoltaic mode. Figure 2.22 shows that there is an open-circuit voltage when the junction receives radiation. The voltage rises logarithmically from 100 to 500 mV as the input radiation increases by a factor of 10^4. This is the principle of the solar cell that is used for direct conversion of the sun's radiation into electric power.

SPECTRAL RESPONSE

All of the aforementioned silicon sensors have the spectral response shown in Figure 2.18(c). There is no response above 1100 nm because the energy of the photons is too low to permit them to jump the band gap. For wavelengths shorter than 900 nm, the response drops off because there are fewer, more-energetic photons per watt. Each photon generates only one hole–electron pair.

Because none of the common sensors is capable of measuring the radiation emitted by the skin (300 K) which has a peak output at 9000 nm, special sensors have been developed, such as the InSb sensor shown in Figure 2.18(c).

2.17 OPTICAL COMBINATIONS

In order to specify the combinations of sources, filters, and sensors, instrument designers require radiometric units that must be weighted according to the response curve of each element. E_e, the total effective irradiance, is found by breaking up the spectral curves into many narrow bands and then multiplying each together and adding the resulting increments (Stimson, 1974). Thus

$$E_e = \Sigma S_\lambda F_\lambda D_\lambda \Delta\lambda \tag{2.30}$$

where

S_λ = relative source output
F_λ = relative filter transmission
D_λ = relative sensor responsivity

Figure 2.18(d) shows several results of this type of calculation. One of the examples shown is an efficient system capable of making measurements in the dark without stimulating the eye. Such a device can be used for tracking eye movements. It can be formed from a tungsten source, a Kodak 87 Wratten filter, and a silicon sensor. If GaAs provides enough output, it can replace both the tungsten source and the Kodak 87 Wratten filter (Young and Sheena, 1988).

PROBLEMS

2.1 For Figure P2.1, plot the ratio of the output voltage to the input voltage v_o/v_i as a function of the displacement x_i of a potentiometer with a total displacement x_t for ranges of R_m, the input resistance of the meter. Show that the maximal error occurs in the neighborhood of $x_i/x_t = 0.67$. What is the value of this maximal error?

2.2 The practical limitation for wire spacing in potentiometer construction is between 20 and 40 turns/mm. Find the resolution limitation for a translational and a rotational potentiometer. Propose a way to increase the resolution of a rotational potentiometer.

2.3 Discuss some of the possible problems involved in elastic-resistance strain-gage sensors and their solutions.

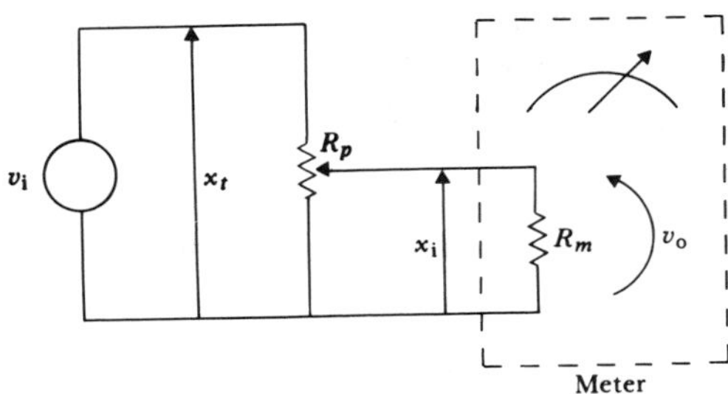

Figure P2.1

2.4 The electromotive force E for a thermocouple is given by (2.21). Calculate and plot E for conditions in which the reference source is at 0 °C and temperature varies from 0 °C to 50 °C. The thermocouple material is copper constantan with $a = 38.7\ \mu\text{V/°C}$ and $b = 0.082\ \mu\text{V/°C}^2$. How significant is the second term in your calculations? Note that these calculated curves are not exactly satisifed in the practical situation. Thus an experimental calibration must be measured over the range of interest.
2.5 Using the results of Problem 2.4, calculate the sensitivity α at 37 °C for the copper-constantan thermocouple.
2.6 Calculate the value of the thermistor temperature coefficient α for $T =$ 300 K and $\beta = 4000$ K.
2.7 For the LVDT shown in Figure 2.6(c), sketch the voltages c–e, d–e, and c–d as the core is displaced through its normal range.
2.8 For a 1-cm^2 capacitance sensor, R is 100 MΩ. What is x, the plate spacing required to pass sound frequencies above 20 Hz?
2.9 For Example 2.1, what size shunting capacitor should be added to extend the low-corner frequency to 0.05 Hz, as is required to detect pulse waveforms? How is the sensitivity changed?
2.10 For a piezoelectric sensor plus cable that has 1-nF capacitance, design a *voltage amplifier* (not a charge amplifier) by using only *one* noninverting amplifier that has a gain of 10. It should handle a charge of 1 μC generated by the carotid pulse without saturation. It should not drift into saturation because of bias currents. It should have a frequency response from 0.05 to 100 Hz. Add the minimal number of extra components to achieve the design specifications.
2.11 Design a charge amplifier for a piezoelectric sensor that has 500-pF capacitance. It should pass frequencies from 0.05 to 100 Hz so that it can detect carotid pulses, and it should not drift into saturation.
2.12 Sketch typical thermistor v–i characteristics with and without a heat sink. Explain why there is a difference.
2.13 Calculate and sketch a curve for the radiant output of the skin at 300 K at 2000, 5000, 10,000, and 20,000 nm.
2.14 Calculate the photon energy for the peak wavelength given off by the skin at 300 K.
2.15 Sketch an optical system using curved mirrors instead of lenses that could replace the system shown in Figure 2.17(b).
2.16 Sketch the circuit for a photo Darlington transistor, which is two cascaded emitter-follower transistors. Estimate its linearity and response time.
2.17 For the solar cell shown in Figure 2.22 what value load resistor would receive the maximal power?
2.18 For the photomultiplier shown in Figure 2.21, when R_L is high enough for adequate sensitivity, the stray capacity–R_L product produces a time constant that is too long. Design a circuit that is 10 times faster and has no loss in sensitivity.
2.19 For Figure 2.18(d), plot the relative combination product for GaP, HbO, CdS.

REFERENCES

Alihanka, J., K. Vaahtoranta, and S.-E Björkqvist, *Apparatus in medicine for the monitoring and/or recording of the body movements of a person on a bed, for instance of a patient*, United States Patent 4,320,766, 1982.

Anonymous, *Manual on the Use of Thermocouples in Temperature Measurement.* Publication 470A. Philadelphia: American Society for Testing and Materials, 1974.

Bowman, L., and J. D. Meindl, "Capacitive sensors," in J. G. Webster (ed.), *Encyclopedia of Medical Devices and Instrumentation.* New York: Wiley, 1988, pp. 551–556.

Carr, K. L., "Thermography," in J. G. Webster (ed.), *Encyclopedia of Medical Devices and Instrumentation.* New York: Wiley, 1988, pp. 2746–2759.

Christensen, D. A., "Thermometry," in J. G. Webster (ed.), *Encyclopedia of Medical Devices and Instrumentation.* New York: Wiley, 1988, pp. 2759–2765.

Cobbold, R. S. C., *Transducers for Biomedical Measurements.* New York: Wiley, 1974.

Conway, J. M., K. H. Norris, and C. E. Bodwell, "A new approach for the estimation of body composition: Infrared interactance." *Am. J. Clin. Nutr.,* 1984, 40, 1123–1130.

Coremans, J., "Thermistors," in J. G. Webster (ed.), *Encyclopedia of Medical Devices and Instrumentation.* New York: Wiley, 1988, pp. 2730–2738.

Dechow, P. C., "Strain gages," in J. G. Webster (ed.), *Encyclopedia of Medical Devices and Instrumentation.* New York: Wiley, 1988, pp. 2715–2721.

Doebelin, E. O., *Measurement Systems: Application and Design,* 4th ed. New York: McGraw-Hill, 1990.

Epstein, M., "Fiber optics in medicine," in J. G. Webster (ed.), *Encyclopedia of Medical Devices and Instrumentation.* New York: Wiley, 1988, pp. 1284–1302.

Fraden, J., "Noncontact temperature measurements in medicine," in D. L. Wise (ed.), *Bioinstrumentation and Biosensors.* New York, Dekker, 1991, pp. 511–550.

Fraden, J., *Handbook of Modern Sensors: Physics, Designs, and Applications.* 2nd ed. Woodbury, NY: American Institute of Physics, 1997.

Geddes, L. A., and L. E. Baker, *Principles of Applied Biomedical Instrumentation,* 3rd ed. New York: Wiley, 1989.

Hennig, E. M., "Piezoelectric sensors," in J. G. Webster (ed.), *Encyclopedia of Medical Devices and Instrumentation.* New York: Wiley, 1988, pp. 2310–2319.

Hokanson, D. E., D. S. Sumner, and D. E. Strandness, Jr., "An electrically calibrated plethysmograph for direct measurement of limb blood flow." *IEEE Trans. Biomed. Eng.,* 1975, BME-22, 25–29.

Korites, B. J., *Microsensors.* Rockland, MA: Kern International, 1987.

Lawton, R. W., and C. C. Collins, "Calibration of an aortic circumference gauge." *J. Appl. Physiol.,* 1959, 14, 465–467.

Lion, K. S., *Elements of Electronic and Electrical Instrumentation.* New York: McGraw-Hill, 1975.

Lion, K. S., *Instrumentation in Scientific Research.* New York: McGraw-Hill, 1959.

Pallás-Areny, R., and J. G. Webster, *Sensors and Signal Conditioning.* New York: Wiley, 1991.

Patel, A., M. Kothari, J. G. Webster, W. J. Tompkins, and J. J. Wertsch, "A capacitance pressure sensor using a phase-locked loop." *J. Rehabil. Res. Devel.,* 1989, 26(2), 55–62.

Peterson, J. I., "Optical sensors," in J. G. Webster (ed.), *Encyclopedia of Medical Devices and Instrumentation.* New York: Wiley, 1988, pp. 2121–2134.

Re. T. J., and M. R. Neuman, "Thermal contact-sensing electronic thermometer." *Biomed. Instrum. Technol.,* 1991, 25, 54–59.

Reddy, N. P., and S. K. Kesavan, "Linear variable differential transformers," in J. G. Webster (ed.), *Encyclopedia of Medical Devices and Instrumentation.* New York: Wiley, 1988, pp. 1800–1806.

Sapoff, M., *Thermistors for Biomedical Use.* Fifth Symposium on Temperature, Proceedings. Washington, DC: Instrument Society of America, 1971, 2109–2121.

Servais, S. B., J. G. Webster and H. J. Monotoye, "Estimating human energy expenditure using an accelerometer device." *J. Clin. Eng.*, 1984, 9, 159–171.

Sheingold, D. H. (ed.), *Transducer Interfacing Handbook.* Norwood, MA: Analog Devices, Inc., 1980.

Sievert, C., Jr., and T. J. Potter, "Endoscopes," in J. G. Webster (ed.), *Encyclopedia of Medical Devices and Instrumentation.* New York: Wiley, 1988, pp. 1203–1211.

Sirohi, R. S., and M. P. Kothiyal, *Optical Components, Systems, and Measurement Techniques.* New York: Marcel Dekker, 1991.

Stimson, A., *Photometry and Radiometry for Engineers.* New York: Wiley, 1974.

Sydenham, P. H. (ed.), *Handbook of Measurement Science. Vol. 1: Theoretical Fundamentals; Vol. 2: Practical Fundamentals.* New York; Wiley, 1982 (vol. 1), 1983 (vol. 2).

Togawa, T., "Thermometry, diode," in J. G. Webster (ed.), *Encyclopedia of Medical Devices and Instrumentation.* New York: Wiley, 1988, pp. 2765–2770.

Tompkins, W. J., and J. G. Webster (eds.), *Interfacing Sensors to the IBM PC.* Englewood Cliffs, NJ: Prentice-Hall, 1988.

Udd, E. (ed.), *Fiber Optic Sensors: An Introduction for Engineers and Scientists.* New York: Wiley, 1991.

van Citters, R. L., "Mutual inductance transducers," in R. F. Rushmer (ed.), *Methods in Medical Research,* Vol. XI. Chicago: Year Book, 1966, 26–30.

Vaughan, M. S., "Temperature measurement in the clinical setting," in J. G. Webster (ed.), *Encyclopedia of Medical Devices and Instrumentation.* New York: Wiley, 1988, pp. 2723–2730.

Webster, J. G. (ed.), *Tactile Sensors for Robotics and Medicine.* New York: Wiley, 1988.

Webster, J. G., *Transducers and Sensors,* An IEEE/EAB Individual Learning Program. Piscataway, NJ: IEEE, 1989.

Welch, A. J., and J. A. Pearce, "Thermocouples," in J. G. Webster (ed.), *Encyclopedia of Medical Devices and Instrumentation.* New York: Wiley, 1988, pp. 2739–2746.

Young, L. R., and D. Sheena, "Eye-movement measurement techniques," in J. G. Webster (ed.), *Encyclopedia of Medical Devices and Instrumentation.* New York: Wiley, 1988, pp. 1259–1269.

3

AMPLIFIERS AND SIGNAL PROCESSING

John G. Webster

Most bioelectric signals are small and require amplification. Amplifiers are also used for interfacing sensors that sense body motions, temperature, and chemical concentrations. In addition to simple amplification, the amplifier may also modify the signal to produce frequency filtering or nonlinear effects. This chapter emphasizes the *operational amplifier* (op amp), which has revolutionized electronic circuit design. Most circuit design was formerly performed with discrete components, requiring laborious calculations, many components, and large expense. Now a 30-cent op amp, a few resistors, and a knowledge of Ohm's law are all that is needed.

3.1 IDEAL OP AMPS

An *op amp* is a high-gain dc differential amplifier. It is normally used in circuits that have characteristics determined by external negative-feedback networks.

The best way to approach the design of a circuit that uses op amps is first to assume that the op amp is ideal. After the initial design, the circuit is checked to determine whether the nonideal characteristics of the op amp are important. If they are not, the design is complete; if they are, another design check is made, which may require additional components.

IDEAL CHARACTERISTICS

Figure 3.1 shows the equivalent circuit for a nonideal op amp. It is a dc differential amplifier, which means that any differential voltage, $v_d = (v_2 - v_1)$, is multiplied by the very high gain A to produce the output voltage v_o.

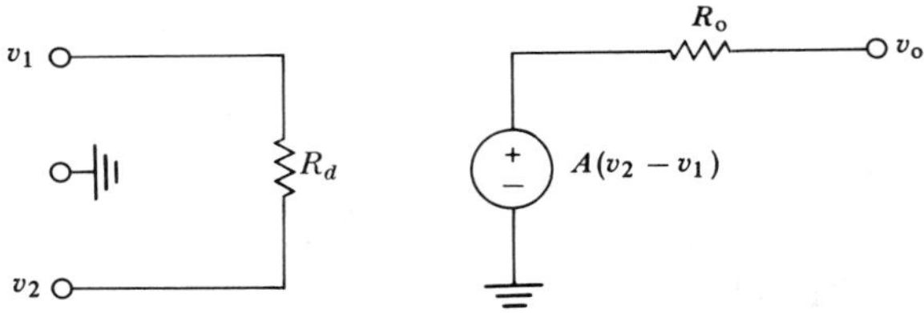

Figure 3.1 Op-amp equivalent circuit The two inputs are v_1 and v_2. A differential voltage between them causes current flow through the differential resistance R_d. The differential voltage is multiplied by A, the gain of the op amp, to generate the output-voltage source. Any current flowing to the output terminal v_o must pass through the output resistance R_o.

To simplify calculations, we assume the following characteristics for an ideal op amp:

1. $A = \infty$ (gain is infinity)
2. $v_o = 0$, when $v_1 = v_2$ (no offset voltage)
3. $R_d = \infty$ (input impedance is infinity)
4. $R_o = 0$ (output impedance is zero)
5. Bandwidth $= \infty$ (no frequency-response limitations) and no phase shift

Later in the chapter we shall examine the effect on the circuit of characteristics that are not ideal.

Figure 3.2 shows the op-amp circuit symbol, which includes two differential input terminals and one output terminal. All these voltages are measured with respect to the ground shown. Power supplies, usually ± 15 V, must be connected to terminals indicated on the manufacturer's specification sheet (Jung, 1986; Horowitz and Hill, 1989).

TWO BASIC RULES

Throughout this chapter we shall use two basic rules (or input terminal restrictions) that are very helpful in designing op-amp circuits.

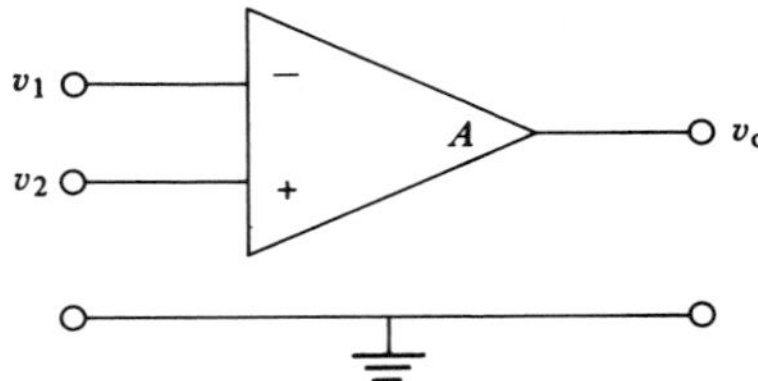

Figure 3.2 Op-amp circuit symbol A voltage at v_1, the inverting input, is greatly amplified and inverted to yield v_o. A voltage at v_2, the noninverting input, is greatly amplified to yield an in-phase output at v_o.

RULE 1 *When the op-amp output is in its linear range, the two input terminals are at the same voltage.*

This is true because if the two input terminals were not at the same voltage, the differential input voltage would be multiplied by the infinite gain to yield an infinite output voltage. This is absurd; most op amps use a power supply of ± 15 V, so v_o is restricted to this range. Actually the op-amp specifications guarantee a linear output range of only ± 10 V, although some saturate at about ± 13 V. A single supply is adequate with some op amps, such as the LM358 (Horowitz and Hill, 1989).

RULE 2 *No current flows into either input terminal of the op amp.*

This is true because we assume that the input impedance is infinity, and no current flows into an infinite impedance. Even if the input impedance were finite, Rule 1 tells us that there is no voltage drop across R_d, so therefore no current flows.

3.2 INVERTING AMPLIFIERS

CIRCUIT

Figure 3.3(a) shows the basic inverting-amplifier circuit. It is widely used in instrumentation. Note that a portion of v_o is fed back via R_f to the negative input of the op amp. This provides the inverting amplifier with the many advantages associated with the use of negative feedback—increased bandwidth, lower output impedance, and so forth. If v_o is ever fed back to the positive input of the op amp, examine the circuit carefully. Either there is a mistake, or the circuit is one of the rare ones in which a regenerative action is desired.

EQUATION

Note that the positive input of the op amp is at 0 V. Therefore, by Rule 1, the negative input of the op amp is also at 0 V. Thus no matter what happens to the rest of the circuit, the negative input of the op amp remains at 0 V, a condition known as a *virtual ground.*

Because the right side of R_i is at 0 V and the left side is v_i, by Ohm's law the current i through R_i is $i = v_i/R_i$. By Rule 2, no current can enter the op amp; therefore i must also flow through R_f. This produces a voltage drop across R_f of iR_f. Because the left end of R_f is at 0 V, the right end must be

$$v_o = -iR_f = -v_i \frac{R_f}{R_i} \qquad \text{or} \qquad \frac{v_o}{v_i} = \frac{-R_f}{R_i} \tag{3.1}$$

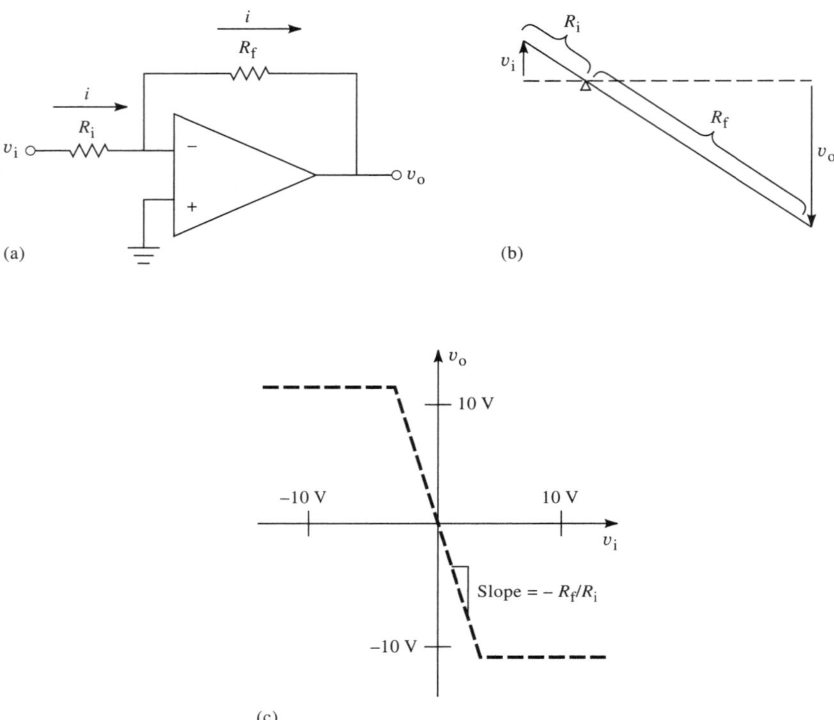

Figure 3.3 (a) An inverting amplifier. Current flowing through the input resistor R_i also flows through the feedback resistor R_f. (b) A lever with arm lengths proportional to resistance values enables the viewer to visualize the input–output characteristics easily. (c) The input–output plot shows a slope of $-R_f/R_i$ in the central portion, but the output saturates at about ±13 V.

Thus the circuit inverts, and the *inverting-amplifier* gain (not the op-amp gain) is given by the ratio of R_f to R_i.

LEVER ANALOGY

Figure 3.3(b) shows an easy way to visualize the circuit's behavior. A lever is formed with arm lengths proportional to resistance values. Because the negative input is at 0 V, the fulcrum is placed at 0 V, as shown. If R_f is three times R_i, as shown, any variation of v_i results in a three-times-bigger variation of v_o. The circuit in Figure 3.3(a) is a voltage-controlled-current-source (VCCS) for any load R_f (Jung, 1986). The current i through R_f is v_i/R_i so v_i controls i. Current sources are useful in electrical impedance plethysmography for passing a fixed current through the body (Section 8.7).

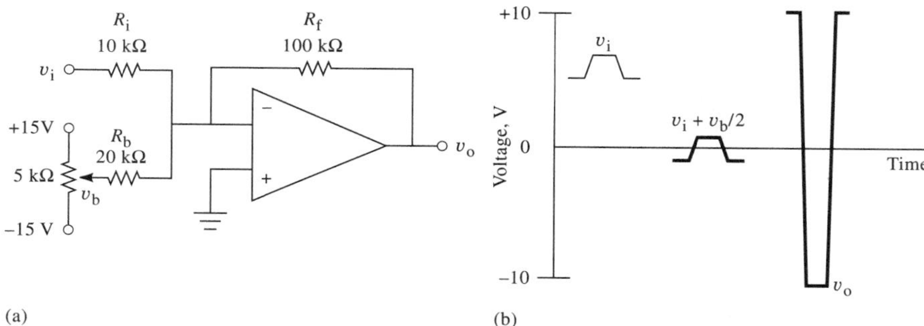

Figure E3.1 (a) This circuit sums the input voltage v_i plus one-half of the balancing voltage v_b. Thus the output voltage v_o can be set to zero even when v_i has a nonzero dc component. (b) The three waveforms show v_i, the input voltage; $(v_i + v_b/2)$, the balanced-out voltage; and v_o, the amplified output voltage. If v_i were directly amplified, the op amp would saturate.

INPUT–OUTPUT CHARACTERISTIC

Figure 3.3(c) shows that the circuit is linear only over a limited range of v_i. When v_o exceeds about ±13 V, it *saturates* (limits), and further increases in v_i produce no change in the output. The linear swing of v_o is about 4 V less than the difference in power-supply voltages. Although op amps usually have power-supply voltages set at ±15 V, reduced power-supply voltages may be used, with a corresponding reduction in the saturation voltages and the linear swing of v_o.

SUMMING AMPLIFIER

The inverting amplifier may be extended to form a circuit that yields the weighted sum of several input voltages. Each input voltage v_{i1}, v_{i2}, . . . , v_{ik} is connected to the negative input of the op amp by an individual resistor the conductance of which $(1/R_{ik})$ is proportional to the desired weighting.

EXAMPLE 3.1 The output of a biopotential preamplifier that measures the electro-oculogram (Section 4.7) is an undesired dc voltage of ±5 V due to electrode half-cell potentials (Section 5.1), with a desired signal of ±1 V superimposed. Design a circuit that will balance the dc voltage to zero and provide a gain of −10 for the desired signal without saturating the op amp.

ANSWER Figure E3.1(a) shows the design. We assume that v_b, the balancing voltage available from the 5-kΩ potentiometer, is ±10 V. The undesired voltage at $v_i = 5$ V. For $v_o = 0$, the current through R_f is zero. Therefore the sum of the currents through R_i and R_b is zero.

$$\frac{v_i}{R_i} + \frac{v_b}{R_b} = 0$$

$$R_b = \frac{-R_i v_b}{v_i} = \frac{-10^4(-10)}{5} = 2 \times 10^4\ \Omega$$

For a gain of -10, (3.1) requires $R_f/R_i = 10$, or $R_f = 100\ \text{k}\Omega$. The circuit equation is

$$v_o = -R_f\left(\frac{v_i}{R_i} + \frac{v_b}{R_b}\right)$$

$$v_o = -10^5\left(\frac{v_i}{10^4} + \frac{v_b}{2 \times 10^4}\right)$$

$$v_o = -10\left(v_i + \frac{v_b}{2}\right)$$

The potentiometer can balance out any undesired voltage in the range ± 5 V, as shown by Figure E3.1(b).

3.3 NONINVERTING AMPLIFIERS

FOLLOWER

Figure 3.4(a) shows the circuit for a unity-gain follower. Because v_i exists at the positive input of the op amp, by Rule 1 v_i must also exist at the negative input. But v_o is also connected to the negative input. Therefore $v_o = v_i$, or the output voltage follows the input voltage. At first glance it seems nothing is gained by using this circuit; the output is the same as the input. However, the circuit is very useful as a *buffer,* to prevent a high source resistance from being loaded down by a low-resistance load. By Rule 2, no current flows into the positive input, and therefore the source resistance in the external circuit is not loaded at all.

NONINVERTING AMPLIFIER

Figure 3.4(b) shows how the follower circuit can be modified to produce gain. By Rule 1, v_i appears at the negative input of the op amp. This causes current $i = v_i/R_i$ to flow to ground. By Rule 2, none of i can come from the negative input; therefore all must flow through R_f. We can then calculate $v_o = i(R_f + R_i)$ and solve for the gain.

$$\frac{v_o}{v_i} = \frac{i(R_f + R_i)}{iR_i} = \frac{R_f + R_i}{R_i} \tag{3.2}$$

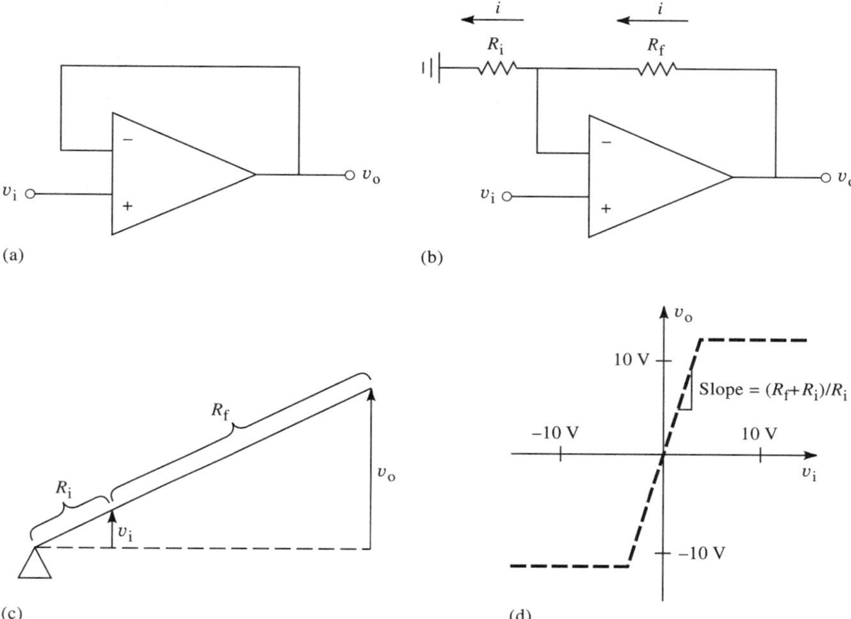

Figure 3.4 (a) A follower, $v_o = v_i$. (b) A noninverting amplifier, v_i appears across R_i, producing a current through R_i that also flows through R_f. (c) A lever with arm lengths proportional to resistance values makes possible an easy visualization of input–output characteristics. (d) The input–output plot shows a positive slope of $(R_f + R_i)/R_i$ in the central portion, but the output saturates at about ± 13 V.

We note that the circuit gain (not the op-amp gain) is positive, always greater than or equal to 1; and that if $R_i = \infty$ (open circuit), the circuit reduces to Figure 3.4(a).

Figure 3.4(c) shows how a lever makes possible an easy visualization of the input–output characteristics. The fulcrum is placed at the left end, because R_i is grounded at the left end. v_i appears between the two resistors, so it provides an input at the central part of the diagram. v_o travels through an output excursion determined by the lever arms.

Figure 3.4(d), the input–output characteristic, shows that a one-op-amp circuit can have a positive amplifier gain. Again saturation is evident.

3.4 DIFFERENTIAL AMPLIFIERS

ONE-OP-AMP DIFFERENTIAL AMPLIFIER

The right side of Figure 3.5(a) shows a simple one-op-amp differential amplifier. Current flows from v_4 through R_3 and R_4 to ground. By Rule 2, no current

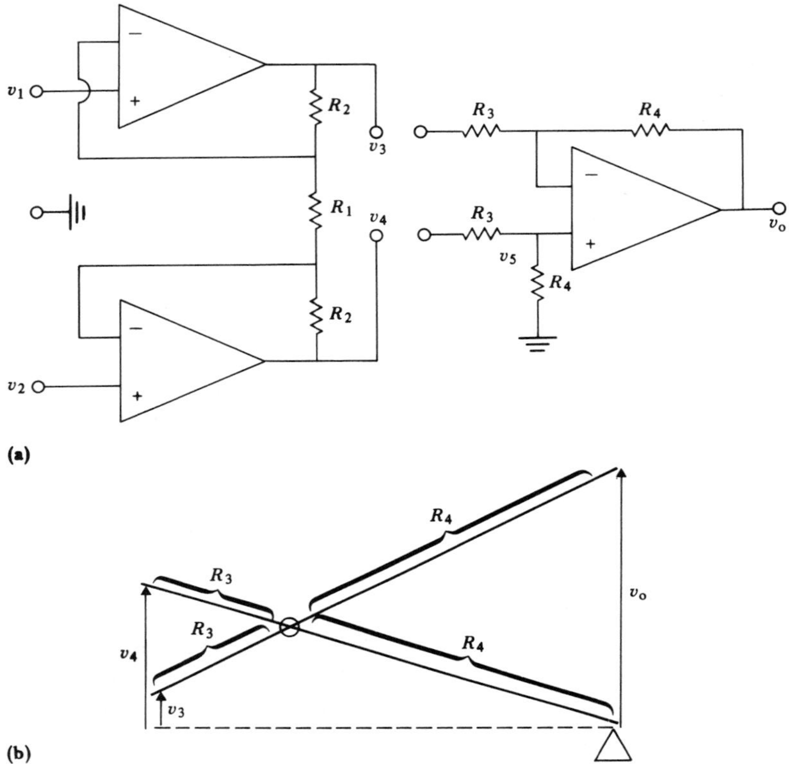

Figure 3.5 (a) The right side shows a one-op-amp differential amplifier, but it has low input impedance. The left side shows how two additional op amps can provide high input impedance and gain. (b) For the one-op-amp differential amplifier, two levers with arm lengths proportional to resistance values make possible an easy visualization of input–output characteristics.

flows into the positive input of the op amp. Hence R_3 and R_4 act as a simple voltage-divider attenuator, which is unaffected by having the op amp attached or by any other changes in the circuit. The voltages in this part of the circuit are visualized in Figure 3.5(b) by the single lever that is attached to the fulcrum (ground).

By Rule 1, whatever voltage appears at the positive input also appears at the negative input. Once this voltage is fixed, the top half of the circuit behaves like an inverting amplifier. For example, if v_4 is 0 V, the positive input of the op amp is 0 V and the v_3–v_o circuit behaves exactly like an inverting amplifier. For other values of v_4, an inverting relation is obtained about some voltage intermediate between v_4 and 0 V. The relationship can be visualized in Figure 3.5(b) by noting that the two levers behave like a pair of scissors. The thumb and finger holes are v_4 and v_3, and the points are at v_o and 0 V.

We solve for the gain by finding v_5.

$$v_5 = \frac{v_4 R_4}{R_3 + R_4} \tag{3.3}$$

Then, solving for the current in the top half, we get

$$i = \frac{v_3 - v_5}{R_3} = \frac{v_5 - v_o}{R_4} \tag{3.4}$$

Substituting (3.3) into (3.4) yields

$$v_o = \frac{(v_4 - v_3)R_4}{R_3} \tag{3.5}$$

This is the equation for a differential amplifier. If the two inputs are hooked together and driven by a common source, with respect to ground, then the *common-mode voltage* v_c is $v_3 = v_4$. Equation (3.5) shows that the ideal output is 0. The differential amplifier-circuit (not op-amp) *common-mode gain* G_c is 0. In Figure 3.5(b), imagine the scissors to be closed. No matter how the inputs are varied, $v_o = 0$.

If on the other hand $v_3 \neq v_4$, then the differential voltage $(v_4 - v_3)$ produces an amplifier-circuit (not op-amp) *differential gain* G_d that from (3.5) is equal to R_4/R_3. This result can be visualized in Figure 3.5(b) by noting that as the scissors open, v_o is geometrically related to $(v_4 - v_3)$ in the same ratio as the lever arms, R_4/R_3.

No differential amplifier perfectly rejects the common-mode voltage. To quantify this imperfection, we use the term *common-mode rejection ratio* (CMRR), which is defined as

$$\text{CMRR} = \frac{G_d}{G_c} \tag{3.6}$$

This factor may be lower than 100 for some oscilloscope differential amplifiers and higher than 10,000 for a high-quality biopotential amplifier.

THREE-OP-AMP DIFFERENTIAL AMPLIFIER

The one-op-amp differential amplifier is quite satisfactory for low-resistance sources, such as strain-gage Wheatstone bridges (Section 2.3). But the input resistance is too low for high-resistance sources. Our first recourse is to add the simple follower shown in Figure 3.4(a) to each input. This provides the required buffering. Because this solution uses two additional op amps, we can also obtain gain from these buffering amplifiers by using a noninverting amplifier, as shown in Figure 3.4(b). However, this solution amplifies the common-mode voltage, as well as the differential voltage, so there is no improvement in CMRR.

A superior solution is achieved by hooking together the two R_i's of the noninverting amplifiers and eliminating the connection to ground. The result is shown on the left side of Figure 3.5(a). To examine the effects of common-mode voltage, assume that $v_1 = v_2$. By Rule 1, v_1 appears at both negative inputs to the op amps. This places the same voltage at both ends of R_1. Hence current through R_1 is 0. By Rule 2, no current can flow from the op-amp inputs. Hence the current through both R_2's is 0, so v_1 appears at both op-amp outputs and the G_c is 1.

To examine the effects when $v_1 \neq v_2$, we note that $v_1 - v_2$ appears across R_1. This causes a current to flow through R_1 that also flows through the resistor string R_2, R_1, R_2. Hence the output voltage

$$v_3 - v_4 = i(R_2 + R_1 + R_2)$$

whereas the input voltage

$$v_1 - v_2 = iR_1$$

The differential gain is then

$$G_d = \frac{v_3 - v_4}{v_1 - v_2} = \frac{2R_2 + R_1}{R_1} \tag{3.7}$$

Since the G_c is 1, the CMRR is equal to the G_d, which is usually much greater than 1. When the left and right halves of Figure 3.5(a) are combined, the resulting three-op-amp amplifier circuit is frequently called an *instrumentation amplifier.* It has high input impedance, a high CMRR, and a gain that can be changed by adjusting R_1. This circuit finds wide use in measuring biopotentials (Section 6.7), because it rejects the large 60 Hz common-mode voltage that exists on the body.

3.5 COMPARATORS

SIMPLE

A comparator is a circuit that compares the input voltage with some reference voltage. The comparator's output flips from one saturation limit to the other, as the negative input of the op amp passes through 0 V. For v_i greater than the comparison level, $v_o = -13$ V. For v_i less than the comparison level, $v_o = +13$ V. Thus this circuit performs the same function as a *Schmitt trigger,* which detects an analog voltage level and yields a logic level output. The simplest comparator is the op amp itself, as shown in Figure 3.2. If a reference voltage

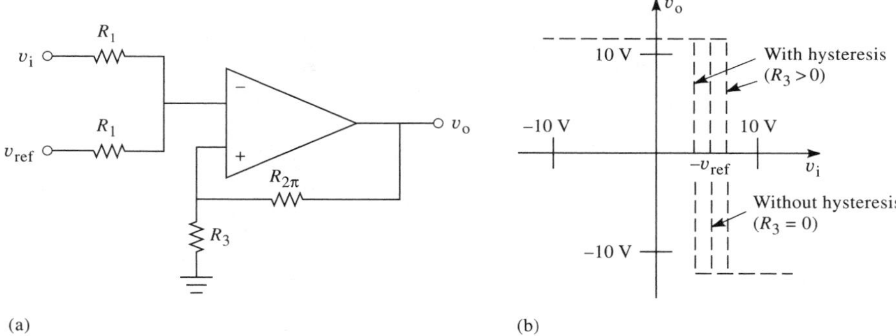

Figure 3.6 (a) Comparator. When $R_3 = 0$, v_o indicates whether $(v_i + v_{ref})$ is greater or less than 0 V. When R_3 is larger, the comparator has hysteresis, as shown in (b), the input–output characteristic.

is connected to the positive input and v_i is connected to the negative input, the circuit is complete. The inputs may be interchanged to invert the output. The input circuit may be expanded by adding the two R_1 resistors shown in Figure 3.6(a). This provides a known input resistance for the circuit and minimizes overdriving the op-amp input. Figure 3.6(b) shows that the comparator flips when $v_i = -v_{ref}$. To avoid building a separate power supply for v_{ref}, we can connect v_{ref} to the −15-V power supply and adjust the values of the input resistors so that the negative input of the op amp is at 0 V when v_i is at the desired positive comparison level. When negative comparison levels are desired, v_{ref} is connected to the +15-V power supply.

WITH HYSTERESIS

For a simple comparator, if v_i is at the comparison level and there is noise on v_i, then v_o fluctuates wildly. To prevent this, we can add hysteresis to the comparator by adding R_2 and R_3, as shown in Figure 3.6(a). The effect of this positive feedback is illustrated by the input–output characteristics shown in Figure 3.6(b). To analyze this circuit, first assume that $v_{ref} = -5$ V and $v_i = +10$ V. Then, because the op amp inverts and saturates, $v_o = -13$ V. Divide v_o by R_2 and R_3 so that the positive input is at, say, −1 V. As v_i is lowered, the comparator does not flip until v_i reaches +3 V, which makes the negative input equal to the positive input, −1 V. At this point, v_o flips to +13 V, causing the positive input to change to +1 V. Noise on v_i cannot cause v_o to flip back, because the negative input must be raised to +1 V to cause the next flip. This requires v_i to be raised to +7 V, at which level the circuit can flip back to its original state. From this example, we see that the width of the hysteresis is four times as great as the magnitude of the voltage across R_3. The width of the hysteresis loop can be varied by replacing R_3 by a potentiometer.

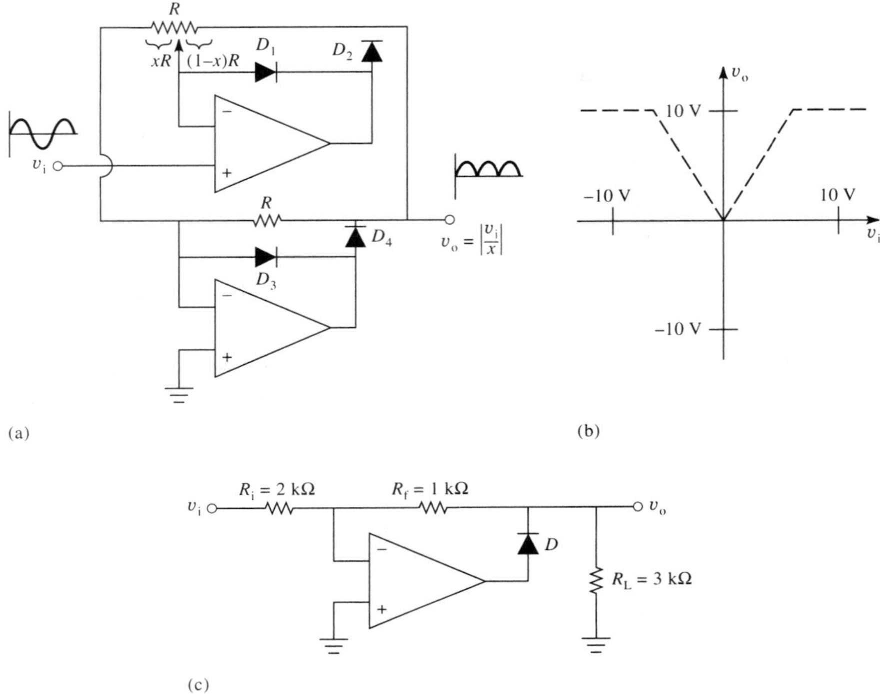

Figure 3.7 (a) Full-wave precision rectifier. For $v_i > 0$, the noninverting amplifier at the top is active, making $v_o > 0$. For $v_i < 0$, the inverting amplifier at the bottom is active, making $v_o > 0$. Circuit gain may be adjusted with a single pot. (b) Input–output characteristics show saturation when $v_o > +13$ V. (Reprinted with permission from *Electronics* Magazine, copyright © December 12, 1974; Penton Publishing, Inc.) (c) One-op-amp full-wave rectifier. For $v_i < 0$, the circuit behaves like the inverting amplifier rectifier with a gain of +0.5. For $v_i > 0$, the op amp disconnects and the passive resistor chain yields a gain of +0.5.

3.6 RECTIFIERS

Simple resistor-diode rectifiers do not work well for voltages below 0.7 V, because the voltage is not sufficient to overcome the forward voltage drop of the diode. This problem can be overcome by placing the diode within the feedback loop of an op amp, thus reducing the voltage limitation by a factor equal to the gain of the op amp.

Figure 3.7(a) shows the circuit for a full-wave precision rectifier (Graeme, 1974b). For $v_i > 0$, D_2 and D_3 conduct, whereas D_1 and D_4 are reverse-biased. The top op amp is a noninverting amplifier with a gain of $1/x$, where x is a fraction corresponding to the potentiometer setting. Because D_4 is not conducting, the lower op amp does not contribute to the output.

For $v_i < 0$, D_1 and D_4 conduct, while D_2 and D_3 are reverse-biased. At the potentiometer wiper v_i serves as the input to the lower op-amp inverting amplifier, which has a gain of $-1/x$. Because D_2 is not conducting, the upper op amp does not contribute to the output. And because the polarity of the gain switches with the polarity of v_i, $v_o = |v_i/x|$.

The advantage of this circuit over other full-wave rectifier circuits (Wait, 1975, p. 173) is that the gain can be varied with a single potentiometer and the input resistance is very high. If only a half-wave rectifier is needed, either the noninverting amplifier or the inverting amplifier can be used separately, thus requiring only one op amp. The perfect rectifier is frequently used with an integrator to quantify the amplitude of electromyographic signals (Section 6.8).

Figure 3.7(c) shows a one-op-amp full-wave rectifier (Tompkins and Webster, 1988). Unlike other full-wave rectifiers, it requires the load to remain constant, because the gain is a function of load.

3.7 LOGARITHMIC AMPLIFIERS

The logarithmic amplifier makes use of the nonlinear volt-ampere relation of the silicon planar transistor (Jung, 1986).

$$V_{BE} = 0.060 \log \left(\frac{I_C}{I_S} \right) \tag{3.8}$$

where

V_{BE} = base-emitter voltage

I_C = collector current

I_S = reverse saturation current, 10^{-13} A at 27 °C

The transistor is placed in the *transdiode* configuration shown in Figure 3.8(a), in which $I_C = v_i/R_i$. Then the output $v_o = V_{BE}$ is logarithmically related to v_i as given by (3.8) over the approximate range 10^{-7} A $< I_C < 10^{-2}$ A. The approximate range of v_o is -0.36 to -0.66 V, so larger ranges of v_o are sometimes obtained by the alternate switch position shown in Figure 3.8(a). The resistor network feeds back only a fraction of v_o in order to boost v_o and uses the same principle as that used in the noninverting amplifier. Figure 3.8(b) shows the input–output characteristics for each of these circuits.

Because semiconductors are temperature-sensitive, accurate circuits require temperature compensation. Antilog (exponential) circuits are made by interchanging the resistor and semiconductor. These log and antilog circuits

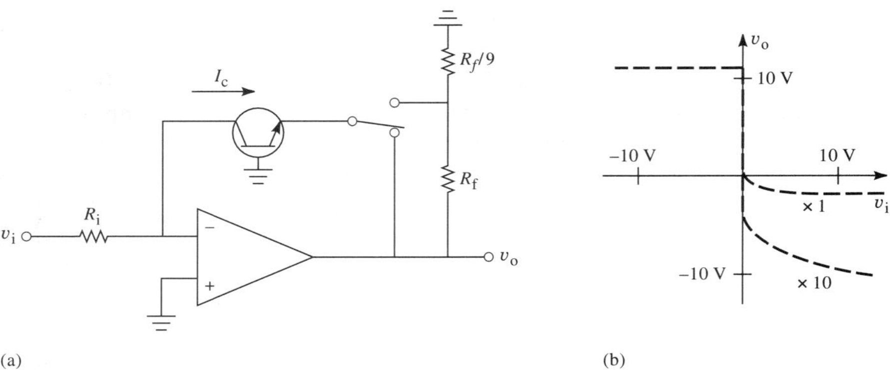

Figure 3.8 (a) A logarithmic amplifier makes use of the fact that a transistor's V_{BE} is related to the logarithm of its collector current. With the switch thrown in the alternate position, the circuit gain is increased by 10. (b) Input–output characteristics show that the logarithmic relation is obtained for only one polarity; ×1 and ×10 gains are indicated.

are used to multiply a variable, divide it, or raise it to a power; to compress large dynamic ranges into small ones; and to linearize the output of devices with logarithmic or exponential input–output relations. In the photometer (Section 11.1), the logarithmic converter can be used to convert transmittance to absorbance.

3.8 INTEGRATORS

So far in this chapter, we have considered only circuits with a flat gain-versus-frequency characteristic. Now let us consider circuits that have a deliberate change in gain with frequency. The first such circuit is the *integrator*. Figure 3.9 shows the circuit for an integrator, which is obtained by closing switch S_1. The voltage across an initially uncharged capacitor is given by

$$v = \frac{1}{C}\int_0^{t_1} i\,dt \tag{3.9}$$

where i is the current through C and t_1 is the integration time. For the integrator, for v_i positive, the input current $i = v_i/R$ flows through C in a direction to cause v_o to move in a negative direction. Thus

$$v_o = -\frac{1}{RC}\int_0^{t_1} v_i\,dt + v_{ic} \tag{3.10}$$

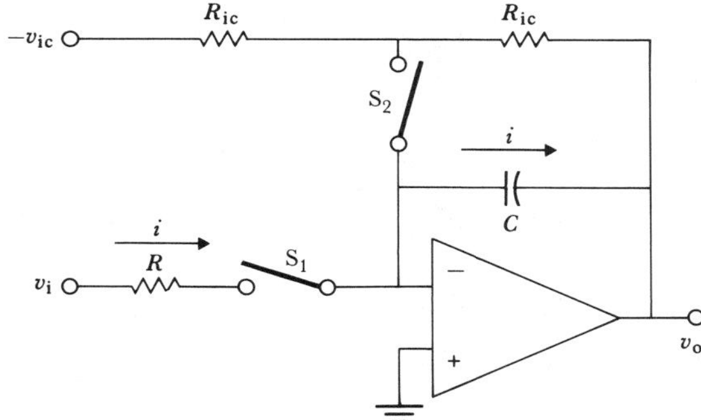

Figure 3.9 A three-mode integrator With S_1 open and S_2 closed, the dc circuit behaves as an inverting amplifier. Thus $v_o = v_{ic}$ and v_o can be set to any desired initial condition. With S_1 closed and S_2 open, the circuit integrates. With both switches open, the circuit holds v_o constant, making possible a leisurely readout.

This shows that v_o is equal to the negative integral of v_i, scaled by the factor $1/RC$ and added to v_{ic}, the voltage due to the initial condition. For $v_{ic} = 0$ and v_i = constant, $v_o = -v_i$ after an integration time equal to RC. Because any real integrator eventually drifts into saturation, a means must be provided to restore v_o to any desired initial condition. If an initial condition of $v_o = 0$ V is desired, a simple switch to short out C is sufficient. For more versatility, S_1 is opened and S_2 closed. This dc circuit then acts as an inverting amplifier, which makes $v_o = v_{ic}$. During integration, S_1 is closed and S_2 open. After the integration, both switches may be opened to hold the output at the final calculated value, thus permitting time for a readout. The circuit is useful for computing the area under a curve, as technicians do when they calculate cardiac output (Section 8.2).

The frequency response of an integrator is easily analyzed because the formula for the inverting amplifier gain (3.1) can be generalized to any input and feedback impedances. Thus for Figure 3.9, with S_1 closed,

$$\frac{V_o(j\omega)}{V_i(j\omega)} = \frac{Z_f}{Z_i} = -\frac{1/j\omega C}{R}$$

$$= -\frac{1}{j\omega RC} = -\frac{1}{j\omega\tau} \tag{3.11}$$

where $\tau = RC$, $\omega = 2\pi f$, and f = frequency. Equation (3.11) shows that the circuit gain decreases as f increases, Figure 3.10 shows the frequency response, and (3.11) shows that the circuit gain is 1 when $\omega\tau = 1$.

EXAMPLE 3.2 The output of the piezoelectric sensor shown in Figure 2.11(b) may be fed directly into the negative input of the integrator shown

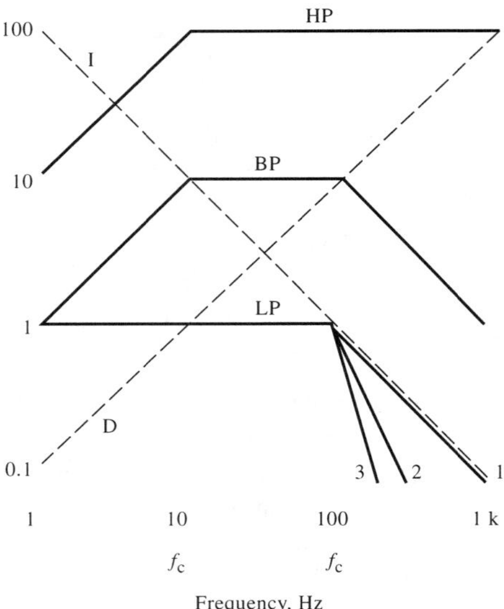

Figure 3.10 Bode plot (gain versus frequency) for various filters Integrator (I); differentiator (D); low pass (LP), 1, 2, 3 section (pole); high pass (HP); bandpass (BP). Corner frequencies f_c for high-pass, low-pass, and bandpass filters.

in Figure 3.9, as shown in Figure E3.2. Analyze the circuit of this charge amplifier and discuss its advantages.

ANSWER Because the FET-op-amp negative input is a virtual ground, $i_{sC} = i_{sR} = 0$. Hence long cables may be used without changing sensor sensitivity or time constant, as is the case with voltage amplifiers. From Figure E3.2, current generated by the sensor, $i_S = K\, dx/dt$, all flows into C, so, using (3.10), we find that v_o is

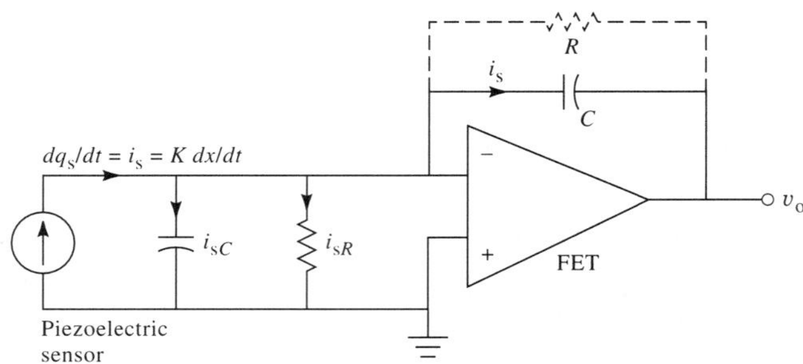

Figure E3.2 The charge amplifier transfers charge generated from a piezoelectric sensor to the op-amp feedback capacitor C.

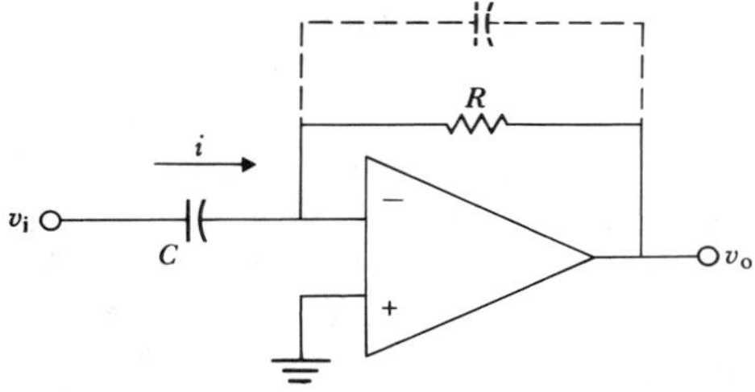

Figure 3.11 A differentiator The dashed lines indicate that a small capacitor must usually be added across the feedback resistor to prevent oscillation.

$$v_o = -v = -\frac{1}{C}\int_0^{t_1} \frac{K\,dx}{dt}\,dt = -\frac{kX}{C}$$

which shows that v_o is proportional to x, even down to dc. Like the integrator, the charge amplifier slowly drifts with time because of bias currents required by the op-amp input. A large feedback resistance R must therefore be added to prevent saturation. This causes the circuit to behave as a high-pass filter, with a time constant $\tau = RC$. It then responds only to frequencies above $f_c = 1/(2\pi RC)$ and has no frequency-response improvement over the voltage amplifier.

3.9 DIFFERENTIATORS

Interchanging the integrator's R and C yields the differentiator shown in Figure 3.11. The current through a capacitor is given by

$$i = C\frac{dv}{dt} \tag{3.12}$$

If dv_i/dt is positive, i flows through R in a direction such that it yields a negative v_o. Thus

$$v_o = -RC\frac{dv_i}{dt} \tag{3.13}$$

The frequency response of a differentiator is given by the ratio of feedback to input impedance.

$$\frac{V_o(j\omega)}{V_i(j\omega)} = -\frac{Z_f}{Z_i} = -\frac{R}{1/j\omega C}$$

$$= -j\omega RC = -j\omega\tau \tag{3.14}$$

Equation (3.14) shows that the circuit gain increases as f increases and that it is equal to unity when $\omega\tau = 1$. Figure 3.10 shows the frequency response.

Unless specific preventive steps are taken, the circuit tends to oscillate. The output also tends to be noisy, because the circuit emphasizes high frequencies. A differentiator followed by a comparator is useful for detecting an event the slope of which exceeds a given value—for example, detection of the R wave in an electrocardiogram.

3.10 ACTIVE FILTERS

LOW-PASS FILTER

Figure 1.9(a) shows a low-pass filter that is useful for attenuating high-frequency noise. A low-pass active filter can be obtained by using the one-op-amp circuit shown in Figure 3.12(a). The advantages of this circuit are that it is capable of gain and that it has a very low output impedance. The frequency response is given by the ratio of feedback to input impedance.

$$\frac{V_o(j\omega)}{V_i(j\omega)} = -\frac{Z_f}{Z_i} = -\frac{\dfrac{(R_f/j\omega C_f)}{[(1/j\omega C_f) + R_f]}}{R_i}$$

$$= -\frac{R_f}{(1 + j\omega R_f C_f)R_i} = -\frac{R_f}{R_i}\frac{1}{1 + j\omega\tau} \tag{3.15}$$

where $\tau = R_f C_f$. Note that (3.15) has the same form as (1.23). Figure 3.10 shows the frequency response, which is similar to that shown in Figure 1.8(d). For $\omega \ll 1/\tau$, the circuit behaves as an inverting amplifier (Figure 3.3), because the impedance of C_f is large compared with R_f. For $\omega \gg 1/\tau$, the circuit behaves as an integrator (Figure 3.9), because C_f is the dominant feedback impedance. The *corner frequency* f_c, which is defined by the intersection of the two asymptotes shown, is given by the relation $\omega\tau = 2\pi f_c \tau = 1$. When a designer wishes to limit the frequency of a wide-band amplifier, it is not necessary to add a separate stage, as shown in Figure 3.12(a), but only to add the correct size C_f to the existing wide-band amplifier.

HIGH-PASS FILTER

Figure 3.12(b) shows a one-op-amp high-pass filter. Such a circuit is useful for amplifying a small ac voltage that rides on top of a large dc voltage, because C_i blocks the dc. The frequency-response equation is

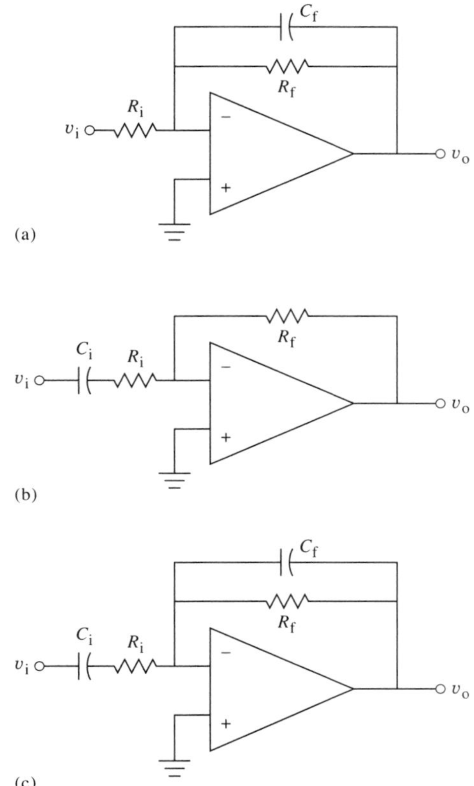

Figure 3.12 Active filters (a) A low-pass filter attenuates high frequencies. (b) A high-pass filter attenuates low frequencies and blocks dc. (c) A bandpass filter attenuates both low and high frequencies.

$$\frac{V_o(j\omega)}{V_i(j\omega)} = -\frac{Z_f}{Z_i} = -\frac{R_f}{1/j\omega C_i + R_i}$$

$$= -\frac{j\omega R_f C_i}{1 + j\omega C_i R_i} = -\frac{R_f}{R_i}\frac{j\omega\tau}{1 + j\omega\tau} \tag{3.16}$$

where $\tau = R_i C_i$. Figure 3.10 shows the frequency response. For $\omega \ll 1/\tau$, the circuit behaves as a differentiator (Figure 3.11), because C_i is the dominant input impedance. For $\omega \gg 1/\tau$, the circuit behaves as an inverting amplifier, because the impedance of R_i is large compared with that of C_i. The corner frequency f_c, which is defined by the intersection of the two asymptotes shown, is given by the relation $\omega\tau = 2\pi f_c \tau = 1$.

BANDPASS FILTER

A series combination of the low-pass filter and the high-pass filter results in a *bandpass filter,* which amplifies frequencies over a desired range and attenuates higher and lower frequencies. Figure 3.12(c) shows that the bandpass function can be achieved with a one-op-amp circuit. Figure 3.10 shows the frequency response. The corner frequencies are defined by the same relations as those for the low-pass and the high-pass filters. This circuit is useful for amplifying a certain band of frequencies, such as those required for recording heart sounds or the electrocardiogram.

3.11 FREQUENCY RESPONSE

Up until now, we have found it useful to consider the op amp as ideal. Now we shall examine the effects of several nonideal characteristics, starting with that of frequency response.

OPEN-LOOP GAIN

Because the op amp requires very high gain, it has several stages. Each of these stages has stray or junction capacitance that limits its high-frequency response in the same way that a simple *RC* low-pass filter reduces high-frequency gain. At high frequencies, each stage has a -1 slope on a log–log plot of gain versus frequency, and each has a $-90°$ phase shift. Thus a three-stage op amp, such as type 709, reaches a slope of -3, as shown by the dashed curve in Figure 3.13. The phase shift reaches $-270°$, which is quite satisfactory for a comparator, because feedback is not employed. For an amplifier, if the gain is greater than 1 when the phase shift is equal to $-180°$ (the closed-loop condition for oscillation), there is undesirable oscillation.

COMPENSATION

Adding an external capacitor to the terminals indicated on the specification sheet moves one of the *RC* filter corner frequencies to a very low frequency. This compensates the uncompensated op amp, resulting in a slope of -1 and a maximal phase shift of $-90°$. This is done with an internal capacitor in the 411, resulting in the solid curve shown in Figure 3.13. This op amp does not oscillate for any amplifier we have described. This op amp has very high dc gain, but the gain is progressively reduced at higher frequencies, until it is only 1 at 4 MHz.

CLOSED-LOOP GAIN

It might appear that the op amp has very poor frequency response, because its gain is reduced for frequencies above 40 Hz. However, an amplifier circuit is never built using the op amp open loop, so we shall therefore discuss only the circuit closed-loop response. For example, if we build an amplifier circuit with a gain of 10, as shown in Figure 3.13, the frequency response is flat up

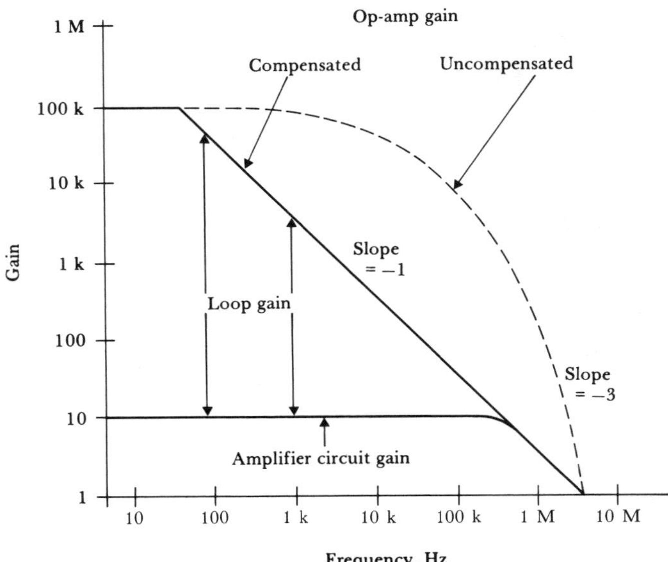

Figure 3.13 Op-amp frequency characteristics Early op amps (such as the 709) were uncompensated, had a gain greater than 1 when the phase shift was equal to −180°, and therefore oscillated unless compensation was added externally. A popular op amp, the 411, is compensated internally, so for a gain greater than 1, the phase shift is limited to −90°. When feedback resistors are added to build an amplifier circuit, the loop gain on this log–log plot is the difference between the op-amp gain and the amplifier-circuit gain.

to 400 kHz and is reduced above that frequency only because the amplifier-circuit gain can never exceed the op-amp gain. We find this an advantage of using negative feedback, in that the frequency response is greatly extended.

LOOP GAIN

The loop gain for an amplifier circuit is obtained by breaking the feedback loop at any point in the loop, injecting a signal, and measuring the gain around the loop. For example, in a unity-gain follower [Figure 3.4(a)] we break the feedback loop and then the injected signal enters the negative input, after which it is amplified by the op-amp gain. Therefore the loop gain equals the op-amp gain. To measure loop gain in an inverting amplifier with a gain of −1 [Figure 3.3(a)], assume that the amplifier-circuit input is grounded. The injected signal is divided by 2 by the attenuator formed of R_f and R_i and is then amplified by the op-amp gain. Thus the loop gain is equal to (op-amp gain)/2.

Figure 3.13 shows the loop-gain concept for a noninverting amplifier. The amplifier-circuit gain is 10. On the log–log plot, the difference between the op-amp gain and the amplifier-circuit gain is the loop gain. At low frequencies, the loop gain is high and the closed-loop amplifier-circuit characteristics are determined by the feedback resistors. At high frequencies, the loop gain is low and the amplifier-circuit characteristics follow the op-amp characteristics. High loop gain is good for accuracy and stability, because the feedback resistors can be made much more stable than the op-amp characteristics.

GAIN-BANDWIDTH PRODUCT

The gain-bandwidth product of the op amp is equal to the product of gain and bandwidth at a particular frequency. Thus in Figure 3.13 the unity-gain-bandwidth product is 4 MHz, a typical value for op amps. Note that along the entire curve with a slope of -1, the gain-bandwidth product is still constant, at 4 MHz. Thus, for any amplifier circuit, we can obtain its bandwidth by dividing the gain-bandwidth product by the amplifier-circuit gain. For higher-frequency applications, op amps such as the OP-37E are available with gain-bandwidth products of 60 MHz.

SLEW RATE

Small-signal response follows the amplifier-circuit frequency response predicted by Figure 3.13. For large signals there is an additional limitation. When rapid changes in output are demanded, the capacitor added for compensation must be charged up from an internal source that has limited current capability I_{max}. The change in voltage across the capacitor is then limited, $dv/dt = I_{max}/C$, and dv_o/dt is limited to a maximal slew rate (15 V/μs for the 411). If this slew rate S_r is exceeded by a large-amplitude, high-frequency sine wave, distortion occurs. Thus there is a limitation on the sine-wave *full-power response,* or maximal frequency for rated output,

$$f_p = \frac{S_r}{2\pi V_{or}} \tag{3.17}$$

where V_{or} is the rated output voltage (usually 10 V). If the slew rate is too slow for fast switching of a comparator, an uncompensated op amp can be used, because comparators do not contain the negative-feedback path that may cause oscillations.

3.12 OFFSET VOLTAGE

Another nonideal characteristic is that of offset voltage. The two op-amp inputs drive the bases of transistors, and the base-to-emitter voltage drop may be slightly different for each. Thus, so that we can obtain $v_o = 0$, the voltage $(v_1 - v_2)$ must be a few millivolts. This offset voltage is usually not important when v_i is 1–10 V. But when v_i is on the order of millivolts, as when amplifying the output from thermocouples or strain gages, the offset voltage must be considered.

NULLING

The offset voltage may be reduced to zero by adding an external nulling pot to the terminals indicated on the specification sheet. Adjustment of this pot increases emitter current through one of the input transistors and lowers it through the other. This alters the base-to-emitter voltage of the two transistors until the offset voltage is reduced to zero.

DRIFT

Even though the offset voltage may be set to 0 at 25 °C, it does not remain there if temperature is not constant. Temperature changes that affect the base-to-emitter voltages may be due to either environmental changes or to variations in the dissipation of power in the chip that result from fluctuating output voltage. The effects of temperature may be specified as a maximal offset voltage change in volts per degree Celsius or a maximal offset voltage change over a given temperature range, say −25 to +85 °C. If the drift of an inexpensive op amp is too high for a given application, tighter specifications (0.1 μV/°C) are available with temperature-controlled chips. An alternative technique modulates the dc as in chopper-stabilized and varactor op amps (Tobey, 1971).

NOISE

All semiconductor junctions generate noise, which limits the detection of small signals. Op amps have transistor input junctions, which generate both noise-voltage sources and noise-current sources. These can be modeled as shown in Figure 3.14. For low source impedances, only the noise voltage v_n is important; it is large compared with the i_nR drop caused by the current noise i_n. The noise is random, but the amplitude varies with frequency. For example, at low frequencies the noise power density varies as $1/f$ (flicker noise), so a large amount of noise is present at low frequencies. At the midfrequencies, the noise is lower and can be specified in rms units of $\text{V} \cdot \text{Hz}^{-1/2}$. In addition, some silicon planar-diffused bipolar integrated-circuit op amps exhibit bursts of noise, called *popcorn noise* (Wait, 1975).

3.13 BIAS CURRENT

Because the two op-amp inputs drive transistors, base or gate current must flow all the time to keep the transistors turned on. This is called *bias current,*

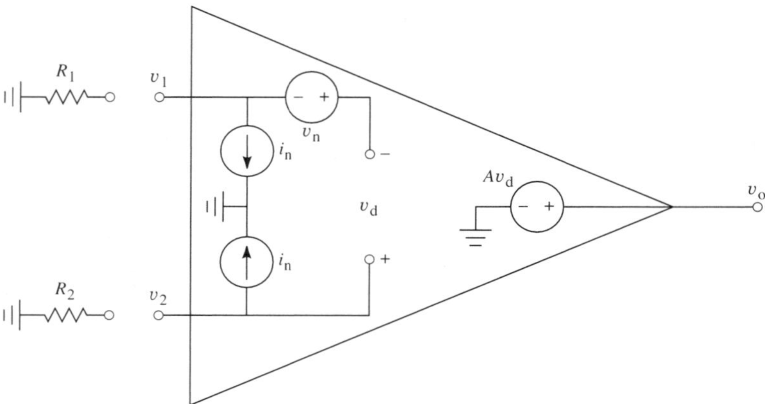

Figure 3.14 Noise sources in an op amp The noise-voltage source v_n is in series with the input and cannot be reduced. The noise added by the noise-current sources i_n can be minimized by using small external resistances.

which for the 411 is about 200 pA. This bias current must flow through the feedback network. It causes errors proportional to feedback-element resistances. To minimize these errors, small feedback resistors, such as those with resistances of 10 kΩ, are normally used. Smaller values should be used only after a check to determine that the current flowing through the feedback resistor, plus the current flowing through all load resistors, does not exceed the op-amp output current rating (20 mA for the 411).

DIFFERENTIAL BIAS CURRENT

The difference between the two input bias currents is much smaller than either of the bias currents alone. A degree of cancelation of the effects of bias current can be achieved by having each bias current flow through the same equivalent resistance. This is accomplished for the inverting amplifier and the noninverting amplifier by adding, in series with the positive input, a compensation resistor the value of which is equal to the parallel combination of R_i and R_f. There still is an error, but it is now determined by the difference in bias current.

DRIFT

The input bias currents are transistor base or gate currents, so they are temperature-sensitive, because transistor gain varies with temperature. However, the changes in gain of the two transistors tend to track together, so the additional compensation resistor that we have described minimizes the problem.

NOISE

Figure 3.14 shows how variations in bias current contribute to overall noise. The noise currents flow through the external equivalent resistances so that the total rms noise voltage is

$$v_t \cong \{[v_n^2 + (i_n R_1)^2 + (i_n R_2)^2 + 4\kappa T R_1 + 4\kappa T R_2]\mathrm{BW}\}^{1/2} \tag{3.18}$$

where

R_1 and R_2 = equivalent source resistances

v_n = mean value of the rms noise voltage, in $\mathrm{V}\cdot\mathrm{Hz}^{-1/2}$, across the frequency range of interest

i_n = mean value of the rms noise current, in $\mathrm{A}\cdot\mathrm{Hz}^{-1.2}$, across the frequency range of interest

κ = Boltzmann's constant (Appendix)

T = temperature, K

BW = noise bandwidth, Hz

The specification sheet provides values of v_n and i_n (sometimes v_n^2 and i_n^2), thus making it possible to compare different op amps. If the source resistances are 10 kΩ, bipolar-transistor op amps yield the lowest noise. For larger source resistances, low-input-current amplifiers such as the *field-effect transistor* (FET) input stage are best because of their lower current noise. Ary (1977) presents design factors and performance specifications for a low-noise amplifier.

For ac amplifiers, the lowest noise is obtained by calculating the characteristic noise resistance $R_n = v_n/i_n$ and setting it equal to the equivalent source resistance R_2 (for the noninverting amplifier). This is accomplished by inserting a transformer with turns ratio $1:N$, where $N = (R_n/R_2)^{1/2}$, between the source and the op amp (Jung, 1986).

3.14 INPUT AND OUTPUT RESISTANCE

INPUT RESISTANCE

The op-amp differential-input resistance R_d is shown in Figures 3.1 and 3.15. For the FET-input 411, it is 1 TΩ, whereas for BJT-input op amps, it is about 2 MΩ, which is comparable to the value of some feedback resistors used. However, we shall see that its value is usually not important because of the benefits of feedback. Consider the follower shown in Figure 3.15. In order to calculate the amplifier-circuit input resistance R_{ai}, assume a change in input voltage v_i. Because this is a follower,

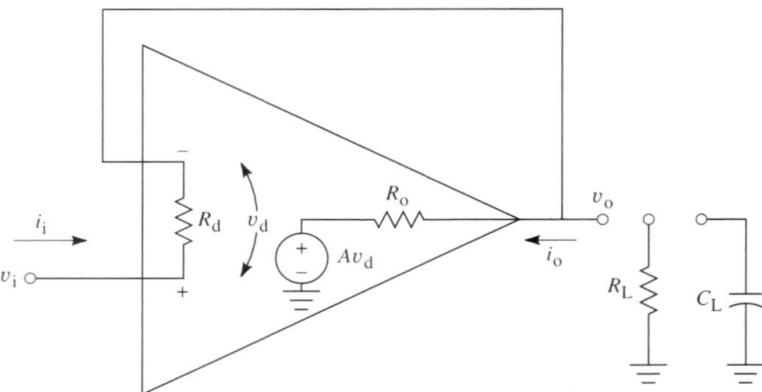

Figure 3.15 The amplifier input impedance is much higher than the op-amp input impedance R_d. The amplifier output impedance is much smaller than the op-amp output impedance R_o.

$$\Delta v_o = A\Delta v_d = A(\Delta v_i - \Delta v_o)$$

$$= \frac{A\Delta v_i}{A + 1}$$

$$\Delta i_i = \frac{\Delta v_d}{R_d} = \frac{\Delta v_i - \Delta v_o}{R_d} = \frac{\Delta v_i}{(A + 1)R_d}$$

$$R_{ai} = \frac{\Delta v_i}{\Delta i_i} = (A + 1)R_d \cong AR_d \tag{3.19}$$

Thus the amplifier-circuit input resistance R_{ai} is about $(10^5) \times (2\ \text{M}\Omega) = 200\ \text{G}\Omega$. This value cannot be achieved in practice, because surface leakage paths in the op-amp socket lower it considerably. In general, all noninverting amplifiers have a very high input resistance, which is equal to R_d times the loop gain. This is not to say that very large source resistances can be used, because the bias current usually causes much larger problems than the amplifier-circuit input impedance. For large source resistances, FET op amps such as the 411 are helpful.

The input resistance of an inverting amplifier is easy to determine. Because the negative input of the op amp is a virtual ground,

$$R_{ai} = \frac{\Delta v_i}{\Delta i_i} = R_i \tag{3.20}$$

Thus the amplifier-circuit input resistance R_{ai} is equal to R_i, the input resistor. Because R_i is usually a small value, the inverting amplifier has small input resistance.

OUTPUT RESISTANCE

The op-amp output resistance R_o is shown in Figures 3.1 and 3.15. It is about 40 Ω for the typical op amp, which may seem large for some applications. However, its value is usually not important because of the benefits of feedback. Consider the follower shown in Figure 3.15. In order to calculate the amplifier-circuit output resistance R_{ao}, assume that load resistor R_L is attached to the output, causing a change in output current Δi_o. Because i_o flows through R_o, there is an additional voltage drop $\Delta i_o R_o$.

$$-\Delta v_d = \Delta v_o = A\,\Delta v_d + \Delta i_o R_o = -A\,\Delta v_o + \Delta i_o R_o$$

$$(A+1)\Delta v_o = \Delta i_o R_o$$

$$R_{ao} = \frac{\Delta v_o}{\Delta i_o} = \frac{R_o}{A+1} \cong R_o/\mathrm{A} \qquad (3.21)$$

Thus the amplifier-circuit output resistance R_{ao} is about $40/(10^5) = 0.0004$ Ω, a value negligible in most circuits. In general, all noninverting and inverting amplifiers have an output resistance that is equal to R_o divided by the loop gain. This is not to say that very small load resistances can be driven by the output. If R_L shown in Figure 3.15 is smaller than 500 Ω, the op amp saturates internally, because the maximal current output for a typical op amp is 20 mA. This maximal current output must also be considered when driving large capacitances C_L at a high slew rate. Then the output current

$$i_o = C_L \frac{dv_o}{dt} \qquad (3.22)$$

The R_o–C_L combination also acts as a low-pass filter, which introduces additional phase shift around the loop and can cause oscillation. The cure is to add a small resistor between v_o and C_L, thus isolating C_L from the feedback loop.

To achieve larger current outputs, the *current booster* is used. An ordinary op amp drives high-power transistors (on heat sinks if required). Then we can use the entire circuit as an op amp by connecting terminals v_1, v_2, and v_o to external feedback networks. This places the booster section within the feedback loop and keeps distortion low.

3.15 PHASE-SENSITIVE DEMODULATORS

Figure 2.7 shows that a linear variable differential transformer requires a phase-sensitive demodulator to yield a useful output signal. A phase-sensitive demodulator does not measure phase but yields a full-wave-rectified output of the in-phase component of a sine wave. Its output is proportional to the amplitude of the input, but it changes sign when the phase shifts by 180°.

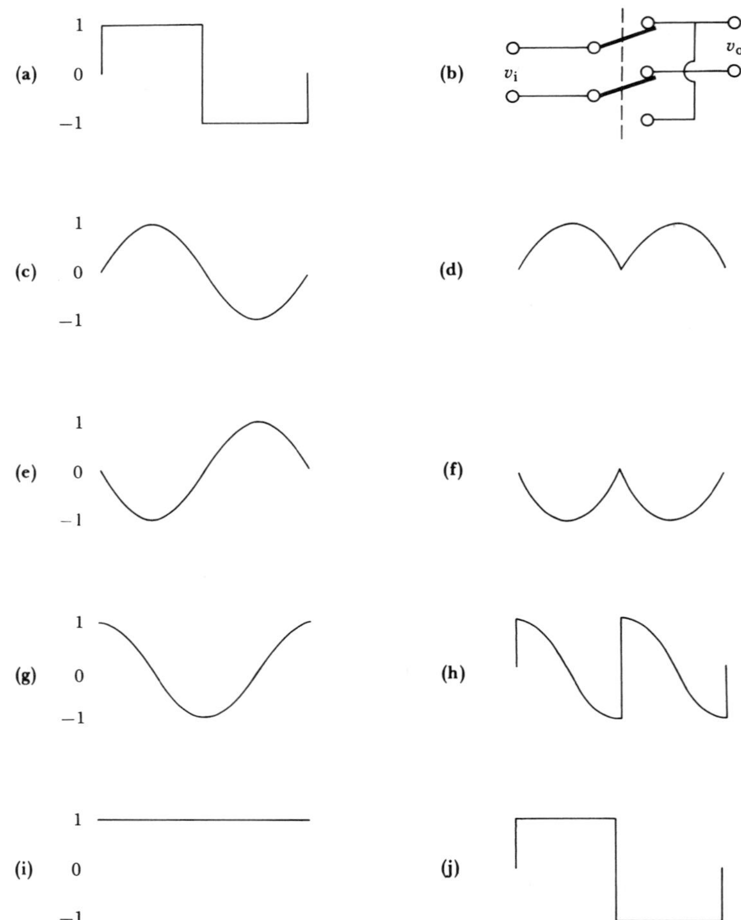

Figure 3.16 Functional operation of a phase-sensitive demodulator (a) Switching function. (b) Switch. (c), (e), (g), (i) Several input voltages. (d), (f), (h), (j) Corresponding output voltages.

Figure 3.16 shows the functional operation of a phase-sensitive demodulator. Figure 3.16(a) shows a switching function that is derived from a *carrier oscillator* and causes the double-pole double-throw switch in Figure 3.16(b) to be in the upper position for +1 and in the lower position for −1. In effect, this multiplies the input signal v_i by the switching function shown in Figure 3.16(a). The in-phase sine wave in Figure 3.16(c) is demodulated by this switch to yield the full-wave-rectified positive signal in Figure 3.16(d). The sine wave in Figure 3.16(e) is 180° out of phase, so it yields the negative signal in Figure 3.16(f).

Amplifier stray capacitance may cause an undesirable *quadrature voltage* that is shifted 90°, as shown in Figure 3.16(g). The demodulated signal in Figure 3.16(h) averages to zero when passed through a low-pass filter and is

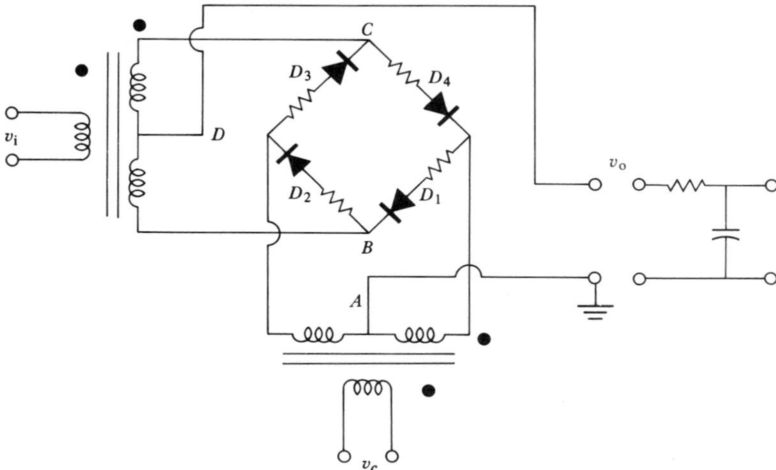

Figure 3.17 A ring demodulator This phase-sensitive detector produces a full-wave-rectified output v_o that is positive when the input voltage v_i is in phase with the carrier voltage v_c and negative when v_i is 180° out of phase with v_c.

rejected. The dc signal shown in Figure 3.16(i) is demodulated to the wave shown in Figure 3.16(j) and is rejected. Any frequency component not locked to the carrier frequency is similarly rejected. Because the phase-sensitive demodulator has excellent noise-rejection capabilities, it is frequently used to demodulate the suppressed-carrier waveforms obtained from LVDTs and the ac-excited strain-gage Wheatstone bridge (Section 2.3). A carrier system and phase-sensitive demodulator are also essential for operation of the electromagnetic blood flowmeter (Section 8.3). The noise-rejection capability may be improved by placing a tuned amplifier before the phase-sensitive demodulator, thus forming a lock-in amplifier (Aronson, 1977).

A practical phase-sensitive demodulator is shown in Figure 3.17. This *ring demodulator* operates with the following action, provided that v_c is more than twice v_i. If the carrier waveform v_c is positive at the black dot, diodes D_1 and D_2 are forward-biased and D_3 and D_4 are reverse-biased. By symmetry, points A and B are at the same voltage. If the input waveform v_i is positive at the black dot, this transforms to a voltage v_{DB} that appears at v_o, as shown in the first half of Figure 3.16(d).

During the second half of the cycle, diodes D_3 and D_4 are forward-biased and D_1 and D_2 are reverse-biased. By symmetry, points A and C are at the same potential. The reversed polarity of v_i yields a positive v_{DC}, which appears at v_o. Thus v_o is a full-wave-rectified waveform. If v_i changes phase by 180°, as shown in Figure 3.16(e), v_o changes polarity. To eliminate ripple, the output is usually low-pass filtered by a filter the corner frequency of which is about one-tenth of the carrier frequency.

The ring demodulator has the advantge of having no moving parts. Also, because transformer coupling is used, v_i, v_c, and v_o can all be referenced to different dc levels. The availability of type 1495 solid-state double-balanced demodulators on a single chip (Jung, 1986) makes it possible to eliminate the bulky transformers but requires more care in biasing v_i, v_c, and v_o at different dc levels.

3.16 MICROCOMPUTERS IN MEDICAL INSTRUMENTATION

The electronic devices that we have described so far in this chapter are useful for acquiring a medical signal and performing some initial processing, such as filtering or demodulation. Microcomputers can frequently replace analog circuits by performing the signal-processing functions of comparator, limiter, rectifier, logarithmic amplifier, integrator, differentiator, active filter, and phase-sensitive demodulator in software (Tompkins, 1993). The generalized instrumentation system shown in Figure 1.1 also indicates additional signal processing, data storage, and control/feedback capability. Traditionally, this additional processing was handled either by using relatively simple digital-electronic circuits or, if a significant amount of processing was required, by connecting the instrument to a computer.

The development of microcomputers has led to the combining of a medical instrument with a signal-processing capability sufficient to perform functions normally done by an operator or a computer. This computing function can certainly be implemented. But from the point of view of medical instrumentation, it is more instructive to view the microcomputer as a microcontroller. The use of a microcomputer generally results in fewer integrated-circuit packages. This reduced complexity, together with the capability for self-calibration and detection of errors, enhances the reliability of the instrument. The most useful applications of microcomputers for medical instrumentation involve this controller function. Microcomputers can provide self-calibration for measurement systems, automatic sequencing of events, and an easy way to enter such patient data as height, weight, and sex for calculating expected or normal performance. All these functions are made possible by the basic structure of the microcomputer system (Tompkins and Webster, 1981; Tompkins and Webster, 1988).

PROBLEMS

3.1 (a) Design an inverting amplifier with an input resistance of 20 kΩ and a gain of −10. (b) Include a resistor to compensate for bias current. (c) Design a summing amplifier such that $v_o = -(10v_1 + 2v_2 + 0.5v_3)$.

3.2 The axon action potential (AAP) is shown in Figure 4.1. Design a *dc-coupled* one-op-amp circuit that will amplify the −100 mV to +50 mV input range to have the maximal gain possible without exceeding the typical guaranteed linear output range.

3.3 Use the circuit shown in Figure E3.1 to design a dc-coupled one-op-amp circuit that will amplify the ± 100-μV EOG to have the maximal gain possible without exceeding the typical guaranteed linear output range. Include a control that can balance (remove) series electrode offset potentials up to ± 300 mV. Give all numerical values.

3.4 Design a noninverting amplifier having a gain of 10 and R_i of Figure 3.4(b) equal to 20 kΩ. Include a resistor to compensate for bias current.

3.5 An op-amp differential amplifier is built using four identical resistors, each having a tolerance of ±5%. Calculate the worst possible CMRR.

3.6 Design a three-op-amp differential amplifier having a differential gain of 5 in the first stage and 6 in the second stage.

3.7 A blood-pressure sensor uses a four-active-arm Wheatstone strain gage bridge excited with 5 V dc. At full scale, each arm changes resistance by ±0.3%. Design an amplifier that will provide a full-scale output over the op amp's full range of linear operation. Use the minimal number of components.

3.8 Design a comparator with hysteresis in which the hysteresis width extends from 0 to +2 V.

3.9 For an inverting half-wave perfect rectifier, sketch the circuit. Plot the input–output characteristics for both the circuit output and the op-amp output, which are not the same point as in most op-amp circuits.

3.10 Using the principle shown in Figure 3.8, design a signal compressor for which an input-voltage range of ±10 V yields an output-voltage range of ±4 V.

3.11 Design an integrator with an input resistance of 1 MΩ. Select the capacitor such that when $v_i = +10$ V, v_o travels from 0 to -10 V in 0.1 s.

3.12 In Problem 3.11, if $v_i = 0$ and offset voltage equals 5 mV, what is the current through R? How long will it take for v_o to drift from 0 V to saturation? Explain how to cure this drift problem.

3.13 In Problem 3.11, if bias current is 200 pA, how long will it take for v_o to drift from 0 V to saturation? Explain how to cure this drift problem.

3.14 Design a differentiator for which $v_o = -10$ V when $dv_i/dt = 100$ V/s.

3.15 Design a one-section high-pass filter with a gain of 20 and a corner frequency of 0.05 Hz. Calculate its response to a step input of 1 mV.

3.16 Design a one-op-amp high-pass active filter with a high-frequency gain of +10 (not −10), a high-frequency input impedance of 10 MΩ, and a corner frequency of 10 Hz.

3.17 Find $V_o(j\omega)/V_i(j\omega)$ for the bandpass filter shown in Figure 3.12(c).

3.18 Figure 6.16 shows that the frequency range of the AAP is 1–10 kHz. Design a one-op-amp active bandpass filter that has a midband input impedance of approximately 10 kΩ, a midband gain of approximately 1, and a frequency response from 1 to 10 kHz (corner frequencies).

3.19 Figure 6.16 shows the maximal single-peak signal and frequency range of the EMG. Design a one-op-amp bandpass filter circuit that will amplify the EMG to have the maximal gain possible without exceeding the typical guaranteed linear output range and will pass the range of frequencies shown.

3.20 Using 411 op amps, explain how an amplifier with a gain of 100 and a bandwidth of 100 kHz can be designed.

3.21 Refer to Figure 3.13. If the amplifier gain is 1000, what is the loop gain at 100 Hz?

3.22 For the differentiator shown in Figure 3.11, ground the input, break the feedback loop at any point, and determine the phase shift in each section. Explain why the circuit tends to oscillate.

3.23 For Problem 3.21, calculate the amplifier input and output resistances at 100 Hz, for inverting and noninverting amplifiers.

3.24 For Figure 3.15, what is the maximal capacitive load C_L that can be connected to a 411 without degrading the normal slew rate (15 V/μs) at the maximal current output (20 mA)?

3.25 (a) For Figure 3.17, assume that the carrier frequency is 3 kHz. Design the *RC* output low-pass filter to have a corner frequency of 20 Hz and a reasonable value capacitor (100 nF). Use (b) a one-section active filter.

3.26 For Figure 3.17, if the forward drop of D_1 is 10% higher than that of the other diodes, what change occurs in v_o?

3.27 Given an oscillator block, design (show the circuit diagram for) an LVDT, phase-sensitive demodulator and a first-order low-pass filter with a corner frequency of 100 Hz. Sketch waveforms at each significant location.

REFERENCES

Aronson, M. H., "Lock-in and carrier amplifiers." *Med. Electron. Data,* 8(3), 1977, C1–C16.

Ary, J. P., "A head-mounted 24-channel evoked potential preamplifier employing low-noise operational amplifiers." *IEEE Trans. Biomed. Eng.,* BME-24, 1977, 293–297.

Franco, S., *Design with Operational Amplifiers and Analog Integrated Circuits.* New York: McGraw-Hill, 1988.

Graeme, J. G., *Applications of Operational Amplifiers.* New York: McGraw-Hill, 1974a.

Graeme, J. G., "Rectifying wide-range signals with precision, variable gain." *Electron.,* Dec. 12, 1974b, 45(25), 107–109.

Horowitz, P., and W. Hill, *The Art of Electronics,* 2nd ed. Cambridge, England: Cambridge University Press, 1989.

Jung, W. G., *IC Op-Amp Cookbook,* 3rd ed. Indianapolis: Howard W. Sams, 1986.

Sheingold, D. H. (ed.), *Analog-Digital Conversion Handbook.* Norwood, MA: Analog Devices, 1972.

Shepard, R. R., "Active filters: Part 12, Short cuts to network design." *Electron.,* Aug. 18, 1969, 42(17), 82–92.

Tobey, G. E., J. G. Graeme, and L. P. Huelsman, *Operational Amplifiers: Design and Application.* New York: McGraw-Hill, 1971.

Tompkins, W. J. (ed.), *Biomedical Digital Signal Processing: C-Language Examples and Laboratory Experiments for the IBM PC.* Englewood Cliffs, NJ: Prentice Hall, 1993.

Tompkins, W. J., and J. G. Webster (eds.), *Design of Microcomputer-Based Medical Instrumentation.* Englewood Cliffs, NJ: Prentice-Hall, 1981.

Tompkins, W. J., and J. G. Webster (eds.), *Interfacing Sensors to the IBM PC.* Englewood Cliffs, NJ: Prentice-Hall, 1988.

Wait, J. V., L. P. Huelsman, and G. A. Korn, *Introduction to Operational Amplifier Theory and Applications.* New York: McGraw-Hill, 1975.

4

THE ORIGIN OF BIOPOTENTIALS

John W. Clark, Jr.

This chapter deals with the genesis of various bioelectric signals that are recorded routinely in modern clinical practice. Given adequate monitoring equipment, the engineer of today can record many forms of bioelectric phenomena with relative ease. These phenomena include the electrocardiogram (ECG), electroencephalogram (EEG), electroneurogram (ENG), electromyogram (EMG), and electroretinogram (ERG).

Engineers generally have a good physical insight into the nature of electromagnetic fields produced by bioelectric sources, and, because of their comprehensive understanding of the physical problem, they may contribute to the solution of biological problems.

This chapter first explains bioelectric phenomena at the cellular level and then discusses volume conductor fields of simple bioelectric sources as well as anatomically complex ones. The volume-conductor-field problem serves as a necessary link between cellular activity and gross, externally recorded biological signals, such as the ECG. We shall describe the functional organization of the peripheral (outside the brain and spinal cord) nervous system and then examine the ENG and EMG. Finally, we shall discuss the ECG, ERG, and EEG.

4.1 ELECTRICAL ACTIVITY OF EXCITABLE CELLS

Bioelectric potentials are produced as a result of electrochemical activity of a certain class of cells, known as *excitable cells,* that are components of nervous, muscular, or glandular tissue. Electrically they exhibit a *resting potential* and, when appropriately stimulated, an *action potential,* as the following paragraphs explain.

THE RESTING STATE

The individual excitable cell maintains a steady electrical potential difference between its internal and external environments. This resting potential of the internal medium lies in the range −50 to −100 mV, relative to the external medium.

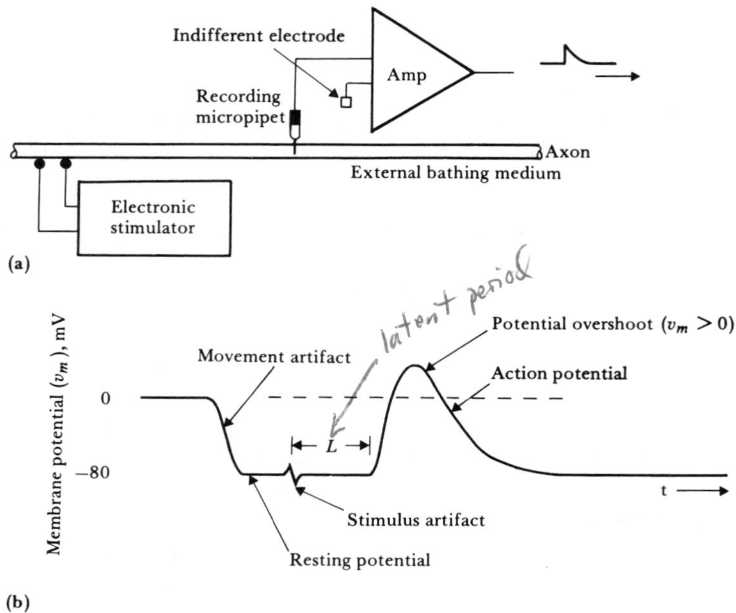

Figure 4.1 Recording of action potential of an invertebrate nerve axon (a) An electronic stimulator supplies a brief pulse of current to the axon, strong enough to excite the axon. A recording of this activity is made at a downstream site via a penetrating micropipet. (b) The movement artifact is recorded as the tip of the micropipet drives through the membrane to record resting potential. A short time later, an electrical stimulus is delivered to the axon; its field effect is recorded instantaneously at downstream measurement site as the stimulus artifact. The action potential proceeds along the axon at a constant propagation velocity. The time period L is the *latent period* or transmission time from stimulus to recording site.

Figure 4.1(a) shows how the resting potential is usually measured. A micromanipulator advances a microelectrode (see Section 5.8) close to the surface of an excitable cell and then, by small movements, pushes it through the cell membrane. For the membrane to seal properly around the penetrating tip, the diameter of the tip must be small relative to the size of the cell in which it is placed. Figure 4.1(b) shows a typical electrical recording from a single nerve fiber, including the dc offset potential (resting potential) that occurs upon penetration of the membrane. It also shows the transient disturbance of membrane potential (the action potential) when an adequate stimulus is given.

The cell membrane is a very thin (7–15 nm) lipoprotein complex that is essentially impermeable to intracellular protein and other organic anions (A^-). The membrane in the resting state is only slightly permeable to Na^+ and rather freely permeable to K^+ and Cl^-. The permeability of the resting membrane to potassium ion (P_K) is approximately 50–100 times larger than its permeability to sodium ion (P_{Na}).

For frog skeletal muscle, the K^+ concentration of the internal media is 140 mmol/liter, whereas that of the external media is 2.5 mmol/liter. Thus the concentration imbalance creates a diffusion gradient that is directed outward across the membrane. The movement of the K^+ along this diffusion gradient (while the nondiffusible anion component stays within the cell) is in such a direction as to make the interior of the cell more negative relative to the outside of the cell (that is, positive charge is removed from the interior). A transmembrane potential difference is thus established. The membrane may then be described electrically as a leaky capacitor. That is, it acts as a charge separator, yet it has a dielectric material (the lipoprotein complex of the membrane itself) that allows a leakage flow of ions across the membrane via pores. The electric field supported by the membrane capacitor is directed inward from positive to negative across the membrane, and it tends to inhibit the outward flow of positively charged ions (such as K^+) and the inward flow of negatively charged ions (such as Cl^-). Thus the diffusional and electrical forces acting across the membrane are opposed to one another, and a steady state is ultimately achieved. The membrane potential at which this steady state exists (considering K^+ to be the main ionic species involved in the resting state; that is, $P_K \gg P_{Na}$) is called the *equilibrium potential* for potassium E_K. It is measured in volts and is calculated from the Nernst equation,

$$E_K = \frac{RT}{nF} \ln \frac{[K]_o}{[K]_i} = 0.0615 \log_{10} \frac{[K]_o}{[K]_i} \quad (V) \tag{4.1}$$

at 37 °C (body temperature). Here n is the valence of K^+, $[K]_i$ and $[K]_o$ are the intracellular and extracellular concentrations of K^+ in moles per liter, R is the universal gas constant (Appendix), T is absolute temperature in K, and F is the Faraday constant (Appendix). Equation (4.1) gives a reasonably good approximation to the potential of the resting membrane, which indicates that the resting membrane is effectively a *potassium membrane.* A more accurate expression for the membrane equilibrium potential E that accounts for the influence of other ionic species in the internal and external media was first developed by Goldman (1943) and later modified by Hodgkin and Katz (1949), who assumed a constant electric field across the membrane.

$$E = \frac{RT}{F} \ln \left\{ \frac{P_K[K]_o + P_{Na}[Na]_o + P_{Cl}[Cl]_i}{P_K[K]_i + P_{Na}[Na]_i + P_{Cl}[Cl]_o} \right\} \tag{4.2}$$

Here E is the equilibrium transmembrane resting potential when net current through the membrane is zero and P_M is the *permeability coefficient* of the membrane for a particular ionic species M.

EXAMPLE 4.1 For frog skeletal muscle, typical values for the intracellular and extracellular concentrations of the major ion species (in millimoles per liter) are as follows.

Species	Intracellular	Extracellular
Na^+	12	145
K^+	155	4
Cl^-	4	120

Assuming room temperature (20 °C) and typical values of permeability coefficient for frog skeletal muscle ($P_{Na} = 2 \times 10^{-8}$ cm/s, $P_K = 2 \times 10^{-6}$ cm/s, and $P_{Cl} = 4 \times 10^{-6}$ cm/s), calculate the equilibrium resting potential for this membrane, using the Goldman equation.

ANSWER From (4.2):

$$E = 0.0581 \log_{10}\left[\frac{P_K(4) + P_{Na}(145) + P_{Cl}(4)}{P_K(155) + P_{Na}(12) + P_{Cl}(120)}\right]$$

$$= 0.0581 \log_{10}\left(\frac{26.9 \times 10^{-6}}{790.24 \times 10^{-6}}\right) = -85.3 \text{ mV}$$

This is close to typical measured values for the resting membrane potential in frog skeletal muscle.

Maintaining steady-state ionic imbalance between the internal and external media of the cell requires the continual active transport of ionic species against their normal electrochemical gradients. The active transport mechanism is located within the membrane and is sometimes referred to as the *sodium–potassium pump.* It actively transports Na^+ out of the cell and K^+ into the cell in the ratio $3Na^+:2K^+$. The associated pump current i_{NaK} is a net outward current that tends to increase the negativity of the intracellular potential. Energy for the pump is provided by a common source of cellular energy, adenosine triphosphate (ATP).

Thus the factors influencing the flow of ions across the membrane are (1) diffusion gradients, (2) the inwardly directed electric field, (3) membrane structure (availability of pores), and (4) active transport of ions against an established electrochemical gradient. The distribution of ions across the cell membrane and the structure of this membrane (P_K, P_{Na}, P_{Cl}) account for the resting potential. Potassium ions diffuse out of the cell according to their concentration gradient, while the nondiffusible organic anion component remains within, creating a potential difference across the membrane. There is thus a slight excess of cations outside the membrane and a slight excess of anions within. The number of ions responsible for the membrane potential, however, is very small relative to the total number present in the internal and external media. The Na^+ influx does not compensate for the K^+ efflux because, in the resting state, $P_{Na} \ll P_K$. Chloride ion diffuses inward down its concentration gradient, but its movement is balanced by the electrical gradient.

THE ACTIVE STATE

Another property of an excitable cell is its ability to conduct an action potential [Figure 4.1(b)] when adequately stimulated. An *adequate stimulus* is one that brings about a depolarization in a membrane that is sufficient to exceed the threshold potential of the membrane and thereby elicit an all-or-none action potential that travels in an unattenuated fashion at a constant conduction velocity along the membrane of the excitable cell. Because of the steady resting potential, the cell membrane is said to be *polarized.* A lessening of the magnitude of this polarization is called *depolarization;* an increase in its magnitude is referred to as *hyperpolarization.* The "all-or-none" property of the action potential means that the membrane potential goes through a very characteristic cycle: a change in potential from the resting level of a certain amount for a fixed duration of time. For a nerve fiber, $\Delta v \cong 120$ mV and the duration is approximately 1 ms. Further increases in intensity or duration of stimulus beyond that required for exceeding the threshold level produce only the same result.

The origin of the action potential lies in the voltage- and time-dependent nature of the membrane permeabilities (or equivalently, in electrical terms, membrane conductivities) to specific ions, notably Na^+ and K^+. As the membrane is depolarized, the membrane permeability to sodium P_{Na} (or, equivalently, the conductance of the membrane to sodium g_{Na}) is significantly increased. As a result, Na^+ rushes into the internal medium of the cell, bringing about further depolarization, which in turn brings about a further increase in g_{Na} (that is, g_{Na} is dependent on the voltage across the membrane). If the membrane threshold is exceeded, this process is self-regenerative and leads to *runaway* depolarization. Under these conditions, the membrane potential tends to approach the Nernst potential of sodium, E_{Na}, which has a value of about +60 mV.

The membrane potential never achieves this level, however, because of two factors: (1) g_{Na} is not only voltage-dependent but also time-dependent, and (as shown in Figure 4.2) it is relatively short-lived compared with the action potential. (2) There is a delayed increase in g_K that acts as a hyperpolarizing influence, tending to return the membrane to resting levels (Figure 4.2). As the membrane potential ultimately returns to the resting level, g_K is still elevated with respect to its resting value and returns slowly along an exponential time course. Because potassium ions continue to leave the cell during this time, the membrane hyperpolarizes and an undershoot is produced in the transmembrane potential waveform (v_m).

The calculated g_{Na} and g_K waveforms of Figure 4.2 are based on *voltage-clamp* data from squid axon. In voltage-clamp experiments, voltage in a mem-

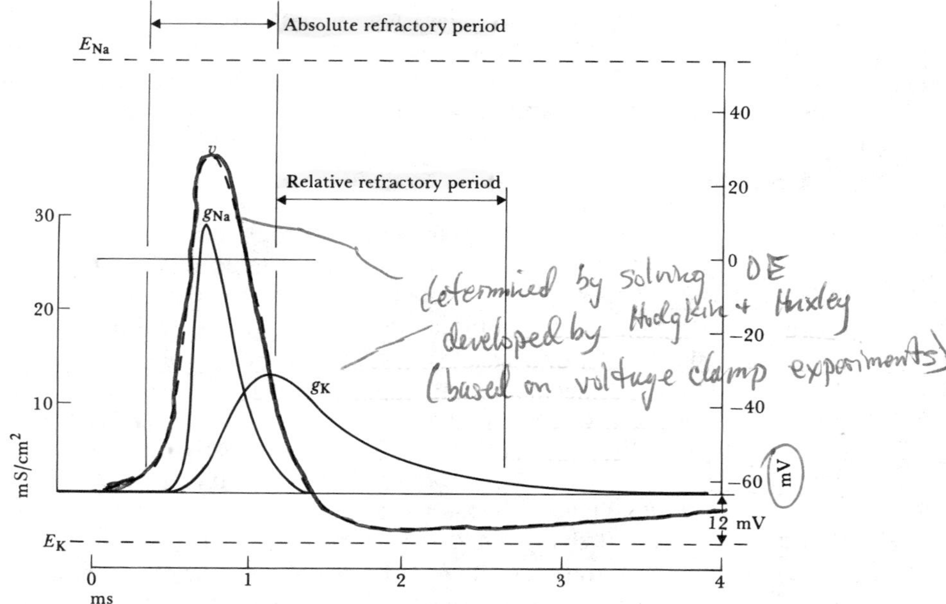

Figure 4.2 Theoretical action potential v and membrane ionic conductance changes for sodium (g_{Na}) and potassium (g_K) are obtained by solving the differential equations developed by Hodgkin and Huxley for the giant axon of the squid at a bathing medium temperature of 18.5 °C. E_{Na} and E_K are the Nernst equilibrium potentials for sodium and potassium across the membrane. (Modified from A. L. Hodgkin and A. F. Huxley, "A Quantitative Description of Membrane Current and Its Application to Conduction and Excitation in Nerve," *Journal of Physiology,* 1952, **117,** p. 530.)

brane is held at prescribed levels via a negative-feedback control circuit. Membrane currents in response to step changes in clamp voltages are studied in order to determine the voltage- and time-dependent nature of g_{Na} and g_K.

Figure 4.3 shows a network equivalent circuit describing the electrical behavior of a small increment of membrane. The entire nerve axon membrane could be characterized in a distributed fashion by utilizing an iterative structure of this same basic form.

When an excitable membrane has an action potential in response to an adequate stimulus, the ability of the membrane to respond to a second stimulus of any sort is markedly altered. During the initial portion of the action potential, the membrane cannot respond to any stimulus, no matter how intense. This interval is referred to as the *absolute refractory period.* It is followed by the *relative refractory period,* wherein an action potential can be elicited by an intense superthreshold stimulus (Figure 4.2). The existence of the refractory period produces an upper limit to the frequency at which an excitable cell may be repetitively discharged. For example, a nerve axon has an absolute

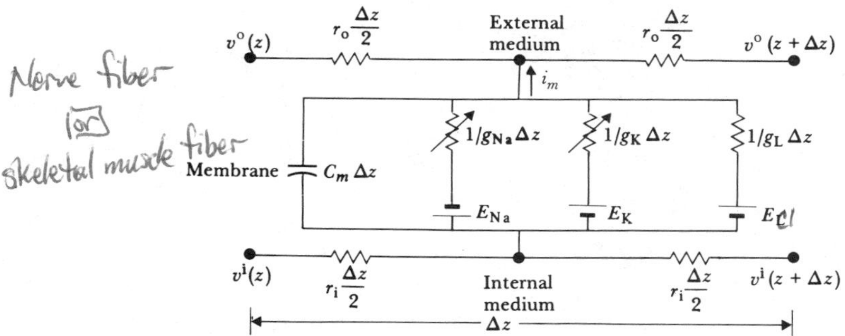

Figure 4.3 Diagram of network equivalent circuit of a small length (Δz) of an unmyelinated nerve fiber or a skeletal muscle fiber The membrane proper is characterized by specific membrane capacitance C_m ($\mu F/cm^2$) and specific membrane conductances g_{Na}, g_K, and g_{Cl} in mS/cm^2 (millisiemens/cm^2). Here an average specific leakage conductance is included that corresponds to ionic current from sources other than Na^+ and K^+ (for example, Cl^-). This term is usually neglected. The cell cytoplasm is considered simply resistive, as is the external bathing medium; these media may thus be characterized by the resistance per unit length r_i and r_o (Ω/cm), respectively. Here i_m is the transmembrane current per unit length (A/cm), and v^i and v^o are the internal and external potentials v at point z, respectively. (Modified from A. L. Hodgkin and A. F. Huxley, "A Quantitative Description of Membrane Current and Its Application to Conduction and Excitation in Nerve," *Journal of Physiology,* 1952, **117,** p. 501.)

refractory period of 1 ms and an upper limit of repetitive discharge of less than 1000 impulses/s.

For an action potential propagating along a single unmyelinated nerve fiber, the region of the fiber undergoing a transition into the active state (the *active region*) at an instant of time is usually small relative to the length of the fiber. Figure 4.4(a) shows schematically the charge distribution along the fiber in the vicinity of the active region. Note that the direction of propagation of the action potential (considered frozen in time) is to the left, and the membrane lying ahead of the active region is polarized, as in the resting state. A reversal of polarity is shown within the active region because of depolarization of the membrane to positive values of potential. The membrane lying behind the active zone is repolarized membrane.

From the indicated charge distribution, *solenoidal* (closed-path) current flows in the pattern shown in Figure 4.4(a). In the region ahead of the active zone, the ohmic potential drop across the membrane caused by this solenoidal current flowing outward through the membrane is of such a polarity as to reduce the magnitude of the transmembrane potential—that is, depolarize the membrane. When the membrane potential is depolarized to the threshold level (about 20 mV more positive than the resting potential), this region becomes activated as well. The same current pattern flowing behind the active

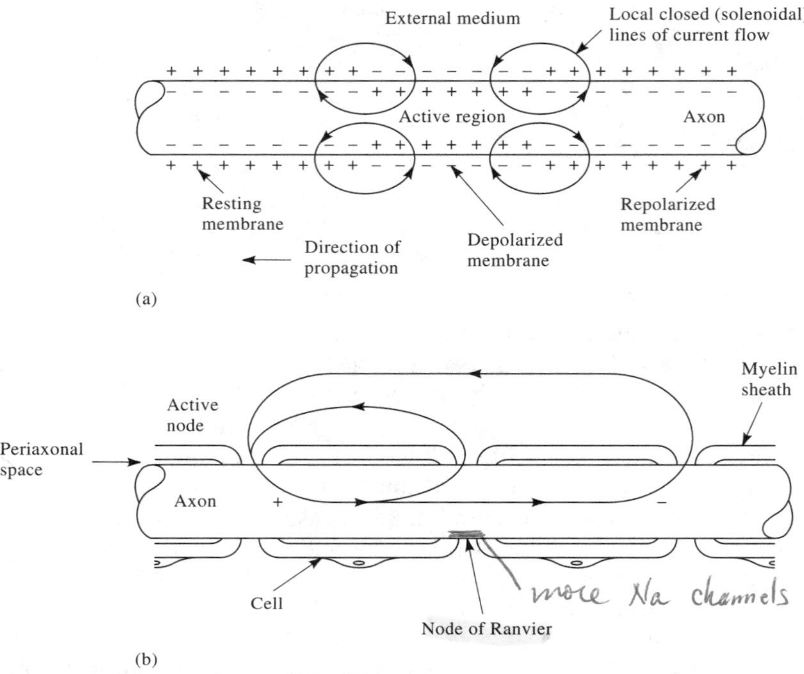

Figure 4.4 (a) Charge distribution in the vicinity of the active region of an unmyelinated fiber conducting an impulse. (b) Local circuit current flow in the myelinated nerve fiber.

region is ineffective in re-exciting the membrane, which is in the refractory state. The nature of this process is therefore self-excitatory, each new increment of membrane being brought to the threshold level by lines of current from the active source region. The membrane stays in the active state for only a brief period of time and ultimately repolarizes completely. In this way, the action potential propagates down the length of the fiber in an unattenuated fashion, the signal being built up at each point along the way.

Most neurons in invertebrates are unmyelinated, but most vertebrate neurons are myelinated. That is, the axon is insulated by a sheath of myelin, a lipoprotein complex formed from successive wrappings of the axon by a special support cell found along nerve fibers. In peripheral nerves—those that lie outside the central nervous system (CNS)—this support cell is known as a *Schwann cell.* In myelinated CNS neurons, this function is served by a special glial cell known as an *oligodendrogliocyte.* The myelin sheath is interrupted at regular intervals (1–2 mm, depending on the species) by nodes of Ranvier; a single Schwann cell thus provides the insulating myelin sheath covering of the axon between two successive nodes of Ranvier [Figure 4.4(b)]. The tightly wrapped membranes of the Schwann cell closely adhere to the axon membrane and effectively increase its thickness by a factor of 100. This substantially decreases the capacitance of the modified membrane and increases the transverse impedance to current flow in the internodal region of the fiber. Sodium ion channels are distributed in a nonuniform manner in myelinated fibers, being densely clustered at the nodes of Ranvier and much more sparely distributed in the internodal region (Shrager, 1987; Meiri *et al.,* 1986). Multiple types of potassium channels (fast-gated, slow-gated) are largely distributed in the paranodal regions lying adjacent to each node of Ranvier. These channels are distributed to a lesser extent throughout the remainder of the internodal region in both amphibian and mammalian species (Grissmer, 1986; Röper and Schwarz, 1989). Once the myelinated nerve fiber is activated, conduction proceeds through a process of local circuit current flow, much as in the case of the unmyelinated nerve fiber described earlier [Figure 4.4(a)]. There are differences, however, in that the sources for action current flow are localized at the nodes of Ranvier and are therefore not uniformly distributed along the axonal membrane, as in the case of the unmyelinated fiber. Myelination of the internode reduces leakage currents and improves the transmission properties of the cable-like myelinated fiber. Local circuit currents emanating from an active node have an exponentially diminishing effect over an axial distance spanning several internodal lengths. Accordingly, they contribute to a drop in nodal potential as current passes outward through a given inactive nodal membrane [Figure 4.4(b)].

Thus conduction in myelinated nerve fibers proceeds via the rapid, sequential activation of the nodes of Ranvier, and local circuit current provides the underlying mechanism for bringing the nodal membrane voltage to threshold. This process is frequently called *saltatory conduction* (from the Latin *saltare,* "to leap or dance"), because action potentials appear to leap from node to node. For an axon of a given diameter, myelination improves the conduction rate by a factor of approximately 20. By its very nature, however, the myelinated nerve fiber represents a more complicated, distributed bioelectric action current source than the unmyelinated nerve fiber.

Mathematical modeling studies of conduction in both unmyelinated and myelinated nerve fibers have appeared in the literature (Moore *et al.,* 1978). Extensions of these fundamental modeling methods have been made in a variety of areas, including (1) fiber branching and tapering (Stockbridge, 1988); (2) internodal potentials and after-potentials (Blight, 1985); (3) distributed parameter models of myelinated nerve fibers (Halter and Clark, 1991); (4) electric stimulation of myelinated nerve fibers (Wood and Cummins, 1985); (5) conduction in demyelinated nerve fibers (Waxman and Brill, 1978); and (6) conduction in unmyelinated, septated nerve fibers (Barach and Wikswo, 1987).

4.2 VOLUME CONDUCTOR FIELDS

A fundamental problem in electrophysiology is the problem of the single active cell immersed in a volume conductor (a salt solution simulating the composition of body fluids). A study of this problem provides considerable insight into other, more complex volume-conductor-field problems, such as the ENG, EMG, ECG, and so forth.

The problem consists of two parts: (1) the bioelectric source and (2) its bathing medium or electrical load. The bioelectric source is the active cell that behaves approximately as a constant-current source, delivering its current to the bathing medium under a large range of loading conditions. The source considered for the present is the single active, unmyelinated nerve fiber. The volume conductor is considered infinite in extent (that is, large relative to the extent of the electric field surrounding the nerve fiber). The lines of flow of the current emanating from the active fiber into the bathing medium of specific resistivity ρ are indicated schematically in Figure 4.4(a). This pattern of current flow is consistent with the charge distribution shown in Figure 4.4(a).

Because the action potential is assumed to be traveling down the fiber at a constant conduction velocity, a temporal waveform $v_m(t)$ can be converted easily to a spatial distribution $v_m(z)$, where z is the axial distance along the fiber. For a simple monophasic action potential, the associated potential waveform at the outer surface of the membrane is (1) triphasic in nature, (2) of greater spatial extent than the action potential, and (3) much smaller in peak-to-peak magnitude. Potentials in the extracellular medium of a single fiber fall off exponentially in magnitude with increasing radial distance from the fiber (potential zero within fifteen fiber radii). The peak to peak magnitude of potential at fiber surface depends on the active surface area (usually on the order of tens of μV).

If, for the case of the infinite volume conductor, ρ is increased, the potential at the field point increases, as it would in the case in which the volume conductor is made smaller (field point within the volume conductor). In each of these cases, the total extracellular resistance to current flow from the constant-current bioelectric source is greater, and (from Ohm's law) potential is increased.

If instead, we consider the source to be an active nerve trunk with its thousands of component nerve fibers simultaneously activated, the extracellular field in an infinite homogeneous bathing medium appears quite similar to that of a single fiber, as shown in Figure 4.5. The extracellular field potential is formed from the contributions of the superimposed electric fields of the component sources within the trunk. The general form of the extracellular response of a nerve trunk to electric stimulation is triphasic, it is in the low-microvolt range in amplitude, and loses both amplitude and high-frequency content at larger radial distances from the nerve trunk.

The sciatic nerve utilized in this experiment is actually a rather complex bioelectric source. It consists of large motor fibers running from the spinal cord to the leg muscles, as well as large and small sensory fibers running from sensory receptors in the leg and skin to the spinal cord. In Figure 4.5(a), the entire nerve trunk (containing both motor and sensory fibers) was simultaneously excited by a brief suprathreshold electric stimulus.

It is possible, however, to excite the motor and sensory components of

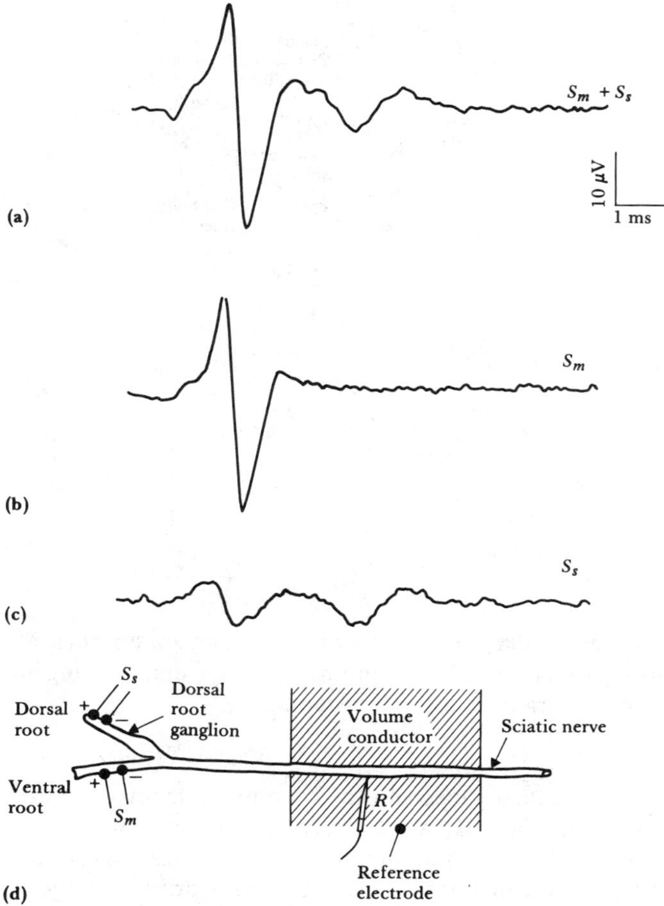

Figure 4.5 Extracellular field potentials (average of 128 responses) were recorded at the surface of an active (1-mm-diameter) frog sciatic nerve in an extensive volume conductor. The potential was recorded with (a) both motor and sensory components excited ($S_m + S_s$), (b) only motor nerve components excited (S_m), and (c) only sensory nerve components excited (S_s).

the trunk separately by isolating the nerve trunk in the vicinity of the spinal cord. Here the nerve trunk divides into a sensory branch (the *dorsal root*) and a motor branch (the *ventral root*). The results of separate motor and sensory stimulation are shown in Figure 4.5(b) and (c). We observe that stimulation of the many large motor fibers in the trunk provides the largest extracellular response. We also observe that stimulation of the sensory root actually excites at least two groups of sensory fibers—a group of large, fast fibers (group I) and a group of smaller, slower fibers (group II). Observing the extracellular waveform produced by combined stimulation [Figure 4.5(a)], we can see the approximate superposition of motor and sensory responses.

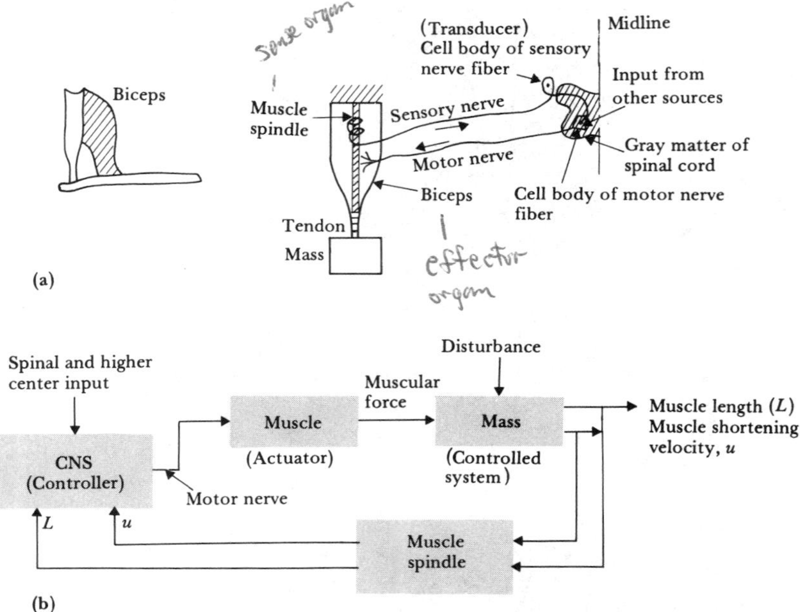

Figure 4.6 Schematic diagram of a muscle-length control system for a peripheral muscle (biceps) (a) Anatomical diagram of limb system, showing interconnections. (b) Block diagram of control system.

The volume conductor load of the active nerve trunk can also be altered and made more complicated. A relatively simple variation is to increase the specific resistivity of the bathing medium or to decrease the radial extent of the volume conductor, or both. These alterations produce larger extracellular potentials.

The foregoing discussion may be considered an explanation of the electrogenesis of the ENG, which is commonly recorded from the surface of an arm, a leg, or the face. The concepts introduced here apply directly to the interpretation of many bioelectric phenomena including those associated with the nervous system (e.g., evoked potential recordings from fiber tracts in the spinal cord and sensory centers in the brain), as well as active skeletal muscle (EMG), cardiac muscle (ECG), and smooth muscle (EGG).

4.3 FUNCTIONAL ORGANIZATION OF THE PERIPHERAL NERVOUS SYSTEM

THE REFLEX ARC

The spinal nervous system is functionally organized on the basis of what is commonly called the *reflex arc* [Figure 4.6(a)]. The components of this arc are as follows: (1) A *sense organ,* consisting of many individual sense receptors that respond preferentially to an environmental stimulus of a particular kind, such as pressure, temperature, touch, or pain. (2) A *sensory nerve,* containing many individual nerve fibers that perform the task of transmitting information (encoded in the form of action potential frequency) from a peripheral sense receptor to other cells lying within the central system (brain and spinal cord). (3) The CNS, which in this case serves as a central integrating station. Here information is evaluated, and, if warranted, a "motor" decision is implemented. That is, action potentials are initiated in motor-nerve fibers associated with the motor-nerve trunk. (4) A *motor nerve,* serving as a communication link between the CNS and peripheral muscle. (5) The *effector organ,* which consists, in this case, of skeletal muscle fibers that contract (shorten) in response to the driving stimuli (action potentials) conducted by motor-nerve fibers.

The simplest example of the behavior of the reflex arc is the knee-jerk reflex, in which the patellar tendon below the knee is given a slight tap that stretches specialized length receptors, called *muscle spindles,* within the muscle and subsequently excites them. This excitation results in action potentials that propagate along the sensory nerve that enters the spinal cord and communicates with CNS cells, specifically motoneurons. The resultant motor activity reflexly brings about contraction of the muscle that was initially stimulated, and the shortening muscle jerks the limb, producing the well-known knee-jerk response. Note that the initial stimulus to the muscle was a stretch, whereas the response was a contraction of the muscle. This simple reflex arc has many of the features of a negative-feedback loop, in which the control variable is muscle length [Figure 4.6(b)]. The CNS acts as the controller, the muscle spindle as a feedback length sensor, and the muscle-limb system as the process to be controlled.

JUNCTIONAL TRANSMISSION

Within the reflex arc there are intercommunicating links between neurons (neuro–neuro junctions) called *synapses.* There are also communicating links between neurons and muscle fibers (*neuromuscular junctions*) at a small specialized region of the muscle fiber referred to as the *end-plate region.* The junctional transmission process in each of these cases is electrochemical in nature. There is a prejunctional fiber involved in the neuromuscular junction that, when depolarized, releases a neurotransmitter substance, *Acetylcholine* (Ach) which diffuses across a very small fluid-filled gap region approximately 20 nm in thickness. The fluid filling the gap is assumed to be ordinary interstitial body fluid. Once the neurotransmitter reaches the postjunctional membrane, it combines with a membrane receptor complex that ultimately leads to a relatively brief transient depolarization of the membrane and to subsequent initiation of an action potential that propagates away from the junctional region. The electrochemical transmission process at the junction involves a time delay on the order of 0.5–1.0 ms. For a more detailed description of

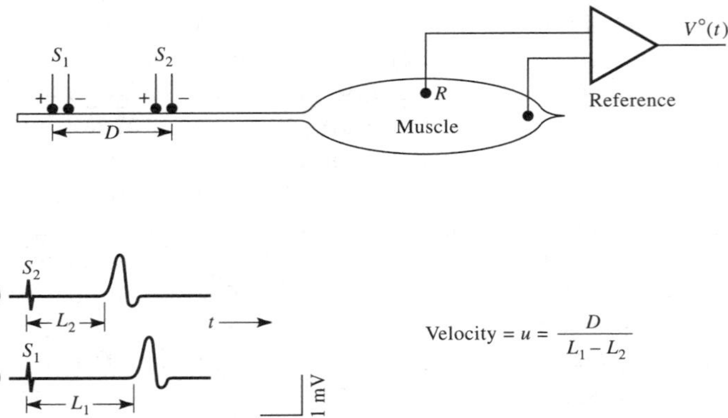

Figure 4.7 Measurement of neural conduction velocity via measurement of latency of evoked electrical response in muscle. The nerve was stimulated at two different sites a known distance D apart.

interneuronal and neuromuscular transmission, see Kandel *et al.* (1991, Chapters 9, 10, and 11).

Another time delay associated with the neuromuscular system is the delay between electrical activation of the musculature and the onset of mechanical contraction. This delay, which is referred to as *excitation-contraction time,* is a property of the muscle itself. When the muscle is repeatedly stimulated, the mechanical response summates. At high stimulation rates, the mechanical responses fuse into one continuous contraction called a *tetanus* (or *tetanic contraction*).

4.4 THE ELECTRONEUROGRAM (ENG)

Conduction velocity in a peripheral nerve is measured by stimulating a motor nerve at two points a known distance apart along its course. Subtraction of the shorter latency from the longer latency (Figure 4.7) gives the conduction time along the segment of nerve between the stimulating electrodes. Knowing the separation distance, we can determine the conduction velocity of the nerve, which has potential clinical value since, e.g., conduction velocity in a regenerating nerve fiber is slowed following nerve injury.

Although field potentials from nerves are of a much smaller amplitude than extracellular potentials from surrounding excitable muscle fibers, such potentials can be recorded with either concentric needle electrodes or surface electrodes (Chap. 5). Nerve field potentials can be evoked by applying stimuli to "mixed" nerves that contain both motor and sensory components (such as the ulnar nerve of the arm), in which case the resultant field potentials are derived from both types of active fibers.

Neural field potentials can also be elicited from a purely sensory nerve (such as the sural nerve in the leg) or from sensory components of a mixed nerve, in which the stimulation is applied in a manner that does not excite the motor components of the nerve. The study of field potentials from sensory nerves in general has been shown to be of considerable value in diagnosing peripheral nerve disorders.

Although conduction velocity and latency are the most generally useful parameters associated with peripheral nerve function, the characteristics of field potentials evoked in muscle supplied by the stimulated nerve are also important. When considering evoked muscle potentials, the duration of the response is frequently of interest, because a slowing of conduction in a few motor nerve fibers may, in fact, lead to late activation of a portion of the muscle, and the resultant field potential may be prolonged and polyphasic. When field potentials of the nerve are measured in such a case, temporal dispersion due to the slowed conduction in some of the fibers may lead to a significant decrease in the amplitude of the signal.

FIELD POTENTIALS OF SENSORY NERVES

Extracellular field responses from sensory nerves can be easily measured from the median or ulnar nerves of the arm by using ring-stimulating electrodes applied to the fingers (Figure 4.8). Recording at two sites a known distance apart along the course of the nerve enables one to compute the conduction velocity of the sensory nerve. In the case of the ulnar nerve (roughly, it supplies the third and fourth fingers), evoked neural potentials can be recorded from different sites along the course of the nerve as high as the armpit. In the case of the median nerve (roughly speaking, it supplies the index and the second fingers), potentials can be recorded from the nerve at and above the elbow.

Long pulses cause muscle contractions, limb movement, and undesired signals (*artifacts*). These are easily avoided by positioning the limb in a comfortable, relaxed posture and applying a brief, intense stimulus (square pulse of approximately 100-V amplitude with a duration of 100–300 μs). Such a stimulus excites the large, rapidly conducting sensory nerve fibers but not small pain fibers or surrounding muscle. To minimize artifacts caused by stimuli, we use a stimulus isolation unit (isolation transformer, diode-bridge circuit, optical coupler, and so on) to isolate the bipolar stimulating electrodes from ground. A patient ground is placed at the wrist between the stimulating and recording electrodes to provide a ground point for the passive electric field coupling from the stimulating electrodes. The skin should be abraded under both the stimulating and the recording electrode (Section 5.5) to reduce skin resistance and ensure good contact.

Clinically, field potentials are recorded using high-gain, high-input-impedance differential preamplifiers with good common-mode rejection capability and low inherent amplifier noise (Section 6.5). Figure 4.8 shows that the measured ENGs are on the order of 10 μV, and power-line interference is sometimes a problem even with good amplifier common-mode properties.

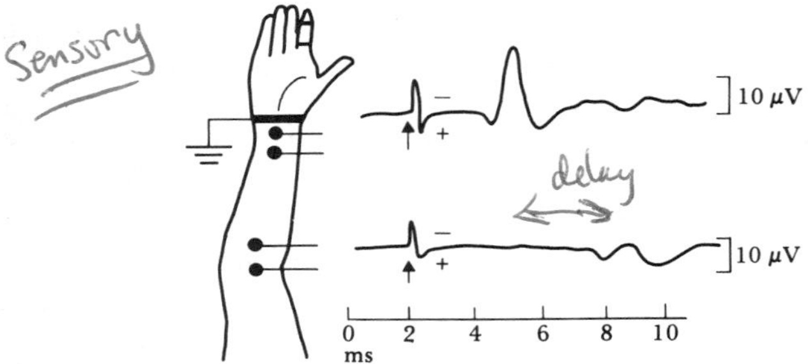

Figure 4.8 Sensory nerve action potentials evoked from median nerve of a healthy subject at elbow and wrist after stimulation of index finger with ring electrodes. The potential at the wrist is triphasic and of much larger magnitude than the delayed potential recorded at the elbow. Considering the median nerve to be of the same size and shape at the elbow as at the wrist, we find that the difference in magnitude and waveshape of the potentials is due to the size of the volume conductor at each location and the radial distance of the measurement point from the neural source. (From J. A. R. Lenman and A. E. Ritchie, *Clinical Electromyography,* 2nd ed., Philadelphia: Lippincott, 1977; reproduced by permission of the authors.)

The input leads should be properly twisted together and shielded. In addition, if warranted, the subject could be placed in an adequately shielded room or cage.

A further step we can use to enhance the signal-to-noise ratio in the presence of random noise (for the most part generated by the amplifier) is to use a *signal averager* (Section 6.8).

MOTOR-NERVE CONDUCTION VELOCITY

In vivo measurement of the conduction velocity of a motor nerve may be obtained as shown in Figure 4.7. For example, the peroneal nerve of the left leg may be stimulated first behind the knee and second behind the ankle. A muscular response is obtained from the side of the foot, using surface or needle electrodes.

REFLEXLY EVOKED FIELD POTENTIALS

When a peripheral nerve is stimulated and an evoked field potential is recorded in the muscle it supplies, it is sometimes possible to record a second potential that occurs later than the initial response. As the neural stimulus site is brought progressively closer to the muscle, the latency of the first response decreases, while the latency of the second response is increased. This behavior of the second reponse indicates that to activate the muscle, the stimulus must travel along the nerve toward the central nervous system (proximally) for some distance before ultimately traveling in the opposite direction (distally). The

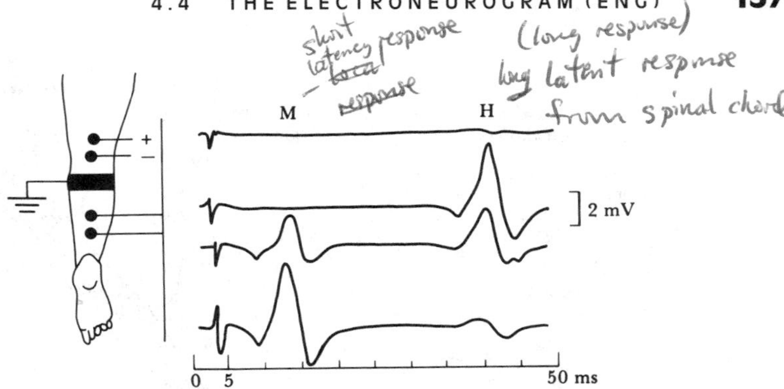

Figure 4.9 The H reflex The four traces show potentials evoked by stimulation of the medial popliteal nerve with pulses of increasing magnitude (the stimulus artifact increases with stimulus magnitude). The later potential or H wave is a low-threshold response, maximally evoked by a stimulus too weak to evoke the muscular response (M wave). As the M wave increases in magnitude, the H wave diminishes. (From J. A. R. Lenman and A. E. Ritchie, *Clinical Electromyography,* 2nd ed., Philadelphia: Lippincott, 1977; reproduced by permission of the authors.)

latency of the second response is such that the activity could have traveled proximally along sensory nerves as far as the spinal cord to elicit a spinal reflex.

If the posterior tibial nerve in the leg is stimulated, a late potential can be evoked from the triceps sural muscle (Figure 4.9). This long latency response has a low threshold and appears at stimulus intensities that are well below the levels required to elicit the conventional (short-latency) *M wave.* This long-latency potential—known as the *H wave*—was discovered by Hoffman (Figure 4.9). Its latency indicates that it is a spinal reflex. It is, in fact, the electrical homolog of the simple "ankle-jerk" reflex.

Thus, when a mixed peripheral nerve such as the posterior tibial nerve is stimulated by a stimulus of low intensity, only fibers of large diameter are stimulated because they have the lowest threshold. These large fibers are sensory fibers from muscle spindles that conduct toward the CNS and ultimately connect with motor fibers in the spinal cord via a single synapse. The motoneurons discharge and produce a response in the gastrocnemius muscle of the leg (the H wave). With a stimulus of medium intensity, smaller motor fibers in the mixed nerve are stimulated in addition to the sensory fibers, producing a direct, short-latency muscle response, the M wave (Figure 4.9). With still stronger stimuli, impulses conducted centrally along the motor fibers may interfere with the production of the H wave (these excited motor fibers are in their refractory period) so that only an M wave is produced (Figure 4.9). The amplitude of the H response depends on the number of motoneurons discharged. Its amplitude is also somewhat variable as a result of fluctuating background neural conditions within the spinal cord. These neural disturbances are provided by the activity of other spinal and higher center neurons impinging on the motoneuron(s) involved in the reflex.

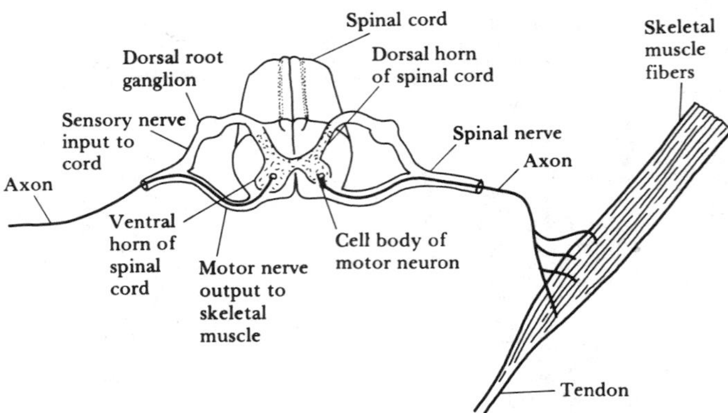

Figure 4.10 Diagram of a single motor unit (SMU), which consists of a single motoneuron and the group of skeletal muscle fibers that it innervates. Length transducers [muscle spindles, Figure 4.6(a)] in the muscle activate sensory nerve fibers whose cell bodies are located in the dorsal root ganglion. These bipolar neurons send axonal projections to the spinal cord that divide into a descending and an ascending branch. The descending branch enters into a simple reflex arc with the motor neuron, while the ascending branch conveys information regarding current muscle length to higher centers in the CNS via ascending nerve fiber tracts in the spinal cord and brain stem. These ascending pathways are discussed in Section 4.8.

4.5 THE ELECTROMYOGRAM (EMG)

Skeletal muscle is organized functionally on the basis of the *motor unit* (see Figure 4.10). The motor unit is the smallest unit that can be activated by a volitional effort, in which case all constituent muscle fibers are activated synchronously. The component fibers of the motor units extend lengthwise in loose bundles along the muscle. In cross section, however, the fibers of a given motor unit are interspersed with fibers of other motor units. Thus the component muscle fibers of the *single motor unit* (SMU) constitute a distributed, unit bioelectric source located in a volume conductor consisting of all other muscle fibers, both active and inactive. The evoked extracellular field potential from the active fibers of an SMU has a triphasic form of brief duration (3–15 ms) and an amplitude of 20–2000 μV, depending on the size of the motor unit. The frequency of discharge usually varies from 6 to 30 per second.

One of the disadvantages of recording the EMG by using the convenient surface electrodes is that they can be used only with superficial muscles and are sensitive to electrical activity over too wide an area. Various types of monopolar, bipolar, and multipolar insertion-type electrodes are commonly used in electromyography for recording from deep muscles and from SMUs. These types of electrodes generally record local activity from small regions within the muscle in which they are inserted. Often a simple fine-tipped monopolar needle electrode can be used to record SMU field potentials even during powerful voluntary contractions. Bipolar recordings are also employed. Various types of electrodes are discussed in Chapter 5.

Figure 4.11 shows motor unit potentials from the normal dorsal interosseus muscle under graded levels of contraction. At high levels of effort, many superimposed motor unit responses give rise to a complicated response (the *interference pattern*) in which individual units can no longer be distinguished. In interpreting Figure 4.11, note that when a muscle contracts progressively under volition, active motor units increase their rate of firing and new (previously inactive) motor units are also recruited.

The shape of SMU potentials is considerably modified by disease. In peripheral neuropathies, partial denervation of the muscle frequently occurs and is followed by regeneration. Regenerating nerve fibers conduct more slowly than healthy axons. In addition, in many forms of peripheral neuropathy, the excitability of the neurons is changed and there is widespread slowing of nerve conduction. One effect of this is that neural impulses are more difficult to initiate and take longer in transit to the muscle, generally causing scatter or desynchronization in the EMG pattern.

A number of mathematical modeling studies of single-fiber and multiple-fiber (single motor unit) action potentials have appeared in the literature (Nandedkar *et al.,* 1985; Ganapathy *et al.,* 1987). Signal processing methods have been employed in EMG analysis (Reucher *et al.,* 1987), as have automatic techniques for the detection, decomposition, and analysis of EMG signals (Mambrito and DeLuca, 1984; Gerber *et al.,* 1984).

4.6 THE ELECTROCARDIOGRAM (ECG)

ANATOMY AND FUNCTION OF THE HEART

The heart serves as a four-chambered pump for the circulatory system (Figure 4.12). The main pumping function is supplied by the ventricles, and the atria are merely antechambers to store blood during the time the ventricles are pumping. The resting or filling phase of the heart cycle is referred to as *diastole.* The contractile or pumping phase is called *systole.* The smooth, rhythmic contraction of the atria and ventricles has an underlying electrical precursor in the form of a well-coordinated series of electrical events that takes place within the heart. That this set of electrical events is intrinsic to the heart itself is well demonstrated when the heart (particularly that of cold-blooded vertebrates such as the frog or turtle) is removed from the body and placed in a nutrient medium (such as glucose-Ringer solution). The heart continues to beat rhythmically for many hours. The coordinated contraction of the atria

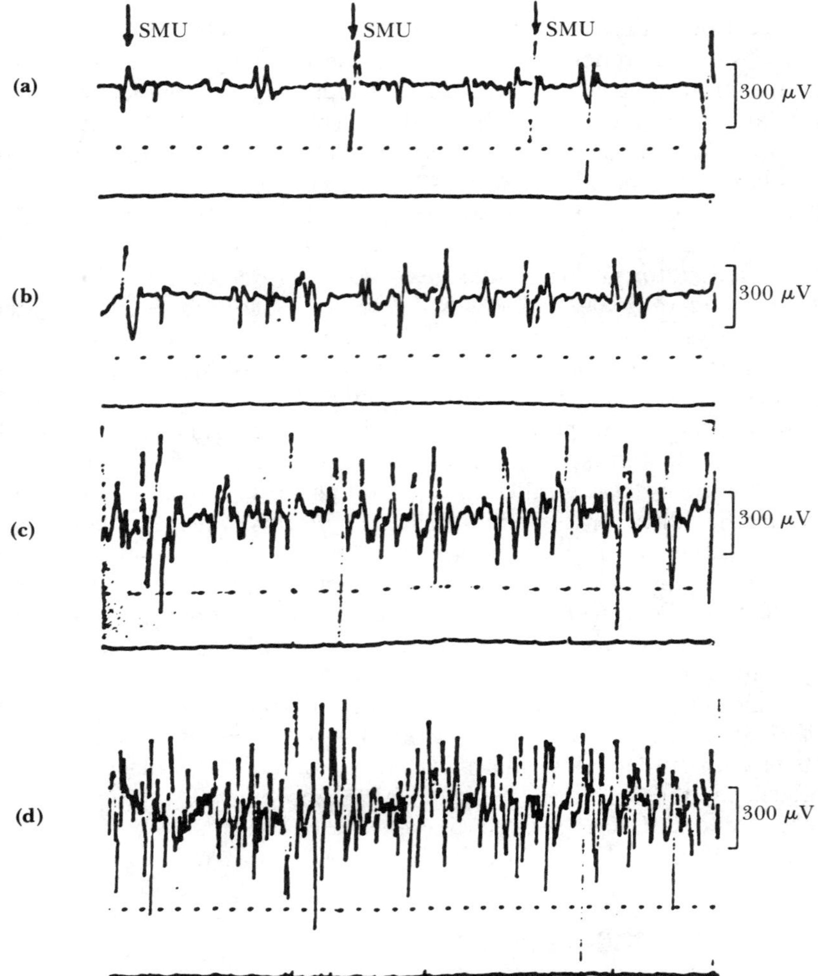

Figure 4.11 Motor unit action potentials from normal dorsal interosseus muscle during progressively more powerful contractions. In the interference pattern (c), individual units can no longer be clearly distinguished. (d) Interference pattern during very strong muscular contraction. Time scale is 10 ms per dot. (From J. A. R. Lenman and A. E. Ritchie, *Clinical Electromyography*, 2nd ed., Philadelphia: Lippincott, 1977; reproduced by permission of the authors.)

and ventricles is set up by a specific pattern of electrical activation in the musculature of these structures. Moreover, the electrical activation patterns in the walls of the atria and ventricles are initiated by a coordinated series of events in the "specialized conduction system" of the heart (Figure 4.12).

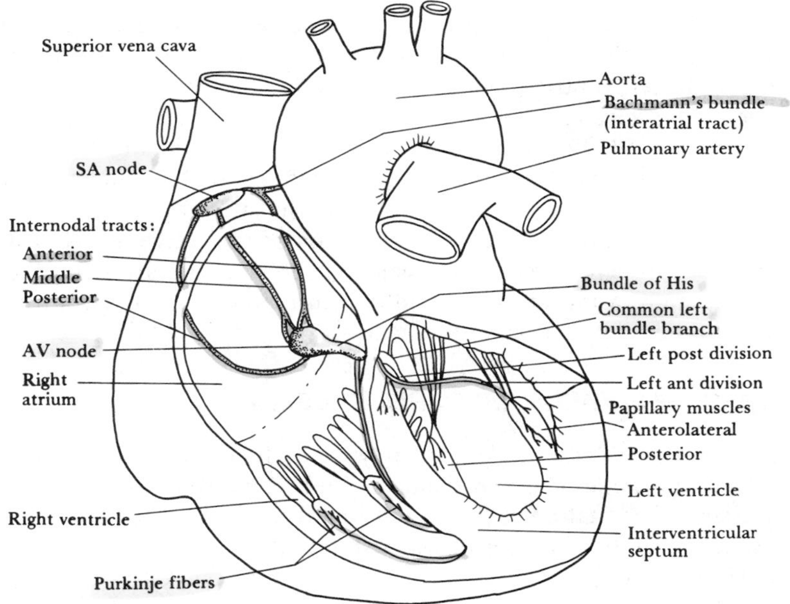

Figure 4.12 Distribution of specialized conductive tissues in the atria and ventricles, showing the impulse-forming and conduction system of the heart The rhythmic cardiac impulse originates in pacemaking cells in the sinoatrial (SA) node, located at the junction of the superior vena cava and the right atrium. Note the three specialized pathways (anterior, middle, and posterior internodal tracts) between the SA and atrioventricular (AV) nodes. Bachmann's bundle (interatrial tract) comes off the anterior internodal tract leading to the left atrium. The impulse passes from the SA node in an organized manner through specialized conducting tracts in the atria to activate first the right and then the left atrium. Passage of the impulse is delayed at the AV node before it continues into the bundle of His, the right bundle branch, the common left bundle branch, the anterior and posterior divisions of the left bundle branch, and the Purkinje network. The right bundle branch runs along the right side of the interventricular septum to the apex of the right ventricle before it gives off significant branches. The left common bundle crosses to the left side of the septum and splits into the anterior division (which is thin and long and goes under the aortic valve in the outflow tract to the anterolateral papillary muscle) and the posterior division (which is wide and short and goes to the posterior papillary muscle lying in the inflow tract). (From B. S. Lipman, E. Massie, and R. E. Kleiger, *Clinical Scalar Electrocardiography.* Copyright © 1972 by Yearbook Medical Publishers, Inc., Chicago. Used with permission.)

In relation to the heart as a whole, the specialized conduction system is very small. It constitutes only a minute portion of the total mass of the heart. The wall of the left ventricle (Figure 4.12) is 2.5–3.0 times as thick as the wall of the right ventricle, and the intraventricular septum is nearly as thick as the left ventricular wall. The major portion of the muscle mass of the ventricle consists of the free walls of the right and left ventricles and the septum. Considering the heart as a bioelectric source, the strength of this source can be expected to be directly related to the mass of the active muscle (that is, to the number of active myocardial cells). Therefore, the atria and the free walls and septum of the ventricles can be considered the major contributors to external potential fields from the heart.

ELECTRICAL BEHAVIOR OF CARDIAC CELLS

The heart comprises several different types of tissues (SA and AV nodal tissue; atrial, Purkinje, and ventricular tissue). Representative cells of each type of tissue differ anatomically to a considerable degree. They are all electrically excitable, and each type of cell exhibits its own characteristic action potential (Figure 4.13).

THE VENTRICULAR CELL

The ventricular myocardium is composed of millions of individual cardiac cells ($15 \times 15 \times 150$ μm long). Figure 4.14 is a drawing of a small section of cardiac muscle as seen under light microscopy. The individual cells are relatively long and thin, and although they run generally parallel to one another, there is considerable branching and interconnecting (*anastamosing*). The cells are surrounded by a plasma membrane that makes end-to-end contact with adjacent cells at a dense structure known as the *intercalated disk* (Figure 4.14). Each fiber contains many contractile *myofibrils* that follow the axis of the cell from one end (intercalated disk) to the other. These myofibrils constitute the "contractile machinery" of the fiber. The component cells of cardiac tissue are in intimate contact at the intercalated disks, both electrically and mechanically, so the heart muscle functions as a unit (a *functional syncytium*).

Prior to excitation, the typical ventricular cell has a resting potential of approximately −90 mV. The initial rapid depolarization phase has a rate of rise that is usually greater than 150 V/s. This phase is followed by an initial rapid repolarization that leads to a maintained depolarizing plateau region lasting approximately 200–300 ms. A final repolarization phase restores membrane potential to the resting level and is maintained for the remainder of the cardiac cycle. The duration of the action potential waveform is collectively referred to as *electrical systole;* the resting phase is referred to as *electrical diastole.*

Most models of membrane excitability that have been used in cardiac electrophysiology are of the Hodgkin–Huxley (HH) type (Hodgkin and Hux-

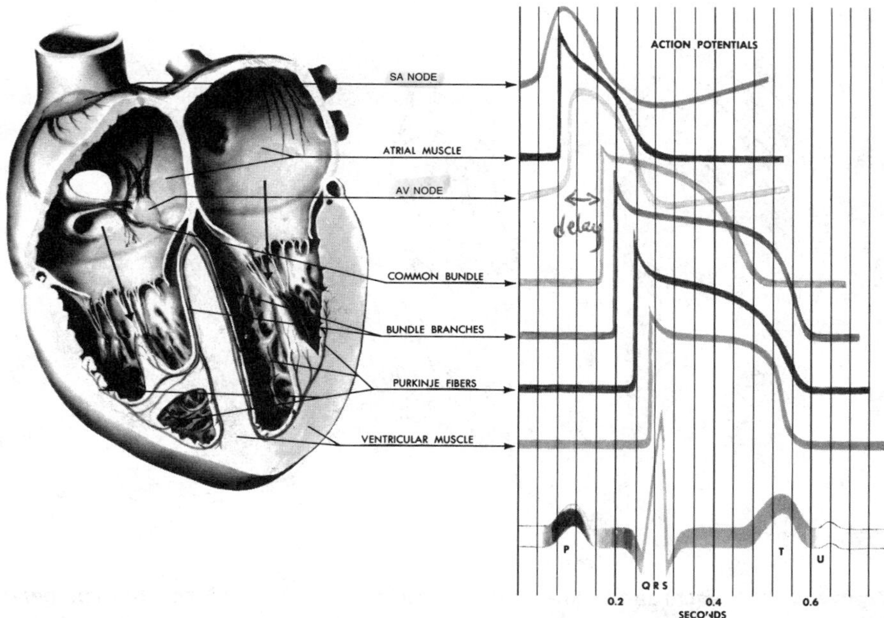

Figure 4.13 Representative electric activity from various regions of the heart The bottom trace is a scalar ECG, which has a typical QRS amplitude of 1–3 mV. (© Copyright 1969 CIBA Pharmaceutical Company, Division of CIBA-GEIGY Corp. Reproduced, with permission, from *The Ciba Collection of Medical Illustrations,* by Frank H. Netter, M.D. All rights reserved.)

ley, 1952). The HH formalism was first applied to Purkinje fibers of the specialized conduction system by Noble (1962). This model was later extensively revised by McAllister, Noble, and Tsien (MNT) (1975), and variations have been used in simulations of the electrophysiological responses of ventricular (Beeler and Reuter, 1977) and sinoatrial (SA) pacemaker cells (Yanagihara *et al.,* 1980). More recent experimental findings have necessitated significant changes in the MNT model, in order to incorporate mathematical descriptions

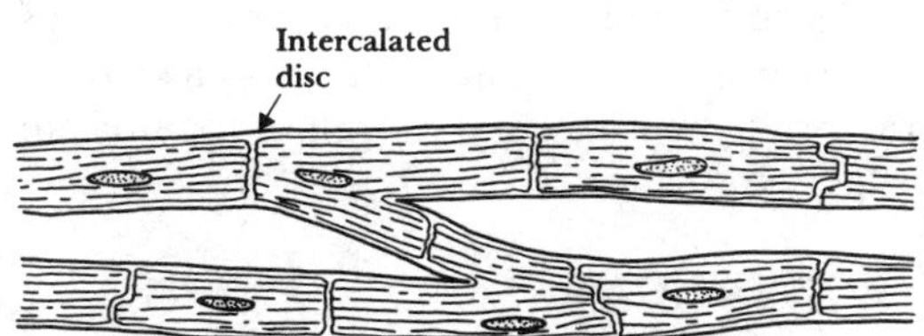

Figure 4.14 The cellular architecture of myocardial fibers Note the centroid nuclei and transverse intercalated disks between cells.

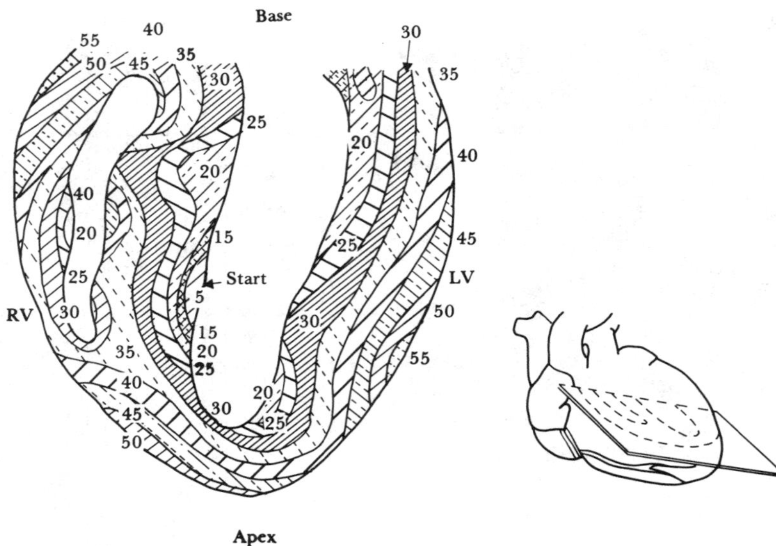

Figure 4.15 Isochronous lines of ventricular activation of the human heart Note the nearly closed activation surface at 30 ms into the QRS complex. (Modified from "The Biophysical Basis for Electrocardiography," by R. Plonsey, in *CRC Critical Reviews in Bioengineering,* **1,** 1, p. 5, 1971, © The Chemical Rubber Co., 1971. Used by permission of The Chemical Rubber Co. Based on data by D. Durrer *et al.,* "Total Excitation of the Isolated Human Heart," 1970, *Circulation,* **41,** 899–912, by permission of the American Heart Association, Inc.)

of the ATP-dependent Na^+–K^+ pump, the Na^+–Ca^{2+} exchanger, and a lumped fluid compartment model for describing intracellular and extracellular concentration changes in Na^+, K^+, and Ca^{2+} [for example, see Coulombe and Coraboeuf (1983)]. A comprehensive Purkinje fiber model incorporating these additional features has been developed by DiFrancesco and Noble (DN) (1985).

This type of HH model has been applied to characterization of the membrane properties of the rabbit SA node (Demir *et al.,* 1994), the rabbit atrium (Earm and Noble, 1990), and the guinea pig ventricle (Nordin, 1993; Luo and Rudy, 1994). These models include membrane pumps and the Na–Ca exchanger, thus expanding the explanatory range of a given cell model, making it possible to characterize the abnormal as well as the normal electrical behavior of a cardiac cell.

VENTRICULAR ACTIVATION

Investigators have conducted studies of ventricular activation on experimental animals using multiple "plunge-type" electrodes inserted into many sites in the heart (Spach and Barr, 1975) (see Figure 4.15). The time of arrival of electrical activation is noted, and *isochronous* (synchronously excited) excitation surfaces can be mapped. Figure 4.15 shows a plot of isochronous lines of activation for the perfused heart of a human who died from a noncardiac condition. Note that activation first takes place on the septal surface of the left ventricle (5 ms into the QRS complex) and that the activity spreads with increasing time in a direction from left to right across the septum. At 20 ms, several regions of the right and left ventricles are simultaneously active. As time increases, excitation spreads and tends to become more confluent. For example, at 30 ms a nearly closed activation surface is seen. Excitation then proceeds in a relatively uniform fashion in an *epicardial* (outside the heart) direction. The apex of the heart is activated roughly in the period 30–40 ms, along with other sites on the right and left ventricular walls where "breakthrough" of activation has occurred. From both Figure 4.15 and data taken in other planes, we can see that the posterior-basal region of the heart is the last region activated.

The isochronous electromotive surface propagates through the myocardium in an outward direction from the *endocardium* (the inside of the heart). The seat of this electromotive surface is, of course, the individual cardiac cell. In a localized region of the heart, however, many of these cells are active simultaneously because of the high degree of electrical interaction between cells. The anatomical substrate for this electrical interaction is the high degree of branching of individual cardiac cells and the low resistance of the intercalated disks at the junctions between cells (Barr *et al.,* 1965).

BODY-SURFACE POTENTIALS

The preceding section dealt with the sequence of events involved in electrical activation of the ventricle. This activation sequence leads to the production of closed-line action currents that flow in the thoracic volume conductor (considered a purely passive medium containing no electric sources or sinks). Potentials measured at the outer surface of this medium—that is, on the body surface—are referred to as *electrocardiograms,* or ECGs.

In the electrocardiographic problem, the heart is viewed as an electrical equivalent generator. A common assumption is that, at each instant of time in the sequence of ventricular activation, the electrical activity of the heart can be represented by a net equivalent current dipole located at a point that we call the *electrical center* of the heart. This center is assumed to lie within the anatomical boundaries of the heart.

Of course, several regions of both ventricles may be active simultaneously (as in Figure 4.15). In this case, the electrical activity of each region at any instant could be thought of as being represented by a current dipole and a net dipolar contribution from all active areas determined at the electrical center. The thoracic medium can be considered the resistive load of this equivalent cardiac generator. There is attenuation of the field with increasing distance from the source, as well as ohmic potential drops measured between surface points (between points A and B in Figure 4.16, for instance) or between

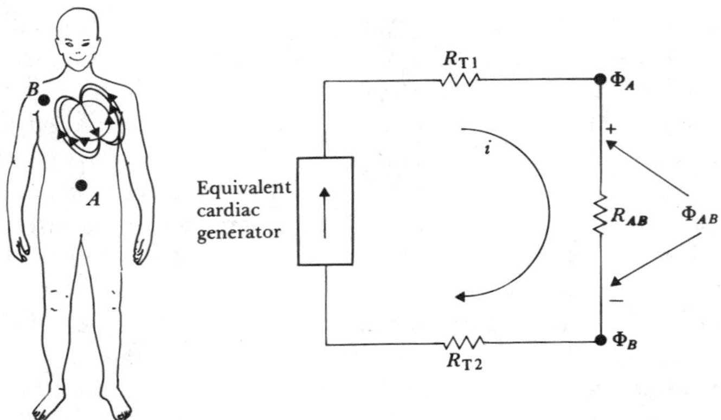

Figure 4.16 The electrocardiographic problem Points A and B are arbitrary observation points on the torso, R_{AB} is the resistance between them, and R_{T1}, R_{T2} are lumped thoracic medium resistances. The bipolar ECG scalar lead voltage is $\Phi_A - \Phi_B$, where these voltages are both measured with respect to an indifferent reference potential.

a single surface point and an assigned reference point. The general volume conductor problem is illustrated in a highly schematic fashion in Figure 4.16.

A scalar lead gives the magnitude of a single body-surface potential difference plotted versus time. A typical scalar electrocardiographic lead is shown in Figure 4.13 (bottom), where the significant features of the waveform are the P, Q, R, S, and T waves, the durations of each wave, and certain time intervals such as the P–R, S–T, and Q–T intervals. Figure 4.13 also shows the temporal relationship between single transmembrane cellular activities in various regions of the heart (atria, ventricles, and specialized conduction system) and this typical ECG waveform.

Clearly the P wave is produced by atrial depolarization, the QRS complex primarily by ventricular depolarization, and the T wave by ventricular repolarization. The manifestations of atrial repolarization are normally masked by the QRS complex. The P–R and S–T intervals are normally at zero potential, the P–R interval being caused mainly by conduction delay in the AV node. The S–T segment is related to the average duration of the plateau regions of individual ventricular cells. A small additional wave, called the U wave, is sometimes recorded temporally after the T wave. It is an inconstant finding, believed to be due to slow repolarization of ventricular papillary muscles.

Section 6.2 describes the 12 standard leads that constitute a diagnostic ECG, so they will not be considered further here.

NORMAL AND ABNORMAL CARDIAC RHYTHMS

Each beat of the normal human heart originates in the SA node. The normal heart rate is approximately 70 beats per minute (bpm). The rate is slowed (*bradycardia*) during sleep and is accelerated (*tachycardia*) by emotion, exercise, fever, and many other stimuli. Detailed aspects of the control that the nervous system has over heart rate are beyond the scope of this book; the reader interested in further discussion is referred to Levy and Berne (1988). Because many parts of the heart possess an inherent rhythmicity (nodal tissue, Purkinje fibers of the specialized conduction system, and atrial tissues, for example), any part under abnormal conditions can become the dominant cardiac pacemaker. This can happen when the activity of the SA node is depressed, when the bundle of His is interrupted or damaged, or when an abnormal (ectopic) focus or site in the atria or in specialized conduction-system tissue in the ventricles discharges at a rate faster than the SA node.

When the bundle of His is interrupted completely, the ventricles beat at their own slow inherent rate (the *idioventricular rhythm*). The atria continue to beat independently at the normal sinus rate, and complete or third-degree block is said to occur [Figure 4.17(a)]. The idioventricular rate in human beings is approximately 30–45 beats/min.

When the His bundle is not completely interrupted, incomplete heart block is present. In the case of *first-degree heart block,* all atrial impulses reach the ventricles, but the P–R interval is abnormally prolonged because of an increase in transmission time through the affected region [Figure 4.17(b)]. In the case of *second-degree heart block,* not all atrial impulses are conducted to the ventricles. There may be, for example, one ventricular beat every second or third atrial beat (2:1 block, 3:1 block, and so on).

In another form of incomplete heart block involving the AV node, the P–R interval progressively lengthens until the atrial impulse fails to conduct to the ventricle (*Wenckebach phenomenon*). The first conducted beat after the pause (or dropped beat) has a shorter P–R interval (sometimes of normal length) than any subsequent P–R interval. Then the process of the lengthening of the P–R interval begins anew, progressing over several cardiac cycles until another beat is dropped. The electrocardiographic sequence starting with the ventricular pause and ending with the next blocked atrial beat constitutes a *Wenckebach period.* The ratio of the number of P waves to QRS complexes determines the block (for example, 6:5 or 5:4 Wenckebach periods).

When one branch of the bundle of His is interrupted, causing right- or left-bundle-branch block, excitation proceeds normally down the intact bundle and then sweeps back through the musculature to activate the ventricle on the blocked side. The ventricular rate is normal, but the QRS complexes are prolonged and deformed.

ARRHYTHMIAS

A portion of the myocardium (or the AV node or specialized conduction system) sometimes becomes "irritable" and discharges independently. This site is then referred to as an *ectopic focus*. If the focus discharges only once, the result is a beat that occurs before the next expected normal beat, and the cardiac rhythm is therefore transiently interrupted. (With respect to atrial,

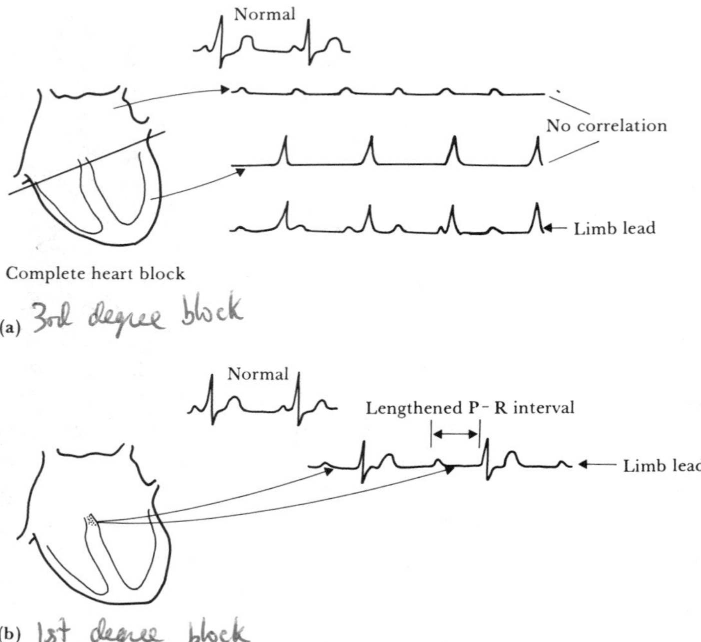

Figure 4.17 Atrioventricular block (a) Complete heart block. Cells in the AV node are dead and activity cannot pass from atria to ventricles. Atria and ventricles beat independently, ventricles being driven by an ectopic (other-than-normal) pacemaker. (b) AV block wherein the node is diseased (examples include rheumatic heart disease and viral infections of the heart). Although each wave from the atria reaches the ventricles, the AV nodal delay is greatly increased. This is first-degree *heart block.* (Adapted from Brendan Phibbs, *The Human Heart,* 3rd ed., St. Louis: The C.V. Mosby Company, 1975.)

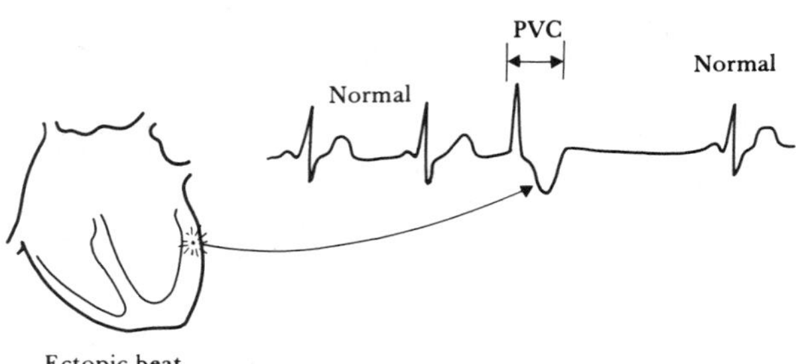

Figure 4.18 Normal ECG followed by an ectopic beat An irritable focus, or *ectopic pacemaker,* within the ventricle or specialized conduction system may discharge, producing an extra beat, or *extrasystole,* that interrupts the normal rhythm. This extrasystole is also referred to as a premature ventricular contraction (PVC). (Adapted from Brendan Phibbs, *The Human Heart,* 3rd ed., St. Louis: The C.V. Mosby Company, 1975.)

nodal, or ventricular *ectopic beat,* see Figure 4.18.) If the focus discharges repetitively at a rate that exceeds that of the SA node, it produces rapid regular tachycardia. [With respect to atrial, nodal, or ventricular paroxysmal tachycardia or atrial flutter, see Figure 4.19(a) and (b).] A rapidly and irregularly discharging focus or, more likely, a group of foci in the atria or ventricles may be the underlying mechanism responsible for atrial or ventricular fibrillation [Figure 4.20(a) and (b)].

Rhythm disturbances can arise from sources other than ectopic foci or competing pacemakers. A feasible alternative is a *circus re-excitation* or *reentrant* mechanism (Allessie *et al.,* 1973). This concept assumes a region of depressed conductivity within the atrium, Purkinje system, or ventricle. It is therefore *ischemic* (deficient in its blood supply) relative to surrounding normal tissue. This brings about pronounced electrophysiological changes in the ischemic zone and a decreased velocity of conduction (see Figure 4.21).

Propagation in this area is slow enough to permit other areas to recover from initial excitation and be re-entered by the slowly emerging impulse. The re-entrant impulse may in turn re-excite the area of slow conduction to complete a circus-movement loop. Intermittent establishment of a re-entrant circuit would result in occasional ectopic beats (*extrasystoles*), and continuous propagation of impulses in the established circuit would underlie an episode of tachyarrhythmia.

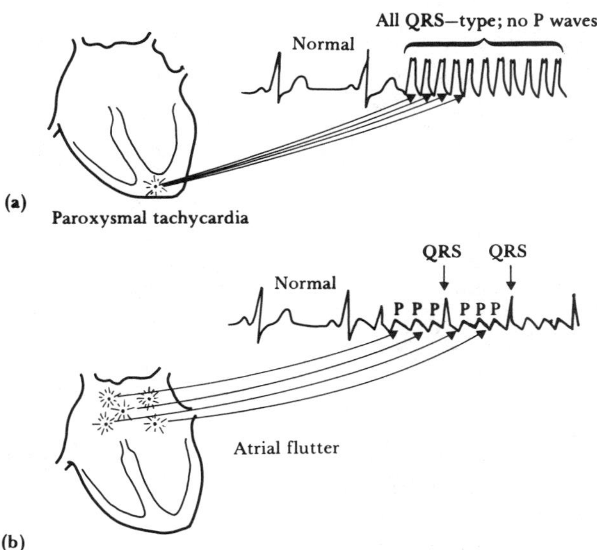

Figure 4.19 (a) Paroxysmal tachycardia. An ectopic focus may repetitively discharge at a rapid regular rate for minutes, hours, or even days. (b) Atrial flutter. The atria begin a very rapid, perfectly regular "flapping" movement, beating at rates of 200 to 300 beats/min. (Adapted from Brendan Phibbs, *The Human Heart,* 3rd ed., St. Louis: The C.V. Mosby Company, 1975.)

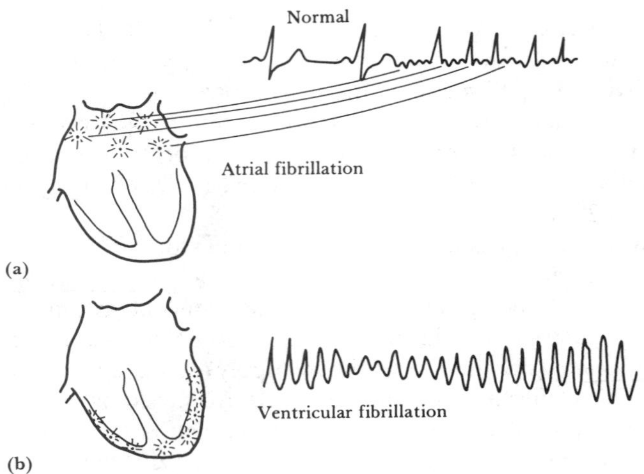

Figure 4.20 (a) Atrial fibrillation. The atria stop their regular beat and begin a feeble, uncoordinated twitching. Concomitantly, low-amplitude, irregular waves appear in the ECG, as shown. This type of recording can be clearly distinguished from the very regular ECG waveform containing atrial flutter. (b) Ventricular fibrillation. Mechanically the ventricles twitch in a feeble, uncoordinated fashion with no blood being pumped from the heart. The ECG is likewise very uncoordinated, as shown. (Adapted from Brendan Phibbs, *The Human Heart,* 3rd ed., St. Louis: The C.V. Mosby Company, 1975.)

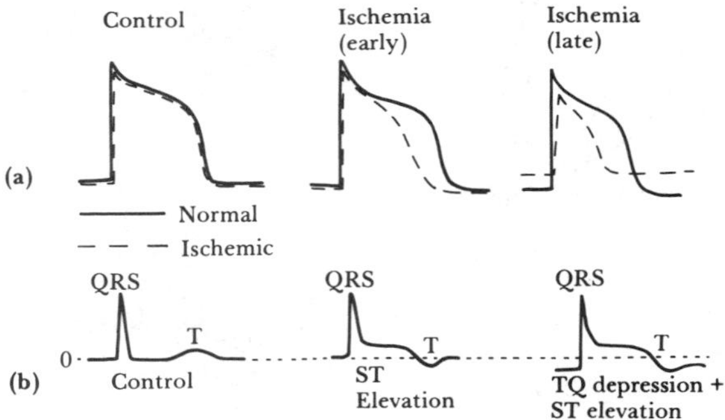

Figure 4.21 (a) Action potentials recorded from normal (solid lines) and ischemic (dashed lines) myocardium in a dog. Control is before coronary occlusion. (b) During the control period prior to coronary occlusion, there is no ECG S–T segment shift; after ischemia, there is such a shift. (From Andrew G. Wallace, "Electrophysiology of the Myocardium," in *Clinical Cardiopulmonary Physiology,* 3rd ed. New York: Grune & Stratton, 1969; used with permission of Grune & Stratton. Based on data by W. E. Sampson and H. M. Scher, "Mechanism of S–T Segment Alteration During Acute Myocardial Injury," 1960, *Circulation Research,* **8,** by permission of The American Heart Association.)

ALTERATION OF POTENTIAL WAVEFORMS IN ISCHEMIA

Of particular interest in Figure 4.21 is the change in the intracellular and extracellular potential waveforms in ischemia. Note particularly that in late ischemia (ischemia that occurs several minutes after induced coronary occlusion), there are decreases in the magnitudes of the resting potential, the velocity of the upstroke, and the height and duration of the action potential. (The decreases in the velocity of the upstroke is indicative of a lowered velocity of conduction of the action-potential wavefront through this ischemic region.) The slope of the potential during the plateau region of the action potential is also altered in ischemia. These changes in the action-potential waveform bring about changes in the extracellular potential fields produced by individual cardiac cells. The field contributions of these cells, as well as those of other normal cells, superimpose in the linear volume-conductor medium to bring about altered forms of the ventricular portion of the ECG (for example, the QRS complex, S–T segment, and T wave, as shown in Figures 4.13 and 4.21).

It is now well known that occlusion of the blood supply to a given myocardial region brings about relatively rapid electrolytic adjustments in this region. Specifically, there is a loss of K^+ and an uptake of Na^+ within the ischemic cell (Ca^{2+} and H^+ also accumulate within the cell, and water shifts inward as well), resulting in a lessening of the magnitude of the resting potential. The shifts in ion concentration are indicative of "depressed" activity of the Na^+–K^+ pump, which is metabolically dependent. Changes in the resting potential and action-potential waveform in ischemia are simply external manifestations of these underlying electrochemical changes brought about by an inadequate oxygen supply.

4.7 THE ELECTRORETINOGRAM (ERG)

ANATOMY OF VISION

The normal eye is an approximately spherical organ about 24 mm in diameter (Figure 4.22). The retina, located at the back of the eye, is the sensory portion of the eye.

The light-transmitting parts of the eye are the cornea, anterior chamber, lens, and vitreous chamber, named in the order in which these structures are traversed by light. A transparent fluid, the *aqueous humor,* is found in the anterior chamber. The vitreous chamber is filled by a transparent gel, the *vitreous body*. The aqueous humor provides a nutrient transport medium, but it is also of further optical significance. It is normally maintained at a pressure (20–25 mm Hg) that is adequate to inflate the eye against its resistive outer coats (the sclera and choroid). This makes possible the precise geometrical configuration of the retina and the optical pathway that is necessary to ensure formation of a clear visual image. In addition, the aqueous humor is the

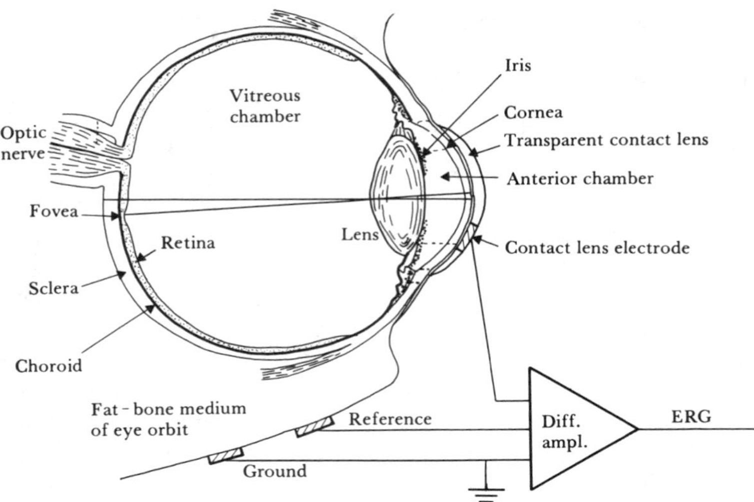

Figure 4.22 The transparent contact lens contains one electrode, shown here on horizontal section of the right eye. Reference electrode is placed on the right temple.

essential link between the circulatory system and the lens and cornea, which themselves lack blood vessels. To satisfy the respiratory and nutritive requirements of these two structures, there is a continual movement of fluid and solute material between the aqueous humor and contiguous blood vessels. Interference with this flow, in pathological conditions, not only leads to damage of the lens and cornea but may also result in the development of pressures within the eye that are high enough to injure the retina. *Glaucoma* is the term applied to this high-pressure condition.

In considering the neural organization of the retina, we need examine only five types of nerve cells: *photoreceptors* and *bipolar, horizontal, amacrine,* and *ganglion* cells. The ganglion cells, the axons of which produce the nerve fibers sweeping across the inner retinal surface to be collected at the optic disk (and which form the greater bulk of the nerve fibers of the optic nerve), are substantially fewer in number than the photoreceptors. There is a convergence in the neural pathways of the retina as a whole. [That is, many photoreceptors terminate on each bipolar cell ($n:1$), and many bipolar cells, in turn, terminate on a single ganglion cell. The degree of convergence varies considerably, being greater in the peripheral parts of the retina and minimal at the fovea (Figure 4.22). That is, the neural chain from photoreceptor to ganglion cell is 1 : 1 in the foveal region.] The synaptic interconnections between photoreceptors and bipolar cells and between bipolar cells and ganglion cells occur in two well-defined regions. The *external plexiform layer* is the region of contact between photoreceptor and bipolar cells, and the *internal plexiform layer* is the region of contact between bipolar and ganglion cells.

Lateral connections are also found in both layers. For example, horizontal cells interconnect rods and cones (defined below) at the level of the external plexiform layer, and amacrine cells provide a second horizontal network at the level of the inner plexiform layer. The retina may thus be considered functionally organized into two parts: an outer sensory layer containing the photoelectric sensors (photoreceptors) and an inner layer responsible for organizing and relaying electrical impulses generated in the photoreceptor layer to the brain.

Two types of photoreceptors occur in the human retina: *rods* (the agents of vision in dim light) and *cones* (the mediators of color vision in brighter light). Both rods and cones are differentiated into outer and inner segments. The inner segments are the major sites of metabolism and contain all the synaptic terminals. Outer segments—typically cylindrical and thin in rods and stout and conical in cones—are sites of visual excitation. The first stage in the transduction of light to neural messages is the absorption of photons by photopigments localized in the outer segments of the retina's photoreceptors (Dartnall, 1962). The photopigment localized in the compact membrane infoldings of the rod's external segment is *rhodopsin.* It is easily isolated and has been extensively studied. The cones in human beings contain one of three photopigments with photospectral absorption characteristics that differ from one another and from the rod pigment rhodopsin. The cone pigments are very difficult to isolate in humans and other vertebrates, and their spectral characteristics have usually had to be measured by indirect means, such as reflection densitometry (Rushton, 1963). All the pigments are *photolabile;* that is, events initiated by light absorption result eventually in breakdown or "bleaching" of the photopigment. The exact process of transduction is not entirely known, but the bleaching of the rhodopsin probably releases transmitter ions that cause a change in graded membrane potentials. These in turn result in ganglion-cell action potentials that are transmitted down the optic nerve. Consult Kandel *et al.* (1991, Chap. 28) as a general text on phototransduction.

ELECTROPHYSIOLOGY OF THE EYE

When the retina is stimulated with a brief flash of light, a characteristic temporal sequence of changes in potential can be recorded between an exploring electrode—placed either on the inner surface of the retina or on the cornea—and an indifferent electrode placed elsewhere on the body (usually the temple, forehead, or earlobe). These potential changes are collectively known as the *electroretinogram* (ERG), and they are clinically recorded with the aid of an Ag/AgCl electrode embedded in a special contact lens used as the exploring electrode. (See, for example, Strong, 1973.) The saline-filled contact lens is in good contact with the cornea, which is very thin and in intimate contact with the aqueous humor and passive fluid medium of the inner eye. The contact lens is usually well-tolerated by the subject and permits long examinations without discomfort. By considering the eye as a fluid-filled sphere and

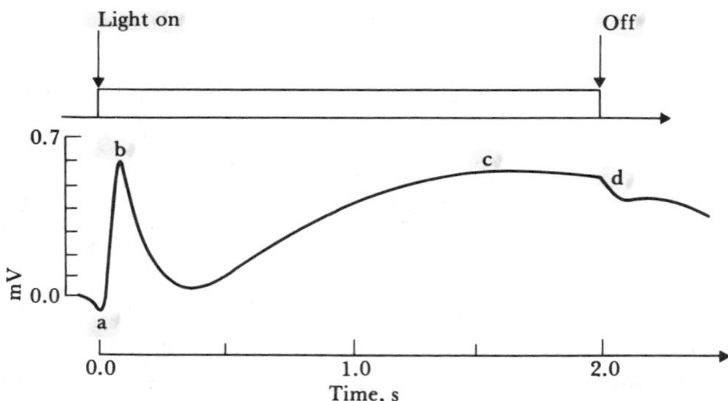

Figure 4.23 Vertebrate electroretinogram

the retina as a thin sheetlike bioelectric source attached to the posterior pole of the sphere (Figure 4.22), we can easily visualize the volume-conductor problem in electroretinography.

Figure 4.23 shows a typical vertebrate ERG waveform in response to a 2-s light flash. The four most commonly identified components of the ERG waveform (the a, b, c, and d waves) are common to most vertebrates, including humans. The first part of the response to a brief light flash is the *early-receptor potential* (ERP) generated by the initial light-induced changes in the photopigment molecules. It appears almost instantaneously with the onset of the light stimulus. The second component, with a latency of 1–5 ms, is the *late-receptor potential* (LRP), which has been found to be maximal near the synaptic endings of the photoreceptors and therefore reflects the outputs of the receptors. Normally the ERP and LRP form the leading edge of the a wave. However, in the absence of the b wave, their entire time course can be studied. The ERP is linear with light intensity; the LRP is already markedly nonlinear, varying approximately logarithmically with intensity.

SPATIAL PROPERTIES OF THE ERG

It is possible to record ERGs from localized areas of the retina in addition to the classical response that we have described in previous sections. [This conventional response is usually elicited from the dark-adapted eye via a brief light flash (flash ERG)]. In a frog, the sum of the ERGs produced by several retinal regions is equal to the single ERG produced when all these regions are stimulated simultaneously.

The spatial properties of the human ERG have also been established [Aiba *et al.* (1967)]. Linear superposition of ERG responses has likewise been confirmed in humans. In applying localized light stimuli to portions of the human retina, we must take precautions to prevent light scattered within the eye from stimulating a much larger area of retina than is intended. Thus relatively high steady background illumination is supplied that illuminates most of the retina, and a localized stimulus is superimposed. The background illumination light causes the retina to adapt and renders it much less sensitive to light scattered from the stimulus region. In general, relatively high-background and low-stimulus intensities are preferred, making these locally generated ERG potentials low in amplitude and detectable with only average response calculations involving large numbers of responses. Without these special precautions, the resultant ERG represents the overall retinal response to light stimulation. Little is known about the actual nature of the light input to a particular retinal locus in the photoreceptive layer.

Despite the anatomical complexities of the retina, the problems of obtaining good records from untrained subjects, and the need for employing averaging techniques in obtaining spatially localized ERGs, the ERG has potential importance in assessing functional retinal behavior. It serves as an objective record of retinal function, is not dependent on the function of the optic nerve or the optic pathways, and is minimally affected by clouding of the optic pathway. For a detailed discussion of the clinical electrophysiology of the retina, consult Heckenlively and Arden (1991) and Ogden (1989).

THE ELECTRO-OCULOGRAM (EOG)

In addition to the transient potential recorded as the ERG, there is a steady corneal-retinal potential. This steady dipole may be used to measure eye position by placing surface electrodes to the left and right of the eye, on the nose and the temple. When the gaze is straight ahead, the steady dipole is symmetrically placed between the two electrodes, and the EOG output is zero. When the gaze is shifted to the left, the positive cornea becomes closer to the left electrode, which becomes more positive. There is an almost linear relationship between horizontal angle of gaze and EOG output up to approximately $\pm 30^\circ$ of arc. Electrodes may also be placed above and below the eye to record vertical eye movements.

The EOG, unlike other bipotentials, requires a dc amplifier. The output is in the microvolt region, so recessed Ag/AgCl electrodes are required to prevent drift. It is necessary to abrade the skin to short out changes in the potential that exists between the inside and the outside of the skin. A noise is present that is compounded of effects from EEG, EMG, and the recording equipment; it is equivalent to approximately 1° of eye movement. Thus EOG data suffer from a lack of accuracy at the extremes. Specifically eye movements of less than 1° or 2° are difficult to record, whereas large eye movements (for example, greater than 30° of arc) do not produce bioelectric amplitudes that are strictly proportional to eye position. For an analysis of the accuracy and precision of electro-oculographic recording, consult North (1965).

The EOG is frequently the method of choice for recording eye movements in sleep and dream research, in recording eye movements from infants and children, and in evaluating reading ability and visual fatigue. For a practical clinical EOG setup, see Dement (1964), who employs EOG recordings to monitor eye movements during sleep.

4.8 THE ELECTROENCEPHALOGRAM (EEG)

The background electrical activity of the brain in unanesthetized animals was described qualitatively in the nineteenth century, but it was first analyzed in a systematic manner by the German psychiatrist Hans Berger, who introduced the term *electroencephalogram* (EEG) to denote the potential fluctuations recorded from the brain. Conventionally, the electrical activity of the brain is recorded with three types of electrodes—scalp, cortical, and depth electrodes. When electrodes are placed on the exposed surface (cortex) of the brain, the recording is called an *electrocorticogram* (ECoG). Thin insulated needle electrodes of various designs may also be advanced into the neural tissue of the brain, in which case the recording is referred to as a *depth recording*. (There is surprisingly little damage to the brain tissue when electrodes of appropriate size are employed.) Whether obtained from the scalp, cortex, or depths of the brain, the recorded fluctuating potentials represent a superposition of the volume-conductor fields produced by a variety of active neuronal current generators. Unlike the relatively simple bioelectric source considered in Section 4.2 (the nerve trunk with its enclosed bundles of circular cylindrical nerve axons), the sources generating the field potentials recorded here are aggregates of neuronal elements with complex interconnections. The neuronal elements mentioned previously are the dendrites, cell bodies (somata), and axons of nerve cells. Moreover, the architecture of the neuronal brain tissue is not uniform from one location to another in the brain. Therefore, prior to undertaking any detailed study of electroencephalography, we must first discuss necessary background information regarding (1) the gross anatomy and function of the brain, (2) the ultrastructure of the cerebral cortex, (3) the potential fields of single neurons leading to an interpretation of extracellular potentials recorded in the cerebral cortex, and (4) typical clinical EEG waveforms recorded via scalp electrodes. We shall next focus on the general volume-conductor problem in electroencephalography and then briefly discuss abnormal EEG waveforms.

INTRODUCTION TO THE ANATOMY AND FUNCTION OF THE BRAIN

The central nervous system (CNS) consists of the spinal cord lying within the bony vertebral column and its continuation, the brain, lying within the skull [Figure 4.24(a)]. The brain is the greatly modified and enlarged portion of the CNS, surrounded by three protective membranes (the *meninges*) and enclosed within the cranial cavity of the skull. The spinal cord is likewise surrounded by downward continuations of the meninges, and it is encased within the protective vertebral column. Both brain and spinal cord are bathed in a special extracellular fluid called *cerebral spinal fluid* (CSF).

Division of the brain into three main parts—*cerebrum, brainstem,* and *cerebellum*—provides a useful basis for the study of brain localization and function. The brainstem is the oldest part of the brain. Its size and functions have changed very little with the evolution of the vertebrates. It is actually a short extension of the spinal cord and serves three major functions: (1) It is a connecting link among the cerebral cortex, the spinal cord, and the cerebellum. (2) It is a center of integration for several visceral functions, such as control of heart rate and respiratory frequency. (3) It is an integration center for various motor reflexes. The diencephalon is the most superior portion of the brainstem; its chief component and largest structure is the *thalamus.* The thalamus serves as a major relay station and integration center for all of the general and special sensory systems, sending information to their respective cortical reception areas. It serves as a gateway to the cerebrum.

The cerebellum is a coordinator in the voluntary (somatic) muscle system and acts in conjunction with the brainstem and cerebral cortex to maintain balance and provide harmonious muscle movements. The cerebrum occupies a special dominant position in the central nervous system, and within the cerebrum are localized the conscious functions of the nervous system.

Within the CNS there are *ascending (sensory)* nerve tracts that run from the spinal cord or brain stem to various areas of the brain, conveying information regarding changes in the external environment of the body that are reported by various peripheral biological sensors. There are a variety of such sensors, including the general sensors of temperature, pain, fine touch, pressure, as well as the special senses of vision, audition, equilibrium, taste, and olfaction. Figure 4.24(b) shows the basic plan associated with the general sense pathways from the periphery to the cortex. A three-neuron chain is involved in conveying information to the cortex where the primary neuron has its cell body in a ganglion outside the CNS and makes synaptic contact with a secondary neuron whose cell body is located in a nucleus within either the spinal cord [e.g., the dorsal horn) or the brain stem (Figure 4.24(a)]. Note from Figure 4.24(b) that the axon of the secondary neuron crosses (decussates) to the other side of the cord and joins a nerve fiber tract bound for the thalamus. The third neuron in the pathway is located in the thalamus, and its axon travels in the thalamocortical radiations to the post-central gyrus which is located just posterior to the central sulcus (Figure 4.25). The post-central gyrus is the cortical projection area for the general senses. The neural pathways for the special senses, particularly audition and vision, follow the same general ground plan; however, there are notable deviations from the scheme shown in Figure 4.24(b):

(i) There are usually more than three neurons involved in the pathway.
(ii) Not all of the "secondary neurons" decussate, i.e., most of the neurons cross to the opposite side of the body, but a significant number ascend to the thalamus on the same side of the body.

The auditory and visual pathways have their own special thalamic relay centers—the medial and lateral geniculate bodies, respectively, as well as their own cortical projection areas (Figure 4.25).

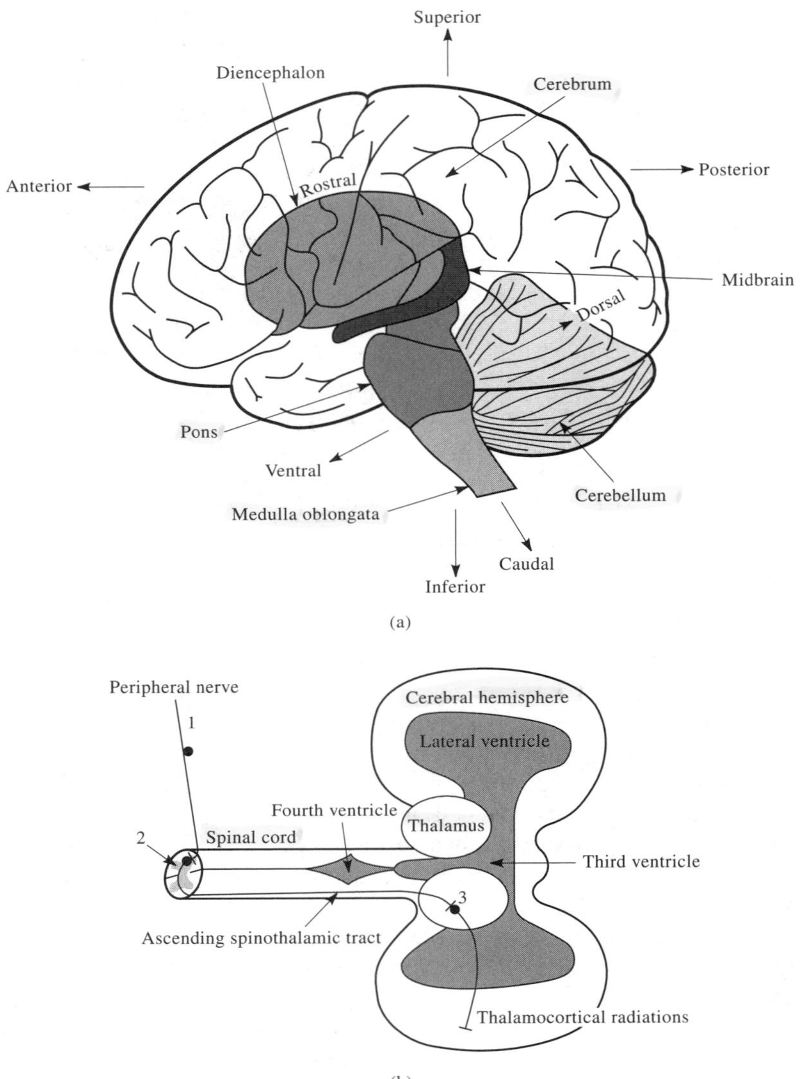

Figure 4.24 (a) Anatomical relationship of brainstem structures (medulla oblongata, pons, midbrain, and diencephalon) to the cerebrum and cerebellum. General anatomic directions of orientation in the nervous system are superimposed on the diagram. Here the terms rostral (toward head), caudal (toward tail), dorsal (back), and ventral (front) are associated with the brainstem; remaining terms are associated with the cerebrum. The terms medial and lateral imply nearness and remoteness, respectively, to or from the central midline axis of the brain. (b) A simplified diagram of the CNS showing a typical general sense pathway from the periphery (neuron 1) to the brain (neuron 3). Note that the axon of the secondary neuron (2) in the pathway

Likewise, within the CNS there are *descending (motor)* nerve tracts that originate in various brain structures such as the cerebrum and cerebellum (Figure 4.24) and terminate ultimately on motor neurons in the ventral horn of the spinal cord (Figure 4.10). These motoneurons, in turn, control the contractile activity of the skeletal musculature. The corticospinal tract is an example, in that axons from the primary motor cortex (Figure 4.25) project directly to motor neurons in the spinal cord. Like the ascending general sense pathway, the descending corticospinal tracts each cross to the opposite side of the body prior to making synaptic contact with the spinal motor neurons.

Thus there exist two-way communication links between the brain and spinal cord that allow higher centers in the brain to control or modify the behavior of the elemental spinal reflex arc at a given spinal level. By means of such communication links, the brain is not only informed of a peripheral event but can also modify the response of the spinal reflex to that environmental stimulus. Information is transmitted to the brain by means of a frequency-modulated train of nerve impulses that, upon reaching specific areas of the brain, stimulates the activity of other neurons there. Similarly, the decision to implement a motor action in response to the initial stimulus is manifested in the activity of cortical neurons from various areas of the brain, depending on the particular motor action to be taken.

Electrical activity in either ascending or descending nerve fiber tracts may be represented to a first approximation by an activation current dipole oriented in the direction of propagation (bioelectric source model). One should be aware that the character of the volume conductor media can change along the length of a particular tract between the spinal cord and the cortex, and an appropriate model adopted based on the particular measurement considered. The volume conductor field potential solutions can be used to both fit and interpret body surface potential measurements obtained clinically. Recording such data noninvasively from relatively small volume action current sources, such as nerve fiber tracts, invariably requires the use of cumulative signal averaging techniques. In Figure 4.8, the median nerve was stimulated and compound action potentials were recorded from the subject's forearm. Although not shown in this figure, sensory fibers in the median nerve thus activated, initiate activity in the general sense pathways to the brain. Averaged field potential recordings can be taken at a variety of points along the ascending pathways [e.g., from spinal cord and brain stem tracts taking note of the crossed nature of the pathway, and finally at the cortex itself (post-central gyrus)]. The field potentials associated with long nerve tracts depends to a large extent on: (a) whether the tract is straight or bent, and (b) the resistance

---

decussates (crosses) to the opposite side of the cord. [Part (a) modified from Harry E. Thomas, *Handbook of Biomedical Instrumentation and Measurement,* 1974, p. 254. Reprinted with permission of Reston Publishing Company, Inc. a Prentice-Hall company, 11480 Sunset Hills Road, VA 22090.]

(geometry and specific conductivity) of the surrounding volume conductor media. The subject of nerve tracts has been discussed previously; however, the activity of both nuclei in the ascending pathway and clusters of cells in the cortex, depends not only on the ensemble of neurons there, but also on the geometry of the ensemble and the different types of synaptic connections involved. This important subject is discussed later; however, for the present, these different types of averaged field potentials are called collectively somatosensory evoked potentials.

Averaged sensory evoked potentials in response to brief auditory "clicks" or flashes of light are also routinely recorded as the auditory evoked response (AER) and the visual evoked response (VER), respectively (Jacobson, 1994; Heckenlively and Arden, 1991). Using an electromagnetic stimulating device held over the primary motor cortex (just anterior to the central sulcus), it is also possible to induce currents that activate the corticospinal tract, making possible the recording of averaged field potentials from the descending motor pathways (York, 1987; Geddes, 1987; Esselle and Stuchly, 1992). The same volume conductor principles are applicable to the analysis of these different types of evoked potential recordings. The cerebrum is a paired structure, with right and left cerebral hemispheres, each relating to the opposite side of the body. That is, voluntary movements of the right hand are "willed" by the left cerebral hemisphere. The surface layer of the hemisphere is called the *cortex;* it receives sensory information from skin, eyes, ears, and other receptors located generally on the opposite side of the body. This information is compared with previous experience and produces movements in response to these stimuli.

Each hemisphere consists of several layers. The outer layer is a dense collection of nerve cells that appear gray in color when examined in a fresh state. It is consequently called gray matter. This outer layer, roughly 1 cm thick, is called the *cerebral cortex.* It has a highly convoluted surface consisting of *gyri* (ridges) and *sulci* (valleys), the deeper sulci being termed *fissures.* The deeper layers of the hemisphere (beneath the cortex) consist of *axons* (or white matter) and collections of cell bodies termed *nuclei.* Some of the integrative functions of the cerebrum can be localized within certain regions of the cortex; others are more diffusely distributed.

A major dividing landmark of the cerebral cortex is the lateral fissure (Figure 4.25), which runs on the lateral (side) surface of the brain from the open end in front, posteriorly and dorsally (backward and upward). The lateral fissure defines a side lobe of cortex inferior to (below) it that is called the *temporal lobe* (Figure 4.25). The superior (upper) part of this lobe contains the primary auditory cortex, which is the part of the cortex that receives auditory impulses via neural pathways leading from the auditory receptors in the inner ear.

The visual system is another example of the projection of the senses onto the cerebral cortex. The *occipital lobe* at the back of the head is the primary visual cortex. Light flashed into the eye evokes large electrical potentials from electrodes placed over this area of the cortex.

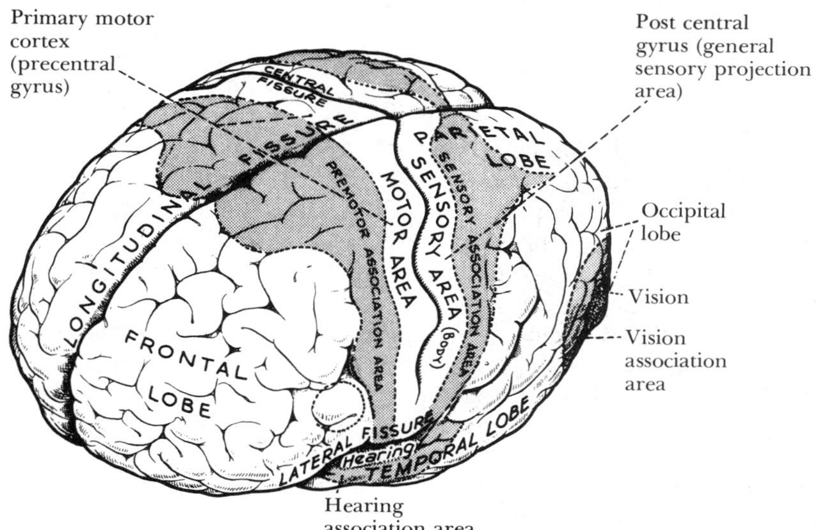

Figure 4.25 The cerebrum, showing the four lobes (frontal, parietal, temporal, and occipital), the lateral and longitudinal fissures, and the central sulcus. (From A. B. McNaught and R. Callander, *Illustrated Physiology,* 3rd ed., 1975. Edinburgh: Churchill Livingstone. Used with permission of Churchill Livingstone.)

Another major landmark of the cerebral cortex is the central sulcus (Figure 4.25). However, it is not so prominent and unvarying an anatomical landmark as the lateral fissure. The central sulcus runs from the medial surface (surface along the midline of the brain) over the convexity of the hemisphere to the lateral fissure. The central sulcus also represents the posterior border of the frontal lobe. The gyrus lying just anterior (forward) to the central sulcus is the *precentral gyrus,* which functions as the primary motor cortex. From this gyrus, nerve signals run down through the brainstem to the spinal cord for control of skeletal muscles via neural control of motoneurons in the ventral horn of the spinal cord (Figure 4.10). Lesions (destruction) of part of the precentral gyrus cause partial paralysis on the opposite side of the body.

Proceeding more anteriorly from the central sulcus, we encounter an area called the *premotor cortex,* where more complex motor movements such as speech are organized. The anterior and inferior portions of the frontal lobe are involved in the control of emotional behavior. For years this part of the cortex has been considered the locus of the higher intellect of the human being, largely because the main difference between the brains of monkeys and the human brain is the great prominence of the human's prefrontal lobes. However, efforts to demonstrate that the prefrontal cortex is more important to higher intellectual functions than other portions of the cortex have not been entirely successful.

Immediately behind the central sulcus lies the *parietal lobe.* Its anterior border is the central sulcus, its ventral boundary is the lateral fissure or a line continuing in the same direction, and its posterior boundary is rather ill defined on the lateral surface. Several component areas of the parietal lobe can be distinguished. Immediately posterior to the central sulcus is the primary *somatosensory cortex,* the *postcentral gyrus.* This region receives impulses from all the general sense receptors from the skin (such as pressure, touch, and pain receptors). Each little area along this gyrus is related to a particular part of the body (for example, the legs on the medial end, the hand in the center, and the face on the end next to the lateral fissure). If a recording electrode is placed appropriately during a neurosurgical procedure, a cortical response can be evoked by tactile stimuli delivered to the *contralateral* (opposite) hand. Likewise, if a stimulus is applied through the same electrode, the subject reports a tingling sensation in the contralateral hand. Higher-order sensory discrimination, such as the ability to recognize a number drawn on the palm of the hand, is organized solely in the parietal lobe. Destruction of the parietal lobe results in a loss of this discriminative ability. For example, a subject may still know that he or she is being touched but cannot tell where or what is being drawn on the palm of the hand. The parietal lobe is also responsible for a person's awareness of the general position of the body and its limbs in space.

ULTRASTRUCTURE OF THE CEREBRAL CORTEX

The functional part of the cerebrum is the cerebral cortex, a relatively thin layer of gray matter (1.5–4.0 mm in thickness) covering the outer surface of the cerebrum, including its intricate convolutions. Because it is the most recent phylogenetic acquisition of the brain, the cerebral cortex has undergone a relatively greater development than other parts of the brain. The greatest advance in relative growth has been the neocortex, which is present on the superior and lateral aspects of the cerebral hemispheres. The distinctly different type of cortex located on the medial surface and base of the brain is known as the *paleocortex.* We shall use the term *cortex* in this chapter to refer specifically to the neocortex.

There are many types of cortical neurons, and they are not randomly distributed along the axis normal to the cortical surface. They show an orderliness, in fact, that affects both the distribution of cell types and their packing density. Relative segregation by depth produces a stratification; each stratum is called a *cortical layer.* The cortex is generally arranged in six layers containing neurons and fiber bundles. The cells are of two main types: *pyramidal* and *nonpyramidal* (many subtypes have been identified). There are also a large number of horizontally oriented layers of nerve fibers that extend between adjacent regions of the cortex, as well as vertically oriented bundles that extend from the cortex to more distant regions of the cortex or downward to the brainstem and spinal cord.

Figure 4.26 shows a schematic drawing of a typical cortical pyramidal cell.

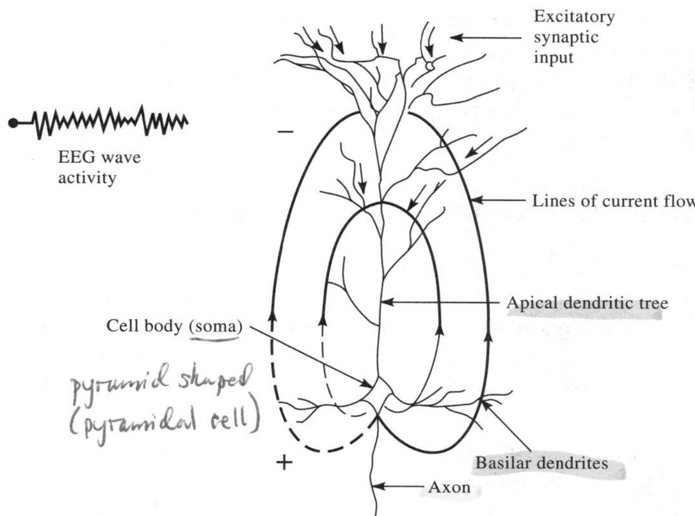

Figure 4.26 Electrogenesis of cortical field potentials for a net excitatory input to the apical dendritic tree of a typical pyramidal cell. For the case of a net inhibitory input, polarity is reversed and the apical region becomes a source (+). Current flow to and from active fluctuating synaptic knobs on the dendrites produces wave-like activity. See text.

The bodies of this type of cell are commonly triangular in shape, with the base down and the apex directed toward the cortical surface. (Pyramidal cell bodies vary greatly in size, from axial dimensions of 15×10 μm up to 120×90 μm or more for the giant pyramids of the motor cortex, which are called *Betz cells* after their discoverer.) These cells usually consist of the following parts: (1) a long apical dendrite (up to 2 mm in length) that ascends from the apex of the cell body through the overlaying cellular layers and that frequently reaches and branches terminally within the outermost layer of the cortex and (2) at the base of the pyramid-shaped cell, dense branching (arborization) occurs, largely horizontally. Axons of pyramidal cells emerge from the cortex as projection fibers to other areas of the cortex or to other structures such as the thalamus, cerebellum, or spinal cord. Frequently these axons send recurrent collateral (feedback) branches back on the cellular regions from which they sprang. Axons of some pyramidal cells turn back toward the cortical surface (never leaving the gray matter) to end via their many branches on the dendrites of other cells.

Nonpyramidal cells of the neocortex differ remarkably from pyramidal cells. Their cell bodies are small, and dendrites spring from them in all directions to ramify in the immediate vicinity of the cell. The axon may arise from a large dendrite; it commonly divides repeatedly to terminate on the cell bodies and dendrites of immediately adjacent cells. The axons of other nonpyramidal cells may turn upward toward the cortical surface, or they may leave the cortex (though this is not common).

For a detailed exposition of the various cells, layers, cellular interconnections, inputs, and outputs of the neocortex, see Kandel, Swartz and Jessell, 1991.

BIOELECTRIC POTENTIALS FROM THE BRAIN

When we make records of potential differences between an exploring electrode resting on the cortical surface and a distant reference electrode, we are in effect recording the resultant field potential at a boundary of a large conductile medium containing an array of active elements. It is evident from what is now known of the electrophysiology of the cortex (information derived from depth and microelectrode recordings) that under normal circumstances, conducted action potentials in axons contribute little to surface cortical records, because they usually occur asynchronously in time in large numbers of axons, which run in many directions relative to the surface. Thus their net influence on potential at the surface is negligible.

An exception occurs, of course, in the case of a response evoked by the simultaneous stimulation of a cortical input, as in the case of direct stimulation of thalamic nuclei or their afferent pathways, which project directly to the cortex via thalamocortical axons. Electrophysiologists have shown that surface records obtained under other circumstances signal principally the net effect of local postsynaptic potentials of cortical cells. These may be of either sign (excitatory or inhibitory) and may occur directly underneath the electrode or at some distance from it. A potential change recorded at the surface is a measure of the net potential (current resistance iR) drop between the surface site and the distant reference electrode. It is obvious, however, that if all the cell bodies and dendrites of cortical cells were randomly arranged in the cortical matrix, the net influence of synaptic currents would be zero. This would result in a "closed field" situation which produces relatively small far-field potentials (Lorente de No, 1947). Any electrical change recorded at the surface must be due to the orderly and symmetric arrangement of some class of cells within the cortex.

Pyramidal cells of the cerebral cortex are oriented vertically, with their long apical dendrites running parallel to one another. Potential changes in one part of the cell relative to another part create "open" potential fields in which current may flow and potential differences can be measured at the cortical surface. Figure 4.26 illustrates this concept in diagrammatic fashion. Synaptic inputs to the apical dendritic tree cause depolarization of the dendritic membrane. As a result, subthreshold current flows in a closed path through the cytoplasmic core of the dendrites and cell body of the pyramidal cell, returning ultimately to the synaptic sites via the extracellular bathing medium. From the indicated direction of the lines of current flow, the extracellular medium about the soma behaves as a *source* (+), while the upper part of the apical dendritic tree behaves as a *sink* (−).

The influence of a particular dendritic *postsynaptic potential* (PSP) on the cortical surface recording depends on its sign [excitatory (−) or inhibitory (+)] orientation, and on its location relative to the measurement site. The effect of each PSP may be regarded as creating a radially oriented dipole. Therefore, continuing synaptic input creates a series of potential dipoles and resulting current flows that are staggered but are overlapped in space and time. Surface potentials of any form can be generated by one population of presynaptic fibers and the cells on which they terminate, depending on the proportion that are inhibitory or excitatory, the level of the postsynaptic cells in the cortex, and so forth.

Nonpyramidal cells in the neocortex, on the other hand, are unlikely to contribute substantially to surface records. Their spatially restricted dendritic trees are radially arranged around their cell bodies such that charge differences between the dendrites and the cell body produce fields of current flow that sum to zero when viewed from a relatively great distance on the cortical surface (closed-field situation).

Thus, to summarize, the apical dendrites of pyramidal cells are a forest of similarly oriented, densely packed units in the superficial layers of the cortex. As excitatory and inhibitory synaptic endings on the dendrites of each cell become active, current flows into and out of these current sinks and sources from the rest of the dendritic processes and the cell body. The cell–dendrite relationship is therefore one of a constantly shifting current dipole, and variations in orientation and strength of the dipole produce wave-like fluctuations in a volume conductor (Figure 4.26). When the sum of dendritic activity is negative relative to the cell, the cell is depolarized and quite excitable. When it is positive, the cell is hyperpolarized and less excitable.

RESTING RHYTHMS OF THE BRAIN

Electric recordings from the exposed surface of the brain or from the outer surface of the head demonstrate continuous oscillating electric activity within the brain. Both the intensity and the patterns of this electric activity are determined to a great extent by the overall excitation of the brain resulting from functions in the brainstem reticular activating system (RAS). The undulations in the recorded electric potentials (Figure 4.27) are called *brain waves,* and the entire record is called an *electroencephalogram* (EEG).

The intensities of the brain waves on the surface of the brain (recorded relative to an indifferent electrode such as the earlobe) may be as large as 10 mV, whereas those recorded from the scalp have a smaller amplitude of approximately 100 μV. The frequencies of these brain waves range from 0.5 to 100 Hz, and their character is highly dependent on the degree of activity of the cerebral cortex. For example, the waves change markedly between states of wakefulness and sleep. Much of the time, the brain waves are irregular and no general pattern can be observed. Yet at other times, distinct patterns do occur. Some of these are characteristic of specific abnormalities of the brain, such as epilepsy (discussed later). Others occur in normal persons and may be classified as belonging to one of four wave groups (*alpha, beta, theta,* and *delta*), which are shown in Figure 4.27(a).

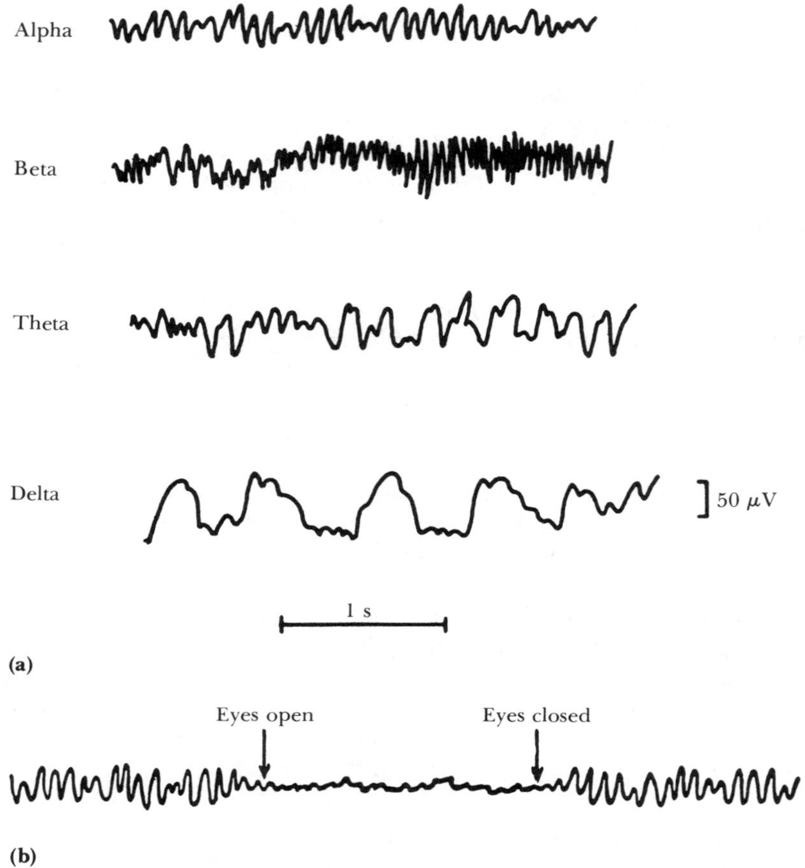

Figure 4.27 (a) Different types of normal EEG waves. (b) Replacement of alpha rhythm by an asynchronous discharge when patient opens eyes. (c) Representative abnormal EEG waveforms in different types of epilepsy. (From A. C. Guyton, *Structure and Function of the Nervous System,* 2nd ed., Philadelphia: W.B. Saunders, 1972; used with permission.)

Alpha waves are rhythmic waves occurring at a frequency between 8 and 13 Hz. They are found in EEGs of almost all normal persons when they are awake in a quiet, resting state of cerebration. These waves occur most intensely in the occipital region but can also be recorded, at times, from the parietal and frontal regions of the scalp. Their voltage is approximately 20–200 μV. When the subject is asleep, the alpha waves disappear completely. When the awake subject's attention is directed to some specific type of mental activity, the alpha waves are replaced by asynchronous waves of higher frequency but lower amplitude. Figure 4.27(b) demonstrates the effect on the alpha waves of simply opening the eyes in bright light and then closing them again. Note

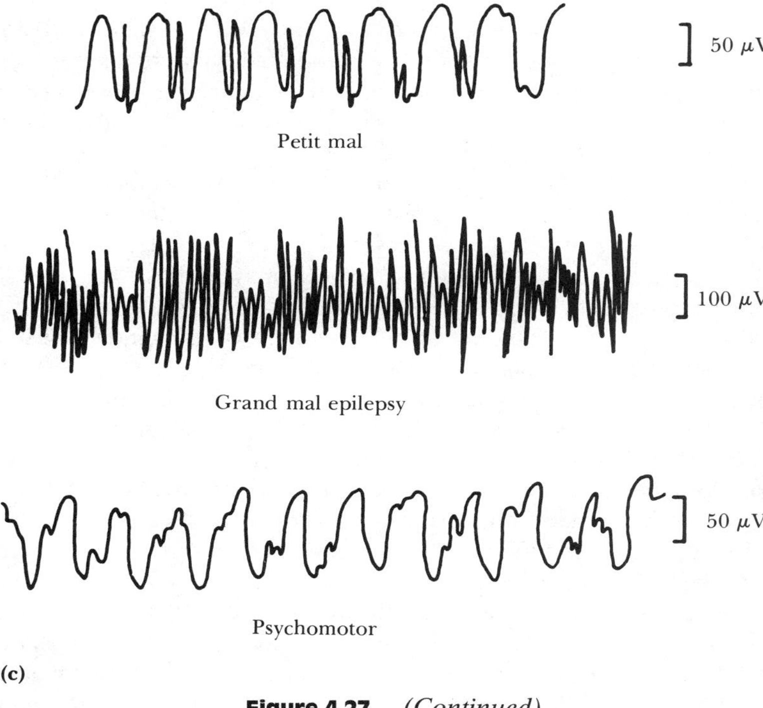

Figure 4.27 *(Continued)*

that the visual sensations cause immediate cessation of the alpha waves; these are replaced by low-voltage, asynchronous waves.

Beta waves normally occur in the frequency range of 14 to 30 Hz, and sometimes—particularly during intense mental activity—as high as 50 Hz. These are most frequently recorded from the parietal and frontal regions of the scalp. They can be divided into two major types: beta I and beta II. The beta I waves have a frequency about twice that of the alpha waves. They are affected by mental activity in much the same way as the alpha waves (they disappear and in their place appears an asynchronous, low-voltage wave). The beta II waves, on the other hand, appear during intense activation of the central nervous system and during tension. Thus one type of beta activity is elicited by mental activity, whereas the other is inhibited by it.

Theta waves have frequencies between 4 and 7 Hz. These occur mainly in the parietal and temporal regions in children, but they also occur during emotional stress in some adults, particularly during periods of disappointment and frustration. For example, they can often be brought about in the EEG of a frustrated person by allowing the person to enjoy some pleasant experience and then suddenly removing the element of pleasure. This causes approximately 20 s of theta waves.

Delta waves include all the waves in the EEG below 3.5 Hz. Sometimes these waves occur only once every 2 or 3 s. They occur in deep sleep, in infancy, and in serious organic brain disease. They can also be recorded from the brains of experimental animals that have had subcortical transections producing a functional separation of the cerebral cortex from the reticular activating system. Delta waves can thus occur solely within the cortex, independent of activities in lower regions of the brain.

A single cortical cell can give rise only to small extracellular current, so large numbers of neurons must be synchronously active to give rise to the potentials recorded from the cerebral surface. The individual waves of the EEG are of long duration (for example, 30–500 ms), and one might well ask how they are produced. They can be long-lasting depolarizations of the cell membranes (for example, of the apical dendrites of pyramidal cells) or a summation of a number of shorter responses. In any event, a sufficiently large number of neurons must discharge together to give rise to these cortical potentials. The term *synchronization* is used to describe the underlying process that acts to bring a group of neurons into unified action. Synaptic interconnections are generally thought to bring about synchronization, although extracellular field interaction between cells has been proposed as a possible mechanism. Rhythmically firing neurons are very sensitive to voltage gradients in their surrounding medium.

Besides the synchronization required for each wave of resting EEG, the series of repeated waves suggests a rhythmic and a trigger or pacemaker process that initiates such rhythmic action. By means of knife cuts below the intact connective-tissue covering (*meningeal layer* or *pia matter*) of the brain, one may prepare *chronic islands* of cortex—with all neuronal connections cut, but with the blood supply via surface vessels intact. Only a low level of EEG activity remains in such islands. Though the isolated islands of cortex may not show spontaneous EEG activity, they still have the ability to respond rhythmically, which may be readily demonstrated by the rhythmic responses that are elicited by applying a single electrical stimulus. The inference is that various regions of the cortex, though capable of exhibiting rhythmic activity, require trigger inputs to excite rhythmicity. The RAS, mentioned earlier, appears to provide this pacemaker function.

THE CLINICAL EEG

The system most often used to place electrodes for monitoring the clinical EEG is the International Federation 10-20 system shown in Figure 4.28. This system uses certain anatomical landmarks to standardize placement of EEG electrodes. The differential amplifier requires a separate ground electrode plus differential inputs to the following three types of electrode connections: (1) Between each member of a pair (bipolar), (2) between one monopolar lead and a distant reference electrode (usually attached to one or both earlobes), and (3) between one monopolar lead and the average of all. In the average reference mode, the system reference is formed by connecting all scalp-recording locations through equal high resistances to a common point.

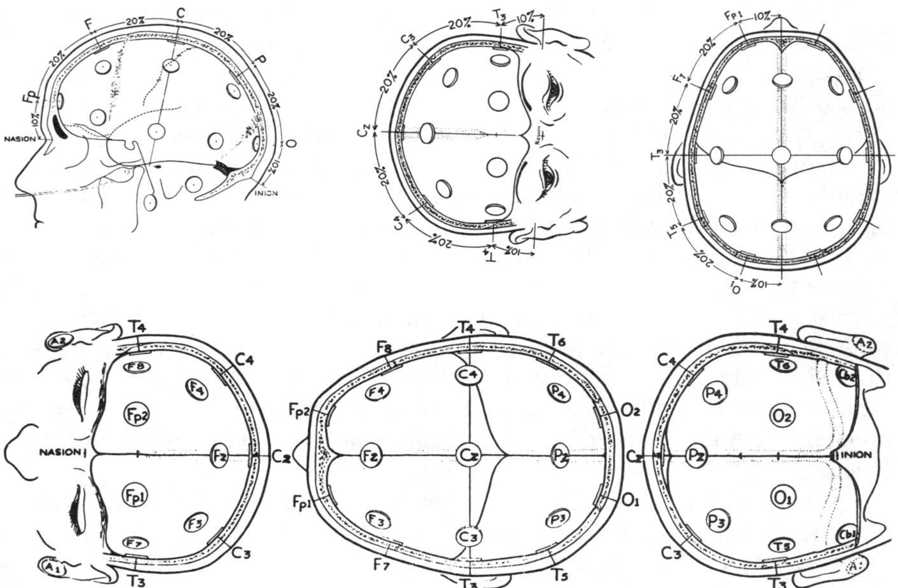

Figure 4.28 The 10-20 electrode system This system is recommended by the International Federation of EEG Societies. (From H. H. Jasper, "The Ten-Twenty Electrode System of the International Federation in Electroencephalography and Clinical Neurophysiology," *EEG Journal,* 1958, **10** (Appendix), 371–375.)

In the bipolar system, differential measurements are made between successive pairs of electrodes. The advantage of using a differential recording between closely spaced electrodes (between successive pairs in the standard system, for example) is cancelation of far-field activity common to both electrodes; one thereby obtains sharp localization of the response. Although the same electric events are recorded in each of the three ways, they appear in a different format in each case. The potential changes that occur are amplified by high-gain, differential, capacitively coupled amplifiers. The output signals are usually displayed via ink-writing strip-chart recorders, a method that usually limits the frequency response to the range of 0.5 to 80 Hz.

In the routine recording of clinical EEGs, the input electrodes are a problem. They must be small, they must be easily affixed to the scalp with minimal disturbance of the hair, they must cause no discomfort, and they must remain in place for extended periods of time. Technicians prepare the surface of the scalp, degrease the recording area by cleaning it with alcohol, apply a conducting paste, and glue nonpolarizable Ag/AgCl electrodes to the scalp with a glue (collodion) or hold them in place with rubber straps.

The EEG is usually recorded with the subject awake but resting recumbent on a bed with eyes closed. With the patient relaxed in such a manner, artifacts from electrode-lead movement are significantly reduced, as are contaminating signals from the scalp. Muscle activity from the face, neck, ears, and so on is perhaps the most subtle contaminant of EEG records in the recording of both spontaneous ongoing activity in the brain and activity evoked by a sensory stimulus *(evoked response).* For example, the frequency spectrum of the field produced by mildly contracted facial muscles contains frequency components well within the nominal EEG range (0.5–100 Hz). After technicians have achieved resting, quiescent conditions in the normal adult subject, the subject's scalp recordings show a dominant alpha rhythm in the parietal-occipital areas, whereas in the frontal areas, there is a low-amplitude, higher-frequency beta rhythm in addition to the alpha rhythm. In the normal subject there is a symmetry between the recordings of the right and left hemispheres. To appreciate the wide range of EEG measurement artifacts, see Hill and Parr (1963).

In general there is a relationship between the degree of cerebral activity and the average frequency of the EEG rhythm: The frequency increases progressively with higher and higher degrees of activity. For example, delta waves are frequently found in stupor, surgical anesthesia, and sleep; theta waves in infants; alpha waves during relaxed states; and beta waves during intense mental activity. However, during periods of mental activity, the waves usually become asynchronous rather than synchronous, so that the magnitude of the summed surface potential recording decreases despite increased cortical activity.

SLEEP PATTERNS

When an individual in a relaxed, inattentive state becomes drowsy and falls asleep, the alpha rhythm is replaced by slower, larger waves (Figure 4.29). In deep sleep, very large, somewhat irregular delta waves are observed. Interspersed with these waves—during moderately deep sleep—are bursts of alpha-like activity called *sleep spindles.* The alpha rhythm and the patterns of the drowsy and sleeping subject are *synchronized,* in contrast with the low-voltage *desynchronized,* irregular activity seen in the subject who is in an alert state.

The high-amplitude, slow waves seen in the EEG of a subject who is asleep are sometimes replaced by rapid, low-voltage irregular activity resembling that obtained in alert subjects. However, the sleep of a subject with this irregular pattern is not interrupted; in fact, the threshold for arousal by sensory stimuli is elevated. This condition has therefore come to be called *paradoxical sleep.* During paradoxical sleep, the subject exhibits rapid, roving eye movements. For this reason, it is also called *rapid-eye-movement* sleep or REM sleep. Conversely, *spindle* or synchronized sleep is frequently called *nonrapid-eye-movement,* NREM, or slow-wave sleep. Human subjects aroused at a time when their EEG exhibits a paradoxical (REM) sleep pattern generally report

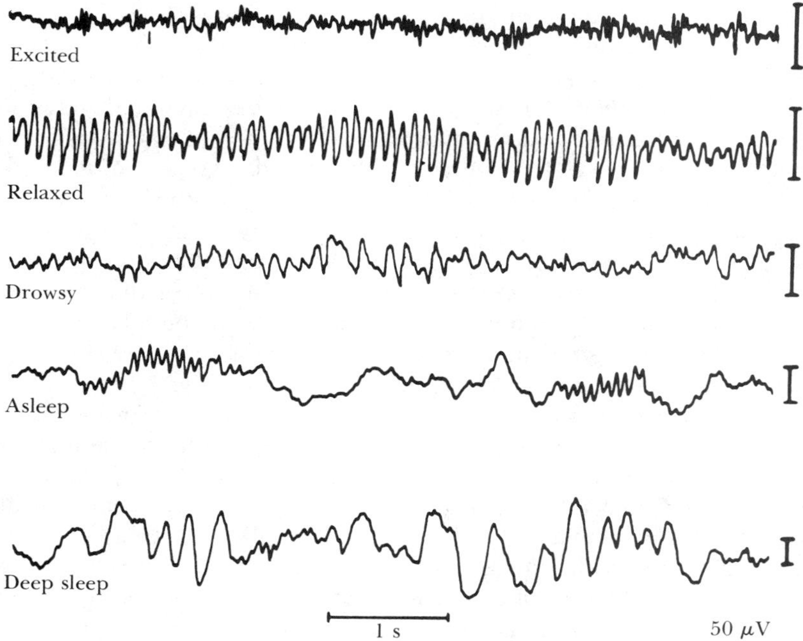

Figure 4.29 The electroencephalographic changes that occur as a human subject goes to sleep The calibration marks on the right represent 50 μV. (From H. H. Japser, "Electrocephalography," in *Epilepsy and Cerebral Localization,* edited by W. G. Penfield and T. C. Erickson. Springfield, Ill.: Charles C. Thomas, 1941.)

that they were dreaming, whereas individuals wakened from spindle sleep do not. This observation and other evidence indicate that REM sleep and dreaming are closely associated. It is interesting that during REM sleep, there is a marked reduction in muscle tone, despite the rapid eye movements.

THE VOLUME-CONDUCTOR PROBLEM IN ELECTROENCEPHALOGRAPHY

Geometrically speaking, the brain approximates a sphere surrounded by concentric shells that differ in impedance and comprise the *meninges* (connective tissue coverings of the brain), cerebral spinal fluid, skull, and scalp. This model is inaccurate to the extent that the brain is not really a true sphere, and its coverings are irregular in shape and thickness. Such irregularities are insignificant for the upper half of the brain, but complications are introduced by the marked departure of the lower parts of the brain from a spherical shape, as well as by variations in impedance produced by the openings (to the spinal column) through the base of the shell. Various cerebral structures differ somewhat in specific resistivity. The resistivity also varies in relation to the predominant direction of the fibers within the white matter. Thus the brain is neither a homogeneous nor an isotropic conducting medium.

In practice, neurological generators do not correspond precisely to simple, one-dimensional dipoles. Any source of activity large enough to manifest itself in the EEG constitutes at least a small area of the cortex the neurons of which are synchronously active. This source may be regarded as a three-dimensional sheet polarized across its thickness. If it is small enough, it may still be conveniently represented as an equivalent dipole per unit volume. A larger area of the cortex may be curved, or even convoluted, and the equivalent dipole then becomes a complex vector sum of the whole. When there are many widely scattered active-current generators, an infinite number of combinations may give rise to the same pattern of surface potentials.

Determining the equivalent dipole of cerebral activity is therefore of practical value only when EEG sources are highly "focal." Fortunately, this condition occurs frequently in the brain's response to sensory stimulation, as well as in pathological conditions. For example, Nunez (1981) considers in some depth the subject of the calculation of field potentials from equivalent current sources in inhomogeneous media. Particularly in Chapter 5 of his book, Nunez provides an introduction to the equivalent source models that have been used in the field of theoretical electroencephalography to interpret scalp potentials. Examples of these models include the simple dipole at the center of a spherical conducting medium, the *radially oriented* dipole not at the center of a sphere (the radially oriented eccentric dipole), the freely oriented eccentric dipole in a sphere, the dipole in a three-concentric spherical shell model, and a dipole current source below a multilayered planar conducting medium.

Considerable interest has arisen in determining the location of intracerebral sources of the potentials that are measured on the scalp. In general, nonuniqueness of this *inverse* problem is well known in that different configurations of sources can lead to the same surface distribution. The usual approach taken to obtaining an approximate solution to the inverse problem is as follows:

1. Assume a model (such as the eccentrically located dipole in a uniform, homogeneous spherical conducting medium. Assume that the electric field is quasi-static).
2. After obtaining a solution to the associated boundary-value problem (the *forward* problem), produce model-generated potential values at measurement points on the cortical surface.
3. Compare these theoretical potential values with particular discrete-time values of EEG waveforms measured at the same surface sites, and form a general least-squares reconstruction error function wherein the error is defined as the difference between predicted and measured potential at several selected cortical measurement sites.
4. Iteratively adjust the EEG dipolar source parameters at each discrete-time instant to obtain the best fit to sampled EEG waveforms in a least-

squares sense. The optimal dipole location is assumed to be the dipole location obtained when the reconstruction error function is so minimized.

The influence of anisotropy on various EEG phenomena has been studied [Henderson, Butler, and Glass (1975)]. These investigations, together with various *in vivo* studies, substantially agree that the presence of tissue anisotropy tends to attenuate and smear the pattern of scalp-recorded EEGs. This type of degradation apparently does not affect the model's ability to predict the locus of the EEG equivalent-dipole generator (although the dipole moment might be underestimated). This is important in the sense that one of the major objectives of electroencephalography is the determination of the sources of cerebral activity—for localized or focal activity—because in evoked cortical potentials and deep-brain pathologies, this concept of the equivalent-dipole generator is of clinical value.

THE ABNORMAL EEG

One of the more important clinical uses of the EEG is in the diagnosis of different types of epilepsy and in the location of the focus in the brain causing the epilepsy. Epilepsy is characterized by uncontrolled excessive activity by either a part or all of the central nervous system. A person predisposed to epilepsy has attacks when the basal level of excitability of all or part of the nervous system rises above a certain critical threshold. However, as long as the degree of excitability is held below this threshold, no attack occurs.

There are two basic types of epilepsy, *generalized epilepsy* and *partial epilepsy*. Generalized epilepsy involves the entire brain at once, whereas partial epilepsy involves a portion of the brain—sometimes only a minute focal spot and at other times a fair amount of the brain. Generalized epilepsy is further divided into *grand mal* and *petit mal* epilepsy.

Grand mal epilepsy is characterized by extreme discharges of neurons originating in the brainstem portion of the RAS. These discharges then spread throughout the cortex, to the deeper parts of the brain, and even to the spinal cord to cause generalized tonic convulsions of the entire body. They are followed near the end of the attack by alternating muscular contractions, called *clonic convulsions*. The grand mal seizure lasts from a few seconds to as long as 3–4 min and is characterized by post-seizure depression of the entire nervous system. The subject may remain in a stupor for 1 min to as long as a day or more after the attack is over.

The middle recording in Figure 4.27(c) shows a typical EEG during a grand mal attack. This response can be recorded from almost any region of the cortex. The recorded potential is of a high magnitude, and the response is synchronous, with the same periodicity as normal alpha waves. The same type of discharge occurs on both sides of the brain at the same time, indicating that the origin of the abnormality is in the lower centers of the brain that control the activity of the cerebral cortex, not in the cortex itself. Electrical recordings from the thalamus and reticular formation of experimental animals during an induced grand mal attack indicate typical high-voltage synchronous activity in these areas, similar to that recorded from the cerebral cortex. Experiments on animals have further shown that a grand mal attack is caused by intrinsic hyperexcitability of the neurons that make up the RAS structures or by some abnormality of the local neural pathways of this system.

Petit mal epilepsy is closely allied to grand mal epilepsy. It occurs in two forms, the *myoclonic* form and the *absence* form. In the myoclonic form, a burst of neuronal discharges, lasting a fraction of a second, occurs throughout the nervous system. These discharges are similar to those that occur at the beginning of a grand mal attack. The person exhibits a single violent muscular jerk involving arms or head. The entire process stops immediately, however, and the attack is over before the subject loses consciousness or stops what he or she is doing. This type of attack often becomes progressively more severe until the subject experiences a grand mal attack. Thus the myoclonic form of petit mal is similar to a grand mal attack, except that some form of inhibitory influence promptly stops it.

The absence type of petit mal epilepsy is characterized by 5–20 s of unconsciousness, during which the subject has several twitchlike contractions of the muscles, usually in the head region. There is a pronounced blinking of the eyes, followed by a return to consciousness and continuation of previous activities. This type of epilepsy is also closely allied to grand mal epilepsy. In rare instances, it can initiate a grand mal attack.

Figure 4.27(c) shows a typical *spike-and-dome* pattern that is recorded during the absence type of petit mal epilepsy. The spike portion of the record is almost identical to the spikes occuring in grand mal epilepsy, but the dome portion is distinctly different. The spike-and-dome pattern can be recorded over the entire cortex, illustrating again that the seizure originates in the RAS.

Partial epilepsy can involve almost any part of the brain, either localized regions of the cerebral cortex or deeper structures of both the cerebrum and brainstem. Partial epilepsy almost always results from some organic lesion of the brain, such as a scar that pulls on the neuronal tissue, a tumor that compresses an area of the brain, or a destroyed region of the brain tissue. Lesions such as these can cause local neurons to fire very rapid discharges. When the rate exceeds approximately 1000 per second, synchronous waves begin spreading over adjacent cortical regions. These waves presumably result from the activity of localized reverberating neuronal circuits that gradually recruit adjacent areas of the cortex into the "discharge," or firing, zone. The process spreads to adjacent areas at rates as slow as a few millimeters per minute to as fast as several centimeters per minute. When such a wave of excitation spreads over the motor cortex, it causes a progressive "march" of muscular contractions throughout the opposite side of the body, beginning perhaps in the leg region and marching progressively upward to the head region, or at other times marching in the opposite direction. This is called Jacksonian epilepsy or *Jacksonian march.*

Another type of partial epilepsy is the so-called *psychomotor seizure,* which may cause (1) a short period of amnesia, (2) an attack of abnormal rage,

(3) sudden anxiety or fear, (4) a moment of incoherent speech or mumbling, or (5) a motor act of rubbing the face with the hand, attacking someone, and so forth. Sometimes the person does not remember his or her activities during the attack; at other times the person is completely aware of, but unable to control, his or her behavior. The bottom tracing of Figure 4.27(c) represents a typical EEG during a psychomotor seizure showing a low-frequency rectangular-wave response with a frequency between 2 and 4 Hz with superimposed 14-Hz waves.

The EEG frequently can be used to locate tumors and also abnormal spiking waves originating in diseased brain tissue that might predispose to epileptic attacks. Once such a focal point is found, surgical excision of the focus often prevents future epileptic seizures.

4.9 THE MAGNETOENCEPHALOGRAM (MEG)

Active bioelectric sources in the brain generate magnetic as well as electric fields. However, the magnitude of the magnetic field associated with active cortex is extremely low. For example, it is estimated that the magnetic field of the alpha wave is approximately 0.1 pT at a distance of 5 cm from the surface of the scalp. By way of comparison, this biomagnetic field associated with the magnetoencephalogram (MEG) is roughly one hundred million times weaker than the magnetic field of the earth ($\sim$ 50 μT). Recent technological advances in the study of superconductivity have made measurement of these extremely low-strength magnetic fields possible. Specifically, the Superconducting Quantum Interference Device (SQUID) magnetometer, which is based on a superconducting effect at liquid helium temperature, has a sensitivity on the order of 0.01 pT. Background fields such as the earth's magnetic field and urban magnetic noise fields ($\sim$ 10–100 nT) can be removed for all practical purposes by using a gradiometer technique.

Using the MEG offers a number of advantages: (1) The brain and overlying tissues can be characterized as a single medium having a constant magnetic permeability μ. Therefore, the magnetic field (unlike the electric field) is not influenced by the shell-like anisotropic inhomogeneities (meninges, fluid layers, skull, muscle layer, and scalp) surrounding the brain. (2) The measurement is indirect in that electrodes are not necessary to record the MEG. That is, the SQUID detector does not need to touch the scalp, because the magnetic field does not disappear in air.

The magnetic vector potential **A** has the same orientation as the equivalent current dipole representing an active region of the brain. For a derivation of the vector potential in terms of the volume current density **J**, see Plonsey (1969). Because the magnetic field lies perpendicular to the vector potential, radially oriented current dipoles produce magnetic fields that are oriented tangentially to the sphere representing the head. Similarly, tangentially oriented brain dipoles produce radially oriented magnetic fields.

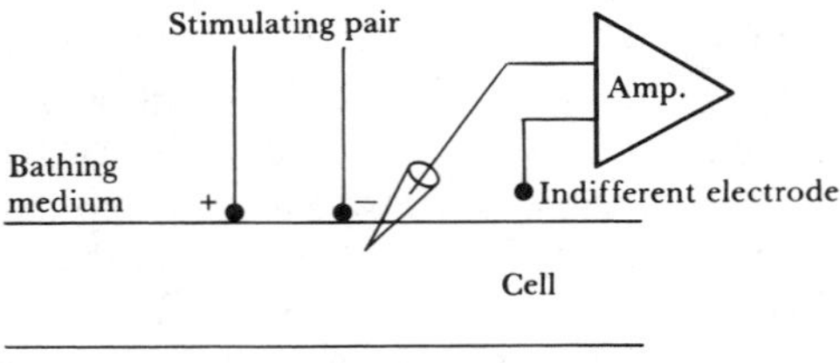

Figure P4.1

The local time dependence of biomagnetic fields can be recorded faithfully with SQUID detection systems, but in order to measure the field distribution over the surface of the scalp, measurements must be made at many locations. This is a time-consuming process. SQUID magnetometer vendors have systems with well over 100 channels (Wiksow, 1995). Advances in material fabrication techniques in the field of superconductivity should yield smaller detector coils for better spatial resolution and, subsequently, more precise localization of intracerebral sources of activity.

PROBLEMS

4.1 What are the four main factors involved in the movement of ions across the cell membrane in the steady-state condition?

4.2 Assume that life on Mars requires an interior cell potential of +100 mV and that the extracellular concentrations of the three major species are as given in Example 4.1. Choose *one* species that has the permeability coefficient given, and assume the other two permeabilities are zero. Design the cell by calculating the intracellular concentration of the chosen species.

4.3 An excitable cell is impaled by a micropipet, and a second extracellular electrode is placed close by at the outer-membrane surface. Brief pulses of current are then passed between these electrodes, which may or may not cause it to conduct an action potential. Explain how the polarity of the stimulating pair influences the membrane potential, and subsequently the activity, of the excitable cell.

4.4 Explain the subthreshold-membrane potential changes that would occur in the immediate vicinity of each of two extracellular stimulating electrodes placed at the outer-membrane surface of an excitable cell. (See Figure P4.1.) Assume that membrane potential is determined by impaling the cell with a micropipet at various points in the vicinity of the stimulating electrodes and recording the potential with respect to an indifferent extracellular electrode.

4.5 Refer to Problem 4.4. Suppose that the electrical properties of an elongated excitable cell of cylindrical geometry (such as a nerve or skeletal muscle fiber) can be modeled fairly accurately with a distributed parameter "cable" model such as that of Figure 4.3. What should the temporal-membrane potential response to brief square pulses of stimulating current look like at some fixed distance from a particular stimulating electrode? As the separation distance between the particular stimulating electrode and the exploring micropipet is progressively increased, in what manner should the amplitude of the subthreshold response change?

4.6 If a stimulus of adequate strength is supplied to the stimulating pair of Problem 4.4, an action potential is generated. Explain by means of the concept of "local-circuit" current flow how the action potential is able to propagate in an unattenuated fashion down the fiber and away from the site of stimulation.

4.7 If an elongated fiber is stimulated in the middle (as opposed to at either end), is an action potential propagated in both directions along the fiber? If so, would you expect any differences in the action-potential response measured at equal distances on either side of the stimulation site?

4.8 Define the following terms: (a) absolute refractory period, (b) relative refractory period, (c) compound nerve-action potential, (d) synapse, (e) neuro–myo junction, (f) motor unit, (g) reflex arc.

4.9 An excised, active nerve trunk serves as a bioelectric source located on the axis of a circular cylindrical volume conductor. Field potentials are recorded at various radial distances from the nerve trunk from an appropriate electrode assembly connected to an amplifier. (a) Describe the behavior of the field potential with increasing radial distance from the nerve (angle and axial distance are fixed). (b) Describe the effect of increasing the specific resistivity ρ of the bathing medium on the magnitude of the field potential, and explain how this change in ρ might be accomplished experimentally. (c) In what manner would changing the radius of the surrounding volume conductor affect the magnitude and waveshape of the extracellular field potential? (d) When can a volume conductor of finite dimensions be considered an essentially "infinite" volume conductor?

4.10 The experimental situation posed in Problem 4.9 is roughly analogous to the problem of recording either surface or intramuscular potentials from the arm of a human subject whose ulnar or median nerve has been stimulated (see Figure 4.8, for instance). Explain in terms of changes in specific resistivity and geometry why potential waveforms recorded at the wrist may differ considerably from those recorded at the level of the forearm (see Figure 4.8).

4.11 Define the M wave and the H reflex.

4.12 In many forms of peripheral neuropathies, the excitability of some neurons is changed and their conduction velocities are consequently altered. Describe the effect that this might have on an EMG recording and on muscular contraction.

4.13 A muscle is paralyzed if its neural connection to, or within, the CNS is interrupted. A disconnection at the level of the motor neuron is called a *lower motoneuron lesion.* A disconnection higher in the spinal cord or brain is called an *upper motoneuron lesion.* In both cases, the contractility of the peripheral skeletal muscle is initially preserved, but after a period of disuse, the muscle atrophies. (Atrophy, however, is much delayed in the case of an upper motoneuron lesion.) Consider Figure P4.2 to represent schematically

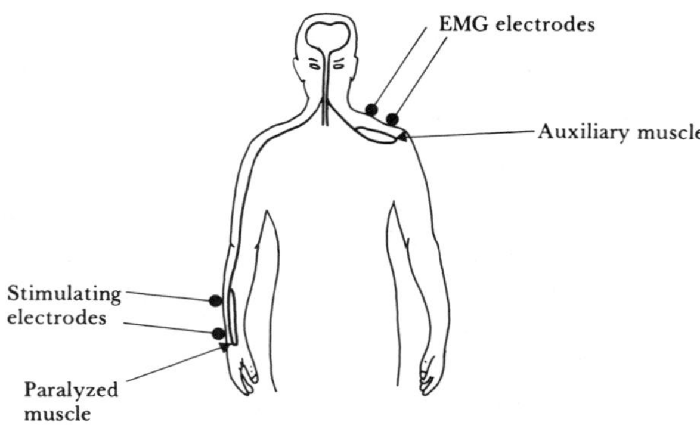

Figure P4.2

a quadriplegic patient with paralyzed extremities as shown. Suggest a scheme for using the EMG from an auxiliary intact muscle (for example, the left trapezius muscle) to aid in the control of stimulation of the paralyzed limb. (The motor nerve supply to the trapezius muscle is assumed to lie above the site of spinal-cord lesion and is therefore under volitional control. Draw a block diagram of the suggested control system. Label the anatomical structures serving as the plant (or controlled system), the controller, the feedback pathway, the actuator, and so forth. [*Hint:* The EMG signal is usually amplified, rectified, and low-pass-filtered before it is used to modulate a stimulator. For further interesting discussions of the work in this area, see Vodovnik *et al.* (1981).]

4.14 Conduct a search of the literature on the subject of the use of electromyography in the study of: (a) The function of ocular muscles. [The EMG yields valuable information regarding the synergistic action of the different ocular muscles, and it is of value in the interpretation of paralytic squint.] (b) Myasthenia gravis and other disorders of neuromuscular transmission.

4.15 Define the following cardiac anatomical terms: (a) internodal tracts, (b) subendocardial layer, (c) intercalated disk, (d) bundle branches, (e) ventricular activation.

4.16 Draw a typical lead II electrocardiogram and label all waves (P, QRS, T) and intervals. Explain what is happening electrically within the heart during each wave or interval.

4.17 The electrical activity of the His bundle normally is not present in the typical ECG recorded at the body surface because of its relatively small tissue mass. However, clinical recordings of His bundle activity could be of considerable importance in the analysis of various disorders of the conduction system. The His bundle signal can be enhanced for such analyses by successively averaging the surface electrocardiogram, or—better yet—by using an invasive technique wherein a small bipolar electrode is introduced into the right atrial chamber via conventional techniques of cardiac catheterization.

Conduct a search of the literature on the topics of noninvasive and invasive methods of recording His bundle activity, as well as the use of this signal in diagnosing various disorders of the conduction system.

4.18 Why is it necessary for the ventricular action potential to have a relatively long absolute refractory period?

4.19 PVCs can be identified because (1) they arrive early, (2) the following beat occurs at the normal time, because it is generated by the SA node, and (3) the QRS width is greater than the normal 80 ms. Design an instrument to detect and count PVCs by using all these criteria. Your *block diagram design* should include a brief description of how each block functions.

4.20 Draw and label a block diagram of the retina considered as a photoelectric sensor. What is the output at the ganglion cell layer? At the photoreceptive layer?

4.21 Explain the components of the ERG in terms of retinal cell activity.

4.22 Discuss, in terms of volume-conductor theory, the production of an ERG signal at a point on the corneal surface of the eye when the retinal bioelectric source is considered an array of current dipole sources per unit volume. Consider the possibility of (experimentally) exciting each of the elements of retinal dipole array individually, one at a time, by applying a localized spot of light superimposed on a background illumination that partially adapts the retina. What special technical considerations are involved?

4.23 Discuss the use of the steady corneal-retinal potential of the eye to measure eye movements. How accurate is this technique? Discuss at least two applications of this method.

4.24 Explain the functional role played by the following CNS structures.

a. The ascending pathways of the general sensory-nerve fibers and the descending pathways of the motor-nerve fibers

b. The ascending reticular formation (RAS)

c. The pre- and post-central gyri

d. The primary auditory and visual cortices

e. The specific and nonspecific thalamic neural fibers to the cortex

4.25 Relate EEG-wave activity recorded at the surface of the cortex to the underlying activity of cortical neurons.

4.26 Discuss in general terms the design of a spectrum analyzer for automatic analysis of EEG waves.

4.27 How might volume-conductor theory aid in the analysis of evoked cortical potentials produced by specific repetitive stimuli (auditory, visual, etc.)?

4.28 Design the switches and resistor networks required, and show all connections between any four electrodes on the scalp (Fig. 4.28) and one differential amplifier, to record the EEG for each of the *three* types of electrode connections (monopolar, bipolar, and "average" potential recordings). See text associated with Figure 4.28.

4.29 Give the block diagram design of a system that would provide nonvisual feedback to a subject who wished to maximize the amplitude of his EEG alpha waves. Explain its operation.

REFERENCES

Aiba, T. S., *et al.,* "The electroretinogram evoked by the excitation of human foveal cones." *J. Physiol.,* 1967, 189, 43–62.

Allessie, M. A., *et al.,* "Circus movement in rabbit atrial muscle as a mechanism of tachycardia." *Circ. Res.,* 1973, 33, 54–62.

Barach, J. P., and J. P. Wikswo, Jr., "Computer simulation of action potential propagation in septated nerve fibers." *Biophys. J.,* 1987, 51, 177–183.

Barr, L., *et al.,* "Propagation of action potentials and the structure of the nexus in cardiac muscle." *J. Gen. Physiol.,* 1965, 48, 797–823.

Beeler, G. W., and H. Reuter, "Reconstruction of the action potential of ventricular myocardial fibres." *J. Physiol.,* 1977, 268, 177–210.

Blight, A. R., "Computer simulation of action potentials and afterpotentials in mammalian myelinated axons." *Neuroscience,* 1985, 15, 13–31.

Clark, J. W., and R. Plonsey, "The extracellular potential field of the single active nerve fiber in a volume conductor." *Biophys. J.,* 1968, 8, 842–864.

Coulombe, A., and E. Coraboeuf, "Simulation of potassium accumulation in clefts of Purkinje fibres: Effect on membrane electrical activity." *J. Theor. Biol.,* 1983, 104, 211–229.

Dartnall, H. J. A., "The photobiology of visual processes," in H. Dawson (ed.), *The Eye,* 1st ed. New York: Academic Press, 1962, Vol. 2, pp. 321–533.

De Luca, C. J., "Electromyography," in J. G. Webster (ed.), *Encyclopedia of Medical Devices and Instrumentation.* New York: Wiley, 1988, pp. 1111–1120.

Dement, W. C., "Eye movements during sleep," in M. B. Bender (ed.), *The Oculomotor System.* New York: Harper, 1964, pp. 366–416.

Demir, S., J. W. Clark, C. Murphey, and W. Giles: A mathematical model of a rabbit sinoatrial node cell. Am. J. Physiol. 1994, 266:C832–C852.

DiFrancesco, D., and D. Noble, "Model of cardiac electrical activity incorporating ionic pumps and concentration changes." *Phil. Trans. R. Soc.,* 1985, B307, 353–398.

Earm, Y., and D. Noble, "A model of the single atrial cell: Relation between calcium current and calcium release," *Proc. Royal Soc.* (London) 1990, B:240, 83–96.

Eberstein, A., "Electroneurography: Nerve conduction studies," in J. G. Webster (ed.), *Encyclopedia of Medical Devices and Instrumentation.* New York: Wiley, 1988, pp. 1120–1126.

Elharrar, V., and D. P. Zipes, "Cardiac electrophysiological alterations during myocardial ischemia." *Amer. J. Physiol.,* 1977, 233, H329–H345.

Esselle, K. P., and M. A. Stuchly, "Neural stimulation with magnetic fields," *IEEE Trans. Biomed. Eng.,* 39, 1992, 39, 693–700.

Ganapathy, N., and J. W. Clark, Jr., "Extracellular currents and potentials of the active myelinated nerve fiber." *Biophys. J.,* 1987, 52, 749–761.

Geddes, L. A., "Optimal stimulus duration for extracranial cortical stimulation," *Neurosurgery,* 1987, 20, 94–99.

Gerber, A., R. M. Studer, R. J. P. deFigueiredo, and G. S. Moschytz, "A new framework and computer program for quantitative EMG signal analysis," in Centennial Issue of the *IEEE Trans. Biomed. Eng.,* 1984, 31, 857–863.

Gevins, A. S., and M. J. Aminoff, "Electroencephalography: Brain electrical activity," in J. G. Webster (ed.), *Encyclopedia of Medical Devices and Instrumentation.* New York: Wiley, 1988, pp. 1084–1107.

Goldman, D. E., "Potential, impedance and rectification in membranes." *J. Gen. Physiol.,* 1943, 27, 37–60.

Grissmer, S., "Properties of potassium and sodium channels in frog internode." *J. Physiol.,* 1986, 381, 119–134.

Halter, J. A., and J. W. Clark, Jr., "A distributed-parameter model of the myelinated nerve fiber," *J. Theo. Biol.,* 1991, 148, 345–382.

Heckenlively, J. R., and G. B. Arden (eds.), *Principles and Practice of Clinical Electrophysiology of Vision.* St. Louis, MO: Mosby Year Book, 1991.

Henderson, C. J., S. R. Butler, and A. Glass, "The localization of equivalent dipoles of EEG sources by the application of electric field theory." *Electroencephalog. Clin. Neurophysiol.,* 1975, 39, 117–130.

Hill, D., and G. Parr, *Encephalography.* London: MacDonald, 1963.

Hodgkin, A. L., and A. F. Huxley, "A quantitative description of membrane current and its application to conduction and excitation in nerve." *J. Physiol.,* 1952, 117, 500–544.

Hodgkin, A. L., and B. Katz, "The effect of sodium ions on the electrical activity of the giant axon of the squid." *J. Physiol.,* 1949, 108, 37–77.

Jacobson, J. T., *Principles and Applications in Auditory Evoked Potentials.* Boston, MA: Allyn and Bacon, 1994.

Jasper, H. H., "The ten-twenty electrode system of the International Federation." *Electroencephalog. Clin. Neurophysiol.,* 1958, 10 (Appendix), 371–375.

Kandel, E. R., J. H. Schwartz, and T. M. Jessell, *Principles of Neural Science,* 3rd ed. New York: Elsevier Science Publishing, 1991.

Lenman, J. A. R., and A. E. Ritchie, *Clinical Electromyography,* 4th ed. Philadelphia: Lippincott, 1987.

Levy, M. N., and R. M. Berne, *Physiology,* 2nd ed. St. Louis: Mosby, 1988.

Lipman, B. S., M. Dunn, and E. Massie, *Clinical Electrocardiography,* 7th ed. Chicago: Year Book, 1984.

Lorente de No, R., "Action potential of the motoneurons of the hypoglossus nucleus," *J. Cell. Comp. Physiol.,* 1947, 29:207–287.

Luo, C. H., and Y. Rudy, "A dynamic model of the cardiac ventricular action potential I: Simulations of ionic currents and concentration changes," *Circ. Res.,* 1994, 74:1097–1113.

Mambrito, B., and C. J. DeLuca, "A technique for the detection, decomposition and analysis of the EMG signal." *Electroencephalog. Clin. Neurophysiol.,* 1984, 58, 175–188.

McAllister, R. E., D. Noble, and R. W. Tsien, "Reconstruction of the electrical activity of cardiac Purkinje fibers." *J. Physiol.,* 1975, 251, 1–59.

Meiri, H., R. Steinberg, and B. Medalion, "Detection of sodium channel distribution in rat sciatic nerve following lysophosphatidylcholine-induced demyelination." *J. Membrane Biol.,* 1986, 92, 47–56.

Moore, J. W., R. W. Joyner, M. H. Brill, S. G. Waxman, and M. Najar-Joa, "Simulations of conduction in uniform myelinated fibers." *Biophys. J.,* 1978, 21, 147–160.

Nandedkar, S. D., E. Stalberg, and D. B. Sanders, "Simulation techniques in electromyography." *IEEE Trans. Biomed. Eng.,* 1985, 32, 775–785.

Noble, D., "A modification of the Hodgkin–Huxley equations applicable to Purkinje fiber action and pacemaker potentials." *J. Physiol.,* 1962, 160, 317–352.

Nordin, C., "Computer model of membrane current and intracellular Ca^{2+} flux in the isolated guinea pig ventricular myocyte." *Am. J. Physiol.,* 1993, 265:H2117–H2136.

North, A. W., "Accuracy and precision of electro-oculographic recording." *Invest. Ophthalmol.,* 1965, 4, 343–348.

Nunez, P. L., *Electric Fields of the Brain: The Neurophysics of the EEG.* New York: Oxford University Press, 1981.

Ogden, T. E., *Retina: Basic Science and Inherited Retinal Disease,* Vol. I. St. Louis: Mosby, 1989, Chap. 19.

Phibbs, B., *The Human Heart,* 4th ed. St. Louis: Mosby, 1979.

Plonsey, R., *Bioelectric Phenomena.* New York: McGraw-Hill, 1969.

Plonsey, R., "The biophysical basis for electrocardiography." *Crit. Rev. Bioeng.,* 1971a, 1(1), 1–48.

Plonsey, R., "Electrocardiography," in J. G. Webster (ed.), *Encyclopedia of Medical Devices and Instrumentation.* New York: Wiley, 1988, pp. 1017–1040.

Plonsey, R., and R. C. Barr, *Bioelectricity.* New York: Plenum, 1988, pp. 149–163.

Reucher, H., G. Rau, and J. Silny, "Spatial filtering of noninvasive multielectrode EMG: Parts I and II." *IEEE Trans. Biomed. Eng.,* 1987, 34, 98–113.

Röper, J., and J. R. Schwarz, "Heterogeneous distribution of fast and slow potassium channels in rat myelinated fibers." *J. Physiol.,* 1989, 416, 93–110.

Rosenfalck, P., "Intra- and extra-cellular potential fields of active nerve and muscle fibers." *Acta Physiol. Scand.,* 1969, 75, suppl. 321, 1–168.

Rushton, W. A. H., "A cone pigment in the protanope." *J. Physiol.,* 1963, 168, 345–359.

Shrager, P., "The distribution of sodium and potassium channels in single demyelinated axons of the frog." *J. Physiol.,* 1987, 392, 587–602.

Spach, M. S., R. C. Barr, G. A. Serwer, *et al.,* "Extracellular potentials related to intracellular action potentials in the dog Purkinje system." *Circ. Res.,* 1972, 30, 505–519.

Stockbridge, N., "Differential conduction at axonal bifurcations. II. Theoretical basis." *J. Neurophys.,* 1988, 59, 1286–1295.

Strong, P., *Biophysical Measurements.* Beaverton, OR: Tektronix, Inc., 1973, pp. 168–170.

Vodovnik, L., T. Bajd, A. Kralj, F. Gracanin, and P. Strojnik, "Functional electrical stimulation for control of locomotor systems." *CRC Crit. Rev. in Bioeng.,* 1981, 6, 63–151.

Waxman, S. G., and M. H. Brill, "Conduction through demyelinated plaques in multiple sclerosis: Computer simulations of facilitation by short internodes." *J. Neurol., Neurosurg. and Psychiat.,* 1978, 41, 408–416.

Wikswo, J. P., Jr., "SQUID magnetometers for biomagnetism and nondestructive testing: important questions and initial answers," *IEEE Trans. Appl. Superconductivity,* 1995, 5, 74–120.

Wood, S. L., and K. L. Cummins, "Bidirectional nerve refractory stimulation." *IEEE Trans. Biomed. Eng.,* 1985, 32, 428–438.

Yanagihara, K., A. Noma, and H. Irisawa, "Reconstruction of sino-atrial node pacemaker potential based on the voltage clamp experiments." *Japan J. Physiol.,* 1980, 30, 841–857.

York, D. H., "Review of descending motor pathways involved with transcranial stimulation." *Neurosurgery,* 1987, 20, 70–73.

5

BIOPOTENTIAL ELECTRODES

Michael R. Neuman

In order to measure and record potentials and, hence, currents in the body, it is necessary to provide some interface between the body and the electronic measuring apparatus. This interface function is carried out by biopotential electrodes. In any practical measurement of potentials, current flows in the measuring circuit for at least a fraction of the period of time over which the measurement is made. Ideally this current should be very small. However, in practical situations, it is never zero. Biopotential electrodes must therefore have the capability of conducting a current across the interface between the body and the electronic measuring circuit.

Our first impression is that this is a rather simple function to achieve and that biopotential electrodes should be relatively straightforward. But when we consider the problem in more detail, we see that the electrode actually carries out a transducing function, because current is carried in the body by ions, whereas it is carried in the electrode and its lead wire by electrons. Thus the electrode must serve as a transducer to change an ionic current into an electronic current. This greatly complicates electrodes and places constraints on their operation. We shall briefly examine the basic mechanisms involved in the transduction process and shall look at how they affect electrode characteristics. We shall next examine the principal electrical characteristics of biopotential electrodes and discuss electrical equivalent circuits for electrodes based on these characteristics. We shall then cover some of the different forms that biopotential electrodes take in various types of medical instrumentation systems, and finally, we shall look at electrodes used for measuring the ECG, EEG, EMG, and intracellular potentials.

5.1 THE ELECTRODE-ELECTROLYTE INTERFACE

The electrode-electrolyte interface is schematically illustrated in Figure 5.1. A net current that crosses the interface, passing from the electrode to the electrolyte, consists of (1) electrons moving in a direction opposite to that of the current in the electrode, (2) cations (denoted by C^+) moving in the same direction as the current, and (3) anions (denoted by A^-) moving in a direction

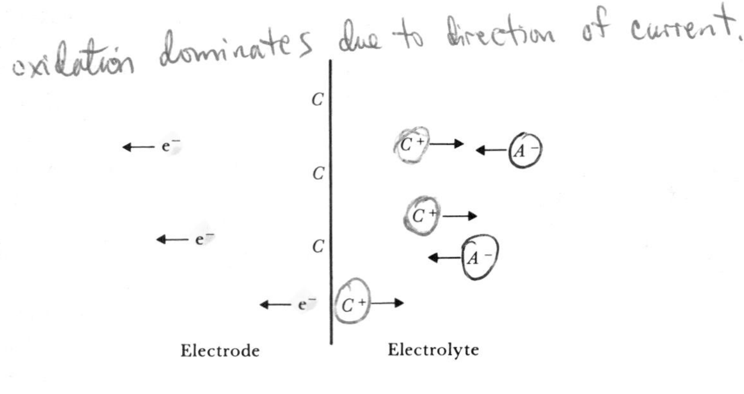

Figure 5.1 Electrode-electrolyte interface The current crosses it from left to right. The electrode consists of metallic atoms C. The electrolyte is an aqueous solution containing cations of the electrode metal C^+ and anions A^-.

opposite to that of the current in the electrolyte. For charge to cross the interface—there are no free electrons in the electrolyte and no free cations or anions in the electrode—something must occur at the interface that transfers the charge between these carriers. What actually occurs are chemical reactions at the interface, which can be represented in general by the following equations:

$$C \rightleftarrows C^{n+} + ne^- \tag{5.1}$$

$$A^{m-} \rightleftarrows A + me^- \tag{5.2}$$

where n is the valence of C and m is the valence of A. Note that in (5.1) we are assuming that the electrode is made up of some atoms of the same material as the cations and that this material in the electrode at the interface can become oxidized to form a cation and one or more free electrons. The cation is discharged into the electrolyte; the electron remains as a charge carrier in the electrode.

The reaction involving the anions is given in (5.2). In this case an anion coming to the electrode–electrolyte interface can be oxidized to a neutral atom, giving off one or more free electrons to the electrode.

Note that both reactions are often reversible and that reduction reactions (going from right to left in the equations) can occur as well. As a matter of fact, when no current is crossing the electrode–electrolyte interface, these reactions often still occur. But the rate of oxidation reactions equals the rate of reduction reactions, so the net transfer of charge across the interface is zero. When the current flow is from electrode to electrolyte, as indicated in Figure 5.1, the oxidation reactions dominate. When the current is in the opposite direction, the reduction reactions dominate.

To further explore the characteristics of the electrode–electrolyte interface, let us consider what happens when we place a piece of metal into a solution containing ions of that metal. These ions are cations, and the solution, if it is to maintain neutrality of charge, must have an equal number of anions. When the metal comes in contact with the solution, the reaction represented by (5.1) begins immediately. Initially, the reaction goes predominantly either to the left or to the right, depending on the concentration of cations in solution and the equilibrium conditions for that particular reaction. The local concentration of cations in the solution at the interface changes, which affects the anion concentration at this point as well. The net result is that neutrality of charge is not maintained in this region. Thus the electrolyte surrounding the metal is at a different electric potential from the rest of the solution. A potential difference known as the *half-cell potential* is determined by the metal involved, the concentration of its ions in solution, and the temperature, as well as other second-order factors. Knowledge of the half-cell potential is important for understanding the behavior of biopotential electrodes.

The distribution of ions in the electrolyte in the immediate vicinity of the metal-electrolyte interface has been of great interest to electrochemists, and several theories have been developed to describe it. Geddes (1972) compares the charges and potential distributions for four of these theories, whereas Cobbold (1974), in a discussion of the half-cell potential, considers the Stern model. Rather than analyze these theories here, we shall accept their general conclusion. Some sort of separation of charges at the metal-electrolyte interface results in an electric double layer, wherein one type of charge is dominant on the surface of the metal and the opposite charge is distributed in excess in the immediately adjacent electrolyte. This charge distribution at the electrode–electrolyte interface can affect electrode performance as we will see in Section 5.3.

It is not possible to measure the half-cell potential of an electrode because—unless we use a second electrode—we cannot provide a connection between the electrolyte and one terminal of the potential-measuring apparatus. Because this second electrode also has a half-cell potential, we merely end up measuring the difference between the half-cell potential of the metal and that of the second electrode. There would of course be a very large number of combinations of pairs of electrodes, so tabulations of such differential half-cell potentials would be very extensive. To avoid this problem, electrochemists have adopted the standard convention that a particular electrode—the hydrogen electrode—is defined as having a half-cell potential of zero under conditions that are readily achievable in the laboratory. We can then measure the half-cell potentials of all other electrode materials with respect to this electrode.

Table 5.1 lists several common materials that are used for electrodes and gives their half-cell potentials. The table also gives the oxidation–reduction reactions that occur at the surfaces of these electrodes and enable us to arrive

Table 5.1 Half-cell Potentials for Common Electrode Materials at 25 °C

The metal undergoing the reaction shown has the sign and potential E^0 when referenced to the hydrogen electrode.

Metal and Reaction	Potential E^0, V
$Al \rightarrow Al^{3+} + 3e^-$	−1.706
$Zn \rightarrow Zn^{2+} + 2e^-$	−0.763
$Cr \rightarrow Cr^{3+} + 3e^-$	−0.744
$Fe \rightarrow Fe^{2+} + 2e^-$	−0.409
$Cd \rightarrow Cd^{2+} + 2e^-$	−0.401
$Ni \rightarrow Ni^{2+} + 2e^-$	−0.230
$Pb \rightarrow Pb^{2+} + 2e^-$	−0.126
$H_2 \rightarrow 2H^+ + 2e^-$	0.000 by definition
$Ag + Cl^- \rightarrow AgCl + e^-$	+0.223
$2Hg + 2Cl^- \rightarrow Hg_2Cl_2 + 2e^-$	+0.268
$Cu \rightarrow Cu^{2+} + 2e^-$	+0.340
$Cu \rightarrow Cu^+ + e^-$	+0.522
$Ag \rightarrow Ag^+ + e^-$	+0.799
$Au \rightarrow Au^{3+} + 3e^-$	+1.420
$Au \rightarrow Au^+ + e^-$	+1.680

SOURCE: Data from *Handbook of Chemistry and Physics,* 55th edition, CRC Press, Cleveland, Ohio, 1974–1975, with permission.

at the potentials. The hydrogen electrode is based on the reaction

$$H_2 \rightleftarrows 2H \rightleftarrows 2H^+ + 2e^- \tag{5.3}$$

where H_2 gas bubbled over a platinum electrode is the source of hydrogen molecules. The platinum also serves as a catalyst for the reaction on the left-hand side of the equation and as an acceptor of the generated electrons.

5.2 POLARIZATION

The half-cell potential of an electrode is described in Section 5.1 for conditions in which no electric current exists between the electrode and the electrolyte. If, on the other hand, there is a current, the observed half-cell potential is often altered. The difference is due to polarization of the electrode. The difference between the observed half-cell potential and the equilibrium zero-current half-cell potential is known as the *overpotential.* Three basic mechanisms contribute to this phenomenon, and the overpotential can be separated into three components: the ohmic, the concentration, and the activation overpotentials.

The *ohmic overpotential* is a direct result of the resistance of the electrolyte. When a current passes between two electrodes immersed in an electrolyte, there is a voltage drop along the path of the current in the electrolyte as a result of its resistance. This drop in voltage is proportional to the current and the resistivity of the electrolyte. The resistance between the electrodes can itself vary as a function of the current. Thus the ohmic overpotential does not necessarily have to be linearly related to the current. This is especially true in electrolytes having low concentrations of ions. This situation, then, does not necessarily follow Ohm's law.

The *concentration overpotential* results from changes in the distribution of ions in the electrolyte in the vicinity of the electrode–electrolyte interface. Recall that the equilibrium half-cell potential results from the distribution of ionic concentration in the vicinity of the electrode–electrolyte interface when no current flows between the electrode and the electrolyte. Under these conditions, reactions (5.1) and (5.2) reach equilibrium, so the rates of oxidation and reduction at the interface are equal. When a current is established, this equality no longer exists. Thus it is reasonable to expect the concentration of ions to change. This change results in a different half-cell potential at the electrode. The difference between this and the equilibrium half-cell potential is the concentration overpotential.

The third mechanism of polarization results in the *activation overpotential.* The charge-transfer processes involved in the oxidation-reduction reaction (5.1) are not entirely reversible. In order for metal atoms to be oxidized to metal ions that are capable of going into solution, the atoms must overcome an energy barrier. This barrier, or *activation energy,* governs the kinetics of the reaction. The reverse reaction—in which a cation is reduced, thereby plating out an atom of the metal on the electrode—also involves an activation energy, but it does not necessarily have to be the same as that required for the oxidation reaction. When there is a current between the electrode and the electrolyte, either oxidation or reduction predominates, and hence the height of the energy barrier depends on the direction of the current. This difference in energy appears as a difference in voltage beween the electrode and the electrolyte, which is known as the *activation overpotential.*

These three mechanisms of polarization are additive. Thus the net overpotential of an electrode is given by

$$V_p = V_r + V_c + V_a \tag{5.4}$$

where

V_p = total overpotential, or polarization potential, of the electrode
V_r = ohmic overpotential
V_c = concentration overpotential
V_a = activation overpotential

When two aqueous ionic solutions of different concentration are separated by an ion-selective semipermeable membrane, an electric potential exists across this membrane. It can be shown (Plonsey, 1969) that this potential is given by the Nernst equation

$$E = -\frac{RT}{nF} \ln \left(\frac{a_1}{a_2} \right) \tag{5.5}$$

where a_1 and a_2 are the activities of the ions on each side of the membrane. [Other terms are defined in (4.1) and the Appendix.] In dilute solutions, ionic activity is approximately equal to ionic concentration. When intermolecular effects become significant, which happens at higher concentrations, the activity of the ions is less than their concentration.

The half-cell potentials listed in Table 5.1 are known as the standard half-cell potentials because they apply to standard conditions. When the electrode-electrolyte system no longer maintains this standard condition, half-cell potentials different from the standard half-cell potential are observed. The differences in potential are determined primarily by temperature and ionic activity in the electrolyte. *Ionic activity* can be defined as the availability of an ionic species in solution to enter into a reaction.

The standard half-cell potential is determined at a standard temperature; the electrode is placed in an electrolyte containing cations of the electrode material having unity activity. As the activity changes from unity (as a result of changing concentration), the half-cell potential varies according to the Nernst equation

$$E = E^0 + \frac{RT}{nF} \ln (a_{c^{n+}}) \tag{5.6}$$

where

E = half-cell potential

E^0 = standard half-cell potential

n = valence of electrode material

$a_{c^{n+}}$ = activity of cation C^{n+}

Equation (5.6) represents a specific application of the Nernst equation to the reaction of (5.1). The more general form of this equation can be written for a general oxidation-reduction reaction as

$$\alpha A + \beta B \rightleftarrows \gamma C + \delta D + ne^- \tag{5.7}$$

where n electrons are transferred. The general Nernst equation for this situation is

$$E = E^0 + \frac{RT}{nF} \ln \left(\frac{a_C^\gamma a_D^\delta}{a_A^\alpha a_B^\beta} \right) \tag{5.8}$$

where the a's represent the activities of the various constituents of the reaction.

An electrode–electrolyte interface is not required for a potential difference to exist. If two electrolytic solutions are in contact and have different concentrations of ions with different ionic mobilities, a potential difference, known as a *liquid-junction potential,* exists between them. For solutions of the same composition but different activities, its magnitude is given by

$$E_j = \frac{\mu_+ - \mu_-}{\mu_+ + \mu_-} \frac{RT}{nF} \ln \left(\frac{a'}{a''} \right) \tag{5.9}$$

where μ_+ and μ_- are the mobilities of the positive and negative ions, and a' and a'' are the activities of the two solutions. Though liquid-junction potentials are generally not so high as electrode–electrolyte potentials, they can easily be of the order of tens of millivolts. For example, two solutions of sodium chloride, at 25 °C, with activities that vary by a factor of 10, have a potential difference of approximately 12 mV. Note that you can generate potentials of the order of some biological potentials by merely creating differences in concentration in an electrolyte. This is a factor to consider when you are examining actual electrode systems used for biopotential measurements.

5.3 POLARIZABLE AND NONPOLARIZABLE ELECTRODES

Theoretically, two types of electrodes are possible: those that are perfectly polarizable and those that are perfectly nonpolarizable. This classification refers to what happens to an electrode when a current passes between it and the electrolyte. *Perfectly polarizable electrodes* are those in which no actual charge crosses the electrode–electrolyte interface when a current is applied. Of course, there has to be current across the interface, but this current is a displacement current, and the electrode behaves as though it were a capacitor. *Perfectly nonpolarizable electrodes* are those in which current passes freely across the electrode–electrolyte interface, requiring no energy to make the transition. Thus, for perfectly nonpolarizable electrodes there are no overpotentials.

Neither of these two electrodes can be fabricated; however, some practical electrodes can come close to acquiring their characteristics. Electrodes made of noble metals such as platinum come closest to behaving as perfectly polarizable electrodes. Because the materials of these electrodes are relatively inert, it is difficult for them to oxidize and dissolve. Thus current passing between the electrode and the electrolyte changes the concentration primarily of ions at the interface, so a majority of the overpotential seen from this type of electrode

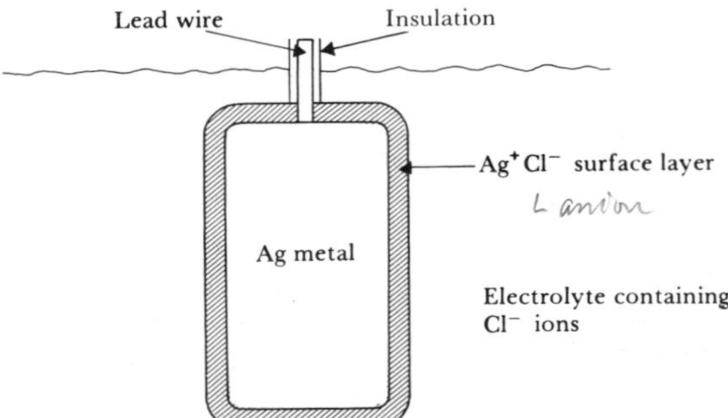

Figure 5.2 A silver/silver chloride electrode, shown in cross section.

is a result of V_c, the concentration overpotential. The electrical characteristics of such an electrode show a strong capacitive effect.

THE SILVER/SILVER CHLORIDE ELECTRODE

The silver/silver chloride (Ag/AgCl) electrode is a practical electrode that approaches the characteristics of a perfectly nonpolarizable electrode and can be easily fabricated in the laboratory. It is a member of a class of electrodes each of which consists of a metal coated with a layer of a slightly soluble ionic compound of that metal with a suitable anion. The whole structure is immersed in an electrolyte containing the anion in relatively high concentrations.

The structure is shown in Figure 5.2. A silver metal base with attached insulated lead wire is coated with a layer of the ionic compound AgCl. (This material—AgCl—is only very slightly soluble in water, so it remains stable.) The electrode is then immersed in an electrolyte bath in which the principal anion of the electrolyte is Cl^-. For best results, the electrolyte solution should also be saturated with AgCl so that there is little chance for any of the surface film on the electrode to dissolve.

The behavior of the Ag/AgCl electrode is governed by two chemical reactions. The first involves the oxidation of silver atoms on the electrode surface to silver ions in solution at the interface.

$$\mathrm{Ag} \rightleftarrows \mathrm{Ag}^+ + \mathrm{e}^- \tag{5.10}$$

$$\mathrm{Ag}^+ + \mathrm{Cl}^- \rightleftarrows \mathrm{AgCl}\downarrow \tag{5.11}$$

The second reaction occurs immediately after the formation of Ag^+ ions. These ions combine with Cl^- ions already in solution to form the ionic compound AgCl. As mentioned before, AgCl is only very slightly soluble in water, so most of it precipitates out of solution onto the silver electrode and contributes to the silver chloride deposit. Silver chloride's rate of precipitation and of returning to solution is a constant K_s known as the *solubility product.*

Under equilibrium conditions the ionic activities of the Ag^+ and Cl^- ions must be such that their product is the solubility product.

$$a_{Ag^+} \times a_{Cl^-} = K_s \tag{5.12}$$

In biological fluids the concentration of Cl^- ions is relatively high, which gives it an activity somewhat less than unity. The solubility product for AgCl, on the other hand, is of the order of 10^{-10}. This means that, when an Ag/AgCl electrode is in contact with biological fluids, the activity of the Ag^+ ion must be very low and of the same order of magnitude as the solubility product.

We can determine the half-cell potential for the Ag/AgCl electrode by writing (5.6) for the reaction of (5.10).

$$E = E^0_{Ag} + \frac{RT}{nF} \ln (a_{Ag^+}) \tag{5.13}$$

By using (5.12), we can rewrite this as

$$E = E^0_{Ag} + \frac{RT}{nF} \ln \left(\frac{K_s}{a_{Cl^-}} \right) \tag{5.14}$$

or

$$E = E^0_{Ag} + \frac{RT}{nF} \ln (K_s) - \frac{RT}{nF} \ln (a_{Cl^-}) \tag{5.15}$$

The first and second terms on the right-hand side of (5.15) are constants; only the third is determined by ionic activity. In this case, it is the activity of the Cl^- ion, which is relatively large and not related to the oxidation of Ag, which is caused by the current through the electrode. The half-cell potential of this electrode is consequently quite stable when it is placed in an electrolyte containing Cl^- as the principal anion, provided the activity of the Cl^- remains stable. Because this is the case in the body, we shall see in later sections of this chapter that the Ag/AgCl electrode is relatively stable in biological applications.

There are several procedures that can be used to fabricate Ag/AgCl electrodes (Janz and Ives, 1968). Two of them are of particular importance in biomedical electrodes. One is the electrolytic process for forming Ag/AgCl electrodes. An electrochemical cell is made up in which the Ag electrode on which the AgCl layer is to be deposited serves as anode and another piece of Ag—having a surface area much greater than that of the anode—serves as cathode. A 1.5-V battery serves as the energy source, and a series resistance limits the peak current, thereby controlling the maximal rate of reaction. A milliammeter can be placed in the circuit to observe the current, which is proportional to the rate of reaction.

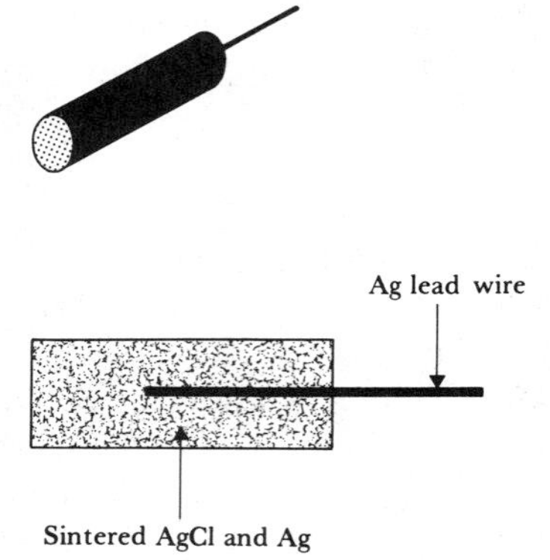

Figure 5.3 **Sintered Ag/AgCl electrode**

The reactions of (5.10) and (5.11) begin to occur as soon as the battery is connected, and the current jumps to its maximal value. As the thickness of the deposited AgCl layer increases, the rate of reaction decreases and the current drops. This situation continues, and the current approaches zero asymptotically. Theoretically, the reaction is not complete until the current drops to zero. In practice this never occurs because of other processes going on that conduct a current. Therefore the reaction can be stopped after a few minutes, once the current has reached a relatively stable low value—of the order of 10 μA for most biological electrodes.

The second process for producing Ag/AgCl electrodes useful in medical instrumentation is a sintering process that forms pellet electrodes, as shown in Figure 5.3. The electrode consists of an Ag lead wire surrounded by a sintered Ag/AgCl cylinder. It is formed by placing the cleaned lead wire in a die that is then filled with a mixture of powdered Ag and AgCl. The die is compressed in an arbor press to form the powdered components into a pellet, which is then removed from the die and baked at 400 °C for several hours. These electrodes tend to have a greater endurance than the electrolytically deposited AgCl electrodes, and they are best applied when repeated usage is necessary. The electrolytically deposited AgCl has a tendency to flake off under mechanical stress, leaving portions of metallic Ag in contact with the electrolyte.

Silver chloride is not a very good conductor of an electric current. If the powder that was compressed to make the sintered electrode consisted only of finely ground silver chloride, this would result in a high-resistance electric connection between the electrolytic solution and the silver wire in the center of the sintered electrode. Electrochemists found that they could increase the conductivity of the silver chloride pellet by including metallic silver powder along with the silver chloride powder. The amount of metallic silver is small enough to make highly unlikely any direct connection from the silver wire to the electrode through silver particles. Instead, there is always some silver chloride beween the silver particles, but the presence of the silver particles makes the current-conduction pathway through the silver chloride much shorter.

A similar situation occurs in the electrolytically prepared silver/silver chloride electrode. Although the silver chloride layer is much thinner in this case than it is for the sintered electrode, it remains a pure silver chloride layer for only a short time after it is deposited. Silver chloride is a silver-halide salt, and these materials are photosensitive. Light striking these salts can cause the silver ions to be reduced to metallic silver atoms. Thus, for all practical purposes, the electrolytically deposited silver chloride layer contains fine silver particles as well. Evidence of their presence appears when the layer is grown and immediately becomes dark gray because of the fine silver particles (pure silver chloride is amber-colored).

In addition to its nonpolarizable behavior, the Ag/AgCl electrode exhibits less electric noise than the equivalent metallic Ag electrodes. Geddes and Baker (1967) showed that electrodes with the AgCl layer exhibited far less noise than was observed when the AgCl layer was removed. Also, a majority of the noise for the purely metallic electrodes was at low frequencies. This would provide the most serious interference for low-frequency recordings, such as the EEG.

A second kind of electrode that has characteristics approaching those of the perfectly nonpolarizable electrode is the *calomel electrode.* It is used primarily as a reference electrode for electrochemical determinations and is frequently applied as the reference electrode when pH is measured (see Section 10.2). The calomel electrode is often constructed as a glass tube with a porous glass plug at its base filled with a paste of mercurous chloride or calomel (Hg_2Cl_2) mixed with a saturated potassium chloride (KCl) solution. Like AgCl, the Hg_2Cl_2 is only slightly soluble in water, so most of it retains its solid form. A layer of elemental mercury is placed on top of the paste layer with an electric lead wire within it. This entire assembly is then positioned in the center of a larger glass tube with a porous glass plug at its base. The tube is filled with a saturated KCl solution so that the Hg_2Cl_2 layer of the inner tube is in contact with this electrolyte through the porous plug of the inner tube. We have a half-cell made up of Hg in intimate contact with an Hg_2Cl_2 layer, which is in contact with the saturated KCl electrolyte. The porous plug at the bottom of the electrode assembly is used to make contact between the internal KCl solution and the solution in which the electrode is immersed. This is actually a liquid-liquid junction that can result in a liquid–liquid junction potential, which will add to the electrode half-cell potential.

Silver/silver chloride electrodes can be fabricated in the same form as the calomel electrode and used for electroanalytical chemical measurements. In this case, the mercury is replaced by silver and AgCl replaces the Hg_2Cl_2 in the electrode structure.

Using the same argument as that used for the Ag/AgCl electrode, we can show that the half-cell potential of this electrode is dependent on the Cl^- activity in the saturated KCl solution. This is stable at a given temperature, because the solution is saturated. In application, the tip of this electrode assembly that contains the porous plug is dipped into the electrolytic solution that it is to contact. In pH measurements, a pH electrode is also dipped into the solution, and the potential difference between the two electrodes is measured.

5.4 ELECTRODE BEHAVIOR AND CIRCUIT MODELS

The electrical characteristics of electrodes have been the subject of much study. Often the current–voltage characteristics of the electrode–electrolyte interface are found to be nonlinear, and, in turn, nonlinear elements are required for modeling electrode behavior. Specifically, the characteristics of an electrode are sensitive to the current passing through the electrode, and the electrode characteristics at relatively high current densities can be considerably different from those at low current densities. The characteristics of electrodes are also waveform-dependent. When sinusoidal currents are used to measure the electrode's circuit behavior, the characteristics are also frequency-dependent.

The characterization of electrode–electrolyte interfacial impedances has been well reviewed by Geddes (1972), Cobbold (1974), Ferris (1974), and Schwan (1963). It is only summarized here. For sinusoidal inputs, the terminal characteristics of an electrode have both a resistive and a reactive component. Over all but the lowest frequencies, this situation can be modeled as a series resistance and capacitance. We should not be surprised to see a capacitance entering into this model, because the half-cell potential described earlier was the result of the distribution of ionic charge at the electrode–electrolyte interface that had been considered a double layer of charge. This, of course, should behave as a capacitor—hence the capacitive reactance seen for real electrodes.

The series resistance–capacitance equivalent circuit breaks down at the lower frequencies, where this model would suggest an impedance going to infinity as the frequency approaches dc. To avoid this problem, we can convert this series *RC* circuit to a parallel *RC* circuit that has a purely resistive impedance at very low frequencies. If we combine this circuit with a voltage source representing the half-cell potential and a series resistance representing the interface effects and resistance of the electrolyte, we can arrive at the biopotential electrode equivalent circuit model shown in Figure 5.4.

In this circuit, R_d and C_d represent the resistive and reactive components just discussed. These components are still frequency- and current-density–dependent. In this configuration it is also possible to assign physical meaning to the components. C_d represents the capacitance across the double layer of charge at the electrode–electrolyte interface. The parallel resistance R_d represents the leakage resistance across this double layer. All the components of this equivalent circuit have values determined by the electrode material,

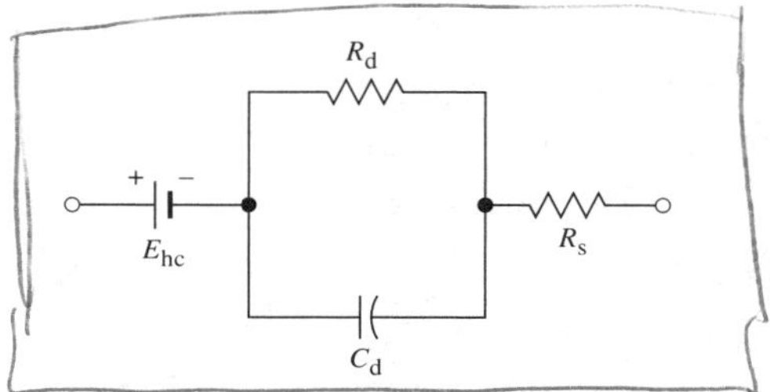

Figure 5.4 Equivalent circuit for a biopotential electrode in contact with an electrolyte E_{hc} is the half-cell potential, R_d and C_d make up the impedance associated with the electrode–electrolyte interface and polarization effects, and R_s is the series resistance associated with interface effects and due to resistance in the electrolyte.

and—to a lesser extent—by the material of the electrolyte and its concentration.

The equivalent circuit of Figure 5.4 demonstrates that the electrode impedance is frequency-dependent. At high frequencies, where $1/\omega C \ll R_d$, the impedance is constant at R_s. At low frequencies, where $1/\omega C \gg R_d$, the impedance is again constant but its value is larger, being $R_s + R_d$. At frequencies between these extremes, the electrode impedance is frequency-dependent.

The impedance of Ag/AgCl electrodes varies significantly from that of a pure silver electrode at frequencies under 100 Hz. This has been demonstrated by Geddes *et al.* (1969), whose data are reproduced in Figure 5.5.

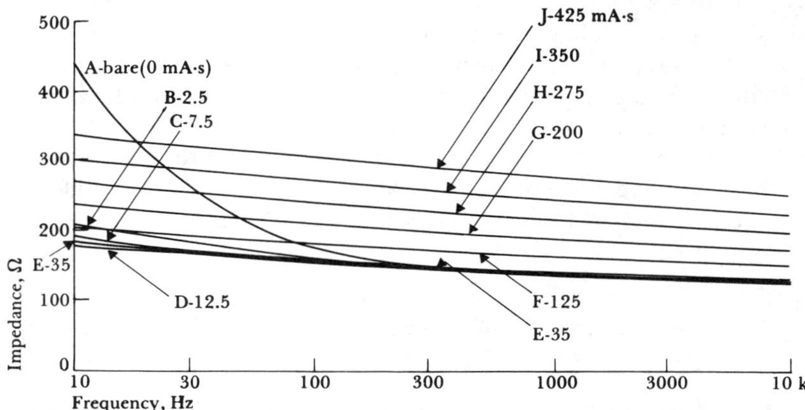

Figure 5.5 Impedance as a function of frequency for Ag electrodes coated with an electrolytically deposited AgCl layer. The electrode area is 0.25 cm². Numbers attached to curves indicate number of mA · s for each deposit. (From L. A. Geddes, L. E. Baker, and A. G. Moore, "Optimum Electrolytic Chloriding of Silver Electrodes," *Medical and Biological Engineering,* 1969, **7,** pp. 49–56.)

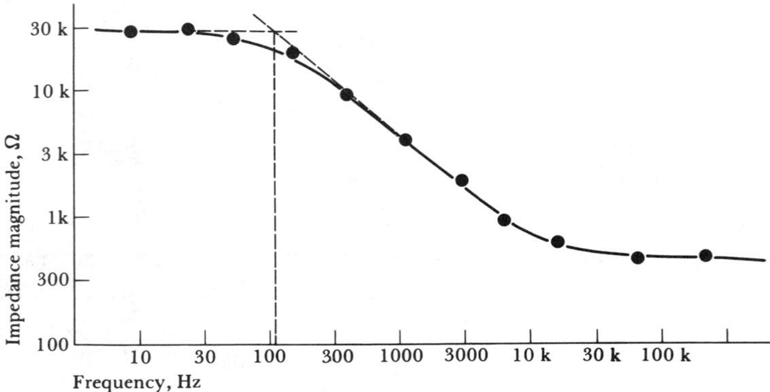

Figure 5.6 Experimentally determined magnitude of impedance as a function of frequency for electrodes.

A metallic silver electrode having a surface area of 0.25 cm² had the impedance characteristic shown by curve A. At a frequency of 10 Hz, the magnitude of its impedance was almost three times the value at 300 Hz. This indicates a strong capacitive component to the equivalent circuit. Electrolytically depositing 2.5 mA · s of AgCl greatly reduced the low-frequency impedance, as reflected in curve B. Depositing thicker AgCl layers had minimal effects until the charge deposited exceeded approximately 100 mA · s. The curves were then seen to shift to higher impedances in a parallel fashion as the amount of AgCl deposited increased. Geddes (1972) points out that depositing an AgCl layer using a charge of between 100 and 500 mA · s/cm² provides the lowest value of electrode impedance. If the current density is maintained at greater than 5 mA/cm², we can adjust current and time to provide the most convenient values for depositing the desired layer.

The impedance decreases with frequency for different electrode materials as shown in Figure 5.6. For 1 cm² at 10 Hz, nickel- and carbon-loaded silicone rubber have an impedance of approximately 30 kΩ, whereas Ag/AgCl has an impedance of less than 10 Ω (Das and Webster, 1980).

5.5 THE ELECTRODE–SKIN INTERFACE AND MOTION ARTIFACT

In Section 5.1 we examined the electrode–electrolyte interface and saw how it influenced the electrical properties that are seen in practical electrodes. When biopotentials are recorded from the surface of the skin, we must consider an additional interface—the interface between the electrode–electrolyte and the skin—in order to understand the behavior of the electrodes. In coupling an electrode to the skin, we generally use a transparent electrolyte gel containing Cl^- as the principal anion to maintain good contact. Alternatively, we

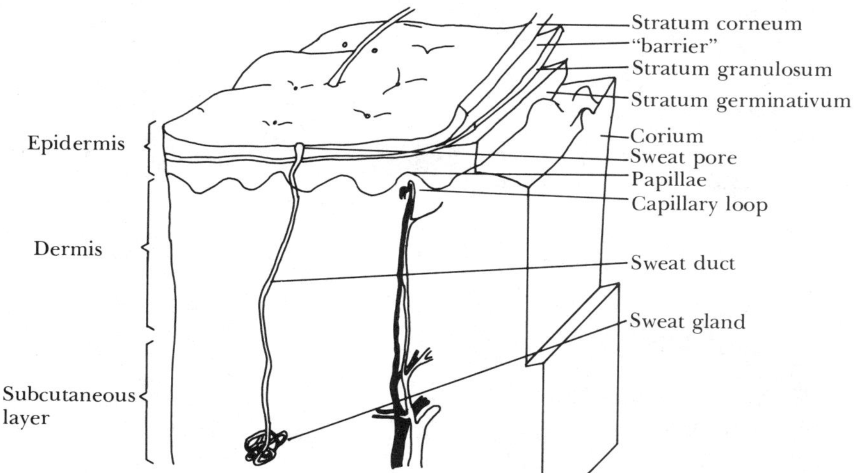

Figure 5.7 Magnified section of skin, showing the various layers (Copyright © 1977 by The Institute of Electrical and Electronics Engineers. Reprinted, with permission, from *IEEE Trans. Biomed. Eng.,* March 1977, vol. BME-24, no. 2, pp. 134–139.)

may use an electrode cream, which contains Cl^- and has the consistency of hand lotion. The interface between this gel and the electrode is an electrode–electrolyte interface, as described above. However, the interface beween the electrolyte and the skin is different and requires some explanation. Before we give this explanation, let us briefly review the structure of the skin.

Figure 5.7 shows a cross-sectional diagram of the skin. The skin consists of three principal layers that surround the body to protect it from its environment and that also serve as appropriate interfaces. The outermost layer, or *epidermis,* plays the most important role in the electrode–skin interface. This layer, which consists of three sublayers, is constantly renewing itself. Cells divide and grow in the deepest layer, the *stratum germinativum,* and are displaced outward as they grow by the newly forming cells underneath them. As they pass through the *stratum granulosum,* they begin to die and lose their nuclear material. As they continue their outward journey, they degenerate further into layers of flat keratinous material that forms the *stratum corneum,* or horny layer of dead material on the skin's surface. These layers are constantly being worn off and replaced at the stratum granulosum by new cells. The epidermis is thus a constantly changing layer of the skin, the outer surface of which consists of dead material that has different electrical characteristics from live tissue.

The deeper layers of the skin contain the vascular and nervous components of the skin as well as the sweat glands, sweat ducts, and hair follicles. These layers are similar to other tissues in the body and, with the exception of the sweat glands, do not bestow any unique electrical characteristics on the skin.

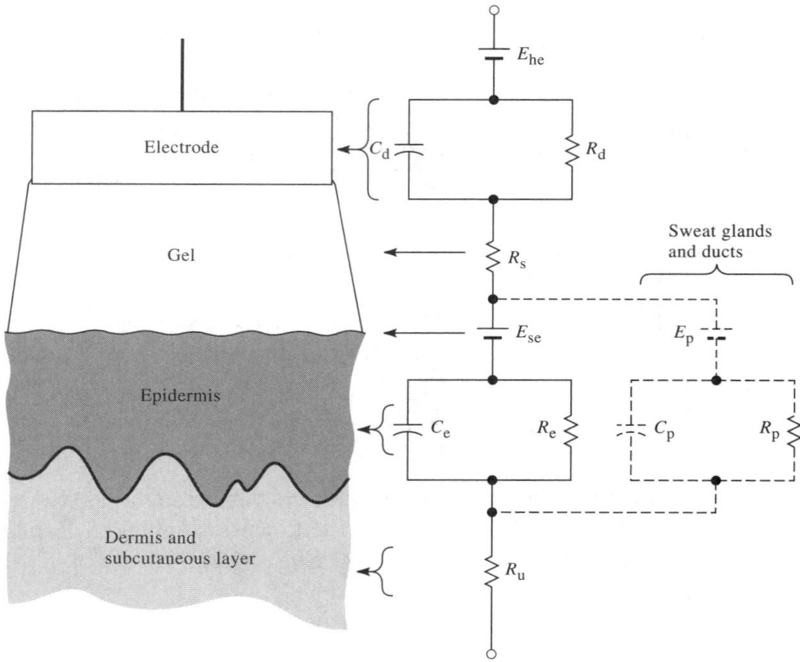

Figure 5.8 A body-surface electrode is placed against skin, showing the total electrical equivalent circuit obtained in this situation. Each circuit element on the right is at approximately the same level at which the physical process that it represents would be in the left-hand diagram.

To represent the electric connection between an electrode and the skin through the agency of electrolyte gel, our equivalent circuit of Figure 5.4 must be expanded, as shown in Figure 5.8. The electrode–electrolyte interface equivalent circuit is shown adjacent to the electrode–gel interface. The series resistance R_s is now the effective resistance associated with interface effects of the gel between the electrode and the skin. We can consider the epidermis, or at least the stratum corneum, as a membrane that is semipermeable to ions, so if there is a difference in ionic concentration across this membrane, there is a potential difference E_{se}, which is given by the Nernst equation. The epidermal layer is also found to have an electric impedance that behaves as a parallel RC circuit, as shown. For 1 cm^2, skin impedance reduces from approximately 200 kΩ at 1 Hz to 200 Ω at 1 MHz (Rosell *et al.,* 1988). The dermis and the subcutaneous layer under it behave in general as pure resistances. They generate negligible dc potentials.

Thus we see that—if the effect of the stratum corneum can be reduced—a more stable electrode will result. We can minimize the effect of the stratum corneum by removing it, or at least a part of it, from under the electrode. There are many ways to do this, ranging from vigorous rubbing with a pad soaked in acetone to abrading the stratum corneum with sandpaper to puncture it. In all cases, this process tends to short out E_{se}, C_e, and R_e, as shown in Figure 5.8, thereby improving the stability of the signal.

A factor that is sometimes important in examination of, for example, psychogenic electrodermal responses or the galvanic skin reflex (GSR), is the contribution of the sweat glands and sweat ducts. The fluid secreted by sweat glands contains Na^+, K^+, and Cl^- ions, the concentrations of which differ from those in the extracellular fluid. Thus there is a potential difference between the lumen of the sweat duct and the dermis and subcutaneous layers. There also is a parallel R_pC_p combination in series with this potential that represents the wall of the sweat gland and duct, as shown by the broken lines in Figure 5.8. These components are often neglected when we consider biopotential electrodes that are not used to measure the electrodermal response or GSR (Boucsein, 1992).

When a polarizable electrode is in contact with an electrolyte, a double layer of charge forms at the interface. If the electrode is moved with respect to the electrolyte, this movement mechanically disturbs the distribution of charge at the interface and results in a momentary change of the half-cell potential until equilibrium can be reestablished. If a pair of electrodes is in an electrolyte and one moves while the other remains stationary, a potential difference appears beween the two electrodes during this movement. This potential is known as *motion artifact* and can be a serious cause of interference in the measurement of biopotentials.

Because motion artifact results primarily from mechanical disturbances of the distribution of charge at the electrode–electrolyte interface, it is reasonable to expect that motion artifact is minimal for nonpolarizable electrodes.

Observation of the motion-artifact signals reveals that a major component of this noise is at low frequencies. Section 6.6 and Figure 6.16 will show that different biopotential signals occupy different portions of the frequency spectrum. Figure 6.16 shows that low-frequency artifact does not affect signals such as the EMG or axon action potential (AAP) nearly so much as it does the ECG, EEG, and EOG. In the former case, filtering can be effectively used to minimize the contribution of motion artifact on the overall signal. But in the latter case, such filtering also distorts the signal. Consequently, it is important in these applications to use a nonpolarizable electrode to minimize motion artifact stemming from the electrode–electrolyte interface.

This interface, however, is not the only source of motion artifact encountered when biopotential electrodes are applied to the skin. The equivalent circuit in Figure 5.8 shows that, in addition to the half-cell potential E_{hc}, the electrolyte gel–skin potential E_{se} can also cause motion artifact if it varies with movement of the electrode. Variations of this potential indeed do represent a major source of motion artifact in Ag/AgCl skin electrodes (Tam and Webster, 1977). They have shown that this artifact can be significantly reduced when the stratum corneum is removed by mechanical abrasion with a fine abrasive paper. This method also helps to reduce the epidermal component of the skin impedance. Tam and Webster (1977) also point out, however, that removal of the body's outer protective barrier makes that region of skin more susceptible to irritation from the electrolyte gel. Therefore, the choice of a gel material is important. Remembering the dynamic nature of the epidermis, note also that the stratum corneum can regenerate itself in as short a time as 24 hours, thereby renewing the source of motion artifact. This is a factor to be taken into account if the electrodes are to be used for chronic recording. A potential between the inside and outside of the skin can be measured (Burbank and Webster, 1978). Stretching the skin changes this skin potential by 5–10 mV, and this change appears as motion artifact. Ten 0.5-mm skin punctures through the barrier layer shortcircuits the skin potential and reduces the stretch artifact to less than 0.2 mV. de Talhouet and Webster (1996) provide a model for the origin of this skin potential and show how it can be reduced by stripping layers of the skin using Scotch tape.

5.6 BODY-SURFACE RECORDING ELECTRODES

Over the years many different types of electrodes for recording various potentials on the body surface have been developed. This section describes the various types of these electrodes and gives an example of each. The reader interested in more extensive examples should consult Geddes (1972).

METAL-PLATE ELECTRODES

One of the most frequently used forms of biopotential sensing electrodes is the metal-plate electrode. In its simplest form, it consists of a metallic conductor in contact with the skin. An electrolyte soaked pad or gel is used to establish and maintain the contact.

Figure 5.9 shows several forms of this electrode. The one most commonly used for limb electrodes with the electrocardiograph is shown in Figure 5.9(a). It consists of a flat metal plate that has been bent into a cylindrical segment. A terminal is placed on its outside surface near one end; this terminal is used to attach the lead wire to the electrocardiograph. A post, placed on this same side near the center, is used to connect a rubber strap to the electrode and hold it in place on an arm or leg. The electrode is traditionally made of German silver (a nickel–silver alloy). Before it is attached to the body, its concave surface is covered with electrolyte gel. Similarly arranged flat metal disks are also used for this type of electrode. Although based upon preceding sections of this chapter one would expect that better electrode designs could be used with electrocardiographs today, these traditional electrodes remain popular and are frequently used.

A second common variety of metal-plate electrode is the metal disk illustrated in Figure 5.9(b). This electrode, which has a lead wire soldered or welded to the back surface, can be made of several different materials. Sometimes the connection between lead wire and electrode is protected by a layer of insulating

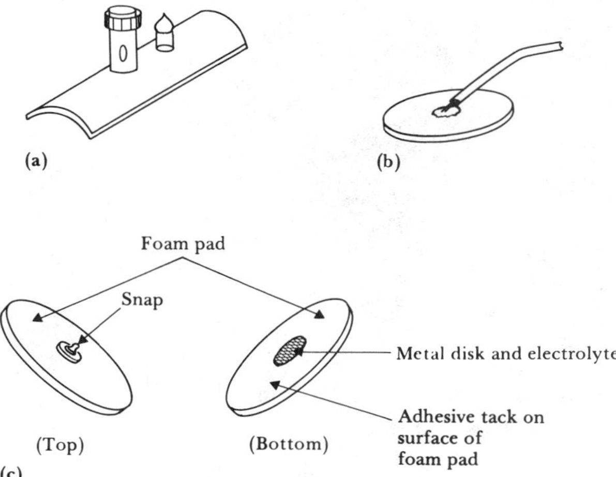

Figure 5.9 Body-surface biopotential electrodes (a) Metal-plate electrode used for application to limbs. (b) Metal-disk electrode applied with surgical tape. (c) Disposable foam-pad electrodes, often used with electrocardiographic monitoring apparatus.

material, such as epoxy or polyvinyl chloride. This structure can be used as a chest electrode for recording the ECG or in cardiac monitoring for long-term recordings. In these applications the electrode is often fabricated from a disk of Ag that may or may not have an electrolytically deposited layer of AgCl on its contacting surface. It is coated with electrolyte gel and then pressed against the patient's chest wall. It is maintained in place by a strip of surgical tape or a plastic foam disk with a layer of tack on one surface.

This style of electrode is also popular for surface recordings of EMG or EEG. In recording EMGs, investigators use stainless steel, platinum, or gold-plated disks to minimize the chance that the electrode will enter into chemical reactions with perspiration or the gel. These materials produce polarizable electrodes, and motion artifact can be a problem with active patients. Electrodes used in monitoring EMGs or EEGs are generally smaller in diameter than those used in recording ECGs. Disk-shaped electrodes such as these have also been fabricated from metal foils (primarily silver foil) and are applied as single-use disposable electrodes. The thinness of the foil allows it to conform to the shape of the body surface. Also, because it is so thin, the cost can be kept relatively low.

Economics necessarily plays an important role in determining what materials and apparatus are used in hospital administration and patient care. In choosing suitable cardiac electrodes for patient-monitoring applications, physicians are more and more turning to pregelled, disposable electrodes with the adhesive already in place. These devices are ready to be applied to the patient and are not cleaned after use. This minimizes the amount of time that personnel must devote to the use of these electrodes.

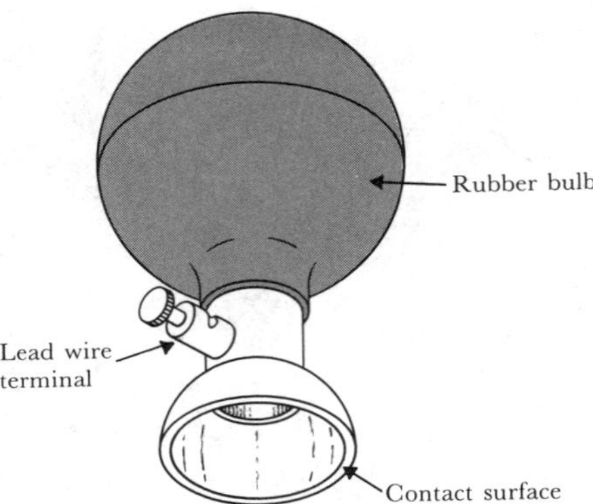

Figure 5.10 A metallic suction electrode is often used as a precordial electrode on clinical electrocardiographs.

A popular type of electrode of this variety is illustrated in Figure 5.9(c). It consists of a relatively large disk of plastic foam material with a silver-plated disk on one side attached to a silver-plated snap similar to that used on clothing in the center of the other side. A lead wire with the female portion of the snap is then snapped onto the electrode and used to connect the assembly to the monitoring apparatus. The silver-plated disk serves as the electrode and may be coated with an AgCl layer. A layer of electrolyte gel covers the disk. The electrode side of the foam is covered with an adhesive material that is compatible with the skin. A protective cover or strip of release paper is placed over this side of the electrode and foam, and the complete electrode is packaged in a foil envelope so that the water component of the gel will not evaporate away. To apply the electrode to the patient, the technician has only to clean the area of skin on which the electrode is to be placed, open the electrode packet, remove the release paper from the tack, and press the electrode against the patient. This procedure is quickly accomplished and no special technique need be learned, such as using the correct amount of gel or cutting strips of adhesive tape to hold the electrode in place.

SUCTION ELECTRODES

A modification of the metal-plate electrode that requires no straps or adhesives for holding it in place is the suction electrode illustrated in Figure 5.10. Such electrodes are frequently used in electrocardiography as the *precordial* (chest) leads, because they can be placed at particular locations and used to take a recording. They consist of a hollow metallic cylindrical electrode that makes contact with the skin at its base. An appropriate terminal for the lead wire is attached to the metal cylinder, and a rubber suction bulb fits over its other base. Electrolyte gel is placed over the contacting surface of the electrode, the bulb is squeezed, and the electrode is then placed on the chest wall. The bulb is released and applies suction against the skin, holding the electrode

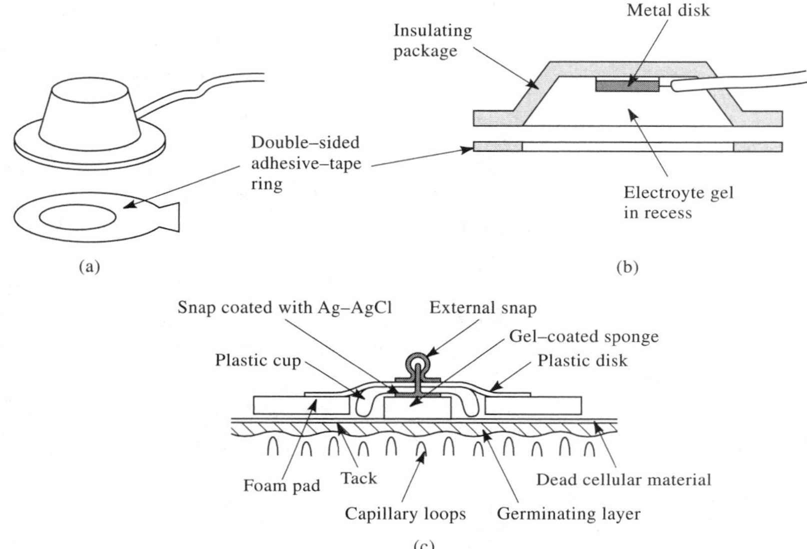

Figure 5.11 Examples of floating metal body-surface electrodes (a) Recessed electrode with top-hat structure. (b) Cross-sectional view of the electrode in (a). (c) Cross-sectional view of a disposable recessed electrode of the same general structure shown in Figure 5.9(c). The recess in this electrode is formed from an open foam disk, saturated with electrolyte gel and placed over the metal electrode.

assembly in place. This electrode can be used only for short periods of time; the suction and the pressure of the contact surface against the skin can cause irritation. Although the electrode itself is quite large, Figure 5.10 shows that the actual contacting area is relatively small. This electrode thus tends to have a higher source impedance than the relatively large-surface-area metal-plate electrodes used for ECG limb electrodes, as shown in Figure 5.9(a).

FLOATING ELECTRODES

In the previous section, we noted that one source of motion artifact in biopotential electrodes is the double layer of charge at the electrode–electrolyte interface. The use of nonpolarizable electrodes, such as the Ag/AgCl electrode, can greatly diminish this artifact. But it still can be present, and efforts to stabilize the interface mechanically can reduce it further. *Floating electrodes* offer a suitable technique to do so.

Figure 5.11 shows examples of these devices. Figure 5.11(a) depicts a floating electrode known as a top-hat electrode; its internal structure is illustrated in cross section in Figure 5.11(b). The principal feature of the electrode is that the actual electrode element or metal disk is recessed in a cavity so that it does not come in contact with the skin itself. Instead, the element is surrounded by electrolyte gel in the cavity. The cavity does not move with respect to the metal disk, so it does not produce any mechanical movement of the double layer of charge. In practice, the electrode is filled with electrolyte gel and then attached to the skin surface by means of a double-sided adhesive-tape ring, as shown in Figure 5.11. The electrode element can be a disk made of a metal such as silver coated with AgCl. Another frequently encountered form of the floating electrode uses a sintered Ag/AgCl pellet instead of a metal disk. These electrodes are found to be quite stable and are suitable for many uses.

A single-use, disposable modification of the floating electrode is shown in cross section in Figure 5.11(c). Its structure is basically the same as that of the disposable metal-plate electrode shown in Figure 5.9(c), but it has one added component—a disk of thin, open-cell foam saturated with electrolyte gel. The foam is firmly affixed to the metal-disk electrode, thereby providing an intermediate electrolyte–gel layer between the electrode and the skin. Because the foam is fixed to the metal disk, the gel contained within it at the disk interface is mechanically stable. The other surface of the foam that is placed against the skin is able to move with the skin, thereby diminishing the motion artifact that sometimes results from differential movement between the skin and the electrolyte gel. Figure 5.11(c) shows that it is possible to lightly abrade through the high-impedance barrier layer of the skin without breaking the deeper capillary loops and drawing blood.

ELECTRODE STANDARDS

During defibrillation, large currents may flow through the electrodes, change the electrode overpotential, and make it difficult to determine whether the defibrillation has been successful. In general Ag/AgCl electrodes are satisfactory, whereas polarizable electrodes are not. Standards for pregelled disposable electrodes (Anonymous, 1984) require face-to-face bench testing to ensure that the offset voltage is less than 100 mV, the noise is less than 150 μV, the 10-Hz impedance is less than 2 kΩ, the defibrillation overload recovery to four 2-mC charges is less than 100 mV, and the bias current tolerance to 200 nA for 8 h yields less than 100 mV offset (Carim, 1988). The defibrillation recovery voltage versus time for 12 electrode materials shows that the optimal recovery occurs for 500 mC/cm^2 of AgCl electrodeposited on Ag (Das and Webster, 1980). Additional tests on humans can assess motion artifact, adhesive tack, and skin irritation (Webster, 1984b).

FLEXIBLE ELECTRODES

The electrodes described so far are solid and either are flat or have a fixed curvature. The body surface, on the other hand, is irregularly shaped and can change its local curvature with movement. Solid electrodes cannot conform to this change in body-surface topography, which can result in additional

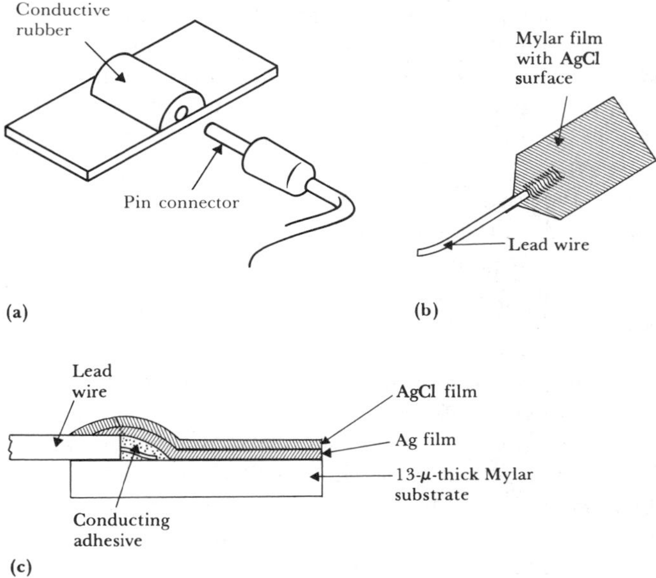

Figure 5.12 Flexible body-surface electrodes (a) Carbon-filled silicone rubber electrode. (b) Flexible thin-film neonatal electrode (after Neuman, 1973). (c) Cross-sectional view of the thin-film electrode in (b). [Parts (b) and (c) are from International Federation for Medical and Biological Engineering. *Digest of the 10th ICMBE,* 1973.]

motion artifact. To avoid such problems, *flexible electrodes* have been developed, examples of which are shown in Figure 5.12.

One type of flexible electrode is a woven, stretchable, nylon fabric impregnated with silver particles. Lead wire bonding is achieved by the use of epoxy. Gel pads are used for short-term monitoring.

Figure 5.12(a) shows another technique employed to provide flexible electrodes. A carbon-filled silicone rubber compound in the form of a thin strip or disk is used as the active element of an electrode. A pin connector is pushed into the lead connector hole, and the electrode is used in the same way as a similar type of metal-plate electrode.

Flexible electrodes are especially important for monitoring premature infants. Electrodes for detecting the ECG and respiration by the impedance technique are attached to the chest of premature infants, who usually weigh less than 2500 g. Conventional electrodes are not appropriate; they cannot conform to the shape of the infant's chest and can cause severe skin ulceration at pressure points. They must also be removed when chest x rays are taken of the infant, because they are opaque and can obstruct the view of significant portions of the thoracic cavity. Neuman (1973) developed flexible, thin-film electrodes for use on newborn infants that minimize these problems. The basic electrode consists of a 13-μm-thick Mylar film on which an Ag and AgCl film have been deposited, as shown in Figure 5.12(b). The actual structure of the electrode is illustrated in cross section in Figure 5.12(c). The flexible lead wire is attached to the Mylar substrate by means of a conducting adhesive, and a silver film approximately 1 μm thick is deposited over this and the Mylar. An AgCl layer is then grown on the surface of the silver film via the electrolytic process.

In addition to the advantage of being flexible and conforming to the shape of the newborn's chest, these electrodes have a layer of silver thin enough to be essentially x-ray-transparent, so they need not be removed when chest x rays of the infant are taken. All that shows up on the x rays is the lead wire. Consequently, the infant's skin is also protected from the irritation caused by removing and reapplying the adhesive tape that holds the electrode in place. This has been demonstrated to reduce the occurrence of skin irritation significantly in nurseries in which flexible, thin-film electrodes have been used.

The flexible electrodes we have described require some type of adhesive tape to hold them in place against the skin. New electrolytic hydrogel materials have been developed that are in the form of a thin, flexible slab of gelatinous material. This substance has a sticky surface that is similar to the adhesive tack on the tape used to hold electrodes in place. By virtue of the mobile ions that it contains, it is also electrically conductive. A piece of this material the same size as the flexible electrode can be secured on the electrode's surface and used to hold it in place against the skin. Because the electrode and this interface material are both flexible, a good, mechanically secure electric contact can be made between the electrode and the skin. One drawback of this material is its relatively high electric resistance, compared to that of the electrolyte gel routinely used with electrodes. Hydrogels are less effective at hydrating the dry epidermal layer (Jossinet and McAdams, 1990). This is not a severe problem anymore, however, because the amplifiers used with these electrodes now have input impedances of the order of 10 MΩ or higher, which is much greater than the resistance of the electrolytic material. Often there is less motion artifact when these electrodes are used.

5.7 INTERNAL ELECTRODES

Electrodes can also be used within the body to detect biopotentials. They can take the form of *percutaneous electrodes,* in which the electrode itself or the lead wire crosses the skin, or they may be entirely *internal electrodes,* in which the connection is to an implanted electronic circuit such as a radiotelemetry transmitter. These electrodes differ from body-surface electrodes in that they do not have to contend with the electrolyte–skin interface and its associated limitations, as described in Section 5.5. Instead, the electrode behaves in the way dictated entirely by the electrode–electrolyte interface. No electrolyte gel is required to maintain this interface, because extracellular fluid is present.

There are many different designs for internal electrodes. An investigator

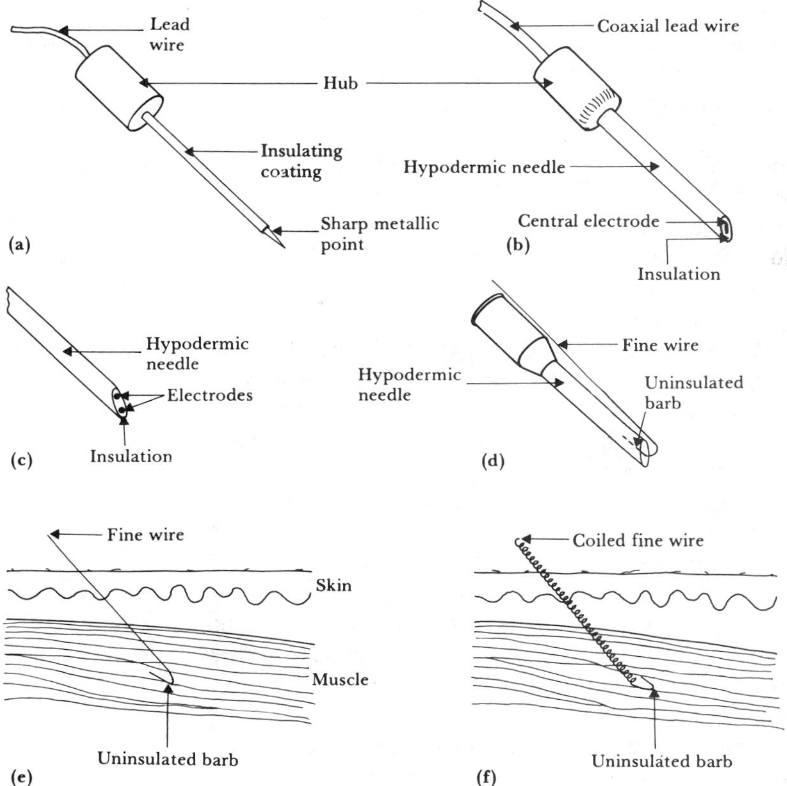

Figure 5.13 Needle and wire electrodes for percutaneous measurement of biopotentials (a) Insulated needle electrode. (b) Coaxial needle electrode. (c) Bipolar coaxial electrode. (d) Fine-wire electrode connected to hypodermic needle, before being inserted. (e) Cross-sectional view of skin and muscle, showing fine-wire electrode in place. (f) Cross-sectional view of skin and muscle, showing coiled fine-wire electrode in place.

studying a particular bioelectric phenomenon by using internal electrodes frequently designs his or her own electrodes for that specific purpose. The following paragraphs describe some of the more common forms of these electrodes and give examples of their application.

Figure 5.13 shows different types of percutaneous needle and wire electrodes. The basic needle electrode consists of a solid needle, usually made of stainless steel, with a sharp point. The shank of the needle is insulated with a coating such as an insulating varnish; only the tip is left exposed. A lead wire is attached to the other end of the needle, and the joint is encapsulated in a plastic hub to protect it. This electrode, frequently used in electromyography, is shown in Figure 5.13(a). When it is placed in a particular muscle, it obtains an EMG from that muscle acutely and can then be removed.

A variation of this type of electrode is used on patients undergoing surgery to monitor the ECG continuously. The electrode consists of stainless steel hypodermic needles placed subcutaneously on each limb. Lead wires with special connectors attached to the needle at the hub connect the electrodes to the cardioscope (Section 6.9). These electrodes remain in place as the patient is manipulated during the surgical procedure. They are away from the surgical field. Electrolyte gel is not necessary.

A shielded percutaneous electrode can be fabricated in the form shown in Figure 5.13(b). It consists of a small-gage hypodermic needle that has been modified by running an insulated fine wire down the center of its lumen and filling the remainder of the lumen with an insulating material such as an epoxy resin. When the resin has set, the tip of the needle is filed to its original bevel, exposing an oblique cross section of the central wire, which serves as the active electrode. The needle itself is connected to ground through the shield of a coaxial cable, thereby extending the coaxial structure to its very tip.

Multiple electrodes in a single needle can be formed as shown in Figure 5.13(c). Here two wires are placed within the lumen of the needle and can be connected differentially so as to be sensitive to electrical activity only in the immediate vicinity of the electrode tip.

The needle electrodes just described are principally for acute measurements, because their stiffness and size make them uncomfortable for long-term implantation. When chronic recordings are required, percutaneous wire electrodes are more suitable. There are many different types of wire electrodes and schemes for introducing them through the skin. [The interested reader should refer to Geddes (1972) for a more detailed review.] The principle can be illustrated, however, with the help of Figure 5.13(d). A fine wire—often made of stainless steel ranging in diameter from 25 to 125 μm—is insulated with an insulating varnish to within a few millimeters of the tip. This noninsulated tip is bent back on itself to form a J-shaped structure. The tip is introduced into the lumen of the needle, as shown in Figure 5.13(d). The needle is inserted through the skin into the muscle at the desired location, to the desired depth. It is then slowly withdrawn, leaving the electrode in place, as shown in Figure 5.13(e). Note that the bent-over portion of wire serves as a barb holding the wire in place in the muscle. To remove the wire, the technician applies a mild uniform force to straighten out the barb and pulls it out through the wire's tract.

A variation on this basic approach has been described by Caldwell and Reswick (1975). Realizing that wire electrodes chronically implanted in active muscles undergo a great amount of flexing as the muscle moves (which can cause the wire to slip as it passes through the skin and increase the irritation and risk of infection at this point, or even cause the wire to break), they developed the helical spiral electrode shown in Figure 5.13(f). It, too, is made from a very fine insulated wire coiled into a tight helix of approximately 150 μm diameter that is placed in the lumen of the inserting needle. The uninsulated barb protrudes from the tip of the needle and is bent back along the needle before insertion. It holds the wire in place in the tissue when the needle is removed from the muscle. Of course, the external end of the electrode now

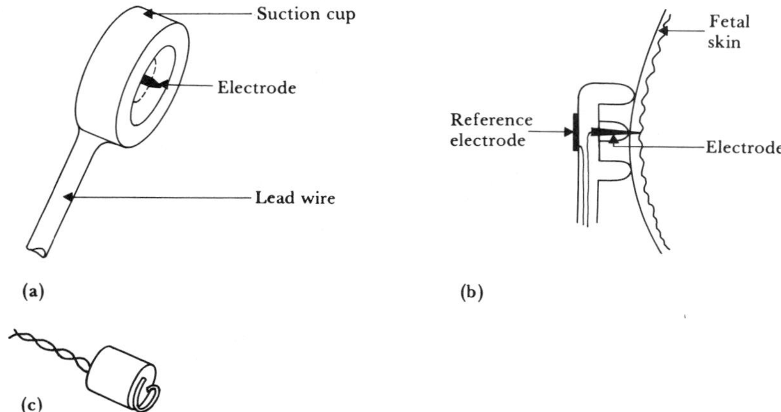

Figure 5.14 Electrodes for detecting fetal electrocardiogram during labor, by means of intracutaneous needles (a) Suction electrode. (b) Cross-sectional view of suction electrode in place, showing penetration of probe through epidermis. (c) Helical electrode, which is attached to fetal skin by corkscrew-type action.

passes through the needle and the needle must be removed—or at least protected—before the electrode is connected to the recording apparatus.

Another group of percutaneous electrodes are those used for monitoring fetal heartbeat. In this case it is desirable to get the electrocardiogram from the fetus during labor by direct connection to the presenting part (usually the head) through the uterine cervix (the mouth of the uterus). The fetus lies in a bath of amniotic fluid that contains ions and is conductive, so surface electrodes generally do not provide an adequate ECG as a result of the shorting effect of the amniotic fluid. Thus electrodes used to obtain the fetal ECG must penetrate the skin of the fetus.

An example of a suction electrode that does this is shown in Figure 5.14(a). A sharp-pointed probe in the center of a suction cup can be applied to the fetal presenting part, as shown in Figure 5.14(b). When suction is applied to the cup after it has been placed against the fetal skin, the surface of the skin is drawn into the cup and the central electrode pierces the stratum corneum, contacting the deeper layers of the epidermis. On the back of the suction electrode is a reference electrode that contacts the fluid, and the signal seen between these two electrodes is the voltage drop across the resistance of the stratum corneum. Thus, although the amniotic fluid essentially places all the body of the fetus at a common potential, the potentials beneath the stratum corneum can be different, and fetal ECGs that have peak amplitudes of the order of 50–700 μV can be reliably recorded.

Another intradermal electrode that is widely applied for detecting fetal ECG during labor is the helical electrode developed by Hon (1972). It consists of a stainless steel needle, shaped approximately like one turn of a helix,

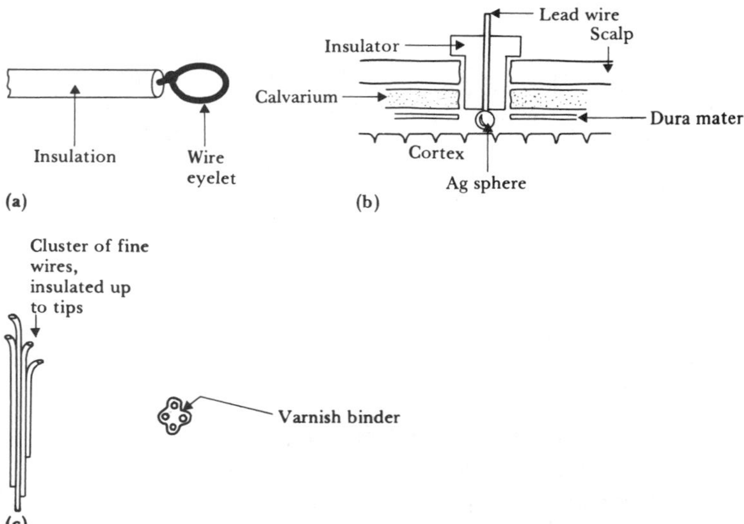

Figure 5.15 Implantable electrodes for detecting biopotentials (a) Wire-loop electrode. (b) Silver-sphere cortical-surface potential electrode. (c) Multielement depth electrode.

mounted on a plastic hub. [See Figure 5.14(c).] The back surface of the hub contains an additional stainless steel reference electrode. When labor has proceeded far enough, this electrode can be attached to the fetal presenting part by rotating it so that the needle twists just beneath the surface of the skin as would a corkscrew shallowly penetrating a cork. This electrode remains firmly attached, and because of the shortness of the helical needle, it does not penetrate deep enough into the skin to cause serious damage. It operates on the same basic principle as the suction electrode.

Often when radio telemetry is used, we want to implant electrodes within the body and not penetrate the skin with the wires. In this case the radio transmitter must be implanted in the body. A wide variety of electrodes are used in this application. Only a few examples are given here.

The simplest electrode for this application is shown in Figure 5.15(a). Insulated multistranded stainless steel wire suitable for implantation has one end stripped so that an eyelet can be formed from the strands of stainless steel. This is best done by individually taking each strand and forming the eyelet either by twisting the wires together one by one at the point at which the insulation stops or by spot-welding each strand to the wire mass at this point. The eyelet can then be sutured to the point in the body at which electric contact is to be established.

Figure 5.15(b) shows another example of an implantable electrode for obtaining cortical-surface potentials from the brain. Critchfield *et al.* (1971) applied this electrode for the radio telemetry of subdural EEGs. The electrode consists of a 2-mm-diameter silver sphere located at the tip of the cylindrical Teflon insulator through which the electrode lead wire passes. The calvarium is exposed through an incision in the scalp, and a burr hole is drilled. A small slit is made in the exposed dura, and the silver sphere is introduced through this opening so that it rests on the surface of the cerebral cortex. The assembly is then cemented in place onto the calvarium by means of a dental acrylic material.

Deep cortical potentials can be recorded from multiple points using the technique described by Delgado (1964), as shown in Figure 5.15(c). This kind of electrode consists of a cluster of fine insulated wires held together by a varnish binder. Each wire has been cut transversely to expose an uninsulated cross section that serves as the active electrode surface. By staggering the ends of the wires as shown, we can produce electrodes located at known differences in depth in an array. The other ends of the electrodes can be attached to appropriate implantable electronic devices or to a connector cemented on the skull to allow connection to an external recording apparatus.

5.8 ELECTRODE ARRAYS

Although implantable electrode arrays can be fabricated one at a time using clusters of fine insulated wires, this technique is both time-consuming and expensive. Furthermore, when such clusters are made individually, each one will be somewhat different from the other. A way to minimize these problems is to utilize microfabrication technology to fabricate identical two- and three-dimensional electrode arrays. Examples of some of the types of structures that are possible are shown in Figure 5.16. One-dimensional linear arrays of six pairs of biopotential recording electrodes have been described by Mastrototaro *et al.* (1992). These probes are illustrated in Figure 5.16(a) consist of square Ag/AgCl electrodes 40 μm on a side on thin-film gold conductors that have been deposited on either flexible polyimide substrates or more robust molybdenum substrate coated with an anodically grown oxide layer to provide the necessary insulation. The probes were typically 10 mm long, 0.5 mm wide, and 125 μm thick. Lead wires were attached to the bonding pads at the proximal end of the probe. These electrode arrays were designed to be used for measuring transmural potential distributions in the beating myocardium. Their flexibility was important to minimize tissue damage as the muscle contracts and relaxes.

Two-dimensional electrode arrays for mapping the electrical potentials across a region of the surface of an organ such as the heart are shown in Figure 5.16(b). These electrodes essentially represent an extension of the approach used for the one-dimensional arrays described above. A pattern of miniature electrodes is formed on a rigid or flexible surface and connected by conductors to the associated instrumentation. This interconnection can be quite a problem because large arrays require many connections. Sock elec-

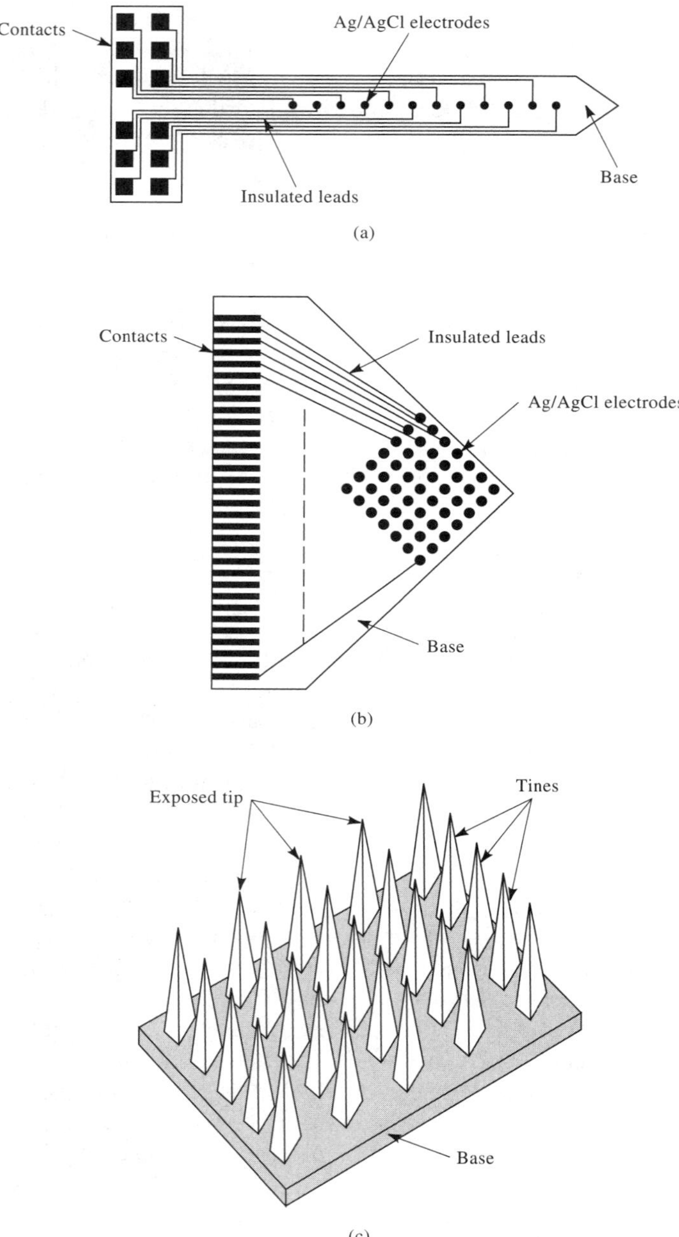

Figure 5.16 Examples of microfabricated electrode arrays. (a) One-dimensional plunge electrode array (after Mastrototaro *et al.,* 1992), (b) Two-dimensional array, and (c) Three-dimensional array (after Campbell *et al.,* 1991).

trodes consisting of individual silver spheres roughly 1 mm in diameter incorporated into a fabric sock that fits snugly over the heart have been used to map epicardial potentials. Each sphere is at the tip of an insulated wire that connects it to the recording apparatus. Needless to say, an array of a large number of electrodes of this type is difficult to build and awkward to use due to the large number of wires coming from the sock.

Ash *et al.* (1992) have shown that this process can be simplified by using a multilayer ceramic integrated circuit package as the electrode array. They have used this structure to map epicardial potentials. These investigators have also used thin-film microfabrication technology to form arrays of 144 miniature Ag/AgCl electrodes on polyimide substrates [Figure 5.16(b)]. The thin gold films serve as conductors as well as the bases for the Ag/AgCl electrodes. The interconnections were completed by using a miniature ribbon cable designed for surface mount microelectronic applications.

Three-dimensional electrode arrays fabricated using silicon microfabrication technology have been described by Campbell *et al.* (1991). Their device has the appearance of a two-dimensional comb [Figure 5.16(c)] with each tine being roughly 1.5 mm long and surrounded with insulating material up to the tip. The exposed tip serves as the electrode, and a wire connection on the base of the structure was needed to make contact with each tine electrode. Although this array is a three-dimensional structure, it really only measures from a two-dimensional array of electrodes because all of the tines are the same length. A truly three-dimensional electrode array can be fabricated by taking a set of one-dimensional electrode array probes such as seen in Figure 5.16(a) or a microelectrode probe as seen in Figure 5.20(b) and assembling them in an array of tines similar in appearance to that of Figure 5.16(c).

5.9 MICROELECTRODES

In studying the electrophysiology of excitable cells, it is often important to measure potential differences across the cell membrane. To be able to do this, we must have an electrode within the cell. Such electrodes must be small with respect to the cell dimensions to avoid causing serious cellular injury and thereby changing the cell's behavior. In addition to being small, the electrode used for measuring intracellular potential must also be strong so that it can penetrate the cell membrane and remain mechanically stable.

Electrodes that meet these requirements are known as *microelectrodes.* They have tip diameters ranging from approximately 0.05 to 10 μm. Microelectrodes can be formed from solid-metal needles, from metal contained within or on the surface of a glass needle, or from a glass micropipet having a lumen filled with an electrolytic solution. Examples of each type are given in the following paragraphs. More detailed descriptions can be found in Geddes (1972), Ferris (1974), and Cobbold (1974).

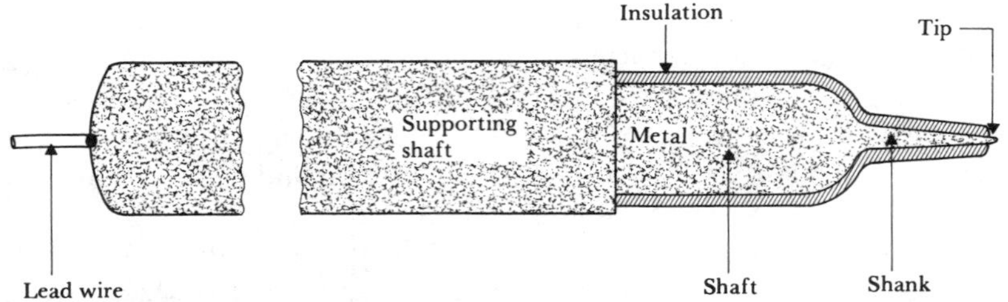

Figure 5.17 The structure of a metal microelectrode for intracellular recordings.

METAL MICROELECTRODES

The metal microelectrode is essentially a fine needle of a strong metal that is insulated with an appropriate insulator up to its tip, as shown in Figure 5.17. A metal needle is prepared in such a way as to produce a very fine tip. This is usually done by electrolytic etching, using a cell in which the metal needle is the anode. The needle is etched as it is slowly withdrawn from the electrolyte solution. Very fine tips can be formed in this way, but a great deal of patience and practice are required to gain the skill to make them. Suitable strong metals for these microelectrodes are stainless steel, platinum–iridium alloy, and tungsten. The compound tungsten carbide is also used because of its great strength.

The etched metal needle is then supported in a larger metallic shaft that can be insulated. This shaft serves as a sturdy mechanical support for the microelectrode and as a means of connecting it to its lead wire. The microelectrode and supporting shaft are usually insulated by a film of some polymeric material or varnish. Only the extreme tip of the electrode remains uninsulated.

SUPPORTED-METAL MICROELECTRODES

The properties of two different materials are used to advantage in supported-metal microelectrodes. A strong insulating material that can be drawn to a fine point makes up the basic support, and a metal with good electrical conductivity constitutes the contacting portion of the electrode.

Figure 5.18 shows examples of supported metal microelectrodes. The classic example of this form is a glass tube drawn to a micropipet structure with its lumen filled with an appropriate metal. Often this type of microelectrode, as shown in Figure 5.18(a), is prepared by first filling a glass tube with a metal that has a melting point near the softening point of the glass. The tube can then be heated to the softening point and pulled to form a narrow constriction. When it is broken at the constriction, two micropipets filled with metal are formed. In this type of structure, the glass not only provides the mechanical support but also serves as the insulation. The active tip is the only metallic area exposed in cross section where the pipet was broken away. Metals such

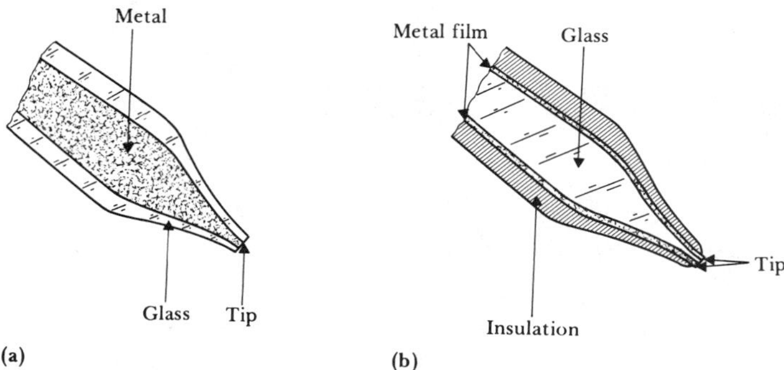

Figure 5.18 Structures of two supported metal microelectrodes (a) Metal-filled glass micropipet. (b) Glass micropipet or probe, coated with metal film.

as silver-solder alloy and platinum and silver alloys are used. In some cases metals with low melting points, such as indium or Wood's metal, are used.

New supported-metal electrode structures have been developed using techniques employed in the semiconductor microelectronics industry. Figure 5.18(b) shows the cross section of the tip of a deposited-metal-film microelectrode. A solid glass rod or glass tube is drawn to form the micropipet. A metal film is deposited uniformly on this surface to a thickness of the order of tenths of a micrometer. A polymeric insulation is then coated over this, leaving just the tip, with the metal film exposed.

MICROPIPET ELECTRODES

Glass micropipet microelectrodes are fabricated from glass capillaries. The central region of a piece of capillary tubing, as shown in Figure 5.19(a), is heated with a burner to the softening point. It is then rapidly stretched to produce the constriction shown in Figure 5.19(b). Special devices, known as microelectrode pullers, that heat and stretch the glass capillary in a uniform reproducible way to fabricate micropipets are commercially available. The two halves of the stretched capillary structure are broken apart at the constriction to produce a pipet structure that has a tip diameter of the order of 1 μm. This pipet is fabricated into the electrode form shown in Figure 5.19(c). It is filled with an electrolyte solution that is frequently 3*M* KCl. A cap containing a metal electrode is then sealed to the pipet, as shown. The metal electrode contacts the electrolyte within the pipet. The electrode is frequently a silver wire prepared with an electrolytic AgCl surface. Platinum or stainless steel wires are also occasionally used.

MICROELECTRODES BASED ON MICROELECTRONIC TECHNOLOGY

The technology used to produce transistors and integrated circuits can also be used to micromachine small mechanical structures. This technique has been

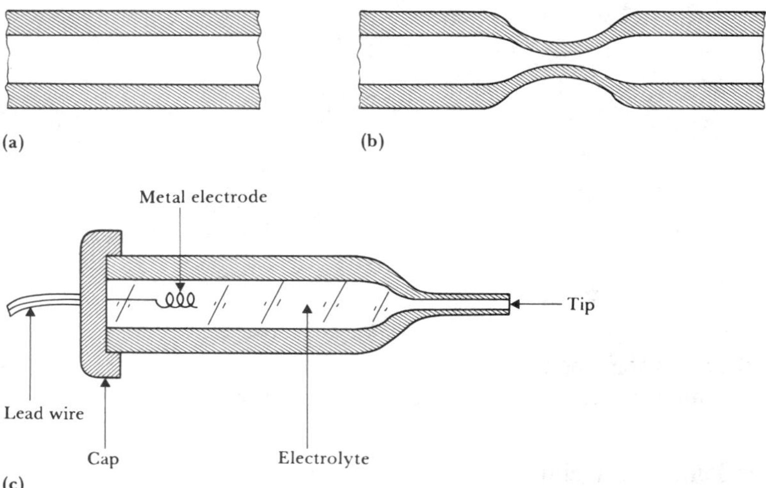

Figure 5.19 A glass micropipet electrode filled with an electrolytic solution (a) Section of fine-bore glass capillary. (b) Capillary narrowed through heating and stretching. (c) Final structure of glass-pipet microelectrode.

used by several investigators to produce metal microelectrodes. The structure shown in Figure 5.20(a) uses the technology for fabricating beam-lead transistors (Wise *et al.,* 1970). The basic structure consists of narrow gold strips deposited on a silicon substrate the surface of which has been first insulated by growing an SiO_2 film. The gold strips are then further insulated by depositing SiO_2 over their surface. The silicon substrate is next etched to a thin, narrow structure that is just wide enough to accommodate the gold strips in the region of the tip. The silicon substrate is etched a millimeter or two back from the tip so that only the gold strips and their SiO_2 insulation remain. The insulation is etched away from the very tip of the gold strips to expose the contacting surface of the electrodes. Although this technology cannot produce tips as small as can be produced with the glass micropipet technique previously described, it is possible to make multielectrode arrays and to maintain very precisely the geometry between individual electrodes in the array. The high reproducibility of microelectronic processing allows many electrodes to be made that all have very similar geometric properties. Thus the characteristics vary little from one electrode array to the next.

Several other designs for sensing and stimulating electrodes have been developed through the use of microelectronic technology. An array of multisite microelectrodes can be grown on a thin silicon probe [Figure 5.20(b)] that can be placed in the cortex of the brain to detect local potentials (Drake *et al.*, 1988). A similar design was utilized by Prohaska *et al.* (1986), but here the actual gold or silver/silver chloride electrode was located in a very small chamber filled with an electrolytic solution such as sodium chloride and was

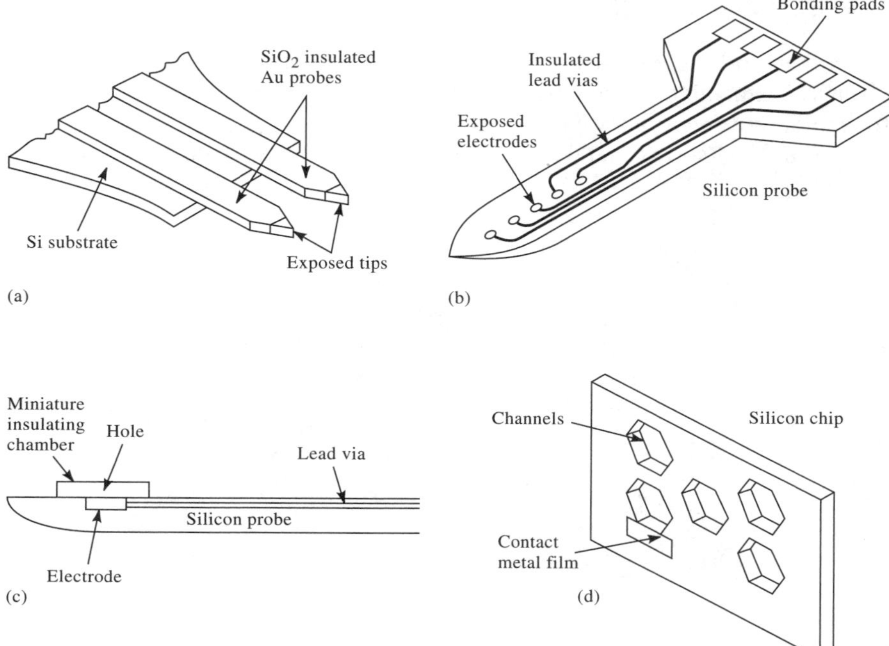

Figure 5.20 Different types of microelectrodes fabricated using microelectronic technology (a) Beam-lead multiple electrode. (Based on Figure 7 in K. D. Wise, J. B. Angell, and A. Starr, "An Integrated Circuit Approach to Extracellular Microelectrodes." Reprinted with permission from *IEEE Trans. Biomed. Eng.,* 1970, BME-17, pp. 238–246. Copyright © 1970 by the Institute of Electrical and Electronics Engineers.) (b) Multielectrode silicon probe after Drake *et al.* (c) Multiple-chamber electrode after Prohaska *et al.* (d) Peripheral-nerve electrode based on the design of Edell.

made from an insulating film with a small hole to allow communication with the nervous tissue in which it was placed [Figure 5.20(c)]. A novel new electrode for sensing signals emitted from peripheral nerves has been developed (Edell, 1986). His electrode consisted of an array of channels etched through a silicon chip [Figure 5.20(d)]. He used these electrodes in animal studies wherein peripheral nerves were transected and each side of the cut nerve was aligned on opposite sides of the silicon chip so that the nerve could regenerate and grow through the channels on the chip to reestablish the connections. Gold metalization on the silicon surface surrounding each channel was used to make electric contact with the nerve fibers that passed through the channels on the silicon chip.

ELECTRICAL PROPERTIES OF MICROELECTRODES

To understand the electrical behavior of microelectrodes, we must derive an electrical equivalent circuit from physical considerations. This circuit differs for metal and micropipet electrodes.

Figure 5.21 shows metal microelectrodes. The microelectrode contributes

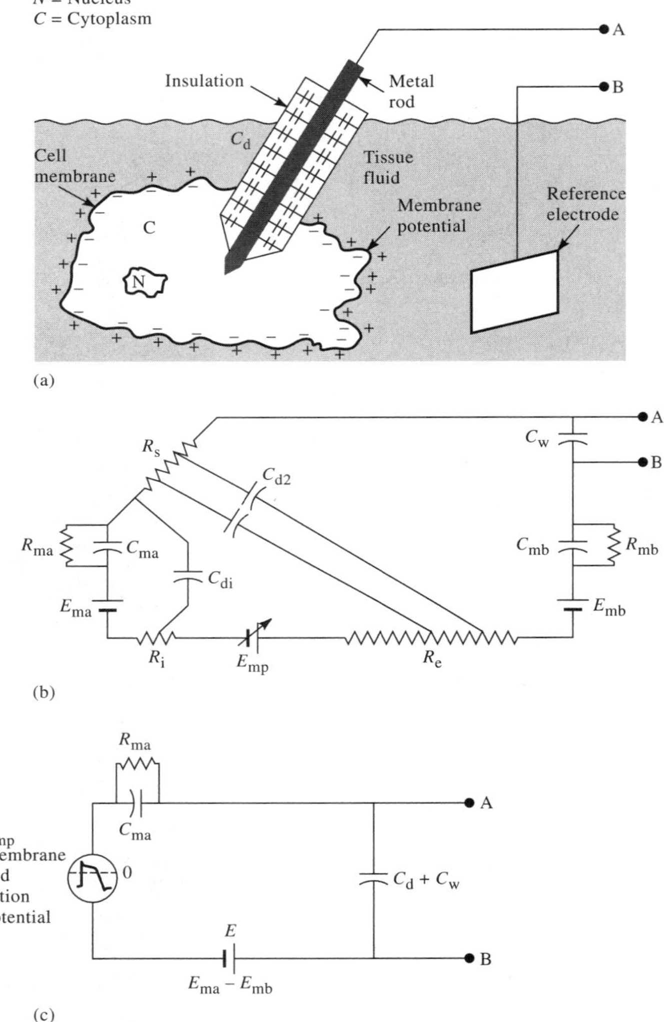

Figure 5.21 Equivalent circuit of metal microelectrode (a) Electrode with tip placed within a cell, showing origin of distributed capacitance. (b) Equivalent circuit for the situation in (a). (c) Simplified equivalent circuit. (From L. A. Geddes, *Electrodes and the Measurement of Bioelectric Events,* Wiley-Interscience, 1972. Used with permission of John Wiley and Sons, New York.)

a series resistance R_s that is due to the resistance of the metal itself. A major contributor to this resistance is the metal in the shank and tip portion of the microelectrode, because the ratio of length to cross-sectional area is much higher in this portion than it is for the shaft. The metal is coated with an insulating material over all but its most distal tip, so a capacitance is set up between the metal and the extracellular fluid. This is a distributed capacitance C_d that we can represent in lumped form by separating the shank and tip from the shaft. In the shank region, we can consider the microelectrode to be a coaxial cylinder capacitor; the capacitance per unit length (F/m) is given by

$$\frac{C_{d1}}{L} = \frac{2\pi\epsilon_r\epsilon_0}{\ln(D/d)} \tag{5.16}$$

where

ϵ_0 = dielectric constant of free space (Appendix A.1)
ϵ_r = relative dielectric constant of insulation material
D = diameter of cylinder consisting of electrode plus insulation
d = diameter of electrode
L = length of shank

Of course, this coaxial-cable approximation is not a very good approximation for the shank region, which is tapered, but it is reasonable for a rough calculation. Because insulation thicknesses are usually on the order of 1 μm in the shank and tip, it is important to consider the structure using the coaxial cylinder analog. However, when we consider the shaft portion of the electrode, if the thickness of the insulation is still approximately 1 μm, the diameter of the metal shaft can be on the order of several millimeters. Here the ratio of diameters would be practically unity, so we can simplify the calculation by unwrapping the circumferential surface of the shaft and considering the system to be a parallel-plate capacitor of area equal to the circumferential surface area and of thickness equal to t, the thickness of the insulation layer. The capacitance per unit length (F/m) is given by

$$\frac{C_{d2}}{L} = \frac{\epsilon_r\epsilon_0\pi d}{t} \tag{5.17}$$

Note that this capacitance comes from only that portion of the electrode shaft that is submerged in the extracellular fluid. Often only the shank is submerged, so C_{d2} is zero.

The other significant contributions to the equivalent circuit from the metal microelectrode are the components contributed by the metal-electrolyte interface, R_{ma}, C_{ma}, and E_{ma}. A similar set of components, C_{mb}, R_{mb}, and E_{mb}, are associated with the reference electrode. Because of the much larger surface area of the reference electrode compared with the tip of the microelectrode, the impedance due to these components is much lower. Of course, the half-cell potential due to the reference electrode is unaffected by the surface area. The tip of the microelectrode is within a cell, so there is a series resistance R_i associated with the electrolyte within the cell membrane and another series resistance R_e due to the extracellular fluid. The cell membrane itself can be modeled simply as a variable potential E_{mp}, but in more detailed analyses an equivalent circuit of greater complexity is required. Some of the distributed capacitance of the shank, C_{d1}, is between the microelectrode and the extracellular fluid, as shown in the equivalent circuit, whereas the remainder of it is beween the microelectrode and the intracellular fluid.

There is also a capacitance associated with the lead wires, C_w. The physical basis for this equivalent circuit is shown in Figure 5.21(a); the actual equivalent circuit is shown in Figure 5.21(b). Often it is acceptable to simplify this equivlent circuit to that shown in Figure 5.21(c), which neglects the impedance of the reference electrode and the series-resistance contribution from the intracellular and extracellular fluid and lumps all the distributed capacitance together. Under circumstances in which the input impedance of the amplifier connected to this electrode is not sufficiently large, we see that this circuit can behave as a high-pass filter and significant waveform distortion can result.

The effective impedance of metal microelectrodes is frequency-dependent and can be of the order of 10 to 100 MΩ. We can, however, lower this impedance by increasing the effective surface area of the tip of the microelectrode through the application of platinum black, as we did in the case of the hydrogen electrode. Impedance reduction of one or two orders of magnitude can be achieved in this way. At lower frequencies, the impedance can be reduced by applying an Ag/AgCl surface to the electrode tip. Care must be taken in doing this, however, because of the mechanically fragile nature of this film and its tendency to flake off.

The equivalent circuit for the micropipet electrode is somewhat more complicated than that of the metal microelectrode. The physical situation is illustrated in Figure 5.22(a), and the resulting equivalent circuit is shown in Figure 5.22(b). The internal electrode in the micropipet gives the metal-electrolyte interface components R_{ma}, C_{ma}, and E_{ma}. In series with this is a resistive element R_t corresponding to the resistance of the electrolyte in the shank and tip region of the microelectrode. Connected to this is the distributed capacitance C_d corresponding to the capacitance across the glass in this region. The distributed capacitance due to the shaft region has been neglected, because the glass wall of the electrode is much thicker in this region and the capacitive contribution is quite small.

There are two potentials associated with the tip of the micropipet micro-electrode. The *liquid-junction potential* E_j corresponds to the liquid junction set up beween the electrolyte in the micropipet and the intracellular fluid. In addition, a potential known as the *tip potential* E_t arises because the thin glass wall surrounding the tip region of the micropipet behaves like a glass membrane and has an associated membrane potential.

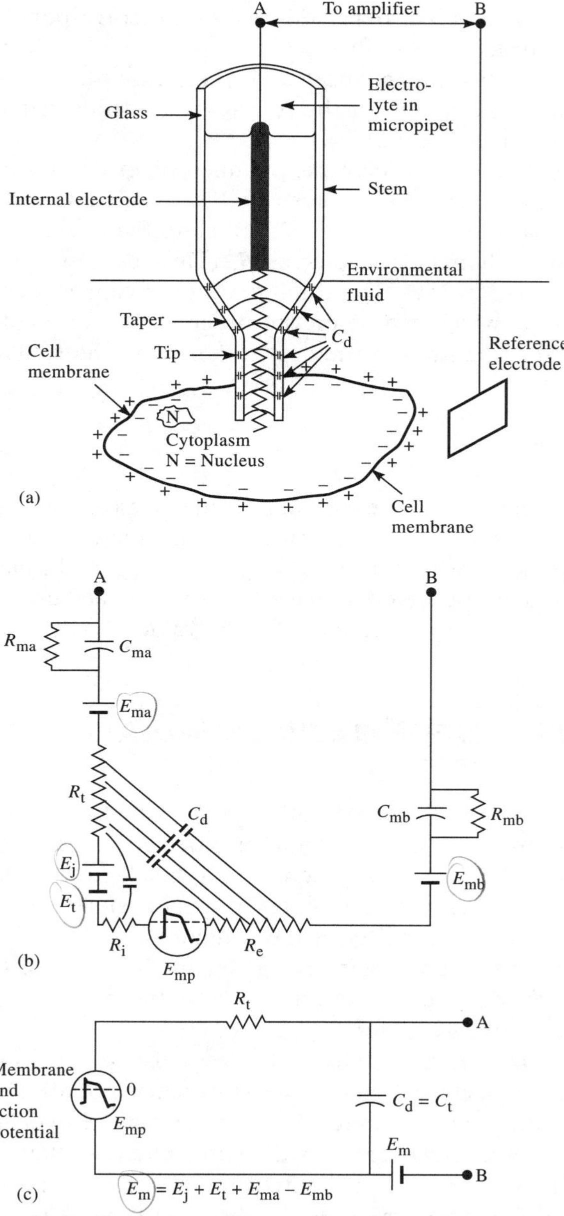

Figure 5.22 Equivalent circuit of glass micropipet microelectrode (a) Electrode with its tip placed within a cell, showing the origin of distributed capacitance. (b) Equivalent circuit for the situation in (a). (c) Simplified equivalent circuit. (From L. A. Geddes, *Electrodes and the Measurement of Bioelectric Events,* Wiley-Interscience, 1972. Used with permission of John Wiley and Sons, New York.)

The equivalent circuit also includes resistances corresponding to the intracellular R_i and extracellular R_e fluids. These are coupled to the microelectrode through the distributive capacitance C_d, as is the case for the metal microelectrode. The equivalent circuit for the reference electrode remains unchanged from that shown in Figure 5.21(b).

Unlike the metal microelectrode, the micropipet's major impedance contribution is resistive. This can be illustrated by approximating the equivalent circuit to give that shown in Figure 5.22(c). Here the overall series resistance of the electrode is lumped together as R_t. This resistance generally ranges in value from 1 to 100 MΩ. The total distributed capacitance is lumped together to form C_t, which can be on the order of tens of picofarads. And all the associated dc potentials are lumped together in the source E_m, which is given by

$$E_m = E_j + E_t + E_{ma} - E_{mb} \tag{5.18}$$

Note that the micropipet-type microelectrode behaves as a low-pass filter. The high series resistance and distributed capacitance cause the electrode output to respond slowly to rapid changes in cell-membrane potential. To reduce this problem, positive-feedback, negative-capacitance amplifiers (see Section 6.6) are used to reduce the effective value of C_t.

5.10 ELECTRODES FOR ELECTRIC STIMULATION OF TISSUE

Electrodes used for the electric stimulation of tissue follow the same general design as those used for the recording of bioelectric potentials. They differ in that currents as large as milliamperes cross the electrode–electrolyte interface in stimulating electrodes. Examples of specific electrodes used in cardiac pacemakers, other functional electric stimulators, and cardiac defibrillators (where currents are even larger) are given in Chapter 13. Other types of stimulating electrodes are of the same form as the potential recording electrodes described in this chapter.

In considering stimulating electrodes, we must bear in mind that the net current across the electrode–electrolyte interface is not always zero. When a biphasic stimulating pulse is used, the average current over long periods of time should be zero. However, over the stimulus cycle, there are periods of time during which the net current across the electrode is in one direction at one time and in the other direction at a different time. Also, the magnitudes of the currents in the two directions may be unequal. In studying the electrical characteristics of the electrode–electrolyte interface under such circumstances, we may well imagine that the equivalent circuit changes as the stimulus progresses. Thus the effective equivalent circuit for the electrode is determined by the stimulus parameters, principally the current and the duration of the stimulus.

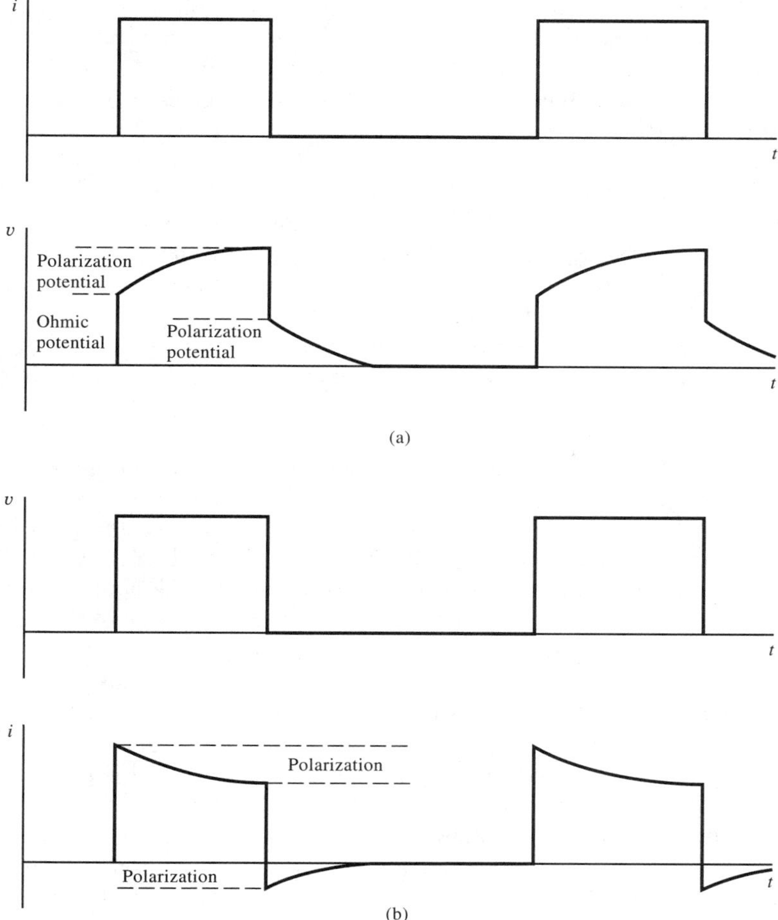

Figure 5.23 Current and voltage waveforms seen with electrodes used for electric stimulation (a) Constant-current stimulation. (b) Constant-voltage stimulation.

Rectangular biphasic or monophasic pulses are frequently used for electric stimulation. However, other waveshapes, such as decaying exponentials in trapezoids, or sine waves, have also been used. Frequently a stimulus is used that is either constant current or constant voltage during the pulse. The response of a typical electrode to this type of stimulus is illustrated in Figure 5.23.

A constant-current stimulus pulse is applied to the stimulating electrodes in Figure 5.23(a), giving the voltage response shown. Note that the resulting voltage pulse is not constant. This is reasonable when we consider that there is a strong reactive component to the electrode–electrolyte interface, or in other words, polarization occurs. The initial rise in voltage corresponding to the leading edge of the current pulse is due to the voltage drop across the resistive components of the electrode–electrolyte interface, but we see that the voltage continues to rise with the constant current. This is due to the establishment of a change in the distribution of charge concentration at the electrode–electrolyte interface—in other words, to a change in polarization resulting from the unidirectional current. As stated in the description of the simplified electrode–electrolyte equivalent circuit of Figure 5.4, this polarization effect can be represented by a capacitor. Again it is important to remember that the size of this capacitor is determined by several factors, one of which is the current density at the interface, so the equivalent circuit for stimulating electrodes changes with the stimulus.

When the current falls back to its low value, the voltage across the electrodes drops, but not back to its initial value. Instead, after the initial steep fall, there is a slower decay corresponding to the dissipation of the polarization charge at the interface.

Constant-voltage stimulation of the electrodes is shown in Figure 5.23(b). In this case the current corresponding to the rising edge of the voltage pulse is seen to jump in a large step and, as the distribution of the polarization charge becomes established, to fall back to a lower steady-state value. When the voltage pulse falls, the current is seen to change direction and then slowly return to its initial zero value. This is a result of the dissipation of the polarization charge built up at the electrode–electrolyte interface.

In choosing materials for stimulating electrodes, we must take into consideration chemical reactions occurring at the electrode–electrolyte interface. If a stimulating current causes the material of the electrode to be oxidized, the electrode is consumed, which limits its lifetime and also increases the concentration of the ions of electrode material in the vicinity of the electrode. This could be toxic to the tissue. When electrodes such as the Ag/AgCl electrode are used, the stimulating current can result in either the formation of additional Cl^- or the reduction of what is already formed, thereby greatly changing the characteristics of the electrode. Thus the best stimulating electrodes are made from noble metals (or at least stainless steel), which undergo only minimal chemical reactions. It is also possible in this case that the chemical reaction involves water in the vicinity of the electrodes. Thus the local activity of hydrogen or hydroxyl ions in the vicinity of the electrode can greatly change. These changes in acidity or alkalinity can produce tissue damage that limits the overall effectiveness of the noble-metal electrodes. Of course, the polarization of these electrodes is large, and the waveforms, such as those shown in Figure 5.23, are less rectangular than they would be with nonpolarizable electrodes. In extreme cases, electrode voltage and current density can result in the electrolysis of water that leads to the evolution of small bubbles of hydrogen or oxygen gas. Clearly, this is a situation that should be avoided.

Carbon-filled silicone rubber electrodes are used for transcutaneous stimulation used in clinical pain management (Carim, 1988). Arzbaecher (1982) describes a pill electrode that can either record a large P wave or pace the heart from the esophagus. A stimulating electrode based upon the iridium/

iridium oxide system [Robblee *et al.* (1983)] has been shown to inject charge into biologic tissue. This type of electrode has been found to provide maximal current density for biphasic stimuli while minimizing chemical changes that could lead to tissue damage. The passage of charge from the electrode to tissue and in the reverse direction results in a chemical change in the iridium oxide chemical composition as opposed to injecting iridium ions into solution or reducing them.

5.11 PRACTICAL HINTS IN USING ELECTRODES

In using metal electrodes for measurement and stimulation, we should understand a few practical points not mentioned elsewhere in this chapter. The first point is the importance of constructing the electrode and any parts of the lead wire that may be exposed to the electrolyte *all of the same material.* Furthermore, a third material such as solder should not be used to connect the electrode to its lead wire unless it is certain that this material will not be in contact with the electrolyte. It is far better either to weld the lead wire to the electrode or at least to form a mechanical bond through crimping or peening. Dissimilar metals should not be used in contact because their half-cell potentials are different. And, because they are connected and in contact with the same electrolyte, it is more than likely that an electrochemical reaction will be set up between them that can result in additional polarization and often in corrosion of one of the metals. This factor also tends to make half-cell potentials less stable, thereby contributing to increased electric noise from the electrode.

When pairs of electrodes are used for measuring differentials, such as in detecting surface potentials on the body or internal potentials within it, it is far better to use the same material for each electrode, because the half-cell potentials are approximately equal. This means that the net dc potential seen at the input to the amplifier connected to the electrodes is relatively small, possibly even zero. This minimizes possible saturation effects in the case of high-gain direct-coupled amplifiers.

Electrodes placed on the skin's surface have a tendency to come off. Frequently, this is due to a loss of effectiveness of the tack on the tape holding them in place. However, the problem need not arise if the electrodes are well designed. Lead wires to these surface electrodes should be extremely flexible yet strong. If they are, only tension on the lead wire can apply a force that is likely to remove the electrode. If the lead wire remains loose, it cannot apply any forces to the electrode because of its high flexibility. It is helpful to provide additional relief from strain by taping the lead wire to the skin a few cm from the electrode with some slack in the wire between the tape and the electrode.

The point at which the lead wire enters the electrode is a point of frequent failure. Even though the insulation appears intact, the wire within may be broken as a result of severe repeated flexing at this point. Well-designed electrodes minimize this problem by providing strain relief at this point, so that there is a gradual transition between the wire and the solid material of the electrode. Using a tapered region of insulation that gradually increases from the diameter of the wire to one closer to that of the electrode often minimizes this problem and distributes the flexing forces over a greater portion of the wire.

Another point to consider is that the insulation of the lead wire and the electrode can also present problems. Electrodes are often in a high-humidity environment or are continually soaked in extracellular fluid or even in cleaning solution (if they are of the reusable type). The insulation of these electrodes is usually made of a polymeric material, so it can absorb water. Some of these materials can become more conductive when they absorb water, and, in the case of implantable electrodes, there may be some high-resistance contact with the lead wire as well as at the electrode itself. If the lead wire is made of a material different from that of the electrode, the problems just described can result, thereby increasing the observed electric noise and possibly leading to a weakening of the lead wires due to corrosion. Thus it is important to understand the insulation material used with the electrode and to make sure that there is a layer of it thick enough to prevent this problem from occurring.

One final point regarding electrodes for measuring biopotentials: In deriving the equivalent circuit for electrodes such as those shown in Figure 5.4, we stressed that for high-fidelity recordings of the measured biopotential, the input impedance of the amplifier to which the electrodes are connected must be much higher than the source impedance represented by the equivalent circuit. If this condition is not met, not only will the amplitude of the recorded signal be less than it should be, but significant distortion also will be introduced into the waveform of the signal. This is demonstrated by Geddes (1966) for electrodes used to record electrocardiograms. Geddes shows how lowering the input impedance of the amplifier causes the recorded signal to take on a more and more biphasic character as well as a reduced amplitude.

PROBLEMS

5.1 A set of biopotential electrodes made of silver are attached to the chest of a patient to detect the electrocardiogram. When current passes through the anode, it causes silver to be oxidized, producing silver ions in solution. There is a 10-μA leakage current between these electrodes. Determine the number of silver ions per second entering the solution at the electrode–electrolyte interface.

5.2 Design a system for electrolytically forming Ag/AgCl electrodes. Give the chemical reactions that occur at *both* electrodes.

5.3 Design an Ag/AgCl electrode that will pass 150 mC (millicoulombs) of charge without removing all the AgCl. Calculate the mass of AgCl required. Show the electrode in cross section and give the active area.

5.4 When electrodes are used to record the electrocardiogram, an electrolyte gel is usually put between them and the surface of the skin. This makes it possible for the metal of the electrode to form metallic ions that move into the electrolyte gel. Often, after prolonged use, this electrolyte gel begins to dry out and change the characteristic of the electrodes. Draw an equivalent circuit for the electrode while the electrolyte gel is fresh. Then discuss and illustrate the way you expect this equivalent circuit to change as the electrolyte gel dries out. In the extreme case where there is no electrolyte gel left, what does the equivalent circuit of the electrode look like? How can this affect the quality of the recorded electrocardiogram?

5.5 A zinc wire and an aluminum wire accidentally come in contact with a part of the body that has been saturated with a physiological saline solution. Is there a potential difference between these two wires? If your answer is yes, quantitatively state the value of this potential under open-circuit conditions.

5.6 Design the electrode of the smallest area that has an impedance of 10 Ω at 100 Hz. State your source of information, describe construction of the electrode, and calculate its area.

5.7 A pair of biopotential electrodes is used to detect the electrocardiogram of an adult male. It has become necessary to determine the equivalent-source impedance of this electrode pair so that a particular experiment can be performed. Describe an experimental procedure that can be used to determine this quantity, using a minimum of test equipment.

5.8 Using test equipment found in most labs, design (show a block diagram and wiring connections) a test facility for measuring the impedance versus frequency of 1-cm^2 electrodes. It should use the largest current density that does not cause a change in the impedance.

5.9 A pair of biopotential electrodes is placed in a saline solution and connected to a stimulator that passes a direct current through the electrodes. It is noted that the offset potentials from the two electrodes are different. Explain why this happens during the passage of current. Sketch the distribution of ions about each electrode while the current is on.

5.10 Electrodes having a source resistance of 4 kΩ each are used in a bipolar configuration with a differential amplifier having an input impedance of 70 kΩ. What will be the percentage reduction in the amplitude of the biopotential signal? How can this distortion of the signal be reduced?

5.11 A nurse noticed that one electrode of a pair of Ag/AgCl cardiac electrodes used on a chronic cardiac monitor was dirty and cleaned it by scraping it with steel wool (Brillo) until it was shiny and bright. The nurse then placed the electrode back on the patient. How did this procedure affect the signal observed from the electrode and electrode impedances?

5.12 A metal microelectrode has a tip that can be modeled as being cylindrical. The metal itself is 1 μm in diameter, and the tip region is 3 mm long. The metal has a resistivity of $1.2 \times 10^{-5}\ \Omega \cdot$cm and is coated over its circumference with an insulation material 0.2 μm thick. The insulation material has a relative dielectric constant of 1.67. Only the base of the cylinder is free of insulation.

a. What is the resistance associated with the tip of this microelectode?

b. What is the area of the surface of the electrode that contacts the electrolytic solution within the cell? The resistance associated with the electrode–electrolyte interface of this material is $10^3\ \Omega$ for 1 cm^2. What is the resistance due to this microelectrode's contact with the electrolyte?

c. What is the capacitance associated with the tip of the microelectrode when the capacitances at the interface of the electrode–electrolytic solution are neglected?

d. Draw an approximate equivalent circuit for the tip portion of this microelectrode.

e. At what frequencies do you expect to see distortions when the electrode is connected to an amplifier having a purely resistive input impedance of 10 MΩ? You may assume that the reference electrode has an impedance low enough so that it will not enter into the answer to this question. If the amplifier's input impedance is raised to 100 MΩ, how does this affect the frequency response of the system? Is this difference significant for most intracellular biological applications?

5.13 A micropipet electrode has a lumenal diameter of 3 μm at its tip. At this point, the glass wall is only 0.5 μm thick and 2 mm long. The resistance of the electrolyte in the tip is 40 MΩ. The glass has a relative dielectric constant of 1.63. Estimate the frequency response of this electrode when it is connected to an infinite-input-impedance amplifier. How can this frequency response be improved?

5.14 A pair of biopotential electrodes are used to monitor a bioelectric signal from the body. The monitoring electronic circuit has a low-input impedance that is of the same order of magnitude as the source impedance in the electrodes.

a. Sketch an equivalent circuit for this situation.

b. Describe qualitatively what you expect the general characteristics of the frequency response of this system to be. It is not necessary to plot an analytic Bode plot.

5.15 A pair of identical stainless steel electrodes is designed to be used to stimulate skeletal muscles. The stimulus consists of a rectangular constant-voltage pulse applied to the electrodes. The pulse has an amplitude of 5 V with a duration of 10 ms. Draw, on the basis of the equivalent circuits of each of the electrodes, an equivalent circuit for the load seen by the constant-voltage pulse generator. Simplify your circuit as much as possible. What is the waveshape of the current at the generator terminals? Remember that a constant-voltage generator has a source impedance of zero. Explain and sketch the resulting current waveform.

5.16 An exotic new animal, recently discovered, has an unusual electrolyte makeup in that its major anion is Br^- rather than Cl^-. Scientists want to measure the EEG of this animal, which is less than 25 μV. Electrodes made

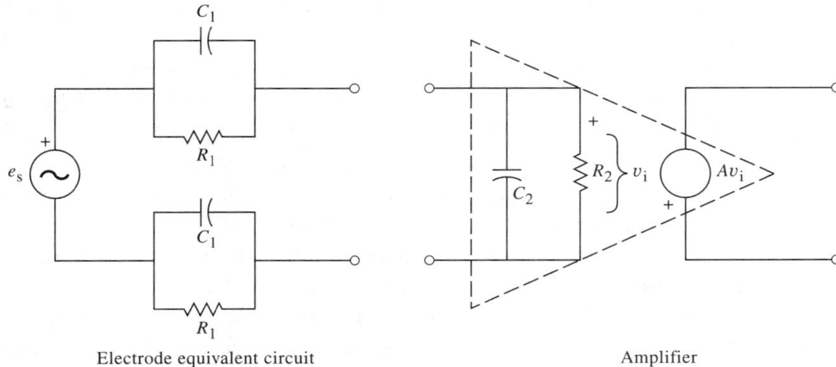

Figure P5.1

of Ag/AgCl seem to be noisy. Can you suggest a better electrode system and explain why it is better?

5.17 Needle-type EMG electrodes are placed directly in a muscle. Figure P5.1 shows their simplified equivalent circuit and also the equivalent circuit of the input stage of an amplifier. The value of the capacitor C_2 in the amplifier may be varied to any desired quantity.

a. Assuming $C_2 = 0$ and the amplifier gain is A, write an equation showing the output voltage of the amplifier as a function of e_s (the signal) and frequency.

b. Determine a value for C_2 that gives electrode–amplifier characteristics that are independent of frequency.

c. What is the amplifier's output voltage in part (b) when the signal is e_s?

5.18 Figure P5.2 shows the equivalent circuit of a biopotential electrode. A pair of these electrodes are tested in a beaker of physiological saline solution. The test consists of measuring the magnitude of the impedance between the electrodes as a function of frequency via low-level sinusoidal excitation so that the impedances are not affected by the current crossing the electrode-electrolyte interface. The impedance of the saline solution is small enough to be neglected. Sketch a Bode plot (log of impedance magnitude versus log of

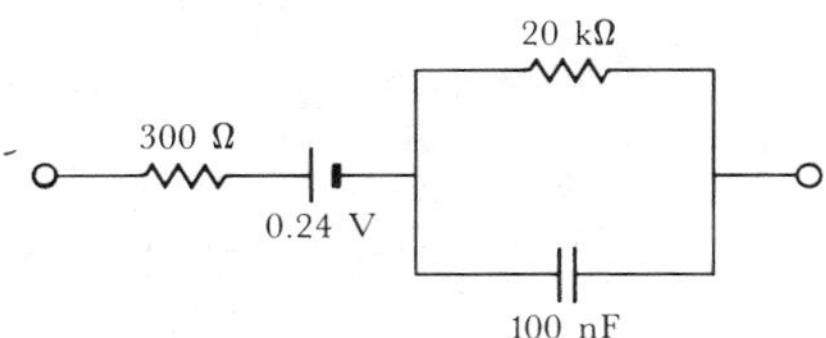

Figure P5.2

frequency) of the impedance between the electrodes over a frequency range of 1–100,000 Hz.

5.19 A pair of biopotential electrodes are implanted in an animal to measure the electrocardiogram for a radio-telemetry system. One must know the equivalent circuit for these electrodes in order to design the optimal input circuit for the telemetry system. Measurements made on the pair of electrodes have shown that the polarization capacitance for the pair is 200 nF and that the half-cell potential for each electrode is 223 mV. The magnitude of the impedance between the two electrodes was measured via sinusoidal excitation at several different frequencies. The results of this measurement are given in the accompanying table. On the basis of all of this information, draw an equivalent circuit for the electrode pair. State what each component in your circuit represents physically, and give its value.

Frequency	Impedance (magnitude), Ω
5 Hz	20,000
10 Hz	19,998
.	.
.	.
.	.
40 kHz	602
50 kHz	600
100 kHz	600

REFERENCES

Anderson, D. J., K. Najafi, S. J. Tanghe, D. A. Evans, K. L. Levy, J. F. Hetke, X. Xue, J. J. Zappia, and K. J. Wise, "Batch-fabricated thin-film electrodes for stimulation of the central auditory system," *IEEE Trans. Biomed. Eng.,* 1989, 36, 693–704.

Anonymous, *American National Standard for Pregelled Disposable Electrodes.* Arlington, VA: Association for the Advancement of Medical Instrumentation, 1984.

Arzbaecher, R., "New applications of the pill electrode for cardiac recording and pacing." *Proc. Annu. Int. Conf. IEEE Eng. Med. Biol. Soc.,* 1982, 4, 508–510.

Ash, R. B., T. A. Johnson, and H. T. Nagle, "Application of a multi-level ceramic integrated circuit package for epicardial mapping," *Proc. Annu. Int. Conf. IEEE Eng. Med. Biol. Soc.,* 1992, 2392.

Atkin, T., K. Najafi, R. H. Smoke, and R. M. Bradley, "A micromachined silicon sieve electrode for nerve regeneration applications," *IEEE Trans. Biomed. Eng.,* 1994, 41, 305–313.

Boucsein, W., *Electrodermal Activity.* New York: Plenum Press, 1992.

Burbank, D. P., and J. G. Webster, "Reducing skin potential motion artefact by skin abrasion," *Med. Biol. Eng. Comput.,* 1978, 16, 31–38.

Caldwell, C. W., and J. B. Reswick, "A percutaneous wire electrode for chronic research use," *IEEE Trans. Biomed. Eng.,* 1975, BME-22, 429–432.

Campbell, P. K., K. E. Jones, R. J. Huber, K. W. Horch, and R. A. Normann, "A silicon-based, three-dimensional neural interface: Manufacturing processes for an intracortical interface," *IEEE Trans. Biomed. Eng.,* 1991, 38, 758–768.

Carim, H. M., "Bioelectrodes," in J. G. Webster (ed.), *Encyclopedia of Medical Devices and Instrumentation.* New York: Wiley, 1988, pp. 195–226.

Cobbold, R. S. C., *Transducers for Biomedical Measurements: Principles and Applications.* New York: Wiley, 1974.

Critchfield, F. H., C. Xinteras, B. Johnson, and M. R. Neuman, "Surgical and engineering techniques for the supra-cephalic mounting of a multichannel EEG telemeter." *Dig. Int. Conf. Med. Biol. Eng.,* 1971, paper number 22-4.

Das, D. P., and J. G. Webster, "Defibrillation recovery curves for different electrode materials." *IEEE Trans. Biomed. Eng.,* 1980, BME-27, 230–233.

de Talhouet, H., and J. G. Webster, "The origin of skin-stretch-caused motion artifacts under electrodes," *Physiol. Meas.,* 1996, 17, 81–93.

Delgado, J. M. R., "Electrodes for extracellular recording and stimulation," in W. L. Nastuk (ed.), *Physical techniques in biological research.* New York: Academic, 1964, Vol. 5A.

Drake, K. L., K. D. Wise, J. Farraye, D. J. Anderson, and S. L. BeMent, "Performance of planar multisite microprobes in recording extracellular single-unit intracortical activity." *IEEE Trans. Biomed. Eng.,* 1988, 35, 719–732.

Edell, D. J., "A peripheral nerve information transducer for amputees: Long-term multi-channel recordings from rabbit peripheral nerves." *IEEE Trans. Biomed. Eng.,* 1986, BME-33, 203–214.

Fein, H., *An Introduction to Microelectrode Technique and Instrumentation.* New Haven, CT: W-P Instruments, 1977.

Ferris, C. D., *Introduction to Bioelectrodes.* New York: Plenum, 1974.

Geddes, L. A., *Electrodes and the Measurement of Bioelectric Events.* New York: Wiley, 1972.

Geddes, L. A., and L. E. Baker, "The relationship between input impedance and electrode area in recording the ECG." *Med. Biol. Eng.,* 1966, 4, 439–450.

Geddes, L. A., L. E. Baker, and A. G. Moore, "Optimum electrolytic chloriding of silver electrodes." *Med. Biol. Eng.,* 1969, 7, 49–56.

Hon, E. H., R. H. Paul, and R. W. Hon, "Electronic evaluation of fetal heart rate. XI: Description of a spiral electrode." *Obstet. Gynecol.,* 1972, 40, 362–363.

Janz, G. J., and D. J. G. Ives, "Silver–silver chloride electrodes." *Ann. N.Y. Acad. Sci.,* 1968, 148, 210–221.

Jossinet, J., and E. McAdams, "Hydrogel electrodes in biosignal recording." *Proc. Annu. Int. Conf. IEEE Eng. Med. Biol. Soc.,* 1990, 12, 1490–1491.

Kovacs, G. T. A., C. W. Storment, M. Halks-Miller, C. R. Belczynski, Jr., C. C. D. Santina, E. R. Lewis, and N. I. Maluf, "Silicon-substrate microelectrode arrays for parallel recording of neural activity in peripheral and cranial nerves," *IEEE Trans. Biomed. Eng.,* 1994, 41, 567–577.

Mastrototaro, J. J., H. Z. Massoud, T. C. Pilkington, and R. E. Ideker, "Rigid and flexible thin-film multielectrode arrays for transmural cardiac recording," *IEEE Trans. Biomed. Eng.,* 1992, 39, 271–279.

Miller, H. A., and D. C. Harrison (eds.), *Biomedical Electrode Technology.* New York: Academic, 1974.

Neuman, M. R., "Flexible thin film skin electrodes for use with neonates." *Dig. Int. Conf. Med. Biol. Eng.,* 1973, paper no. 35.11.

Plonsey, R., *Bioelectric Phenomena.* New York: McGraw-Hill, 1969.

Prohaska, O. J., F. Olcaytug, P. Pfundner, and H. Dragaun, "Thin-film multiple electrode probes: Possibilities and limitations." *IEEE Trans. Biomed. Eng.,* 1986, BME-33, 223–229.

Robinson, D. A., "The electrical properties of metal microelectrodes." *Proc. IEEE,* 1968, 56, 1065–1071.

Robblee, R. S., J. L. Lefko, and S. B. Brummer, "Activated iridium: an electrode suitable for reversible charge injection in saline solution," *J. Electrochem. Soc.,* 1983, 1130, 731–733.

Rosell, J., J. Colominas, P. Riu, R. Pallas-Areny, and J. G. Webster, "Skin impedance from 1 Hz to 1 MHz." *IEEE Trans. Biomed. Eng.,* 1988, 35, 649–651.

Schwan, H. P., "Determination of biological impedances," in W. L. Nastuk (ed), *Physical Techniques in Biological Research.* New York: Academic, 1963, pp. 323–407.

Tam, H. W., and J. G. Webster, "Minimizing electrode motion artifact by skin abrasion." *IEEE Trans. Biomed. Eng.,* 1977, BME-24, 134–139.

Webster, J. G., "Reducing motion artifacts and interference in biopotential recording." *IEEE Trans. Biomed. Eng.,* 1984a, BME-31, 823–826.

Webster, J. G., "What is important in biomedical electrodes?" *Proc. Annu. Conf. Eng. Med. Biol.,* 1984b, 26, 96.

Wise, K. D., K. Najafi, J. Ji, J. F. Hetke, S. J. Tanghe, A. Hoogerwerf, D. J. Anderson, S. L. BeMent, M. Ghazzi, W. Baer, T. Hull, and Y. Yang. "Micromachined silicon microprobes for CNS recording and stimulation." *Proc. Annu. Int. Conf. IEEE Eng. Med. Biol. Soc.,* 1990, 12, 2334–2335.

6

BIOPOTENTIAL AMPLIFIERS

Michael R. Neuman

Amplifiers are an important part of modern instrumentation systems for measuring biopotentials. Such measurements involve voltages that often are at low levels, have high source impedances, or both. Amplifiers are required to increase signal strength while maintaining high fidelity. Amplifiers that have been designed specifically for this type of processing of biopotentials are known as *biopotential amplifiers.* In this chapter we examine some of the basic features of biopotential amplifiers and also look at specialized systems.

6.1 BASIC REQUIREMENTS

The essential function of a biopotential amplifier is to take a weak electric signal of biological origin and increase its amplitude so that it can be further processed, recorded, or displayed. Usually such amplifiers are in the form of voltage amplifiers, because they are capable of increasing the voltage level of a signal. Nonetheless, voltage amplifiers also serve to increase power levels, so they can be considered power amplifiers as well. In some cases, biopotential amplifiers are used to isolate the load from the source. In this situation, the amplifiers provide only current gain, leaving the voltage levels essentially unchanged.

To be useful biologically, all biopotential amplifiers must meet certain basic requirements. They must have high input impedance, so that they provide minimal loading of the signal being measured. The characteristics of biopotential electrodes can be affected by the electric load they see, which, combined with excessive loading can result in distortion of the signal. Loading effects are minimized by making the amplifier input impedance as high as possible, thereby reducing this distortion. Modern biopotential amplifiers have input impedances of at least 10 MΩ.

The input circuit of a biopotential amplifier must also provide protection to the organism being studied. Any current or potential appearing across the amplifier input terminals is capable of affecting the biological potential being measured. In clinical systems, electric currents produced by the biopotential amplifier and seen at its input terminals can result in microshocks or macroshocks in the patient being studied—a situation that can have grave consequences. To avoid these problems, the amplifier should have isolation and protection circuitry, so that the current through the electrode circuit can be kept at safe levels and any artifact generated by such current can be minimized.

The output circuit of a biopotential amplifier does not present so many critical problems as the input circuit. Its principal function is to drive the amplifier load, usually an indicating or recording device, in such a way as to maintain maximal fidelity and range in this readout. Therefore, the output impedance of the amplifier must be low with respect to the load impedance, and the amplifier must be capable of supplying the power required by the load.

Biopotential amplifiers must operate in that portion of the frequency spectrum in which the biopotentials that they amplify exist. Because of the low level of such signals, it is important to limit the bandwidth of the amplifier so that it is just great enough to process the signal adequately. In this way, we can obtain optimal signal-to-noise ratios. Biopotential signals usually have amplitudes of the order of a few millivolts or less. Such signals must be amplified to levels compatible with recording and display devices. This means that most biopotential amplifiers must have high gains—of the order of 1000 or greater.

Very frequently biopotential signals are obtained from bipolar electrodes. These electrodes are often symmetrically located, electrically, with respect to ground. Under such circumstances, the most appropriate biopotential amplifier is a differential one. Because such bipolar electrodes frequently have a common-mode voltage with respect to ground that is much larger than the signal amplitude, and because the symmetry with respect to ground can be distorted, such biopotential differential amplifiers must have high common-mode rejection ratios to minimize artifact due to the common-mode signal.

A final requirement for biopotential amplifiers that are used both in medical applications and in the laboratory is that they make quick calibration possible. In recording biopotentials, the scientist and clinician need to know not only the waveforms of these signals but also their amplitudes. To provide this information, the gain of the amplifier must be well calibrated. Frequently biopotential amplifiers have a standard signal source that can be momentarily connected to the input, at the push of a button, to check the calibration. Biopotential amplifiers that need to have adjustable gains usually have a switch by which different, carefully calibrated fixed gains can be selected, rather than having a continuous control (such as the volume control of an audio amplifier) for adjusting the gain. Thus the gain is always known, and there is no chance of its being accidentally varied by someone bumping the gain control.

Biopotential amplifiers have additional requirements that are application-specific and that can be ascertained from an examination of each application. To illustrate some of these, let us first consider the electrocardiogram (ECG), the most frequently used application of biopotential amplifiers.

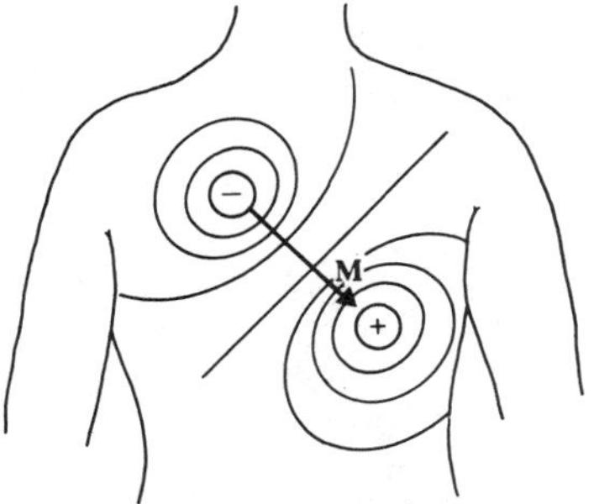

Figure 6.1 Rough sketch of the dipole field of the heart when the R wave is maximal The dipole consists of the points of equal positive and negative charge separated from one another and denoted by the dipole moment vector **M**.

6.2 THE ELECTROCARDIOGRAPH

To learn more about biopotential amplifiers, we shall examine a typical clinical electrocardiograph. First, let us review the ECG itself.

THE ECG

As we learned in Section 4.6, the beating heart generates an electric signal that can be used as a diagnostic tool for examining some of the functions of the heart. This electric activity of the heart can be approximately represented as a vector quantity. Thus we need to know the location at which signals are detected, as well as the time-dependence of the amplitude of the signals. Electrocardiographers have developed a simple model to represent the electric activity of the heart. In this model, the heart consists of an electric dipole located in the partially conducting medium of the thorax. Figure 6.1 shows a typical example.

This particular field and the dipole that produces it represent the electric activity of the heart at a specific instant. At the next instant the dipole can change its magnitude and its orientation, thereby causing a change in the electric field. Once we accept this model (it is an oversimplification), we need not draw a field plot every time we want to discuss the dipole field of the heart. Instead, we can represent it by its dipole moment, a vector directed from the negative charge to the positive charge and having a magnitude proportional to the amount of charge (either positive or negative) multiplied by the separation of the two charges. In electrocardiography this dipole moment, known as the *cardiac vector,* is represented by **M**, as shown in Figure 6.1. As we progress through a cardiac cycle, the magnitude and direction of **M** vary because the dipole field varies.

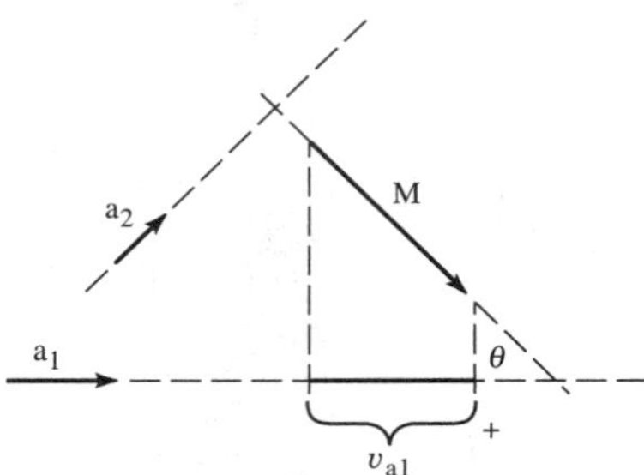

Figure 6.2 Relationships between the two lead vectors $\mathbf{a}_1$ and $\mathbf{a}_2$ and the cardiac vector **M**. The component of **M** in the direction of $\mathbf{a}_1$ is given by the dot product of these two vectors and denoted on the figure by v_{a1}. Lead vector $\mathbf{a}_2$ is perpendicular to the cardiac vector, so no voltage component is seen in this lead.

The electric potentials generated by the heart appear throughout the body and on its surface. We determine potential differences by placing electrodes on the surface of the body and measuring the voltage between them, being careful to draw little current (ideally there should be no current at all, because current distorts the electric field that produces the potential differences). If the two electrodes are located on different equal-potential lines of the electric field of the heart, a nonzero potential difference or voltage is measured. Different pairs of electrodes at different locations generally yield different voltages because of the spatial dependence of the electric field of the heart. Thus it is important to have certain standard positions for clinical evaluation of the ECG. The limbs make fine guideposts for locating the ECG electrodes. We shall look at this in more detail later.

In the simplified dipole model of the heart, it would be convenient if we could predict the voltage, or at least its waveform, in a particular set of electrodes at a particular instant of time when the cardiac vector is known. We can do this if we define a *lead vector* for the pair of electrodes. This vector is a unit vector that defines the direction a constant-magnitude cardiac vector must have to generate maximal voltage in the particular pair of electrodes. A pair of electrodes, or combination of several electrodes through a resistive network that gives an equivalent pair, is referred to as a *lead.*

For a cardiac vector **M**, as shown in Figure 6.2, the voltage induced in a lead represented by the lead vector $\mathbf{a}_1$ is given by the component of **M** in the direction of $\mathbf{a}_1$. In vector algebra, this can be denoted by the dot product

$$v_{a1} = \mathbf{M} \cdot \mathbf{a}_1 \qquad \text{or} \qquad v_{a1} = |\mathbf{M}| \cos \theta \tag{6.1}$$

where v_{a1} is the scalar voltage seen in the lead that has the vector $\mathbf{a}_1$. Let us consider another lead, represented by the lead vector $\mathbf{a}_2$, as seen in Figure 6.2. In this case, the vector is oriented in space so as to be perpendicular to

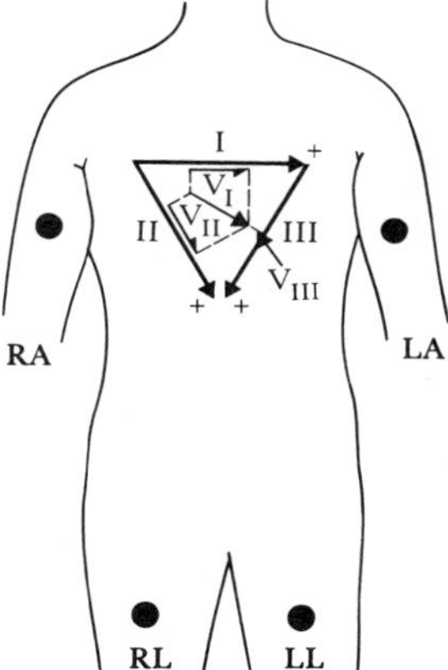

Figure 6.3 Cardiologists use a standard notation such that the direction of the lead vector for lead I is 0°, that of lead II is 60°, and that of lead III is 120°. An example of a cardiac vector at 30° with its scalar components seen for each lead is shown.

the cardiac vector **M**. The component of **M** along the direction of $\mathbf{a}_2$ is zero, so no voltage is seen in this lead as a result of the cardiac vector. If we measured the ECG generated by **M** using one of the two leads shown in Figure 6.2 alone, we could not describe the cardiac vector uniquely. However, by using two leads with different lead vectors, both of which lie in the same plane as the cardiac vector such as $\mathbf{a}_1$ and $\mathbf{a}_2$, we can describe **M**.

In clinical electrocardiography, more than one lead must be recorded to describe the heart's electric activity fully. In practice, several leads are taken in the *frontal plane* (the plane of your body that is parallel to the ground when you are lying on your back) and the *transverse plane* (the plane of your body that is parallel to the ground when you are standing erect).

Three basic leads make up the *frontal-plane* ECG. These are derived from the various permutations of pairs of electrodes when one electrode is located on the right arm (RA in Figure 6.3), the left arm (LA), and the left leg (LL). Very often an electrode is also placed on the right leg (RL) and grounded or connected to special circuits, as shown in Figure 6.15. The resulting three leads are lead I, LA to RA; lead II, LL to RA; and lead III, LL to LA. The lead vectors that are formed can be approximated as an equilateral triangle, known as *Eindhoven's triangle,* in the frontal plane of the body, as shown in Figure 6.3. Because the scalar signal on each lead of Eindhoven's triangle can be represented as a voltage source, we can write Kirchhoff's voltage law for the three leads.

$$\mathrm{I} - \mathrm{II} + \mathrm{III} = 0 \tag{6.2}$$

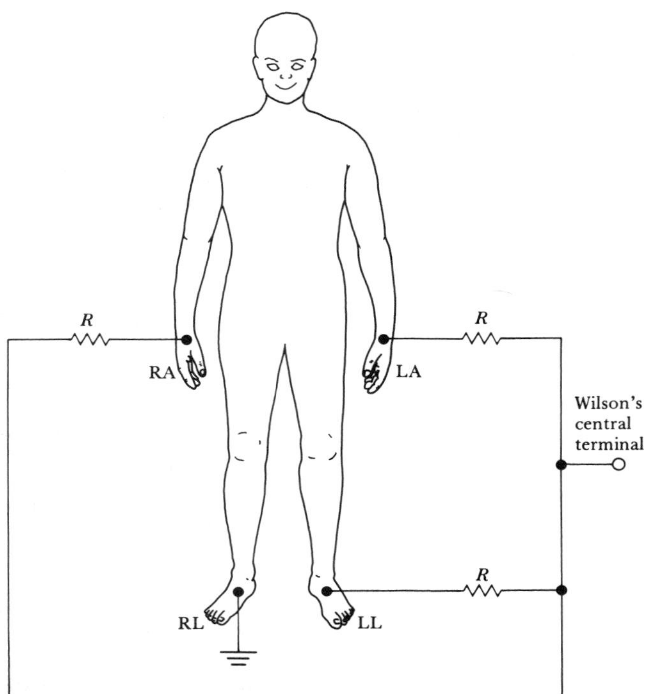

Figure 6.4 Connection of electrodes to the body to obtain Wilson's central terminal

The components of a particular cardiac vector can be determined easily by placing the vector within the triangle and determining its projection along each side. The process can also be reversed, which enables us to determine the cardiac vector when we know the components along the three lead vectors, or at least two of them. It is this latter problem that usually concerns the electrocardiographer.

Three additional leads in the frontal plane—as well as a group of leads in the transverse plane—are routinely used in taking clinical ECGs. These leads are based on signals obtained from more than one pair of electrodes. They are often referred to as *unipolar leads,* because they consist of the potential appearing on one electrode taken with respect to an equivalent reference electrode, which is the average of the signals seen at two or more electrodes.

One such equivalent reference electrode is the *Wilson central terminal,* shown in Figure 6.4. Here the three-limb electrodes just described are connected through equal-valued resistors to a common node. The voltage at this node, which is the Wilson central terminal, is the average of the voltages at each electrode. In practice, the values of the resistors should be at least 5 MΩ so that the loading of any particular lead will be minimal. Thus, a more

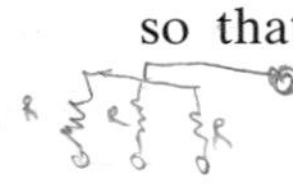

practical approach is to use buffers (voltage followers, see Section 3.3) between each electrode and the equal-valued resistors. The signal beween LA and the central point is known as VL, that at RA as VR, and that at the left foot as VF. Note that for each of these leads, one of the resistances R shunts the circuit between the central terminal and the limb electrode. This tends to reduce the amplitude of the signal observed, and we can modify these leads to *augmented leads* by removing the connection between the limb being measured and the central terminal. This does not affect the direction of the lead vector but results in a 50% increase in amplitude of the signal.

The augmented leads—known as aVL, aVR, and aVF—are illustrated in Figure 6.5, which also illustrates their lead vectors, along with those of leads I, II, and III. Note that when the negative direction for aVR is considered with the other five, all six vectors are equally spaced, by 30°. It is thus possible for the cardiologist looking at an ECG consisting of these six leads to estimate the position of the cardiac cycle by seeing which of the six leads has the greatest signal amplitude at that point in the cycle.

When physicians look at the ECG in the transverse plane, they use *precordial* (chest) leads. They place an electrode at various anatomically defined positions on the chest wall, as shown in Figure 6.6. The potential between this electrode and Wilson's central terminal is the electrocardiogram for that particular lead. Figure 6.6 also shows the lead-vector positions. Physicians can obtain ECGs from the posterior side of the heart by means of an electrode placed in the esophagus. This structure passes directly behind the heart, and the potential between the esophageal electrode and Wilson's central terminal gives a posterior lead.

SPECIFIC REQUIREMENTS OF THE ELECTROCARDIOGRAPH

Because the electrocardiograph is widely used as a diagnostic tool and there are several manufacturers of this instrument, standardization is necessary. Standard requirements for electrocardiographs have been developed over the years (Bailey *et al.* 1990; Anonymous, 1991). Table 6.1 gives a summary of performance requirements from the most recent of these (Anonymous, 1991). These recommendations are a part of a voluntary standard. The Food and Drug Administration is planning to develop mandatory standards for frequently employed instruments such as the electrocardiograph.

FUNCTIONAL BLOCKS OF THE ELECTROCARDIOGRAPH

Figure 6.7 shows a block diagram of a typical clinical electrocardiograph. To understand the overall operation of the system, let us consider each block separately.

1. *Protection circuit* This circuit includes protection devices so that the high voltages that may appear across the input to the electrocardiograph under certain conditions do not damage it.

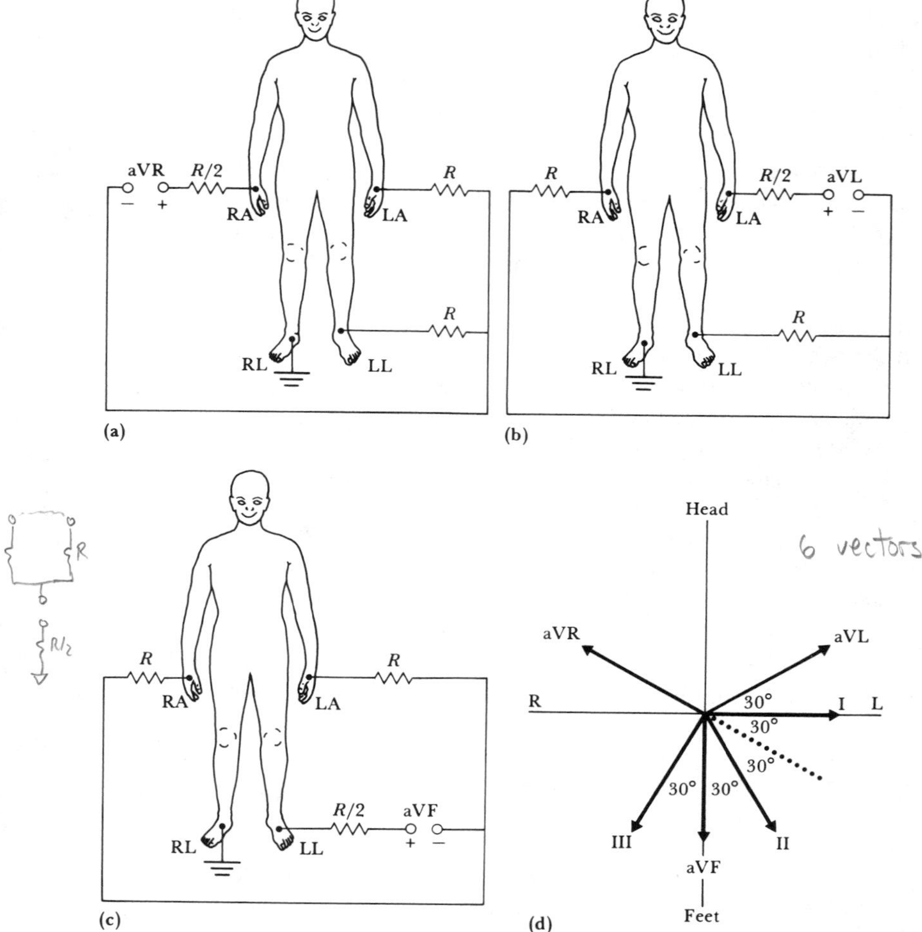

Figure 6.5 (a), (b), (c) Connections of electrodes for the three augmented limb leads. (d) Vector diagram showing standard and augmented lead-vector directions in the frontal plane.

2. *Lead selector* Each electrode connected to the patient is attached to the lead selector of the electrocardiograph. The function of this block is to determine which electrodes are necessary for a particular lead and to connect them to the remainder of the circuit. It is this part of the electrocardiograph in which the connections for the central terminal are made. This block can be controlled by the operator or by the microcomputer of the electrocardiograph when it is operated in automatic mode. It selects one or more leads to be recorded. In automatic mode, each of the 12 standard leads is recorded for a short duration such as 10 s.

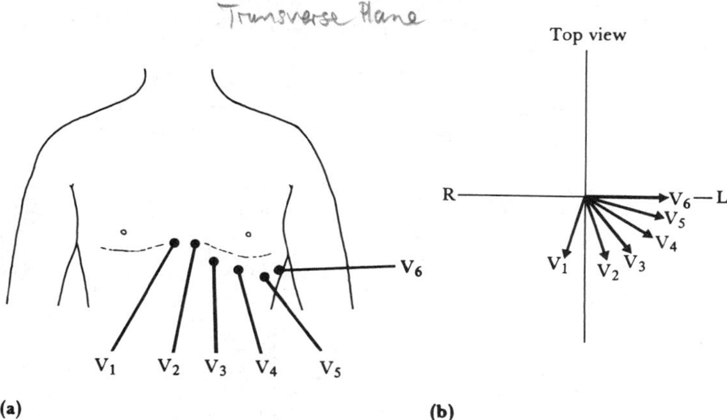

Figure 6.6 (a) Positions of precordial leads on the chest wall. (b) Directions of precordial lead vectors in the transverse plane.

3. *Calibration signal* A 1-mV calibration signal is momentarily introduced into the electrocardiograph for each channel that is recorded.
4. *Preamplifier* The input preamplifier stage carries out the initial amplification of the ECG. This stage should have very high input impedance and a high common-mode-rejection ratio (CMRR). A typical preamplifier stage is the differential amplifier that consists of three operational amplifiers, shown in Figure 3.5. A gain-control switch is often included as a part of this stage.
5. *Isolation circuit* The circuitry of this block contains a barrier to the passage of current from the power line (50 or 60 Hz). For example, if the patient came in contact with a 120-V line, this barrier would prevent dangerous currents from flowing from the patient through the amplifier to the ground of the recorder or microcomputer.
6. *Driven right leg circuit* This circuit provides a reference point on the patient that normally is at ground potential. This connection is made to an electrode on the patient's right leg. Details on this circuit are given in Section 6.5.
7. *Driver amplifier* Circuitry in this block amplifies the ECG to a level at which it can appropriately record the signal on the recorder. Its input should be ac-coupled so that offset voltages amplified by the preamplifier are not seen at its input. These dc voltages, when amplified by this stage, might cause it to saturate. This stage also carries out the bandpass filtering of the electrocardiograph to give the frequency characteristics described in Table 6.1. Also it often has a zero-offset control that is used to position the signal on the chart paper. This control adjusts the dc level of the output signal.
8. *Memory system* Many modern electrocardiographs store electrocardiograms in memory as well as printing them out on a paper chart. The

Table 6.1 Summary of Performance Requirements for Electrocardiographs (Anonymous, 1991)

Section	Requirement Description	Min/max	Units	Min/max Value
3.2.1	Operating Conditions:			
	line voltage	range	V rms	104 to 1127
	frequency	range	Hz	60 ± 1
	temperature	range	°C	25 ± 10
	relative humidity	range	%	50 ± 20
	atmospheric pressure	range	Pa	7×10^4 to 10.6×10^4
3.2.2	Lead Definition (number of leads):	NA	NA	Table 3
	single-channel	min	NA	7
	three-channel	min	NA	12
3.2.3	Input Dynamic Range:			
	range of linear operations of input signal	min	mV	±5
	slew rate change	max	mV/s	320
	dc offset voltage range	min	mV	±300
	allowed variation of amplitude with dc offset	max	%	±5
3.2.4	Gain Control, Accuracy, and Stability:			
	gain selections	min	mm/mV	20, 10, 5
	gain error	max	%	5
	manual override of automatic gain control	NA	NA	NA
	gain change rate/minute	max	%/min	±0.33
	total gain change/hour	max	%	±3
3.2.5	Time Base Selection and Accuracy:			
	time base selections	min	mm/s	25, 50
	time base error	max	%	±5
3.2.6	Output Display:			
	general	NA	NA	per 3.2.3
	width of display	min	mm	40
	trace visibility (writing rates)	max	mm/s	1600
	trace width (permanent record only)	max	mm	1
	departure from time axis alignment	max	mm	0.5
		max	ms	10
	preruled paper division	min	div/cm	10
	error of rulings	max	%	±2
	time marker error	max	%	±2
3.2.7	Accuracy of Input Signal Reproduction:			
	overall error for signals	max	%	±5
	up to ± 5 mV and 125 mV/s	max	μV	±40
	upper cut-off frequency (3 dB)	min	Hz	150
	response to 20 ms, 1.5 mV triangular input	min	mm	13.5
	response after 3 mV, 100 ms impulse	max	mV	0.1
		max	mV/s	0.30
	error in lead weighting factors	max	%	5

Table 6.1 *(Continued)*

Section	Requirement Description	Min/max	Units	Min/max Value
	hysteresis after 15-mm deflection from baseline	max	mm	0.5
3.2.8	Standardizing Voltage:			
	nominal value	NA	mV	1.0
	rise time	max	ms	1
	decay time	min	s	100
	amplitude error	max	%	±5
3.2.9	Input Impedance at 10 Hz (each lead)	min	megohms	2.5
3.2.10	DC Current (any input lead)	max	μA	0.1
	DC Current (any patient electrode)	max	μA	1.0
3.2.11	Common-Mode Rejection:			
	allowable noise with 20 V, 60 Hz and ± 300 mV dc and 51-kilohm	max	mm	10
	imbalance	max	mV	1
3.2.12	System Noise:			
	RTI, *p-p*	max	μV	30
	multichannel crosstalk	max	%	2
3.2.13	Baseline Control and Stability:			
	return time after reset	max	s	3
	return time after lead switch	max	s	1
	Baseline Stability:			
	baseline drift rate RTI	max	μV/s	10
	total baseline drift RTI (2-min period)	max	μV	500
3.2.14	Overload Protection:			
	no damage from differential voltage, 60-Hz, 1-V *p-p*, 10-s application	min	V	1
	no damage from simulated defibrillator discharges:			
	overvoltage	N/A	V	5000
	energy	N/A	J	360
	recovery time	max	s	8
	energy reduction by defibrillator shunting	max	%	10
	transfer of charge through defibrillator chassis	max	μC	100
	ECG display in presence of pacemaker pulses:			
	amplitude	range	mV	2 to 250
	pulse duration	range	ms	0.1 to 2.0
	rise time	max	μs	100
	frequency	max	pulses/min	100
3.2.15	Risk Current (Isolated Patient Connection)	max	μA	10
		as per Applicable Document 2.1.1		
3.2.16	Auxiliary Output (if provided):			
	no damage from short circuit risk current (isolated patient connection)	max	μA	10
		as per Applicable Document 2.1.1		

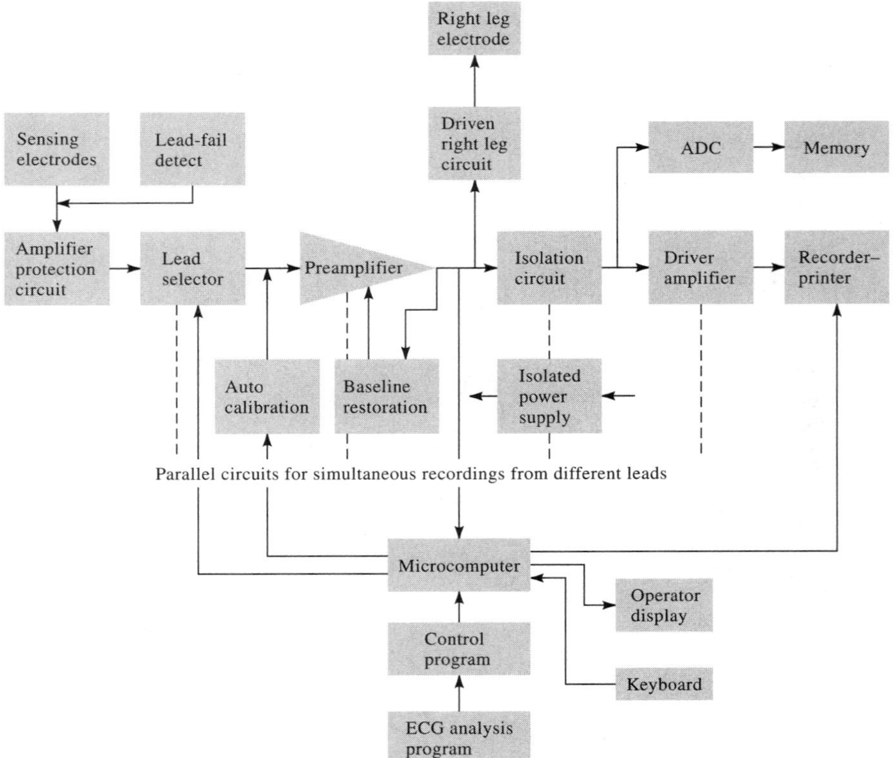

Figure 6.7 Block diagram of an electrocardiograph

signal is first digitized by an analog-to-digital converter (ADC), and then samples from each lead are stored in memory. Patient information entered via the keyboard is also stored. The microcomputer controls this storage activity.

9. *Microcomputer* The microcomputer controls the overall operation of the electrocardiograph. The operator can select several modes of operation by invoking a particular program. For example, she or he can ask the microcomputer to generate the standard 12-lead electrocardiogram by selecting three simultaneous 10-s segments of the six frontal plane leads followed by three 10-s segments of the six transverse plane leads. The microcomputer in some machines can also perform a preliminary analysis of the electrocardiogram to determine the heart rate, recognize some types of arrhythmia, calculate the axes of various features of the electrocardiogram, and determine intervals between these features. A keyboard and an alphanumeric display enable the operator to communicate with the microcomputer.

10. *Recorder-printer* This block provides a hard copy of the recorded ECG signal. It also prints out patient identification, clinical information entered by the operator, and the results of the automatic analysis of the electrocardiogram. Although analog oscillograph-type recorders were employed for this function in the past, modern electrocardiographs make use of thermal or electrostatic recording techniques in which the only moving part is the paper being transported under the print head (Vermariën, 1988). Digitized electrocardiograms can also be stored in permanent memory such as magnetic disks or tape.

6.3 PROBLEMS FREQUENTLY ENCOUNTERED

There are many factors that must be taken into consideration in the design and application of the electrocardiograph as well as other biopotential amplifiers. These factors are important not only to the biomedical engineer, but also to the individual who operates the instrument. In the following paragraphs, we shall describe a few of the more common problems encountered and shall indicate some of their causes.

FREQUENCY DISTORTION

The electrocardiograph does not always meet the frequency-response standards we have described. When this happens, frequency distortion is seen in the ECG.

High-frequency distortion rounds off the sharp corners of the waveforms and diminishes the amplitude of the QRS complex.

An instrument that has a frequency response of 1–150 Hz yields *low-frequency distortion.* The baseline is no longer horizontal, especially immediately following any event in the tracing. Monophasic waves in the ECG appear to be more biphasic.

SATURATION OR CUTOFF DISTORTION

High offset voltages at the electrodes or improperly adjusted amplifiers in the electrocardiograph can produce saturation or cutoff distortion that can greatly modify the appearance of the ECG. The combination of input-signal amplitude and offset voltage drives the amplifier into saturation during a portion of the QRS complex (Section 3.2). The peaks of the QRS complex are cut off because the output of the amplifier cannot exceed the saturation voltage.

In a similar occurrence, the lower portions of the ECG are cut off. This can result from negative saturation of the amplifier. In this case only a portion of the S-wave may be cut off. In extreme cases of this type of distortion even the P and T waves may be below the cutoff level such that only the R wave appears.

GROUND LOOPS

Patients who are having their ECGs taken on either a clinical electrocardiograph or continuously on a cardiac monitor are often connected to other pieces of electric apparatus. Each electric device has its own ground connection either through the power line or, in some cases, through a heavy ground wire attached to some ground point in the room.

A *ground loop* can exist when two machines are connected to the patient. Both the electrocardiograph and a second machine have a ground electrode attached to the patient. The electrocardiograph is grounded through the power line at a particular socket. The second machine is also grounded through the power line, but it is plugged into an entirely different outlet across the room, which has a different ground. If one ground is at a slightly higher potential than the other ground, a current from one ground flows through the patient to the ground electrode of the electrocardiograph and along its lead wire to the other ground. In addition to this current's presenting a safety problem, it can elevate the patient's body potential to some voltage above the lowest ground to which the instrumentation is attached. This produces common-mode voltages on the electrocardiograph that, if it has a poor common-mode-rejection ratio, can increase the amount of interference seen.

OPEN LEAD WIRES

Frequently one of the wires connecting a biopotential electrode to the electrocardiograph becomes disconnected from its electrode or breaks as a result of excessively rough handling, in which case the electrode is no longer connected to the electrocardiograph. Relatively high potentials can often be induced in the open wire as a result of electric fields emanating from the power lines or other sources in the vicinity of the machine. This causes a wide, constant-amplitude deflection of the pen on the recorder at the power-line frequency, as well as, of course, signal loss. Such a situation also arises when an electrode is not making good contact with the patient. A circuit for detecting poor electrode contact is described in Section 6.9.

ARTIFACT FROM LARGE ELECTRIC TRANSIENTS

In some situations in which a patient is having an ECG taken, cardiac defibrillation may be required (Section 13.2). In such a case, a high-voltage high-current electric pulse is applied to the chest of the patient so that transient potentials can be observed across the electrodes. These potentials can be several orders of magnitude higher than the normal potentials encountered in the ECG. Other electric sources can cause similar transients. When this situation occurs, it can cause an abrupt deflection in the ECG, as shown in

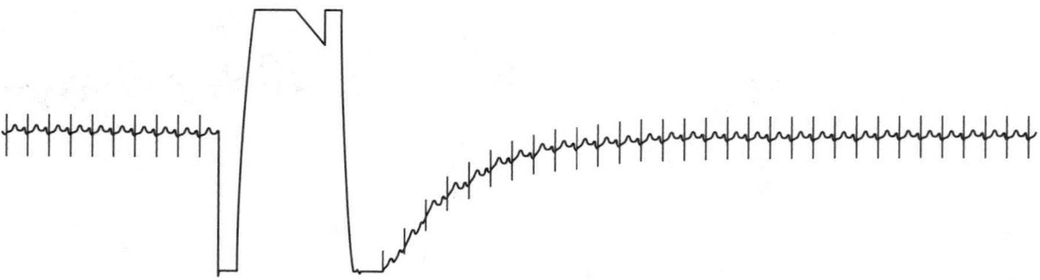

Figure 6.8 Effect of a voltage transient on an ECG recorded on an electrocardiograph in which the transient causes the amplifier to saturate, and a finite period of time is required for the charge to bleed off enough to bring the ECG back into the amplifier's active region of operation. This is followed by a first-order recovery of the system.

Figure 6.8. This is due to the saturation of the amplifiers in the electrocardiograph caused by the relatively high-amplitude pulse or step at its input. This pulse is sufficiently large to cause the buildup of charge on coupling capacitances in the amplifier, resulting in its remaining saturated for a finite period of time following the pulse and then slowly drifting back to the original baseline with a time constant determined by the low corner frequency of the amplifier. The slowly recovering waveform is shown in Figure 6.8.

Transients of the type just described can be generated by means other than defibrillation. Serious artifact caused by motion of the electrodes can produce variations in potential greater than ECG potentials. Another source of artifact is the patient's encountering a built-up static electric charge that can be partially discharged through the body. Older electrocardiographs exhibit a similar transient when they are switched manually from one lead to another, because there are different offset potentials at each electrode. This is usually not seen on newer machines that switch leads automatically, because voltages due to excess charge are discharged during the switching process.

This problem is greatly alleviated by reducing the source of the artifact. Because we do not have time to disconnect an electrocardiograph when a patient is being defibrillated, we can include electronic protection circuitry, such as that described in Section 6.4, in the machine itself. In this way, we can limit the maximal input voltage across the ECG amplifier so as to minimize the saturation and charge buildup effects due to the high-voltage input signals. This results in a more rapid return to normal operation following the transient. Such circuitry is also important in protecting the electrocardiograph from any damage that might be caused by these pulses.

Artifact caused by static electric charge on personnel can be lessened noticeably by reducing the buildup of static charge through the use of conductive clothing, shoes, and flooring, as well as by having personnel touch the bed before touching the patient. Motion artifact from the electrodes can be decreased by using the techniques described in Chapter 5.

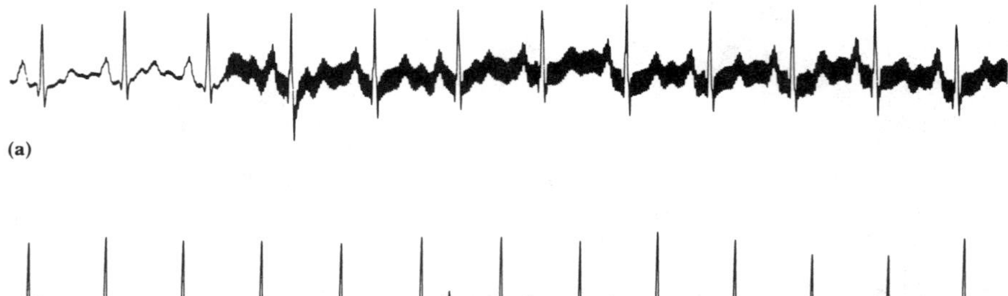

(a)

(b)

Figure 6.9 (a) 60-Hz power-line interference. (b) Electromyographic interference on the ECG. Severe 60-Hz interference is also shown on the bottom tracing in Figure 4.13.

INTERFERENCE FROM ELECTRIC DEVICES

A major source of interference when one is recording or monitoring the ECG is the electric-power system. Besides providing power to the electrocardiograph itself, power lines are connected to other pieces of equipment and appliances in the typical hospital room or physician's office. There are also power lines in the walls, floor, and ceiling running past the room to other points in the building. These power lines can affect the recording of the ECG and introduce interference at the line frequency in the recorded trace, as illustrated in Figure 6.9(a). Such interference appears on the recordings as a result of two mechanisms, each operating singly or, in some cases, both operating together.

Electric-field coupling between the power lines and the electrocardiograph and/or the patient is a result of the electric fields surrounding main power lines and the power cords connecting different pieces of apparatus to electric outlets. These fields can be present even when the apparatus is not turned on, because current is not necessary to establish the electric field. These fields couple into the patient, the lead wires, and the electrocardiograph itself. It is almost as though small capacitors joined these entities to the power lines, as shown by the crude model in Figure 6.10.

The current through the capacitance C_3 coupling the ungrounded side of the power line and the electrocardiograph itself flows to ground and does not cause interference. C_1 represents the capacitance between the power line and one of the leads. Current i_{d1} does not flow into the electrocardiograph because of its high input impedance, but rather through the skin–electrode impedances Z_1 and Z_G to ground. Similarly, i_{d2} flows through Z_2 and Z_G to ground. Body impedance, which is about 500 Ω, can be neglected when compared with the other impedances shown. The voltage amplified is that appearing between inputs A and B, $v_A - v_B$.

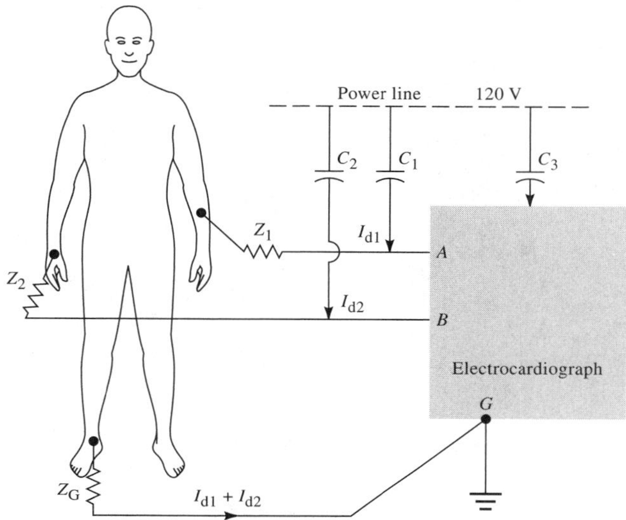

Figure 6.10 A mechanism of electric-field pickup of an electrocardiograph resulting from the power line. Coupling capacitance between the hot side of the power line and lead wires causes current to flow through skin–electrode impedances on its way to ground.

$$v_A - v_B = i_{d1}Z_1 - i_{d2}Z_2 \tag{6.3}$$

Huhta and Webster (1973) suggest that if the two leads run near each other, $i_{d1} \cong i_{d2}$. In this case,

$$v_A - v_B = i_{d1}(Z_1 - Z_2) \tag{6.4}$$

Values measured for 9-m cables show that $i_d \cong 6$ nA, although this value will be dependent on the room and the location of other equipment and power lines. Skin-electrode impedances may differ by as much as 20 kΩ. Hence

$$v_A - v_B = (6\ \text{nA})(20\ \text{k}\Omega) = 120\ \mu\text{V} \tag{6.5}$$

which would be an objectionable level of interference. This can be minimized by shielding the leads and grounding each shield at the electrocardiograph. This is done, in fact, in most modern electrocardiographs. Lowering skin–electrode impedances is also helpful.

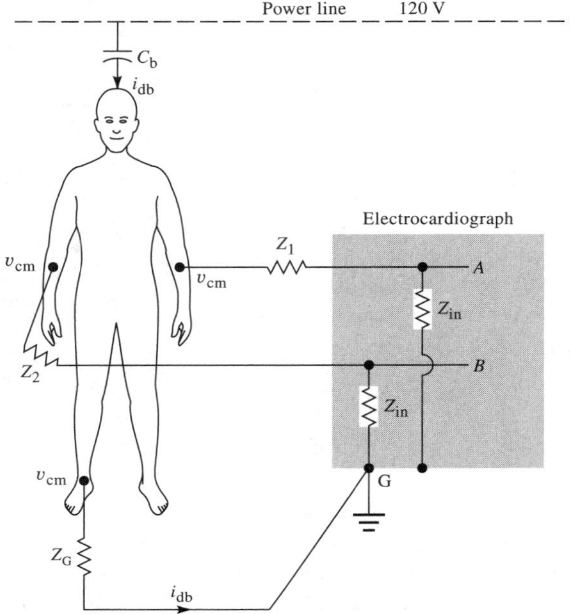

Figure 6.11 Current flows from the power line through the body and ground impedance, thus creating a common-mode voltage everywhere on the body. Z_{in} is not only resistive but, as a result of RF bypass capacitors at the amplifier input, has a reactive component as well.

Figure 6.11 shows that current also flows from the power line into the body. This displacement current i_{db} flows through the ground impedance Z_G to ground. The resulting voltage drop causes a common-mode voltage v_{cm} to appear throughout the body.

$$v_{cm} = i_{db} Z_G \tag{6.6}$$

Substituting typical values yields

$$v_{cm} = (0.2\ \mu\text{A})(50\ \text{k}\Omega) = 10\ \text{mV} \tag{6.7}$$

In poor electrical environments in which $i_{db} > 1\ \mu\text{A}$, v_{cm} can be greater than 50 mV. For a perfect amplifier, this would cause no problem, because a differential amplifier rejects common-mode voltages (Section 3.4). However, real amplifiers have finite input impedances Z_{in}. Thus v_{cm} is decreased because of the attenuator action of the skin–electrode impedances and Z_{in}. That is,

$$v_A - v_B = v_{cm}\left(\frac{Z_{in}}{Z_{in} + Z_1} - \frac{Z_{in}}{Z_{in} + Z_2}\right) \quad (6.8)$$

Because Z_1 and Z_2 are much less than Z_{in},

$$v_A - v_B = v_{cm}\left(\frac{Z_2 - Z_1}{Z_{in}}\right) \quad (6.9)$$

Substituting typical values yields

$$v_A - v_B = (10\,\text{mV})(20\,\text{k}\Omega/5\,\text{M}\Omega) = 40\,\mu\text{V} \quad (6.10)$$

which would be noticeable on an ECG and would be very objectionable on an EEG. This can be minimized by lowering skin–electrode impedance and raising amplifier input impedance.

Thus we see that the difference between the skin–electrode impedances is an important consideration in the design of biopotential amplifiers. Some common-mode voltage is always present, so the input imbalance and Z_{in} are critical factors determining the common-mode rejection, no matter how good the differential amplifier itself is.

The other source of interference from power lines is magnetic induction. Current in power lines establishes a *magnetic field* in the vicinity of the line. Magnetic fields can also sometimes originate from transformers and ballasts in fluorescent lights. If such magnetic fields pass through the effective single-turn coil produced by the electrocardiograph, lead wires, and the patient, as shown in Figure 6.12, a voltage is induced in this loop. This voltage is proportional to the magnetic-field strength and the area of the effective single-turn coil. It can be reduced (1) by reducing the magnetic field through the use of shielding, (2) by keeping the electrocardiograph and leads away from potential magnetic-field regions (both of which are rather difficult to achieve in practice), or (3) by reducing the effective area of the single-turn coil. This last approach can be achieved easily by twisting the lead wires together over as much as possible of the distance between the electrocardiograph and the patient.

OTHER SOURCES OF ELECTRIC INTERFERENCE

Electric interference from sources other than the power lines can also affect the electrocardiograph. *Electromagnetic interference* from nearby high-power radio, television, or radar facilities can be picked up and rectified by the *p–n* junctions of the transistors in the electrocardiograph and sometimes even by the electrode–electrolyte interface on the patient. The lead wires and the patient serve as an antenna. Once the signal is detected, the demodulated signal appears as interference on the electrocardiogram (Anonymous, 1979).

Electromagnetic interference can also be generated by high-frequency generators in the hospital itself. Electrosurgical and diathermy (Section 13.9)

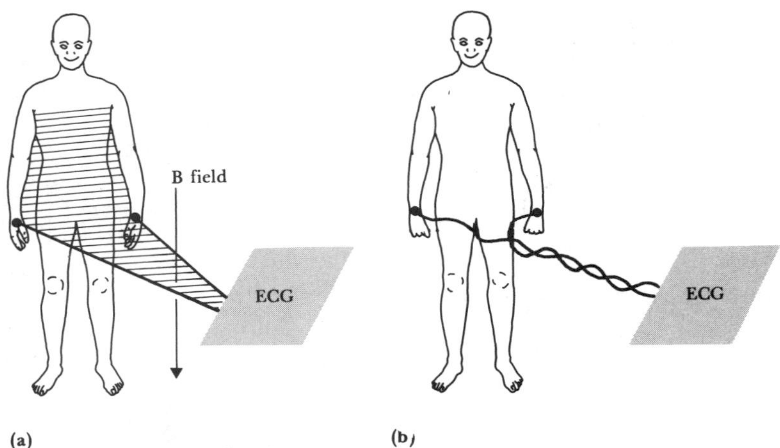

Figure 6.12 Magnetic-field pickup by the electrocardiograph (a) Lead wires for lead I make a closed loop (shaded area) when patient and electrocardiograph are considered in the circuit. The change in magnetic field passing through this area induces a current in the loop. (b) This effect can be minimized by twisting the lead wires together and keeping them close to the body in order to subtend a much smaller area.

equipment is a frequent offender. Grobstein and Gatzke (1977) show both the proper use of electrosurgical equipment and the design of an ECG amplifier required to minimize interference. Electromagnetic radiation can be generated from x-ray machines or switches and relays on heavy-duty electric equipment in the hospital as well. Even arcing in a fluorescent light that is flickering and in need of replacement can produce serious interference.

Electromagnetic interference can usually be minimized by shunting the input terminals to the electrocardiograph amplifier with a small capacitor of approximately 200 pF. The reactance of this capacitor is quite high over the frequency range of the ECG, so it does not appreciably lower the input impedance of the electrocardiograph. However, with today's modern high-input-impedance machines, it is important to make sure that this is really the case. At radio frequencies, its reactance is low enough to cause effective shorting of the electromagnetic interference picked up by the lead wires and to keep it from reaching the transistors in the amplifier.

There is also a source of electric interference located within the body itself that can have an effect on ECGs. There is always muscle located between the electrodes making up a lead of the electrocardiograph. Any time this muscle is contracting, it generates its own electromyographic signal that can be picked up by the lead along with the ECG and can result in interference on the ECG, as shown in Figure 6.9(b). When we look only at the ECG and not at the patient, it is sometimes difficult to determine whether interference

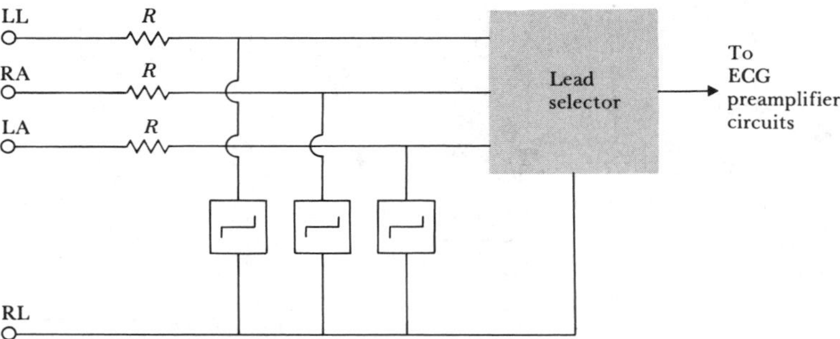

Figure 6.13 A voltage-protection scheme at the input of an electrocardiograph to protect the machine from high-voltage transients. Circuit elements connected across limb leads on left-hand side are voltage-limiting devices.

of this type is muscle interference or the result of electromagnetic radiation. However, while the ECG is being taken, we can easily separate the two sources, because the EMG interference is associated with the patient's muscle contractions.

6.4 TRANSIENT PROTECTION

The isolation circuits described in Section 14.9 are primarily for the protection of the patient in that they eliminate the hazard of electric shock resulting from interaction among the patient, the electrocardiograph, and other electric devices in the patient's environment. There are also times when other equipment attached to the patient can present a risk to the machine. For example, in the operating suite, patients undergoing surgery usually have their ECGs continuously monitored during the procedure. If the surgical procedure involves the use of an electrosurgical unit (Section 13.9), it can introduce onto the patient relatively high voltages that can enter the electrocardiograph or cardiac monitor through the patient's electrodes. If the ground connection to the electrosurgical unit is faulty or if higher-than-normal resistance is present, the patient's voltage with respect to ground can become quite high during coagulation or cutting. These high potentials enter the electrocardiograph or cardiac monitor and can be large enough to damage the electronic circuitry. They can also cause severe transients, of the type shown in Figure 6.8.

Ideally, cardiac monitors and electrocardiographs should be designed so that they are unaffected by such transients. Unfortunately, this cannot be achieved completely. However, it is possible to reduce the effects of these electric transients and to protect the equipment from serious damage. Figure 6.13 shows the basic arrangement of such protective circuits. Two-terminal voltage-limiting devices are connected between each patient electrode and electric ground.

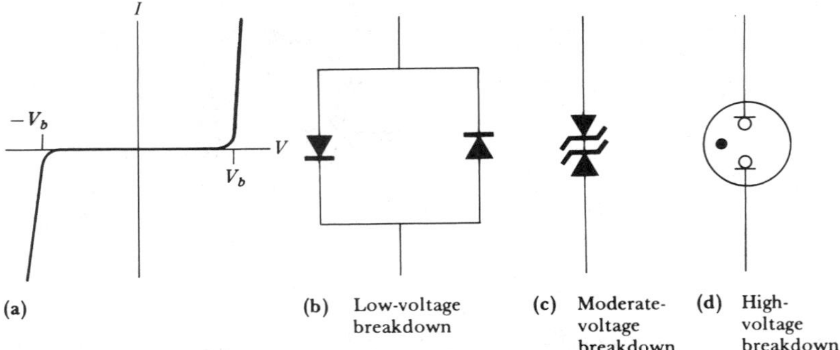

Figure 6.14 Voltage-limiting devices (a) Current-voltage characteristics of a voltage-limiting device. (b) Parallel silicon-diode voltage-limiting circuit. (c) Back-to-back silicon Zener-diode voltage-limiting circuit. (d) Gas-discharge tube (neon light) voltge-limiting circuit element.

Figure 6.14(a) shows the typical current–voltage characteristic of such a device. At voltages less than V_b, the breakdown voltage, the device allows very little current to flow and ideally appears as an open circuit. Once the voltage across the device attempts to exceed V_b, the characteristics of the device sharply change, and current passes through the device to such an extent that the voltage cannot exceed V_b as a result of the voltage drop across the series resistors R (in Figure 6.13). Under these conditions, the device appears to behave as a short circuit in series with a constant-voltage source of magnitude V_b.

In practice, there are several wys to achieve a characteristic approaching this idealized characteristic. Figure 6.14 indicates three of these. *Parallel silicon diodes,* as shown in Figure 6.14(b), give a characteristic with a breakdown voltage of approximately 600 mV. The diodes are connected such that the terminal voltage on one has a polarity opposite that on the other. Thus, when the voltage reaches approximately 600 mV, one of the diodes is forward-biased. And even though the other is reverse-biased, its bias voltage is limited to the forward voltage drop. When the voltage across the network is reversed, the roles of the two diodes are reversed, again limiting the voltage across the network to approximately 600 mV. The transition from nonconducting state to conducting state, however, is not so sharp as shown in the characteristic curve, and signal distortion can begin to appear from these diodes at voltages of approximately 300 mV. Although the ECG itself does not approach such a voltage, it is possible under extreme conditions for dc-offset potentials of that order of magnitude to result from faulty electrodes. The main advantage of this circuit is its low breakdown voltage; the maximal transients at the amplifier input are only approximately 600 mV peak amplitude.

Because the breakdown voltage of this circuit is too small, it is usually increased simply by connecting two or three diodes in series instead of using single diodes in each branch. This has the advantage of not only increasing the breakdown voltage by multiplying the initial 600 mV by the number of diodes in series, but also increasing the resistance of the circuit, both in the conducting and the nonconducting state.

When we want higher breakdown voltages, we can use the circuit of Figure 6.14(c). This circuit consists of two silicon diodes, usually *Zener diodes,* connected back to back. When a voltage is connected across this circuit, one of the diodes is biased in the forward direction and the other in the reverse direction. The breakdown voltage in the forward direction is approximately 600 mV, but that in the reverse direction is much higher. It generally covers the range of 2 to 20 V. Thus this circuit does not conduct until its terminal voltage exceeds the reverse breakdown of the diode by approximately 600 mV. Again, when the polarity of the circuit terminal voltage is reversed, the roles of the two diodes are interchanged.

A device that gives an even higher breakdown voltage is the *gas-discharge tube* illustrated in Figure 6.14(d). This device appears as an open circuit until it reaches its breakdown voltage. It then switches to the conducting state and maintains a voltage that is usually several volts less than the breakdown voltage. Breakdown voltages ranging from 50 to 90 V are typical for this device. This breakdown voltage is considered high for the input to most electrocardiographic amplifiers. Thus it is important to include a circuit element such as a resistor between the gas-discharge tube and the amplifier input to limit the amplifier's input current.

Designers of biopotential amplifiers often use miniature neon lamps as voltage limiters. They are essentially gas discharge tubes and are very inexpensive and have a symmetric characteristic, requiring only a single device per electrode pair. Their resistance in the nonconducting state is nearly infinite, so there is no loading effect on the electrodes—a feature that is most desirable when the biopotential amplifier has very high input impedance.

6.5 COMMON-MODE AND OTHER INTERFERENCE-REDUCTION CIRCUITS

As we noted earlier, common-mode voltages can be responsible for much of the interference in biopotential amplifiers. Although having an amplifier with a high common-mode-rejection ratio minimizes the effects of common-mode voltages, a better approach to this problem is to discover the source of the voltage and try to eliminate it. In this section, we shall look at some of the sources of this and other types of interference to discover ways in which they can be minimized.

ELECTRIC- AND MAGNETIC-FIELD PICKUP

As we saw in Section 6.3, electric interference can be introduced in systems of biopotential measurement through capacitive coupling and magnetic induction. We can minimize these interfering signals by trying to eliminate the sources of the signals via shielding techniques. Electrostatic shielding is accomplished by placing a grounded conducting lane between the source of the electric field and the measurement system. The measurement of very-low-level biopotentials, such as the EEG, has traditionally been carried out in a shielded enclosure containing either continuous solid-metal panels or at least grounded copper screening to minimize interference.

This type of shielding is ineffective for magnetic fields unless the metal panels have a high permeability (such as sheet steel). In other words, the panels must be good magnetic conductors as well as good electric conductors. Such rooms are available to provide magnetic shielding, but a much less expensive way of achieving a reduction of magnetically induced signals is to reduce the effective surface area between the differential inputs to the biopotential amplifier, in the case of differential signals, and between the inputs and ground, in the case of common-mode signals. Something as simple as a twisted pair of lead wires, as illustrated in Figure 6.12(b), may greatly improve the situation.

DRIVEN-RIGHT-LEG SYSTEM

In many modern electrocardiographic systems, the patient is not grounded at all. Instead, the right-leg electrode is connected (as shown in Figure 6.15) to the output of an auxiliary op amp. The common-mode voltage on the body is sensed by the two averaging resistors R_a, inverted, amplified, and fed back to the right leg. This negative feedback drives the common-mode voltage to a low value. The body's displacement current flows not to ground but rather to the op-amp output circuit. This reduces the pickup as far as the ECG amplifier is concerned and effectively grounds the patient.

The circuit can also provide some electric safety. If an abnormally high voltage should appear between the patient and ground as a result of electric leakage or other cause, the auxiliary op amp in Figure 6.15 saturates. This effectively ungrounds the patient, because the amplifier can no longer drive the right leg. Now the parallel resistances R_f and R_o are between the patient and ground. They can be several megohms in value—large enough to limit the current. These resistances do not protect the patient, however, because 120 V on the patient would break down the op-amp transistors of the ECG amplifier, and large currents would flow to ground.

EXAMPLE 6.1 Determine the common-mode voltage v_{cm} on the patient in the driven-right-leg circuit of Figure 6.15 when a displacement current i_d flows to the patient from the power lines. Choose appropriate values for the resistances in the circuit so that the common-mode voltage is minimal and there is only a high-resistance path to ground when the auxiliary operational amplifier saturates. What is v_{cm} for this circuit when $i_d = 0.2\ \mu A$?

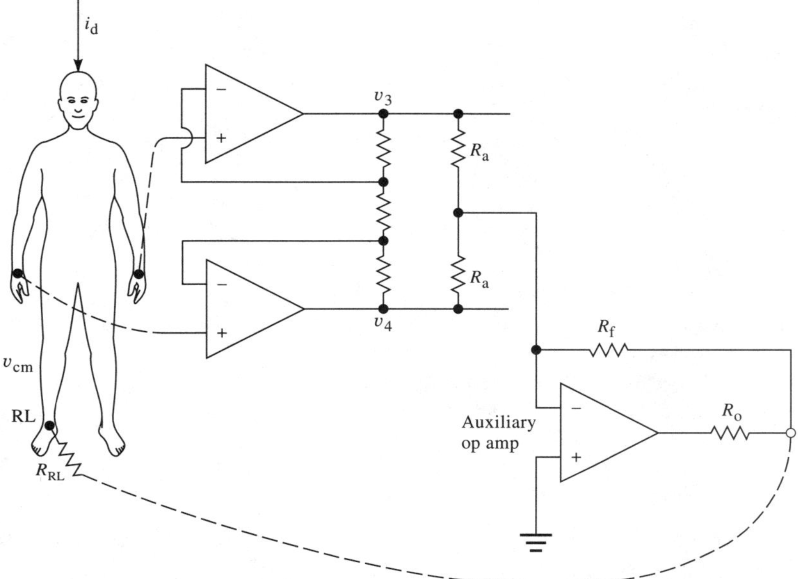

Figure 6.15 Driven-right-leg circuit for minimizing common-mode interference The circuit derives common-mode voltage from a pair of averaging resistors connected to v_3 and v_4 in Figure 3.5. The right leg is not grounded but is connected to output of the auxiliary op amp.

ANSWER The equivalent circuit for the circuit of Figure 6.15 is shown in Figure E6.1. Note that because the common-mode gain of the input stage is 1 (Section 3.4), and because the input stage as shown has a very high input impedance, v_{cm} at the input is isolated from the output circuit. R_{RL} represents the resistance of the right-leg electrode. Summing the currents at the negative input of the operational amplifier, we get

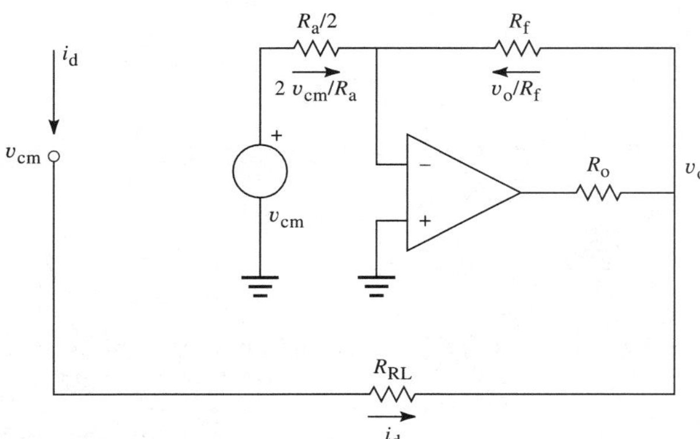

Figure E6.1 Equivalent circuit of driven-right-leg system of Figure 6.19.

$$\frac{2v_{\rm cm}}{R_{\rm a}} + \frac{v_{\rm o}}{R_{\rm f}} = 0 \qquad \text{(E6.1)}$$

This gives

$$v_{\rm o} = -\frac{2R_{\rm f}}{R_{\rm a}} v_{\rm cm} \qquad \text{(E6.2)}$$

but

$$v_{\rm cm} = R_{\rm RL} i_{\rm d} + v_{\rm o} \qquad \text{(E6.3)}$$

Thus, substituting (E6.2) into (E6.3) yields

$$v_{\rm cm} = \frac{R_{\rm RL} i_{\rm d}}{1 + 2R_{\rm f}/R_{\rm a}} \qquad \text{(E6.4)}$$

The effective resistance between the right leg and ground is the resistance of the right-leg electrode divided by 1 plus the gain of the auxiliary operational-amplifier circuit. When the amplifier saturates, as would occur during a large transient $v_{\rm cm}$, its output appears as the saturation voltage $v_{\rm s}$. The right leg is now connected to ground through this source and the parallel resistances $R_{\rm f}$ and $R_{\rm o}$. To limit the current, $R_{\rm f}$ and $R_{\rm o}$ should be large. Values as high as 5 MΩ are used.

When the amplifier is not saturated, we would like $v_{\rm cm}$ to be as small as possible or, in other words, to be an effective low-resistance path to ground. This can be achieved by making $R_{\rm f}$ large and $R_{\rm a}$ relatively small. $R_{\rm f}$ can be equal to $R_{\rm o}$, but $R_{\rm a}$ can be much smaller.

A typical value of $R_{\rm a}$ would be 25 kΩ. A worst-case electrode resistance $R_{\rm RL}$ would be 100 kΩ. The effective resistance between the right leg and ground would then be

$$\frac{100\ \text{k}\Omega}{1 + \dfrac{2 \times 5\ \text{M}\Omega}{25\ \text{k}\Omega}} = 249\ \Omega$$

For the 0.2-μA displacement current, the common-mode voltage is

$$v_{\rm cm} = 249\ \Omega \times 0.2\ \mu\text{A} = 50\ \mu\text{V}$$

6.6 AMPLIFIERS FOR OTHER BIOPOTENTIAL SIGNALS

Up to this point we have stressed biopotential amplifiers for the ECG. Amplifiers for use with other biopotentials are essentially the same. However, other

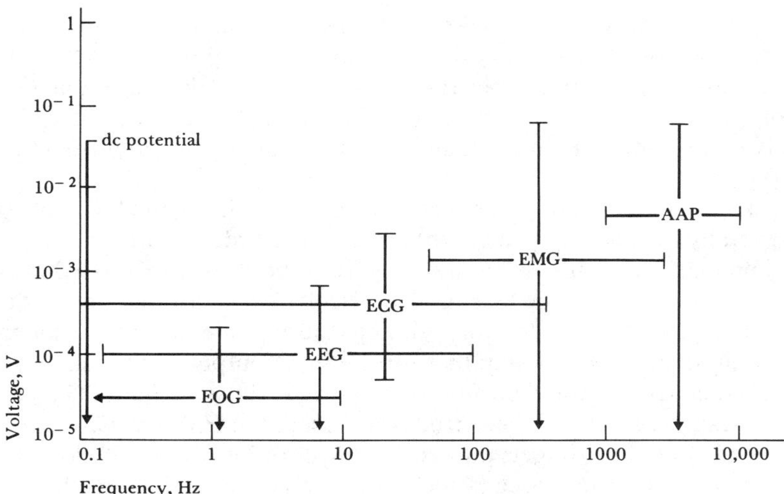

Figure 6.16 Voltage and frequency ranges of some common biopotential signals; dc potentials include intracellular voltages as well as voltages measured from several points on the body. EOG is the electrooculogram, EEG is the electroencephalogram, ECG is the electrocardiogram, EMG is the electromyogram, and AAP is the axon action potential. (From J. M. R. Delgado, "Electrodes for Extracellular Recording and Stimulation," in *Physical Techniques in Biological Research,* edited by W. L. Nastuk, New York: Academic Press, 1964.

signals do put different constraints on some aspects of the amplifier. The frequency content of different biopotentials covers different portions of the spectrum. Some biopotentials have higher amplitudes than others. Both these facts place gain and frequency-response constraints on the amplifiers used. Figure 6.16 shows the ranges of amplitudes and frequencies covered by several of the common biopotential signals. Depending on the signal, frequencies range from dc to about 10 kHz. Amplitudes can range from tens of microvolts to approximately 100 mV. The amplifier for a particular biopotential must be designed to handle that potential and to provide an appropriate signal at its output.

The electrodes used to obtain the biopotential place certain constraints on the amplifier input stage. To achieve the most effective signal transfer, the amplifier must be matched to the electrodes. Also, the amplifier input circuit must not promote the generation of artifact by the electrode, as could occur with excessive bias current. Let us look at a few requirements placed on different types of biopotential amplifiers by the measurement being made.

EMG AMPLIFIER

Figure 6.16 shows that electromyographic signals range in frequency from 25 Hz to several kilohertz. Signal amplitudes range from 100 μV to 90 mV, depending on the type of signal and electrodes used. Thus EMG amplifiers must have a wider frequency response than ECG amplifiers, but they do not have to cover so low a frequency range as the ECGs. This is desirable because motion artifact contains mostly low frequencies that can be filtered more effectively in EMG amplifiers than in ECG amplifiers without affecting the signal.

If skin-surface electrodes are used to detect the EMG, the levels of signals are generally low, having peak amplitudes of the order of 0.1 to 1 mV. Electrode impedance is relatively low, ranging from about 200 to 5000 Ω, depending on the type of electrode, the electrode–electrolyte interface, and the frequency at which the impedance is determined. Thus the amplifier must have somewhat higher gain than the ECG amplifier for the same output-signal range, and its input characteristics should be almost the same as those of the ECG amplifier. When intramuscular needle electrodes are used, the EMG signals can be an order of magnitude stronger, thus requiring an order of magnitude less gain. Furthermore, the surface area of the EMG needle electrode is much less than that of the surface electrode, so its source impedance is higher. Therefore, a higher amplifier input impedance is desirable for quality signal reproduction.

AMPLIFIERS FOR USE WITH GLASS MICROPIPET INTRACELLULAR ELECTRODES

Intracellular electrodes or microelectrodes that can measure the potential across the cell membrane generally detect potentials on the order of 50 to 100 mV. Their small size and small effective surface-contact area give them a very high source impedance, and their geometry results in a relatively large shunting capacitance. These features place on the amplifier the constraint of requiring an extremely high input impedance. Furthermore, the high shunting capacitance of the electrode itself affects the frequency-response characteristics of the system. Often positive-feedback schemes are used in the biopotential amplifier to provide an effective negative capacitance that can compensate for the high shunt capacitance of the source.

The frequency response of microelectrode amplifiers must be quite wide. Intracellular electrodes are often used to measure the dc potential difference across a cell membrane, so the amplifier must be capable of responding to dc signals. When excitable cell-membrane potentials are to be measured, such as in muscle cells and nerve cells, rise times can contain frequencies of the order of 10 kHz, and the amplifiers must be capable of passing these, too. The fact that the potentials are relatively high means that the voltage gain of the amplifier does not have to be so high as in previous examples.

A preamplifier circuit that is especially useful with microelectrodes is the negative-input-capacitance amplifier shown in Figure 6.17. The basic circuit consists of a low-gain, very-high-input-impedance, noninverting amplifier with a capacitor C_f providing positive feedback to the input. If we look at the equivalent circuit for this amplifier [Figure 6.17(b)], we can relate the input voltage and current:

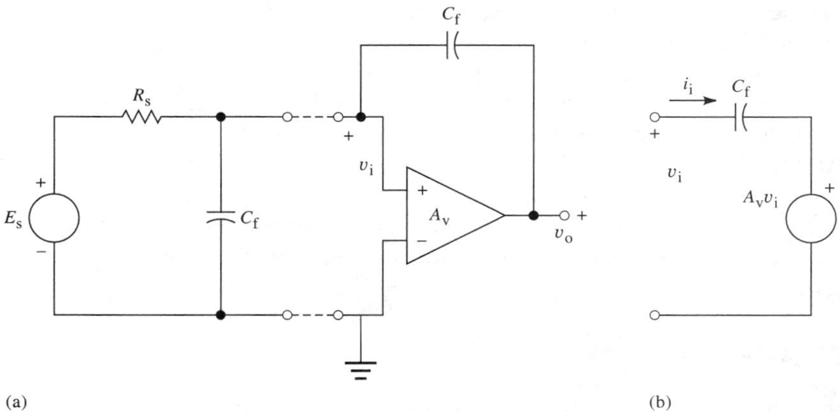

Figure 6.17 (a) Basic arrangement for negative-input capacitance amplifier. Basic amplifier is on the right-hand side; equivalent source with lumped series resistance R_s and shunt capacitance C_s is on the left. (b) Equivalent circuit of basic negative-input capacitance amplifier.

$$v_i = \frac{1}{C_f}\int i_i\,dt + A_v v_i \tag{6.11}$$

where A_v is the amplifier gain, provided the operational amplifier itself draws no current. This equation can be rearranged as follows:

$$v_i = \frac{1}{(1 - A_v)C_f}\int i_i\,dt \tag{6.12}$$

Thus the equivalent capacitance at the amplifier input is $(1 - A_v)C_f$. If A_v is greater than unity, this equivalent capacitance is negative. The amplifier is connected to the microelectrode, with its high source resistance R_s. The shunt capacitance from the electrode and cable is C_s. The total circuit capacitance is

$$C = C_s + (1 - A_v)C_f \tag{6.13}$$

which is zero when

$$C_s = (A_v - 1)C_f \tag{6.14}$$

This condition can be met by adjusting either the amplifier gain A_v or the feedback capacitance C_f.

In practical negative-input-capacitance amplifiers, the idealized condition of (6.14) cannot be met, because the gain of any amplifier has some frequency dependence. There are thus frequencies where the input capacitance of the amplifier does not cancel the source capacitance, and the circuit does not have an ideal transient response. The amplifier employs positive feedback and does not have an ideal frequency response, so it is possible for the conditions of oscillation to be met at some frequency, and the amplifier will then become unstable. Thus it is important that the amplifier be carefully adjusted to meet the condition if (6.14) as closely as possible without becoming unstable. Another consequence of the positive feedback is that the amplifier tends to be noisy. This is not a serious problem, however, because the voltages from microelectrodes are usually relatively high.

EEG AMPLIFIERS

Figure 6.16 shows that the EEG requires an amplifier with a frequency response of from 0.1 to 100 Hz. When surface electrodes are used, as in clinical electroencephalography, amplitudes of signals range from 25 to 100 μV. Thus amplifiers with relatively high gain are required. These electrodes are smaller than those used for the ECG, so they have somewhat higher source impedances, and a high input impedance is essential in the EEG amplifier. Because the signal levels are so small, common-mode voltages can have more serious effects. Therefore more stringent efforts must be made to reduce common-mode interference, as well as to use amplifiers with higher common-mode-rejection ratios and low noise.

6.7 EXAMPLE OF A BIOPOTENTIAL PREAMPLIFIER

As we have seen, biopotential amplifiers can be used for a variety of signals. The gain and frequency response are two important variables that relate the amplifier to the particular signal. An important factor common to all amplifiers is the first stage, or preamplifier. This stage must have low noise, because its output must be amplified through the remaining stages of the amplifier, and any noise is amplified along with the signal. It must also be coupled directly to the electrodes (no series capacitors) to provide optimal low-frequency response as well as to minimize charging effects on coupling capacitors from input bias current. Of course, every attempt should be made to minimize this current. Even without coupling capacitors it can polarize the electrodes, resulting in polarization overpotentials that produce a large dc offset voltage at the amplifiers' input. This is why preamplifiers often have relatively low voltage gains. The offset potential is coupled directly to the input, so it could saturate high-gain preamplifiers, cutting out the signal altogether. To eliminate the saturating effects of this dc potential, the preamplifier can be capacitor-coupled to the remaining amplifier stages. A final consideration is that the preamplifier must have a very high input impedance, because it represents the load on the electrodes (Thakor, 1988a).

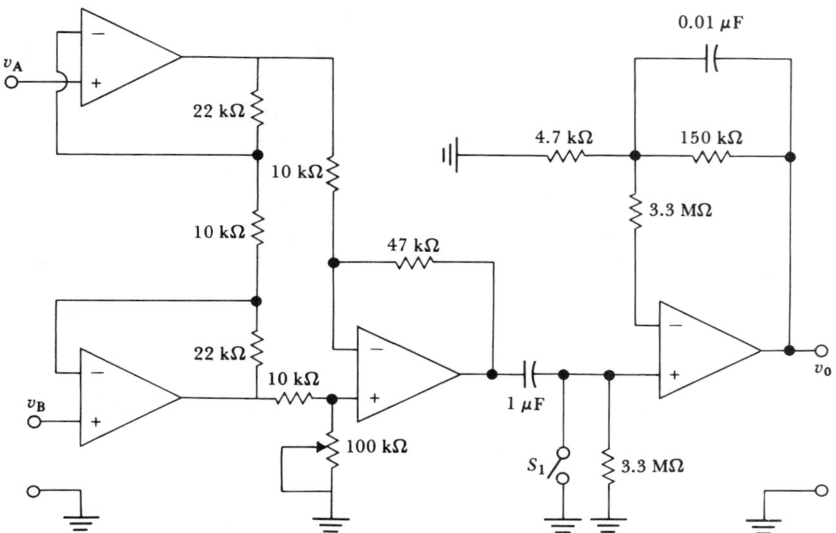

Figure 6.18 This ECG amplifier has a gain of 25 in the dc-coupled stages. The high-pass filter feeds a noninverting-amplifier stage that has a gain of 32. The total gain is 25 × 32 = 800. When μA 776 op amps were used, the circuit was found to have a CMRR of 86 dB at 100 Hz and a noise level of 40 mV peak to peak at the output. The frequency response was 0.04–150 Hz for ±3 dB and was flat over 4–40 Hz.

Often, for safety reasons, the preamplifier either is electrically isolated from the remaining amplifier stages (and hence from the power lines) (Section 14.9) or is located near the signal source to minimize interference pickup on the high-impedance lead wires. In the latter case, we can use for the circuit a battery-powered preamplifier with low power consumption or a power supply that is electrically isolated.

Figure 6.18 shows the circuit of an ECG amplifier. The instrumentation amplifier of Figure 3.5 is used to provide very high input impedance. High common-mode rejection is achieved by adjusting the potentiometer to about 47 kΩ. Electrodes may produce an offset potential of up to 0.3 V. Thus, to prevent saturation, the dc-coupled stages have a gain of only 25. Coupling capacitors are not placed at the input because this would block the op-amp bias current. Adding resistors to supply the bias current would lower the Z_{in}. Coupling capacitors placed after the first op amps would have to be impractically large. Therefore, the single 1-μF coupling capacitor and the 3.3-MΩ resistor form a high-pass filter. The resulting 3.3-s time constant passes all frequencies above 0.05 Hz. The output stage is a noninverting amplifier that has a gain of 32 (Section 3.3).

A second 3.3-MΩ resistor is added to balance bias-current source impedances. The 150-kΩ and 0.01-μF low-pass filter attenuates frequencies above

100 Hz. Switch S_1 may be momentarily closed to decrease the discharge time constant when the output saturates. This is required after defibrillation or lead switching to charge the 1-μF capacitor rapidly to the new value and return the output to the linear region. We do *not* discharge the capacitor voltage to zero. Rather, we want the right end to be at 0 V when the left end is at the dc voltage determined by the electrode offset voltage. Switch closure may be automatic, via a circuit that detects when the output is in saturation, or it may be manual. Although the 741 op amp is satisfactory in this circuit, an op amp such as the 411, which has lower bias current, may be preferred.

6.8 OTHER BIOPOTENTIAL SIGNAL PROCESSORS

CARDIOTACHOMETERS

A cardiotachometer is a device for determining heart rate. The signal most frequently used is the ECG. However, circuitry for deriving heart rate from signals such as the arterial pressure waveform or heart sounds has also been developed.

There are two basic kinds of cardiotachometers. The averaging cardiotachometer determines average heart rate by counting the pulse over a known period of time. The beat-to-beat cardiotachometer, on the other hand, determines the reciprocal of the time interval between heartbeats for each beat and presents it as the heart rate for that particular interval. Any slight variability in the interval between beats shows up as a variation in the instantaneous heart rate determined by this method.

Figure 6.19 shows a beat-to-beat cardiotachometer. The ECG initially passes through a bandpass filter, which passes QRS complexes while reducing artifact and most of the P and T waves. The threshold detector triggers the first 10-μs monostable multivibrator, which produces pulse P_1, as shown in the timing diagram of Figure 6.20.

The falling edge of pulse P_1 triggers a second monostable multivibrator that also produces a 10-μs duration pulse P_2. This pulse occurs 10 μs after the initiation of pulse P_1, as shown in the timing diagram. These two pulses control a NOR (not OR) circuit, which has output P_3. This signal is high during the interval when P_1 and P_2 are not occurring. Immediately following the initiation of a QRS complex, pulse P_3 goes low for a total of 20 μs and is then returned to its initial level. The signal controls an AND gate so that a 1-kHz clock signal is allowed to enter a counting register whenever P_3 is high. Because P_3 is high during the interval between QRS complexes, the 1-ms pulses coming from the clock (P_4) accumulate in register 1 during this period. If the register is initially at zero, the number of pulses in the register by the time the next QRS complex arrives equals the number of milliseconds in the interval between this QRS complex and the previous one. Once the gate prohibits additional clock pulses from entering register 1, pulse P_1 enables

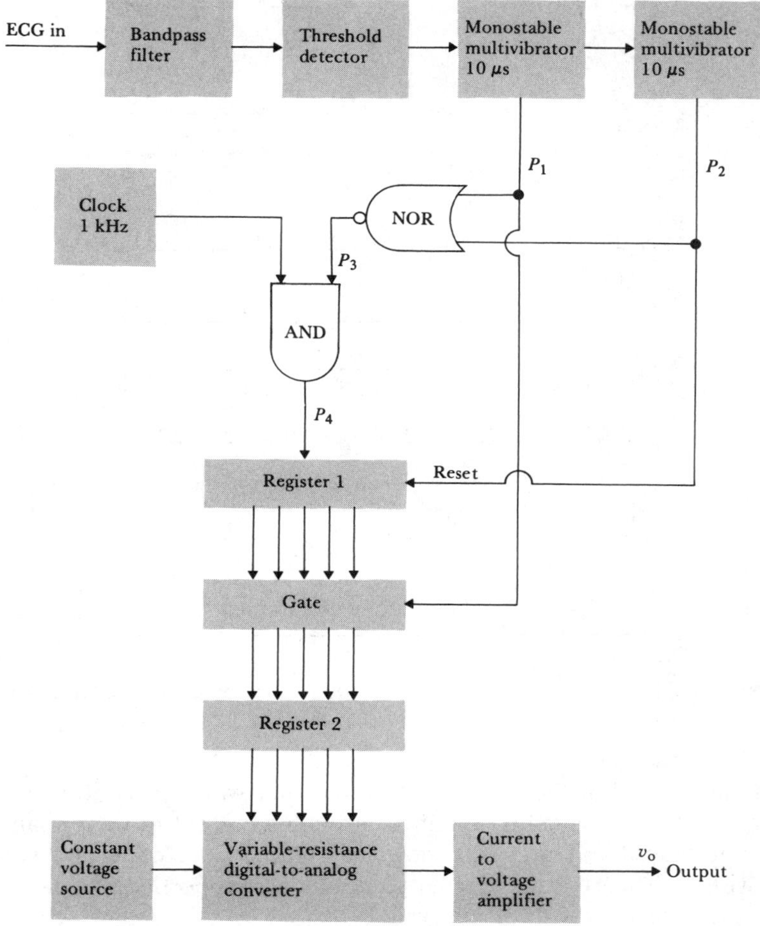

Figure 6.19 Block diagram of a beat-to-beat instantaneous cardiotachometer.

the signal in this register to be stored in a second register, which serves as a memory. The second register is connected to a digital-to-analog converter of the type that produces a resistance proportional to the digital signal at its input. The resistance is proportional to the time interval between QRS complexes. This resistance is connected across a constant-voltage source, yielding a current given by

$$i = \frac{v}{R} = \frac{k}{T_R} \tag{6.15}$$

where k is a constant and T_R is the interval between QRS complexes. We see that the current in the circuit is proportional to the reciprocal of the beat-to-

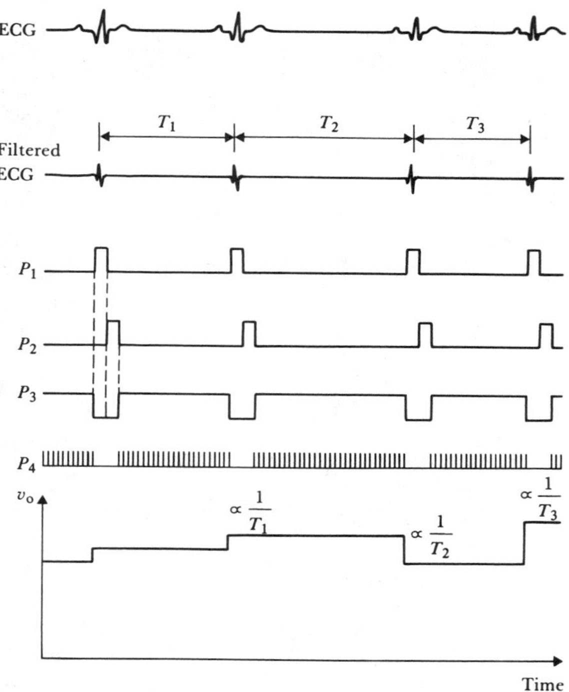

Figure 6.20 Timing diagram for beat-to-beat cardiotachometer in Figure 6.19.

beat time interval of the original ECG; in other words, it is proportional to the heart rate. The signal is amplified through a current-to-voltage amplifier, giving the output voltage v_o, as shown in Figure 6.20. Note that this voltage shifts with each heart beat and that its amplitude is proportional to the duration of the previous beat-to-beat interval.

It is important to note that the contents of register 1 must be zero before the gate is opened to let the clock pulses enter that register. It is for this reason that, once the contents of register 1 have been transferred to register 2, pulse P_2 enters the reset terminal of register 1, sets it to zero, and prepares it to count the next beat-to-beat interval.

Alarm circuits can also be used with this type of cardiotachometer. Often these circuits consist of digital rather than analog comparators, which compare the signal in register 1 to determine whether an interval of longer than a preset value has occurred (this could happen if the heart rate were too low). An additional comparator can monitor the signal in register 2 to determine whether it is less than a preset value, a situation that would occur if the heart rate were too high. In either case, the comparators can then be used to activate appropriate alarms.

Ludwig (1977) presents the design of a beat-to-beat cardiotachometer. Taylor (1975) gives a design for a highly accurate digital cardiotachometer

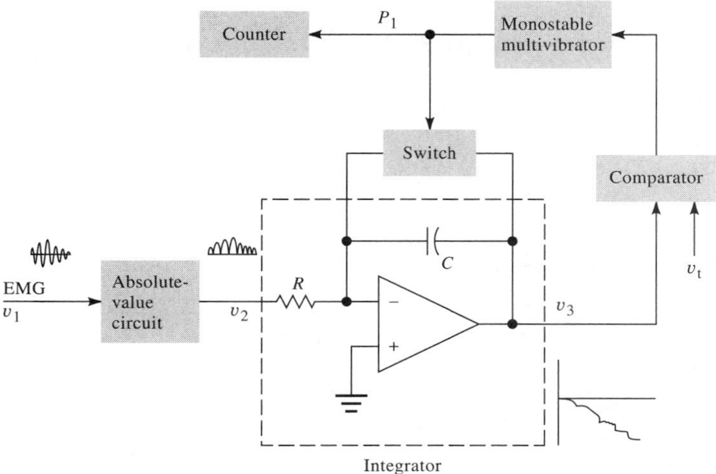

Figure 6.21 **Block diagram of an integrator for EMG signals**

with digital readout. Hartley (1976) presents a method of getting an analog output from a digital rate-determining circuit. Fichtenbaum (1976) shows a simplified design based on a divide-by-N counter.

ELECTROMYOGRAM INTEGRATORS

It is frequently of interest to quantify the amount of EMG activity measured by a particular system of electrodes. Such quantification often assumes the form of taking the absolute value of the EMG and integrating it, as illustrated by the block diagram of the system shown in Figure 6.21.

The raw EMG, amplified appropriately, is fed to an absolute-value circuit or full-wave rectifier (Section 3.6). As indicated in the waveform of Figure 6.22, only positive-going signals (v_2) result following this block. The negative-going portions of the signal have been inverted, making them positive. The signal is then integrated in the operational-amplifier integrator (Section 3.8). The feedback capacitor is charged according to the integral of the incoming waveform. Once the integrator output has exceeded a preset threshold level, a comparator at the output of the integrator fires a monostable multivibrator. The pulse emanating from the multivibrator closes a switch that discharges the integrator's capacitor. The duration of the pulse from the multivibrator must be at least five times greater than the time constant of the capacitor and closed-switch circuit to ensure nearly complete discharge of the capacitor. The integrator then reinitiates integration of the EMG until the cycle repeats itself.

We can view the output from the integrator in two ways. The actual voltage output from the integrator can be recorded on a conventional strip-chart recorder or computer to give the actual integral at any instant. The total integral necessary to reset the integrator is known, so at any instant the integral

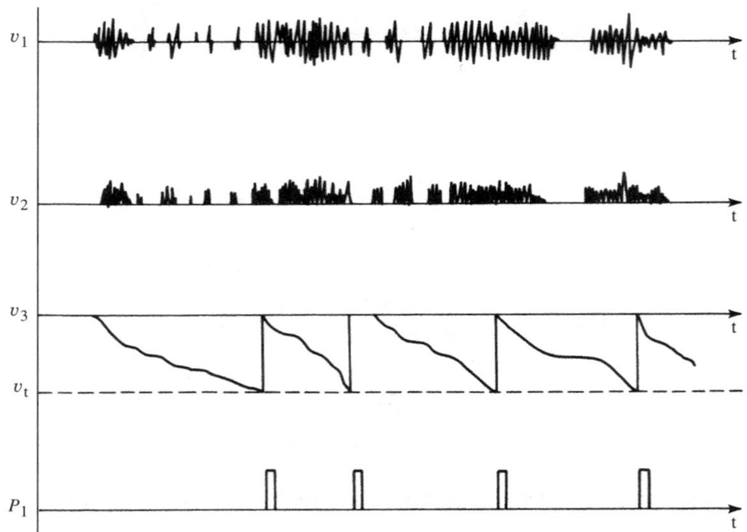

Figure 6.22 The various waveforms for the EMG integrator circuit shown in Figure 6.21.

equals the number of times the integrator has been reset, multiplied by this calibration constant, plus whatever is recorded as being in the integrator at that time. Another way to view the output of the integrator is to count the number of reset pulses, as shown in the block diagram, by passing the reset pulses into a counter. We then determine the approximate integral by determining the number of resets over a specific time interval and calculating total activity, as described earlier.

EVOKED POTENTIALS AND SIGNAL AVERAGERS

Often in neurophysiology we are interested in looking at the neurological response to a particular stimulus. This response is electric in nature, and it frequently represents a very weak signal with a very poor signal-to-noise ratio (SNR). When the stimulus is repeated, the same or a very similar response is repeatedly elicited. This is the basis for biopotential signal processors that can obtain an enhanced response by means of repeated application of the stimulus (Childers, 1988).

Figure 6.23 shows how signal averaging works. The response to each stimulus is recorded. The time at which each stimulus occurs is considered the reference time, and the values for each response at this reference time are summed to get the total response at the reference time. This process is repeated for the values of the responses sampled immediately after the reference time, and the sum is determined for this point in time after the stimulus. The process is then repeated for each sample point after the reference time so that a waveform that is the sum of the individual responses can be displayed,

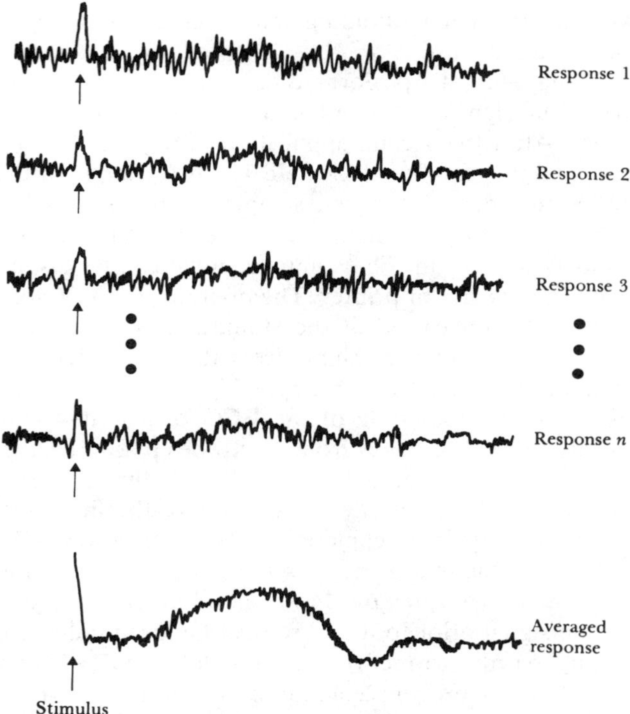

Figure 6.23 Signal-averaging technique for improving the SNR in signals that are repetitive or respond to a known stimulus.

as shown at the bottom of Figure 6.23. The only limits on the number of samples that can be summed are the available memory for storing the responses and the time required to collect the data. Practical signal averagers can process more than 1000 repeated responses.

The noise on the individual responses is random with respect to the stimulus. This means that if a large enough sample is taken, some positive-going noise pulses at a particular instant after the stimulus partially cancel some negative-going noise spikes at the same instant. Thus the net sum of the noise at any instant following the stimulus increases as $\sqrt{n}$, where n is the number of responses. The evoked response, on the other hand, follows the same time course after each stimulus. Thus there is no cancellation in this signal as the individual responses are summed. Instead, the amplitude of the evoked response increases in direct proportion to n. By repetitive summing, one is thus able to enhance the SNR by the factor $n/\sqrt{n} = \sqrt{n}$.

This technique is frequently used with the EEG and ERG. As stated earlier in this chapter, EEGs obtained from surface electrodes are very weak and consequently can have a high noise component. When a repetitive stimulus (such as electric shock, flashing light, or repeating sound) is applied to the test subject, it is difficult to ascertain the response in a directly recorded EEG.

However, if we apply this signal summing or averaging technique, it is possible to obtain the evoked response.

Signal averaging is usually performed on a computer. The basic scheme involves digitizing the signal and then locating the stimulus. The response is stored in memory. After the second application of the stimulus, the signal is digitized, stored, and the stimulus is located. The first sample of the response after the stimulus is added to the first sample of the response to the first stimulus, and the sum remains in memory. The second samples taken of each response are added, and so on. The summed signal can be displayed on an oscilloscope, a chart recorder or printer. The operator of the system can look at the sum after each application of the stimulus to determine how many stimuli are necessary to extract the signal from the noise adequately.

This technique can be used without applying the external stimulus. One example of its use is the recording of the ECG of a fetus. Although it is possible to record the fetal R waves from electrodes placed on the abdomen of the mother, artifacts generated by the ECG of the mother and other biopotentials, as well as by electrode noise, obscure the finer details of the fetal ECG. A signal-averaging technique similar to that we have described can be applied by using the fetal R wave in the same capacity as the stimulus. In this case the computer locates the R wave and averages several hundred milliseconds of the signal prior to it and several hundred milliseconds of the signal following it, in order to recover the complete P-QRS-T configuration of the fetal ECG. Such averaging techniques do not always work, however, because the various intervals of the fetal ECG, as well as the waveforms themselves, may change slightly from one beat to the next. The sum is an average of all the recorded ECG configurations and might provide a waveform that does not indicate the single-beat ECG of the fetal heart.

FETAL ELECTROCARDIOGRAPHY

As we have said, physicians can determine the ECG of a fetus from a pair of biopotential-sensing electrodes placed on the abdomen of the mother. Often it is necessary to try several different placements to get the best signal. Once the best placement is determined, we obtain a recording such as that shown in the top trace of Figure 6.24. For comparison, Figure 6.24 also shows a direct ECG of the same fetus and a direct ECG of the same mother. The fetal ECG signal is usually quite weak; it generally has an amplitude of around 50 μV or less. This makes it extremely difficult to record the heartbeat of the fetus by using electrodes attached to the abdomen of the mother during labor, when the mother is restless and motion artifact as well as EMG interfere. There is also considerable interference from the ECG of the mother (Neuman, 1988).

Note that the QRS complexes of the mother are much stronger than those of the fetus, which makes it difficult to determine the fetal heart rate electronically from recordings of this type. This information can be obtained by hand, however, by measuring the fetal R–R interval on the chart and converting it to heart rate.

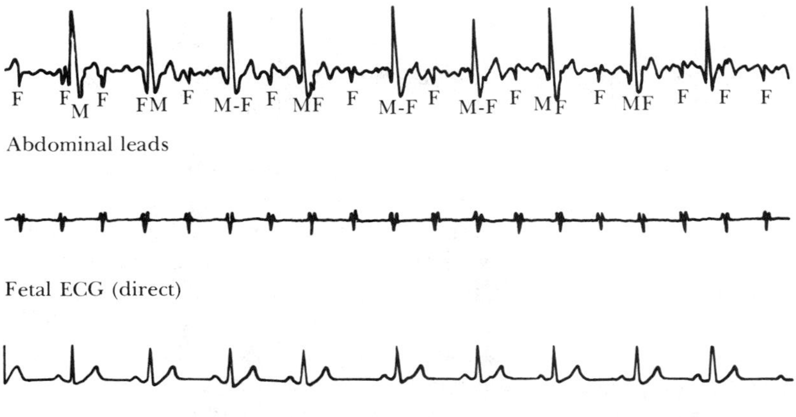

Figure 6.24 Typical fetal ECG obtained from maternal abdomen F represents fetal QRS complexes; M represents maternal QRS complexes. Maternal ECG and fetal ECG (recorded directly from the fetus) are included for comparison. (From "Monitoring of Intrapartum Phenomena," by J. F. Roux, M. R. Neuman, and R. C. Goodlin, in *CRC Critical Reviews in Bioengineering,* **2,** pp. 119–158, January 1975, © CRC Press. Used by permission of CRC Press, Inc.)

Several methods have been devised for improving the quality of fetal ECGs obtained by attaching electrodes to the mother's abdomen. In addition to the signal-averaging technique, physicians have applied various forms of anticoincidence detectors to eliminate the maternal QRS complexes (Offnet and Moisand, 1966; Walden and Birnbaum, 1964). This method, as shown in the block diagram of Figure 6.25, uses at least three electrodes: one on the

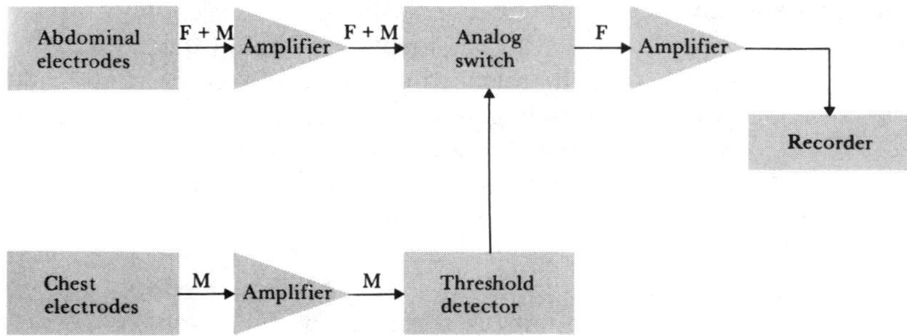

Figure 6.25 Block diagram of a scheme for isolating fetal ECG from an abdominal signal that contains both fetal and maternal ECGs. (From "Monitoring of Intrapartum Phenomena," by J. F. Roux, M. R. Neuman, and R. C. Goodlin, in *CRC Critical Reviews in Bioengineering,* **2,** pp. 119–158, January 1975, © CRC Press. Used by permission of CRC Press, Inc.)

mother's chest, one at the upper part or fundus of the uterus, and one over the lower part of the uterus. The ECG of the mother is obtained from the top two electrodes, and the fetal-plus-maternal signal is obtained from the bottom two. The center electrode is common to both. A threshold detector determines the mother's QRS complexes and uses this information to turn off an analog switch between the electrodes recording the fetal ECG and the recording apparatus. Therefore, whenever a maternal QRS complex is detected, the signal from the abdominal leads is temporarily blocked until the end of the QRS complex, thereby eliminating it from the abdominal recording. Note that this technique also eliminates any fetal QRS complexes that occur simultaneously with the maternal ones. Modern systems incorporate computing circuits to recognize the absence of this fetal signal and to compensate for it when determining the fetal heart rate. One must always be cautious in using such a system since it can indeed count a blocked beat.

THE VECTORCARDIOGRAPH

In Section 6.2 we looked at the basis of the ECG and defined the cardiac vector. The ensuing description of the electrocardiograph showed how a particular component of the cardiac vector could be recorded. Such scalar ECGs are the type that are usually taken. However, we can obtain more information from a *vectorcardiogram* (VCG). A VCG shows a three-dimensional—or at least a two-dimensional—picture of the orientation and magnitude of the cardiac vector throughout the cardiac cycle. It is difficult for practical machines to display the VCG in three dimensions, but it is relatively simple to display it in two dimensions—or, in other words, in a particular plane of the body.

Special lead systems have been developed that can provide the x, y, and z components of the ECG. Any two of these can be fed into the circuit to arrive at the VCG for the plane defined by the axes. The signal from the lead for one axis is connected to input 1, and that for the other enters input 2. These signals, amplified by appropriate identical amplifiers, are fed to the x and y deflection circuits of a cathode-ray oscilloscope or computer. For each heartbeat, a vector loop representing the locus of the tip of the cardiac vector when its tail is at the origin is then traced on the screen or plotted.

This type of display does not give any information on how the ECG signals are changing with respect to time. The scalar ECG is best able to provide this information, but it is of interest to see the time course of the vector loop as well. For this reason, the z (intensity) axis is modulated by a precision clock. Each segment of the vector loop corresponds to a known time interval. The viewer can tell the direction of the vector loop, as well as its time course, because one end of the comet-shaped segment will be at low intensity while the other is at high intensity. Triggering circuits are often included in the vectorcardiograph so that only one cardiac cycle is displayed at a time, thereby avoiding overlap.

Because of the complexities of obtaining the vectorcardiogram and the difficulty that arises in interpreting the patterns, this technique is generally

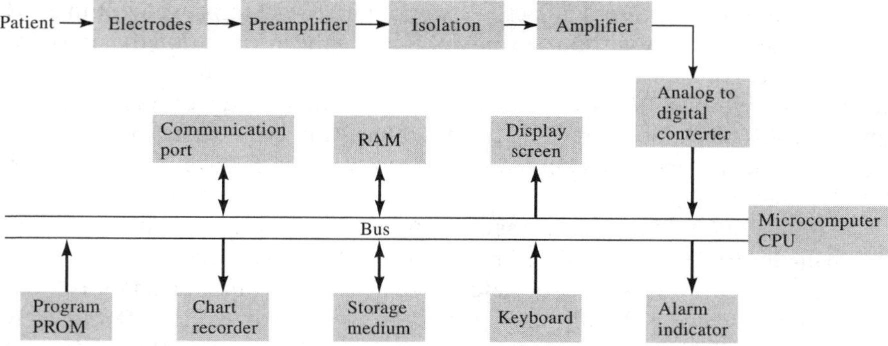

Figure 6.26 Block diagram of a cardiac monitor.

limited to special studies at tertiary-medical-care facilities or to use as a research tool. The scalar 12-lead electrocardiogram is employed for routine clinical studies.

6.9 CARDIAC MONITORS

There are several clinical situations in which continuous observation of the ECG and heart rate is important to the care of the patient. Continuous observation of the ECG during the administration of anesthesia helps doctors monitor the patient's condition while he or she is on the operating table and during recovery from anesthesia. Constant monitoring of the ECG and heart rate of the myocardial-infarction patient during the danger period of several days following the initial incident has made possible the early detection of life-threatening cardiac arrhythmias. Continuous monitoring of the fetal heart rate during labor may help in the early detection of fetal distress.

These and other clinical applications of continuous monitoring of the ECG and heart rate are made possible by *cardiac monitors* and *cardioscopes*. Figure 6.26 shows the basic cardiac monitor in block-diagram form. Its front-end circuitry is similar to that of the electrocardiograph. A pair of electrodes, usually located on the anterior part of the chest, pick up the ECG and are connected by lead wires to the input circuit of the monitor. The input circuit contains protection circuitry, as described in Section 6.4, to protect the monitor from high-voltage transients that can occur during defibrillation procedures.

The next stage of the monitor is a standard biopotential amplifier designed to amplify the ECG. Although it is best to have the frequency-response characteristics described in Section 6.2, cardiac monitors often have a slightly narrower frequency response than would be acceptable for a standard electrocardiograph. The reason for this is that much of the motion-artifact signal seen during movement of the patient is at very low frequencies. By filtering out some of these low frequencies, we can obtain a vast improvement in SNR and recording stability without seriously affecting the information that pertains to cardiac rhythm in the ECG. Frequency response should be from 0.67 to 40 Hz (Anonymous, 1992). Cardiac monitors should not trigger on pacemaker spikes, which continue even when the heart has stopped. And, to avoid double counting, cardiac monitors should not trigger on tall T waves (Anonymous, 1990).

Patient isolation circuitry (Section 6.2) is usually found in the circuit following an ECG preamplifier. This is followed by an additional amplifier to raise the signal to levels appropriate for display, recording, or analysis.

The output from the amplifier can take several routes. It is fed directly to an analog display screen (cathode ray tube), which shows the ECG as it would appear on an electrocardiograph.

If the cardiac monitor is required to show only the ECG, the three blocks are all that are necessary. A monitor of this type is called a *cardioscope*. It is frequently used in an operating room or ambulance to monitor the patients during surgery and critical transport.

In most modern cardiac monitors, the amplified ECG signal is digitized by an analog-to-digital converter, and the remaining processing is carried out by a computer as shown in Figure 6.26. The computer CPU and its peripherals carry out the functions described in the following paragraphs, although they can be performed by dedicated analog or digital circuits as well.

Often a physician wants to have a permanent record of the ECG being monitored. For this reason, many cardiac monitors have a small chart recorder or graphic printer built into them that can be switched on (S_1) by the operator to record a particularly interesting ECG as it appears on the screen.

It is often desirable to have a record of the events in the ECG that lead up to a serious arrhythmia. Such a record can be made if the amplifier output is fed first to a memory loop, which delays the ECG signal by about 15 s. The output from the memory loop can then be fed to the chart recorder via S_2. Thus, when the operator of the monitor sees an interesting ECG waveform, she or he can switch on the chart recorder through the memory loop and obtain a record of the events that led up to that particular pattern.

The ECG is used to determine heart rate either by the computer or a cardiotachometer. The output is displayed on a rate meter so that the operator can immediately tell the patient's heart rate. Alarm circuitry to warn of high and low heart rate is also associated with this system. Frequently, the alarm circuit automatically turns on the chart recorder and connects it to the output of the memory loop so that a recording of what precipitated the alarm is produced. This can be a valuable aid to clinicians in selecting appropriate therapy for the alarm-producing event.

Cardiac monitors can be used with individual patients on a ward by bringing a portable monitor to that patient. Most hospitals also utilize them in an organized system called an intensive-care unit. In such units, there are often individual monitors at each patient's bedside that consist of a cardioscope and a cardiotachometer with a rate meter and alarms. These individual monitors are connected to a central unit located at the nursing station. The unit contains a cardioscope, which shows the ECGs for all patients being monitored, and a rate meter and slave alarm* unit for each patient. Memory loops and a chart recorder are also located at the central display. The chart recorder can be activated either at the central station or by remote control from the individual monitors at the patient's bedside.

Special-purpose computers that can recognize cardiac arrhythmias and record the frequency of their occurrence are applied in intensive-care units. The machines can also prepare hard-copy charts showing trends in the patient's monitored parameters and can keep records of various therapeutic measures taken by the clinical staff. The computer can also be a big help in the intensive-care unit by carrying out many observational and secretarial functions, thereby freeing the clinical staff to care for the patient (Thakor, 1988b).

The availability of microcomputers has made it possible to combine the monitoring of ambulatory patients with detection of cardiac arrhythmias. Portable tape recorders or electronic memories connected to ECG amplifiers can collect data from ambulatory patients; these data are analyzed later by a computer (Jurgen, 1976).

Microcomputers in cardiac monitors perform two basic functions, data management and data analysis. In the former case, the microcomputer controls the various components of the system and directs the transport of data from one block to another along the bus. Carrying out the second function involves the actual analysis of the electrocardiogram. It includes filtering and artifact reduction, identification of the various components of the electrocardiogram, determination of the heart rate, and identification of arrhythmias. More than one microcomputer can be used in a monitor system to carry out these functions. The microcomputer is under the control of a program that is usually stored in read-only memory (ROM). This makes it possible to update the monitor by replacing ROM rather than modifying any hardware of the instrument.

Random-access memory (RAM) is provided for temporarily storing the signals, and an alternative medium such as an optical disk or a magnetic tape cartridge is used to archive selected incidents or the entire monitored data. There is also a staff interface to the system that consists of a keyboard and a display monitor.

Computerized cardiac monitors can be integrated into other hospital information systems. Frequently these monitors also have a network connection that enables them to interact with other information systems or to transmit data to physicians' offices located away from the intensive-care unit.

Ambulatory cardiac monitors are often used in the diagnosis and treatment of heart disease. The most frequently applied ambulatory monitor—the Holter monitor—includes a miniature magnetic tape recorder that the patient wears. These devices consist of a battery-powered ECG amplifier and tape

*An alarm placed at the patient's bedside is "master" to a "slave" alarm that is placed at the nurse's station.

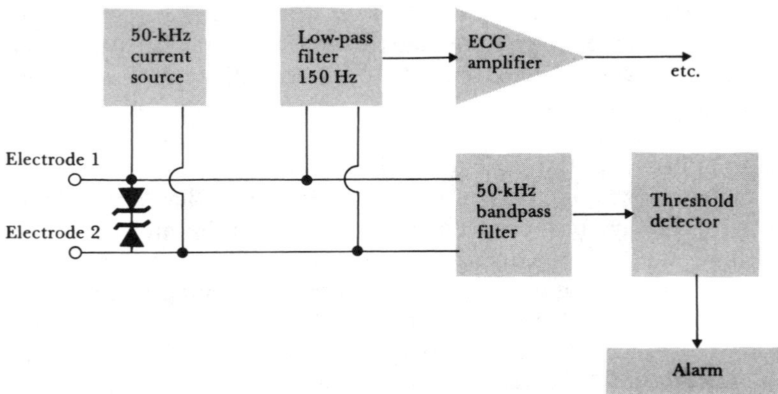

Figure 6.27 Block diagram of a system used with cardiac monitors to detect increased electrode impedance, lead wire failure, or electrode fall-off.

recorder that are connected to electrodes placed on the patient's chest. The instrument is sufficiently small to allow the patient to wear it on a belt or harness, and the tapes can hold from 24 to 48 h of continuous ECG recording. Some recorders can collect data from three leads simultaneously so that vectorcardiograms can be stored. Special computerized playback units rapidly analyze the tapes for cardiac arrhythmias and display these portions of the electrocardiogram on a computer screen or generate a hard-copy printout of them. The playback units also summarize the total recording in a report that indicates variables such as heart rate, variability in heart rate, type and number of arrhythmias, and amount of artifact.

In situations in which cardiac monitors are used to observe a patient's ECG over a long period of time, artifact and failure of the monitor can occur as a result of a poor electrode–patient interface. The longer the electrodes remain on the patient, the more often this occurs. In intensive-care units, electrodes are routinely changed—sometimes once a shift, sometimes once a day—to ensure against this type of breakdown. Most cardiac monitors also have alarm circuits that indicate when electrodes fall off the patient or the electrode-patient connection degenerates.

Figure 6.27 is a block diagram of a typical lead fall-off alarm. A 50-kHz high-impedance source is connected across the electrodes. Peak amplitudes of the current can be as great as 100–200 μA without any risk to the patient, because the microshock hazard to excitable tissue decreases as the frequency increases above 50 Hz (Figure 14.3). The current passes through the body between the electrodes, and, as long as there is good electrode contact, the voltage drop is relatively small. If the electrode connections become poor, as can happen when the electrolyte gel begins to dry, or if one of the electrodes falls off or the wire breaks, the impedance between the electrodes jumps considerably. This causes the voltage produced by the 50-kHz source to rise. The high-frequency signal is separated from the ECG by the filtering scheme, as shown. The ECG passes through a low-pass filter with approximately a 150-Hz corner frequency, and is processed in the usual way. A bandpass filter with a 50-kHz center frequency passes the voltage resulting from the current source to a threshold detector. This detector sets off an alarm when the voltage exceeds a certain threshold, which would correspond to poor electrode contact. When an electrode falls off the patient, the interelectrode impedance should increase to infinity, resulting in the possibility of 50-kHz voltages high enough to cause some damage to the electronic devices. For this reason, a high-voltage protection circuit, such as that described in Section 6.4, is frequently connected across the input terminals to the monitor. In the case shown in Figure 6.27, back-to-back zener diodes are used.

6.10 BIOTELEMETRY

Biopotential and other signals are often processed by radiotelemetry, a technique that provides a wireless link between the patient and the majority of the signal-processing components. By using a miniature radio transmitter attached to the patient to broadcast the information over a limited range, clinicians can monitor a patient or study a research animal while the subject has full mobility. This technique also provides the best method of isolating the patient from the recording equipment and power lines. For a single-channel system of biopotential radiotelemetry, a miniature battery-operated radio transmitter is connected to the electrodes on the patient. This transmitter broadcasts the biopotential over a limited range to a remotely located receiver, which detects the radio signals and recovers the signal for further processing. In this situation there is obviously negligible connection or stray capacitance between the electrode circuit connected to the radio transmitter and the rest of the instrumentation system. The receiving system can even be located in a room separate from the patient's. Hence the patient is completely isolated, and the only risk of electric shock that the patient runs is due to the battery-powered transmitter itself. Thus, if the transmitter power supply is kept at a low voltage, there is negligible risk to the patient.

Many types of radiotelemetry systems are used in biomedical instrumentation (Mackay, 1970). The basic configuration of the system, however, is pretty much the same for all. Figure 6.28 shows a single-channel radiotelemetry system for the ECG. A preamplifier amplifies the ECG signal to a level at which it can modulate the radiofrequency (RF) carrier generated by an oscillator. Frequency modulation is often applied on single-channel systems, but more recent designs use pulse-duration or pulse-position modulation with pulse-code modulation applied where ultrahigh reliability is required, such as in clinical monitoring. The modulated RF signal can be either directly applied to the radiating antenna (which in some cases is merely a coil around the

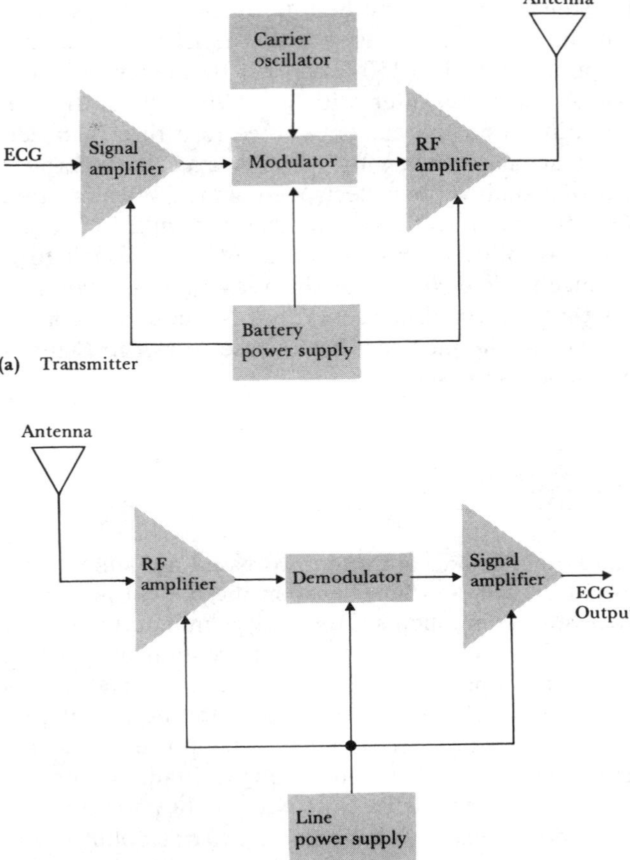

Figure 6.28 Block diagram of a single-channel radiotelemetry system

transmitter package) or amplified through an RF amplifier to provide higher signal levels. The entire transmitter is powered by a small battery pack. It is carried by the patient and usually attached by means of a special harness. Ultraminiature radio transmitters can be attached by surgical tape directly to the patient's skin. In research with experimental animals, experimenters can surgically implant the tiny transmitters within the bodies of the animals so that no external connections or wires are required.

In the receiving system, a pickup antenna receives the modulated RF signal, which is amplified through a tuned RF amplifier. The signal is then demodulated to recover the original information from the carrier. The signal can be further amplified to provide a usable output. The receiver system is generally powered directly from the power line, because it is in a permanent location and is not attached to the patient in any way.

Multichannel radiotelemetry systems consist of two or more channels of data transmitted over a single carrier wave. The method of combining the various channels of information into a single signal is known as *multiplexing.* There are two basic methods of multiplexing. *Frequency-division multiplexing* makes use of continuous-wave subcarrier frequencies. Consider a simple three-channel frequency-division multiplex radiotelemetry system. In the transmitter the signals frequency-modulate three different subcarrier oscillators, respectively, each being at such a frequency that its modulated signal does not overlap the frequency spectra of the other two modulated signals. The frequency-modulated signals from all channels are added together through a summing amplifier to give a composite signal in which none of the parts overlap in frequency. This signal then modulates the RF carrier of the transmitter and is broadcast.

At the receiver for the frequency-division multiplexing system, the RF signal is amplified and detected to give a signal equivalent to the sum of the three modulated subcarriers. Each of these is next separated by a bandpass filter tuned to the frequencies that it contains. The separated signal is then demodulated and sent to the appropriate recording device.

The bandwidth of signal information that can be contained in any one channel depends on the frequency deviation of the subcarrier modulator and on the bandwidth of the subcarrier channel itself. Of course, as more subcarrier channels are used, they have to occupy higher and higher frequencies so that they do not overlap. And the RF channel has to have a wider bandwidth to handle the information.

The second multiplexing scheme that is used in multichannel radiotelemetry is *time-division multiplexing.* Figure 6.29 shows a three-channel time-division multiplexing system. In this case, the three signals are amplified and applied to a commutator circuit. This circuit is an electronic switch that is rapidly scanning the three signals and a fourth "frame reference signal" in succession. For the sake of this description, let us consider the three input signals (v_1, v_2, and v_3) to be constant voltages. The frame reference signal v_4 is already at a constant voltage. An oscillator drives the commutator circuit so that it samples each voltage for an instant of time, thereby giving a pulse-train sequence for v_1, v_2, and v_3, as shown in Figure 6.29(b). When the commutator reaches the frame reference signal v_4, it samples this signal for a longer time to give a wider pulse. This is done to make it easy to recognize the frame-reference signal in the receiving system.

The receiver consists of a conventional radio receiver and a demodulator that recovers the signal, as shown in Figure 6.29(b). This signal must be demultiplexed by another commutator, which is driven by an oscillator in the receiver. The frame-reference pulse is used to achieve synchronization between the commutators in the transmitter and the receiver. The frame-pulse-detector circuit recognizes the longer frame-reference pulse and tells the commutator oscillator to set the commutator so that the next pulse it receives is v_1, which is then followed by v_2 and finally by v_3. In this way the receiving commutator follows the transmitting commutator. Each separated

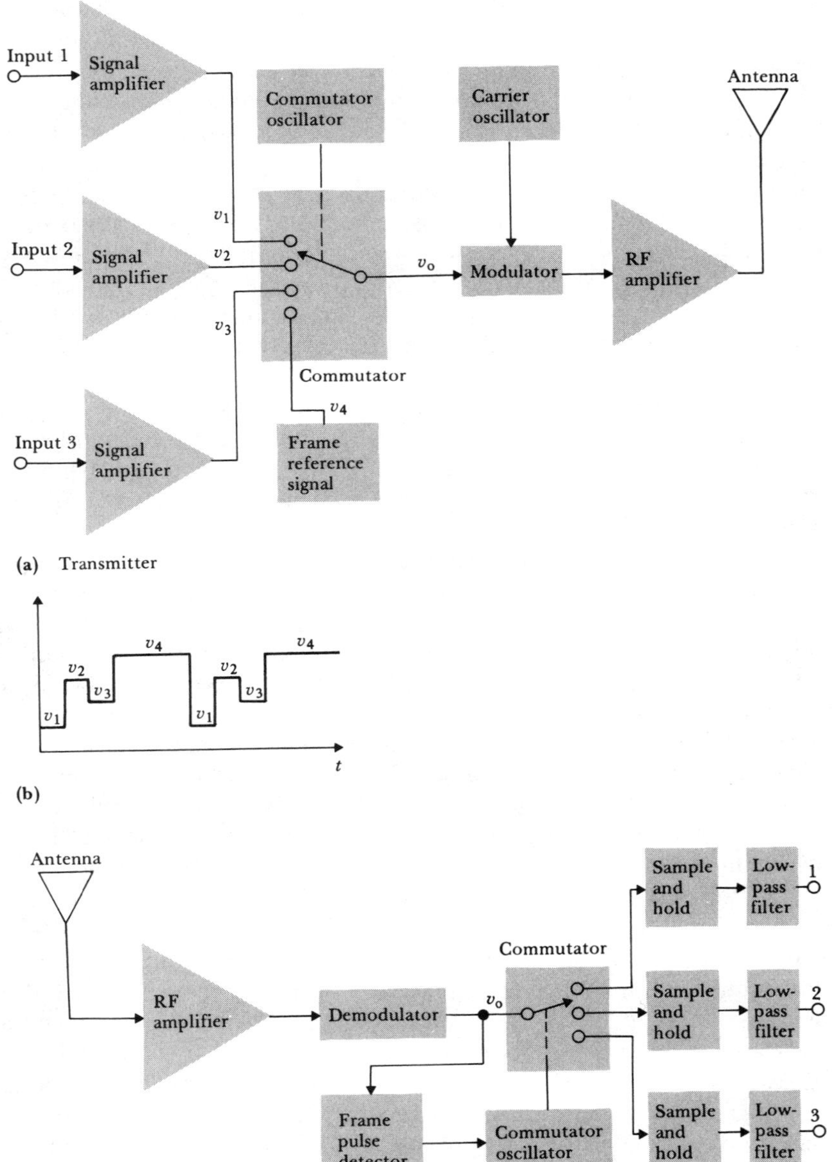

Figure 6.29 Block diagram of a three-channel time-division multiplexed radio-telemetry system (a) Transmitter. (b) Example of output waveform from commutator in transmitter. (c) Receiver.

pulse at the receiver is then applied to a sample-and-hold circuit that holds its amplitude until the next pulse is received. A low-pass filter follows this sample-and-hold circuit and removes the artifact generated by the commutating technique.

The bandwidth of each signal channel of this system is determined by the rate at which it is sampled. Ideally, the rate of sampling should be at least twice that of the highest-frequency component to be transmitted, but in practical circuits the rate of sampling is usually at least five times that of the highest-frequency component.

It is important to note that, although radiotelemetry systems provide ideal isolation with no patient ground required, they are not completely immune to problems of electric noise. Because coupling is achieved by a radiated electromagnetic signal, other electromagnetic signals at similar frequencies can interfere and cause artifacts. In extreme cases, these other signals can even bring about complete loss of signal.

In addition, the relative orientation between transmitting and receiving antennas is important. There can be orientations in which none of the signals radiated from the transmitting antenna are picked up by the receiving antenna. In such cases, there is no transmission of signals. In high-quality radiotelemetry systems, it is therefore important to have a means of indicating when signal interference or signal dropout is occurring. Such a signal makes it possible to take steps to rectify this problem and informs the clinical staff that the information being received is noise and should be disregarded.

Telemetry over short distances has been carried out using a near-infrared signal as the carrier. Weller (1977), Šantić (1984), and Kimmich (1988) have described infrared telemetry systems in which the ECG information modulates the pulsing frequency of infrared light-emitting diodes that are driven with pulses to conserve energy. These systems were found to be very reliable when used in an enclosed area such as a hospital room. They were not susceptible to the type of interference that affects radiotelemetry.

PROBLEMS

6.1 What position of the cardiac vector at the peak of the R wave of an electrocardiogram gives the greatest sum of voltages for leads I, II, and III?

6.2 What position of the cardiac vector during the R wave gives identical signals in leads II and III? What does the ECG seen in lead I look like for this orientation of the vector?

6.3 An ECG has a scalar magnitude of 1 mV on lead II and a scalar magnitude of 0.5 mV on lead III. *Calculate* the scalar magnitude on lead I.

6.4 Design a system that has as inputs the *scalar* voltages of lead II and lead III and as output the *scalar* voltage of the cardiac vector **M**.

6.5 Design the lead connections for the VF and aVF leads. For each, choose minimal resistor values that meet the requirements for input impedance given in Table 6.1.

6.6 A student designs a new lead system by inverting Eindhoven's triangle. She places one electrode on each hip and one on the neck. For this new system, design a resistor network (show the circuit and give resistor values) to yield conventional lead aVF (show polarity). Explain the reason for each resistor.

6.7 Design an electrocardiograph with an input-switching system such that we can record the six frontal-plane leads by means of changing the switch.

6.8 Discuss the factors that enter into choosing a resistance value for the three resistors used to establish the Wilson central terminal. Describe the advantages and disadvantages of having this resistance either very large or very small.

6.9 The central-terminal requirements for an electrocardiograph that meets the recommendations of Table 6.1 sets the minimal value of the resistances at 1.7 MΩ. Show that this value is a result of the specification given in Table 6.1.

6.10 A student attempts to measure his own ECG on an oscilloscope having a differential input. For Figure 6.11, Z_{in} = 1 MΩ, Z_1 = 20 kΩ, Z_2 = 10 kΩ, Z_G = 30 kΩ, and i_{db} = 0.5 μA. Calculate the power-line interference the student observes.

6.11 Design a driven-right-leg circuit and show all resistor values. For 1 μA of 60-Hz current flowing through the body, the common-mode voltage should be reduced to 2 mV. The circuit should supply no more than 5 μA when the amplifier is saturated at ±13 V.

6.12 An engineer sees no purpose for $R/2$ in Figure 6.5(a) and replaces it with a wire in order to simplify the circuit. What is the result?

6.13 An ECG lead is oriented such that its electrodes are placed on the body in positions that pick up an electromyogram from the chest muscles as well as the electrocardiogram. Design a circuit that separates these two signals as well as possible, and discuss the limitations of such a circuit.

6.14 A cardiac monitor is found to have 1 mV p-p of 60-Hz interference. Describe a procedure that you could use to determine whether this is due to an electric field or a magnetic field pickup.

6.15 Assume zero skin-electrode impedance, and design (give component values for) simple *filters* that will attenuate incoming 1-MHz radiofrequency interference to 0.001 of its former value. Sketch the placement of these filters (show all connections) to prevent interference from entering an ECG amplifier. Then calculate the 60-Hz interference that they cause for common-mode voltage of 10 mV and skin-electrode impedances of 50 kΩ and 40 kΩ.

6.16 You design an ECG machine using FETs such that the Z_{in} of Figure 6.11 exceeds 100 MΩ. Because of RF interference, you wish to add equal shunt capacitors at the two Z_{in} locations. For v_{cm} = 10 mV, Z_2 100 kΩ, and Z_1 = 80 kΩ, calculate the maximal capacitance so 60-Hz $|v_A - v_B|$ = 10 μV. Calculate the result using Equation (6.9).

6.17 Silicon diodes having a forward resistance of 2 Ω are to be used as voltage-limiting devices in the protection circuit of an electrocardiograph. They are connected as shown in Figure 6.14(b). The protection circuit is shown in Figure 6.13. If voltage transients as high as 500 V can appear at the electrocardiograph input during defibrillation, what is the minimal value of R that the designer can choose so that the voltage at the preamplifier input does not exceed 800 mV? Assume that the silicon diodes have a breakdown voltage of 600 mV.

6.18 For Figure 6.15, assume $i_d = 500$ nA and RL skin impedance is 100 kΩ. Design (give component values for) a *driven-right-leg* circuit to achieve $v_{cm} =$ 10 μV.

6.19 For Figure 6.17, assume $C_s = 10$ pF and $C_f = 20$ pF. Design the amplifier circuit to replace the triangle containing A_v. Use an op amp and passive components to achieve an ideal negative-input capacitance amplifier. Show the circuit diagram and connections to other components that appear in Figure 6.17.

6.20 Design a technique for automatically calibrating an electrocardiograph at the beginning of each recording. The calibration can consist of a 1-mV standardizing pulse.

6.21 A student decides to remove the switch across the 3.3 MΩ resistor in Figure 6.18 and place it across the 1-μF capacitor to "discharge the capacitor after defibrillation." Sketch what the typical output looks like before, during, and after defibrillation and switch closure, and explain why it looks that way.

6.22 Redesign Figure 6.18 by placing a capacitor in series with the 10-kΩ resistor between the two inverting inputs. Eliminate the last op amp, and adjust other components to keep the same gain, corner frequencies, and ability to use a switch to return the output to the linear region.

6.23 Design a biopotential preamplifier that is battery-powered and isolated in such a way that there is less than 0.5-pF coupling capacitance between the input and output terminals. The amplifier should have a nominal gain of 10 and an input impedance greater than 10 MΩ differentially and greater than 10 GΩ with respect to ground. The output impedance should be less than 100 Ω and single-ended.

6.24 Design a circuit that uses one op amp plus other passive components that will detect QRS complexes of the ECG even when the amplitude of the T wave exceeds that of the QRS complex and provides output signals suitable for counting these complexes on a counter.

6.25 Design an automatic reset circuit for an electrocardiograph.

6.26 When each R wave occurs, register 2 in the beat-to-beat instantaneous cardiotachometer yields a number equal to the number of milliseconds since the previous R wave. Design a variable-resistance digital-to-analog converter that yields heart rate for the three-bit binary output.

6.27 For Figure 6.19, design the circuit diagram for the variable-resistance digital-to-analog converter. An op-amp *summer* with digitally controlled switches that short resistors or open branches may be useful to achieve Equation (6.15).

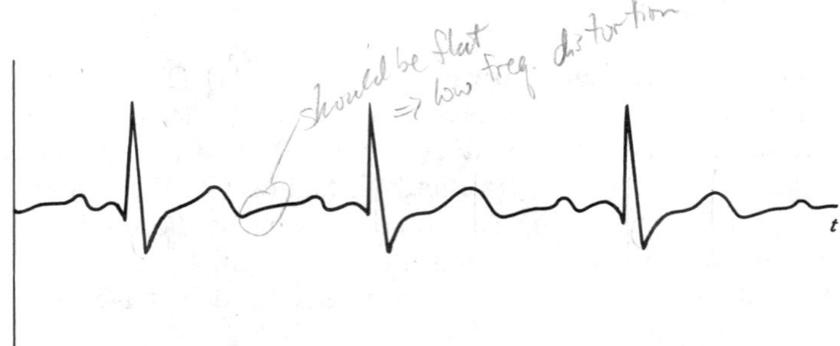

Figure P6.1

6.28 Design an arrhythmia-detection system for detecting and counting the PVCs shown in Figure 4.18. Note that PVCs occur earlier than expected, but the following beat occurs at the normal time, because it is generated by the SA node. Show a block diagram and describe the operation of the system.

6.29 Contrast averaging and beat-to-beat cardiotachometers in terms of what would happen to the output signal of each if they were both connected to a patient who suddenly went into cardiac arrest.

6.30 Design a full-wave rectifier circuit that provides a signal corresponding to the absolute value of an electromyogram. The electromyogram coming from the electrodes has a maximal peak amplitude of 1 mV. What are the limitations of such a rectifier?

6.31 In an evoked-response experiment in which the EEG is studied after a patient is given the stimulus of a flashing light, the experimenter finds that the response has approximately the same amplitude as the random noise of the signal. If a signal averager is used, how many samples must be averaged to get an SNR of 10:1? If we wanted an SNR of 100:1, would it be practical to use this technique?

6.32 A single-channel radiotelemetry system is used to continuously monitor a patient's ECG. Occasionally, because of improper orientation of the antenna or the transmitter's getting out of range, there is a loss of signal. Design a system to indicate this signal loss at the telemetry receiver.

6.33 A physician wishes to obtain two simultaneous ECGs in the frontal plane from leads that have lead vectors at right angles. The signal will be used to generate a VCG. Describe how you would go about obtaining these two signals, and suggest a test to determine whether the leads are truly orthogonal.

6.34 The ECG shown in Figure P6.1 is distorted as a result of an instrumentation problem. Discuss possible causes of this distortion, and suggest means of correcting the problem.

6.35 Figure P6.2 shows ECGs from simultaneous leads I and II. Sketch the vector loop for this QRS complex in the frontal plane.

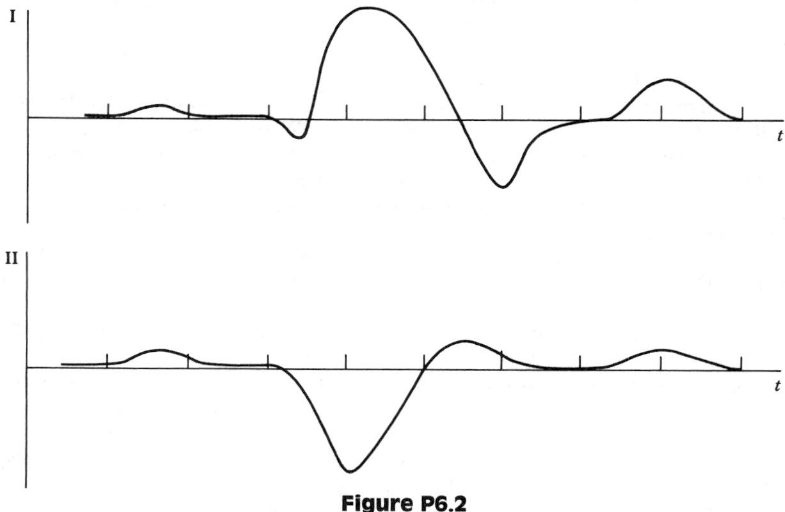

Figure P6.2

REFERENCES

Anonymous, "Ambulatory ECG monitors." *Health Devices,* 1989, 18, 295–321.

Anonymous, "Diagnostic electrocardiographic devices" ANSI/AAMI EC11-1991. Arlington, VA: Association for the Advancement of Medical Instrumentation, 1991.

Anonymous, "Cardiac monitors, heart rate meters and alarms" ANSI/AAMI EC13-1992. Arlington, VA: Association for the Advancement of Medical Instrumentation, 1992.

Bailey, J. J. et al., "Recommendations for standardization and specifications in automated electrocardiography: Bandwidth and digital signal processing." *Circulation,* 1990, 81(2), 730–739.

Childers, D. G., "Evoked potentials," in J. G. Webster (ed.), *Encyclopedia of Medical Devices and Instrumentation.* New York: Wiley, 1988, pp. 1245–1254.

Cobold, R. S. C., *Transducers for Biomedical Measurements: Principles and Applications.* New York: Wiley, 1974.

Delgado, J. M. R., "Electrodes for extracellular recording and stimulation," in W. L. Nastuk, *Physical Techniques in Biological Research.* New York: Academic, 1964, pp. 88–139.

Fichtenbaum, M., "Counter inverts period to measure low frequency." *Electron.,* March 4, 1976, 49, 100.

Geddes, L. A., and L. E. Baker, *Principles of Applied Biomedical Instrumentation,* 3rd ed. New York: Wiley, 1989.

Grobstein, S. R., and R. D. Gatzke, "A battery-powered ECG monitor for emergency and operating room environments." *Hewlett-Packard J.,* September 1977, 29(1), 26–32.

Hartley, R., "Analogue-display rate meter built around digital switching elements." *Med. Biol. Eng.,* 1976, 14, 107–108.

Huhta, J. C., and J. G. Webster, "60-Hz interference in electrocardiography." *IEEE Trans. Biomed. Eng.,* 1973, BME-20, 91–101.

Jacobsen, N. K., and J. L. Stuart, "A field-portable, microprocessor-controlled, data processing and storing cardiotachometer." *Biotelem. Patient Monit.,* 1982, 9, 80–88.

Jongsma, H. W., H. P. van Geijn, K. J. Dalton, and R. J. Parsons (eds.), "Fetal electro and phonocardiography." *Clin. Phys. Physiol. Meas.,* 1989, 10 (suppl. B), 1–78.

Jurgen, R. K., "Software (and hardware) for the 'medics.'" *IEEE Spectrum,* April 1976, 13(4), 40–43.

Kimmich, H. P., "Biotelemetry," in J. G. Webster (ed.), *Encyclopedia of Medical Devices and Instrumentation.* New York: Wiley, 1988, pp. 409–425.

Ludwig, H., "Heart- or respiration-rate calculator." *Med. Biol. Eng. Comput.,* 1977, 15, 700–702.

Mackay, R. S., *Bio-medical Telemetry,* 2nd ed. New York: Wiley, 1970.

Neuman, M. R., "Neonatal monitoring," in J. G. Webster (ed.), *Encyclopedia of Medical Devices and Instrumentation.* New York: Wiley, 1988, pp. 2015–2034.

Offner, F., and B. Moisand, "A coincidence technique for fetal electrocardiography." *Amer. J. Obstet. Gynecol.,* 1966, 95, 676.

Plonsey, R., *Bioelectric Phenomena.* New York: McGraw-Hill, 1969.

Plonsey, R., "The biophysical basis for electrocardiography." *CRC Crit. Rev. Bioeng.,* 1971, 1, 1–48.

Roux, J. F., M. R. Neuman, and R. Goodlin, "Monitoring intrapartum phenomena." *CRC Crit. Rev. Bioeng.,* January 1975, 119–158.

Santic, A., and M. R. Neuman, "A low-power infrared biotelemetry system," in H. P. Kimich and H.-J. Klewe (eds.), *Biotelemetry VIII.* Nijmegen, Netherlands: Kimmich/Klewe, 1984, pp. 147–150.

Smith, M. S., and E. L. Pritchett, "Electrocardiographic monitoring in ambulatory patients with cardiac arrhythmias." *Cardiol. Clin.,* 1983, 1, 293–304.

Svetz, P., and N. Duane, "The α β γ of bioelectric measurements." *Electron. Des.,* August 2, 1975, 23(16), 68.

Taylor, K., and M. Mandelberg, "Precision digital instrument for calculation of heart rate and R–R interval." *IEEE Trans. Biomed. Eng.,* May 1975, BME-22, 255–257.

Thakor, N. V., "Electrocardiographic monitors," in J. G. Webster (ed.), *Encyclopedia of Medical Devices and Instrumentation.* New York: Wiley, 1988a, pp. 1002–1017.

Thakor, N. V., "Electrocardiography, computers in," in J. G. Webster (ed.), *Encyclopedia of Medical Devices and Instrumentation.* New York: Wiley, 1988b, pp. 1040–1061.

van Oosterom, A., "Lead systems for the abdominal fetal electrocardiogram." *Clin. Phys. Physiol. Meas.,* 1989, 10 (Suppl. B), 21–26.

Vermariën, H., "Recorders, graphic," in J. G. Webster (ed.), *Encyclopedia of Medical Devices and Instrumentation.* New York: Wiley, 1988, pp. 2495–2511.

Walden, W. D., and S. J. Birnbaum, "Fetal electrocardiography with cancellation of maternal complexes." *Amer. J. Obstet. Gynecol.,* 1964, 94, 596.

Weller, C., "Electrocardiography by infrared telemetry." *J. Physiol.* (London), 1977, 267, 11–12.

Winter, B. B., and J. G. Webster, "Driven-right-leg circuit design." *IEEE Trans. Biomed. Eng.,* 1983, BME-30, 62–66.

7

BLOOD PRESSURE AND SOUND

Robert A. Peura

Determining an individual's blood pressure is a standard clinical measurement, whether taken in a physician's office or in the hospital during a specialized surgical procedure. Blood-pressure values in the various chambers of the heart and in the peripheral vascular system help the physician determine the functional integrity of the cardiovascular system. A number of direct (invasive) and indirect (noninvasive) techniques are being used to measure blood pressure in the human. The accuracy of each should be established, as well as its suitability for a particular clinical situation.

Fluctuations in pressure recorded over the frequency range of hearing are called *sounds.* The sources of heart sounds are the vibrations set up by the accelerations and decelerations of blood.

The function of the blood circulation is to transport oxygen and other nutrients to the tissues of the body and to carry metabolic waste products away from the cells. In Section 4.6 we pointed out that the heart serves as a four-chambered pump for the circulatory system. This is illustrated in Figure 4.12. The heart is divided into two pumping systems, the right side of the heart and the left side of the heart. These two pumps and their associated valves are separated by the pulmonary circulation and the systemic circulation. Each pump has a filling chamber, the atrium, which helps to fill the ventricle, the stronger pump. Figure 7.15 is a diagram that shows how the electric and mechanical events are related during the cardiac cycle. The four heart sounds are also indicated in this diagram.

Figure 7.1 is a schematic diagram of the circulatory system. The left ventricle ejects blood through the aortic valve into the aorta, and the blood is then distributed through the branching network of arteries, arterioles, and capillaries. The resistance to blood flow is regulated by the arterioles, which are under local, neural, and endocrine control. The exchange of the nutrient material takes place at the capillary level. The blood then returns to the right side of the heart via the venous system. Blood fills the right atrium, the filling chamber of the right heart, and flows through the tricuspid valve into the right ventricle. The blood is pumped from the right ventricle into the pulmonary artery through the pulmonary valve. It next flows through the pulmonary arteries, arterioles, capillaries, and veins to the left atrium. At the pulmonary capillaries, O_2 diffuses from the lung alveoli to the blood, and CO_2 diffuses

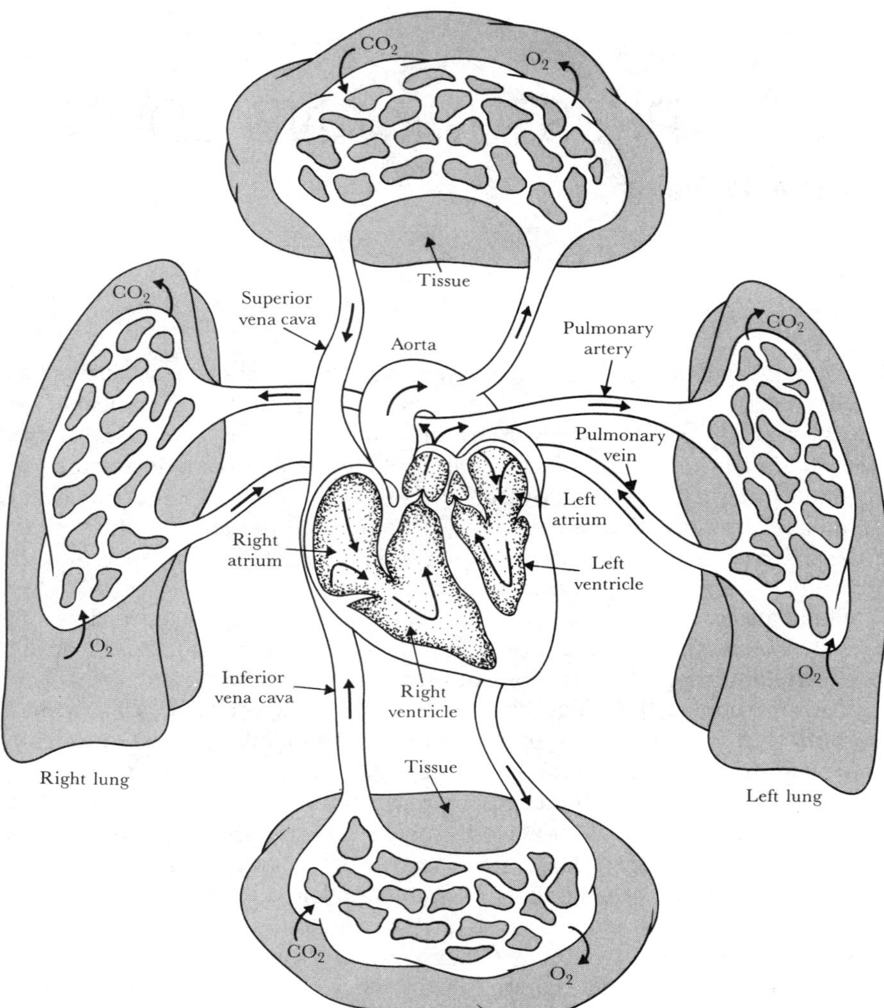

Figure 7.1 The left ventricle ejects blood into the systemic circulatory system. The right ventricle ejects blood into the pulmonary circulatory system.

from the blood to the alveoli. The blood flows from the left atrium, the filling chamber of the left heart, through the mitral valve into the left ventricle. When the left ventricle contracts in response to the electric stimulation of the myocardium (discussed in detail in Section 4.6), blood is pumped through the aortic valve into the aorta.

The pressures generated by the right and left sides of the heart differ somewhat in shape and in amplitude (see Figure 7.2). As we noted in Section 4.6, cardiac contraction is caused by electric stimulation of the cardiac muscle.

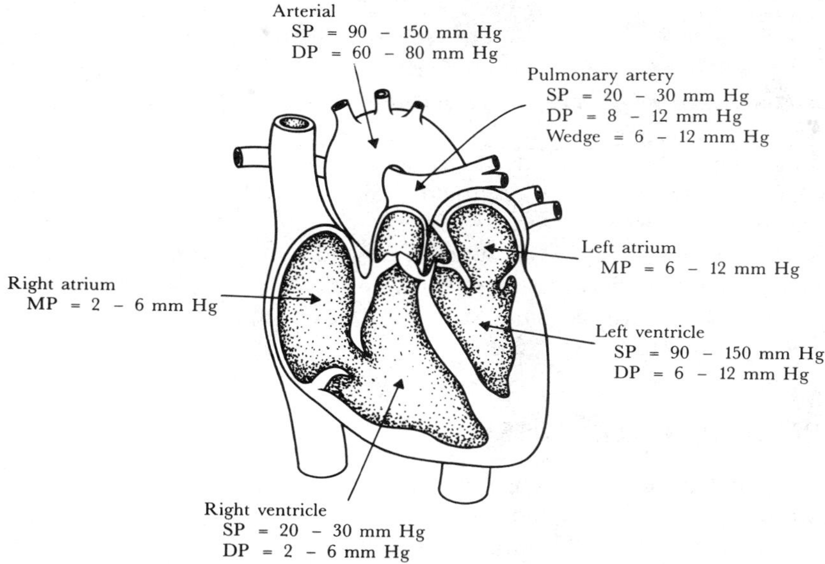

Figure 7.2 Typical values of circulatory pressures SP is the systolic pressure, DP the diastolic pressure, and MP the mean pressure. The wedge pressure is defined in Section 7.13.

An electric impulse is generated by specialized cells located in the sinoatrial node of the right atrium. This electric impulse quickly spreads over both atria. At the junction of the atria and ventricles, the electric impulse is conducted after a short delay at the atrial-ventricular node. Conduction quickly spreads over the interior of both ventricles by means of a specialized conduction system, the His bundle, and the Purkinje system. Conduction then propagates throughout both ventricles. This impulse causes mechanical contraction of both ventricles. Mechanical contraction of the ventricular muscle generates ventricular pressures that force blood through the pulmonary and aortic valves into the pulmonary circulation and the systemic circulation, causing pressures in each. Section 7.9 describes the correlation of the four heart sounds with the electric and mechanical events of the cardiac cycle. Briefly, the heart sounds are associated with the movement of blood during the cardiac cycle. Murmurs are vibrations caused by the turbulence in the blood moving rapidly through the heart.

7.1 DIRECT MEASUREMENTS

Blood-pressure sensor systems can be divided into two general categories according to the location of the sensor element. The most common clinical method for directly measuring pressure is to couple the vascular pressure to an *external* sensor element via a liquid-filled catheter. In the second general category, the liquid coupling is eliminated by incorporating the sensor into the tip of a catheter that is placed in the vascular system. This device is known as an *intravascular pressure sensor.*

A number of different kinds of sensor elements may be used; they include strain gage, linear-variable differential transformer, variable inductance, variable capacitance, optoelectronic, piezoelectric, and semiconductor devices (King, 1988). Cobbold (1974) compares the significant electric and mechanical properties of commercial pressure sensors. This section describes the principles of operation of an extravascular and an intravascular system. For a description of other sensors, see Chapter 2.

EXTRAVASCULAR SENSORS

The extravascular sensor system is made up of a catheter connected to a three-way stopcock and then to the pressure sensor (Figure 7.3). The catheter-sensor system, which is filled with a saline-heparin solution, must be flushed with the solution every few minutes to prevent blood from clotting at the tip.

The physician inserts the catheter either by means of a *surgical cut-down,* which exposes the artery or vein, or by means of *percutaneous insertion,* which involves the use of a special needle or guide-wire technique. Blood pressure is transmitted via the catether liquid column to the sensor and, finally, to the diaphragm, which is deflected. Figure 2.2(a) shows an early pressure sensor, in which the displacement of the diaphragm is transmitted to a system composed of a moving armature and an unbonded strain gage. Figures 7.3 and 14.15 show modern disposable blood-pressure sensors.

INTRAVASCULAR SENSORS

Catheter-tip sensors have the advantage that the hydraulic connection via the catheter, between the source of pressure and the sensor element, is eliminated. The frequency response of the catheter–sensor system is limited by the hydraulic properties of the system. Detection of pressures at the tip of the catheter without the use of a liquid-coupling system can thus enable the physician to obtain a high frequency response and eliminate the time delay encountered when the pressure pulse is transmitted in a catheter–sensor system.

A number of basic types of sensors are being used commercially for the detection of pressure in the catheter tip. These include various types of strain-gage systems bonded onto a flexible diaphragm at the catheter tip. Gages of this type are available in the F 5 catheter (1.67 mm OD) size. In the French scale (F), used to denote the diameter of catheters, each unit is approximately equal to 0.33 mm. Smaller-sized catheters may become available as the technology improves and the problems of temperature and electric drift, fragility, and nondestructive sterilization are solved more satisfactorily. A disadvantage

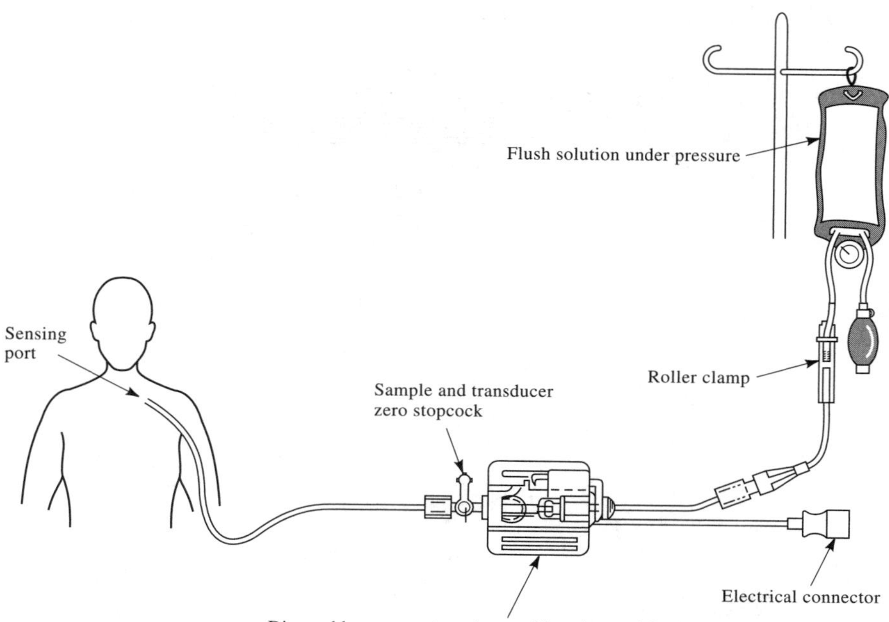

Figure 7.3 Extravascular pressure-sensor system A catheter couples a flush solution (heparinized saline) through a disposable pressure sensor with an integral flush device to the sensing port. The three-way stopcock is used to take blood samples and zero the pressure sensor.

of the catheter-tip pressure sensor is that it is more expensive than others and may break after only a few uses, further increasing its cost per use.

The fiber-optic intravascular pressure sensor can be made in sizes comparable to those described above, but at a lower cost. The fiber-optic device measures the displacement of the diaphragm optically by the varying reflection of light from the back of the deflecting diaphragm. (Recall that Section 2.14 detailed the principles of transmission of light along a fiber bundle.) These devices are inherently safer electrically, but unfortunately they lack a convenient way to measure relative pressure without an additional lumen either connected to a second pressure sensor or vented to the atmosphere.

A fiber-optic microtip sensor for *in vivo* measurements inside the human body is shown in Figure 7.4(a) in which one leg of a bifurcated fiber bundle is connected to a light-emitting diode (LED) source and the other to a photodetector (Hansen, 1983). The pressure-sensor tip consists of a thin metal membrane mounted at the common end of the mixed fiber bundle. External pressure causes membrane deflection, varying the coupling between the LED source and the photodetector. Figure 7.4(b) shows the output signal versus membrane deflection. Optical fibers have the property of emitting and ac-

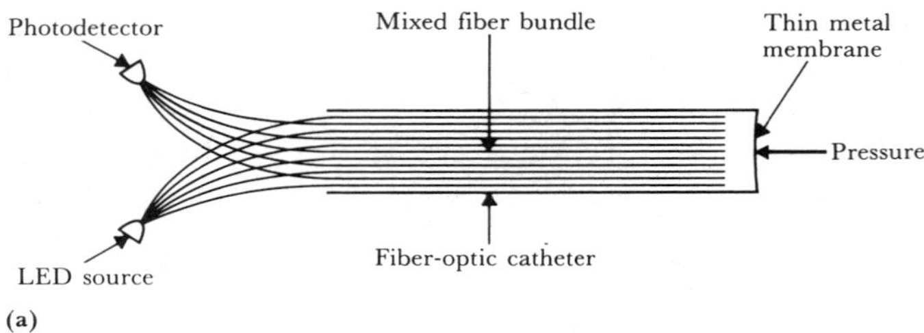

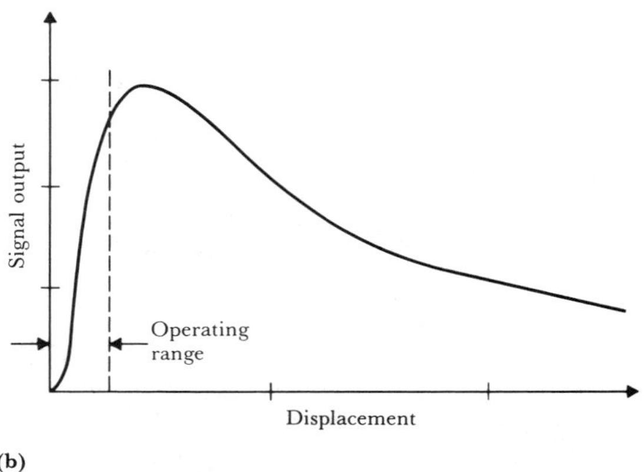

Figure 7.4 (a) Schematic diagram of an intravascular fiber-optic pressure sensor. Pressure causes deflection in a thin metal membrane that modulates the coupling between the source and detector fibers. (b) Characteristic curve for the fiber-optic pressure sensor.

cepting light within a cone defined by the acceptance angle θ_A, which is equal to the fiber numerical aperture, N_A (Section 2.14). The coupling between LED source and detector is a function of the overlap of the two acceptance angles on the pressure-sensor membrane. The operating portion of the curve is the left slope region where the characteristic is steepest.

Roos and Carroll (1985) describe a fiber-optic pressure sensor for use in magnetic resonance imaging (MRI) fields in which a plastic shutter assembly modulates the light transversing a channel between source and detector. Neuman (1988) has described a fiber-optic pressure sensor for intracranial pressure measurements in the newborn. Figure 7.5 shows a schematic of the device, which is applied to the anterior fontanel. Pressure is applied with the sensor such that the curvature of the skin surface is flattened. When this applanation occurs, equal pressure exists on both sides of the membrane, which consists of soft tissue between the scalp surface and the dura. Monitoring of the probe pressure determines the dura pressure. The membrane position is determined

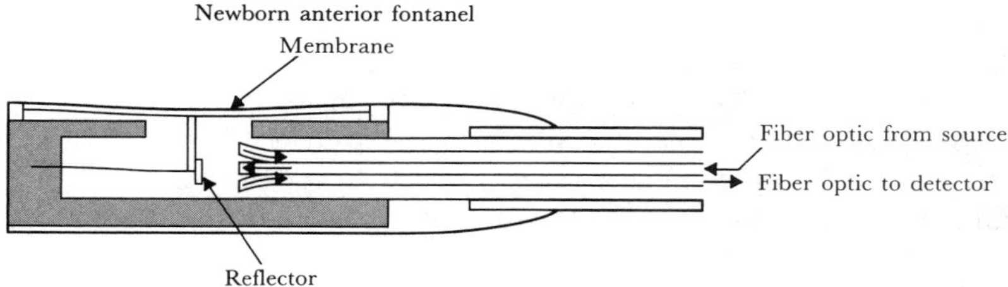

Figure 7.5 Fiber-optic pressure sensor for intracranial pressure measurements in the newborn. The sensor membrane is placed in contact with the anterior fontanel of the newborn.

by a reflector that is attached to the membrane and varies the amount of light coupling between the source and detector fibers.

Air pressure from a pneumatic servo system controls the air pressure within the pressure sensor, which is adjusted such that diaphragm—and thus the fontanel tissue—is flat, indicating that the sensor air pressure and the fontanel or intracranial pressure are equal.

Similar pressure-unloading techniques are used in thin, compliant sensors that measure interface pressures between the skin and support structures such as seat cushions (Webster, 1991).

Silicon fusion bonding is used to fabricate micro silicon pressure-sensor chips (Howe *et al.,* 1990). A wedge-shaped cavity is etched in a silicon wafer to form a diaphragm. Piezoresistive strain gages are implanted and a metal connection is made in order to create a sensor for a catheter tip micropressure sensor.

DISPOSABLE PRESSURE SENSORS

Traditionally, physiological pressure sensors have been reusable devices, but most modern hospitals have adopted inexpensive, disposable pressure sensors in order to lower the risk of patient cross-contamination and reduce the amount of handling of pressure sensors by hospital personnel. Because reusable pressure sensors are subject to the abuses of reprocessing and repeated user handling, they tend to be less reliable than disposable sensors.

By micromachining silicon, a pressure diaphragm is etched and piezoresistive strain gages are diffused into the diaphragm for measuring its displacement. This process results in a small, integrated, sensitive, and relatively inexpensive pressure sensor. This silicon chip is incorporated into a disposable pressure-monitoring tubing system. The disposable pressure sensor system also contains a thick-film resistor network that is laser-trimmed to remove offset voltages and set the same sensitivity for similar disposable sensors. In addition, a thick-film thermistor network is usually incorporated for temperature compensation. The resistance of the bridge elements is usually high in order to reduce self-heating, which may cause erroneous results. This results in a high output impedance for the device. Thus, a high-input impedance monitor must be used with disposable pressure sensors.

Pressure sensors can monitor blood pressure in postsurgical patients as part of a closed-loop feedback system. Such a system injects controlled amounts of the drug nitroprusside to stabilize the blood pressure (Sheppard and Jannett, 1988).

Each February issue of *Medical Electronics* lists the manufacturers of blood-pressure instruments and their types.

7.2 HARMONIC ANALYSIS OF BLOOD-PRESSURE WAVEFORMS

The basic sine-wave components of any complex time-varying periodic waveform can be dissected into an infinite sum of properly weighted sine and cosine functions of the proper frequency that, when added, reproduce the original complex waveform. It has been shown that researchers can apply techniques of Fourier analysis when they want to characterize the oscillatory components of the circulatory and respiratory systems, because two basic postulates for Fourier analysis—periodicity and linearity—are usually satisfied (Attinger *et al.,* 1970).

Cardiovascular physiologists and some clinicians have been employing Fourier-analysis techniques in the quantification of pressure and flow since this method was established in the 1950s. Early Fourier analysis used bandpass filters. More recent analysts have used computer techniques to obviate the need for special hardware. The advantage of the technique is that it allows for a quantitative representation of a physiological waveform; thus it is quite easy to compare corresponding harmonic components of pulses.

O'Rourke (1971) points out that the physician who turns to a standard medical textbook for assistance in interpreting the arterial pulse is likely to be confused, misled, and disappointed. He further indicates that in recent years, analysis of the frequency components of the pulse appears to have yielded more information on arterial properties than any other approach. He proposes that the arterial pulse be represented in terms of its frequency components.

The blood-pressure pulse can be divided into its fundamental component (of the same frequency as the blood-pressure wave) and its significant harmonics. Figure 7.6 shows the first six harmonic components of the blood-pressure wave and the resultant sum. When we compare the original waveform and the waveform reconstructed from the Fourier components, we find that they agree quite well, indicating that the first six harmonics give a fairly good reproduction. Note that the amplitude of the sixth harmonic is approximately 12% of the fundamental. We can achieve more faithful reproduction of the original waveform by adding higher harmonic components.

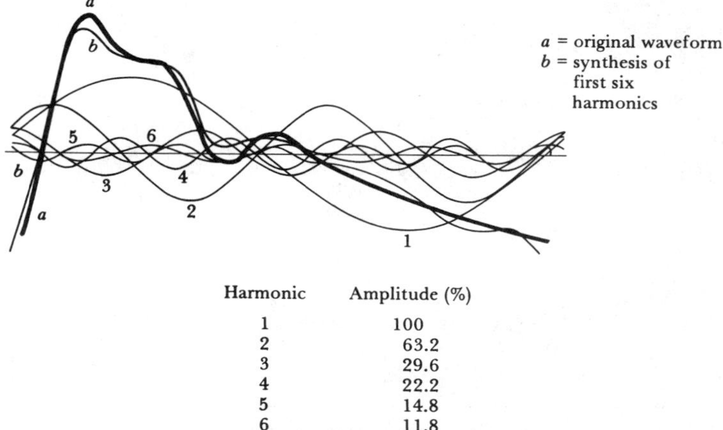

Harmonic	Amplitude (%)
1	100
2	63.2
3	29.6
4	22.2
5	14.8
6	11.8

Figure 7.6 The first six harmonics of the blood-pressure waveform The table gives relative values for amplitudes. (From T. A. Hansen, "Pressure Measurement in the Human Organism," *Acta Physiologica Scandinavica,* 1949, 19, Suppl. 68, 1–227. Used with permission.)

7.3 DYNAMIC PROPERTIES OF PRESSURE-MEASUREMENT SYSTEMS

An understanding of the dynamic properties of a pressure-measurement system is important if we wish to preserve the dynamic accuracy of the measured pressure. Errors in measurement of dynamic pressure can have serious consequences in the clinical situation. For instance, an underdamped system can lead to overestimation of pressure gradients across *stenotic* (narrowed) heart valves. The liquid-filled catheter sensor is a hydraulic system that can be represented by either distributed- or lumped-parameter models. Distributed-parameter models are described in the literature (Fry, 1960) which gives an accurate description of the dynamic behavior of the catheter–sensor system. However, distributed-parameter models are not normally employed, because the single-degree-of-freedom (lumped-parameter) model is easier to work with, and the accuracy of the results obtained by using these models is acceptable for the clinical situation.

ANALOGOUS ELECTRIC SYSTEMS

The modeling approach taken here develops a lumped-parameter model for the catheter and sensor separately and shows how, with appropriate approximations, it reduces to the lumped-parameter model for a second-order system.

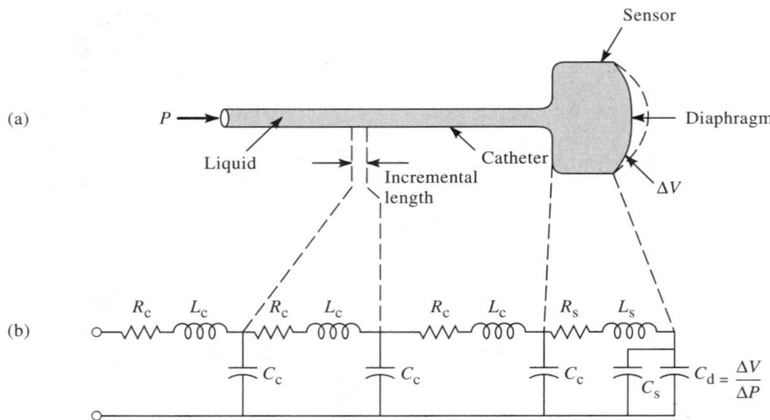

Figure 7.7 (a) Physical model of a catheter–sensor system. (b) Analogous electric system for this catheter–sensor system. Each segment of the catheter has its own resistance R_c, inertance L_c, and compliance C_c. In addition, the sensor has resistance R_s, inertance, L_s, and compliance C_s. The compliance of the diaphragm is C_d.

Figure 7.7 shows the physical model of a catheter–sensor system. An increase in pressure at the input of the catheter causes a flow of liquid to the right from the catheter tip, through the catheter, and into the sensor. This liquid shift causes a deflection of the sensor diaphragm, which is sensed by an electromechanical system. The subsequent electric signal is then amplified.

A liquid catheter has inertial, frictional, and elastic properties represented by inertance, resistance, and compliance, respectively. Similarly, the sensor has these same properties, in addition to the compliance of the diaphragm. Figure 7.7(b) shows an electric analog of the pressure-measuring system, wherein the analogous elements for hydraulic inertance, resistance, and compliance are electric inductance, resistance, and capacitance, respectively.

The analogous circuit in Figure 7.7(b) can be simplified to that shown in Figure 7.8(a). The compliance of the sensor diaphragm is much larger than that of the liquid-filled catheter or sensor cavity, provided that the saline solution is bubble-free and the catheter material is relatively noncompliant. The resistance and inertance of the liquid in the sensor can be neglected compared to those of the liquid in the catheter. Let us now derive equations relating the resistance and inductance to the properties of the system.

The liquid resistance R_c of the catheter is due to friction between shearing molecules flowing through the catheter. It can be represented by the equation

$$R_c = \frac{\Delta P}{F} \ (\text{Pa} \cdot \text{s/m}^3) \tag{7.1}$$

or

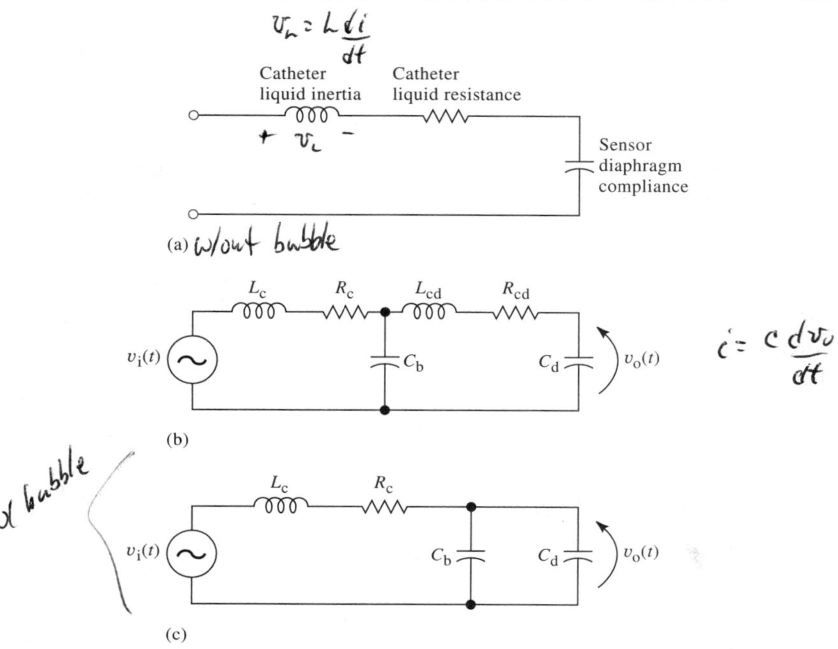

Figure 7.8 (a) Simplified analogous circuit. Compliance of the sensor diaphragm is larger than compliance of catheter or sensor cavity for a bubble-free, noncompliant catheter. The resistance and inertance of the catheter are larger than those of the sensor, because the catheter has longer length and smaller diameter. (b) Analogous circuit for catheter–sensor system with a bubble in the catheter. Catheter properties proximal to the bubble are inertance L_c and resistance R_c. Catheter properties distal to the bubble are L_{cd} and R_{cd}. Compliance of the diaphragm is C_d; compliance of the bubble is C_b. (c) Simplified analogous circuit for catheter–sensor system with a bubble in the catheter, assuming that L_{cd} and R_{cd} are negligible with respect to R_c and L_c.

$$R_c = \frac{\Delta P}{\bar{u}A}$$

where

P = pressure difference across the segment in Pa (pascal = N/m^2)
F = flow rate, m^3/s
$\bar{u}$ = average velocity, m/s
A = cross-sectional area, m^2

Poiseuille's equation enables us to calculate R_c when we are given the values of catheter length L, in meters; radius r, in meters; and liquid viscosity η, in pascal-seconds. The equation applies for laminar or Poiseuille flow. It is

$$R_c = \frac{8\eta L}{\pi r^4} \tag{7.2}$$

The liquid inertance L_c of the catheter is due primarily to the mass of the liquid. It can be represented by the equation

$$L_c = \frac{\Delta P}{dF/dt} \text{ (Pa} \cdot \text{s}^2/\text{m}^3) \tag{7.3}$$

or

$$L_c = \frac{\Delta P}{aA}$$

where a = acceleration, m/s^2.

This equation reduces further to

$$L_c = \frac{m}{A^2}$$

or

$$L_c = \frac{\rho L}{\pi r^2} \tag{7.4}$$

where m = mass of liquid (kg) and ρ = density of liquid (kg/m^3).

Equations (7.2) and (7.4) show that we can neglect the resistive and inertial components of the sensor with respect to those of the liquid catheter. The reason for this is that the liquid-filled catheter is longer than the cavity of the sensor and of smaller diameter. Geddes (1970) develops a more refined model of fluid inertance—one based on kinetic energy considerations—in which he considers that the effective mass is four-thirds times that of the fluid in the catheter.

The compliance C_d of the sensor diaphragm is given by the equation

$$C_d = \frac{\Delta V}{\Delta P} = \frac{1}{E_d}$$

where E_d is the volume modulus of elasticity of the sensor diaphragm.

We can find the relationship between the input voltage v_i, analogous to applied pressure, and the output voltage v_o, analogous to pressure at the diaphragm, by using Kirchhoff's voltage law. Thus,

$$v_i(t) = \frac{L_c C_d d^2 v_o(t)}{dt^2} + \frac{R_c C_d dv_o(t)}{dt} + v_o(t) \tag{7.5}$$

Table 7.1 Mechanical Characteristics of Fluids

Parameter	Substance	Temperature	Value
η	water	20 °C	0.001 Pa · s
η	water	37 °C	0.0007 Pa · s
η	air	20 °C	0.000018 Pa · s
ρ	air	20 °C	1.21 kg/m^3
$\Delta V/\Delta P$	water	20 °C	0.53×10^{-15} m^5/N per milliliter volume
η	blood	all	$\cong 4 \times \eta$ for water

Using the general form of a second-order system equation derived in Section 1.10, we can show that the natural undamped frequency ω_n is $1/(L_c C_d)^{1/2}$ and that the damping ratio ζ is $(R_c/2)(C_d/L_c)^{1/2}$. For the hydraulic system under study, by substituting (7.2) and (7.4) into the expressions for ω_n and ζ, we can show that

$$f_n = \frac{r}{2}\left(\frac{1}{\pi \rho L}\frac{\Delta P}{\Delta V}\right)^{1/2} \tag{7.6}$$

and

$$\zeta = \frac{4\eta}{r^3}\left(\frac{L(\Delta V/\Delta P)}{\pi \rho}\right)^{1/2} \tag{7.7}$$

Table 7.1 lists a number of useful relationships and pertinent constants.

We can study the transient response and the frequency response of the catheter-sensor system by means of the analogous electric circuit. In addition, we can study the effects of changes in the hydraulic system by adding appropriate elements to the circuit. For example, an air bubble in the liquid makes the system more compliant. Thus its effect on the system is the same as that caused by connecting an additional capacitor in parallel to that representing the diaphragm compliance. Example 7.1 illustrates how the analogous circuit is used.

EXAMPLE 7.1 A 5-mm-long air bubble has formed in the rigid-walled catheter connected to a Statham P23Dd sensor. The catheter is 1 m long, 6 French diameter, and filled with water at 20 °C. (The isothermal compression of air $\Delta V/\Delta P$ is 1 ml per cm of water pressure per liter of volume.) Plot the frequency-response curve of the system with and without the bubble. (Internal radius of the catheter is 0.46 mm; volume modulus of elasticity of the diaphragm is 0.49×10^{15} N/m^5.)

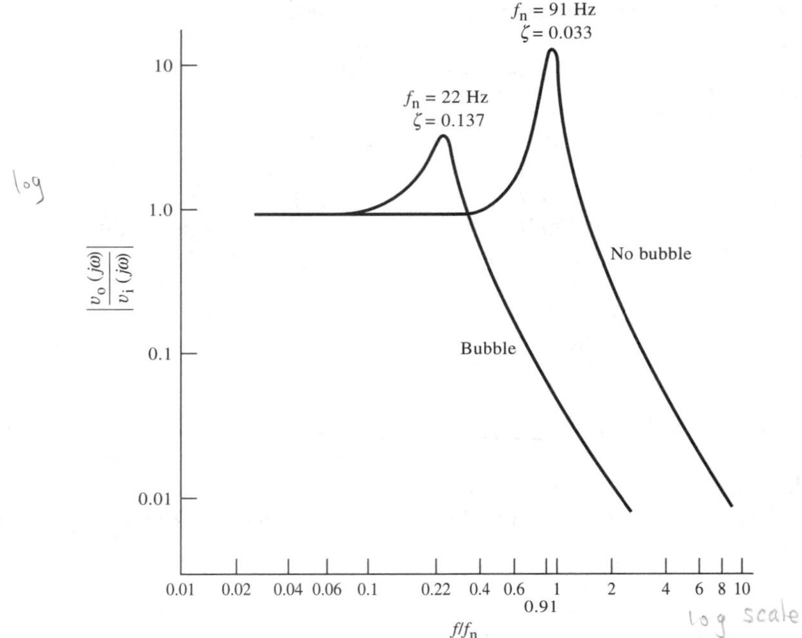

Figure 7.9 Frequency-response curves for catheter–sensor system with and without bubbles. Natural frequency decreases from 91 Hz to 22 Hz and damping ratio increases from 0.033 to 0.137 with the bubble present.

ANSWER The analogous circuit for the hydraulic system with and without the bubble is shown in Figure 7.8(b) and (c). We can calculate the values of the natural frequency f_n and the damping ratio ζ without the bubble by using (7.6) and (7.7). That is,

$$f_n = \frac{r}{2}\left(\frac{1}{\pi L}\frac{\Delta P}{\rho \Delta V}\right)^{1/2}$$

$$= \frac{0.046 \times 10^{-2}}{2}\left(\frac{1}{\pi(1)}\frac{0.49 \times 10^{15}}{1 \times 10^{3}}\right)^{1/2} = 91 \text{ Hz}$$

$$\zeta = \frac{4\eta}{r^3}\left(\frac{L}{\pi\rho}\frac{\Delta V}{\Delta P}\right)^{1/2}$$

$$= \frac{4(0.001)}{(0.046 \times 10^{-2})^3}\left(\frac{1}{\pi}\frac{1}{(1 \times 10^{3})(0.49 \times 10^{15})}\right)^{1/2} = 0.033$$

The frequency response for the catheter-sensor system is shown in Figure 7.9.

The next step is to calculate the new values of ζ and f_n for the case in which a bubble is present. Because the two capacitors are in parallel, the total capacitance for the circuit is equal to the sum of these two. That is,

$$C_t = C_d + C_b \tag{7.8}$$

or

$$C_t = \frac{\Delta V}{\Delta P_d} + \frac{\Delta V}{\Delta P_b}$$

The value of $\Delta V/\Delta P_d = 1/E_d = 2.04 \times 10^{-15}\ \text{m}^5/\text{N}$. The volume of the bubble is

$$\pi r^2 l = 3.33 \times 10^{-9}\ \text{m}^3 = 3.33 \times 10^{-6}\ \text{l}.$$

One centimeter of water pressure is 98.5 N/m^2. Thus

$$\Delta V/\Delta P_b = \frac{3.33 \times 10^{-9}\,\text{l} \times (1 \times 10^{-3}\ \text{m}^3/\text{l})}{98.5\ \text{N/m}^2}$$

$$= 3.38 \times 10^{-14}\ \text{m}^5/\text{N}$$

Consequently, $C_t = 3.38 \times 10^{-14}\ \text{m}^5/\text{N}$. We can find the new values for f_n and ζ by referring to (7.6) and (7.7) and assuming that the only parameter that changes is the value of $\Delta V/\Delta P$. Thus

$$f_{n,\text{bubble}} = f_{n,\text{no bubble}} \left(\frac{\Delta P \Delta V_{\text{total}}}{\Delta P / \Delta V_{\text{no bubble}}} \right)^{1/2}$$

or

$$f_{n,\text{bubble}} = 92 \left(\frac{2.04 \times 10^{-15}}{3.38 \times 10^{-14}} \right)^{1/2} = 22\ \text{Hz}$$

and

$$\zeta_{\text{bubble}} = \zeta_{\text{no bubble}} \left(\frac{\Delta V / \Delta P_{\text{total}}}{\Delta V / \Delta P_{\text{no bubble}}} \right)^{1/2} = 0.137$$

The frequency response for the system with the bubble present is shown in Figure 7.9. Note that the bubble lowers f_n and increases ζ. This lowering of f_n may cause distortion problems with the higher harmonics of the blood-pressure waveform.

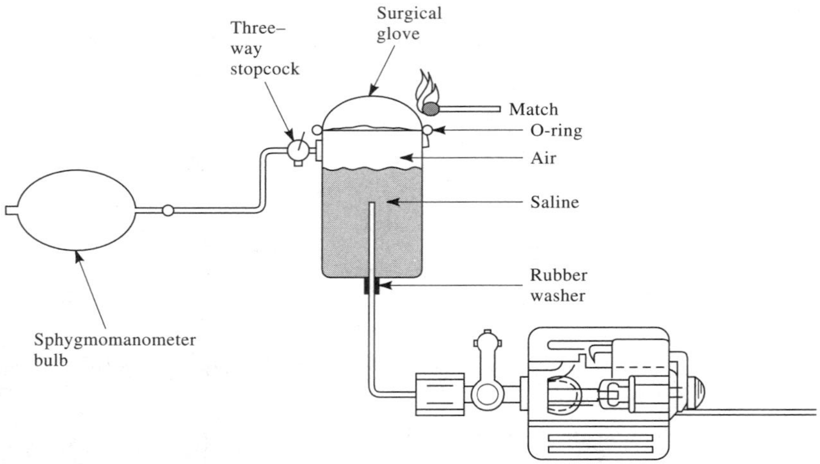

Figure 7.10 Transient-response technique for testing a pressure-sensor–catheter-sensor system.

7.4 MEASUREMENT OF SYSTEM RESPONSE

The response characteristics of a catheter-sensor system can be determined by two methods. The simplest and most straightforward technique involves measuring the transient step response for the system. A potentially more accurate method—but a more complicated one because it requires special equipment—involves measuring the frequency response of the system.

TRANSIENT STEP RESPONSE

The basis of the transient-response method is to apply a sudden step input to the pressure catheter and record the resultant damped oscillations of the system. This is also called the *pop* technique, for reasons that will become evident in the following discussion. The transient response can be found by the method shown in Figure 7.10.

The catheter, or needle, is sealed in a tube by a screw adaptor that compresses a rubber washer against the insert. Flushing of the system is accomplished by passing excess liquid out of the three-way stopcock. The test is performed by securing a rubber membrane over the tube by means of an O ring. Surgical-glove material is an excellent choice for the membrane. The technician pressurizes the system by squeezing the sphygmomanometer bulb, punctures the balloon with a burning match or hot soldering iron, and observes the response. The response should be observed on a recorder running at a speed that makes it possible to distinguish the individual oscillations. However,

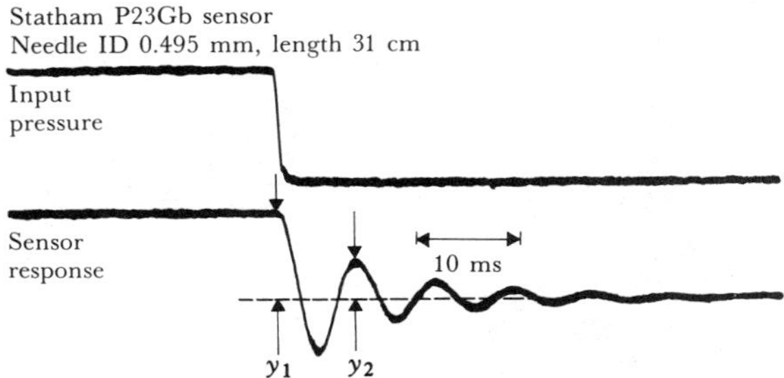

Figure 7.11 Pressure-sensor transient response Negative-step input pressure is recorded on the top channel; the bottom channel is sensor response for a Statham P23Gb sensor connected to a 31-cm needle (0.495 mm ID). (From I. T. Gabe, "Pressure Measurement in Experimental Physiology," in D. H. Bergel, ed., *Cardiovascular Fluid Dynamics,* vol. I, New York: Academic Press, 1972.)

if the frequency bandwidth of the recorder is inadequate, the technician can use a storage oscilloscope or data acquisition system.

Figure 7.11 shows an example of the transient response. In this case the response represents a second-order system. The technician can measure the amplitude ratio of successive positive peaks and determine the logarithmic decrement Λ. Equation (1.38) yields the damping ratio ζ. The observer can measure T, the time between successive positive peaks, and determine the undamped natural frequency from $\omega_n = 2\pi/[T(1 - \zeta^2)^{1/2}]$.

SINUSOIDAL FREQUENCY RESPONSE

As we noted before, the sinusoidal frequency-response method is more complex because it requires more specialized equipment. Figure 7.12 is a schematic diagram of a sinusoidal pressure-generator test system. A pump produces sinusoidal pressures that are normally monitored at the pressure source by a pressure sensor with known characteristics. This is used because the amplitudes of the source-pressure waveforms are not normally constant for all frequencies. The source pressure is coupled to the catheter sensor under test by means of bubble-free saline. The air is removed by boiling the liquid.

We can find an accurate model for the catheter–sensor system by determining the amplitude and phase of the output as a function of frequency without the constraint of the second-order system model required in the transient-response case. In some cases, resonance at more than one frequency may be present.

7.5 EFFECTS OF SYSTEM PARAMETERS ON RESPONSE

We have shown in our discussion of the model of the catheter-sensor system that the values of the damping ratio ζ and natural frequency ω_n are functions

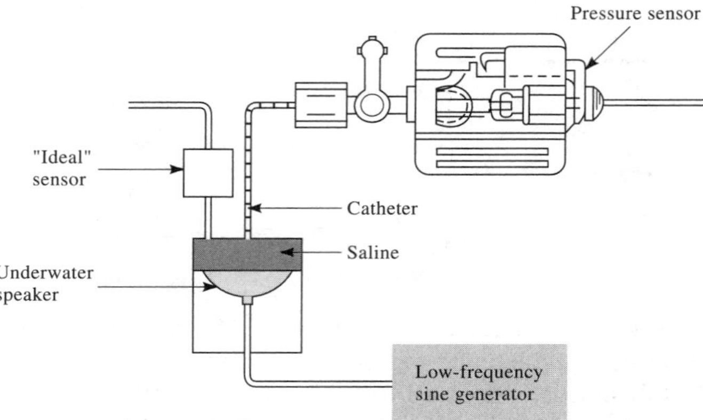

Figure 7.12 A sinusoidal pressure-generator test system A low-frequency sine generator drives an underwater-speaker system that is coupled to the catheter of the pressure sensor under test. An "ideal" pressure sensor, with a frequency response from 0 to 100 Hz, is connected directly to the test chamber housing and monitors input pressure.

of the various system parameters. This section reports on experimental verification of these theoretical derivations (Shapiro and Krovetz, 1970). By using step transient-response and sinusoidal pressure-generation techniques similar to those described above, these investigators determined the effects on the performance of the catheter-sensor system of deaerating water and of using various catheter materials and connectors. They found that even minute air bubbles, which increase the compliance of the catheter manometer system, drastically decreased the damped natural frequency $\omega_d = 2\pi/T$. For a PE-190 catheter of lengths 10 to 100 cm, the damped natural frequency decreased by approximately 50 to 60% for an unboiled-water case compared to a boiled-water case. Length of the catheter was shown to be inversely related to the damped natural frequency for Teflon and polyethylene catheters for the diameters tested (0.58 to 2.69 mm). The theoretical linear relationship between the damped natural frequency and $1/(\text{catheter length})^{1/2}$ seemed to hold within experimental errors. A linear relationship was found to exist between the inner diameter of the catheter and the damped natural frequency for both polyethylene and Teflon catheters, as predicted by the model equations.

In comparing the effect of catheter material on frequency response, Shapiro and Krovetz (1970) found that because Teflon is slightly stiffer than polyethylene, it has a slightly higher frequency response at any given length. As expected, the increased compliance of silicone rubber tubing caused a marked decrease in frequency response. The authors concluded that silicone rubber is a poor material for determining parameters other than mean pressure.

They examined the effect of connectors on the system response by inserting—in series with the catheter—various connecting needles that added little to the overall length of the system. They found that the damped natural frequency was linearly related to the needle bore for needles of the same length. The connector serves as a simple series hydraulic damper that decreases the frequency response. They suggested that the fewest possible number of connectors be used and that all connectors be tight-fitting and have a water seal.

In further tests, they found that coils and bends in the catheter cause changes in the resonant frequency. However, the magnitude of these changes was insignificant compared with changes caused by factors that affected compliance.

7.6 BANDWIDTH REQUIREMENTS FOR MEASURING BLOOD PRESSURE

When we know the representative harmonic components of the blood-pressure waveform—or, for that matter, any periodic waveform—we can specify the bandwidth requirements for the instrumentation system. As with all biomedical measurements, bandwidth requirements are a function of the investigation.

For example, if the mean blood pressure is the only parameter of interest, it is of little value to try to achieve a wide bandwidth system. It is generally accepted that harmonics of the blood-pressure waveform higher than the tenth may be ignored. As an example, the bandwidth requirements for a heart rate of 120 beats/min (or 2 Hz) would be 20 Hz.

For a perfect reproduction of the original waveform, there should be no distortion in the amplitude or phase characteristics. The waveshape can be preserved, however, even if the phase characteristics are not ideal. This is the case if the relative amplitudes of the frequency components are preserved but their phases are displaced in proportion to their frequency. Then the synthesized waveform gives the original waveshape, except that it is delayed in time, depending on the phase shift.

Measurements of the derivative of the pressure signal increase the bandwidth requirements, because the differentiation of a sinusoidal harmonic increases the amplitude of that component by a factor proportional to its frequency. As with the original blood-pressure waveform, the bandwidth requirements for the derivative of the blood pressure can be estimated by a Fourier analysis of the derivative signal. The amplitude-versus-frequency characteristics of any catheter-manometer system used for the measurement of ventricular pressures that are subsequently differentiated must remain flat to within 5%, up to the twentieth harmonic (Gersh *et al.*, 1971).

7.7 TYPICAL PRESSURE-WAVEFORM DISTORTION

Accurate measurements of blood pressure are important in both clinical and physiological research. This section gives examples of typical types of distor-

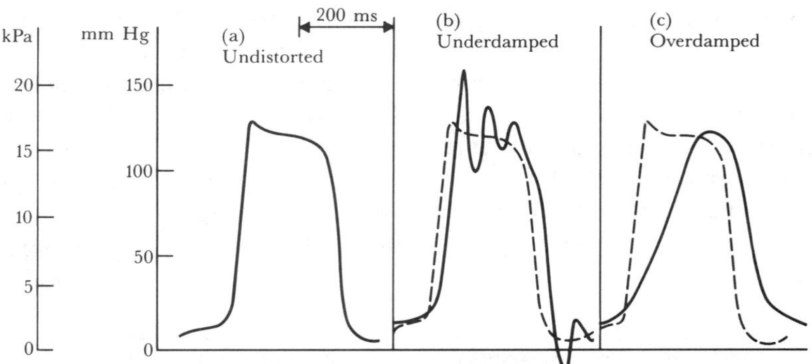

Figure 7.13 Pressure-waveform distortion (a) Recording of an undistorted left-ventricular pressure waveform via a pressure sensor with bandwidth dc to 100 Hz. (b) Underdamped response, where peak value is increased. A time delay is also evident in this recording. (c) Overdamped response that shows a significant time delay and an attenuated amplitude response.

tion of blood-pressure waveform that are due to an inadequate frequency response of the catheter–sensor system. There may be serious consequences when an underdamped system leads to overestimation of the presure gradients across a stenotic (narrowed) heart valve.

Figure 7.13 shows examples of distortion of pressure waveform. The actual blood-pressure waveform [Figure 7.13(a)] was recorded with a high-quality pressure sensor with a bandwidth from dc to 100 Hz. Note that in the underdamped case, the amplitude of the higher-frequency components of the pressure wave are amplified, whereas for the overdamped case these higher-frequency components are attenuated. The actual peak pressure [Figure 7.13(a)] is approximately 130 mm Hg (17.3 kPa). The underdamped response [Figure 7.13(b)] has a peak pressure of about 165 mm Hg (22 kPa), which may lead to a serious clinical error if this peak pressure is used to assess the severity of aortic-valve stenosis. The minimal pressure is in error, too; it is −15 mm Hg (−2 kPa) and the actual value is 5 mm Hg (0.7 kPa). There is also a time delay of approximately 30 ms in the underdamped case.

The overdamped case [Figure 7.13(c)] shows a significant time delay of approximately 150 ms and an attenuated amplitude of 120 mm Hg (16 kPa); the actual value is 130 mm Hg (17.3 kPa). This type of response can occur in the presence of a large air bubble or a blood clot at the tip of the catheter.

An underdamped catheter–sensor system can be transformed to an overdamped system by pinching the catheter. This procedure increases the damping ratio ζ, and has little effect on the natural frequency. (See Problem 7.6.)

Another example of distortion in blood-pressure measurements is known as *catheter whip.* Figure 7.14 shows these low-frequency oscillations that appear in the blood-pressure recording. This may occur when an aortic ventricular catheter, in a region of high pulsatile flow, is bent and whipped about by the

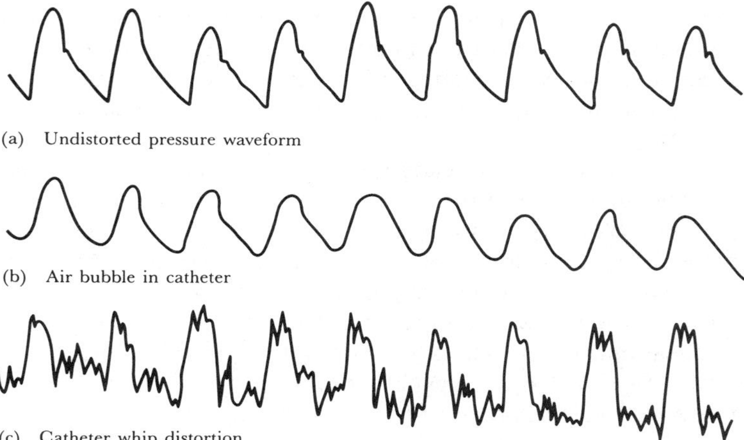

Figure 7.14 Distortion during the recording of arterial pressure The bottom trace is the response when the pressure catheter is bent and whipped by accelerating blood in regions of high pulsatile flow.

accelerating blood. This type of distortion can be minimized by the use of stiff catheters or by careful placement of catheters in regions of low flow velocity.

7.8 SYSTEMS FOR MEASURING VENOUS PRESSURE

Measurements of venous pressure are an important aid to the physician for determining the function of the capillary bed and the right side of the heart. The pressure in the small veins is lower than the capillary pressure and reflects the value of the capillary pressure. The intrathoracic venous pressure determines the diastolic filling pressure of the right ventricle (Rushmer, 1970). The central venous pressure is measured in a central vein or in the right atrium. It fluctuates above and below atmospheric pressure as the subject breathes, whereas the extrathoracic venous pressure is 2 to 5 cm H_2O (0.2 to 0.5 kPa) above atmospheric. The reference level for venous pressure is at the right atrium.

Central venous pressure is an important indicator of myocardial performance. It is normally monitored on surgical and medical patients to assess proper therapy in cases of heart dysfunction, shock, hypovolemic or hypervolemic states, or circulatory failure. It is used as a guide to determine the amount of liquid a patient should receive.

Physicians usually measure steady-state or mean venous pressure by making a percutaneous venous puncture with a large-bore needle, inserting a catheter through the needle into the vein, and advancing it to the desired position. The needle is then removed. A plastic tube is attached to the intravenous catheter by means of a stopcock, which enables clinicians to administer drugs or fluids as necessary. Continuous dynamic measurements of venous pressure can be made by connecting to the venous catheter a high-sensitivity pressure sensor with a lower dynamic range than that necessary for arterial measurements.

Problems in maintaining a steady baseline occur when the patient changes position. Errors may arise in the measurements if the catheter is misplaced or if it becomes blocked by a clot or is impacted against a vein wall. It is normal practice to accept venous-pressure values only when respiratory swings are evident. Normal central venous pressures range widely from 0 to 12 cm H_2O (0 to 1.2 kPa), with a mean pressure of 5 cm H_2O (0.5 kPa).

Esophageal manometry uses a similar low-pressure catheter system (Weihrauch, 1988). A hydraulic capillary infusion system infuses 0.6 ml/min to prevent sealing of the catheter orifice in the esophagus.

7.9 HEART SOUNDS

The auscultation of the heart gives the clinician valuable information about the functional integrity of the heart. More information becomes available when clinicians compare the temporal relationships between the heart sounds and the mechanical and electric events of the cardiac cycle. This latter approach is known as *phonocardiography*.

There is a wide diversity of opinion concerning the theories that attempt to explain the origin of heart sounds and murmurs. More than 40 different mechanisms have been proposed to explain the first heart sound. A basic definition shows the difference between heart sounds and murmurs (Rushmer, 1970). *Heart sounds* are vibrations or sounds due to the acceleration or deceleration of blood, whereas *murmurs* are vibrations or sounds due to blood turbulence.

MECHANISM AND ORIGIN

Figure 7.15 shows how the four heart sounds are related to the electric and mechanical events of the cardiac cycle. The first heart sound is associated with the movement of blood during ventricular systole (Rushmer, 1970). As the ventricles contract, blood shifts toward the atria, closing the atrioventricular valves with a consequential oscillation of blood. The first heart sound further originates from oscillations of blood between the descending root of the aorta and ventricle and from vibrations due to blood turbulence at the aortic and pulmonary valves. Splitting of the first heart sound is defined as an asynchronous closure of the tricuspid and mitral valves. The second heart sound is a low-frequency vibration associated with the deceleration and reversal of flow in the aorta and pulmonary artery and with the closure of the semilunar valves (the valves situated between the ventricles and the aorta or the pulmonary

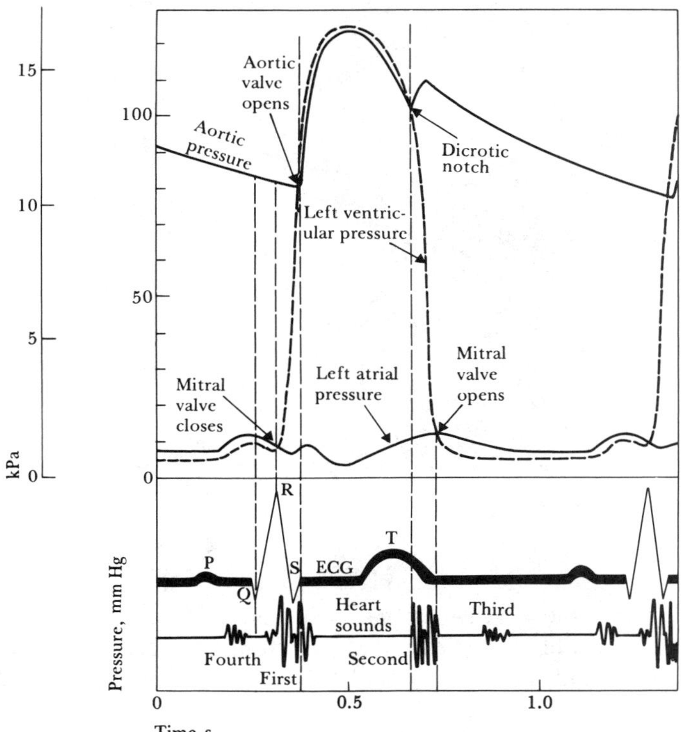

Figure 7.15 Correlation of the four heart sounds with electric and mechanical events of the cardiac cycle.

trunk). This second heart sound is coincident with the completion of the T wave of the ECG.

The third heart sound is attributed to the sudden termination of the rapid-filling phase of the ventricles from the atria and the associated vibration of the ventricular muscle walls, which are relaxed. This low-amplitude, low-frequency vibration is audible in children and in some adults.

The fourth or atrial heart sound—which is not audible but can be recorded by the phonocardiogram—occurs when the atria contract and propel blood into the ventricles.

The sources of most murmurs, developed by turbulence in rapidly moving blood, are known. Murmurs during the early systolic phase are common in children, and they are normally heard in nearly all adults after exercise. Abnormal murmurs may be caused by stenoses and insufficiencies (leaks) at the aortic, pulmonary, and mitral valves. They are detected by noting the time of their occurrence in the cardiac cycle and their location at the time of measurement.

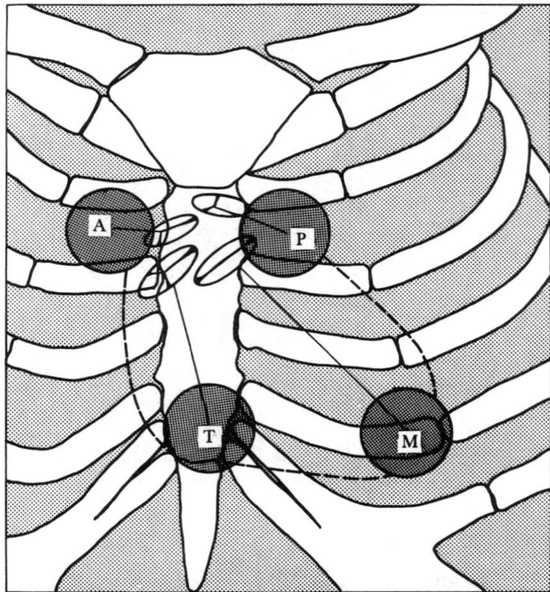

Figure 7.16 Auscultatory areas on the chest A, aortic; P, pulmonary; T, tricuspid; and M, mitral areas. (From A. C. Burton, *Physiology and Biophysics of the Circulation,* 2nd ed. Copyright © 1972 by Year Book Medical Publishers, Inc., Chicago. Used by permission.)

AUSCULTATION TECHNIQUES

Heart sounds travel through the body from the heart and major blood vessels to the body surface. Because of the acoustical properties of the transmission path, sound waves are attenuated and not reflected. The largest attenuation of the wave-like motion occurs in the most compressible tissues, such as the lungs and fat layers.

There are optimal recording sites for the various heart sounds, sites at which the intensity of sound is the highest because the sound is being transmitted through solid tissues or through a minimal thickness of inflated lung. There are four basic chest locations at which the intensity of sound from the four valves is maximized (Figure 7.16).

Heart sounds and murmurs have extremely small amplitudes, with frequencies from 0.1 to 2000 Hz. Two difficulties may result. At the low end of the spectrum (below about 20 Hz), the amplitude of heart sounds is below the threshold of audibility. The high-frequency end is normally quite perceptible to the human ear, because this is the region of maximal sensitivity. However, if a phonocardiogram is desired, the recording device must be carefully selected for high frequency-response characteristics. That is, a light-beam, ink-jet, or digital-array recorder would be adequate, whereas a standard pen strip-chart recorder would not.

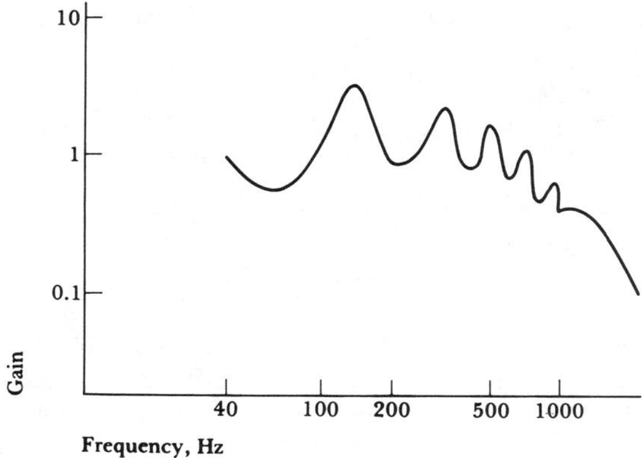

Figure 7.17 The typical frequency-response curve for a stethoscope can be found by applying a known audiofrequency signal to the bell of a stethoscope by means of a headphone–coupler arrangement. The audio output of the stethoscope earpiece was monitored by means of a coupler microphone system. (From P. Y. Ertel, M. Lawrence, R. K. Brown, and A. M. Stern, *Stethoscope Acoustics* I, "The Doctor and his Stethoscope." *Circulation* 34, 1966; by permission of American Heart Association.)

Because heart sounds and murmurs are of low amplitude, extraneous noises must be minimized in the vicinity of the patient. It is standard procedure to record the phonocardiogram for nonbedridden patients in a specially designed, acoustically quiet room. Artifacts from movements of the patient appear as baseline wandering.

STETHOSCOPES

Stethoscopes are used to transmit heart sounds from the chest wall to the human ear. Some variability in interpretation of the sounds stems from the user's auditory acuity and training. Moreover, the technique used to apply the stethoscope can greatly affect the sounds perceived.

Ertel *et al.* (1966a) have investigated the acoustics of stethoscope transmission and the acoustical interactions of human ears with stethoscopes. They found that stethoscope acoustics reflected the acoustics of the human ear. Younger individuals revealed slightly better responses to a stethoscope than their elders. The mechanical stethoscope amplifies sound because of a standing-wave phenomenon that occurs at quarter-wavelengths of the sound. Figure 7.17 is a typical frequency-response curve for a stethoscope; it shows that the mechanical stethoscope has an uneven frequency response, with many resonance peaks.

These investigators emphasized that the critical area of the performance of a stethoscope (the clinically significant sounds near the listener's threshold of hearing) may be totally lost if the stethoscope attenuates them as little as

3 dB. A physician may miss, with one instrument, sounds that can be heard with another.

When the stethoscope chest piece is firmly applied, low frequencies are attenuated more than high frequencies. The stethoscope housing is in the shape of a bell. It makes contact with the skin, which serves as the diaphragm at the bell rim. The diaphragm becomes taut with pressure, thereby causing an attenuation of low frequencies.

Loose-fitting earpieces cause additional problems, because the leak that develops reduces the coupling between the chest wall and the ear, with a consequent decrease in the listener's perception of heart sounds and murmurs.

Many types of electronic stethoscopes have been proposed by engineers. These devices have selectable frequency-response characteristics ranging from the "ideal" flat-response case and selected bandpasses to typical mechanical-stethoscope responses. Physicians, however, have not generally accepted these electronic stethoscopes, mainly because they are unfamiliar with the sounds heard with them. Their size, portability, convenience, and resemblance to the mechanical stethoscope are other important considerations.

7.10 PHONOCARDIOGRAPHY

A phonocardiogram is a recording of the heart sounds and murmurs (Vermariën, 1988). It eliminates the subjective interpretation of these sounds and also makes possible an evaluation of the heart sounds and murmurs with respect to the electric and mechanical events in the cardiac cycle. In the clinical evaluation of a patient, a number of other heart-related variables may be recorded simultaneously with the phonocardiogram. These include the ECG, carotid arterial pulse, jugular venous pulse, and apex cardiogram. The indirect carotid, jugular, and apex-cardiogram pulses are recorded by using a microphone system with a frequency response from 0.1 to 100 Hz. The cardiologist evaluates the results of a phonocardiograph on the basis of changes in waveshape and in a number of timing parameters (Tavel, 1972).

7.11 CARDIAC CATHETERIZATION

The cardiac-catheterization procedure is a combination of several techniques that are used to assess hemodynamic function and cardiovascular structure. Cardiac catheterization is performed in virtually all patients in whom heart surgery is contemplated. This procedure yields information that may be crucial in defining the timing, risks, and anticipated benefit for a given patient (Grossman, 1974). Catheterization procedures are performed in specialized laboratories outfitted with x-ray equipment for visualizing heart structures and the position of various pressure catheters. In addition, measurements are made

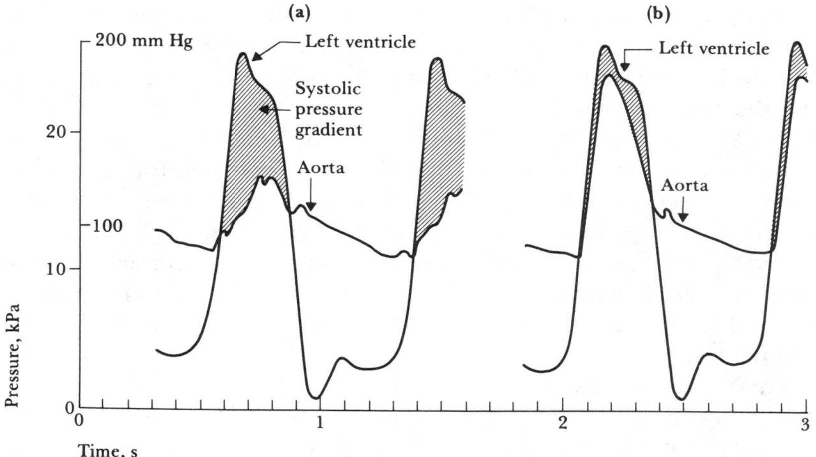

Figure 7.18 (a) Systolic pressure gradient (left ventricular-aortic pressure) across a stenotic aortic valve. (b) Marked decrease in systolic pressure gradient with insertion of an aortic ball valve.

of cardiac output, blood and respiratory gases, blood-oxygen saturation, and metabolic products. The injection of radiopaque dyes into the ventricles or aorta makes it possible for the clinician to assess ventricular or aortic function. In a similar fashion, injection of radiopaque dyes into the coronary arteries makes possible a clinical evaluation of coronary-artery disease. In the following paragraphs, we shall discuss a number of specific procedures carried out in a catheter laboratory.

Clinicians can measure pressures in all four chambers of the heart and in the great vessels by positioning catheters, during fluoroscopy, in such a way that they can recognize the characteristic pressure waveforms. They measure pressures across the four valves to determine the valves' pressure gradients.

An example of a patient with aortic stenosis will help illustrate the procedure. Figure 7.18(a) shows the pressures of the stenotic patient before the operation: Note the pressures in the left ventricle and in the aorta, and the systolic pressure gradient. Figure 7.18(b) reflects the situation after the operation: Note the marked decrease in the pressure gradient brought about by the insertion of a ball-valve aortic prosthesis. These pressures may be measured by using a two-lumen catheter positioned such that the valve is located between the two catheter openings. The clinician can find the various time indices that describe the injection and filling periods of the heart directly from the recordings of blood pressure in the heart.

Clinicians can also use balloon-tipped, flow-directed catheters without fluoroscopy (Ganz and Swan, 1974). An inflated balloon at the catheter tip is carried by the bloodstream from the intrathoracic veins through the right atrium, ventricle, and pulmonary artery and into a small pulmonary artery—

where it is wedged, blocking the local flow. The wedge pressure in this pulmonary artery reflects the mean pressure in the left atrium, because a column of stagnant blood on the right side of the heart joins the free-flowing blood beyond the capillary bed.

This catheter is also commonly used to measure cardiac output using the principle of thermodilution. Cardiac output is valuable for assessing the pumping function of the heart and can also be measured using dye dilution, the Fick method, and impedance cardiography (Sections 8.1, 8.2 and 8.7).

Blood samples can be drawn from within the various heart chambers and vessels where the catheter tip is positioned. These blood samples are important in determining the presence of shunts between the heart chambers or great vessels. For example, a shunt from the left to the right side of the heart is indicated by a higher-than-normal O_2 content in the blood in the right heart in the vicinity of the shunt. The O_2 content is normally determined by an oximeter (Section 10.3). Cardiac blood samples are also used to assess such metabolic end products as lactate, pyruvate, CO_2, and such injected substances as radioactive materials and colored dyes.

Angiographic visualization is an essential tool used to evaluate cardiac structure. Radiopaque dye is injected rapidly into a cardiac chamber or blood vessel, and the hemodynamics are viewed and recorded on x-ray film, movie film, or videotape. (In Section 12.6 we will discuss the principles of radiography and fluoroscopy.) Specially designed catheters and power injectors are used in order that a bolus of contrast material can be delivered rapidly into the appropriate vessel or heart chamber. Standard angiographic techniques are employed, where indicated, in the evaluation of the left and right ventricles *(ventriculography),* the coronary arteries *(coronary arteriography),* the pulmonary artery *(pulmonary angiography),* and the aorta *(aortography).* During heart catheterization, ectopic beats and/or cardiac fibrillation frequently occur. These are usually caused by a mechanical stimulus from the catheter or from a jet of contrast material. For this reason, clinicians must have a functional defibrillator (Section 13.2) readily available in the catheterization laboratory.

The percutaneous translumenal coronary angioplasty (PTCA) catheter is used to enlarge the lumen of stenotic coronary arteries, thereby improving distal flow and relieving symptoms of ischemia and signs of myocardial hypoperfusion. After initial coronary angiography is performed and the coronary lesions are adequately visualized, a guiding catheter is introduced and passed around the aortic arch. The PTCA catheter is then placed over the guidewire and connected to a manifold (for pressure recording and injections) and to the inflation device. The guidewire is generally advanced into the coronary artery, across from and distal to the lesion to be dilated. A balloon catheter is advanced over the wire and placed across the stenosis. The pressure gradient across the stenosis is measured by using the pressure lumens on the PTCA catheter. This measurement is done to determine the severity of the stenosis. The balloon is repeatedly inflated—usually for 30 to 60 s each time—until the stenosis is fully expanded. Test injections are performed to determine whether the coronary artery flow has been improved. (Clinicians observe the distal runoff by using a radiopaque dye.)

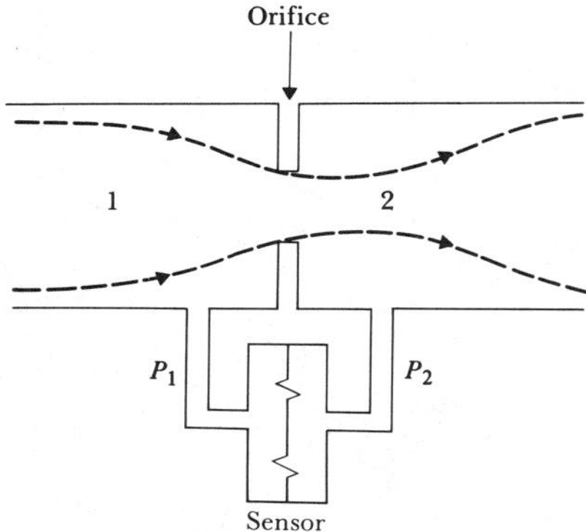

Figure 7.19 Model for deriving equation for heart-valve orifice area P_1 and P_2 are upstream and downstream static pressures. Velocity u is calculated for minimal flow area A at location 2.

A successful PTCA is an alternative to coronary by-pass surgery for a large proportion of patients with coronary artery disease. It avoids the morbidity associated with thoracotomy, cardiopulmonary by-pass, and general anesthesia. In addition, the hospital stay is much shorter. The patient can be discharged about two days after the procedure. Restenosis following coronary angioplasty has been shown to recur in 15 to 35% of cases.

Areas of a valve orifice can be calculated from basic fluid-mechanics equations (Herman *et al.,* 1974). Physicians can assess valvular stenosis by measuring the pressure gradient across the valve of interest and the flow through it.

Bernoulli's equation for frictionless flow (Burton, 1972) is

$$P_t = P + \rho g h + \frac{\rho u^2}{2} \qquad (7.8)$$

where

P_t = fluid total pressure
P = local fluid static pressure
ρ = fluid density
g = acceleration of gravity (Appendix A.1)
h = height above reference level
u = fluid velocity

We first assume frictionless flow for the model shown in Figure 7.19 and equate total pressures at locations 1 and 2. We assume that the difference in heights is negligible and that the velocity at location 1 is negligible compared with u, the velocity at location 2. Then (7.8) reduces to

$$P_1 - P_2 = \frac{\rho u^2}{2} \tag{7.9}$$

from which

$$u = \left[\frac{2(P_1 - P_2)}{\rho}\right]^{1/2}$$

At location 2, the flow $F = Au$, where A is the area. Hence

$$A = \frac{F}{u} = F\left[\frac{\rho}{2(P_1 - P_2)}\right]^{1/2} \tag{7.10}$$

In practice, there are losses due to friction, and the minimal flow area is smaller than the orifice area. Hence (7.10) becomes

$$A = \frac{F}{c_d}\left[\frac{\rho}{2(P_1 - P_2)}\right]^{1/2} \tag{7.11}$$

where c_d is a discharge coefficient. It has been found empirically that for semilunar valves, septal defects, and patent ductus, $c_d = 0.85$, whereas for mitral valves, $c_d = 0.6$ (Yellin *et al.*, 1975).

EXAMPLE 7.3 Calculate the approximate area of the aortic valve for the patient with the aortic and left-ventricular pressures shown in Figure 7.18(a). The patient's cardiac output was measured by thermodilution as 6400 ml/min and the heart rate as 78 beats/min. Blood density is 1060 kg/m^3.

ANSWER From Figure 7.18(a), the ejection period is 0.31 s and the average pressure drop is 7.33 kPa. During the ejection period, the flow (in SI units) is

$$F = (6.4 \times 10^{-3}\ \text{m}^3/\text{min})(1/78\ \text{min/beat})(1/0.31\ \text{beat/s})$$

$$= 264 \times 10^{-6}\ \text{m}^3/\text{s}$$

From (7.11) we have

$$A = \frac{264 \times 10^{-6}}{0.85}\left[\frac{1060}{2(7330)}\right]^{1/2}$$

$$= 83 \times 10^{-6}\ \text{m}^2 = 83\ \text{mm}^2$$

7.12 EFFECTS OF POTENTIAL AND KINETIC ENERGY ON PRESSURE MEASUREMENTS

In certain situations, the effects of potential- and kinetic-energy terms in the measurement of blood pressure may yield inaccurate results.

Bernoulli's equation (7.8) shows that the total pressure of a fluid remains constant in the absence of dissipative effects. The static pressure P of the fluid is the desired pressure; it is measured in a blood vessel when the potential- and kinetic-energy terms are zero.

We first examine the effect of the potential-energy term on the static pressure of the fluid. When measurements of blood pressure are taken with the patient in a supine (on-the-back) position and with the sensor so placed that it is at heart level, no corrections need be made for the potential-energy term. However, when the patient is sitting or standing, the long columns of blood in the arterial and venous pressure systems contribute a hydrostatic pressure, ρgh.

For a patient in the erect position, the arterial and venous pressure both increase to approximately 85 mm Hg (11.3 kPa) at the ankle. When the arm is held above the head, the pressure in the wrist becomes about 40 mm Hg (5.3 kPa). The sensor diaphragm should be placed at the same level as the pressure source. If this is not possible, the difference in height must be accounted for. For each 1.3-cm increase in height of the source, 1.0 mm Hg (133 Pa) must be added to the sensor reading.

The kinetic-energy term $\rho u^2/2$ becomes important when the velocity of blood flow is high. When a blood-pressure catheter is inserted into a blood vessel or into the heart, two types of pressures can be determined—side (static) and end (total) pressures. "Side pressure" implies that the end of the catheter has openings at right angles to the flow. In this case, the pressure reading is accurate because the kinetic-energy term is minimal. However, if the catheter pressure port is in line with the flow stream, then the kinetic energy of the fluid at that point is transformed into pressure. If the catheter pressure port faces *up*stream, the recorded pressure is the side pressure plus the additional kinetic-energy term $\rho u^2/2$. On the other hand, if the catheter pressure port faces *down*stream, the value is approximately $\rho u^2/2$ less than the side pressure. When the catheter is not positioned correctly, artifacts may develop in the pressure reading.

The data given in Table 7.2 demonstrate the relative importance of the kinetic-energy term in different parts of the circulation (Burton, 1972). As Table 7.2 shows, there are situations in the aorta, venae cavae, and pulmonary artery in which the kinetic-energy term is a substantial part of the total pressure. For the laminar-flow case, this error decreases as the catheter pressure port is moved from the center of the vessel to the vessel wall, where the average velocity of flow is less. The kinetic-energy term could also be important in a disease situation in which an artery becomes narrowed.

7.13 INDIRECT MEASUREMENTS OF BLOOD PRESSURE

Indirect measurement of blood pressure is an attempt to measure intraarterial pressures noninvasively. The most standard manual techniques employ either

Table 7.2 Relative Importance of the Kinetic-Energy Term in Different Parts of the Circulation

Vessel	Vel (cm/s)	KE (mm Hg)	Systolic (mm Hg)	Systolic (kPa)	% KE of Total
Aorta (systolic)					
At rest	100	4	120	(16)	3
Cardiac output at 3 × rest	300	36	180	(24)	17
Brachial artery					
At rest	30	0.35	110	(14.7)	0.3
Cardiac output at 3 × rest	90	4	120	(16)	3
Venae cavae					
At rest	30	0.35	2	(0.3)	12
Cardiac output at 3 × rest	90	3.2	3	(0.4)	52
Pulmonary artery					
At rest	90	3	20	(2.7)	13
Cardiac output at 3 × rest	270	27	25	(3.3)	52

SOURCE: From A. C. Burton, *Physiology and Biophysics of the Circulation.* Copyright 1972 by Year Book Medical Publishers, Inc., Chicago. Used by permission.

the palpation or the auditory detection of the pulse distal to an occlusive cuff. Figure 7.20 shows a typical system for indirect measurement of blood pressure. It employs a sphygmomanometer consisting of an inflatable cuff for occlusion of the blood vessel, a rubber bulb for inflation of the cuff, and either a mercury or an aneroid manometer for detection of pressure.

Blood pressure is measured in the following way. The occlusive cuff is inflated until the pressure is above systolic pressure and then is slowly bled off (2–3 mm Hg/s) (0.3–0.4 kPa/s). When the systolic peaks are higher than the occlusive pressure, the blood spurts under the cuff and causes a palpable pulse in the wrist *(Riva-Rocci method).* Audible sounds *(Korotkoff sounds)* generated by the flow of blood and vibrations of the vessel under the cuff are heard through a stethoscope. The manometer pressure at the first detection of the pulse indicates the systolic pressure. As the pressure in the cuff is decreased, the audible Korotkoff sounds pass through five phases (Geddes, 1970). The period of transition from muffling (phase IV) to silence (phase V) brackets the diastolic pressure.

In employing the palpation and auscultatory techniques, you should take several measurements, because normal respiration and vasomotor waves modulate the normal blood-pressure levels. These techniques also suffer from the disadvantage of failing to give accurate pressures for infants and hypotensive patients.

Using an occlusive cuff of the correct size is important if the clinician is to obtain accurate results. The pressure applied to the artery wall is assumed to be equal to that of the external cuff. However, the cuff pressure is transmitted via interposed tissue. With a cuff of sufficient width and length, the cuff

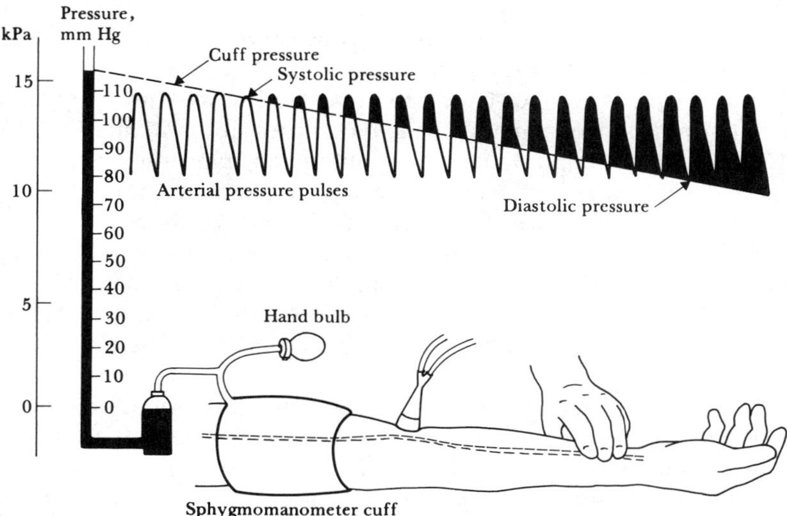

Figure 7.20 Typical indirect blood-pressure measurement system The sphygmomanometer cuff is inflated by a hand bulb to pressures above the systolic level. Pressure is then slowly released, and blood flow under the cuff is monitored by a microphone or stethoscope placed over a downstream artery. The first Korotkoff sound detected indicates systolic pressure, whereas the transition from muffling to silence brackets diastolic pressure. (From R. F. Rushmer, *Cardiovascular Dynamics,* 3rd ed., 1970. Philadelphia: W. B. Saunders Co. Used with permission.)

pressure is evenly transmitted to the underlying artery. It is generally accepted that the width of the cuff should be about 0.40 times the circumference of the extremity. However, no general agreement appears to exist about the length of the pneumatic cuff (Geddes, 1970). If a short cuff is used, it is important that it be positioned over the artery of interest. A longer cuff reduces the problem of misalignment. The cuff should be placed at heart level to avoid hydrostatic effects.

The auscultatory technique is simple and requires a minimum of equipment. However, it cannot be used in a noisy environment, whereas the palpation technique can. The hearing acuity of the user must be good for low frequencies from 20 to 300 Hz, the bandwidth required for these measurements. Bellville and Weaver (1969) have determined the energy distribution of the Korotkoff sounds for normal patients and for patients in shock. When there is a fall in blood pressure, the sound spectrum shifts to lower frequencies. The failure of the auscultation technique for hypotensive patients may be due to low sensitivity of the human ear to these low-frequency vibrations (Geddes, 1970).

There is a common misconception that normal human blood pressure is 120/80, meaning that the systolic value is 120 mm Hg (16 kPa) and that the diastolic value is 80 mm Hg (10.7 kPa). This is not the case. A careful study (by Master *et al.,* 1952) showed that the age and sex of an individual determine the "normal value" of blood pressure.

A number of techniques have been proposed to measure automatically and indirectly the systolic and diastolic blood pressure in humans (Cobbold, 1974). The basic technique involves an automatic sphygmomanometer that inflates and deflates an occlusive cuff at a predetermined rate. A sensitive detector is used to measure the distal pulse or cuff pressure. A number of kinds of detectors have been employed, including ultrasonic, piezoelectric, photoelectric, electroacoustic, thermometric, electrocardiographic, rheographic, and tissue-impedance devices (Greatorex, 1971; Visser and Muntinga, 1990). Three of the commonly used automatic techniques are described in the following paragraphs.

The first technique employs an automated auscultatory device wherein a *microphone* replaces the stethoscope. The cycle of events that takes place begins with a rapid (20–30 mm Hg/s) (2.7–4 kPa/s) inflation of the occlusive cuff to a preset pressure about 30 mm Hg higher than the suspected systolic level. The flow of blood beneath the cuff is stopped by the collapse of the vessel. Cuff pressure is then reduced slowly (2–3 mm Hg/s) (0.3–0.4 kPa/s). The first Korotkoff sound is detected by the microphone, at which time the level of the cuff pressure is stored. The muffling and silent period of the Korotkoff sounds is detected, and the value of the diastolic pressure is also stored. After a few minutes, the instrument displays the systolic and diastolic pressures and recycles the operation. Design considerations for various types of automatic indirect methods of measurement of blood pressure can be found in the literature (Greatorex, 1971).

The *ultrasonic* determination of blood pressure employs a transcutaneous Doppler sensor that detects the motion of the blood-vessel walls in various states of occlusion. Figure 7.21 shows the placement of the compression cuff over two small transmitting and receiving ultrasound crystals (8 MHz) on the arm (Stegall *et al.,* 1968). The Doppler ultrasonic transmitted signal is focused on the vessel wall and the blood. The reflected signal (shifted in frequency) is detected by the receiving crystal and decoded (Section 8.4). The difference in frequency, in the range of 40 to 500 Hz, between the transmitted and received signals is proportional to the velocity of the wall motion and the blood velocity. As the cuff pressure is increased above diastolic but below systolic, the vessel opens and closes with each heartbeat, because the pressure in the artery oscillates above and below the applied external pressure in the cuff. The opening and closing of the vessel are detected by the ultrasonic system.

As the applied pressure is further increased, the time between the opening and closing decreases until they coincide. The reading at this point is the *systolic pressure.* Conversely, when the pressure in the cuff is reduced, the time between opening and closing increases until the closing signal from one pulse coincides with the opening signal from the next. The reading at this point is the *diastolic pressure,* which prevails when the vessel is open for the complete pulse.

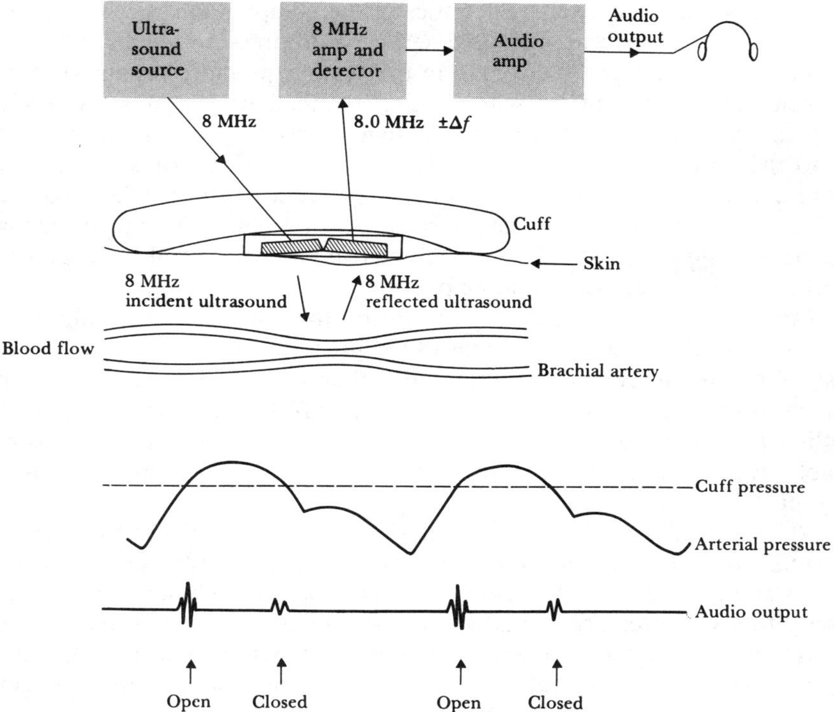

Figure 7.21 Ultrasonic determination of blood pressure A compression cuff is placed over the transmitting (8 MHz) and receiving (8 MHz $\pm \Delta f$) crystals. The opening and closing of the blood vessel are detected as the applied cuff pressure is varied. (From H. F. Stegall, M. B. Kardon, and W. T. Kemmerer, "Indirect Measurement of Arterial Blood Pressure by Doppler Ultrasonic Sphygmomanometry," *J. Appl. Physiol.*, 1968, 25, 793–798. Used with permission.)

The advantages of the ultrasonic technique are that it can be used with infants and hypotensive individuals and in high-noise environments. A disadvantage is that movements of the subject's body cause changes in the ultrasonic path between the sensor and the blood vessel. Complete reconstruction of the arterial-pulse waveform is also possible via the ultrasonic method. A timing pulse from the ECG signal is used as a reference. The clinician uses the pressure in the cuff when the artery opens versus the time from the ECG R wave to plot the rising portion of the arterial pulse. Conversely, the clinician uses the cuff pressure when the artery closes versus the time from the ECG R wave to plot the falling portion of the arterial pulse.

The oscillometric method, a noninvasive blood pressure technique, measures the amplitude of oscillations that appear in the cuff pressure signal which are created by expansion of the arterial wall each time blood is forced through the artery. The uniqueness of the oscillometric method, a blood-pressure cuff technique, is that specific characteristics of the compression cuff's entrained air volume are used to identify and sense blood-pressure values. The cuff-pressure signal increases in strength in the systolic pressure region, reaching a maximum when the cuff pressure is equal to mean arterial pressure. As the cuff pressure drops below this point, the signal strength decreases proportionally to the cuff air pressure bled rate. There is no clear transition in cuff-pressure oscillations to identify diastolic pressure since arterial wall expansion continues to happen below diastolic pressure while blood is forced through the artery (Geddes, 1984). Thus, oscillometric monitors employ proprietary algorithms to estimate the diastolic pressure.

Ramsey (1991) has indicated that, using the oscillometric method, the mean arterial pressure is the single blood-pressure parameter which is the most robust measurement, as compared with systolic and diastolic pressure, because it is measured when the oscillations of cuff pressure reach the greatest amplitude. This property usually allows mean arterial pressure to be measured reliably even in case of hypotension with vasoconstriction and diminished pulse pressure.

When the cuff pressure is raised quickly to pressures higher than systolic pressure it is observed that the radial pulse disappears. Cuff pressures above systolic cause the underlying artery to be completely occluded. However, at suprasystolic cuff pressures, small amplitude pressure oscillations occur in the cuff pressure due to artery pulsations under the upper edge of the cuff, which are communicated to the cuff through the adjacent tissues. With slow cuff-pressure reductions, when the cuff pressure is just below systolic pressure, blood spurts through the artery and the cuff-pressure oscillations become larger. Figure 7.22 illustrates the ideal case in which the cuff pressure is monitored by a pressure sensor connected to a strip chart recorder. A pressure slightly above systolic pressure is detected by determining the shift from small-amplitude oscillations at cuff pressure slightly above systolic pressure and when the cuff pressure begins to increase amplitude (Point 1). As the cuff continues to deflate, the amplitude of the oscillations increases reaching a maximum, and then decreases as the cuff pressure is decreased to zero. Point 2 in Figure 7.22 is the maximum cuff-pressure oscillation which is essentially true mean arterial pressure. Since there is no apparent transition in the oscillation amplitude as cuff pressure passes diastolic pressure, algorithmic methods are used to predict diastolic pressure.

The system description begins with the blood-pressure cuff which compresses a limb and its vasculature by the encircling inflatable compression cuff pressures (Ramsey, 1991). The cuff is connected to a pneumatic system (see Figure 7.23). A solid-state pressure sensor senses cuff pressure, and the electric signal proportional to pressure is processed in two different circuits. One circuit amplifies and corrects the zero offset of the cuff-pressure signal before the analog-to-digital digitization. The other circuit high-pass filters and amplifies the cuff-pressure signal. Cuff pressure is controlled by a microcomputer which activates the cuff inflation and deflation systems during the measurement cycle.

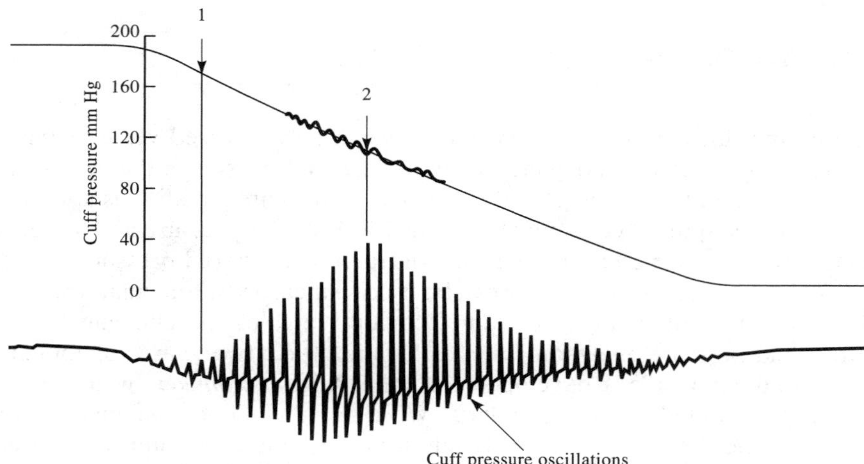

Figure 7.22 The oscillometric method A compression cuff is inflated above systolic pressure and slowly deflated. Systolic pressure is detected (Point 1) where there is a transition from small amplitude oscillations (above systolic pressure) to increasing cuff-pressure amplitude. The cuff-pressure oscillations increase to a maximum (Point 2) at the mean arterial pressure.

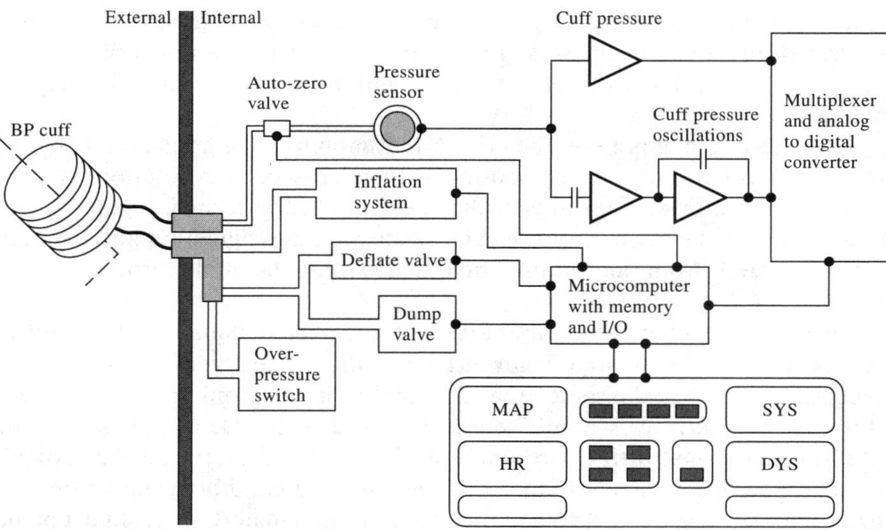

Figure 7.23 Block diagram of the major components and subsystems of an oscillometric blood-pressure monitoring device, based on the Dinamap unit, I/O = input/output; MAP = mean arterial pressure; HR = heart rate; SYS = systolic pressure; DYS = diastolic pressure. From Ramsey M III. Blood pressure monitoring: automated oscillometric devices, *J. Clin. Monit.* 1991, 7, 56–67.

7.14 TONOMETRY

The basic principle of tonometry is that, when a pressurized vessel is partly collapsed by an external object, the circumferential stresses in the vessel wall are removed and the internal and external pressures are equal. This approach has been used quite successfully to measure intraocular pressure and has been used with limited success to determine intraluminal arterial pressure.

The force-balance technique can be used to measure intraocular pressure. Based on the Imbert-Fick law, the technique enables the clinician to find intraocular pressure by dividing the applanation force by the area of applanation. Goldmann (1957) developed an *applanation tonometer,* which is the currently accepted clinical standard. With this technique, the investigator measures the force required to flatten a specific optically determined area. Mackay and Marg (1960) developed a sensor probe that is applied to the corneal surface; the cornea is flattened as the probe is advanced. The intraocular pressure is detected by a force sensor in the center of an annular ring, which unloads the bending forces of the cornea from the sensor.

Forbes *et al.* (1974) developed an applanation tonometer that measures intraocular pressure without touching the eye. An air pulse of linearly increasing force deforms and flattens the central area of the cornea, and it does so within a few milliseconds. The instrument consists of three major components. The first is a pneumatic system that delivers an air pulse the force of which increases linearly with time. As the air pulse decays, it causes a progressive reduction of the convexity of the cornea and, finally, a return to its original shape.

The second component, the system that monitors the applanation, determines the occurrence of applanation with microsecond resolution by continuously monitoring the status of the curvature of the cornea. Figure 7.24(a) and (b) shows the systems of optical transmission and detection and the light rays reflected from an undisturbed and an applanated cornea, respectively.

Two obliquely oriented tubes are used to detect applanation. Transmitter tube T directs a collimated beam of light at the corneal vertex; a telecentric receiver R observes the same area. The light reflected from the cornea passes through the aperture A and is sensed by the detector D. In the case of the undisturbed cornea, little or no light is received by the detector. As the cornea's convexity is progressively reduced to the flattened condition, the amount of light detected is increased. When the cornea is applanated, it acts like a plano mirror with a resulting maximal detected signal. When the cornea becomes concave, a sharp reduction in light detection occurs. The current source for the pneumatic solenoid is immediately shut off when applanation is detected in order to minimize further air-pulse force impinging on the cornea. A direct linear relationship has been found between the intraocular pressure and the time interval to applanation.

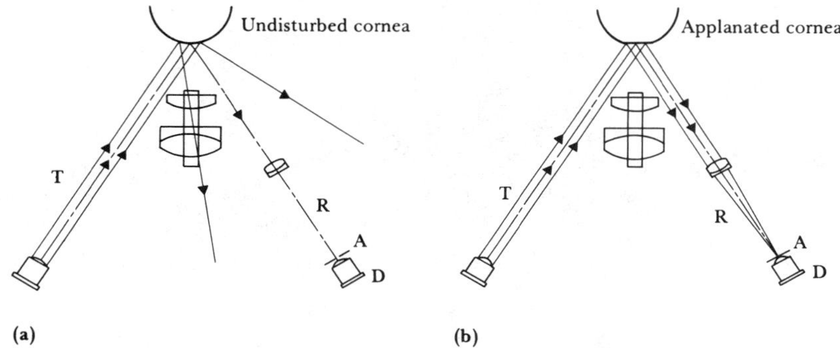

Figure 7.24 Monitoring system for noncontact applanation tonometer (From M. Forbes, G. Pico, Jr., and B. Grolman, "A Noncontact Applanation Tonometer, Description and Clinical Evaluation," *J. Arch. Ophthalmology,* 1975, 91, 134–140. Copyright © 1975, American Medical Association. Used with permission.)

The principles of operation of the arterial tonometry are very similar to those for the ocular tonometry, discussed above. The arterial tonometer measures dynamic arterial blood pressure, i.e., it furnishes continuous measurements of arterial pressure throughout the total heart cycle (Eckerle, 1988). The instrument sensor is placed over a superficial artery which is supported from below by bone. The radial artery at the wrist is a convenient site for arterial tonometer measurements. The arterial tonometer suffers from relatively high cost when compared to a conventional sphygmomanometer. One significant advantage of the arterial tonometer is its ability to make noninvasive, nonpainful, continuous measurements for long periods of time.

Figure 7.25 shows an arterial tonometer model that depicts system operation in which the arterial blood pressure, P, from a superficial artery and the force, F, is measured by a tonometer sensor. The artery wall is represented by a flat ideal membrane, M. A free-body diagram is used to describe the force balances. The ideal membrane only transmits a tensile force, T, without any bending moment. Vertical force balance shows that the tension vector, T, is perpendicular to the pressure vector. Thus, the force, F, is in quadrature to and independent of T and only depends on the blood pressure and the area of the frictionless piston, A. Thus, measurement of the force, F, permits direct measurement of the intraarterial pressure.

Eckerle (1988) indicates that several conditions must be met by the tonometer sensor and an appropriate superficial artery for proper system operation:

1. A bone provides support for the artery, opposite to the applied force.
2. The hold-down force flattens the artery wall at the measurement site without occluding the artery.
3. Compared to artery diameter, the skin thickness over the artery is insignificant.

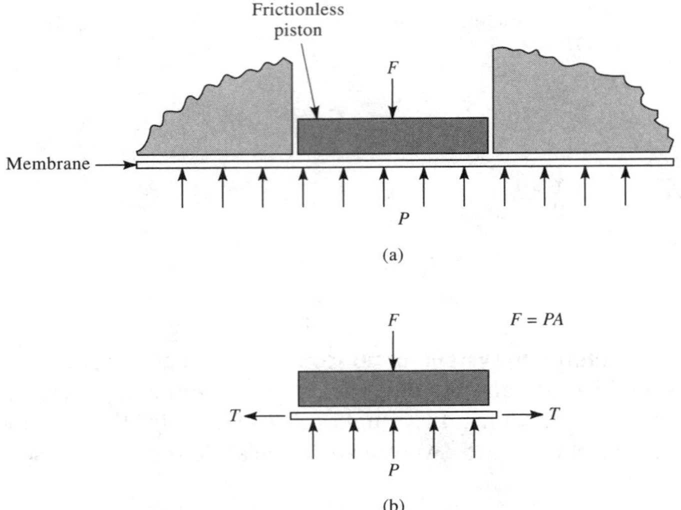

Figure 7.25 Idealized model for an arterial tonometer. (a) a flattened portion of an arterial wall (membrane). P is the blood pressure in a superficial artery, and F is the force measured by a tonometer transducer. (b) a free-body diagram for the idealized model of (a) in which T is the membrane tensile force perpendicular to both F and P. From Eckerle, J. D., "Tonometry, arterial," in J. G. Webster (ed.), *Encyclopedia of Medical Devices and Instrumentation.* New York: Wiley, 1988, pp. 2770–2776.

4. The artery wall has the properties of an ideal membrane.
5. The arterial rider, positioned over the flattened area of the artery, is smaller than the artery.
6. The force transducer spring constant K_T, is larger than the effective spring constant of the artery.

When all these conditions hold, it has been shown on a theoretical basis that the electrical output signal of the force sensor is directly proportional to the intra-arterial blood pressure (Pressman and Newgard, 1963). However, a major practical problem with the above approach, using a single arterial tonometer, is that the arterial rider must be precisely located over the superficial artery. A solution to this problem is the use of an arterial tonometer with multiple element sensors. Figure 7.26 shows a linear array of force sensors and arterial riders positioned such that at least one element of the array is centered over the artery. A computer algorithm is used to automatically select from the multiple sensors, the sensor element which is positioned over the artery. One approach uses two pressure distribution characteristics in the vicinity of the artery in which an element-selection algorithm searches for a

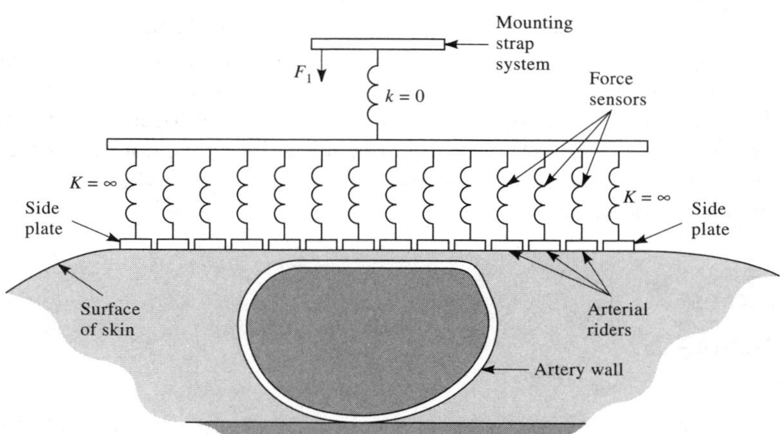

Figure 7.26 Multiple-element arterial tonometer. The multiple element linear array of force sensors and arterial riders are used to position the system such that some element of the array is centered over the artery. From Eckerle, J. D., "Tonometry, arterial," in J. G. Webster (ed.), *Encyclopedia of Medical Devices and Instrumentation.* New York: Wiley, 1988, pp. 2770–2776.

(spatial) local minimum in diastolic pressure in a region near the maximum pulse amplitude (Eckerle, 1988). The sensor with these characteristics is assumed to be centered over the artery, and the blood pressure from this sensor is measured with this element.

In addition to positioning the sensor over the artery, the degree of arterial flattening is another important factor for accurate tonometric pressure measurements. The hold-down force F_1, (in Figure 7.26), which causes arterial flattening, is a function of the interaction of anatomical factors. The hold-down force for each subject must be determined before tonometric readings can be taken. The hold-down force is gradually increased (or decreased) while recording the tonometer sensor output.

Multiple-element tonometer sensors have been manufactured from a monolithic silicon substrate using anisotropic etching to define pressure-sensing diaphragms (10 μm thick in the silicon). Piezoresistive strain gages in the diaphragms are fabricated using integrated-circuit (IC) processing techniques. The strain gage's resistance is used to determine the pressure exerted on each sensor element.

Note that the radial artery is not the only measurement site at which a tonometer may be applied. Other possible sites for tonometric measurements include the brachial artery at the inner elbow (the antecubital fossa), the temporal artery in front of the ear, and the dorsalis pedis artery on the upper foot (Eckerle, 1988). Arterial tonometers have not been commercially successful because of inaccuracy caused by wrist movement, tendons overlying arteries, etc.

Gizdulich and Wesseling (1990) measure the arterial pressure in the finger continuously and indirectly by using the Peñás method. They apply counter-pressure just sufficient to hold the arteries under a pressure cuff at their unstressed diameter at zero transmural pressure monitored by an infrared plethysmograph. Because the method occludes the veins, continuous use for more than 20 min causes discomfort and swelling. Thus, the pressure must be relieved periodically.

PROBLEMS

7.1 Compare the transient-step and sinusoidal-frequency methods for determining the response characteristics of a catheter–sensor system.

7.2 Find (a) the damping ratio, (b) the undamped frequency, and (c) the frequency-response curve of the pressure sensor for which the transient response to a step change in pressure is shown in Figure 7.11.

7.3 Find the frequency-response curve of the sensor in Problem 7.2, given that its chamber is filled with the whole blood at body temperature (37 °C). The original data in Problem 7.2 were obtained with water at 20 °C.

7.4 What happens to the frequency response of a P23Dd sensor, 6 F, 1-m, water-filled catheter system (at 20 °C) when a tiny pinhole leak occurs at the junction of the catheter and sensor? The leak allows a 0.40-ml/min flow for a pressure head of 100 mm Hg (13.3 kPa). Plot frequency-response curves for the system with and without leak. (An intentional leak is often desirable to permit constant flushing of the catheter and thus inhibit the formation of clots.)

7.5 By changing only the *radius* of the catheter, redesign the (no-bubble) catheter of Figure 7.9 to achieve the damping ratio $\zeta = 1$. Calculate the resulting natural frequency f_n.

7.6 A low-pass filter is added to the catheter of a pressure-sensor–catheter system by pinching the catheter. The system consists of a Statham P23Dd sensor and a 1-m, 6 F, polyethylene catheter. The pinch effectively reduces the diameter of the catheter to 25% of its original diameter.

- **a.** How long must the pinch be for the system's damping factor to be equal to 0.7?
- **b.** Sketch the frequency response for the system with and without the pinch.
- **c.** Sketch the time response for the system with and without the pinch when it is excited by a 100-mm-Hg step input.
- **d.** Discuss how faithfully the two systems will reproduce the blood-pressure waveform for humans, dogs, and shrews with heart-rate variations of 1 to 3.3 beats/s, 1.5 to 5 beats/s, and 12 to 22 beats/s, respectively.

7.7 A heart murmur has a frequency of 300 Hz. Give the block diagram and sketch waveforms for the special instrumentation that enables us to show the occurrence of this murmur on a 0–80-Hz pen recorder.

7.8 Name the two basic causes of abnormal heart murmurs. For each type, give an example and show on a sketch when it occurs relative to systole and diastole.

7.9 In block-diagram form, show the elements required for an automatic indirect system for measuring blood pressure.

7.10 Design a portable system for indirectly measuring blood pressure every 5 min on ambulatory subjects. It should operate without attention from the subject for 24 h. Show a block diagram and describe the system's operation, including power source, sensor, storage, and algorithm.

7.11 A patient who has been vomiting for several days is dehydrated. Liquid is infused through a venous catheter at the rate of 250 ml/h. Sketch the resulting central venous pressure versus time, and explain any large change in the slope of the curve. How does the jugular venous pulse change during this procedure?

7.12 Determine whether the kinetic-energy term is significant for measurements of pressure in the human descending aorta. Assume that the peak velocity of flow in the center of the aorta is approximately 1.5 m/s and that the density of the blood ρ is 1060 kg/m^3.

7.13 Design a noninvasive (no breaks in the skin) system for measuring the velocity of propagation of a blood-pressure wave from the aortic valve in the heart to the radial artery on the wrist. Name and describe the sensors required, show their placement, sketch the expected waveforms, and indicate the times required to measure velocity of propagation.

7.14 One of the problems of tonometers is that each operator pushes with a different force. Sketch the block diagram of a system that would apply a *ramp* input of force from low to high and pick out the maximal pulse pressure (systolic minus diastolic). Sketch the expected output versus time.

REFERENCES

Attinger, E. O., A. Anne, and D. A. McDonald, "Use of Fourier series for the analysis of biological systems." *Biophys. J.* 1970, 6, 291–304.

Bahr, D. E., and J. Petzke, "The automatic arterial tonometer." *Proc. Annu. Conf. Eng. Med. Biol.,* 1973, 26, 259.

Bellville, J. W., and C. S. Weaver, *Techniques in Clinical Physiology.* New York: Macmillan, 1969.

Burton, A. C., *Physiology and Biophysics of the Circulation,* 2nd ed. Chicago: Year Book, 1972.

Cobbold, R. S. C., *Transducers for Biomedical Measurements: Principles and Applications.* New York: Wiley, 1974.

Collins, C. C., "Biomedical transensors: A review." *J. Biomed. Syst.,* 1970, 1, 23–39.

Eckerle, J. S., "Tonometry, arterial," in J. G. Webster (ed.), *Encyclopedia of Medical Devices and Instrumentation.* New York: Wiley, 1988, pp. 2770–2776.

Ertel, P. Y., M. Lawrence, R. K. Brown, and A. M. Stern, "Stethoscope acoustics I. The doctor and his stethoscope." *Circ.,* 1966a, 34, 889–898.

Ertel, P. Y., M. Lawrence, R. K. Brown, and A. M. Stern, "Stethoscope acoustics II. Transmission and filtration patterns." *Circ.,* 1966b, 34, 899–908.

Fleming, D. G., W. H. Ko, and M. R. Neuman (eds.), *Indwelling and Implantable Pressure Transducers.* Cleveland: CRC Press, 1977.

Forbes, M., G. Pico, Jr., and B. Grolman, "A noncontact applanation tonometer, description and clinical evaluation." *J. Arch. Ophthal.,* 1974, 91, 134–140.

Fry, D. L., "Physiologic recording by modern instruments with particular reference to pressure recording." *Physiol. Rev.,* 1960, 40, 753–788.

Ganz, W., and H. J. C. Swan, "Balloon-tipped flow-directed catheters," in W. Grossman (ed.), *Cardiac Catheterization and Angiography.* Philadelphia: Lea & Febiger, 1974.

Geddes, L. A., *The Direct and Indirect Measurement of Blood Pressure.* Chicago: Year Book, 1970.

Geddes, L. A., *Cardiovascular Devices and Their Applications.* New York: Wiley, 1984.

Gersh, B. J., C. E. W. Hahn, and C. P. Roberts, "Physical criteria for measurement of left ventricular pressure and its first derivative." *Cardiovasc. Res.,* 1971, 5, 32–40.

Gizdulich, P., and K. H. Wesseling, "Reconstruction of brachial arterial pulsation from finger arterial pressure." *Proc. Annu. Int. Conf. IEEE Eng. Med. Biol. Soc.,* 1990, 12, 1046–1047.

Goldmann, H., "Applanation tonometry," in F. W. Newell (ed.), *Glaucoma: Transactions of the Second Conference, December 1956, Princeton, N.J.* Madison, NJ: Madison Printing, 1957, pp. 167–220.

Greatorex, C. A., "Indirect methods of blood-pressure measurement," in B. W. Watson (ed.), *IEE Medical Electronics Monographs 1–6.* London: Peter Peregrinus, 1971.

Grossman, W., *Cardiac Catheterization and Angiography.* Philadelphia: Lea & Febiger, 1974.

Gupta, R., J. W. Miller, A. P. Yoganathan, B. M. Kim, F. E. Udwadia, and W. H. Corcoran, "Spectral analysis of arterial sounds: A noninvasive method of studying arterial disease." *Med. Biol. Eng.,* 1975, 13, 700–705.

Hansen, A. T., "Fiber-optic pressure transducers for medical application." *Sensors and Actuators,* 1983, 4, 545–554.

Hansen, A. T., "Pressure measurement in the human organism." *Acta Physiol. Scand.,* 1949, 19 (Suppl. 68), 1–227.

Herman, M. V., P. F. Cohn, and R. Gorlin, "Resistance to blood flow by stenotic valves: Calculation of orifice area," in W. Grossman (ed.), *Cardiac Catheterization and Angiography.* Philadelphia: Lea & Febiger, 1974.

Howe, R. T., R. S. Muller, K. J. Gabriel, and W. S. N. Trimmer, "Silicon micromechanics: Sensors and actuators on a chip." *IEEE Spectrum,* 1990, 27 (7), 29–35.

King, G. E., "Blood pressure measurement," in J. G. Webster (ed.), *Encyclopedia of Medical Devices and Instrumentation.* New York: Wiley, 1988, pp. 467–482.

Kingsley, B., and B. L. Segal, "Cardiovascular vibratory phenomena," in C. Ray (ed.), *Medical Engineering.* Chicago: Year Book, 1974.

Loudon, R. G., and R. P. Baughman, "Lung sounds," in J. G. Webster (ed.), *Encyclopedia of Medical Devices and Instrumentation.* New York: Wiley, 1988, pp. 1825–1831.

Mackay, R. S., and E. Marg, "Fast automatic ocular pressure measurement based on an exact theory." *IRE Trans. Med. Electron.,* 1960, ME-7, 61–67.

Master, A. M., C. I. Garfield, and M. B. Walters, *Normal Blood Pressure and Hypertension.* Philadelphia: Lea & Febiger, 1952.

Neuman, M. R., "Neonatal monitoring," in J. G. Webster (ed.), *Encyclopedia of Medical Devices and Instrumentation.* New York: Wiley, 1988, pp. 2015–2034.

O'Rourke, P. L., "The arterial pulse in health and disease." *Amer. Heart J.,* 1971, 82, 687–702.

Pressman, G. L., and P. M. Newgard, "A transducer for the continuous external measurement of arterial blood pressure," *IEEE Trans. Biomed. Electron.,* 1963, 10, 73–81.

Probhaker, G., and W. Carr, "Endoradiosonde pressure-sensor system for chronic biomedical monitoring." *Proc. Annu. Int. Conf. IEEE Eng. Med. Biol. Soc.,* 1990, 12, 514–515.

Ramsey, M. III., "Blood pressure monitoring: automated oscillometric devices." *J. Clin. Monit.,* 1991, 7, 56–67.

Roos, C. F., and F. E. Carroll, Jr., "Fiber-optic pressure transducer for use near MR magnetic fields." *Radiology,* 1985, 156, 548.

Rushmer, R. F., *Cardiovascular Dynamics,* 3rd ed. Philadelphia: Saunders, 1970.

Shapiro, G. G., and L. J. Krovetz, "Damped and undamped frequency responses of underdamped catheter manometer systems." *Amer. Heart J.,* 1970, 80, 226–236.

Sheppard, L. C., and T. C. Jannett, "Blood pressure, automatic control of," in J. G. Webster (ed.), *Encyclopedia of Medical Devices and Instrumentation.* New York: Wiley, 1988, pp. 460–466.

Stegall, H. F., M. B. Kardon, and W. T. Kemmerer, "Indirect measurement of arterial blood pressure by Doppler ultrasonic sphygmomanometry." *J. Appl. Physiol.,* 1968, 25, 793–798.

Vermariën, H., "Phonocardiography," in J. G. Webster (ed.), *Encyclopedia of Medical Devices and Instrumentation.* New York: Wiley, 1988, pp. 2265–2277.

Visser, K. R., and J. H. J. Muntinga, "Blood pressure estimation investigated by electric impedance measurement." *Proc. Annu. Int. Conf. IEEE Eng. Med. Biol. Soc.,* 1990, 12, 691–692.

Webster, J. G. (ed.), *Prevention of Pressure Sores: Engineering and Clinical Aspects.* Bristol, England: Adam Hilger, 1991.

Weihrauch, T. R., "Esophageal manometry," in J. G. Webster (ed.), *Encyclopedia of Medical Devices and Instrumentation.* New York: Wiley, 1988, pp. 1236–1245.

Yellin, E. L., R. W. M. Frater, and C. S. Peskin, "The application of the Gorlin equation to the stenotic mitral valve," in A. C. Bell and R. M. Nerem (eds.), *1975 Advances in Bioengineering.* New York: Am. Soc. Mech. Engr., 1975.

8

MEASUREMENT OF FLOW AND VOLUME OF BLOOD

John G. Webster

One of the primary measurements the physician would like to acquire from a patient is that of the concentration of O_2 and other nutrients in the cells. Such quantities are normally so difficult to measure that the doctor is forced to accept the second-class measurements of blood flow and changes in blood volume, which usually correlate with concentration of nutrients. If blood *flow* is difficult to measure, the physician may settle for the third-class measurement of blood *pressure,* which usually correlates adequately with blood flow. If blood pressure cannot be measured, the physician may fall back on the fourth-class measurement of the ECG, which usually correlates adequately with blood pressure.

Note that the measurement of blood flow—the main subject of this chapter—is the one that most closely reflects the primary measurement of concentration of O_2 in the cells. However, measurement of blood flow is usually more difficult to make and more invasive than measurement of blood pressure or of the ECG.

Commonly used flowmeters, such as the orifice or turbine flowmeters, are unsuitable for measuring blood flow because they require cutting the vessel and can cause formation of clots. The specialized techniques described in this chapter have therefore been developed. Each April issue of *Medical Electronics* lists the manufacturers and types of blood flowmeters.

8.1 INDICATOR-DILUTION METHOD THAT USES CONTINUOUS INFUSION

The indicator-dilution methods described in this chapter do not measure instantaneous pulsatile flow but, rather, flow averaged over a number of heartbeats.

CONCENTRATION

When a given quantity m_0 of an indicator is added to a volume V, the resulting concentration C of the indicator is given by $C = m_0/V$. When an additional quantity m of indicator is then added, the incremental increase in concentration is $\Delta C = m/V$. When the fluid volume in the measured space is continuously removed and replaced, as in a flowing stream, then in order to maintain a fixed change in concentration, the clinician must continuously add a fixed quantity of indicator per unit time. That is, $\Delta C = (dm/dt)/(dV/dt)$. From this equation, we can calculate flow (von Reth and Versprille, 1988).

$$F = \frac{dV}{dt} = \frac{dm/dt}{\Delta C} \tag{8.1}$$

EXAMPLE 8.1 Derive (8.1), using principles of mass transport.

ANSWER The rate at which indicator enters the vessel is equal to the indicator's input concentration C_i times the flow F. The rate at which indicator is injected into the vessel is equal to the quantity per unit time, dm/dt. The rate at which indicator leaves the vessel is equal to the indicator's output concentration C_o times F. For steady state, $C_iF + dm/dt = C_oF$ or $F = (dm/dt)/(C_o - C_i)$.

FICK TECHNIQUE

We can use (8.1) to measure *cardiac output* (blood flow from the heart) as follows (Capek and Roy, 1988).

$$F = \frac{dm/dt}{C_a - C_v} \tag{8.2}$$

where

$$\begin{aligned} F &= \text{blood flow, liters/min} \\ dm/dt &= \text{consumption of } O_2\text{, liters/min} \\ C_a &= \text{arterial concentration of } O_2\text{, liters/liter} \\ C_v &= \text{venous concentration of } O_2\text{, liters/liter} \end{aligned}$$

Figure 8.1 shows the measurements required. The blood returning to the heart from the upper half of the body has a different concentration of O_2 from the blood returning from the lower half, because the amount of O_2 extracted by the brain is different from that extracted by the kidneys, muscles, and so forth. Therefore, we cannot accurately measure C_v in the right atrium. We must measure it in the pulmonary artery after it has been mixed by the pumping action of the right ventricle. The physician may float the catheter

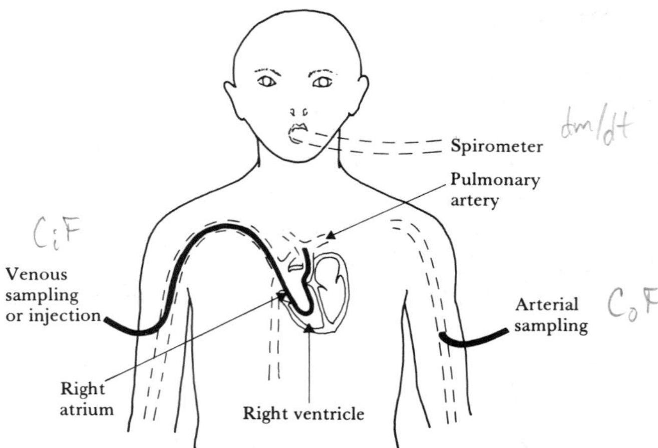

Figure 8.1 Several methods of measuring cardiac output In the Fick method, the indicator is O_2; consumption is measured by a spirometer. The arterial-venous concentration difference is measured by drawing samples through catheters placed in an artery and in the pulmonary artery. In the dye-dilution method, dye is injected into the pulmonary artery and samples are taken from an artery. In the thermodilution method, cold saline is injected into the right atrium and temperature is measured in the pulmonary artery.

into place by temporarily inflating a small balloon surrounding the tip. This is done through a second lumen in the catheter.

As the blood flows through the lung capillaries, the subject adds the indicator (the O_2) by breathing in pure O_2 from a spirometer (see Figure 9.6). The exhaled CO_2 is absorbed in a soda-lime canister, so the consumption of O_2 is indicated directly by the net gas-flow rate.

The clinician can measure the concentration of the oxygenated blood C_a in any artery, because blood from the lung capillaries is well mixed by the left ventricle and there is no consumption of O_2 in the arteries. An arm or leg artery is generally used.

EXAMPLE 8.2 Calculate the cardiac output, given the following data: spirometer O_2 consumption 250 ml/min; arterial O_2 content, 0.20 ml/ml; venous O_2 content, 0.15 ml/ml.

ANSWER From (8.2),

$$F = \frac{dm/dt}{C_a - C_v}$$

$$= \frac{0.25 \text{ liter/min}}{(0.20 \text{ liter/liter}) - (0.15 \text{ liter/liter})}$$

$$= 5 \text{ liters/min}$$

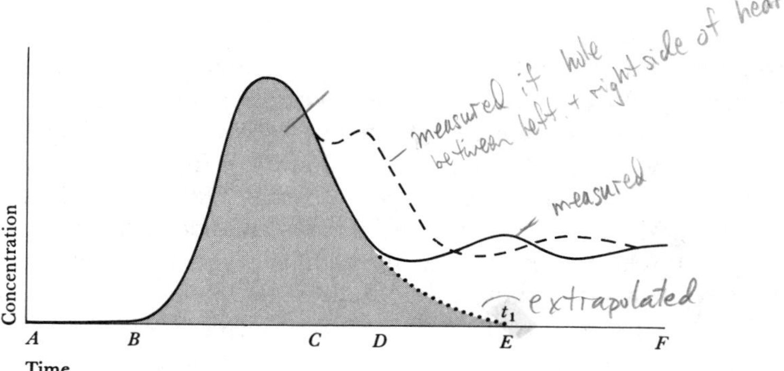

Figure 8.2 Rapid-injection indicator-dilution curve After the bolus is injected at time A, there is a transportation delay before the concentration begins rising at time B. After the peak is passed, the curve enters an exponential decay region between C and D, which would continue decaying along the dotted curve to t_1 if there were no recirculation. However, recirculation causes a second peak at E before the indicator becomes thoroughly mixed in the blood at F. The dashed curve indicates the rapid recirculation that occurs when there is a hole between the left and right sides of the heart.

The units for the concentrations of O_2 represent the volume of O_2 that can be extracted from a volume of blood. This concentration is very high for blood, because large quantities of oxygen can be bound to hemoglobin. It would be very low if water were flowing through the vessels, even if the Po_2 were identical in both cases.

The Fick technique is nontoxic, because the indicator (O_2) is a normal metabolite that is partially removed as blood passes through the systemic capillaries. The cardiac output must be constant over several minutes so that the investigator can obtain the slope of the curve for O_2 consumption. The presence of the catheter causes a negligible change in cardiac output.

8.2 INDICATOR-DILUTION METHOD THAT USES RAPID INJECTION

EQUATION

The continuous-infusion method has been largely replaced by the rapid-injection method, which is more convenient. A bolus of indicator is rapidly injected into the vessel, and the variation in downstream concentration of the indicator versus time is measured until the bolus has passed. The solid line in Figure 8.2 shows the fluctuations in concentration of the indicator that occur after the injection. The dotted-line extension of the exponential decay shows the curve that would result if there were no recirculation. For this case we can calculate the flow as outlined in the following paragraphs.

An increment of blood of volume dV passes the sampling site in time dt. The quantity of indicator dm contained in dV is the concentration $C(t)$ times the incremental volume. Hence $dm = C(t)dV$. Dividing by dt, we obtain $dm/dt = C(t)dV/dt$. But $dV/dt = F_i$, the instantaneous flow; therefore $dm = F_iC(t)dt$. Integrating over time through t_1, when the bolus has passed the downstream sampling point, we obtain

$$m = \int_0^{t_1} F_i C(t)\, dt \tag{8.4}$$

where t_1 is the time at which all effects of the first pass of the bolus have died out (point E in Figure 8.2). Minor variations in the instantaneous flow F_i produced by the heartbeat are smoothed out by the mixing of the bolus and the blood within the heart chambers and the lungs. Thus we can obtain the average flow F from

$$F = \frac{m}{\int_0^{t_1} C(t)\, dt} \tag{8.5}$$

The integrated quantity in (8.5) is equal to the shaded area in Figure 8.2, and we can obtain it by counting squares or using a planimeter. Small, special-purpose computers that extrapolate the dotted line in real time and compute the flow are also available.

If the initial concentration of indicator is not zero—as may be the case when there is residual indicator left over from previous injections—then (8.5) becomes

$$F = \frac{m}{\int_0^{t_1} [\Delta C(t)]\, dt} \tag{8.6}$$

DYE DILUTION

A common method of clinically measuring cardiac output is to use a colored dye, *indocyanine green* (cardiogreen). It meets the necessary requirements for an indicator in that it is (1) inert, (2) harmless, (3) measurable, (4) economical, and (5) always intravascular. In addition, its optical absorption peak is 805 nm, the wavelength at which the optical absorption coefficient of blood is independent of oxygenation. The dye is available as a liquid that is diluted in isotonic saline and injected directly through a catheter, usually into the pulmonary artery. About 50% of the dye is excreted by the kidneys in the first 10 min, so repeat determinations are possible.

The plot of the curve for concentration versus time is obtained from a constant-flow pump, which draws blood from a catheter placed in the femoral or brachial artery. Blood is drawn through a colorimeter cuvette (Figure 2.17), which continuously measures the concentration of dye, using the principle of absorption photometry (Section 11.1). The 805-nm channel of a two-channel blood oximeter can be used for measuring dye-dilution curves. The clinician calibrates the colorimeter by mixing known amounts of dye and blood and drawing them through the cuvette.

The shape of the curve can provide additional diagnostic information. The dashed curve in Figure 8.2 shows the result when a left-right shunt (a hole between the left and right sides of the heart) is present. Blood recirculates faster than normal, resulting in an earlier recirculation peak. When a right-left shunt is present, the delay in transport is abnormally short, because some dye reaches the samping site without passing through the lung vessels.

THERMODILUTION

The most common method of measuring cardiac output is that of injecting a bolus of cold saline as an indicator. A special four-lumen catheter (Trautman and Newbower, 1988) is floated through the brachial vein into place in the pulmonary artery. A syringe forces a gas through one lumen; the gas inflates a small, doughnut-shaped balloon at the tip. The force of the flowing blood carries the tip into the pulmonary artery. The cooled saline indicator is injected through the second lumen into the right atrium. The indicator is mixed with blood in the right ventricle. The resulting drop in temperature of the blood is detected by a thermistor located near the catheter tip in the pulmonary artery. The third lumen carries the thermistor wires. The fourth lumen, which is not used for the measurement of thermodilution, can be used for withdrawing blood samples. The catheter can be left in place for about 24 h, during which time many determinations of cardiac output can be made, something that would not be possible if dye were being used as the indicator. Also, it is not necessary to puncture an artery.

We can derive the following equation, which is analogous to (8.6).

$$F = \frac{Q}{\rho_b c_b \int_0^{t_1} \Delta T_b(t)\, dt} \ (\mathrm{m^3/s}) \tag{8.7}$$

where

Q = heat content of injectate, J $(= V_i \, \Delta T_i \rho_i c_i)$

ρ_b = density of blood, kg/m^3

c_b = specific heat of blood, J/(kg · K)

When an investigator uses the thermodilution method, there are a number of problems that cause errors. (1) There may be inadequate mixing between the injection site and the sampling site. (2) There may be an exchange of heat between the blood and the walls of the heart chamber. (3) There is heat

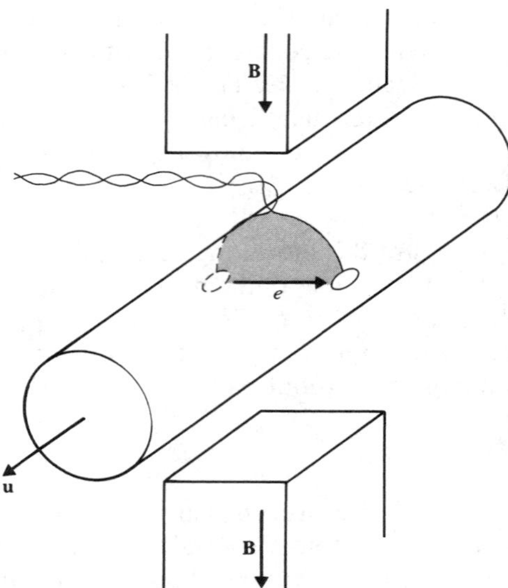

Figure 8.3 Electromagnetic flowmeter When blood flows in the vessel with velocity **u** and passes through the magnetic field **B,** the induced emf *e* is measured at the electrodes shown. When an ac magnetic field is used, any flux lines cutting the shaded loop induce an undesired transformer voltage.

exchange through the catheter walls before, during, and after injection. However, the instrument can be calibrated by simultaneously performing dye-dilution determinations and applying a correction factor that corrects for several of the errors.

8.3 ELECTROMAGNETIC FLOWMETERS

The electromagnetic flowmeter measures instantaneous pulsatile flow of blood and thus has a greater capability than indicator-dilution methods, which measure only average flow. It operates with any conductive liquid, such as saline or blood.

PRINCIPLE

The electric generator in a car generates electricity by induction. Copper wires move through a magnetic field, cutting the lines of magnetic flux and inducing an emf in the wire. This same principle is exploited in a commonly used blood flowmeter, shown in Figure 8.3. Instead of copper wires, the flowmeter depends on the movement of blood, which has a conductance similar to that of saline. The formula for the induced emf is given by Faraday's law of induction.

$$e = \int_0^{L_1} \mathbf{u} \times \mathbf{B} \cdot d\mathbf{L}$$

where

$\mathbf{B}$ = magnetic flux density, T
$\mathbf{L}$ = length between electrodes, m
$\mathbf{u}$ = instantaneous velocity of blood, m/s

For a uniform magnetic field B and a uniform velocity profile u, the induced emf is

$$e = BLu \tag{8.8}$$

where these three components are orthogonal.

Let us now consider real flowmeters, several of which exhibit a number of divergences from this ideal case. If the vessel's cross section were square and the electrodes extended the full length of two opposite sides, the flowmeter would measure the correct average flow for any flow profile. The electrodes are small, however, so velocities near them contribute more to the signal than do velocities farther away.

Figure 8.4 shows the weighting function that characterizes this effect for circular geometry. It shows that the problem is less when the electrodes are located outside the vessel wall. The instrument measures correctly for a uniform flow profile. For axisymmetric nonuniform flow profiles, such as the parabolic flow profile resulting from laminar flow, the instrument measurement is correct if u is replaced by $\bar{u}$, the average flow velocity. Because we usually know the cross-sectional area A of the lumen of the vessel, we can multiply A by $\bar{u}$ to obtain F, the volumetric flow. However, in many locations of blood vessels in the body, such as around the curve of the aorta and near its branches, the velocity profile is asymmetric, so errors result.

Other factors can also cause error.

1. Regions of high velocity generate higher incremental emf's than regions of low velocity, so circulating currents flow in the transverse plane. These currents cause varying drops in resistance within the conductive blood and surrounding tissues.
2. The ratio of the conductivity of the wall of the blood vessel to that of the blood varies with the *hematocrit* (percentage of cell volume to blood volume), so the shunting effects of the wall cause a variable error.
3. Fluid outside the wall of the vessel has a greater conductivity than the wall, so it shunts the flow signal.

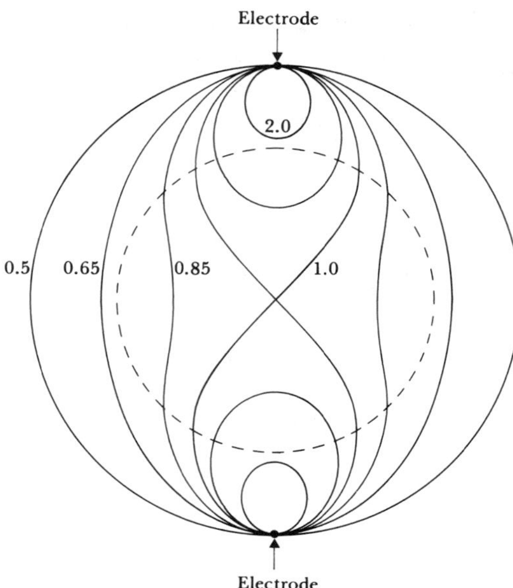

Figure 8.4 Solid lines show the weighting function that represents relative velocity contributions (indicated by numbers) to the total induced voltage for electrodes at the top and bottom of the circular cross section. If the vessel wall extends from the outside circle to the dashed line, the range of the weighting function is reduced. (Adapted from J. A. Shercliff, *The Theory of Electromagnetic Flow Measurement,* © 1962, Cambridge University Press.)

4. The magnetic-flux density is not uniform in the transverse plane; this accentuates the problem of circulating current.
5. The magnetic-flux density is not uniform along the axis, which causes circulating currents to flow in the axial direction.

To minimize these errors, most workers recommend calibration for animal work by using blood from the animal—and, where possible, the animal's own vessels also. Blood or saline is usually collected in a graduated cylinder and timed with a stopwatch.

DC FLOWMETER

The flowmeter shown in Figure 8.3 can use a dc magnetic field, so the output voltage continuously indicates the flow. Although a few early dc flowmeters were built, none were satisfactory, for the following three reasons. (1) The voltage across the electrode's metal-to-solution interface is in series with the flow signal. Even when the flowmeter has nonpolarizable electrodes, the random drift of this voltage is of the same order as the flow signal, and there is no way to separate the two. (2) The ECG has a waveform and frequency content similar to that of the flow signal; near the heart, the ECG's waveform is much larger than that of the flow signal and therefore causes interference. (3) In the frequency range of interest, 0 to 30 Hz, $1/f$ noise in the amplifier is large, which results in a poor SNR.

AC FLOWMETER

The clinician can eliminate the problems of the dc flowmeter by operating the system with an ac magnet current of about 400 Hz. Lower frequencies require bulky sensors, whereas higher frequencies cause problems due to stray capacitance. The operation of this carrier system results in the ac flow voltage shown in Figure 8.5 When the flow reverses direction, the voltage changes phase by 180°, so the phase-sensitive demodulator (described in Section 3.15) is required to yield directional output.

Although ac operation is superior to dc operation, the new problem of *transformer voltage* arises. If the shaded loop shown in Figure 8.3 is not exactly parallel to the B field, some ac magnetic flux intersects the loop and induces a transformer voltage proportional to dB/dt in the output voltage. Even when the electrodes and wires are carefully positioned, the transformer voltage is usually many times larger than the flow voltage, as indicated in Figure 8.5. The amplifier voltage is the sum of the transformer voltage and the flow voltage.

There are several solutions to this problem. (1) It may be eliminated at the source by use of a *phantom electrode.* One of the electrodes is separated into two electrodes in the axial direction. Two wires are led some distance from the electrodes, and a potentiometer is placed between them. The signal from the potentiometer wiper yields a signal corresponding to a "phantom" electrode, which can be moved in the axial direction. The shaded loop in Figure 8.3 can thus be tilted forward or backward or placed exactly parallel to the B field. (2) Note in Figure 8.5 that we can sample the composite signal when the transformer voltage is zero. At this time the flow voltage is at its maximum, and the resulting *gated signal* measures only the flow voltage. However, if undesired phase shifts cause the gating to be done even a few degrees away from the proper time, large errors and drifts result. (3) The best method for reducing the effects of transformer voltage is to use the *quadrature-suppression* circuit shown in Figure 8.6 (Wyatt, 1971).

The magnitude of the voltage in the transformer at the amplifier output is detected by the quadrature demodulator, which has a full-wave-rectified output. This is low-pass-filtered to yield a dc voltage, which is then modulated by the quadrature generator to produce a signal proportional to the transformer voltage. The signal is fed to a balancing coil on the input transformer, thus balancing out the transformer voltage at the input. With enough gain in this negative-feedback loop, the transformer voltage at the amplifier output is reduced by a factor of 50. This low transformer voltage prevents overloading of the in-phase demodulator, which extracts the desired in-phase flow signal shown in Figure 8.5. By choosing low-noise FETs for the amplifier input stage,

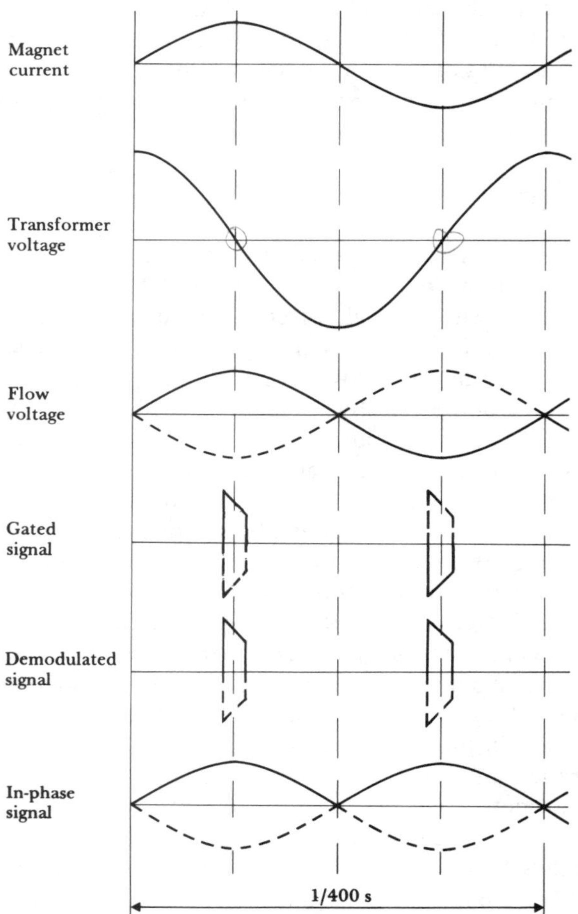

Figure 8.5 Electromagnetic flowmeter waveforms The transformer voltage is 90° out of phase with the magnet current. Other waveforms are shown solid for forward flow and dashed for reverse flow. The gated signal from the gated-sine-wave flowmeter includes less area than the in-phase signal from the quadrature-suppression flowmeter.

the proper turns ratio on the step-up transformer (Section 3.13), and full-wave demodulators, we can obtain an excellent SNR.

Some flowmeters, unlike the sine-wave flowmeters described previously, use *square-wave excitation.* In this case the transformer voltage appears as a very large spike, which overloads the amplifier for a short time. After the amplifier recovers, the circuit samples the square-wave flow voltage and processes it to obtain the flow signal. To prevent overload of the amplifier, *trapezoidal excitation* has also been used.

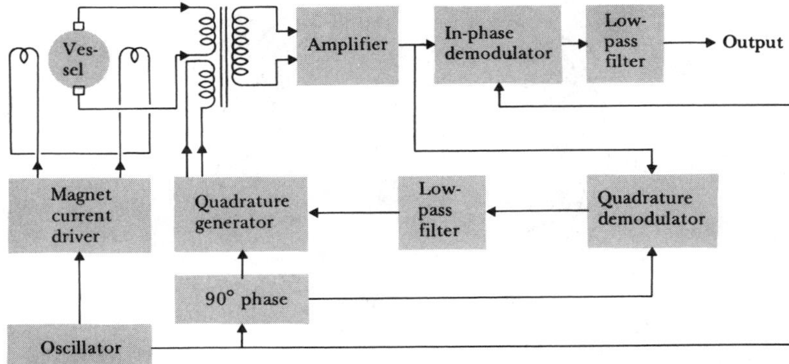

Figure 8.6 The quadrature-suppression flowmeter detects the amplifier quadrature voltage. The quadrature generator feeds back a voltage to balance out the probe-generated transformer voltage.

PROBE DESIGN

A variety of probes to measure blood flow have been used (Cobbold, 1974). The electrodes for these probes are usually made of platinum. Best results are obtained when the electrodes are platinized (electrolytically coated with platinum) to provide low impedance and are recessed in a cavity to minimize the flow of circulating currents through the metal. When the electrodes must be exposed, bright platinum is used, because the platinized coating wears off anyway. Bright platinum electrodes have a higher impedance and a higher noise level than platinized ones.

Some probes do not use a magnetic core, but they have lower sensitivity. A common *perivascular probe* is shown in Figure 8.7, in which a toroidal laminated Permalloy core is wound with two oppositely wound coils. The resulting magnetic field has low leakage flux. To prevent capacitive coupling between the coils of the magnet and the electrodes, an electrostatic shield is placed between them. The probe is insulated with a potting material that has a very high resistivity and impermeability to salt water (blood is similar to saline).

The open slot on one side of the probe makes it possible to slip it over a blood vessel without cutting the vessel. A plastic key may be inserted into the slot so that the probe encircles the vessel. The probe must fit snugly during diastole so that the electrodes make good contact. This requires some constriction of an artery during systole, when the diameter of the artery is about 7% greater. Probes are made in 1-mm increments in the range of 1 to 24 mm to ensure a snug fit on a variety of sizes of arteries. To be able to measure any size of artery requires a considerable expenditure for probes: Individual probes typically cost $500 each. The probes do not operate satisfactorily on veins, because the electrodes do not make good contact when the

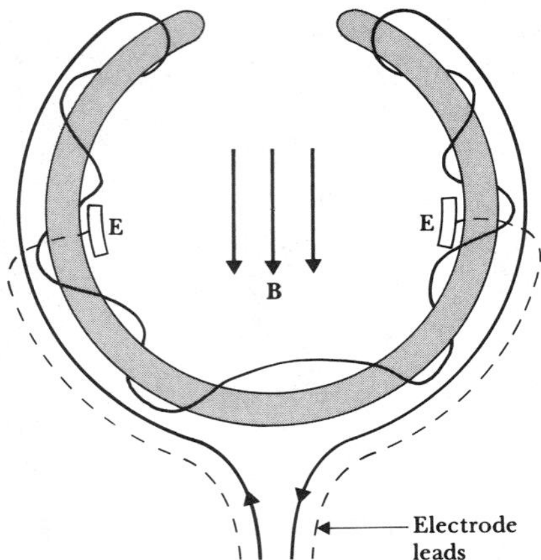

Figure 8.7 The toroidal-type cuff probe has two oppositely wound windings on each half of the core. The magnetic flux thus leaves the top of both sides, flows down in the center of the cuff, enters the base of the toroid, and flows up through both sides.

vein collapses. Special flow-through probes are used outside the body for measuring the output of cardiac-bypass pumps.

8.4 ULTRASONIC FLOWMETERS

The ultrasonic flowmeter, like the electromagnetic flowmeter, can measure instantaneous flow of blood. The ultrasound can be beamed through the skin, thus making transcutaneous flowmeters practical. Advanced types of ultrasonic flowmeters can also measure flow profiles. These advantages are making the ultrasonic flowmeter the subject of intensive development. Let us examine some aspects of this development.

TRANSDUCERS

For the transducer to be used in an ultrasonic flowmeter, we select a piezoelectric material (Section 2.6) that converts power from electric to acoustic form (Christensen, 1988). Lead zirconate titanate is a crystal that has the highest conversion efficiency. It can be molded into any shape by melting. As it is cooled through the Curie temperature, it is placed in a strong electric field to polarize the material. It is usually formed into disks that are coated on opposite faces with metal electrodes and driven by an electronic oscillator. The resulting electric field in the crystal causes mechanical constriction. The piston-like movements generate longitudinal plane waves, which propagate into the tissue. For maximal efficiency, the crystal is one-half wavelength thick.

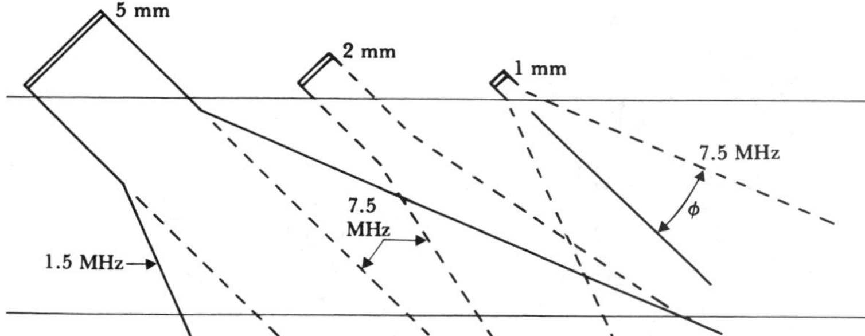

Figure 8.8 Near and far fields for various transducer diameters and frequencies. Beams are drawn to scale, passing through a 10-mm-diameter vessel. Transducer diameters are 5, 2, and 1 mm. Solid lines are for 1.5 MHz, dashed lines for 7.5 MHz.

Any cavities between the crystal and the tissue must be filled with a fluid or watery gel in order to prevent the high reflective losses associated with liquid–gas interfaces.

Because the transducer has a finite diameter, it will produce diffraction patterns, just as an aperture does in optics. Figure 8.8 shows the outline of the beam patterns for several transducer diameters and frequencies. In the *near field,* the beam is largely contained within a cylindrical outline and there is little spreading. The intensity is not uniform, however: There are multiple maximums and minimums within this region, caused by interference. The near field extends a distance d_{nf} given by

$$d_{nf} = \frac{D^2}{4\lambda} \tag{8.9}$$

where D = transducer diameter and λ = wavelength.

In the *far field* the beam diverges, and the intensity is inversely proportional to the square of the distance from the transducer. The angle of beam divergence ϕ, shown in Figure 8.8, is given by

$$\sin \phi = \frac{1.2\lambda}{D} \tag{8.10}$$

Figure 8.8 indicates that we should avoid the far field because of its lower spatial resolution. To achieve near-field operation, we must use higher frequencies and larger transducers.

To select the operating frequency, we must consider several factors. For a beam of constant cross section, the power decays exponentially because of absorption of heat in the tissue. The absorption coefficient is approximately proportional to frequency, so this suggests a low operating frequency. However, most ultrasonic flowmeters depend on the power scattered back from

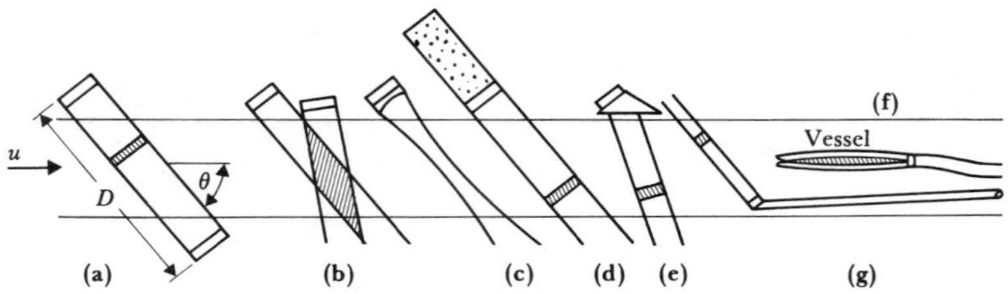

Figure 8.9 Ultrasonic transducer configurations (a) A transit-time probe requires two transducers facing each other along a path of length D inclined from the vessel axis at an angle θ. The hatched region represents a single acoustic pulse traveling between the two transducers. (b) In a transcutaneous probe, both transducers are placed on the same side of the vessel, so the probe can be placed on the skin. Beam intersection is shown hatched. (c) Any transducer may contain a plastic lens that focuses and narrows the beam. (d) For pulsed operation, the transducer is loaded by backing it with a mixture of tungsten powder in epoxy. This increases losses and lowers Q. Shaded region is shown for a single time of range gating. (e) A shaped piece of Lucite on the front loads the transducer and also refracts the beam. (f) A transducer placed on the end of a catheter beams ultrasound down the vessel. (g) For pulsed operation, the transducer is placed at an angle.

moving red blood cells. The back-scattered power is proportional to f^4, which suggests a high operating frequency. The usual compromise dictates a frequency between 2 and 10 MHz.

TRANSIT-TIME FLOWMETER

Figure 8.9(a) shows the transducer arrangement used in the transit-time ultrasonic flowmeter (Christensen, 1988). The effective velocity of sound in the vessel is equal to the velocity of sound, c, plus a component due to $\hat{u}$, the velocity of flow of blood averaged along the path of the ultrasound. For laminar flow, $\hat{u} = 1.33\overline{u}$; and for turbulent flow, $\hat{u} = 1.07\overline{u}$, where $\overline{u}$ is the velocity of the flow of blood averaged over the cross-sectional area. Because the ultrasonic path is along a single line rather than averaged over the cross-sectional area, $\hat{u}$ differs from $\overline{u}$. The transit time in the downstream (+) and upstream (−) directions is

$$t = \frac{\text{distance}}{\text{conduction velocity}} = \frac{D}{c \pm \hat{u}\cos\theta} \tag{8.11}$$

The difference between upstream and downstream transit times is

$$\Delta t = \frac{2\,D\hat{u}\cos\theta}{(c^2 - \hat{u}^2\cos^2\theta)} \cong \frac{2\,D\hat{u}\cos\theta}{c^2} \tag{8.12}$$

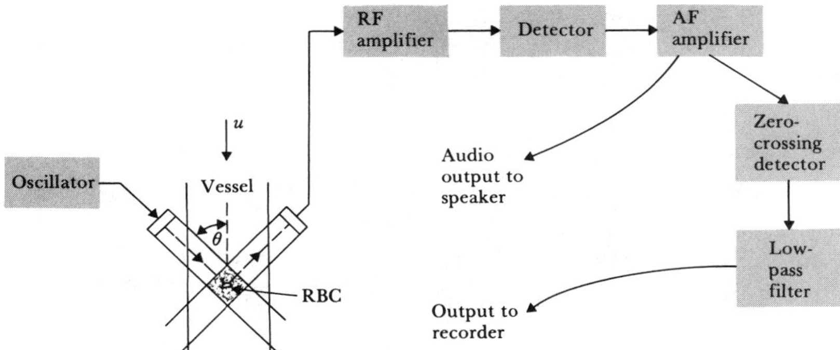

Figure 8.10 Doppler ultrasonic blood flowmeter In the simplest instrument, ultrasound is beamed through the vessel walls, back-scattered by the red blood cells, and received by a piezoelectric crystal.

and thus the average velocity $\hat{u}$ is proportional to Δt. A short acoustic pulse is transmitted alternately in the upstream and downstream directions. Unfortunately, the resulting Δt is in the nanosecond range, and complex electronics are required to achieve adequate stability. Like the electromagnetic flowmeter, the transit-time flowmeter and similar flowmeters using a phase-shift principle can operate with either saline or blood as a fluid, because they do not require particulate matter for scattering. However, they do require invasive surgery to expose the vessel.

CONTINUOUS-WAVE DOPPLER FLOWMETER

When a target recedes from a fixed source that transmits sound, the frequency of the received sound is lowered because of the Doppler effect. For small changes, the fractional change in frequency equals the fractional change in velocity.

$$\frac{f_{\mathrm{d}}}{f_0} = \frac{u}{c} \tag{8.13}$$

where

f_{d} = Doppler frequency shift

f_0 = source frequency

u = target velocity

c = velocity of sound

The flowmeter shown in Figure 8.10 requires particulate matter such as blood cells to form reflecting targets. The frequency is lowered twice. One shift occurs between the transmitting source and the moving cell that receives the signal. The other shift occurs between the transmitting cell and the receiving transducer.

$$\frac{f_d}{f_0} = \frac{2u}{c+u} \cong \frac{2u}{c} \tag{8.14}$$

The approximation is valid, because $c \cong 1500$ m/s and $u \cong 1.5$ m/s. The velocities do not all act along the same straight line, so we add an angle factor

$$f_d = \frac{2f_0 u \cos\theta}{c} \tag{8.15}$$

where θ is the angle between the beam of sound and the axis of the blood vessel, as shown in Figure 8.10. If the flow is not axial, or the transducers do not lie at the same angle, such as in Figure 8.9(b), we must include additional trigonometric factors.

Figure 8.10 shows the block diagram of a simple continuous-wave flowmeter. The oscillator must have a low output impedance to drive the low-impedance crystal. Although at most frequencies the crystal transducer has a high impedance, it is operated at mechanical resonance, where the impedance drops to about 100 Ω. The ultrasonic waves are transmitted to the moving cells, which reflect the Doppler-shifted waves to the receiving transducer. The receiving transducer is identical to the transmitting transducer. The amplified RF (radio frequency) signal plus carrier signal is detected to produce an AF (audio frequency) signal at a frequency given by (8.15).

Listening to the audio output using a speaker, we get much useful qualitative information. A simple *frequency-to-voltage converter* provides a quantitative output to a recorder. The *zero-crossing detector* emits a fixed-area pulse each time the audio signal crosses the zero axis. These pulses are low-pass-filtered to produce an output proportional to the velocity of the blood cells.

Although the electromagnetic blood flowmeter is capable of measuring both forward and reverse flow, the simple ultrasonic-type flowmeter full-wave rectifies the output, and the sense of direction of flow is lost. This results because—for either an increase or a decrease in the Doppler-shifted frequency—the beat frequency is the same. Examination of the field intersections shown in Figure 8.10 suggests that the only received frequency is the Doppler-shifted one. However, the received carrier signal is very much larger than the desired Doppler-shifted signal. Some of the RF carrier is coupled to the receiver by the electric field from the transmitter. Because of side lobes in the transducer apertures, some of the carrier signal travels a direct acoustic path to the receiver. Other power at the carrier frequency reaches the receiver after one or more reflections from fixed interfaces. The resulting received signal is composed of a large-amplitude signal at the carrier frequency plus the very low (approximately 0.1%) amplitude Doppler-shifted signal.

The Doppler-shifted signal is not at a single frequency, as implied by (8.15), for several reasons.

1. Velocity profiles are rarely blunt, with all cells moving at the same velocity. Rather, cells move at different velocities, producing different shifts of the Doppler frequency.
2. A given cell remains within the beam-intersection volume for a short time. Thus the signal received from one cell is a pure frequency multiplied by some time-gate function, yielding a band of frequencies.
3. Acoustic energy traveling within the main beam, but at angles to the beam axis, plus energy in the side lobes, causes different Doppler-frequency shifts due to an effective change in θ.
4. Tumbling of cells and local velocities resulting from turbulence cause different Doppler-frequency shifts.

All these factors combine to produce a band of frequencies. The resulting spectrum is similar to band-limited random noise, and from this we must extract flow information.

We would like to have high gain in the RF amplifier in order to boost the low-amplitude Doppler-frequency components. But the carrier is large, so the gain cannot be too high or saturation will occur. The RF bandwidth need not be wide, because the frequency deviation is only about 0.001 of the carrier frequency. However, RF-amplifier bandwidths are sometimes much wider than required, to permit tuning to different transducers.

The detector can be a simple square-law device such as a diode. The output spectrum contains the desired difference (beat) frequencies, which lie in the audio range, plus other undesired frequencies.

EXAMPLE 8.3 Calculate the maximal audio frequency of a Doppler-ultrasonic blood flowmeter that has a carrier frequency of 7 MHz, a transducer angle of 45°, a blood velocity of 150 cm/s, and an acoustic velocity of 1500 m/s.

ANSWER Substitute these data into (8.15).

$$f_d = \frac{2(7 \times 10^6\ \text{Hz})(1.5\ \text{m/s})\cos(45°)}{1500\ \text{m/s}} \cong 10\ \text{kHz} \tag{8.16}$$

The dc component must be removed with a high-pass filter in the AF amplifier. We require a corner frequency of about 100 Hz in order to reject large Doppler signals due to motion of vessel walls. Unfortunately, this high-pass filter also keeps us from measuring slow cell velocities (less than 1.5 cm/s), such as occur near the vessel wall. A low-pass filter removes high frequencies and also noise. The corner frequency is at about 15 kHz, which includes all frequencies that could result from cell motion, plus an allowance for spectral spreading.

In the simplest instruments, the AF output drives a power amplifier and speaker or earphones. The output is a band of frequencies, so it has a whooshing sound that for steady flows sounds like random noise. Venous flow sounds like a low-frequency rumble and may be modulated when the subject breathes. Arterial flow, being pulsatile, rises to a high pitch once each beat and may be followed by one or more smaller, easily heard waves caused by the underdamped flow characteristics of arteries. Thus this simple instrument can be used to trace and qualitatively evaluate blood vessels within 1 cm of the skin in locations in the legs, arms, and neck. We can also plot the spectrum of the AF signal versus time to obtain a more quantitative indication of velocities in the vessel.

The function of the *zero-crossing detector* is to convert the AF input frequency to a proportional analog output signal. It does this by emitting a constant-area pulse for each crossing of the zero axis. The detector contains a comparator (a Schmitt trigger), so we must determine the amount of hysteresis for the comparator. If the input were a single sine wave, the *signal-to-hysteresis ratio* (SHR) could be varied over wide limits, and the output would indicate the correct value. But the input is band-limited random noise. If the SHR is low, many zero crossings are missed. As the SHR increases, the indicated frequency of the output increases. A SHR of 7 is a good choice, because the output does not vary significantly with changes in SHR. Automatic gain control can be used to maintain this ratio. Very high SHRs are not desirable; noise may trigger the comparator. The signal increases and decreases with time because of the beating of the signal components at the various frequencies. Thus the short-term SHR fluctuates, and for a small portion of the time the signal is too low to exceed the hysteresis band.

The output of the zero-crossing detector is a series of pulses. These pulses are passed through a low-pass filter to remove as many of the high-frequency components as possible. The filter must pass frequencies from 0 to 25 Hz in order to reproduce the frequencies of interest in the flow pulse. But the signal is similar to band-limited random noise. Thus the pulses are not at uniform intervals, even for a fixed flow velocity, but are more like a Poisson process. Hence the output contains objectionable noise. The low-pass filter must therefore be chosen as a compromise between the high corner frequency desired to reproduce the flow pulse and the low corner frequency desired for good filtering of noise.

A major defect of the detector used in simple flowmeters is that it cannot detect the direction of flow. The recorded output looks as it would if the true velocity had been full-wave-rectified. Compared with the electromagnetic flowmeter, this is a real disadvantage, because reverse flow occurs frequently in the body. A first thought might be to translate the Doppler-shifted frequencies not to the region about dc, but to the region about 20 kHz. Forward flow might thus be 30 kHz, and reverse flow 10 kHz. The difficulty with this approach is that the high-amplitude carrier signal is translated to 20 kHz. The Doppler signals are so small that considerable effort is required to build any reasonable frequency-to-voltage converter that is not dominated by the 20-kHz signal.

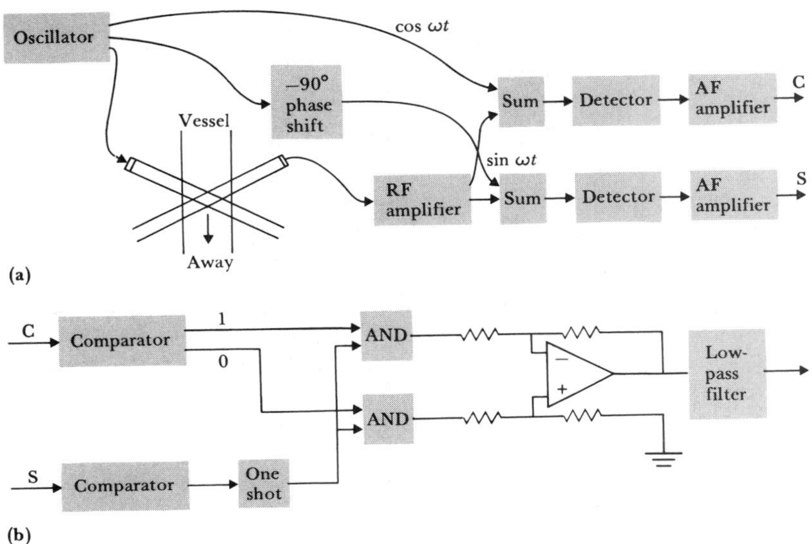

Figure 8.11 Directional Doppler block diagram (a) Quadrature-phase detector. Sine and cosine signals at the carrier frequency are summed with the RF output before detection. The output C from the cosine channel then leads (or lags) the output S from the sine channel if the flow is away from (or toward) the transducer. (b) Logic circuits route one-shot pulses through the top (or bottom) AND gate when the flow is away from (or toward) the transducer. The differential amplifier provides bidirectional output pulses that are then filtered.

A better approach is to borrow a technique from radar technology, which is used to determine not only the speed at which an aircraft is flying but also its direction. This is the quadrature-phase detector.

Figure 8.11(a) shows the analog portion of the *quadrature-phase detector* (McLeon, 1967). A phase-shift network splits the carrier into two components that are in quadrature, which means that they are 90° apart. These reference cosine and sine waves must be several times larger than the RF-amplifier output, as shown in Figure 8.12(a). The reference waves and the RF-amplifier output are linearly summed to produce the RF envelope shown in Figure 8.12(b). We assume temporarily that the RF-amplifier output contains no carrier.

If the flow of blood is in the same direction as the ultrasonic beam, we consider the blood to be flowing away from the transducer, as shown in Figure 8.11(a). For this direction, the Doppler-shift frequency is lower than that of the carrier. The phase of the Doppler wave lags behind that of the reference carrier, and the Doppler vector [see Figure 8.12(a)] rotates clockwise. In Figure 8.12(b), for time 1, the carrier and the Doppler add, producing a larger sum in the cosine channel. The sine channel is unchanged. For time 2, the carrier and the Doppler add, producing a larger sum in the sine channel.

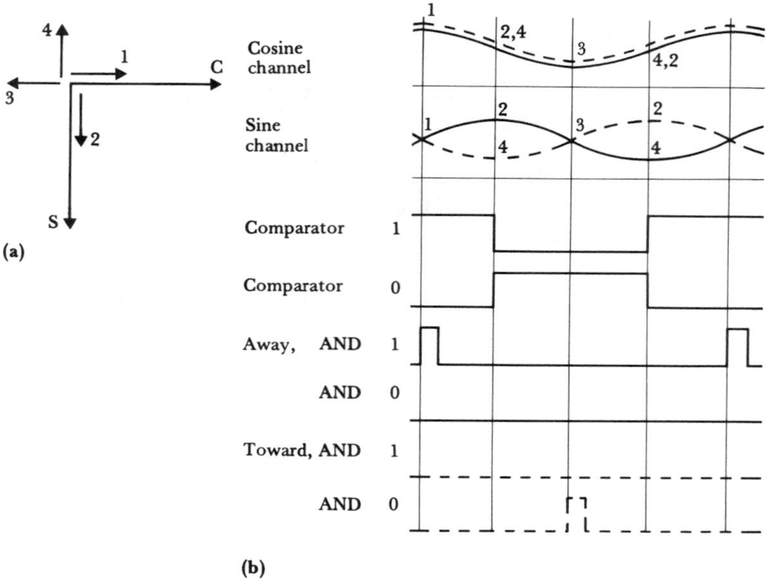

Figure 8.12 Directional Doppler signal waveforms (a) Vector diagram. The sine wave at the carrier frequency lags the cosine wave by 90°. If flow is away from the transducer, the Doppler frequency is lower than the carrier. The short vector represents the Doppler signal and rotates clockwise, as shown by the numbers 1, 2, 3, and 4. (b) Timing diagram. The top two waves represent the single-peak envelope of the carrier plus the Doppler before detection. Comparator outputs respond to the cosine channel audio signal after detection. One-shot pulses are derived from the sine channel and are gated through the correct AND gate by comparator outputs. The dashed lines indicate flow toward the transducer.

Similar reasoning produces the rest of the wave for times 3 and 4. Note that the sine channel lags behind the cosine channel.

If the flow of blood is toward the transducer, the Doppler frequency is higher than the carrier frequency, and the Doppler vector rotates counterclockwise. This produces the dashed waves shown in Figure 8.12(b), and the phase relation between the cosine and sine channels is reversed. Thus, by examining the sign of the phase, we measure direction of flow. The detector produces AF waves that have the same shape as the RF envelope.

Figure 8.11(b) shows the logic that detects the sign of the phase. The cosine channel drives a comparator, the digital output of which, shown in Figure 8.12(b), is used for gating and does not change with direction of flow of blood. The sine channel triggers a one-shot the width of which must be short. Depending on the direction of flow, this one-shot is triggered either at the beginning of or halfway through the period, as shown in Figure 8.12(b).

The AND gates then gate it into the top or bottom input of the differential amplifier, thus producing a bidirectional output.

The preceding discussion is correct for a sinusoidal RF signal. Our RF signal is like band-limited random noise, however, so there is some time shifting of the relations shown in Figure 8.12(b). Also, a large fixed component at the carrier frequency is present, which displaces the Doppler vectors away from the position shown. As long as the reference cosine and sine waves are more than twice the amplitude of the total RF output, time shifting of the gating relations is not excessive. These time shifts are not problems in practice; a short one-shot pulse can shift almost $\pm 90°$ before passing out of the correct comparator gate.

It is possible to add another one-shot and several logic blocks to obtain pulse outputs on both positive and negative zero crossings. This doubles the frequency of the pulse train and reduces the fluctuations in the output to 0.707 of their former value.

PULSED DOPPLER

Continuous-wave flowmeters provide little information about flow profile. Therefore, several instruments have been built (Christensen, 1988) that operate in a radar-like mode. The transmitter is excited with a brief burst of signal. The transmitted wave travels in a single packet, and the transmitter can also be used as a receiver, because reflections are received at a later time. The delay between transmission and reception is a direct indication of distance, so we can obtain a complete plot of reflections across the blood vessel. By examining the Doppler shift at various delays, we can obtain a velocity profile across the vessel.

To achieve good range resolution, the transmitted-pulse duration should ideally be very short. To achieve a good SNR and good velocity discrimination, it should be long. The usual compromise is an 8-MHz pulse of 1-μs duration, which produces a traveling packet 1.5 mm long, as shown in Figure 8.9(d). The intensity of this packet is convolved with the local velocity profile to produce the received signal. Thus, the velocity profile of the blood vessel is smeared to a larger-than-actual value. Because of this problem, and also because the wave packet arrives at an angle to normal, the location of the vessel walls is indistinct. It is possible, however, to mathematically "deconvolve" the instrument output to obtain a less-smeared representation of the velocity profile.

There are two constraints on pulse repetition rate f_r. First, to avoid *range ambiguities,* we must analyze the return from one pulse before sending out the next. Thus

$$f_r < \frac{c}{2R_m} \tag{8.17}$$

where R_m is the maximal useful range. Second, we must satisfy the *sampling theorem,* which requires that

$$f_r > 2f_d \tag{8.18}$$

Combining (8.17) and (8.18) with (8.15) yields

$$u_m(\cos\theta)R_{max} < \frac{c^2}{8f_0} \tag{8.19}$$

which shows that the product of the range and the maximal velocity along the transducer axis is limited. In practice, measurements are constrained even more than indicated by (8.19) because of (1) spectral spreading, which produces some frequencies higher than those expected, and (2) imperfect cutoff characteristics of the low-pass filters used to prevent *aliasing* (generation of fictitious frequencies by the sampling process).

Because we cannot easily start and stop an oscillator in 1 μs, the first stage of the oscillator operates continuously. The transmitter and the receiver both use a common piezoelectric transducer, so a *gate* is required to turn off the signal from the transmitter during reception. A one-stage gate is not sufficient to isolate the large transmitter signals from the very small received signals. Therefore, two gates in series are used to turn off the transmitter.

The optimal transmitted signal is a pulse-modulated sine-wave carrier. Although it is easy to generate this burst electrically, it is difficult to transduce this electric burst to a similar acoustic burst. The crystal transducer has a high Q (narrow bandwidth) and therefore rings at its resonant frequency long after the electric signal stops. Therefore, the transducer is modified to achieve a lower Q (wider bandwidth) by adding mass to the back [Figure 8.9(d)] or to the front [Figure 8.9(e)]. The Q is not lowered to a desirable value of about 2 to 5, because this would greatly decrease both the efficiency of the transmission and the sensitivity of the reception. The Q is generally 5 to 15, so some ringing still exists.

When we generate a short sine-wave burst, we no longer have a single frequency. Rather, the pulse train of the repetition rate is multiplied by the carrier in time, producing carrier sidebands in the frequency domain. This spectrum excites the transducer, producing a field that is more complex than that for continuous-wave excitation. This causes spectral spreading of the received signal.

LASER DOPPLER BLOOD FLOWMETER

In a laser Doppler blood flowmeter, a 5-mW He-Ne laser beams 632.8-nm light through fiber optics into the skin (Holloway, 1988). Moving red blood cells in the skin frequency shift the light and cause spectral broadening. Reflected light is carried by fiber optics to a photodiode. Filtering, weighting, squaring, and dividing are necessary for signal processing. Capillary blood flow has been studied in the skin and many other organs.

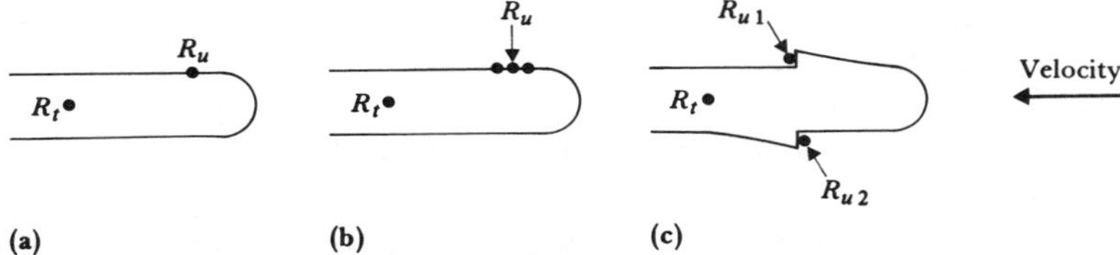

Figure 8.13 Thermal velocity probes (a) Velocity-sensitive thermistor R_u is exposed to the velocity stream. Temperature-compensating thermistor R_t is placed within the probe. (b) Thermistors placed down- and upstream from R_u are heated or not heated by R_u, thus indicating velocity direction. (c) Thermistors exposed to and shielded from flow can also indicate velocity direction.

8.5 THERMAL-CONVECTION VELOCITY SENSORS

PRINCIPLE

The thermodilution methods described in Sections 8.1 and 8.2 depend on the mixing of the heat indicator into the entire flow stream. In contrast, thermal velocity sensors depend on convective cooling of a heated sensor and are therefore sensitive only to local velocity.

Figure 8.13(a) shows a simple probe. The thermistor R_u is heated to a temperature difference ΔT above blood temperature by the power W dissipated by current passing through R_u. Experimental observations (Grahn *et al.*, 1969) show that these quantities are related to the blood velocity u by

$$\frac{W}{\Delta T} = a + b \log u \tag{8.20}$$

where a and b are constants. Thus the method is nonlinear, with a large sensitivity at low velocities and a small sensitivity at high velocities.

PROBES

Catheter-tip probes are designed with two types of sensors (Cobbold, 1974). The first type uses the thermistors shown in Figure 8.13 and provides a high sensitivity and reasonable resistance values. Because the thermistor shown in Figure 8.13(a) is cooled equally for both directions of velocity, the output of the instrument is a full-wave-rectified replica of the true velocity. To overcome this limitation, the probe shown in Figure 8.13(b) has two additional thermistors located a few tenths of a millimeter downstream and upstream from R_u. Depending on the direction of velocity, one or the other is heated by the heat carried through the blood from the thermistor R_u. These two additional thermistors are placed in a bridge that is balanced for zero velocity. A comparator detects the bridge unbalance and switches the output from positive to negative. The probe shown in Figure 8.13(c) uses two velocity sensors arranged so that one is exposed to the fluid velocity while the other is shielded from the fluid velocity.

The second type of sensor uses a glass bead with a thin strip of platinum deposited on its surface. The platinum may be painted on and then fired in a furnace, or it may be *sputtered* (deposited by electric discharge in a vacuum). A disadvantage of platinum-film sensors is their low resistance (a few ohms) and low sensitivity.

A real question arises about what is actually being measured. When a catheter is inserted into a blood vessel, the sensor may be centered and thus measure maximal velocity, or it may be against the wall of the vessel and thus measure a low velocity. One way of ensuring that the sensor is not against the wall is to rotate the catheter, searching for the maximal output. Catheters are also sensitive to radial velocity of blood, as well as to radial vibrations of the catheter (catheter whip). Thus, in addition to any errors due to measuring velocity, errors in trying to estimate flow can arise from lack of knowledge about location of the sensor. Either type of probe (if it is made sufficiently small) can be placed at the end of a hypodermic needle and inserted perpendicular to the vessel for measuring velocity profiles.

CIRCUIT

A *constant-current* sensor circuit cannot be used for two reasons. First, the time constant of the sensor embedded in the probe is a few tenths of a second—much too long to achieve the desired frequency response of 0 to 25 Hz. Second, to achieve a reasonable sensitivity at high velocities, the sensor current must be so high that when the flow stops, lack of convection cooling increases the sensor temperature more than 5 °C above the blood temperature and fibrin coats the sensor.

The *constant-temperature* sensor circuit shown in Figure 8.14 overcomes both of these problems. The circuit is initially unbalanced by adjusting R_1. The unbalance is amplified by the high-gain op amp, and its output is fed back to power the resistance bridge. Operation of the circuit is as follows: Assume that thermistor R_u is 5 °C higher than blood temperature because of self-heating. If the velocity increases, R_u cools and its resistance increases. A more positive voltage enters the noninverting op-amp terminal, so v_b increases. This increases bridge power and R_u heats up, thus counteracting the original cooling. The system uses high-gain negative feedback to keep the bridge always in balance. Thus R_u remains nearly constant, and therefore its temperature remains nearly constant. The high-gain negative feedback divides the sensor time constant by a factor equal to the loop gain, so frequency response is greatly improved. In effect, if the sensor becomes slightly cooled, the op amp can provide a large quantity of power to rapidly heat it back to the desired temperature.

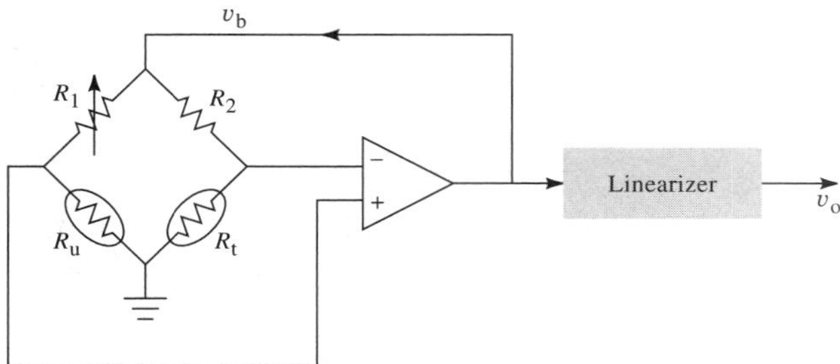

Figure 8.14 Thermal velocity meter circuit A velocity increase cools R_u, the velocity-measuring thermistor. This increases voltage to the noninverting op-amp input, which increases bridge voltage v_b and heats R_u. R_t provides temperature compensation.

The circuit operates satisfactorily with only one sensor, R_u, provided that the blood temperature is constant. Should the blood temperature vary, a temperature-compensating thermistor R_t is added to keep the bridge in balance. So that its rise in temperature is very small, R_t must have a much lower resistance-temperature coefficient than R_u, to ensure that R_t is a sensor of temperature and not of velocity. The thermal resistance of R_t can be lowered by making it large in size, by using a heat sink, or by placing it within the probe so that the effective cooling area is much larger. Another solution is to increase the resistance values for R_2 and R_t so that their power dissipation is much lower.

A linearizer is required to solve (8.20). We may square v_b to obtain W and then use an antilog converter to obtain v_o. For the directional probe shown in Figure 8.17(b), a unity-gain inverting amplifier and switch may be used to yield the direction of flow.

Calibration can be accomplished by using a sinusoidal-flow pump or a cylindrical pan of liquid rotating on a turntable.

The main use of thermal-velocity sensors is to measure the velocity of blood and to compile velocity profiles in studies of animals, although such sensors have also been regularly used to measure velocity and acceleration of blood at the aortic root in human patients undergoing diagnostic catheterization (Roberts, 1972). The same principle has also been applied to the measurement of the flow of air in lungs by installing a heated platinum wire in a breathing tube.

8.6 CHAMBER PLETHYSMOGRAPHY

Plethysmographs measure changes in volume. The only accurate way to measure changes in volume of blood in the extremities noninvasively is to use a

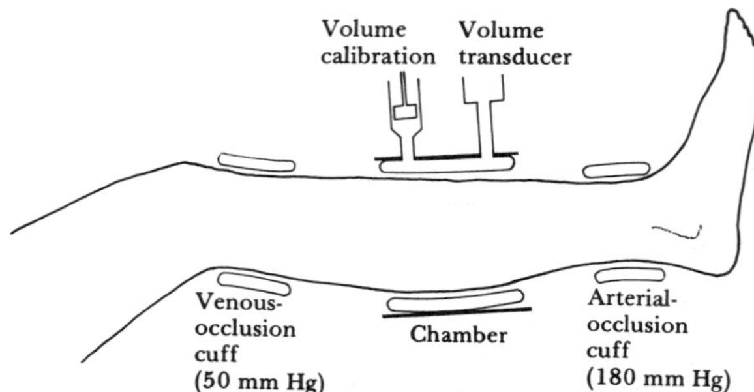

Figure 8.15 In chamber plethysmography, the venous-occlusion cuff is inflated to 50 mm Hg (6.7 kPa), stopping venous return. Arterial flow causes an increase in volume of the leg segment, which the chamber measures. The text explains the purpose of the arterial-occlusion cuff.

chamber plethysmograph. By timing these volume changes, we can measure flow by computing $F = dV/dt$. A cuff is used to prevent venous blood from leaving the limb—hence the name *venous-occlusion plethysmography* (Seagar *et al.,* 1984).

EQUIPMENT

Figure 8.15 shows the equipment used in a venous-occlusion plethysmograph. The chamber has a rigid cylindrical outer container and is placed around the leg. As the volume of the leg increases, the leg squeezes some type of bladder and decreases its volume. If the bladder is filled with water, the change in volume may be measured by observing the water rising in a calibrated tube. For recording purposes, some air may be introduced above the water and the change in air pressure measured. Water-filled plethysmographs are temperature-controlled to prevent thermal drifts. Because of their hydrostatic pressure, they may constrict the vessels in the limb and cause undesirable physiological changes.

Air may be used in the bladder and the resulting changes in pressure measured directly. Some systems do not use a bladder. They attempt to seal the ends of a rigid chamber to the limb, but then leaks may be a problem. One device uses a pneumotachometer to measure the flow of air into and out of the chamber. This flow is then integrated to yield changes in volume. This equipment is designed to accommodate a variety of limb sizes, so the chambers and bladders are made in a family of sizes. Alternatively, a single chamber may be used for several sizes of limb. Devices that are capable of doing this are made with iris diaphragms that form the ends of the chamber and close down on the limb.

METHOD

Figure 8.16 shows the sequence of operations that yields a measurement of flow (Raines and Darling, 1976). A calibration may be marked on the record

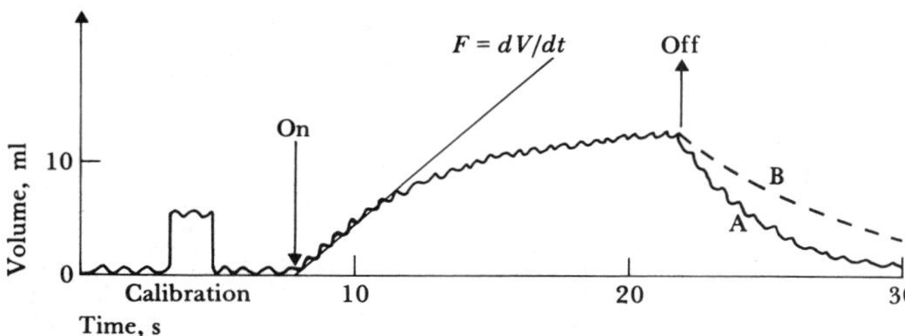

Figure 8.16 After venous-occlusion cuff pressure is turned on, the initial volume-versus-time slope is caused by arterial inflow. After the cuff is released, segment volume rapidly returns to normal (A). If a venous thrombosis blocks the vein, return to normal is slower (B).

by injecting into the chamber a known volume of fluid, using the volume-calibration syringe. The venous-occlusion cuff is then applied to a limb and pressurized to 50 mm Hg (6.7 kPa), which prevents venous blood from leaving the limb. Arterial flow is not hindered by this cuff pressure, and the increase in volume of blood in the limb per unit time is equal to the arterial inflow. If the chamber completely encloses the limb distal to the cuff, the arterial flow into the limb is measured. If the chamber encloses only a segment of a limb, as shown in Figure 8.15, an arterial-occlusion cuff distal to the chamber must be inflated to 180 mm Hg (24 kPa) to ensure that the changes in chamber volume measure only arterial flow entering the segment of the limb.

A few seconds after the cuffs are occluded, the venous pressure exceeds 50 mm Hg (6.7 kPa), venous return commences, and the volume of blood in the limb segment plateaus. When the clinician releases the pressure of the venous-occlusion cuff, the volume of blood in the limb segment rapidly returns to normal (Figure 8.16, curve A). If a venous thrombosis (vein clot) partially blocks the return of venous blood, the volume of blood in the veins returns to normal more slowly (Figure 8.16, curve B). This technique is a useful noninvasive test for venous thrombosis.

Semmlow (1988) notes that the measurement of erection, or penile tumescence, is the only physiological response that reliably differentiates male sexual arousal from other emotional states. Early water- or air-filled chamber plethysmographs for measuring tumescence have been replaced by less bulky circular metal bands and elastic strain gages.

8.7 ELECTRIC-IMPEDANCE PLETHYSMOGRAPHY

It is simple to attach electrodes to a segment of tissue and measure the resulting impedance of the tissue. As the volume of the tissue changes in response to pulsations of blood (as happens in a limb) or the resistivity changes in response

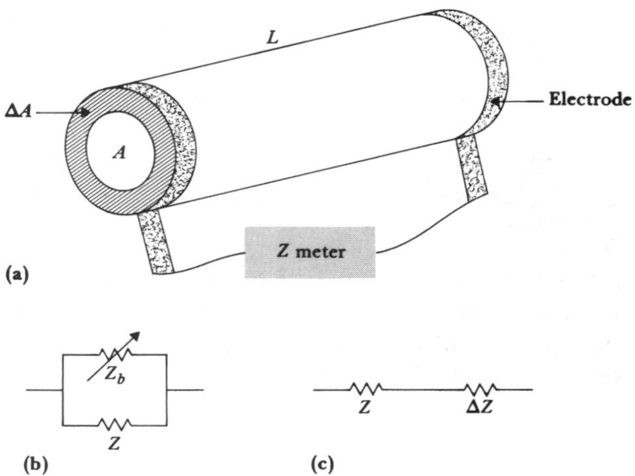

Figure 8.17 (a) A model for impedance plethysmography. A cylindrical limb has length L and cross-sectional area A. With each pressure pulse, A increases by the shaded area ΔA. (b) This causes impedance of the blood, Z_b, to be added in parallel to Z. (c) Usually ΔZ is measured instead of Z_b.

to increased air in the tissue (as happens in the lung), the impedance of the tissue changes (Anderson, 1988).

Electric-impedance plethysmography has been used to measure a wide variety of variables, but in many cases the accuracy of the method is poor or unknown.

PRINCIPLE

In the early 1950s, Nyboer (1970) developed the equations used in impedance plethysmography. However, we shall follow Swanson's (1976) derivation, which is conceptually and mathematically simpler. Figure 8.17 shows Swanson's model of a cylindrical limb. The derivation requires three assumptions: (1) The expansion of the arteries is uniform. This assumption is probably valid in healthy vessels, but it may not be valid in diseased ones. (2) The resistivity of blood, ρ_b, does not change. In fact, ρ_b decreases with velocity because of alignment of the cells with flow streamlines and movement of cells toward the axis. Also, ρ_b is real for dc but has a small reactive component at higher frequencies. (3) Lines of current are parallel to the arteries. This assumption is probably valid for most limb segments, but not for the knee.

The shunting impedance of the blood, Z_b, is due to the additional blood volume ΔV that causes the increase in cross-sectional area ΔA.

$$Z_b = \frac{\rho_b L}{\Delta A} \tag{8.21}$$

$$\Delta V = L\,\Delta A = \frac{\rho_b L^2}{Z_b} \tag{8.22}$$

But we must replace the Z_b of Figure 8.21(b) in terms of the normally measured $\Delta Z = [(Z_b \parallel Z) - Z]$ of Figure 8.21(c). Now

$$\Delta Z = \frac{ZZ_b}{Z + Z_b} - Z = \frac{-Z^2}{Z + Z_b} \tag{8.23}$$

and because $Z \ll Z_b$,

$$\frac{1}{Z_b} \cong \frac{-\Delta Z}{Z^2} \tag{8.24}$$

Substituting (8.24) in (8.22) yields

$$\Delta V = \frac{-\rho_b L^2 \Delta Z}{Z^2} \tag{8.25}$$

If the assumptions are valid, (8.25) shows that we can calculate ΔV from ρ_b (Geddes and Baker, 1989) and from other quantities that are easily measured.

Although (8.25) is valid at any frequency, there are several considerations that suggest the use of a frequency of about 100 kHz.

1. It is desirable to use a current greater than 1 mA in order to achieve adequate SNR. At low frequencies this current causes an unpleasant shock. But the current required for perception increases with frequency (Section 14.2). Therefore, frequencies above 20 kHz are used to avoid perception of the current.
2. The skin-electrode impedance decreases by a factor of about 100 as the frequency is increased from low values up to 100 kHz. High frequencies are therefore used to decrease both the skin-electrode impedance and the undesirable changes in this impedance that result from motion of the patient.
3. If a frequency much higher than 100 kHz is used, the low impedances of the stray capacitances make design of the instrument difficult.

TWO OR FOUR ELECTRODES

For reasons of economy and ease in application, some impedance plethysmographs use two electrodes, as shown in Figure 8.18. The current *i* flows through the same electrodes used to measure the voltage *v*. This causes several problems.

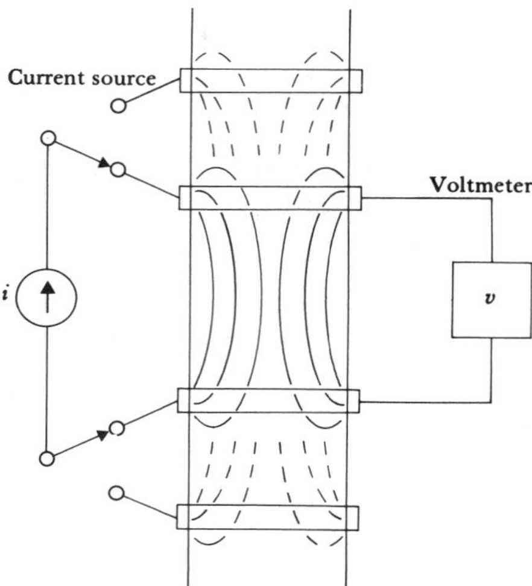

Figure 8.18 In two-electrode impedance plethysmography, switches are in the position shown, resulting in a high current density (solid lines) under voltage-sensing electrodes. In four-electrode impedance plethysmography, switches are thrown to the other position, resulting in a more uniform current density (dashed lines) under voltage-sensing electrodes.

1. The current density is higher near the electrodes than elsewhere in the tissue. This causes the measured impedance, $Z = v/i$, to weight impedance of the tissue more heavily near the electrodes than elsewhere in the tissue.
2. Pulsations of blood in the tissue cause artifactual changes in the skin–electrode impedance, as well as changes in the desired tissue impedance. Because the skin–electrode impedance is in series with the desired tissue impedance, it is impossible to separate the two and determine the actual change in impedance of the tissue.
3. The current density is not uniform in the region of interest, so (8.26) cannot be used.

To solve these problems, clinicians use the four-electrode impedance plethysmograph shown in Figure 8.18. The current flows through the two outer electrodes, so the current density is more uniform in the region sensed by the two inner voltage electrodes. Variations in skin–electrode impedance cause only a second-order error.

CONSTANT CURRENT SOURCE

Figure 8.19 shows the circuit of a four-electrode impedance plethysmograph. Ideally, the current source i causes a constant current to flow through Z, regardless of changes in ΔZ or other impedances. In practice, however, a

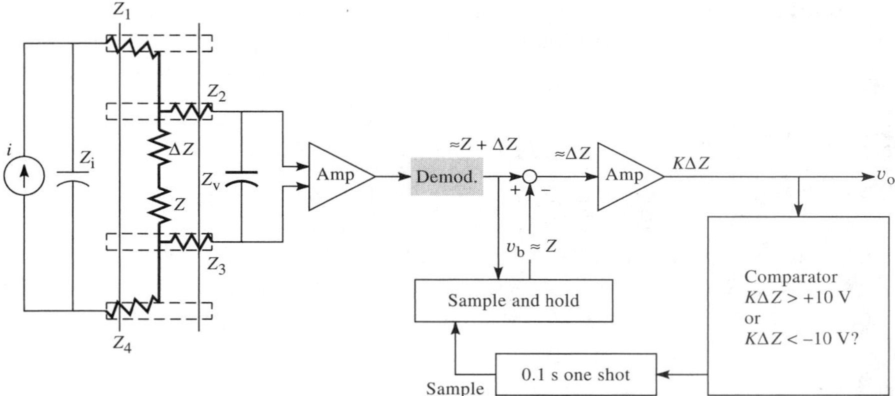

Figure 8.19 In four-electrode impedance plethysmography, current is injected through two outer electrodes, and voltage is sensed between two inner electrodes. Amplification and demodulation yield $Z + \Delta Z$. Normally, a balancing voltage v_b is applied to produce the desired ΔZ. In the automatic-reset system, when saturation of v_o occurs, the comparator commands the sample and hold to sample $Z + \Delta Z$ and hold it as v_b. This resets the input to the final amplifier and v_o zero. Further changes in ΔZ cause changes in v_o without saturation.

shunting impedance Z_i results from stray and cable capacitance. At 100 kHz, 15 pF of stray capacitance causes an impedance of about 100 kΩ. Thus changes in Z_1, ΔZ, and Z_4 cause the constant current to divide between Z and Z_i in a changing manner. In practice this is not a problem, because changes in Z_1, ΔZ, and Z_4 are small, and careful design can keep Z_i large enough. Also, Z and Z_i are close to 90° out of phase, which reduces the effects of the problem. Frequently the constant current is supplied through a low-capacity transformer to prevent ground-loop problems.

VOLTAGE-SENSING AMPLIFIER

Figure 8.19 shows that electrodes Z_2 and Z_3 are used to sense the voltage. Ideally, the voltage amplifier has an input impedance sufficiently high that no current flows through Z_2 and Z_3. In practice, however, a shunting impedance Z_v results from stray, cable, and amplifier capacitance. Thus changes in Z_2 and Z_3 cause the desired voltage to be attenuated in a changing manner. In practice this is not a problem, because changes in Z_2 and Z_3 are small, and careful design can keep Z_v large enough. Also, Z_2 and Z_3 are 90° out of phase with Z_v, which reduces the effects of the problem. Not shown in Figure 8.19 are common-mode impedances from each amplifier input to ground. These impedances can convert common-mode voltages to erroneous differential voltages unless the instrument is carefully designed. Frequently, the voltage is sensed through a low-capacity transformer, which greatly reduces common-mode and ground-loop problems. The amplifier requires only modest gain, because a typical voltage sensed is $v = iZ = (0.004)(4) = 0.16$ V.

DEMODULATION

The output of the amplifier is a large 100-kHz signal, amplitude-modulated a small amount by $i\Delta Z$. This $i\Delta Z$ may be demodulated by any AM detector, such as a diode followed by a low-pass filter. The phase-sensitive detector described in Section 3.15 is a superior demodulator because it is insensitive to the noise and 60 Hz interference that simpler demodulators detect.

METHODS OF BALANCE

The demodulator produces an output $Z + \Delta Z$. Frequently ΔZ contains the useful information, but it may be only 1/1000th of Z. One approach is to use a high-pass filter to pass frequencies above 0.05 Hz and extract ΔZ. This is satisfactory for measuring pulsatile arterial changes, but not venous or respiratory changes. To build a dc-responding instrument, we subtract a balancing voltage v_b from the demodulated signal to yield ΔZ, as shown in Figure 8.19. We may derive v_b from an adjustable dc source, but then slight changes in i produce artifactual changes in ΔZ. A better technique is to derive v_b from a rectified signal from the master oscillator that generates i. Then the system behaves like a Wheatstone bridge: A change in excitation voltage does not unbalance the bridge.

But still there is a problem. When the electrodes are first applied or when the patient moves, Z changes by an amount much larger than ΔZ. The operator must manually adjust v_b to keep ΔZ small, which is necessary if the operator is to be able to amplify and display ΔZ adequately. An *automatic-reset system* has been developed to eliminate the bother of manual adjustment. Whenever ΔZ saturates its amplifier, a sample-and-hold circuit makes $v_b = Z + \Delta Z$, which momentarily resets ΔZ to zero. The sudden vertical-reset trace is easily distinguished from the slower-changing physiological data. A detailed design of an automatically balancing electric-impedance plethysmograph is available (Shankar and Webster, 1985).

APPLICATIONS

Electric-impedance plethysmography is used to measure a wide variety of changes in the volume of tissue (Geddes and Baker, 1989). Electrodes placed on both legs provide an indication of whether pulsations of volume are normal (Shankar and Webster, 1991). If the pulsatile waveform in one leg is much smaller than that in the other, this indicates an obstruction in the first leg. If pulsatile waveforms are reduced in both legs, this indicates an obstruction in their common supply. A clinically useful noninvasive method for detecting venous thrombosis in the leg is venous-occlusion plethysmography.

When impedance plethysmography measures the changes in volume shown in Figure 8.16, this approach replaces the cumbersome chamber shown in Figure 8.15

Electrodes on each side of the thorax provide an excellent indication of rate of ventilation, but they give a less accurate indication of volume of ventilation. Such transthoracic electrical impedance monitoring is widely used for infant apnea monitoring to prevent sudden infant death syndrome (SIDS). Computer algorithms use pattern recognition techniques such as threshold crossing, adaptive threshold, and peak detection to reject cardiogenic and movement artifacts (Neuman, 1988).

Electrodes around the neck and around the waist cause current to flow through the major vessels connected to the heart. The resulting changes in impedance provide a rough estimate of beat-by-beat changes in cardiac output (Kubicek *et al.,* 1970). An extensive review of this *impedance cardiography* is available (Mohapatra, 1988). The impedance-cardiographic outputs from the neck, upper thorax, and lower thorax during supine, sitting, and bicycle exercise have been measured (Patterson *et al.,* 1991). Arrangements of spot electrodes do not duplicate band electrodes and do not yield good estimates of cardiac output but estimate only regional flow. Band electrodes yield good estimates of cardiac output for normals, but they may fail to give reasonable predictions on very sick patients.

Although Nyboer (1970) and others claim that flow of blood in the limbs can be measured, Swanson (1976) shows their techniques to be poor predictors of flow.

An eight-electrode catheter in the left ventricle injects current through band electrodes 1 and 8, and measures voltages from all the electrodes in between (Valentinuzzi and Spinelli, 1989). The change in impedance yields change in ventricular volume and, from this, cardiac output. Plots of pressure–volume diagrams and their area yield stroke work.

Some systems claim to measure body water and body fat by measuring the electric impedance between the limbs. But a wrist-to-ankle measurement is influenced mostly by the impedance of the arm and leg and less than 5% of the total impedance is contributed by the trunk, which has half the body mass. Separate measurements of the arms, legs, and trunk might improve the prediction (Patterson, 1989).

The number of independent measurements from N electrodes is equal to $N(N - 1)/2$. If we place 16 electrodes around the thorax, we can obtain 120 independent measurements and can use these data to compute a two-dimensional image of resistivity distribution within the thorax. A review describes methods of injecting current patterns, measuring electrode voltages, and optimizing reconstruction algorithms to create these images (Webster, 1990). The spatial resolution is only about 10%, but this *electric-impedance tomography* may be useful for monitoring the development of pneumonia, measuring stomach emptying, or monitoring ventilation.

The advantages of electric-impedance plethysmography are that it is noninvasive and that it is relatively simple to use. The disadvantages are

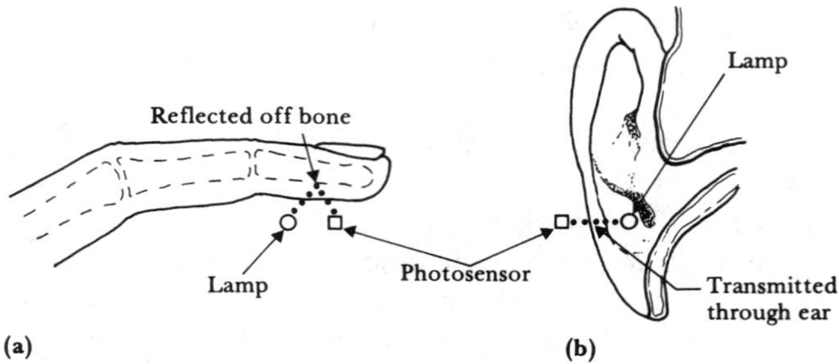

Figure 8.20 (a) Light transmitted into the finger pad is reflected off bone and detected by a photosensor. (b) Light transmitted through the aural pinna is detected by a photosensor.

that it is not sufficiently accurate for many of the attempted applications and that even the cause of the changes in impedance is not clear in some cases.

8.8 PHOTOPLETHYSMOGRAPHY

Light can be transmitted through a capillary bed. As arterial pulsations fill the capillary bed, the changes in volume of the vessels modify the absorption, reflection, and scattering of the light. Although photoplethysmography is simple and indicates the timing of events such as heart rate, it provides a poor measure of changes in volume, and it is very sensitive to motion artifact.

LIGHT SOURCES

Figure 8.20 shows two photoplethysmographic methods, in which sources generate light that is transmitted through the tissue (Geddes and Baker, 1989). A miniature tungsten lamp may be used as the light source, but the heat generated causes vasodilation, which alters the system being measured. This may be considered desirable, however, because a larger pulse is produced. A less bulky unit may be formed using a GaAs LED (Lee *et al.,* 1975), which produces a narrow-band source with a peak spectral emission at a wavelength of 940 nm [Figure 2.18(a)].

PHOTOSENSORS

Photoconductive cells have been used as sensors, but they are bulky and present a problem in that prior exposure to light changes the sensitivity of the cell. In addition, a filter is required to restrict the sensitivity of the sensor

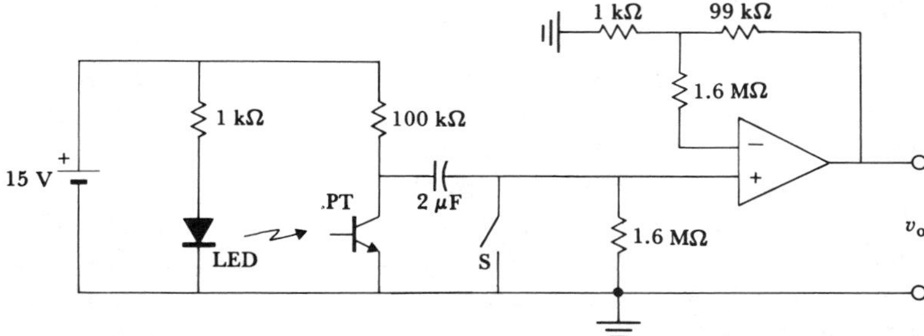

Figure 8.21 In this photoplethysmograph, the output of a light-emitting diode is altered by tissue absorption to modulate the phototransistor. The dc level is blocked by the capacitor, and switch S restores the trace. A noninverting amplifier can drive low impedance loads, and it provides a gain of 100.

to the near-infrared region so that changes in blood O_2 content that are prominent in the visible-light region will not cause changes in sensitivity. A less bulky unit can be formed using an Si phototransistor. A filter that passes only infrared light is helpful for all types of sensors to prevent 120 Hz signals from fluorescent lights from being detected. This does not prevent dc light from tungsten lights or daylight from causing baseline shifts, so lightproof enclosures are usually provided for these devices.

CIRCUITS

The output from the sensor represents a large value of transmittance, modulated by very small changes due to pulsations of blood. To eliminate the large baseline value, frequencies above 0.05 Hz are passed through a high-pass filter. The resulting signal is greatly amplified to yield a sufficiently large waveform. Any movement of the photoplethysmograph relative to the tissue causes a change in the baseline transmittance that is many times larger than the pulsation signal. These large artifacts due to motion saturate the amplifier; thus it is a good thing to have a means of quickly restoring the output trace.

EXAMPLE 8.4 Design the complete circuit for a solid-state photoplethysmograph.

ANSWER A typical LED requires a forward current of 15 mA. Using a 15-V supply would require a series resistor of $R_L = v/i = 15/0.015 = 1$ kΩ. A typical phototransistor passes a maximum of 150 μA. To avoid saturation, choose a series resistor $R_p = v/i = 15/0.00015 = 100$ kΩ. The largest convenient paper capacitor is 2 μF. The output resistor $R_o = 1/(2\pi f_0 C) = 1/[2\pi(0.05)(2 \times 10^{-6})] = 1.6$ MΩ. Figure 8.21 shows the circuit.

APPLICATIONS

For a patient who remains quiet, the photoplethysmograph can measure heart rate. It offers an advantage in that it responds to the pumping action of the heart and not to the ECG. When properly shielded, it is unaffected by the use of electrosurgery, which usually disables the ECG. However, when the patient is in a state of shock, vasoconstriction causes peripheral flow to be greatly reduced, and the resulting small output may make the device unusable. To prevent this problem, the device has been used to transmit light through the nasal septum (Groveman *et al.*, 1966). This technique monitors terminal branches of the internal carotid artery and yields an output that correlates with cerebral blood flow.

PROBLEMS

8.1 *Clearance* is defined as the minimal volume of blood entering an organ per unit time required to supply the amount of indicator removed from the blood per unit time during the blood's passage through the organ. Derive a formula for renal clearance, given the arterial concentration of the indicator PAH, all of which is excreted by the kidneys into the urine. Give units.
8.2 In Figure 8.2, the final concentration at time F is higher than the initial concentration at time A. Write a formula that yields the circulating blood volume from the information obtained during an indicator dilution test. Give units.
8.3 In the decaying exponential portion of Figure 8.2, the concentrations at times C and D are given. Calculate the shaded area under the dotted curve between times C and E. Give units.
8.4 A physician is using the rapid-injection thermodilution method of finding a patient's cardiac output. Calculate the cardiac output (in milliliters per second and in liters per minute) from the following data:

$$V_i = 10\text{ ml}, \Delta T_i = -30\text{ K}$$
$$\rho_i = 1005\text{ kg/m}^3, c_i = 4170\text{ J/(kg}\cdot\text{K)}$$
$$\rho_b = 1060\text{ kg/m}^3, c_b = 3640\text{ J/(kg}\cdot\text{K)}$$
$$\int_0^{t_1} \Delta T_b\, dt = -5.0\text{ s}\cdot\text{K}$$

8.5 Name the indicator-dilution technique for measuring cardiac output that does *not* require arterial puncture. Give the equation for calculating cardiac output, and define all terms.
8.6 For cardiac catheterization, describe the characteristics of the dye used to improve *visualization.* Describe the characteristics of the dye used for measuring *cardiac output.*
8.7 The maximal average velocity of blood in a dog, 1 m/s, occurs in the dog's aorta, which is 0.015 m in diameter. The magnetic flux density in an

electromagnetic blood flowmeter is 0.03 T. What is the voltage at the electrodes?

8.8 In order to determine the frequency response of an electromagnetic flowmeter, the clinician can transiently shortcircuit the magnet current by using a microswitch. For steady flow, sketch the resulting output of the flowmeter. Describe the mathematical steps you could implement on a computer in order to convert the resulting transient wave to the flowmeter's frequency response.

8.9 On a common time scale, sketch the waveforms for the magnet current, flow signal, and transformer voltage for the following electromagnetic flowmeters: (1) gated sine wave, (2) square wave, (3) trapezoidal. Indicate the best time for sampling each flow signal.

8.10 For Figure 8.6, design a simpler electromagnetic flowmeter without quadrature suppression. Show the block diagram and show all connections for a ring demodulator.

8.11 For the Doppler ultrasonic flowmeter shown in Figure 8.9(b), suppose that the two transducers are inclined at angles θ and ϕ to the axis. Derive a formula for f_d, the Doppler frequency shift.

8.12 Use information from Section 3.5 to design a comparator with a SHR of 7, as required for the Doppler zero-crossing detector (Section 8.4).

8.13 For Figure 8.11, show how to add another one-shot block and several logic blocks to obtain pulse outputs on both positive and negative zero crossings.

8.14 A pulsed Doppler flowmeter has $f_r = 15$ kHz, $f_0 = 8$ MHz, and $\theta = 45°$. Calculate R_m and u_m.

8.15 Expand Figures 8.14 and 8.13(b) to show a complete block diagram of a directionally sensitive thermal velocity meter and probe.

8.16 The chamber plethysmograph shown in Figure 8.15 has a volume of 200 ml. Calculate the rapid change in tissue volume that produces a 120 Pa change in chamber pressure. Assume an adiabatic process: $P(V)^{1.4} =$ constant.

8.17 Calculate the arterial inflow for the test shown in Figure 8.16.

8.18 For Figure 8.19, assume $Z + \Delta Z = Z_2 = Z_3 = 100\ \Omega$ and $Z_v = -j2000\ \Omega$ (capacitive). How large is the error caused by a 5-Ω change in Z_2? Is an error of this magnitude important?

8.19 Design a circuit that uses the same two electrodes (plus one ground electrode) to monitor ventilation by impedance and the conventional ECG, with no cross interference.

REFERENCES

Anderson, F. A., Jr., "Impedance plethysmography," in J. G. Webster (ed.), *Encyclopedia of Medical Devices and Instrumentation.* New York: Wiley, 1988, pp. 1632–1643.

Baker, D. W., "Pulsed ultrasonic Doppler blood-flow sensing." *IEEE Trans. Sonics Ultrason.*, 1970, SU-17, 170–185.

Bergel, D. H., and U. Gessner, "The electromagnetic flowmeter," in R. F. Rushmer (ed.), *Methods in Medical Research.* Chicago: Year Book, 1966, Vol. XI.

Calvert, M. H., B. R. Pullan, and D. E. Bone, "A simple method for measuring the frequency response of an electromagnetic flowmeter." *Med. Biol. Eng.,* 1975, 13, 592–594.

Capek, J. M., and R. J. Roy, "Fick techniques," in J. G. Webster (ed.), *Encyclopedia of Medical Devices and Instrumentation.* New York: Wiley, 1988, pp. 1302–1314.

Christensen, D. A., *Ultrasonic Bioinstrumentation.* New York: Wiley, 1988.

Cobbold, R. S. C., *Transducers for Biomedical Measurements: Principles and Applications.* New York: Wiley, 1974.

Cooley, W. L., and R. L. Longini, "A new design for an impedance pneumograph," *J. Appl. Physiol.,* 1968, 25, 429–432.

Geddes, L. A., and Baker, L. E., *Principles of Applied Biomedical Instrumentation,* 3rd ed. New York: Wiley, 1989.

Grahn, A. R., M. H. Paul, and H. U. Wessel, "A new direction-sensitive probe for catheter-tip thermal velocity measurements." *J. Appl. Physiol.,* 1969, 27, 407–412.

Groveman, J., D. D. Cohen, and J. B. Dillon, "Rhinoplethysmography: Pulse monitoring at the nasal septum." *Anesth. Analg.,* 1966, 45, 63.

Holloway, G. A., Jr., "Cutaneous blood flow, laser Doppler measurement of," in J. G. Webster (ed.), *Encyclopedia of Medical Devices and Instrumentation.* New York: Wiley, 1988, pp. 908–915.

Hoskins, P. R., "Measurement of arterial blood flow by Doppler ultrasound." *Clin. Phy. Physiol. Meas.,* 1990, 11, 1–26.

Kubicek, W. G., A. H. L. From, R. P. Patterson, D. A. Witsoe, A. Castenda, R. G. Lilleki, and R. Ersek, "Impedance cardiography as a noninvasive means to monitor cardiac function." *J. Assoc. Adv. Med. Instrum.,* 1970, 4, 79–84.

Lee, A. L., A. J. Tahmoush, and J. R. Jennings, "An LED-transistor photoplethysmograph." *IEEE Trans. Biomed. Eng.,* 1975, BME-22, 243–250.

McCutcheon, E. P., *Chronically Implanted Cardiovascular Instrumentation.* New York: Academic, 1973.

McLeod, F. D., "A directional Doppler flowmeter." *Dig. Int. Conf. Med. Biol. Eng.,* Stockholm, 1967, 213.

Mohapatra, S. N., "Impedance cardiography," in J. G. Webster (ed.), *Encyclopedia of Medical Devices and Instrumentation.* New York: Wiley, 1988, pp. 1622–1632.

Neuman, M. R., "Neonatal monitoring," in J. G. Webster (ed.), *Encyclopedia of Medical Devices and Instrumentation.* New York: Wiley, 1988, pp. 2015–2034.

Nyboer, J., *Electrical Impedance Plethysmography,* 2nd ed. Springfield, IL: C. C. Thomas, 1970.

Patterson, R., "Body fluid determinations using multiple impedance measurements." *IEEE Eng. Med. Biol. Magazine,* 1989, 8 (1), 16–18.

Patterson, R. P., L. Wang, and B. Raza, "Impedance cardiography using band and regional electrodes in supine, sitting, and during exercise." *IEEE Trans. Biomed. Eng.,* 1991, 38, 393–400.

Raines, J., and R. C. Darling, "Clinical vascular laboratory: Criteria, procedures, instrumentation." *Med. Electronics Data,* 1976, 7 (1), 33–52.

Seagar, A. D., J. M. Gibbs, and F. M. David, "Interpretation of venous occlusion plethysmographic measurements using a simple model." *Med. Biol. Eng. Comput.,* 1984, 22, 12–18.

Semmlow, J. L., "Sexual instrumentation," in J. G. Webster (ed.), *Encyclopedia of Medical Devices and Instrumentation.* New York: Wiley, 1988, pp. 2609–2621.

Shankar, R., and J. G. Webster, "Noninvasive measurement of compliance in leg arteries." *IEEE Trans. Biomed. Eng.,* 1991, 38, 62–67.

Shankar, T. M. R., and J. G. Webster, "Design of an automatically balancing electrical impedance plethysmography." *J. Clin. Eng.,* 1984, 9, 129–134.

Shercliff, J. A., *The Theory of Electromagnetic Flow Measurement.* Cambridge: Cambridge University Press, 1962.

Swanson, D. K., "Measurement errors and origin of electrical impedance changes in the limbs." Ph.D. dissertation, Department of Electrical and Computer Engineering, University of Wisconsin, Madison, Wisconsin, 1976.

Trautman, E. D., and R. S. Newbower, "Cardiac output, thermodilution mesurement of," in

J. G. Webster (ed.), *Encyclopedia of Medical Devices and Instrumentation.* New York; Wiley, 1988, pp. 584–592.

Valentinuzzi, M. E., and J. C. Spinelli, "Intracardiac measurements with the impedance technique." *IEEE Eng. Med. Biol. Magazine,* 1989, 8 (1), 27–34.

von Reth, E. A., and A. Versprille, "Cardiac output, indicator dilution measurement of," in J. G. Webster (ed.), *Encyclopedia of Medical Devices and Instrumentation.* New York: Wiley, 1988, pp. 578–584.

Webster, J. G. (ed.), *Electrical Impedance Tomography.* Bristol, England: Adam Hilger, 1990.

9

MEASUREMENTS OF THE RESPIRATORY SYSTEM

Frank P. Primiano, Jr.

This chapter deals with the processes in the lungs that are involved in the exchange of gases between the blood and the atmosphere. Measurement of variables associated with these processes enables the physician to perform two clinically relevant tasks: assess the functional status of the respiratory system (lungs, airways, and chest wall) and intervene in its function.

The objective assessment of respiratory function is performed clinically on two time scales. One is relatively long, involving discrete observations, usually in the form of *pulmonary function tests* (PFT), at intervals on the order of days to years. In pulmonary function testing, a subject's parameter values are compared to those expected from specific populations—either normal populations or ones with documented diseases (Primiano, 1981). The required parameters of respiratory function are evaluated using well-defined computational procedures operating on variables measured under specified experimental conditions. Tests of pulmonary function are used to (1) screen the general population for disease; (2) serve as part of periodic physical examinations, especially of individuals with chronic pulmonary conditions; (3) evaluate acute changes during episodes of disease; and (4) follow up after treatment.

The second time scale on which respiratory function is assessed is very short; observations are made either continuously or at intervals on the order of minutes to hours. This activity comes under the heading of *patient monitoring* and is performed in a hospital setting, usually in an intensive-care unit (ICU). It is warranted in crisis situations such as might result from trauma, drug overdose, major surgery, or disease. (See Section 13.6.)

Therapeutic modification of respiratory function can be achieved through surgery, the use of drugs, or physical intervention with respiratory-assist devices. Except for extreme, acute circumstances—such as cardiac surgery, in which the lungs are completely by-passed and blood is arterialized in an extracorporeal oxygenator (Section 13.3)—these approaches attempt to control arterial blood gases by manipulating the composition and distribution of pulmonary gas and the distribution of the flow of pulmonary blood. The same variables used to evaluate lung function can be monitored to provide objective feedback information for this external control of the system.

There are a myriad of instruments that have been used to measure variables associated with respiration. Consequently, we shall limit our discussion to clinically applicable devices that yield accurate, quantitative measures suitable for the computation of parameters routinely evaluated for pulmonary function tests and for pulmonary status assessment during mechanical ventilation. This eliminates from discussion those devices that have primary applications as patient-monitoring or physical-diagnosis tools. Examples include imaging techniques (x-ray films, CT, MRI, PET) and a large group of instruments used to estimate or detect lung-volume change (see the review by Sackner, 1980), such as fluoroscopes; magnetometers; various electrical, mechanical and pneumatic devices placed on or around the torso; nasal temperature sensors; and force plates to detect body movements associated with breathing.

The literature in respiratory physiology suffers from a very poor system of notation. This system evolved, to some extent, from an effort to accommodate the clinically-oriented audience. The symbols used in this chapter are a compromise between those found in the respiratory literature (Macklem, 1986; Miller *et al.,* 1987) and those found in the physical sciences literature. For example, in the respiratory literature, V represents the "gas volume" (Miller *et al.,* 1987) within a container, such as the lungs, and $\dot{V}$ represents "flow of gas" (Macklem, 1986). However, because the gas in the lungs can be expanded and compressed, the rate of change of lung volume is not necessarily equal to the volume flow rate of gas entering the lungs through the nose and mouth, although, in many circumstances, it is well approximated by this flow. Nevertheless, to emphasize the distinction between these two physical entities, we will use $\dot{V}$ for the rate of change of volume of a container having volume V, and Q for the rate of flow of a fluid (gas or liquid) into it, as is commonly done in physics and fluid mechanics. We could not use F for flow, as is done in the rest of this book, because F is routinely used in respiratory physiology to denote molar fraction (fractional concentration) of a gas species in a mixture. Other symbols will be defined as they are introduced.

9.1 MODELING THE RESPIRATORY SYSTEM

Which of the many respiratory variables is to be measured depends on the type of behavior under study as well as on a concept of how the system functions. Ideas about how the respiratory system functions are usually formalized in abstract (that is, verbal or mathematical) models. Not only are the variables that are to be measured specified by models of the respiratory sytsem, but such models also define characteristic parameters of respiratory function and are the basis for the design of experiments to evaluate these parameters. In addition, they motivate control strategies and devices that are used to produce effective respiratory assistance. The definitions and discussions of lung physiology are based on models of the lungs (Primiano and Chatburn, 1988). Therefore, before we attempt measurements, we should understand the essential features of the respiratory system and some approaches to modeling that we can use not only for the respiratory system but also for the measurement devices themselves.

Because it is the respiratory function of living individuals that is to be evaluated, measurements must be minimally invasive, cause minimal discomfort, and be acceptable for use in a clinical environment. This greatly limits the number and types of measurements that can be made and leads to the use of lumped-parameter models. For the sake of discussion, it is convenient to divide respiratory function into two categories: (1) *gas transport* in the lungs (including extrapulmonary airways and pulmonary capillaries) and (2) *mechanics* of the lungs and chest wall. The models describing gas transport deal primarily with changes in concentrations of gas species and volume flow of gas, whereas the models dealing with mechanics primarily relate pressure, lung volume, and rate of change of lung volume. Bear in mind that these two categories are highly interrelated, and the models and measurements from one complement those from the other (Ligas and Primiano, 1988).

GAS TRANSPORT

Models of gas transport, both in the gas phase and across the alveolar-capillary membrane into the blood, are developed from mass balances for the pulmonary system depicted as a set of compartments. What can be considered as a basic gas-transport unit of the lungs is shown in Figure 9.1(a). It consists of a variable-volume alveolar compartment, with its contents well mixed by diffusion; a well-mixed, flow-through blood compartment that exchanges gases with the alveolar compartment by diffusion; and a constant-volume dead space. Gas moves by convection through the dead space, which acts only as a time-delay conduit between its outer opening and its associated alveolar volume. A *pair* of normal lungs during quiet breathing may be represented satisfactorily by the system shown in Figure 9.1(a). Lungs undergoing maximal volume changes, subjected to very high ($\gg 1$ Hz) ventilatory frequencies, or afflicted with diseases that produce abnormal gas transport may require more complicated models comprising combinations of such units, in parallel or in series or both.

A dynamic mass balance can be written for any chemical species X or set of species in the breathed gas mixture. If the production of X by chemical reaction in a system were negligible, a species mass balance could be written as

$$\begin{array}{l}\text{Rate of mass}\\ \text{accumulation}\\ \text{of X in the}\\ \text{system}\end{array} = \sum_{i=1}^{n} \begin{array}{l}\text{Rate of mass}\\ \text{convection}\\ \text{of X through}\\ \text{port } i\end{array} - \begin{array}{l}\text{Net rate of}\\ \text{diffusion out}\\ \text{of the system}\end{array} \tag{9.1}$$

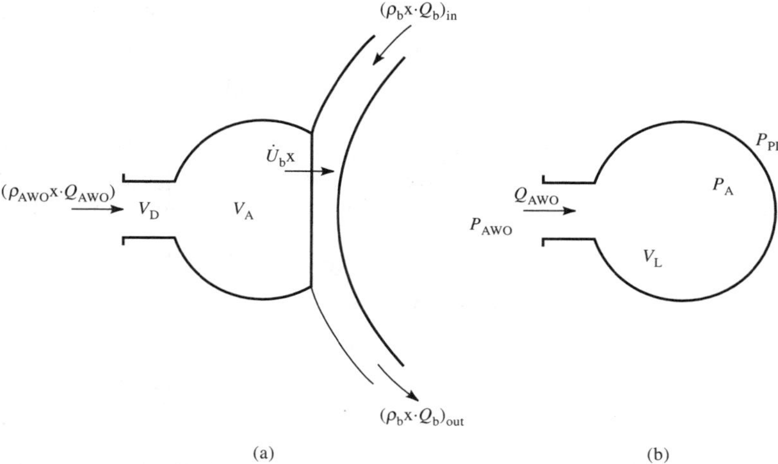

Figure 9.1 Models of the lungs (a) Basic gas-transport unit of the pulmonary system. Here $(\rho x \cdot Q)$ is the molar flow of X through the airway opening, AWO, and the pulmonary capillary blood network, b. $\dot{U}_b x$ is the net rate of molar uptake—that is, the net rate of diffusion of X into the blood. V_D and V_A are the dead-space volume and alveolar volume, respectively. (b) A basic mechanical unit of the pulmonary system. P_A is the pressure inside the lung—that is, in the alveolar compartment. P_{PL} and P_{AWO} are the pressures on the pleural surface of the lungs and at the airway opening, respectively. V_L is the volume of the gas space within the lungs, including the airways; Q_{AWO} is the volume flow of gas into the lungs measured at the airway opening.

This can also be written as a molar balance, because the number of moles N is the ratio of the mass of X to its molecular weight (in mass units). Define $\rho_{AWO}X$ as the mole density (moles per unit volume) of species X and Q_{AWO} as its volume flow (volume per unit time), each measured at the airway opening. Then a molar balance for X in the gas phase in the model of the lungs in Figure 9.1(a) would be

$$\frac{d(N_L x)}{dt} = (\rho_{AWO} x \cdot Q_{AWO}) - \dot{U}_b x \tag{9.2}$$

in which $\dot{U}_b x$ is the net molar rate of uptake of X by the blood. The number of moles of X in the lungs, $N_L x$, is the sum of the moles in the dead-space volume, $N_D x$, and the alveolar compartment, $N_A x$.

MECHANICS

We can conveniently model the mechanical behavior of the respiratory system as a combination of pneumatic and mechanical elements (Chatburn and Primiano, 1988). Figure 9.1(b) shows an idealized mechanical unit of the lungs. It

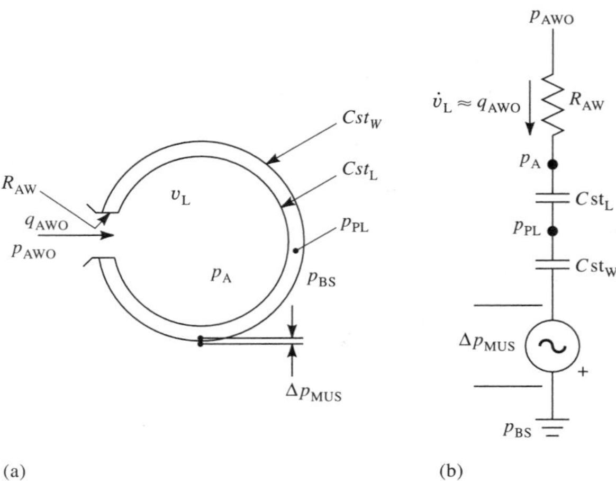

Figure 9.2 Models of normal ventilatory mechanics for small-amplitude, low-frequency (normal lungs, resting) breathing (a) Lung mechanical unit enclosed by chest wall. (b) Equivalent circuit for model in Figure 9.2(a).

consists of a deformable pressure vessel made of a material that exhibits both elastic and plastic behavior and a nonrigid airway that has a variable resistance to convective flow. The system contains a saturated mixture of ideal gases that exhibits inertia during acceleration through the airway and that undergoes an isothermal process during changes of state.

Even though each of the millions of alveoli and terminal airways potentially could act as a separate mechanical unit, it has been found that the mechanics of a *pair* of normal lungs during quiet breathing can be represented by the single unit of Figure 9.1(b). However, at high rates of breathing, normal and abnormal pulmonary systems may require models containing combinations of such units. Note that only in rare instances, such as when there is only a single gas space in each model, do compartments of the gas-transport units [Figure 9.1(a)] correspond one-to-one to mechanical units [Figure 9.1(b)] in the same lungs.

An additional deformable pressure vessel representing the chest wall surrounding the lungs has been added to the system in Figure 9.2(a). The chest wall includes all extrapulmonary structures, such as ribs, respiratory muscles, and abdominal contents, that can undergo motions as a result of breathing. The gap between the lung unit and the chest wall represents the liquid-filled interpleural space.

The mechanics of the respiratory system are described by the relationships between pressure differences across the various subsystems and the changes in volume and flow of gas through them. The subsystems are defined between points in the system at which representative pressures can be computed or measured. Consequently, the difference in pressure across the entire system can be expressed as the algebraic sum of pressure differences across subsystems. Mass balances can be used to follow the path of gas flow through the subsystems, and geometrical constraints determine the distribution of volume changes.

MODEL OF NORMAL RESPIRATORY MECHANICS DURING QUIET BREATHING

When the flows, changes of volume, or their respective time derivatives are large, the equations describing the mechanical behavior of the respiratory system are highly nonlinear. However, for small flows and changes of volume such as those that occur during resting breathing, linear approximations adequately describe the respiratory system. These linear approximations define the familiar properties—such as compliance, resistance, and inertance—that are evaluated for pulmonary function tests.

The following conventions will be applied throughout this chapter to facilitate writing linearized equations. Lower-case letters for a variable will indicate small changes of that variable about an operating point or reference level.

$$y = Y - \hat{Y}$$

where $\hat{Y}$ indicates some fixed reference value for Y. All linearized equations, therefore, are written in lower-case variables. A delta, Δ, will indicate differences between two spatial points.

$$\Delta Y = Y_i - Y_j$$

where i and j indicate different positions, such as AWO and PL. Therefore, the change in the pressure difference across the lungs (the transpulmonary pressure difference) would be

$$(P_{\text{AWO}} - P_{\text{PL}}) - (\hat{P}_{\text{AWO}} - \hat{P}_{\text{PL}}) = \Delta P_{\text{L}} - \Delta \hat{P}_{\text{L}} = \Delta p_{\text{L}}$$

If the alveoli and chest wall exhibit predominantly elastic behavior, the following set of linear equations can be used as a simple model of the mechanics of the respiratory system for normal tidal breathing in the atmosphere [Figure 9.2(a)].

$$p_{\text{AWO}} - p_{\text{A}} = R_{\text{AW}} q_{\text{AWO}} \tag{9.3a}$$

$$p_{\text{A}} - p_{\text{PL}} = \frac{1}{\text{Cst}_{\text{L}}} v_{\text{L}} \tag{9.3b}$$

$$\Delta p_{\text{MUS}} + (p_{\text{PL}} - p_{\text{BS}}) = \frac{1}{\text{Cst}_{\text{W}}} v_{\text{L}} \tag{9.3c}$$

in which lower-case letters are used to designate changes in the following variables with respect to an operating point.

P_{AWO} Hydrostatic pressure at the airway opening

P_A Representative pressure within the lungs (alveolar pressure)

P_{PL} Representation of the average force per unit area acting on the pleural surfaces (interpleural pressure)

ΔP_{MUS} Representation of the average force per unit area on the chest wall, which would cause the same movements produced by the active contraction of the respiratory muscles during breathing (muscle pressure difference)

P_{BS} Hydrostatic pressure acting on the body surface, except at the airway opening

Q_{AWO} Volume flow of gas at the airway opening

V_L Volume of the gas space in the system, assumed to be entirely within the lungs and airways

Three mechanical properties are included in (9.3): airway resistance R_{AW}, pulmonary static compliance Cst_L, and chest-wall static compliance Cst_W. We can evaluate these parameters by applying general definitions of flow resistance through a conduit and compliance of a deformable structure. The resistance to flow of a gas through a conduit is the ratio of the change in pressure drop along the conduit to the change of flow through it while the change in volume of the conduit is zero.

$$R \equiv \frac{\partial(\Delta P)}{\partial Q} \tag{9.4}$$

The "static" compliance of a structure is the ratio of the change in volume of the structure to the change in pressure difference across it while all flows and derivatives of volume are zero.

$$Cst \equiv \frac{\partial V}{\partial(\Delta P)} \tag{9.5}$$

In (9.4) and (9.5), ΔP is the pressure difference across the system under study. Therefore, for the airway,

$$R_{AW} = \frac{\partial(P_{AWO} - P_A)}{\partial Q_{AWO}} \tag{9.6}$$

The partial derivatives in (9.4) through (9.6) are used to indicate that all other variables must be constant when these parameters are evaluated. In particular, Cst can be evaluated experimentally only when the system is at static equilibrium—that is, when all flows and rates of change of volume and pressure in the system are zero. In this situation $P_{AWO} - P_A = 0$, and $(P_A - P_{PL})$ can be measured as $(P_{AWO} - P_{PL})$. Thus pulmonary static compliance can be evaluated as

$$Cst_L = \frac{V_L(t_2) - V_L(t_1)}{\partial P_L(t_2) - \partial P_L(t_1)} \tag{9.7}$$

in which

$$\Delta P_L = (P_{AWO} - P_{PL}) \tag{9.8}$$

is the *transpulmonary* pressure difference, and t_2 and t_1 are two instants in time at which the system is completely motionless and at two different volumes.

It is impossible to measure the muscle pressure difference, ΔP_{MUS}, directly. Consequently, chest-wall static compliance can be evaluated only when $\Delta P_{MUS} = 0$. This occurs, by definition, when the respiratory muscles are completely relaxed. Defining the difference in pressure across the chest wall as

$$\Delta P_W = P_{PL} - P_{BS} \tag{9.9}$$

we obtain chest-wall static compliance from

$$Cst_W = \frac{V_L(t_4) - V_L(t_3)}{\Delta P_W(t_4) - \Delta P_W(t_3)} \tag{9.10}$$

in which t_4 and t_3 are two instants at which the system is static and at two different volumes, *and* the respiratory muscles are completely relaxed.

As the lungs change volume and lose or gain gas through the airway opening, the gas inside is compressed or expanded transiently. During fast changes in volume, this can produce an inequality between the instantaneous rate of volume change, $\dot{V}_L$, and the volume flow of gas at the mouth, Q_{AWO}. However, for normal, tidal breathing, this effect can be neglected and Q_{AWO} can be taken as a good approximation to $\dot{V}_L$. Therefore, (9.3a) and (9.3b) can be combined and rewritten as

$$p_{AWO} - p_{PL} = \frac{1}{Cst_L} v_L + R_{AW} \dot{v}_L \tag{9.11}$$

Equations (9.3c) and (9.11), which describe Figure 9.2(a), can also be represented by the analogous equivalent circuit in Figure 9.2(b).

MEASURABLE VARIABLES IN THE RESPIRATORY SYSTEM

Even though a number of variables are included in the simple models of gas transport and mechanics shown in Figures 9.1 and 9.2, only a very limited subset can be measured directly. These include volume flow of gas through the mouth and nose (and, equivalently, a measure of its integral, the volume of gas breathed); pressure at the mouth and nose and body surface; partial pressures or concentrations of various gases in gas mixtures passing through the airway opening, and in discrete samples of blood *in vitro;* and temperature, including body-core temperature. The values of all other variables in the preceding equations cannot be measured directly but must be inferred from measurements of other variables. A notable example is change in lung volume, which is routinely obtained from gas flow or gas volume measured at the airway opening.

9.2 MEASUREMENT OF PRESSURE

Two noteworthy characteristics of respiratory pressure measurements are the manner in which pressure is measured and the fact that most of the measurements involve pressure differences. All the pressures included in the respiratory models given in Section 9.1 are described in the respiratory literature as "lateral pressure" or "side pressure"; that is, the pressure is measured at the wall of a vessel with the plane of the measurement port parallel to the direction of any flow that might exist. This defines the static or hydrostatic pressure of the fluid dynamicist. The total, or stagnation, pressure and the dynamic pressure at a point are never used in these models even though they may be used in instruments to measure flow (Hänninen, 1991). Confusion can arise, however, because of the way in which the terms *static* and *dynamic* are used with respect to pressure in the respiratory literature. It is standard practice to express the difference in hydrostatic pressure between two points as the sum of two time-varying components, the static component and the dynamic component. For example, the transpulmonary pressure difference can be written as

$$\Delta P_L = (\Delta P_L)\text{st} + (\Delta P_L)\text{dyn} \tag{9.12}$$

The static component is defined as a function of only the volume change in the system. [For example, in (9.11), the first term on the right-hand side has been referred to as the static component of the transpulmonary pressure difference.] The dynamic component, which is the hydrostatic pressure difference minus its static component, is related only to flows, rates of change of volume, and their derivatives. The static component of a pressure difference, therefore, can be measured as the hydrostatic pressure difference when all flows, rates of volume change, and their derivatives are zero, that is, when the system is static.

PRESSURE SENSORS

We can conveniently perform dynamic measurements of respiratory pressures using an electronic strain-gage pressure sensor with a tube or catheter as a probe. Section 7.3 described the characteristics of such systems for the circulatory system in which the catheter and sensor are filled with liquid. The same type of analysis can be used for gas-filled systems, except that the acoustic compliance of the gas may be of the same order of magnitude as—or even high than—the compliance of the sensor diaphragm. Therefore, an appropriate shunt capacitor must be included in the equivalent circuit for the device [Figure 7.8(a)].

An additional point must be considered, however, when a pressure difference is measured. Such measurements are usually accomplished with differential pressure sensors that have two chambers separated by a diaphragm connected to strain-sensitive elements. Gas is introduced into each chamber through a catheter. Therefore, the circuit of Figure 7.8(a) represents the mechanical or pneumatic transfer function for only one side of a differential pressure sensor. Note that the time-varying pressures that exist in the chambers on each side of the diaphragm are influenced by the transfer characteristics of the mechanical–pneumatic circuits between their respective pressure sources and the strain-sensing diaphragm. Thus it is extremely important that the frequency response of the transmission pathways on both sides of the sensor be matched over the frequency range of interest. This becomes critical when high-frequency changes in pressure are to be measured with sensors having chambers of unequal volumes on each side of the diaphragm.

INTRAESOPHAGEAL PRESSURE

Computing pulmonary mechanical properties—for instance, pulmonary static compliance—from (9.7) and (9.8) requires a measure of the spatially averaged pressure acting on the pleural surfaces. Direct measurements of the pressure on the visceral pleural surface made by puncturing the thoracic wall and introducing a catheter into the interpleural space are not clinically applicable. Such measurements have shown, however, that a gravity-related pressure gradient exists in the thin liquid film in the interpleural space surrounding the lungs. This nonuniform pressure, which is lowest in the uppermost part of the chest, makes the point at which a representative pressure can be measured uncertain. Fortunately, the determination of mechanical properties from linearized equations such as (9.3) involves only changes in pressure about an operating point. This relaxes the requirement on the measurement of the absolute pressure.

A significant advance in clinical testing of pulmonary function was the development of a method of estimating changes in the average pressure on the visceral pleural surface from measurements of changes in the pressure in a bolus of fluid introduced into the esophagus. The most commonly used technique involves passing an air-filled catheter with a small latex balloon on its end through the nose into the esophagus (Macklem, 1974).

The esophagus, which is normally a flaccid, collapsed tube, is subjected to the pressure in the interpleural space (acting through the parietal pleura) and to the weight of other thoracic structures, primarily the heart. The pressure in the air trapped in a small balloon situated within the thoracic esophagus depends on the compression, or expansion, caused by these sources. Although the mean intraesophageal pressure does not equal the mean pressure on the pleural surface measured directly by catheter in the interpleural space, under certain conditions the *changes* in the pressure in the esophageal balloon reflect the *changes* in pressure on the pleural surface. The mechanical properties of the balloon and esophagus have minimal effect on the changes in pressure in the balloon if the amount of air in the balloon is sufficiently small that the balloon remains unstressed and the esophageal wall does not undergo motions large enough to influence the transmission of pressure into the balloon.

The balloon should be so located that pressure changes due to motions of other organs competing for space in the thoracic cavity are minimized. The largest noise signal comes from the heartbeat, which usually has a fundamental frequency (on the order of 1 Hz) much higher than that of resting breathing. A region below the upper third of the thoracic esophagus gives low cardiac interference and provides pressure variations that correspond well in magnitude and phase with directly measured representative changes in pleural pressure. The correspondence decreases as lung volume approaches the minimum achievable (the residual) volume. The frequency response of an esophageal balloon pressure-measurement system depends on the mechanical properties and dimensions of the pressure sensor, the catheter, the balloon, and the gas within the system. The use of helium instead of air can extend the usable frequency range of these systems.

9.3 MEASUREMENT OF GAS-FLOW RATE

When the lungs change volume during breathing, a mass of gas is transported through the airway opening by convective flow. Measurement of variables associated with the movement of this gas is of major importance in studies of the respiratory system. The volume-flow rate and the time integral of volume-flow rate are used to estimate rate of change of lung volume and changes of lung volume, respectively. Even though the devices we will describe here are calibrated and are used to measure volume-flow rate or to estimate its time integral, the primary physical process involved is mass flow. Volume-flow rate equals the mass-flow rate divided by the density of the gas at the measurement site. The instruments used to measure volume-flow rate are referred to as *volume flowmeters.* The volume occupied by a given mass (number of moles) of gas under known conditions of temperature and pressure is usually determined by using a spirometer (Section 9.4).

Even though breathing movements are cyclic by nature and involve alternating (bidirectional) gas flow, some tests of pulmonary function, such as those involving the single-breath washout, the forced expiratory vital-capacity maneuver, and the maximal voluntary ventilation, require the measurement of flow in only one direction. In addition, the precision and accuracy demanded of flow measurements vary greatly, depending on the settings in which the measurements are performed, from physiology and clinical function laboratories to mass screening centers to intensive care units. Consequently, there are a variety of instruments that can produce measurements useful in particular applications.

REQUIREMENTS FOR RESPIRATORY GAS-FLOW MEASUREMENTS

Measurement of the motion of material passing through a system requires that the sensor be placed at a position traversed by a known fraction of the material. In respiratory experiments, especially those involving measurement of breathed gas, the usual practice is to have the entire flow stream pass through or into the instrument. This produces several potential problems. Any pressure imposed at the airway during measurements at the airway opening—for example, by a mechanical ventilator—must be withstood by the sensor without damage, distortion, or leakage. Also, the device should not obstruct breathing or produce a back pressure during flow that might affect respiratory performance. For example, the American Thoracic Society recommends that volume-flow measuring devices used for maximal expiratory efforts have a resistance to flow of less than 1.5 cm H_2O/(liter/s) (Anonymous, 1995a).

As with any instrument, the gas-flow-measuring sensor must have a stable baseline (reference output) and sensitivity so that measurements are accurate. However, changes in composition and temperature of gas can affect the calibration factors of various flowmeters. These changes occur between inspired and expired gas and during an expiration. Furthermore, inspired particles of dust, dirt, and medication and expired aerosolized organic particles from the respiratory system can deposit on sensitive parts of the sensors and contaminate them. This not only affects calibration but can also transmit disease. Therefore, the sensor must be either sterilizable or disposable.

One of the major sensor contaminants in expired gas is water. Unless the sensor is heated to near or above body temperature, it can act as a condenser for the water vapor in the saturated expirate. The resulting liquid can foul delicate sensors and change the effective cross-sectional area through which the gas must pass.

The measurement procedure must not alter inspired air by adding excessive heat or toxic substances. Such techniques as laser anemometry (which requires reflecting particles in the stream) and ion anemometry (which produces ozone) are not suitable for measurements at the airway opening. If the sensor is to monitor breathing continuously for a number of breaths, then its dead space becomes important. Carbon dioxide must be flushed out and O_2 replenished if the conduit tubing in the system is of such a volume that the patient will experience excessive rebreathing of expired gas.

The performance characteristics required of respiratory flowmeters depend upon the specific measurements to be made. These can be as different as the flow during quiet breathing of an infant to that during a maximal, forced expiration by an adult athlete. The amplitude ranges, measurement accuracies, and frequency responses necessary for several clinical applications have been presented by Sullivan *et al.* (1984).

Commonly used respiratory volume flowmeters fall into one of four categories: rotating-vane, ultrasonic, thermal-convection, and differential pressure flowmeters.

ROTATING-VANE FLOWMETERS

This type of sensor has a small turbine in the flow path. The rotation of the turbine can be related to the volume flow of gas. Mechanical linkages have been used to display parameters of the flow (such as peak flow and integral over an expiration) on indicator dials on the instrument. Interruption of a light beam by the turbine has also been sensed and converted to voltages proportional to flow and/or its integral, to be recorded photographically or displayed continuously. In devices such as this, the mass of the moving parts and the friction between them combine to prevent high-frequency motions of the turbine in response to accelerating flows. This precludes their use in the measurement of alternating bidirectional flows and makes them primarily suitable for clinical screening.

ULTRASONIC FLOWMETERS

Section 8.4 described the operation and application of ultrasonic sensors in the measurement of blood flow. For respiratory measurements, investigators measure the effect of the flowing gas on the transit time of the ultrasonic signal. The transmitter-receiver crystal pair is mounted either externally and obliquely to the axis of the tube through which the gas flows or internally and coaxially with the flow. The transit time between the transmitter and receiver depends not only on the velocity of the gas between them but also on the composition and temperature of the gas.

In another approach, a rod is placed in the flow stream to produce a pattern of vortices. An ultrasonic transmitter and receiver are mounted diametrically opposite each other in the walls of the tube. The intensity of the ultrasonic signals passing perpendicular to the flow is modulated by the vortices. The modulating frequency is detected and calibrated in units of volume-flow rate. Ultrasonic flowmeters measure unidirectional flows and are suitable for clinical monitoring.

THERMAL-CONVECTION FLOWMETERS

Thermal-convection flowmeters employ sensing elements such as metal wires, metal films, and thermistors, the electrical resistances of which change with temperature. When operated in the self-heated mode, in which sufficient current is passed through them to maintain an average temperature above that of the surrounding fluid, these elements lose heat at a rate that depends on the local mass flow, temperature, specific heat, kinematic viscosity, and thermal conductivity of the fluid. If a feedback circuit is used to operate the primary sensing element at a constant temperature, then a second, unheated element can be included in the circuit to compensate for heat loss due to local, ambient temperature changes. (The thermal response time of the unheated element can be affected by condensation of water vapor from humid gas mixtures.) The details and operation of such devices and circuits are described in Section 8.5.

For situations in which gas properties are sufficiently constant, the output voltage from these circuits is a nonlinear function of mass-flow rate only. Analog and digital implementations of piecewise-linear or polynomial approximations of this function have been developed to provide a linear mass-flow–voltage relationship.

Flowmeters using a single, temperature-compensated, heated wire (hot-wire anemometer) with a linearizing circuit provide unidirectional flow measurements that are satisfactory for testing pulmonary function. If a single hot wire is to be used to obtain volume-flow rate continuously, several conditions must be fulfilled. For a gas of constant density, volume-flow rate through a cross section is proportional to the average mass flow (averaged over the cross section). The mass-flow sensing wire is very small (on the order of 5 μm in diameter and 1 to 2 mm in length) to satisfy heat-transfer and frequency-response requirements. Consequently, mass flow is measured only locally in a correspondingly small region of the flow stream. The duct in which the sensor is located must be designed so that the position at which the measurement is made yields a value representative of the average mass flow through the entire cross section of the duct at every instant of time. At the low Mach numbers involved in respiratory flows, this requires that the velocity profile be well defined at all flows of interest. The variations in cross-sectional area upstream and downstream from the sensor can be optimized for this condition.

Although a single hot-wire sensor provides an output of the same polarity independent of the direction of flow, limiting its use to unidirectional flow, multiple sensors located at separate points along the flow path can be used with the appropriate circuitry to provide directional sensitivity, as described in Section 8.5. However, respiratory measurements involve a changing mixture of gases in the flow stream that affects the heat transfer from the heated wire. The significant variations in composition that occur between inspiration and expiration could invalidate the use of a single calibration factor. During a multibreath N_2 washout (Section 9.4) the N_2–O_2 ratio in the lung changes from approximately 4 to 1 on the first breath to nearly zero at the end of the test. Fortunately, the differences in thermal properties and densities of N_2 and O_2 offset each other sufficiently well that a linearized, temperature-compensated hot-wire anemometer can be used with a constant calibration factor for volume-flow measurements during the successive expirations of a multibreath N_2 washout. In general, however, the sensor should be calibrated for the particular gas mixture to which it is exposed.

The hot-wire anemometer has a number of features that are advantageous in respiratory applications. It has an appropriate frequency response (the sensor itself can respond to frequencies into the kilohertz range). And, because the sensing element in the flow stream is extremely small, the only back pressure produced is that caused by the flow through the duct. In unidirectional applications rebreathing does not occur, so dead space is irrelevant. Accurate readings can be made at low as well as high flow rates if the output is adequately linearized. Special circuits can be provided to overheat the sensor to burn off contaminants as necessary. The main disadvantage is the limitation to unidirectional flow. The relatively high cost of overcoming this with a paired hot-wire system may not be justified.

DIFFERENTIAL PRESSURE FLOWMETERS

Convective flow occurs as a result of a difference in pressure between two points. From the relationship between pressure difference and volume-flow rate through a system, measurement of the difference in pressure yields an estimate of flow. Flowmeters based on this idea have incorporated several mechanisms to establish the relationship between pressure drop and flow. These include the venturi, orifice, and flow resistors of various types. A flowmeter based on a modified Pitot tube has also been developed.

Venturis and Orifices Venturis and orifices with fixed-sized openings have inherently nonlinear pressure-flow relationships and require calibration charts, special circuitry, or digitally implemented algorithms to be useful as flow sensors. A passive, mechanically linearized orifice flowmeter has been produced in which an elastic flap moved by the pressure of the flowing gas impinging on it increases or decreases the orifice size (Sullivan, 1984). However, sensors with computer-controlled orifices that can measure and/or control flow are also in use.

A two-stage, fixed-orifice system has been used to measure flow during forced expiratory vital-capacity maneuvers (Jones, 1990). At high flows, the gas stream passes through a large orifice. When the sensed pressure drop indicates that the flow has decreased to below 2 lps, a solenoid is activated, producing a step decrease in the opening and thus increasing the sensor's sensitivity to low flows. Parameters of the measured waveshape are presented on a digital readout.

A sophisticated extension of this concept has been used in the feedback control of the flow from a mechanical ventilator (Anonymous, 1991). Flow is controlled by a variable orifice in line with an upstream, regulated, constant-pressure gas source. A stepping motor is used to change the open area of the orifice in extremely small increments. The resulting aperture size and the pressure drop across the orifice are continuously monitored. A microcomputer with access to a stored calibration table that relates orifice area, pressure drop, and flow iteratively adjusts the opening to produce the desired instantaneous flow.

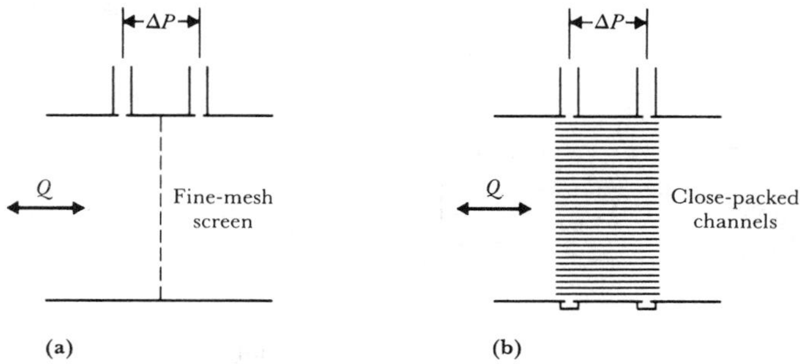

Figure 9.3 Pneumotachometer flow-resistance elements (a) Screen. (b) Capillary tubes or channels.

Pneumotachometers The flow sensors that have historically been, and continue to be, the mainstay of the respiratory laboratory utilize flow resistors with approximately linear pressure–flow relationships. These devices are usually referred to as pneumotachometers (Buess *et al.,* 1988). (In general, the term *pneumotachometer* is synonymous with *gas volume-flowmeter.*) Flow-resistance pneumotachometers are easy to use and can distinguish the directions of alternating flows. They also have sufficient accuracy, sensitivity, linearity, and frequency response for most clinical applications. In addition, they use the same differential pressure sensors and amplifiers required for other respiratory measurements. The following discussion primarily concerns these instruments.

Even though other flow-resistance elements have been incorporated in pneumotachometers, those most commonly used consist of either one (Silverman and Whittenberger, 1950) or more (Sullivan *et al.,* 1984) fine mesh screens [Figure 9.3(a)] placed perpendicular to flow or a tightly packed bundle of capillary tubes or channels [Figure 9.3(b)] with its axis parallel to flow (Fleisch, 1925). These physical devices exhibit, for a wide range of unsteady flows, a nearly linear pressure-drop–flow relationship, with pressure drop approximately in phase with flow.

In practice, the element is mounted in a conduit of circular cross section. The pressure drop is measured across the resistance element at the wall of the conduit (within the boundary layer of the flow). The pressure tap on each side of the resistance is either a single hole through the conduit wall or multiple holes from a circumferential channel within the wall, which is connected to a common external tap.

Because the pressure drop is measured at a single radial distance from the center of the conduit, it is assumed that this pressure drop is representative of the pressure drop governing the total flow through the entire conduit cross section. Thus these flowmeters rely on the flow-resistive element to establish a consistent—though not uniform—velocity profile on each side of the element in the neighborhood of the pressure measurement. This, however, cannot be achieved independent of the ductwork in which the pneumotachometer is placed (Kreit and Sciurba, 1996). Therefore, the placement of the pressure

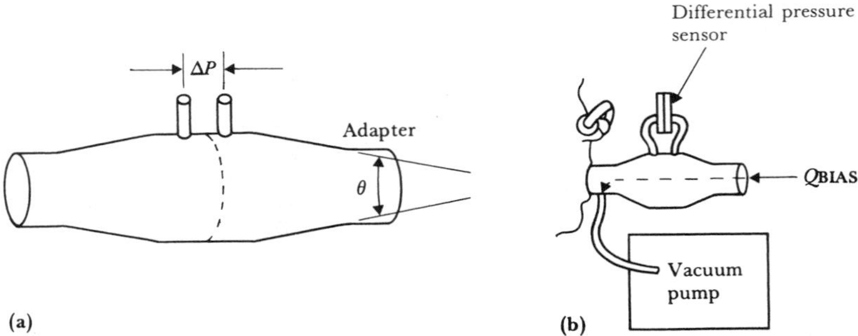

Figure 9.4 Pneumotachometer for measurements at the mouth (a) Diameter adapter that acts as a diffuser. (b) An application in which a constant flow is used to clear the dead space.

ports and the configuration of the tubing that leads from the subject to the pneumotachometer and from the pneumotachometer to the remainder of the system are critical in determining the pressure-drop–flow relationship. This is especially important when alternating and/or high-frequency flow patterns are involved.

A number of trade-offs exist in the design and use of these sensors. The ΔP–Q relationship may be more linear for steady flow when there is a large axial separation between pressure ports than when there is a smaller separation. But during unsteady flow with high-frequency content, the pressure drop for a larger separation may be more influenced by inertial forces. If the sensor is to avoid excessive formation of vortices at high flow rates, the cross-sectional area of the conduit at the flow-resistance element must be large enough to reduce the velocity through the element. This area may be several times that of the mouth of the subject from which the gas originates, requiring an adapter or diffuser between the mouthpiece and the resistance element. If flow separation and turbulence are to be avoided as the cross-sectional area changes, the adapter should have a shallow internal angle θ [Figure 9.4(a)] not exceeding 15°. The shallower the angle, however, the greater the distance between the mouth and the flow-resistance element. Symmetric drops in pressure for the same flow rate in either direction require that the geometries of the conduit on both sides of the resistance element be matched. The volume within the adapters and conduit represents a dead space for cyclic breathing. The designer, when designing the sensor, must balance the effects of decreased angle of the diffuser against the tolerable volume of dead space.

A bias flow can be used to clear this dead space. Air is drawn through the pneumotachometer from a side hole [Figure 9.4(b)] through a long tube connected to a vacuum pump. This causes a constant bias pressure drop across the pneumotachometer if the bias flow created is constant during the breathing of the patient. This approach can work well when high-frequency breathing patterns are of interest. But for low frequencies, such as those of tidal breathing, a regulating device may be required to prevent variation of the bias flow as the patient breathes.

The usable frequency range for capillary pneumotachometers is typically smaller than that for the screen type. Depending on its design, the screen-type pneumotachometer can exhibit a constant-amplitude ratio of pressure difference to flow, and zero phase shift up to as high as 70 Hz (Peslin *et al.*, 1972).

In air, the amplitude ratio of the Fleisch (capillary bundle) pneumotachometer is nearly constant at its steady-flow value up to approximately 10 Hz, increasing by 5% at 20 Hz. The phase angle between ΔP and Q increases linearly with frequency to approximately 8.5° at 10 Hz, which corresponds to an approximate 2-ms time delay between flow and pressure difference. These values change with the kinematic viscosity of the gas (Finucane *et al.*, 1972). Peslin *et al.* (1972) modeled the Fleisch pneumotachometer for frequencies up to 70 Hz by an equation of the form

$$\Delta P = RQ + L\dot{Q} \tag{9.13}$$

where L is the inertance (related to mass) of the gas in one capillary tube, defined by (7.4), and R is the flow resistance for each capillary tube, defined by (7.2). Equation (9.13) can be used as a computational algorithm to compensate the Fleisch pneumotachometer for precise measurements of a gas of constant composition at constant temperature.

An appropriately designed screen pneumotachometer may not require such compensation. But it may instead be subject to equipment-generated, high-frequency noise in clinical applications. An additional point should be stressed: The frequency response of a pneumotachometer is no better than that of its associated differential pressure measurement system. It is essential that the pneumatic (acoustic) impedances, including those of the tubes and connectors between the pressure sensor and the pneumotachometer, on each side of the differential pressure sensor be balanced. This is most easily achieved by ensuring that the geometries and dimensions of the pneumatic pathways from the pneumotachometer to each side of the pressure sensor diaphragm are identical.

Equation (7.2) indicates that the resistance of the Fleisch pneumotachometer is proportional to the viscosity of the flowing gas mixture. The resistance of a screen pneumotachometer, though not computable from (7.2), is also proportional to the viscosity of the gas. The viscosity of a gas mixture depends on its composition and temperature (Turney *et al.*, 1973). When inertance effects are negligible,

$$Q = \frac{\Delta P}{R(T,[F\text{x}])} \tag{9.14}$$

where Q is the flow measured by the pneumotachometer for a gas mixture with species molar fractions $[F\text{x}] = [N_1/N, N_2/N, \ldots, N_\text{x}/N]$ at absolute temperature T. Pneumotachometers are routinely calibrated for steady flow with a single calibration factor being used during experiments. But instantaneous values of T and $[Fx]$ are not constant during a single expiration, and their mean values change from expiration to inspiration. In particular, changes in viscosity of 10 to 15% occur from the beginning to the end of an experiment in which N_2 is washed out of the lungs by pure O_2. A continuous correction in calibration should be made when accurate results are desired.

The prevention of water-vapor condensation in a pneumotachometer is of particular importance. The capillary tubes and screen pores are easily blocked by liquid water, which decreases the effective cross-sectional area of the flow element and causes a change in resistance. Also, as water condenses, the composition of the gas mixture changes. To circumvent these problems, a common practice is to heat the pneumotachometer element, especially when more than a few consecutive breaths are to be studied. The Fleisch pneumotachometer is usually provided with an electrical resistance heater; the screen of the screen pneumotachometer can be heated by passing a current through it. In addition, heated wires can be placed inside, or heating tape or other electrical heat source can be wrapped around any conduit that carries expired gas.

Pitot Tubes The difference between stagnation pressure (measured head-on into the flow) and static pressure (measured perpendicular to the flow) in a flowing gas is the dynamic pressure, and is related to the density and the square of the velocity of the gas. An instrument based on this relationship and using two pressure ports, one facing upstream and the other, downstream, has been developed (Hänninen, 1991). When flow is in one direction, one port measures stagnation pressure and the other assesses static pressure. When flow reverses, the roles of the ports also reverse permitting the alternating flows of breathing to be measured. During a breath, the inspired and expired gas compositions continuously change with concomitant changes in density. Simultaneous measurement of gas composition from the same location at which the pressure measurements are made permits compensation for variations in density. Flow resistance of 1.0 cm H_2O/(liter/s), dead space of 9.5 ml, volume accuracy (from the integral of flow) of $\pm 6\%$ and flow sensitivity of 0.07 liter/s have been reported.

9.4 LUNG VOLUME

The most commonly used indices of the mechanical status of the ventilatory system are the absolute volume and changes of volume of the gas space in the lungs achieved during various breathing maneuvers. Observe Figure 9.5 and assume that a subject's airway opening and body surface are exposed to atmospheric pressure. Then the largest volume to which the subject's lungs can be voluntarily expanded is defined as the *total lung capacity*, TLC. The

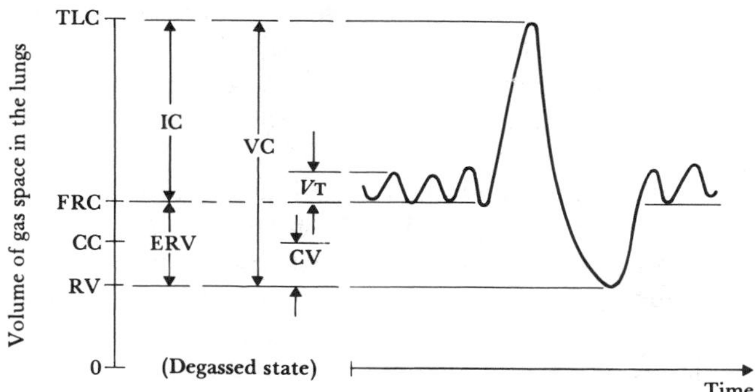

Figure 9.5 Volume ranges of the intact ventilatory system (with no external loads applied). TLC, FRC, and RV are measured as absolute volumes. VC, IC, ERV, and V_T are volume changes. Closing volume (CV) and closing capacity (CC) are obtained from a single-breath washout experiment.

smallest volume to which the subject can slowly deflate his or her lungs is the *residual volume,* RV. And the volume of the lungs at the end of a quiet expiration when the respiratory muscles are relaxed is the *functional residual capacity,* FRC. The difference between TLC and RV is the *vital capacity,* VC, which defines the maximal change in volume the lungs can undergo during voluntary maneuvers. The vital capacity can be divided into the *inspiratory capacity* (IC = TLC − FRC) and the *expiratory reserve volume* (ERV = FRC − RV). The peak-to-peak volume change during a quiet breath is the *tidal volume,* V_T (Petrini, 1988).

CHANGES IN LUNG VOLUME: SPIROMETRY

The measurement of changes in lung volume has been approached in two ways. One is to measure the changes in the volume of the gas space within the body during breathing by using plethysmographic techniques (discussed in Section 9.5). The second approach, referred to as *spirometry,* involves measurements of the gas passing through the airway opening. The latter measurements can provide accurate, continuous estimates of changes in lung volume only when compression of the gas in the lungs is sufficiently small. The flow rate of moles of gas at the airway opening can be expressed as

$$\dot{N}_{\mathrm{AWO}} = \rho_{\mathrm{AWO}} Q_{\mathrm{AWO}} \cong \rho_{\mathrm{L}} \dot{V}_{\mathrm{L}} \tag{9.15}$$

if we neglect the net rate of diffusion into the pulmonary capillary blood. This equation can be rearranged and, if the densities are essentially constant, integrated from some initial time t_0, as follows:

$$\frac{\rho_{\mathrm{AWO}}}{\rho_{\mathrm{L}}} \int_{t_0}^{t} Q_{\mathrm{AWO}}\, dt \cong \int_{t_0}^{t} \dot{V}_{\mathrm{L}}\, dt = V_{\mathrm{L}}(t) - V_{\mathrm{L}}(t_0) \equiv v_{\mathrm{L}} \tag{9.16}$$

in which v_L is, according to the convention used here, the change in the volume of the lungs relative to the reference volume $V(t_0)$. The density ratio accounts for differences in mean temperature, pressure, and composition that may exist between the gas mixture inside the lungs and that in the measurement sensor external to the body.

For purposes of testing pulmonary function, (9.16) is frequently implemented directly by electronically integrating the output of a flowmeter placed at a subject's mouth (with the nose blocked). However, the most common procedure for estimating v_L—in use since the nineteenth century—is to continuously collect the gas passing through the airway opening and to compute the volume it occupied within the lungs. This represents a physical integration of the flow at the mouth; it is performed by a device called a *spirometer.* The widespread and historical use of this device has given rise to use of the term *spirometry* to mean the measurement of changes in lung volume for testing of pulmonary function, regardless of whether a spirometer, flowmeter plus integrator, or plethysmographic technique is used. Consequently, performance recommendations have been published by the American Thoracic Society (Anonymous, 1995a) for spirometry systems in general, regardless of the primary variable measured (volume change or flow) or the type of sensor used. The recommendations address volume range and accuracy, flow range, time interval for which data are to be collected, respiratory load imposed on the subject, and calibration standards to be met by such systems for various pulmonary function tests.

A spirometer is basically an expandable compartment consisting of a movable, statically counterbalanced, rigid chamber or "bell," a stationary base, and a dynamic seal between them (Figure 9.6). The seal is often water, but dry seals of various types have been used. Changes in internal volume of the spirometer, V_S, are proportional to the displacement of the bell. This motion is traditionally recorded on a rotating drum (kymograph) through direct mechanical linkage, but any displacement sensor can be used. A simple approach is to attach a single-turn, precision linear potentiometer to the shaft of the counterweight pulley and use it as a voltage divider. The electric output can then be processed or displayed.

The mouthpiece of the spirometer (Figure 9.6) is placed in the mouth of the subject, whose nose is blocked. As gas moves into and out of the spirometer, the pressure P_S of the gas in the spirometer changes, causing the bell to move. Analysis of the dynamic mechanics of a spirometer indicates that these variations in pressure are reduced by minimizing (1) the mass of the bell and counterweight and the moment of inertia of the pulley, (2) the gas space in the spirometer and tubing, (3) the surface area of the liquid seal exposed to P_S, (4) the viscous and frictional losses by appropriate choice of lubricants and type of dynamic seal, and (5) the flow resistances of any inlet and outlet tubing and valves. During resting breathing, the pressure changes in the gas within the spirometer can be considered negligible. Therefore, only the temperature, average ambient pressure, and change in volume are needed to estimate the amount of gas exchanged with the spirometer. To use the spirome-

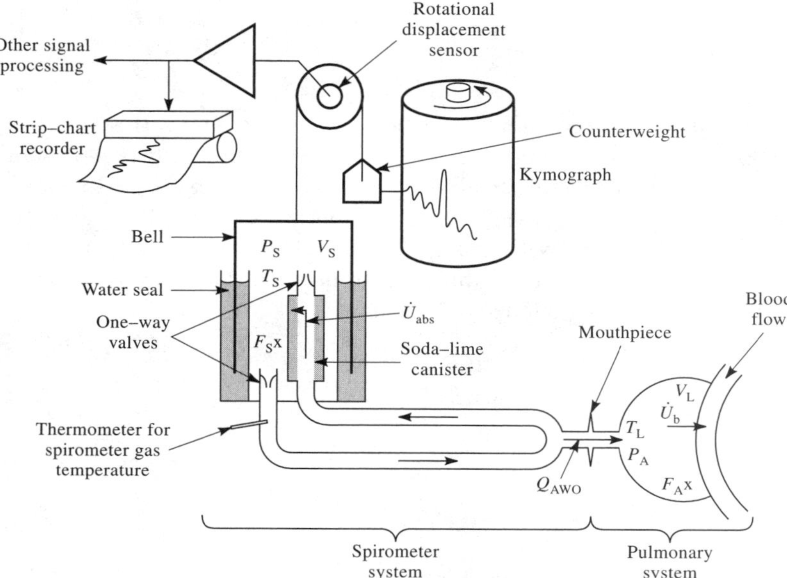

Figure 9.6 A water-sealed spirometer set up to measure slow lung-volume changes. The soda-lime and one-way-valve arrangement prevent buildup of CO_2 during rebreathing.

ter for estimates of change in lung volume during breathing patterns of higher frequency (> 1 Hz) requires, in addition to the variables just mentioned, a knowledge of the acoustic compliance of the gas in the spirometer (see Problem 9.3) and continuous measurement of the change of the spirometer pressure relative to ambient pressure.

The system—lungs plus spirometer—can be modeled as two gas compartments connected such that the number of moles of gas lost by the lungs through the airway opening is equal and opposite to the number gained by the spirometer. For rebreathing experiments, most spirometer systems have a chemical absorber (soda lime) to prevent buildup of CO_2. When compression of gas in the lungs and in the spirometer is neglected, mass balances on the system yield

$$\rho_L \dot{V}_L + \dot{U}_b = -\rho_S \dot{V}_S - \dot{U}_{abs} \tag{9.17}$$

The net rates of uptake from the system by the pulmonary capillary blood, $\dot{U}_b$, and the absorber, $\dot{U}_{abs}$, can be assumed constant during steady breathing. Therefore, when (9.17) is integrated with respect to time, the combined effect of these uptakes is a change in volume essentially proportional to time. This can be approximated as a linear baseline drift easily separable from the breathing pattern. Consequently, rearrangement and integration of (9.17) yield

$$v_L \cong -\frac{\rho_S}{\rho_L}(v_S - \text{drift}) = -\frac{\rho_S}{\rho_L} v_{S'} \tag{9.18}$$

from which it can be seen that the change in lung volume is approximately proportional to $v_{S'}$, the volume change of the spirometer corrected for drift.

The constant of proportionality, $(-\rho_S/\rho_L)$, can be expressed in terms of the measurable quantities, pressure and temperature, by applying an equation of state to the system. With the exception of water vapor in a saturated mixture, all gases encountered during routine respiratory experiments obey the ideal-gas law during changes of state:

$$P = \frac{N}{V}\mathbf{R}T = \rho \mathbf{R}T \tag{9.19}$$

where

$\mathbf{R}$ = universal gas constant

T = absolute temperature

ρ = mole density, which, for a well-mixed compartment, equals the ratio of moles of gas, N, in the compartment to the compartment volume, V

This relationship holds for an entire gas mixture and for an individual gas species X in the mixture. For the latter, the partial pressure Px, number of moles Nx, and density ρx $= N$x$/V$ are substituted into (9.19).

A difficulty in directly substituting (9.19) into (9.18) arises from the presence of water vapor in the system. The gas in the lungs is saturated with water vapor at body-core temperature. The gas in the spirometer is also saturated, even in those with dry seals, after only a few exhalations from the warmer lungs into the cooler spirometer. Water vapor in a saturated mixture does not follow (9.19) during changes of state. Instead, its partial pressure is primarily a function of temperature alone.

Processes in the lungs are approximately isothermal; the change in temperature in the spirometer during most pulmonary function tests is assumed small. Therefore we can compute the partial pressure of the ideal dry gases—the total gas mixture excluding water vapor—as the mean total pressure (taken to be atmospheric pressure P_{atm} for both the spirometer and the lungs) minus the partial pressure of water vapor saturated at the appropriate temperature. Equation (9.18) can then be evaluated as

$$v_L \cong -\left[\frac{(P_{atm} - P_S H_2O)}{(P_{atm} - P_A H_2O)} \frac{T_L}{T_S}\right] v_{S'} \tag{9.20}$$

in which $P_{A}H_2O$ and $P_{S}H_2O$ are the saturated partial pressures of water vapor in the lung [47 mm Hg (6.27 kPa) at $T_L = 37$ °C] and in the spirometer (at the measured temperature in the spirometer, T_S), respectively.

ABSOLUTE VOLUME OF THE LUNG

Because of the complex geometry and inaccessibility of the lungs, we cannot compute their volume accurately either from direct spatial measurements or from two- or three-dimensional pictures provided by various imaging techniques. Three procedures have been developed, however, that can give accurate estimates of the volume of gas in normal lungs. Two are based on static mass balances and involve the washout or dilution of a test gas in the lungs. The test gas must have low solubility in the lung tissue. That is, movement of the gas from the alveoli by diffusion into the parenchyma (tissue) and blood must be much less than that occurring by convection through the airways during the experiment. The third procedure is a total body plethysmographic technique employing dynamic mass balances and gas compression in the lungs (see Section 9.5). These estimates of lung volume provide a static baseline value of absolute lung volume that can be added to a continuous measure of the change of lung volume to provide a continuous estimate of absolute lung volume.

The computational formulas for the test-gas procedures described below are routinely derived in the literature in terms of a volume fraction of the test gas in a mixture. The volume fraction of a gas X is an alternative expression of, and is numerically equal to, its molar fraction, $F\text{X}$. When the ideal-gas law [(9.19)] is applied to X alone and to the total mixture containing X, we can express $F\text{X}$ in terms of the partial pressure of X, $P\text{X}$.

$$F\text{X} = \frac{N\text{X}}{N} = \left(\frac{P\text{X}V}{\mathbf{R}T}\right)\left(\frac{\mathbf{R}T}{PV}\right) = \frac{P\text{X}}{P} \tag{9.21}$$

This follows from Dalton's law of partial pressures, for which all gases in a mixture are visualized as having the same temperature and occupying the same volume. On the other hand, the concept of volume fraction lends itself to the use of the spirometer to measure the volume occupied by a mass of gas at a given pressure and temperature. Assume that $N\text{X}$ moles of X at temperature T occupy a volume $V\text{X}$ when subjected to a pressure P. These $N\text{X}$ moles of X are then added to an X-free gas mixture so that the total moles of mixture is N, and the mixture is allowed to occupy volume V at pressure P. Applying the ideal-gas law to X before addition to the mixture and to the mixture after the addition of X, we can evaluate $F\text{X}$ for the final mixture as

$$F\text{X} = \frac{N\text{X}}{N} = \left(\frac{PV\text{X}}{\mathbf{R}T}\right)\left(\frac{\mathbf{R}T}{PV}\right) = \frac{V\text{X}}{V} \tag{9.22}$$

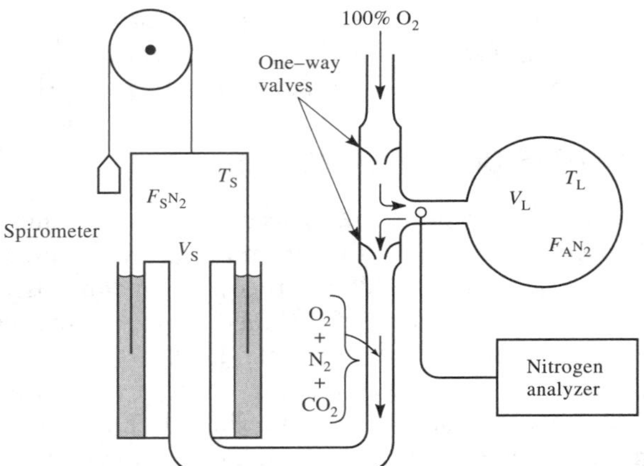

Figure 9.7 Diagram of an N_2 washout experiment The expired gas can be collected in a spirometer, as shown here, or in a rubberized-canvas or plastic Douglas bag. N_2 content is then determined off-line. An alternative is to measure expiratory flow and nitrogen concentration continuously to determine the volume flow of expired nitrogen, which can be integrated to yield an estimate of the volume of nitrogen expired.

Thus the molar fraction $F\text{x}$ can be thought of either as a partial-pressure fraction (9.21) or as an equivalent-volume fraction (9.22), in which the volumes are those that the components of the gas or the gas mixture would occupy if all exhibited the *same* temperature and total pressure. However, in some experiments the temperature and/or total pressure changes from one measurement to another. Remember that in the mass balances describing such experiments, $F\text{x}$ represents molar fraction and must be evaluated as such to account for the observed changes in temperature and pressure. Instruments capable of measuring molar fractions and concentrations of gases in a mixture are described in Section 9.7.

NITROGEN-WASHOUT ESTIMATE OF LUNG VOLUME

Figure 9.7 is a diagram of an apparatus that can be used during a multibreath N_2 washout. The subject inhales only an N_2-free gas mixture (for this example, O_2) because of the one-way valves, but she or he exhales N_2, O_2, CO_2, and water vapor. This experiment is routinely performed to measure FRC, so the subject is switched into the apparatus at the end of a quiet expiration following normal breathing of atmospheric air. He or she is allowed to breathe the N_2-free mixture in a relaxed manner around FRC with relatively constant tidal volumes for a fixed period of time (7 to 10 min) or until the N_2 molar fraction in the expirate is sufficiently near zero ($<2\%$).

A static mass (molar) balance on the N_2 in the lungs from before to after the washout yields an estimate of the lung volume at which the first inspiration of O_2 began. Assume that during the experiment, negligible amounts of N_2 diffuse into the alveolar gas from lung tissue and pulmonary capillary blood. Therefore, the change in the number of moles of N_2 in the lungs as a result of the washout is just the number lost during each expiration (and gained by the spirometer). Assuming that measurements of N_2 fraction are made on a wet-gas basis and that the spirometer contained no N_2 at the start of the experiment, we find that a static mass balance yields

$$F_{AN_2}(t_1)\frac{V_L(t_1)}{T_L} - F_{AN_2}(t_2)\frac{V_L(t_2)}{T_L} = F_{SN_2}(t_2)\frac{V_S(t_2)}{T_S}$$

If the lung volume at the beginning of the washout, t_1, is the same as that at the finish, t_2, then the equation can be rearranged as follows:

$$V_L = \frac{T_L}{T_S}\left[\frac{F_{SN_2}(t_2)V_S(t_2)}{F_{AN_2}(t_1) - F_{AN_2}(t_2)}\right] \tag{9.23}$$

When this volume is that at the end of a quiet expiration, V_L is the FRC.

All variables on the right-hand side of (9.23) are measurable. The numerator in the brackets in (9.23) represents the equivalent expired N_2 volume for the conditions in the measuring device (the spirometer in Figure 9.7). However, if the nitrogen molar fractions in (9.23) were to be measured on a dry-gas basis rather than on a wet basis as assumed here, the right-hand side would have to be multiplied by the dry-gas partial-pressure ratio $(P_{atm} - P_{SH_2O})/(P_{atm} - P_{AH_2O})$.

A washout experiment of this type gives an estimate of the volume of the gas space in the lungs that freely communicates with the airway opening. This is the entire alveolar volume in normal young, adult lungs. In diseased lungs, in which airways are totally or partially obstructed (by mucus, edema, bronchospasm, or tumors, for instance), this estimate of the gas volume in the lungs at FRC can be low.

HELIUM-DILUTION ESTIMATE OF LUNG VOLUME

The procedure described in the previous paragraph involves the removal from the lungs of gas normally resident there in known concentration. An alternative approach is to add a measured amount of a nontoxic, insoluble *tracer* gas to the inspirate, and—after it is uniformly distributed in the lungs—determine its concentration and compute lung volume. Helium is often used as the tracer gas for this experiment, though Ar and Ne are also acceptable.

The lungs communicate directly with the spirometer, as shown in Figure 9.6. At the beginning of the experiment, a fixed amount of He is added to the spirometer. The amount of He added is routinely measured as an equivalent volume at the initial conditions in the spirometer. The change in volume of the spirometer due to addition of pure helium, V_{SHe}, divided by the total spirometer volume after the addition of the helium, $V_S(t_1)$, is $F_{SHe}(t_1)$, the molar fraction of helium on a wet-gas basis in the spirometer at the beginning of the experiment [see (9.22)].

The subject is typically allowed to begin rebreathing from the spirometer when the lung volume is at FRC. The subject breathes at his or her resting rate and tidal volume until the concentration of He in the lungs is in equilibrium with that in the spirometer—that is, until $F_{\mathrm{A}}\mathrm{He}(t_2) = F_{\mathrm{S}}\mathrm{He}(t_2)$. The tracer-gas molar fraction in the expirate is continuously measured by withdrawing a small stream of gas from the mouthpiece, passing it through a He analyzer, and returning it to the spirometer. Equilibration is judged to have occurred when the change in the He fraction from one breath to the next is arbitrarily small. As equilibration approaches, the mean spirometer volume is maintained at its original volume $V_{\mathrm{S}}(t_1)$ by the addition of O_2 if required. The experiment is terminated at the end of a quiet expiration (FRC).

The He is redistributed between spirometer and lungs during rebreathing, but the total amount remains essentially constant because no appreciable quantities are lost by diffusion into the tissues. The average total number of moles of dry gas in the system is kept constant by chemically removing, via the soda-lime canister, the CO_2 added from the blood, and replenishing the O_2 taken up. Therefore, the system can be considered closed with respect to the test gas only, and a static mass (molar) balance for He may be written as

$$F_{\mathrm{S}}\mathrm{He}(t_1)\,\frac{V_{\mathrm{S}}(t_1)}{T_{\mathrm{S}}(t_1)} = F_{\mathrm{S}}\mathrm{He}(t_2)\,\frac{V_{\mathrm{S}}(t_2)}{T_{\mathrm{S}}(t_2)} + F_{\mathrm{A}}\mathrm{He}(t_2)\,\frac{V_{\mathrm{L}}(t_2)}{T_{\mathrm{L}}}$$

for molar fractions measured on a wet basis. This can be rearranged and evaluated when the He fractions in the lungs and spirometer are equal.

$$V_{\mathrm{L}} = \frac{V_{\mathrm{S}}(t_1)}{F_{\mathrm{S}}\mathrm{He}(t_2)}\left[\frac{T_{\mathrm{L}}}{T_{\mathrm{S}}(t_1)}\,F_{\mathrm{S}}\mathrm{He}(t_1) - \frac{T_{\mathrm{L}}}{T_{\mathrm{S}}(t_2)}\,F_{\mathrm{S}}\mathrm{He}(t_2)\right] \tag{9.24}$$

V_{L} is an estimate of FRC for the experiment performed as we have described. Equation (9.24) is frequently rewritten in terms of the equivalent volume of He originally added to the system, $V_{\mathrm{S}}\mathrm{He}$.

$$V_{\mathrm{L}} = \frac{T_{\mathrm{L}}}{T_{\mathrm{S}}(t_1)}\left(\frac{V_{\mathrm{S}}\mathrm{He}}{F_{\mathrm{S}}\mathrm{He}(t_2)}\right) - \frac{T_{\mathrm{L}}}{T_{\mathrm{S}}(t_2)}\left(\frac{V_{\mathrm{S}}\mathrm{He}}{F_{\mathrm{S}}\mathrm{He}(t_1)}\right) \tag{9.25}$$

If the molar fractions are measured on a dry basis, the temperature ratios in (9.24) and (9.25) must be multiplied by their corresponding dry-gas partial-pressure ratios: $[P_{\mathrm{atm}} - P_{\mathrm{S}}\mathrm{H_2O}(t_1)]/(P_{\mathrm{atm}} - P_{\mathrm{A}}\mathrm{H_2O})$ and $[P_{\mathrm{atm}} - P_{\mathrm{S}}\mathrm{H_2O}(t_2)]/(P_{\mathrm{atm}} - P_{\mathrm{A}}\mathrm{H_2O})$, respectively.

The volume computed from (9.24) and (9.25) is a measure of the volume of gas space in the lungs for which the final He molar fraction in the spirometer, $F_{\mathrm{S}}\mathrm{He}(t_2)$, is a representative value. If some parts of the lung do not communicate freely with the airway opening, as in obstructive lung disease, then these equations can provide a low estimate of the FRC.

9.5 RESPIRATORY PLETHYSMOGRAPHY

The term *plethysmography* refers, in general, to measurement of the volume or change in volume of a portion of the body. In respiratory applications, plethysmography has been approached in two ways: by inferring changes in thoracic-cavity volume from geometrical changes at discrete locations on the torso, and by measuring the effects of changes in thoracic volume on variables associated with the gas within a total-body plethysmograph.

THORACIC PLETHYSMOGRAPHY

Several devices have been used to measure continuously the kinematics (motions) of the chest wall that are associated with changes in thoracic volume (Sackner, 1980). The electrical impedance of the thoracic cavity changes with breathing movements and can be sensed (Section 8.7) in order to monitor ventilatory activity. Impedance pneumographs are used for apnea detection and sleep studies in which the presence or absence (relative magnitude and frequency) of breathing movements rather than their actual volume changes are important. The application of magnetometers, strain gages, and variable-inductance sensors requires the simultaneous measurement of motion at two locations on the chest wall. During most breathing patterns, the chest wall behaves as though it has two predominant degrees of freedom corresponding to movements of the ribcage and the diaphragm. The weighted sum of the displacements of these two structures, with the movement of the abdomen taken as a measure of diaphragmatic motion, can yield an estimate of the volume change of the thoracic cavity.

Magnetometers and other linear-displacement sensors measure ribcage and abdominal diameters. *Strain gages* wrapped around the torso measure local perimeter changes during breathing. A small-diameter, mercury-filled, silicone rubber tube is often used as a ventilatory strain gage. The *respiratory inductive plethysmograph* employs a pair of wires, each attached in a zig-zag pattern to its own highly compliant belt. One belt is placed around the ribcage and the other around the abdomen, so that each wire forms a single loop, and the pair is excited by a low-level radio-frequency signal. Changes in a loop's cross-sectional area produce corresponding changes in self-inductance. After demodulation, an output is obtained proportional to the local cross-sectional area of the segment of the chest wall that is encircled by the loop. Respiratory inductive plethysmographs are used in sleep laboratories to provide noninvasive, continuous (8-h or more) estimates of tidal volume and with other measures help to diagnose sleep apnea and disordered breathing during sleep (Broughton, 1993; Kryger, 1994).

Each of these devices measures a different primary geometrical variable (diameter, perimeter, or area), but each is used to estimate changes in thoracic volume as well as changes in relative volume between ribcage and diaphragm (via abdominal motion). The accuracy of these estimates varies among the techniques (there is generally a 5 to 10% error relative to spirometrically evaluated volume changes). Measurement artifacts are caused by body motion (changes in posture and torso shape) and extreme variations in breathing amplitude. However, with suitable calibration, the sensitivity of the inductive plethysmograph to such disturbances can be reduced (Verschakelen, 1989).

TOTAL-BODY PLETHYSMOGRAPHY

The *total-body plethysmograph* (TBP) is a rigid, constant-volume box in which the subject is completely enclosed. Clinically it is used primarily to evaluate the absolute volume of the lungs, and to provide a continuous estimate of alveolar pressure, from which airway resistance R_{AW} can be computed. There are three types or configurations of TBPs: pressure, volume-displacement, and flow-displacement TBPs. These names correspond to the primary plethysmographic variable that is measured and used to compute other variables associated with the lungs. Even though these names provide a means of identifying a particular configuration, they are misleading because a number of measurable plethysmographic variables change in all these systems in response to changes in respiratory variables (Primiano and Greber, 1975).

The pressure plethysmograph is a box that acts as though it were closed or gastight at the frequencies at which pressure changes are measured. The volume-displacement plethysmograph and the flow-displacement plethysmograph are referred to as open, because each has an opening through which gas is intended to enter and leave. A spirometer or a volume flowmeter such as a pneumotachometer is placed at the opening, and pressure changes within these plethysmographs can be kept small by allowing movement of gas between the box and these measuring devices. Consequently, open boxes are suitable for maneuvers in which large changes in lung volume occur. For small-volume-amplitude maneuvers, such as panting, any of the boxes can be employed.

GENERAL EQUATION FOR BREATHING WITHIN A TOTAL-BODY PLETHYSMOGRAPH

Absolute volume of the lungs and changes in alveolar pressure can be inferred from measurements in a TBP in which the subject can breathe within the box. The following analysis is restricted to the *pressure plethysmograph* (Figure 9.8), which is probably the most commonly used configuration. For simplicity, assume that the subject has a normal pulmonary system. Then a single mechanical unit [Figure 9.1(b)] can be used to model the lungs; the representative alveolar pressure P_A is well defined as the pressure within the alveolar compartment (Figure 9.8). The airway exhibits a flow resistance so that, during breathing, P_A does not equal the pressure at the airway opening, P_{AWO}. The gas

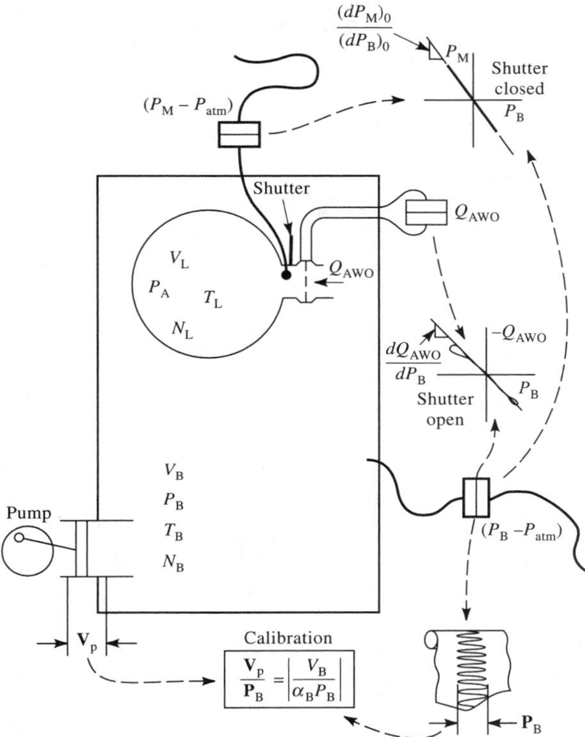

Figure 9.8 A pressure-type total-body plethysmograph is used with the shutter closed to determine lung volume and with the shutter open to determine changes in alveolar pressure. Airway resistance can also be computed if volume flow of gas is measured at the airway opening. Because atmospheric pressure is constant, changes in the pressures of interest can be obtained from measurements made relative to atmospheric pressure.

space within the box outside the subject can be considered a single, well-mixed compartment containing an unsaturated gas mixture with its own variables, P_B, V_B, N_B, and T_B.

The volume of the plethysmograph V_P is occupied by the tissues of the body V_{TIS}, the gas space in the lungs V_L, and the gas space in the box around the subject, V_B.

$$V_P = V_{TIS} + V_L + V_B \tag{9.26}$$

The tissues of the body, composed of liquids and solids, can be considered incompressible when compared with the gas in the lungs. During breathing movements, the tissues change shape, not volume. Also, because the volume of the plethysmograph is a constant (except during calibration procedures), changes in V_P are zero. Consequently, the change in volume of the gas in the lungs is equal and opposite to the changes in volume of the gas space in the box.

$$dV_L = -dV_B \tag{9.27}$$

These equal and opposite changes in volume produce changes in pressure within the lungs and within the gas space in the box as a result of thermodynamic processes. For the lungs and the box, each modeled as a well-mixed compartment, the volume, pressure, and number of moles of the ideal gases undergoing these processes can be related by a barotropic relation of the form

$$P\left(\frac{V}{N}\right)^{\alpha} = K \tag{9.28a}$$

in which α and K are constants. For an isothermal process, such as occurs in the lungs, $\alpha = 1$; for an adiabatic process, $\alpha = 1.4$ (for a diatomic gas such as air). For any other process, α will be between these two limits. In the box, α approaches the adiabatic limit during rapid breathing movements.

The total derivative of (9.28a) relates the changes in V, P, and N.

$$dV = -\frac{V}{\alpha P}dP + \frac{1}{\rho}dN \tag{9.28b}$$

The coefficient of dP is the *acoustic compliance,* C_g, of the gas in the container. This is a measure of its absolute compressibility.

Assume that the gas within the plethysmograph is unsaturated and therefore acts as a mixture of ideal gases. Then, evaluating (9.28b) for the total mixture in the box and the dry gas only in the lungs and substituting into (9.27) yields

$$-\frac{V_L}{P_{ADRY}}dP_{ADRY} + \frac{dN_{LDRY}}{\rho_{LDRY}} = \frac{V_B}{\alpha_B P_B}dP_B - \frac{dN_B}{\rho_B} \tag{9.29a}$$

The box is sealed, and we assume that the net uptake of gas by the pulmonary capillary blood is negligible. We perform separate mass balances on the lungs and box, which yield

$$dN_{LDRY} = \rho_{AWODRY}\, Q_{AWO}\, dt$$
$$dN_B = -\rho_{AWO}\, Q_{AWO}\, dt$$

These can be substituted into (9.29a) to produce, after rearrangement,

$$\frac{V_L}{P_{ADRY}}dP_{ADRY} = -\left[\frac{V_B}{\alpha_B P_B}dP_B + \left(\frac{\rho_{AWO}}{\rho_B} - \frac{\rho_{AWODRY}}{\rho_{LDRY}}\right) Q_{AWO}\, dt\right] \tag{9.29b}$$

The ideal-gas law (9.19) can be used to express the densities in (9.29b) in terms of temperatures and pressures. The *mean* (hydrostatic) pressures of the total gas mixtures at the airway opening, within the lungs, and within the box are all equal to atmospheric pressure: $P_{\mathrm{AWO}} = P_{\mathrm{A}} = P_{\mathrm{B}} = P_{\mathrm{atm}}$. Because the partial pressure of water vapor in the saturated mixture within the lungs is primarily a function of temperature, and because the processes within the lungs are essentially isothermal, the changes in the total alveolar pressure are simply the changes in partial pressure of the dry gases within the lungs: $dP_{\mathrm{A}} = dP_{\mathrm{ADRY}}$. Consequently (9.29b) can be rewritten as

$$\frac{V_{\mathrm{L}}}{(P_{\mathrm{atm}} - P_{\mathrm{AH_2O}})} dP_{\mathrm{A}}$$

$$= -\left[\frac{V_{\mathrm{B}}}{\alpha_{\mathrm{B}} P_{\mathrm{B}}} dP_{\mathrm{B}} + \left(\frac{T_{\mathrm{B}}}{T_{\mathrm{AWO}}} - \frac{(P_{\mathrm{atm}} - P_{\mathrm{AWOH_2O}})}{(P_{\mathrm{atm}} - P_{\mathrm{AH_2O}})} \frac{T_{\mathrm{L}}}{T_{\mathrm{AWO}}}\right) Q_{\mathrm{AWO}}\, dt\right] \tag{9.30}$$

Equation (9.30) represents the governing equation for the total-body plethysmograph in which a subject breathes within the box. It may be used in separate experiments—one with the airway occluded, the other with it open—to compute an estimate of absolute volume of the lungs and a continuous estimate of alveolar pressure, respectively.

VOLUME OF GAS WITHIN THE THORACIC CAVITY BY TOTAL-BODY PLETHYSMOGRAPH

If a subject were to attempt breathing movements within the TBP with his or her airway opening blocked, no gas would flow at that location and Q_{AWO} would be equal to zero. Then (9.30) could be rewritten as

$$\frac{(dP_{\mathrm{A}})_0}{(dP_{\mathrm{B}})_0} = -\frac{(P_{\mathrm{atm}} - P_{\mathrm{AH_2O}})}{V_{\mathrm{L}}} \frac{V_{\mathrm{B}}}{\alpha_{\mathrm{B}} P_{\mathrm{B}}} \tag{9.31}$$

in which the subscript 0 designates zero flow at the airway opening. Equation (9.31) contains V_{L}, which represents the volume of the gas space within the thoracic cavity at the instant when the airway was blocked. We can compute this volume of thoracic gas, designated V_{TG}, from (9.31) if we know $V_{\mathrm{B}}/\alpha_{\mathrm{B}} P_{\mathrm{B}}$, the acoustic compliance of the gas in the box, and if we know the changes in alveolar and box pressure during the blocked breathing movements. A two-step clinical procedure can be performed to evaluate these terms. One step involves a calibrating pump, and the other requires the subject to attempt to pant.

During airway occlusion, the lungs and the box can both be considered closed systems. Then $dN_{\mathrm{B}} = 0$ and (9.28b) evaluated for the box indicates that $V_{\mathrm{B}}/\alpha_{\mathrm{B}} P_{\mathrm{B}}$ can be computed as the ratio of a known change in volume in the gas space in the box to the change in box pressure it causes. The known change in volume is produced by a motor-driven, valveless piston pump mounted in the side of the box (Figure 9.8), which changes the volume V_P of the plethysmograph itself. If the subject stops breathing and blocks his or her nose and mouth while the piston oscillates, the change in volume of the gas space in the box will be only the volume displacement of the pump, $\mathbf{V}_p$. The *amplitude* of the resulting box pressure change, $\mathbf{P}_B$, can be measured. If the frequency of the pump is approximately that of the breathing movements of the subject during the remainder of the experiment, then α_B will have essentially the same value in the two situations, and the ratio $\mathbf{V}_p/\mathbf{P}_B$ will yield the acoustic compliance of the gas in the box around the subject.

When the airway opening is blocked—by a shutter, for example (Figure 9.8)—we assume that, during attempted breathing movements, only compression and expansion of gas occur within the lungs and that no internal flows take place. Then the pressure at every point in the pulmonary gas space is in equilibrium, and a measurement at any accessible point in communication with the lungs (for instance, the mouth) reflects the changes in pressure occurring at every point in the pulmonary system. Therefore, changes in mouth pressure, P_M, behind the closed shutter during the subject's efforts to breathe are assumed equal to changes in pressure in the alveolar region resulting from the same maneuver; that is, $dP_M = dP_A$ when only compression and expansion of the gas occur.

For these conditions, (9.31) can be rearranged to yield V_{TG} as an estimate of the absolute volume of the lungs.

$$V_{TG} = -(P_{atm} - P_{AH_2O}) \frac{\mathbf{V}_p}{\mathbf{P}_B} \frac{(dP_B)_0}{(dP_M)_0} \tag{9.32}$$

We can evaluate the ratio of change in box pressure to change in mouth pressure by displaying P_M versus P_B simultaneously on the two axes of an oscilloscope (Figure 9.8). Then $(dP_B)_0/(dP_M)_0$ represents the inverse of the average slope of the resulting curve, which for rapid panting movements against a closed shutter should be a straight line that retraces itself.

If the airway is occluded at the end of a quiet expiration, V_{TG} can be used as an estimate of FRC. However, V_{TG} is a measure of the total gas space within the body that is compressed and expanded during the closed-airway panting maneuver (DuBois *et al.,* 1956a). Normally this reflects just the gas space contained entirely within the nose, mouth, large airways, and thoracic cavity [plus any volume between the shutter and airway opening which must be subtracted from the volume obtained from (9.32)], because abdominal gas usually represents a negligible fraction of the gas space within the body. However, this approach cannot distinguish between gas within the lungs and any nondissolved gas within the pleural space (pneumothorax). In a subject with no pneumothorax, V_{TG} is an estimate of the total volume of all gas spaces in the pulmonary system, including those that may not communicate freely with the airway opening. Consequently, for normal subjects, V_{TG} measured at the end of expiration corresponds closely to the FRC estimated by the gas-washout and dilution procedures described in Section 9.4. However, in subjects with obstructive lung disease, VTG is a more accurate estimate of FRC than the lower values calculated from the washout and dilution experiments.

CHANGES IN ALVEOLAR PRESSURE BY TOTAL-BODY PLETHYSMOGRAPH

For the computation of a continuous estimate of changes in alveolar pressure during flow of gas through the airways, DuBois *et al.* (1965b) proposed making measurements during a voluntarily produced, high-frequency low-amplitude pant. This results in two primary advantages related to (9.30). First, it improves the validity of the assumption that the effects of gas exchange between the alveoli and pulmonary capillaries are negligible. Because the net uptake by the capillaries can be considered to occur at a nearly constant rate with only slight variation with breathing frequency, its major (low-frequency) component is easily separable from the changes in P_B caused by respiratory movements. At panting frequencies the effects of nonzero net uptake can be assumed to appear primarily as a baseline drift.

Second it reduces the size of the term involving Q_{AWO}. If low-volume amplitudes are produced even at high voluntary frequencies, Q_{AWO} will not be large. The coefficient of Q_{AWO} in (9.39) is the difference between two ratios, both of which are near unity. As the conditions at the airway opening approach either those of the gas space in the box or those in the lungs, the difference between these ratios departs from zero. The extent of this departure can limit the application of the procedure of breathing into the box for the estimation of alveolar pressure to small-volume-amplitude maneuvers unless the subject, during the measurements, breathes a gas mixture the density of which is very close to that of alveolar gas. During panting, the term involving Q_{AWO} in (9.30) is routinely neglected.

High-frequency maneuvers also decrease the effects of leaks in the box. The combination of a small leak that has a high resistance to flow and the acoustic compliance of the gas in the large gas space in the box (about 10^3 liters) acts as a high-pass filter to differences in pressure and as a low-pass filter to the flow between the inside and outside of the plethysmograph. Thus the higher the frequency of the changes in pressure in the box, the less degradation of the breathing-related signal due to leaks to the atmosphere. Note, however, that the high-pass-filter effects of small, controlled leaks in the box, or of leaky ballast chambers on the reference side of plethysmograph pressure sensors, are sometimes intentionally employed to eliminate slow drift in box-pressure readings. This drift can be related to increases in temperature of the gas in the box—increases produced by the subject—or to changes in ambient pressure such as those caused by the movement of elevators or the closing and opening of doors.

With these considerations in mind, we can simplify (9.30) to yield

$$dP_A \cong -\left(\frac{P_{atm} - P_{AH_2O}}{V_L}\right)\left(\frac{V_B}{\alpha_B P_B}\right) dP_B \tag{9.33}$$

This states that for high-frequency low-amplitude breathing, the change in alveolar pressure will be approximately proportional to changes in pressure in the box, given that the coefficient of dP_B is constant. The dry partial pressure in the lungs, the pressure in the box, and the volume of the gas space in the box change very little for an individual as a result of breathing movements. However, α_B depends on the frequency of the maneuver, and it asymptotically approaches the adiabatic limit as frequency increases. Also the volume of the lung can change manyfold from RV to TLC, so a low-volume-amplitude maneuver is mandatory for V_L to be nearly constant.

Note from the discussion of (9.31) that the coefficient of dP_B in (9.33) can be evaluated from simultaneous measurements of mouth pressure and box pressure during an occluded-airway panting maneuver.

9.6 SOME TESTS OF RESPIRATORY MECHANICS

Pulmonary function tests can be divided into two groups: gas-transport tests, concerned with the movement of gas molecules between the atmosphere and blood (see Section 9.8), and mechanics tests that deal primarily with the relationships among lung volume, gas flow, and pressure differences. One of the ultimate objectives of mechanics tests is to determine whether the defect(s) that produce abnormal pressure–volume–flow relationships can be identified as intrinsic to the airways (lumen and/or walls), to the lung parenchyma surrounding and supporting the airways, or to extrapulmonary structures.

A distinction is often made, in pulmonary medicine, between obstructive and restrictive disease processes. This distinction is not, in general, equivalent to asking whether the defect is in the airways or not; it is based more on the functional impairment caused by a disease. *Obstruction* connotes dynamic mechanics, being associated with abnormal rates of change of volume or gas flows in the lungs during breathing movements. *Restriction,* on the other hand, connotes abnormal static mechanics. It is used to refer not only to the lungs but also to extrapulmonary structures (such as the chest wall, muscles, and the abdominal contents). This condition is indicated when the volume attained by the ventilatory system is inappropriate for the difference in pressure applied or when the differences in pressure that the respiratory muscles are capable of producing are abnormally low. Even though restriction and obstruction may not be completely separable in a given disease state, the presence of either or both can ideally be determined by making measurements on the respiratory system under two sets of conditions: static, when all flows and rates of change of all variables are zero, and dynamic, when some or all of these are nonzero.

STATIC MECHANICS

When flows and rates of change of flow and volume are zero, the transpulmonary pressure difference, $\Delta P_L = P_{AWO} - P_{PL}$, as indicated by (9.3a) and (9.3b),

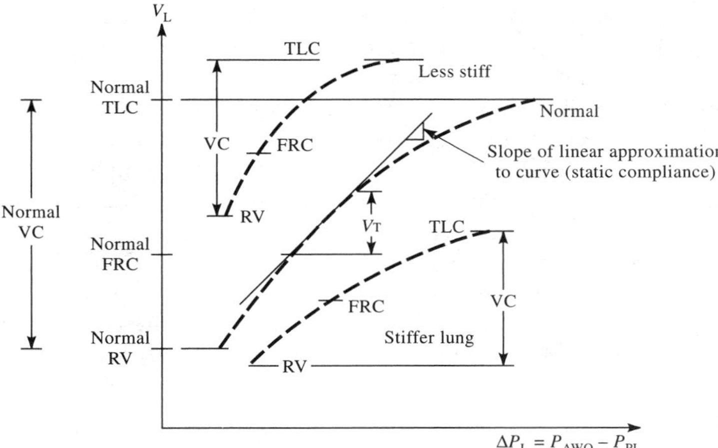

Figure 9.9 Idealized statically determined expiratory pressure–volume relations for the lung. The positions and slopes for lungs with different elastic properties are shown relative to scales of absolute volume and pressure difference.

reduces to a function of volume alone. Therefore, the static mechanical characteristics of the lungs can be obtained by measuring simultaneously the lung volume and the transpulmonary pressure difference, at various lung volumes, while no motions exist in the system. The pulmonary system exhibits a static, or plastic, hysteresis when cycled through a set of static lung volumes. Consequently, to define the ΔP–V relation for the lungs, we must standardize or note the initial volume and direction of change to subsequent volumes. In practice, the expiratory portion of the statically determined ΔP–V curve is routinely used to characterize the lungs.

Static Compliance To produce an expiratory ΔP–V curve, the subject is instructed to inspire to TLC and then to exhale to successively smaller volumes, holding each volume while the pressure in an esophageal balloon is measured relative to pressure at the airway opening. This pressure difference is used as an estimate of transpulmonary pressure (Section 9.2). The lung volume corresponding to each measurement is obtained by spirometer in terms of a change from a reference volume (for example, FRC) determined independently either by gas washout (Section 9.4) or plethysmograph (Section 9.5).

Figure 9.9 shows a facsimile of statically determined expiratory ΔP–V curves for a normal subject, for a subject who has restricted, or stiff, lungs (from diseases such as pulmonary fibrosis or pneumonia), and for a subject who has very distensible lungs (from pulmonary emphysema, for example). The curves extend from TLC to RV for the corresponding individual and are seen to be nonlinear. Such relations are not routinely described for the entire vital capacity. Instead, for small changes in volume, on the order of a tidal volume, straight-line approximations about a volume operating point are used. The slope of the linear approximation to the statically determined ΔP–V curve is the static compliance, as defined by (9.7).

Assuming that the compliance of each of the curves in Figure 9.9 is computed at the respective FRCs, we find, as would be expected, that the stiffer lung has a lower compliance and the less stiff lung a higher compliance than normal. Therefore, we can identify a restricted, or stiff, lung by computing lung compliance at a standardized lung volume. However, even though an increased compliance can be associated in certain cases with obstructive pulmonary diseases, such as emphysema, lungs affected by other obstructive diseases, such as those involving partial blockage of the lumen of large airways, frequently exhibit a normal compliance. Consequently, even when compliance is normal, the dynamic behavior of the system must also be studied to rule out obstruction.

Static Lung Volumes Although pulmonary compliance is a desirable parameter to obtain when the physician suspects restrictive lung disease, it is not routinely measured because it requires the use of an esophageal balloon, which is very unpleasant for the subject. Instead, the observation (made very early in the history of the study of ventilatory mechanics) that TLC, RV, and the difference between them, VC, are reproducible and easily attained is the basis of a simple test to rule out restrictive disease in the absence of measureents of changes in esophageal pressure.

Figure 9.9 shows that the vital capacity of the restricted subject, which can conveniently be estimated by spirometry, is reduced compared with that of the normal individual. Therefore, the observation of a normal VC is a contraindication of a restrictive disease in the pulmonary system. However, a low VC does not imply stiff lungs. A decreased VC can result from a number of circumstances, including obstructive diseases affecting the small airways (emphysema, asthma) and abnormalities of the chest wall and respiratory neuromuscular system (as a result of scoliosis or poliomyelitis, for example).

We can infer an obstructive process from a reduced vital capacity if we also know some measure of absolute lung volume such as TLC or RV. In most cases of obstructive disease, the TLC is approximately normal or increased and the vital capacity tends to be reduced (Figure 9.9) as the residual volume increases, because the obstructed patient is unable to empty his or her lungs effectively. However, some obstructive processes, such as those in the larger airways or the initial phases of small airways disease, may not lead to a reduced vital capacity. Consequently, the dynamic behavior of the system should also be studied.

Maximal Inspiratory and Expiratory Static Pressures When the vital capacity is reduced and a restrictive process is suspected, the clinician can investigate extrapulmonary abnormality by measuring the maximal static inspiratory and expiratory pressures that can be produced at the airway opening at a given lung volume by forceful efforts against a blocked airway opening. By combining (9.3a), (9.3b), and (9.3c), we note that, for no flows in the system,

$$p_{\mathrm{AWO}} - p_{\mathrm{BS}} = \left[\frac{1}{Cst_{\mathrm{L}}} + \frac{1}{Cst_{\mathrm{W}}}\right] v_{\mathrm{L}} - \Delta p_{\mathrm{MUS}} \tag{9.34}$$

where the pressure on the body surface is usually taken to be atmospheric. If a patient is instructed to try to exhale or inhale maximally against a closed nose and mouthpiece while the pressure changes in the mouth are measured, changes in lung volume v_{L} should be negligible, and the measured changes in mouth pressure will reflect the forces produced by the respiratory muscles during these efforts. If the respiratory muscles, the nerve cells controlling them, and the kinematics of the chest wall are not impaired, then $p_{\mathrm{AWO}} - p_{\mathrm{BS}}$ should approximate that attained by a normal person at the corresponding point in his or her vital capacity (usually FRC) at which the test is performed.

DYNAMIC MECHANICS DURING SMALL VOLUME CHANGES AND FLOWS

Equations (9.3a) and (9.3b) indicate that, during resting breathing, changes in transpulmonary pressure can be expressed as a function of the changes in lung volume and gas volume flow rate that would occur if the alveoli were purely elastic structures. However, the alveoli are not purely elastic, in that they are able to exhibit a viscoelastic type of behavior. Inclusion of this in (9.3a) and (9.3b) yields

$$(p_{\mathrm{AWO}} - p_{\mathrm{PL}}) = \frac{1}{Cst_{\mathrm{L}}} v_{\mathrm{L}} + R_{\mathrm{LT}}\dot{v}_{\mathrm{L}} + R_{\mathrm{AW}} q_{\mathrm{AWO}} \tag{9.35}$$

in which R_{LT} is the resistance of the lung tissue and R_{AW} is the airway resistance defined by (9.6). R_{AW} represents a direct measure of obstruction in the airways. Because of the viscoelastic behavior of the alveolar tissue, the computation of R_{AW} from (9.35) is not convenient. Instead, we compute R_{AW} directly from (9.6). The pressure and flow at the airway opening required by (9.6) are easily measured. However, estimating changes in a representative alveolar pressure while there is flow in the airways requires the total-body plethysmograph, as discussed in Section 9.5.

Airway Resistance As an example of a TBP procedure for estimating R_{AW}, consider a subject situated inside a pressure plethysmograph (Figure 9.8). The individual (with nose blocked) breathes within the box through a mouthpiece and flowmeter with an associated shutter that can be closed to block flow at the mouth. While the shutter is open, the pressure at the airway opening (external to the mouthpiece and flowmeter) is the pressure within the plethysmograph P_{B}. That portion of the pressure drop between the alveoli and box that results from the mouthpiece assembly is taken into account by expressing the total alveolar-to-box resistance [given by (9.6) for this arrangement] as the sum of R_{AW} and the mouthpiece assembly resistance R_{MP}. For a panting maneuver, (9.33) indicates that changes in the representative alveolar pressure are proportional to changes in P_B. Consequently, (9.6) can be rewritten

$$R_{AW} + R_{MP} = \left[1 + \frac{(P_{atm} - P_{AH_2O})\,V_B}{V_L \alpha_B P_B}\right] \frac{\partial P_B}{\partial Q_{AWO}} \tag{9.36}$$

The fraction within the brackets in (9.36) is much greater than unity, because V_B, the volume of the gas space in the box, is much larger ($>100:1$) than V_L, the volume within the lungs [$(P_{atm} - P_{AH_2O})/\alpha_B P_B$ being of order 1]. Consequently, as shown by (9.31), a closed-shutter panting maneuver, in which mouth pressure (behind the shutter) and box pressure are measured simultaneously, can be used to evaluate the square-bracketed term in (9.36). We can obtain an approximation to the partial derivative of box pressure P_B with respect to gas flow rate at the airway opening Q_{AWO} by displaying these two variables, measured simultaneously with the shutter open, on orthogonal axes of an oscilloscope (Figure 9.8). For very small changes in volume (a condition of the panting maneuver) and the accelerations associated with panting, normal lungs exhibit a relatively straight P_B–Q_{AWO} relationship with very slight looping. Changes in box pressure are then essentially a function of flow in the airways alone. Thus $\partial P_B/\partial Q_{AWO}$ can be evaluated as the inverse of the slope of the oscilloscope figure, dP_B/dQ_{AWO}, and R_{AW} can be computed as

$$R_{AW} = \left[-\frac{(dP_M)_0}{(dP_B)_0}\right] \frac{dP_B}{dQ_{AWO}} - R_{MP} \tag{9.37}$$

for which R_{MP} can be evaluated independently.

In some instances, however, the P_B–Q_{AWO} plot produced on the oscilloscope screen is highly nonlinear and can display exaggerated looping (dynamic hysteresis). Because of this, it is difficult to determine a representative slope for the plot. A convention has therefore been adopted for the evaluation of R_{AW}. The representative slope is that which corresponds to a straight line drawn between the points on the plot at which flow is +0.5 liter/s and −0.5 liter/s in the region corresponding to the end of inspiration and the beginning of expiration. This straddles the point $Q_{AWO} = 0$; it is usually fairly straight and involves small changes in volume. The representative slope and the parameter R_{AW} computed from it are the only pieces of information from this procedure that are routinely used for the clinical evaluation of airway mechanics.

At least three phenomena associated with the respiratory system can cause exaggerated looping and/or nonlinearity of the P_B–Q_{AWO} relationship in this experiment. (1) Equation (9.33) may not be a good approximation to (9.30) for the conditions of the experiment. (2) The pulmonary system being tested may not act as a single mechanical unit. Hence the concept of a single representative alveolar pressure in phase with flow may be inappropriate. (3) The mechanical properties of the lungs and airways may be so abnormal that even during a panting maneuver, significant changes in dimension (primarily in diameter) of the airways occur, so that the P_B–Q_{AWO} relationship becomes quite bizarre (nonlinear). The latter two situations can be evaluated by additional independent pulmonary function tests.

The aggregate resistance of the smaller airways that play a major role in the final distribution of gas to the alveoli is small compared with that of the larger, upper airways. Consequently, R_{AW} easily reflects obstruction in the larger airways. However, only when the smaller airways are so affected by pulmonary disease that their aggregate resistance is drastically increased do they significantly affect R_{AW}. Thus the parameter R_{AW} is not a sensitive indicator of the initial phases of diseases of the small airways.

Pulmonary Mechanics Evaluated during Breathing For a normal pulmonary system that can be represented by a first-order linear differential equation such as (9.11), mechanical properties (the coefficients C and R) can be estimated by direct substitution of pressure difference, flow, and volume. Measuring the pressure difference across the pulmonary system, i.e., between the airway opening and the thoracic esophagus, and the flow at the airway opening, and integrating this flow (as a measure of volume change) for breathing at voluntarily achieved frequencies permits calculation of the total pulmonary (lung) properties. For example, total lung resistance can be calculated from

$$R_L = \frac{\Delta p_L(t_2) - \Delta p_L(t_1)}{q_{AWO}(t_2) - q_{AWO}(t_1)} \tag{9.38}$$

if t_2 and t_1 are successive instants at which $V_L(t_2) = V_L(t_1)$. (R_L generally would be expected to be greater than R_{AW} because of the resistive component of the viscoelastic lung tissue.) Similarly,

$$C_L = \frac{v_L(t_4) - v_L(t_3)}{\Delta p_L(t_4) - \Delta p_L(t_3)} \tag{9.39}$$

for which t_4 and t_3 are successive instants at which flow at the airway opening, Q_{AWO}, is zero. (In general, when flow at the airway opening is instantaneously zero, there still can be motions and flows at other places within the pulmonary system so that the system may not be truly "static").

If the lungs have a uniform distribution of mechanical properties so that all parts act in unison (exhibit a single degree of freedom), then R_L and C_L will be constants, independent of the breathing pattern the patient produces and frequency at which he or she breathes. In this case, C_L will equal the static compliance, Cst_L, (see Figure 9.9) at all frequencies and the combination of both lungs can be characterized by a single viscoelastic unit and a single time constant, $\tau_L = R_L Cst_L$.

However, if the lungs have a nonuniform distribution of mechanical properties, the values computed for R_L and C_L from (9.38) and (9.39), respectively, will decrease from their static (zero frequency) values as breathing frequency increases. In this case, a single degree-of-freedom pulmonary system can be ruled out, and the lungs can be represented by a combination of viscoelastic units, each with its own time constant. Because C_L is calculated from data obtained while the subject is breathing and does not necessarily have the same value as Cst_L, it is given the name *dynamic compliance* and the symbol $Cdyn_L$.

A test that is sensitive to the nonuniform mechanical changes caused by the onset of small airways disease is based on the values of $Cdyn_L$ obtained at several breathing frequencies. If Cst_L and R_{AW} or R_L are normal, and $Cdyn_L$ obtained for breathing frequencies between 2 and 4 br/s is less than 80% of Cst_L, then small airways disease is indicated.

DYNAMIC MECHANICS DURING LARGE VOLUME CHANGES AND FLOWS

Several features of respiratory mechanics are exhibited for large changes in volume and flows. These characteristics arise because the airways that must distribute the gas within the thoracic cavity are subjected to essentially the same effective forces that the respiratory muscles exert to empty and fill the alveolar regions of the lung.

Specific Airway Resistance The extrapulmonary intrathoracic airways are directly exposed to local interpleural pressure. The airways within the lungs are subjected to the increased and decreased tensile forces produced by the stretching of their parenchymal attachments as the lung regions in which they are embedded inflate and deflate. Consequently, as the alveolar regions are expanded and compressed, the length and cross-sectional area of each airway generation undergo corresponding changes. One manifestation of this is an approximately inverse relationship between plethysmographically determined airway resistance and the lung volume at which it is measured. The larger the volume of the lung, the more expanded the airways and the lower the resistance to flows produced during a panting maneuver, and vice versa. Hence

$$R_{AW} \cong \frac{(SR_{AW})}{V\text{TG}} \tag{9.40}$$

where SR_{AW} is a constant of proportionality referred to as *specific airway resistance* and $V\text{TG}$ is volume of thoracic gas. SR_{AW} represents a characteristic of a subject's airways normalized for his or her lung volume.

Flow Limitation Another manifestation of the effects of changes in airway dimensions during ventilatory maneuvers is the *flow limitation* exhibited during forced expirations. If it were possible to keep the volume of the lung constant and vary the pressure drop and flow through the airways, characteristic *isovolume pressure–flow curves* would result, depending on the volume of

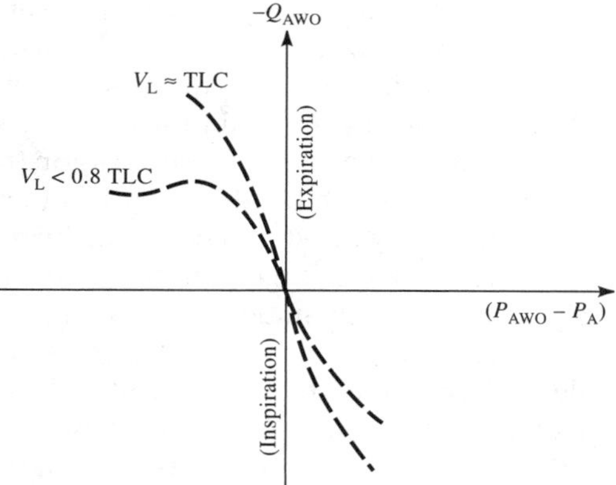

Figure 9.10 Idealized isovolume pressure–flow curves for two lung volumes for a normal respiratory system. Each curve represents a composite from numerous inspiratory–expiratory cycles, each with successively increased efforts. The pressure and flow values measured as the lungs passed through the respective volumes of interest are plotted and connected to yield the corresponding curves.

the lung for which the curve was obtained. Figure 9.10 gives idealized examples of such curves for normal lungs. In practice these curves are obtained by cycling the lungs through a succession of increasing volume amplitudes and flows. The pressure–flow curve for each cycle is plotted, and points on each curve corresponding to the same volume are connected.

Three features generally appear: (1) Inspiratory flow continually increases as the difference in pressure is increased at any lung volume. (2) For high lung volumes (near TLC), expiratory flow increases as the difference in pressure becomes more negative. (3) For lung volumes below about 80% of the TLC, the expiratory flow reaches a value that is never exceeded, even though the difference in pressure becomes more negative. This condition is referred to as *flow limitation* in which the flow is *effort independent.* That is, expiratory flow cannot be increased, no matter how much expiratory efforts are increased.

The flow limit that can be achieved at any given lung volume is dependent on the characteristics of the intrathoracic pulmonary system and the physical properties (viscosity, density) of the gas it contains. Two basic flow-limiting mechanisms appear to be operating (Hyatt, 1986). Central to both is the viscoelastic behavior of the pulmonary airways by which the cross-sectional area of the airways changes in response to changes in transmural pressure.

The first flow-limiting mechanism, accounted for by what is referred to as the *wavespeed theory,* involves the coupling of airway compliance to the pressure drop due to the inertial properties of the gas and convective acceleration of flow in the airways. This coupling determines the wavespeed in the compliant airways—that is, the speed at which a small pressure disturbance, or wave, can propagate along the airway. Flow in a compliant tube cannot exceed that at which the local fluid velocity equals the local wavespeed at any point in the tube. The second mechanism, termed *viscous-flow limitation,* involves the coupling between airway compliance and viscous-flow losses; that is, under appropriate conditions, laminar and turbulent dissipation dominate the pressure distribution that causes airway compression.

At larger lung volumes, for which the driving pressure and local flows are high and airways are held open by the stretched parenchyma, the viscous limit for the peripheral, smaller airways is greater than the wavespeed for the central, larger airways. Consequently in normal lungs, the wavespeed mechanism causes the choke point, or the site at which flow is limited, to be in the central airways while approximately the upper two-thirds of the vital capacity is being exhaled. At lower lung volumes, the viscous-flow limit in the peripheral airways is less than the wavespeed limit anywhere else in the lungs, so flow is limited by the viscous mechanism, and the choke point is in the smaller airways.

Regardless of which mechanism is active, the maximal flow is not greatly influenced by the properties of any airways larger than those that limit flow. Thus the behavior of the lung exhibited while the expiratory flow is effort independent, especially at lower lung volumes, reflects the status of the more peripheral parts of the lungs; the smaller airways in which obstructive disease is thought to produce its initial lesions and the alveolar parenchyma that provide tensile forces that tend to keep these airways expanded.

The Forced Expiratory Vital Capacity Maneuver; the MEFV Curve and Timed Vital Capacity Spirogram Flow limitation phenomena can be displayed in two clinically useful ways, both based on measures of only volume flow of gas at the airway opening during a forced expiration from TLC to RV (referred to as a forced expiratory vital capacity maneuver). The equivalent volume of gas expired during this maneuver is the *forced vital capacity* (FVC). Two alternative (and equivalent) methods of displaying the events within a forced expiration involve either (1) plotting volume flow of gas at the airway opening against its integral (or volume change in a spirometer) subtracted from FVC, or (2) plotting the integral of expired-gas volume flow (or spirometer volume change) subtracted from FVC against time (Figure 9.11). The first method produces the *maximal expiratory flow volume* (MEFV) curve, and the second corresponds to the spirogram of the *timed vital capacity* (TVC) routinely performed as part of standard spirometry tests.

The normal MEFV curve reaches a maximal flow at a volume slightly below TLC; and as the forced expiration continues below about 25% of FVC (below TLC), the expired flow rate decreases nearly linearly with decreasing volume. This linear region corresponds to effort-independent flow; it is reproducible and characteristic of the state of the lungs. The MEFV curve represents the relationship between a variable ($\mathrm{FVC} - \int Q_{\mathrm{AWO}}\, dt$) and its derivative ($-Q_{\mathrm{AWO}}$). When this relationship is a straight line through the origin, it represents a homogeneous, linear, first-order differential equation:

$$Q_{\mathrm{AWO}} = -K\left(\mathrm{FVC} - \int Q_{\mathrm{AWO}}\, dt\right) \qquad (9.41)$$

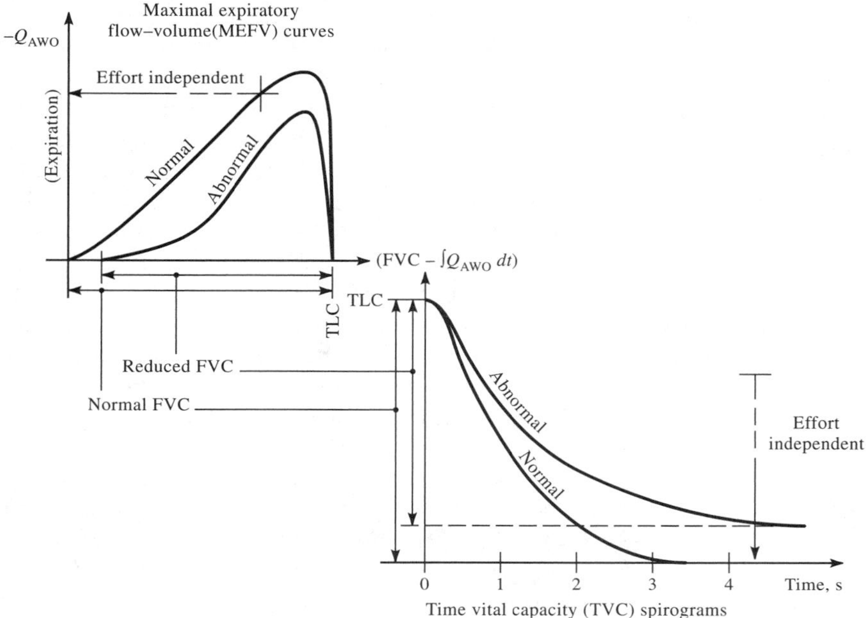

Figure 9.11 Alternative methods of displaying data produced during a forced vital capacity expiration. Equivalent information can be obtained from each type of curve; however, reductions in expiratory flow are subjectively more apparent on the MEFV curve than on the timed spirogram.

Because the coefficient (slope of the linear part of the MEFV curve, $-K$) is negative, it corresponds to an exponential decay from the initial value of FVC $- \int Q_{AWO}\,dt$ at which the relationship became linear. Thus the latter part of the TVC spirogram from 25% FVC below TLC to RV is approximately exponential, corresponding to the effort-independent region of the MEFV curve, and is reproducible and unique for a given subject. The region of each curve between TLC and approximately 25% of the FVC below TLC is effort-dependent. It provides information about the larger, upper airways and the extrapulmonary parts of the ventilatory system (chest wall, respiratory muscles, and so on). The effort-independent region reflects mechanics of the smaller airways and parenchyma of the lungs. We can make comparable inferences about the state of these various parts of the respiratory system from either the MEFV curve or the TVC spirogram.

Forced Expiratory Flows and Forced Expiratory Volume The MEFV curve has gained popularity because it provides a dramatic visual display compatible with subjective evaluation of lung impairment. We can easily recognize decreases in maximal flows or the concave effort-independent region characteristic of even minimal small airways (obstructive) disease. These are harder to detect visually in a spirogram (Figure 9.11). The parameters used to describe the MEFV curve are basically the maximal flow at a given lung volume—for example, *forced expiratory flow* after 50 or 75% of the forced vital capacity has been expired, $FEF_{50\%}$ and $FEF_{75\%}$, respectively. It is especially useful when studying children to normalize these flows to a measure of lung volume to compensate for the size of the individual. An MEFV display normalized by dividing both axes by FVC yields these data directly. Normalized flows such as $FEF_{50\%}/FVC$ are related to the average slope of the effort-independent portion of the MEFV curve.

A number of parameters have been proposed and used to describe the forced expiratory spirogram (Miller *et al.,* 1987). Besides the *forced vital capacity* itself (FVC), only two others are described here: the *forced expiratory volume in one second* (FEV_1) and the *mean forced expiratory flow during the middle half of the FVC* ($FEF_{25-75\%}$) formerly called the maximal midexpiratory flow rate (MMEF).

One characteristic of advanced obstructive disease is a reduction of the maximal volume that can be expired forcefully, compared with that which can be expired slowly, from TLC. Therefore, even though the (slow) VC may be reduced in both restrictive and obstructive diseases, the FVC of an obstructed patient can be smaller than his or her (slow) VC.

FEV_1 is the equivalent volume of gas expired during the initial 1 s of the forced expiratory maneuver. Its value depends partially on data from the effort-dependent part of the curve and is therefore not a good index of small-airways disease; it is correlated more closely with R_{AW}. If the entire TVC spirogram were a perfect exponential, then FEV_1 would be related to the log decrement for the curve. FEV_1/FVC and, equivalently, FEV_1/VC have been found to be nearly independent of body size for normal individuals. A decreased FEV_1/FVC can suggest obstruction, and FEV_1/VC can be reduced even more than FEV_1/FVC in obstructive disease.

The $FEF_{25-75\%}$ represents an average flow produced during the middle half of the forced vital capacity. It is computed as follows:

$$FEF_{25-75\%} = \frac{0.5\ FVC}{t_{25\%FVC} - t_{75\%FVC}} \tag{9.42}$$

where $t_{25\%FVC}$ and $t_{75\%FVC}$ are the times at which 25 and 75% of the forced vital capacity, respectively, have been exhaled. It is an estimate of the maximal flow at 50% of the FVC, that is, $FEF_{50\%}$—which is obtained from the MEFV curve. It is therefore no less valuable as an indicator of small-airways disease. Dividing $FEF_{25-75\%}$ by FVC is a useful normalization for differences in size among individuals.

Peak Expiratory Flow and Maximal Voluntary Ventilation Two other parameters that yield information about obstruction in the larger airways are the *peak expiratory flow* (PEF) and the *maximal voluntary ventilation* (MVV).

The PEF is the highest instantaneous flow at the airway opening that a subject can produce during a maximally forced expiration from TLC. The MVV is the average expiratory flow produced by a subject who is instructed to continually inhale and exhale as deeply and rapidly as possible. Usually this maneuver is performed for only 12 to 15 s to prevent a large reduction in arterial P_{CO_2}. The gas expired during this time period is collected, and its volume is measured by spirometer. Using (9.20), the clinician converts this volume to the volume that the gas would occupy at the temperature and pressure existing within the lungs and routinely expresses the result on a per-minute basis. Like R_{AW}, the PEF and MVV reflect small airway obstruction only when it is major. However, unlike R_{AW}, these parameters can also be affected by neuromuscular impairment.

9.7 MEASUREMENT OF GAS CONCENTRATION

Analysis of the composition of gas mixtures is one of the primary methods of obtaining information about lung function. In respiratory studies, the concentration of a component in a gas mixture is not routinely expressed as mass per volume. It is most frequently given in terms of partial pressure or molar fraction [which can be expressed as an equivalent volume fraction, as shown by (9.22)].

Discrete samples of gas are all that is required in certain circumstances. However, as analytical devices with fast response times have been developed, continuous measurement of intrabreath events has become possible and desirable.

Input systems for continuous sampling usually consist of a thin (capillary) tube or catheter and an input connector (which may include a valve or a mixing chamber). Depending on the particular instrument and the application, the transport pathway can vary from centimeters to meters. This introduces transit delays and mixing within the sample and requires that the characteristics of the catheter and connector be matched to those of the instrument. Adjustment of the delay time, which depends on the mean velocity of the sample through the catheter, requires consideration of the following: the pressure drop along the catheter, the pressure within the instrument, the length and diameter of the catheter, the geometry of the input connector, the composition of the sampled gas, and the volume flow of sample required by the instrument for accurate results. Some minimal time delay is always produced by catheter sampling systems, so electrical or numerical signal processing is required if a gas analyzer's output is to be synchronized with the output from another, independent instrument, such as a flowmeter.

Water vapor presents a major problem in sampling respiratory gas by catheter. Investigators have used thin, flexible, stainless steel inlet tubing, carrying sufficient current through its length to heat its wall above body temperature, to prevent changes of gas composition and plugging of its lumen by water condensation. However, heating of the sample within the catheter increases axial diffusion, which can effectively filter out high-frequency components of fast-changing concentration waveshapes. An additional difficulty in measurement is introduced by the tendency of water vapor to be adsorbed and desorbed from surfaces within the system. The establishment of an equilibrium between the water molecules on these surfaces and in the moving sample lags behind changes in partial pressures in the sample. Therefore, the gas delivered to the sensor does not represent the gas entering the inlet tube until after that equilibrium is reached. For precision measurements in some applications, the sampled gas is passed through a tube filled with a drying agent. This effectively eliminates water vapor but prolongs the overall response time of the measurement system.

The pH and partial pressures of O_2 and CO_2 dissolved in blood are routinely measured clinically *in vitro* in discrete samples withdrawn from the circulatory system. The instruments used are specifically adapted electrode systems such as those discussed in Section 10.2. Indwelling, intravascular monitoring systems designed to provide continuous measurements of blood gases and pH using electro-optical techniques, as described in Section 10.3, have been introduced. Their commercial and clinical viability has yet to be demonstrated, at least in part, due to the cost of the disposable sensor-tipped catheters.

Devices for measurements in the gas phase vary in complexity and capabilities from those that can continuously detect and analyze several gas species simultaneously (such as the mass spectrometer) to those that can be modified to test for a few gases individually (such as the infrared analyzer) to those sensitive to a particular property possessed by only one gas that is important to the respiratory system (such as the paramagnetic O_2 sensor). Increasing versatility usually entails increasing cost.

MASS SPECTROSCOPY

A *mass spectrometer* is an apparatus that produces a stream of charged particles (ions) from a substance being analyzed, separates the ions into a spectrum according to their mass-to-charge ratios, and determines the relative abundance of each type of ion present. Medical mass-spectrometer systems include the following elements: a sample-inlet assembly, an ionization chamber, a dispersion chamber, and an ion-detection (collector) system (Figure 9.12) (Sodal *et al.*, 1988).

The sample-inlet assembly consists of a heated or unheated capillary tube (approximately 0.25 mm ID) and sample-inlet chamber. Gas is drawn through this system by a rotary pump that reduces the pressure in the inlet chamber to about 10–20 mm Hg (1.3–1.7 kPa) absolute. A small amount of gas in the inlet chamber leaks by diffusion through a porous plug into the ionization chamber, which, along with the dispersion chamber, is evacuated to approximately 10^{-7} mm Hg (10^{-5} Pa) by a high-vacuum, high-capacity pump. A stream of electrons traveling between a heated filament and an anode bombards the

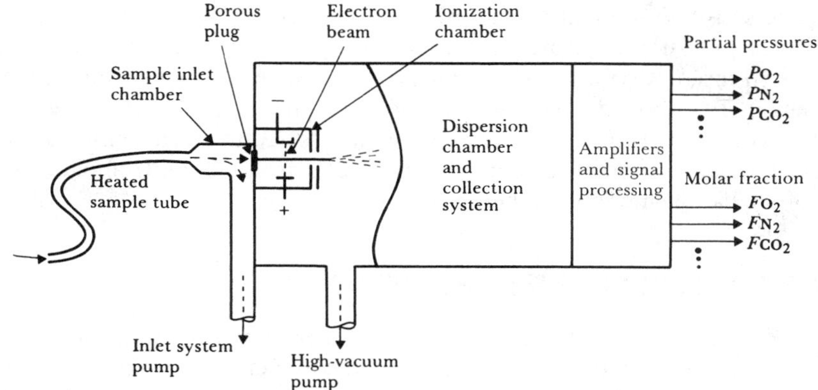

Figure 9.12 **Essential elements of a medical mass spectrometer.**

gas entering the ionization chamber and causes the molecules to lose electrons, thereby producing positive ions. These ions are focused into a beam and accelerated by an electric field into the dispersion chamber, where the ion beam is sorted into its components on a molecular mass basis.

Dispersion techniques incorporated into commercially available systems include a magnetic field, a quadrupole electric field, or measurement of time of flight. The separated ion beams fall on the collector system that produces the output signal of the instrument. A mass spectrometer is of real value in respiratory applications only if it simultaneously produces separate continuous outputs for the several chemical species of interest. This has been achieved in two ways. One method uses a single collector that is swept sequentially by the component beams at a high repetition rate. Individual sample-and-hold circuits, each corresponding to a particular species, register the ion current as the beam for their respective species falls on the collector. The second approach employs multiple collectors, the positions of which can be adjusted so that each continuously receives the ions of only one of the components of interest.

The ion current measured by the collector is proportional to the partial pressure of the corresponding component in the gas mixture. On-line signal processing, in conjunction with an appropriate calibration procedure, makes it possible for the output to be expressed in terms of molar fractions.

The range of molecular weight that is adequate for most respiratory measurements extends from 4 (He) to 44 (CO_2) atomic mass units. Expanded ranges make possible the monitoring of gases such as sulfur hexafluoride (146) and halothane (196). Several important gases produce outputs for the same atomic mass units. In particular, O_2 and CO_2 cannot be measured in the presence of N_2O. Also, CO interferes with N_2. CO and N_2O can be quickly and easily measured with infrared instruments, as described in the following paragraphs.

The sensitivity, linearity, and SNR quoted by manufacturers are appropriate for precision measurements. The response time (2 to 90%) for a step change in input is typically less than 100 ms (except for water vapor, which is longer). Transport delays associated with 1.3-m to 1.6-m inlet catheters are on the order of 200 ms (with sample flowrates of 10 to 30 ml/min). Systems that can measure numerous component gases are available. Compact (desk-top or shelf-top) models are being developed.

THERMAL-CONDUCTIVITY DETECTORS

One of the properties of a gas mixture that changes with composition is its *thermal conductivity.* In general, the thermal conductivity of a gas is inversely related to its molecular weight. For example, H_2 and He have thermal conductivities approximately 6.5 times greater than those of N_2 and O_2. As we noted in Section 9.3, heat transfer between a stationary heated body and a fluid moving past it is related to, among other factors, the velocity, the thermal conductivity, and the temperature of the fluid.

Thermal-conductivity detectors (TCD) have been developed for use in gas chromatography and in instruments designed to analyze gas mixtures for He or H_2. In both applications, heated sensing elements operated in the constant-current mode (Section 8.5) are connected in a Wheatstone bridge. The heated elements can be either thermistors or coiled wires made of a metal with a high temperature coefficient of resistance (such as platinum, tungsten, or nickel). Thermal-conductivity detectors incorporating heated wires are called *katharometers.*

INFRARED SPECTROSCOPY

Various chemical species, whether in the gas phase or in liquid solution, absorb power from specific ranges of the electromagnetic radiation spectrum. The infrared region, spanning wavelengths from 3 to 30 μm, is very useful in the study of gases. This is because most gases absorb infrared "light" and do so only at distinct, highly characteristic wavelengths, thus yielding what has been called a molecular fingerprint (Lord, 1987). The power absorbed is transformed into heat and increases the temperature of the absorbing gas. However, infrared light is absorbed only by molecules made up of dissimilar atoms, because only such molecules possess an electric dipole moment with which the electromagnetic wave can interact. CO_2, CO, N_2O, H_2O, and volatile anesthetic agents are examples of such molecules. Symmetric molecules (such as O_2, N_2, and H_2) and the noble gases (such as He and Ne) do not have an electric dipole moment and do not absorb infrared radiation.

In general, when light of wavelengths characteristic of a particular gas falls on a sample of that gas, only some is absorbed. The remainder is transmitted through the gas. For light of a specific wavelength, the power per unit area transmitted P_t by the sample relative to that entering the sample, P_0, is

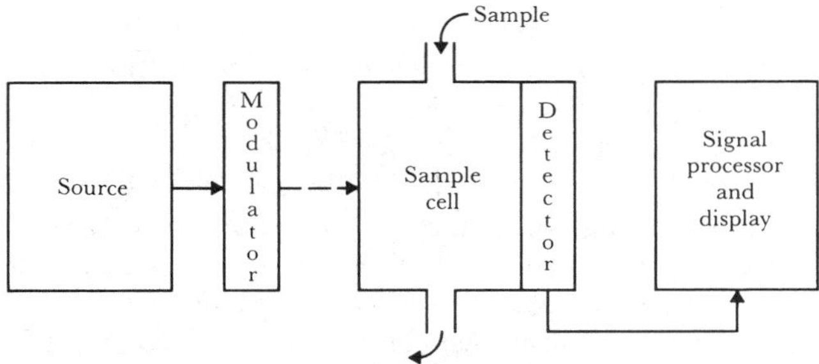

Figure 9.13 General arrangement of the components of an infrared spectroscopy system.

given by Beer's law (we use P for power per unit area to be consistent with Section 11.1):

$$P_t = P_0 e^{-aLC} \tag{9.43}$$

where a is the absorption coefficient, L is the length of the light path through the gas, and C is the concentration of the absorbing gas. Consequently, we can measure the concentration of the components of a gas mixture by determining the power that is either *absorbed* or *transmitted* by the mixture. Instruments based on each of these approaches have been developed to measure gases important in respiration—most notably CO_2, but also CO, water vapor, and anesthetic agents.

The conventional technique that we will refer to here as transmission analysis measures the power *transmitted* at wavelengths corresponding to the substances under study. In contrast, photoacoustic principles have been utilized to measure directly the power *absorbed* by a sample. Both types of instruments have employed wide-band (black-body) sources to irradiate the sample. However, neither produces an entire spectrum of absorption or transmission by which to identify the components of the sample, as "dispersive" spectroscopy does (Lord, 1987). Instead, they measure the behavior at only a well-defined set of wavelengths chosen to maximize the response for the substance of interest and to minimize interference with other substances. Such instruments fall into the category of nondispersive infrared (NDIR) analyzers.

Both transmission systems and photoacoustic systems have a minimum of five essential components, as shown in Figure 9.13: (1) a source of radiation of the required wavelengths; (2) a means, usually a mechanical "chopper," of periodically varying the power and/or wavelength of the source radiation; (3) a sample cell; (4) a detector; and (5) signal processing and display equipment. The relative positions of the components vary from one instrument to another.

Transmission Analysis An NDIR system used to analyze a gas mixture for the presence of a single species of test gas has two identical intermittently interrupted (10 to 90 times per second, depending on the particular instrument) IR beams. The IR power pulses produced travel two parallel paths, one of which includes a test cell. A sample of the gas mixture to be analyzed is continuously drawn through the test cell from a sampling catheter. The second path includes an interference filter, either a thin film filter that transmits selected wavelengths or, in older systems, a reference cell that has windows exactly the same as those of the test cell but containing a gas mixture free of the test gas.

A detector measures the difference between the powers transmitted through the reference pathway and through the test pathway for each pulse of the IR beams. The rms power difference between the two pathways is *approximately* proportional to the concentration of the absorbing gas in the test cell. The output of the detector circuit is demodulated and processed to produce a signal proportional to the concentration (molar density) of the test gas. At flow rates through the sampling catheter on the order of 0.5 to 1 liter/min, a 90% full-scale step-response time of approximately 100 ms can be achieved. However, newer devices are sufficiently small to be used in-line with the gas stream during breathing (Coombes and Halsall, 1988). IR transmission instruments have been developed to measure gases with the following full-scale molar fractions: 10% CO_2, 0.3% CO, 100% N_2O, 7.5% halothane, enflurane, isoflurane and sevoflurane, and 20% desflurane.

Photoacoustic Analysis West *et al.* (1983) define the photoacoustic effect as the process of sound generation in a gas that results from the absorption of photons. They point out that Alexander Graham Bell, among others, described this phenomenon in 1880. Bell was able to produce sound by repeatedly interrupting a beam of sunlight that was focused on a test tube filled with tobacco smoke. The sound pressure waves were caused by the expansion of gas resulting from the absorption of the incident infrared radiation and by the interspersed contraction of the gas when the light source was blocked. This observation has formed the basis for gas analysis, because the radiant energy absorbed by a gas is approximately proportional to the concentration of that gas. Consequently, the higher the gas concentration, the louder the sound for the same input of light.

Figure 9.13 shows the general scheme for a photoacoustic gas analyzer. Broad-band infrared light is modulated by a mechanical chopper and filtered to permit pulses of light of selected wavelengths to be focused on the gas mixture in the test cell. The resulting pressure fluctuations recur at the mechanical chopper's repetition rate, which is in the audio frequency range. An extremely sensitive, stable capacitance microphone detects the sound generated.

The photoacoustic analyzer measures the IR energy absorbed by a test gas by sensing the sound pressure waves produced. The IR energy absorbed by a gas is very small compared to that transmitted. Transmission analyzers determine absorbed energy as a small difference between two large quantities, such as the energy transmitted through the reference cell and that transmitted through the test cell. Consequently, the signal-to-noise ratio of transmission analyzers is inherently lower than that of the photoacoustic analyzer.

Gas analyzers that utilize photoacoustic sensing have found applications in anesthesia monitoring. Instruments have been developed that photoacoustically measure three gases simultaneously: CO_2, N_2O, and one of several volatile anesthetic agents. This is accomplished by using a chopper wheel with three concentric rows of apertures, each row having openings of different spacing and size. Thus, for the same rotational velocity of the wheel, three beams are created, each interrupted at its own frequency. Each beam is filtered so that it contains only wavelengths that will be absorbed by a particular gas of interest. The beams are focused into the cell and simultaneously excite specific constituents of the mixture. Three sounds are produced, each with a characteristic pitch corresponding to one chopping frequency and with an amplitude approximately proportional to the concentration of the gas producing it. The concentration of each gas component of interest can be continuously determined by filtering the sensing microphone's output for its Fourier components at the chopper frequencies and demodulating the resulting amplitude-modulated signals.

Such instruments claim remarkable stability (calibration at 1- to 3-month intervals), 1-min warm-up, high accuracy (less than 1% FS error), and 10–90% response time of 250–300 ms at a 90-ml/min sample flow rate (Møllgaard, 1989).

EMISSION SPECTROSCOPY

Figure 9.14 depicts a device used to detect the concentration of a single gas species in a mixture by measuring the intensity of the light in a given wavelength range produced when the gas mixture is ionized at very low pressures. The respiratory gas routinely measured by such a device is N_2 (East and East, 1988). The system is evacuated by a high-capacity vacuum pump, and the pressure [1 to 4 mm Hg (150 to 550 Pa)] is regulated by a needle valve that allows a small flow of gas to be drawn through the ionization chamber. Respiratory gases ionized by the voltage difference (600 to 1500 V dc) between the electrodes in the ionization chamber emit light in the range of 310 to 480 nm. Reflecting surfaces direct the light through a selective optical filter that absorbs unwanted wavelengths. A photoelectric tube produces a current proportional to the intensity of the light passed by the filter. For a fixed ionization-chamber geometry, vacuum, gas flow, and current, the amplified output of the phototube is a nonlinear function of the molar fraction of the gas species of interest. The output is processed to produce a signal e proportional to this molar fraction.

Spectroscopic N_2 analyzers yield the N_2 molar fraction in a gas mixture on a wet basis. Calibration procedures must take into account the humidity of the sample. The needle valve and conduit from which the sample is with-

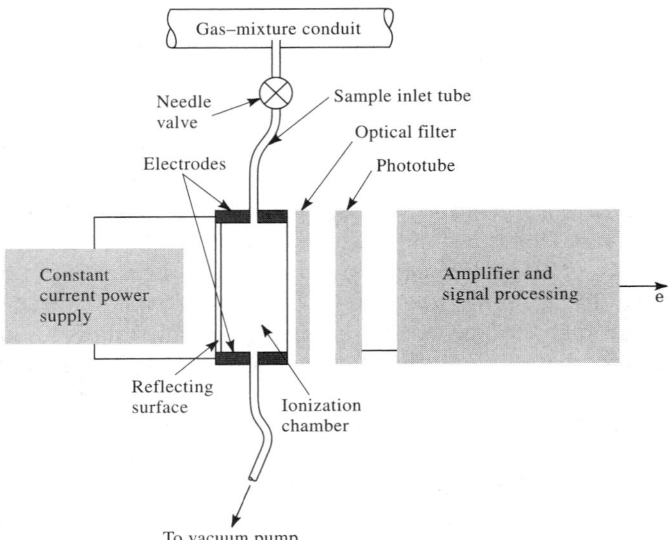

Figure 9.14 N_2 analyzer employing emission spectroscopy.

drawn may have to be heated to prevent plugging or loss of water vapor by condensation. The high vacuum within the system prevents condensation of water vapor in the transport tubing that leads from the needle valve to the ionization chamber, but it may not prevent adsorption and desorption during changes in the partial pressure of the water vapor.

Commercially available N_2 analyzers can produce a steady-state output that has as little as 0.5% rms error over the range of 0 to 80% N_2. In response to a step change in N_2 molar fraction, the transport delay and rise time (between 10% and 90% of the steady-state output) can each be on the order of 40 ms. Variation in output to a constant input can be less than 1.5% over 24 h. Deposition of electrode material on the glass walls of the ionization chamber over a long period of time eventually degrades the instrument's performance. O_2 and CO_2 do not interfere with the accuracy of N_2 determinations. However, He and Ar can produce errors. A mixture containing 40% N_2 and 5% He can produce an output corresponding to up to 43% N_2 from the analyzer.

MEASUREMENT OF OXYGEN CONCENTRATION

Emission and IR absorption instruments cannot measure the concentration of oxygen in gas mixtures. However, oxygen can be measured by mass spectrometers and by fuel cells and galvanic and polargraphic sensors (described in Section 10.2). An additional class of oxygen analyzers is based on the magnetic properties of oxygen.

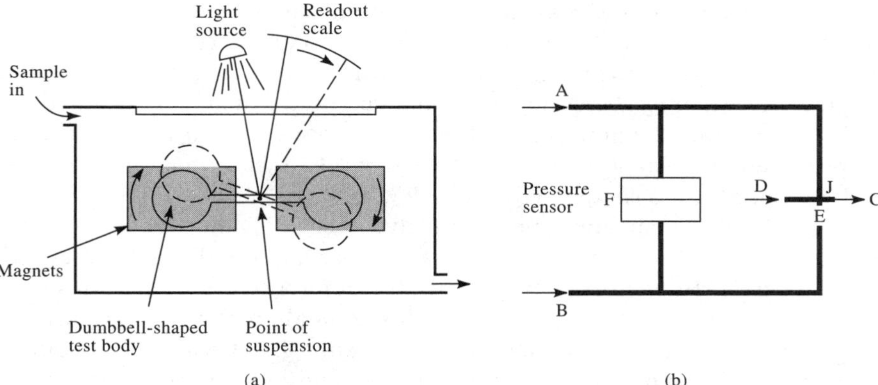

Figure 9.15 Oxygen analyzers (a) Diagram of the top view of a balance-type paramagnetic oxygen analyzer. The test body either is allowed to rotate (as shown) or is held in place by countertorque, which is measured to determine the oxygen concentration in the gas mixture. (b) Diagram of a differential pressure and a magnetoacoustic oxygen analyzer (see text for descriptions).

Paramagnetic Oxygen Sensors Oxygen is unusual among gases in that it is attracted by a magnetic field. This property, referred to as paramagnetism, is characterized by a positive magnetic susceptibility. The high magnetic susceptibility of iron, ferromagnetism, is a special case of paramagnetism (Kocache, 1988). Most gases are diamagnetic, being repulsed by a magnetic field, and thus exhibit negative susceptibilities.

Several instruments exploit the paramagnetic property of oxygen. The first to have been developed is a *balance-type* device that has a diamagnetic, nitrogen-filled, hollow, thin-walled, glass test body situated in a nonuniform magnetic field within a gas-sample chamber. The test body is shaped like a dumbbell [Figure 9.15(a)] and is suspended by a taut fiber attached to the midpoint of the bar connecting its two globes. The fiber acts a a torsional spring and permits the dumbbell to rotate in a plane perpendicular to the magnetic field. When at undisturbed equilibrium, the globes of the test body are positioned in the region of highest magnetic-field concentration—that is, between the poles of the magnet producing the field. When a gas mixture with no paramagnetic components is introduced into the sample chamber, the density of the mixture remains uniform within the chamber, and the dumbbell maintains its equilibrium position. When a paramagnetic gas such as oxygen is in the mixture, however, it is attracted by the field, causing the local gas density between the poles of the magnet to increase. This causes the test body to be displaced in much the same way as a fishing bobber is forced to the surface of the much denser water into which it is cast. The more oxygen, the larger the region of increased density, the higher the density gradient, and the more the force on the test body tending to displace it.

The concentration of oxygen can thus be determined either by measuring the deflection of the test body about its suspension axis (via the reflection of a light beam by a mirror placed on the test body at its center of rotation [Figure 9.15(a)] or by measuring the torque required to maintain the dumbbell in its oxygen-free equilibrium position. This type of system is very accurate (repeatability and linearity are both to within 0.1%), but it has a long (on the order of 10 s) response time (Kocache, 1988).

Another paramagnetic approach involves measuring the pressure required to maintain the flow of an oxygen-containing gas mixture through a magnetic field. In one instrument [Figure 9.15(b)], two reference gas streams meet and mix at a junction point J and exit through a common pathway C. A pulsating magnetic field is focused on one of the reference gas streams at its point of entry E into the junction. The sample to be analyzed is introduced through D into the junction, where it mixes with the reference gases and flows through the exit C. Any oxygen in the gas sample is attracted by the magnetic field at E and increases the local density in the outlet of that reference pathway, retarding flow by effectively increasing flow resistance and causing an increased back-pressure in that branch. This produces a pressure difference between the two reference pathways, and that difference is sensed by a differential pressure sensor at F. The output of the sensor is proportional to the difference in Po_2 between the reference gas and the sample gas (East, 1990). Their performance is slightly inferior to that of paramagnetic balance-type analyzers, but the *differential pressure-type analyzers* display a response time of under 200 ms, which is adequate for breath-by-breath measurements (Kocache, 1988).

A variation of the differential pressure analyzer is used in a commercial instrument in conjunction with the photoacoustic spectrometer we have described (Møllgaard, 1989). Termed *magnetoacoustic spectroscopy,* this technique employs the same general configuration shown in Figure 9.15(b), with a few modifications. The sample is introduced directly into one of the pathways, such as at A, that leads to the mixing junction J. (D is not used.) An alternating magnetic field with a frequency in the audio range is imposed across the junction J, not just across the end of one of the pathways. Consequently, the gas from both pathways is exposed to the same magnetic field. If the sample introduced at A contains some oxygen, then when it reaches J, the oxygen portion of the mixture is alternately expanded and compressed, causing acoustic waves to propagate through the two gas pathways. The amplitude of these sound pressure waves is proportional to the concentration of oxygen in the mixture. Finally, instead of using a single differential pressure sensor to measure the pressure difference between the two pathways, this system has two capacitance microphones to measure the sound in the sample gas and that in the reference gas separately. The sample-gas microphone is the same one used to measure the photoacoustic disturbances caused by the absorption of IR energy by other components in the sample gas. The alternating magnetic-field frequency is different from the frequencies of the modulated IR beams exciting these other gas components. Thus the magnetically generated sound pressure waves can be digitally bandpass-filtered from the output of the sample-gas microphone and compared to the signal emanating from the reference gas to determine the absolute concentration of oxygen in the sample gas. The resulting instrument is stable (gain drift: $< 2\%$ full scale/30 days, temperature drift $< 0.1\%$ of reading/°C); is accurate (better than 1% full scale or 2% of reading) and has a 10 to 90% rise time of less than 250 ms at a sample flow rate of 90 ml/min.

9.8 SOME TESTS OF GAS TRANSPORT

The pulmonary function tests to be discussed in this section are concerned primarily with gas-phase transport beween the airway opening and the alveoli and with interphase (or membrane) transport between alveolar gas and pulmonary capillary blood. Gas-transport tests are designed to achieve one or more of the following objectives: (1) Determine the homogeneity of the distribution of inspired gas (ventilation). (2) Determine the matching of ventilation to perfusion. (3) Evaluate the ability of the alveolar membrane to allow gas transfer. Because of the architecture of the respiratory system, it is usually impossible to isolate the processes involved or to make direct measurements at sites of interest. Access is available only at the boundaries of the system—that is, at the airway opening and the systemic circulation.

An overall evaluation of gas transport by the lungs can be obtained from a measurement of the partial pressures of O_2 and CO_2 in a sample of systemic arterial blood drawn while the subject is breathing air. If the gas-exchange ratio $\dot{V}CO_2/\dot{V}O_2$ is approximately 0.8, then the sum of the arterial partial pressures, P_aO_2 and P_aCO_2, should be approximately 140 mm Hg (18.7 kPa). If this sum is below 120 mm Hg (16 kPa), it is considered abnormal. Because the acquisition of these data requires puncturing an artery, it is not used routinely as a pulmonary-function test, especially for children.

Further discussion of arterial blood gases is limited here to the observation that they can be used in a procedure to distinguish between two possible causes of abnormally low partial pressures: (a) the shunting of blood past the gas-exchange regions in the lung and (b) a mismatch between local alveolar ventilation and blood perfusion. Besides a sample of arterial blood gas, the procedure requires an estimate of the mean PO_2 in the alveoli. This is obtained from an end-expired gas sample collected as the subject exhales to RV. If the steady-state alveolar–arterial (A–a) difference in O_2 partial pressure does not appreciably change when the subject breathes 100% O_2 instead of air, then a shunt is assumed to be present. If the A–a difference decreases, then a ventilation–perfusion mismatch is implied.

GAS-PHASE TRANSPORT

Tests of gas-phase transport are concerned with questions arising from two of the objectives mentioned above: (1) How is the inspired gas (ventilation) distributed in the lungs? (2) What is the equivalent volume of inspired gas that is not taking part in the exchange of gas with the blood? In other words, what is the effective dead space of the lung?

We can assess the distribution of gas in the lungs from multibreath and single-breath maneuvers by using a tracer gas that is insoluble in the pulmonary tissues and blood. The assumption can then be made that the tracer gas can enter or leave the lung only through the airway opening. Gases that fulfill this requirement are He, N_2, Ar, Ne, and Xe.

Multibreath N_2 Washout The multibreath He-dilution and N_2-washout procedures used to estimate lung volume (FRC) as described in Section 9.4 can also provide information about the efficiency of gas mixing in the lungs: whether the gas in all ventilated alveoli is diluted at the same rate by gas inspired during resting breathing. As an example, consider the washout of N_2 from the lungs during resting breathing of 100% O_2. A set of one-way valves, as shown in Figure 9.7, is used to prevent mixing of the inspired O_2 with the expirate. If a pulmonary system were to act as a single, well-mixed compartment ventilated at a constant breathing rate and tidal volume, a time plot of the end-tidal expired-nitrogen molar fraction $F\mathrm{N}_2$ would exhibit an exponential decay. Normal lungs exhibit a washout curve that can be approximated rather closely by a single exponential. The shape of the washout curve is not routinely used as a test of abnormal ventilation because it is difficult to compensate for the effects of variations in tidal volume and breathing frequency on the shape of the curve. However, abnormality is indicated if the $F\mathrm{N}_2$ of a sample of gas obtained at the end of an expiration to RV after 7 min of O_2 breathing has not been reduced to 0.02. This procedure is obviously highly dependent on cooperation, because it is sensitive to changes in the tidal volume and frequency of breathing of the subject.

Single-Breath N_2 Washout The single-breath N_2 washout can yield as many as three useful pieces of information: (1) an estimate of the anatomical dead space (approximately the volume of the conducting airways); (2) a measure of the distribution of ventilation—that is, the relative local rates of filling and emptying the regions of the lungs; and (3) an index of small-airway mechanical function, the closing volume CV.

The procedure requires that, after having reached a steady state while breathing air, the subject inspires 100% O_2 from RV to TLC. A setup similar to Figure 9.7 can be used for this experiment. After a momentary pause at TLC, the individual is instructed to exhale very slowly to residual volume. The $F\mathrm{N}_2$ in the expired gas, $F_E\mathrm{N}_2$, and the volume of the expired gas are continuously measured and displayed against each other on an x–y plot. The interpretation of these plots requires a discussion of the events that occur in the lungs during both a vital-capacity inspiration and a slow vital-capacity expiration.

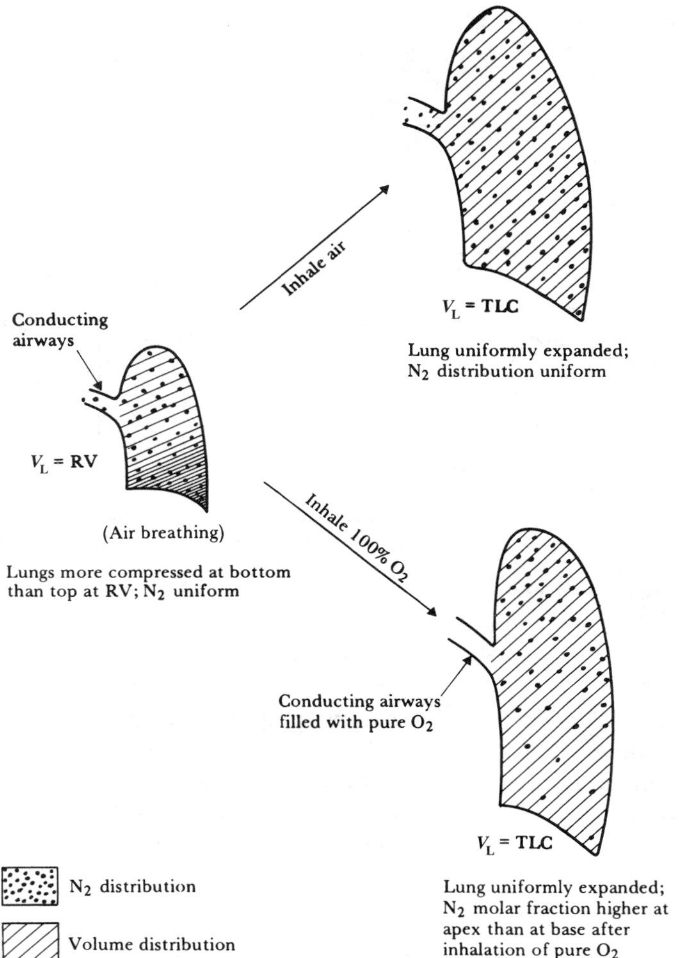

Figure 9.16 Distributions of volume and gas species at RV and TLC for a vital-capacity inspiration of air or pure oxygen.

Consider the idealized normal upright lung (subject sitting or standing erect) (Figure 9.16) At residual volume, the *dependent* parts of the lungs (those lowest with respect to gravity) are less expanded than the upper regions. Thus there is a distribution of local alveolar volumes corresponding roughly to the distribution of the pressure on the interpleural surface of the lungs. That is, there is lower pressure at the top of the lungs (larger alveoli) and higher pressure at the base (smaller alveoli).

After the lung has been expanded to and held at TLC, the lung is more uniformly inflated, with all regions having the same relative local volume. This means that the top parts of the lungs must have changed their volumes

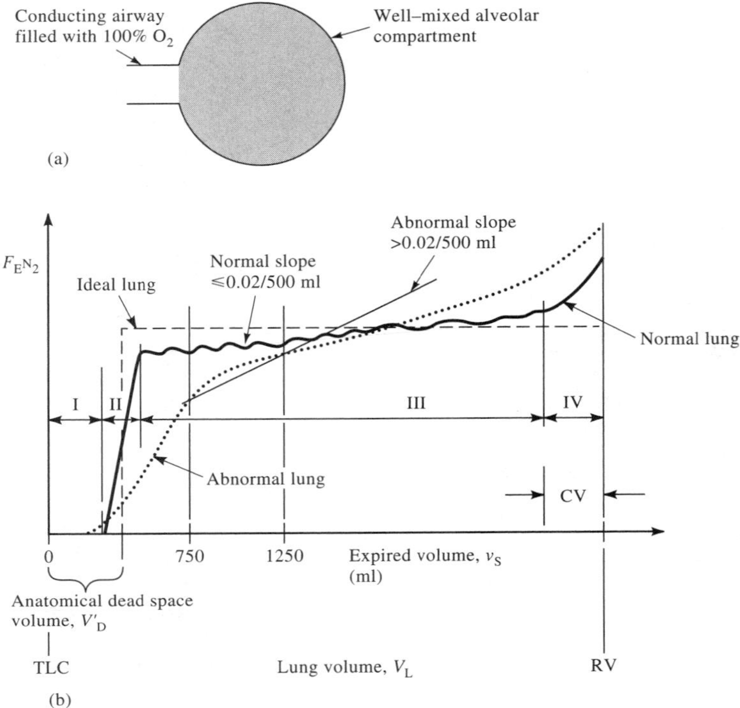

Figure 9.17 Single-breath nitrogen-washout maneuver (a) An idealized model of a lung at the end of a vital-capacity inspiration of pure O_2, preceded by breathing of normal air. (b) Single-breath N_2-washout curves for idealized lung, normal lung, and abnormal lung. Parameters of these curves include anatomical dead space, slope of phase III, and closing volume.

less during a vital-capacity inspiration of 100% O_2 than did the lower, dependent parts in order for the two regions to reach approximately the same final volume at TLC. Consequently, less inspired O_2 was added to the upper regions than to the lower. If the subject had achieved steady state while breathing air prior to performing this test, then it can be assumed that the F_{N_2} in the gas within the lungs was the same throughout. Thus, because less diluting inspired O_2 is added to the upper regions than to the lower during a VC inspiration of pure O_2, the F_{N_2} in the upper regions is higher than in the lower regions at the end of the inspiration. Also, at TLC, the conducting airways (anatomical dead space) are filled with pure O_2.

For a slow vital-capacity expiration, following the inspiration of pure O_2, normal lungs exhibit four identifiable regions or phases on the F_{EN_2}-versus-volume plot [Figure 9.17(b)]. During phase I, lung volume changes but F_{EN_2} is zero as pure O_2 is emptied from the conducting airways past the N_2 sensor. Phase III is a (sloping) plateau corresponding to the emptying of the mixed alveolar gas from all parts of the lungs. Oscillations of $F_{\mathrm{E}}\mathrm{N}_2$ during phase III have been attributed to local emptying of areas of different $F\mathrm{N}_2$ caused by the changes in volume associated with the heartbeat. Phase II is the transition between the emptying of the anatomical dead space and the arrival of mixed alveolar gas at the N_2 sensor. Phase IV signifies a drastic change in the rates of emptying of the dependent parts of the lungs. It is dramatically evident if the lung is normal or has only minimal small-airways disease.

A system consisting of a single, perfectly mixed compartment and a transport dead space [Figure 9.17(a)] would exhibit a washout curve like that in Figure 9.17(b) (dashed line). By analogy, the anatomical dead space V'_{D} is taken as the volume emptied to the point at which the $F_{\mathrm{E}}\mathrm{N}_2$ reaches one-half of a representative mixed alveolar value. For conditions in the spirometer in Figure 9.7, we can compute

$$V'_{\mathrm{D}} = \mathrm{VC} - \frac{\int_0^{\mathrm{VC}} F_{\mathrm{E}}\mathrm{N}_2 dv_{\mathrm{S}}}{\hat{F}_{\mathrm{E}}\mathrm{N}_2(\mathrm{III})} \tag{9.44}$$

where

VC = expired vital capacity

v_{S} = volume change of spirometer used to estimate the integral of expired flow

$\hat{F}_{\mathrm{E}}\mathrm{N}_2(\mathrm{III})$ = mean value for $F_{\mathrm{E}}\mathrm{N}_2$ during phase III

Normal lungs exhibit an average slope during phase III that does not exceed a 0.02-increase in $F_{\mathrm{E}}\mathrm{N}_2$ per 500 ml of expired volume for an interval of expired volume between 750 and 1250 ml below TLC. A greater increase than this is taken to indicate a nonuniformity of local alveolar emptying rates.

The volume above RV at which phase IV begins is designated the *closing volume* CV. The absolute volume at which phase IV begins is the *closing capacity,* CC = RV + CV. As the lung volume approaches RV, the dependent gas spaces in the lung begin to cease emptying at the same rate at which they emptied during phase III. Therefore, the expirate at the airway opening contains a higher proportion of gas from the superior parts of the lungs and less from the dependent portions. Because the upper parts of the lung had the higher $F\mathrm{N}_2$ after the vital-capacity inspiration of pure O_2, the expirate passing the N_2 sensor at the airway opening has an increasing $F_{\mathrm{E}}\mathrm{N}_2$ as an increasing number of dependent lung units decrease their emptying rate as the lung approaches RV.

Inferences about the mechanical state of the small airways in the dependent parts of the lungs have been made from the size of the CV relative to the VC and of the CC relative to TLC. Closing volume varies with body position, age, and disease. In the initial phases of small-airways disease, CV increases. However, as the disease progresses, the slope of phase III increases and the transition between phases III and IV becomes less distinct [Figure

9.17(b)]. The resulting ambiguity in determining the onset of phase IV limits the use of the CV, except as an indicator of the beginnings of abnormalities of the small airways.

EFFECTIVE (PHYSIOLOGIC) DEAD SPACE DURING ALVEOLAR–CAPILLARY GAS EXCHANGE

When inspired gas cannot reach alveoli with well-perfused capillaries, and when blood cannot perfuse capillaries of alveoli that are ventilated with inspired gas, gas exchange is impaired. A tracer gas that is not diffusion-limited by the alveolar membrane can be used to evaluate the matching of ventilation and perfusion. The CO_2 added to the gas in the lungs from the pulmonary capillary blood satisfies this requirement. It diffuses so freely that it can be assumed that the alveoli and capillaries are in equilibrium during the entire expiration: that their respective CO_2 partial pressures are equal.

In the absence of significant right-to-left heart shunt, systemic arterial CO_2 partial pressure provides an excellent estimate of an average pulmonary capillary P_{CO_2}. Consequently, a conceptual *alveolar volume* V_A can be defined as the equivalent volume of gas in the lung that communicates with the airway opening and satisfies two conditions: (1) It contains all the CO_2 in the lungs at the end of inspiration. (2) This CO_2 exists at a partial pressure equal to that of systemic arterial blood. The *effective dead space* V_D, sometimes called the *physiologic dead space,* is the remainder of the gas space in the lungs that is in excess of the alveolar volume. If we consider that none of the CO_2 expired during a breath is contributed by the dead space—that all of it is coming from the alveolar volume—then we can obtain an estimate of the volume of the effective dead space as a fraction of the expired tidal volume V_T from a mass balance of expired CO_2. Thus

$$\frac{V_D}{V_T} = 1 - \frac{\hat{P}_{ECO_2}}{\hat{P}_{aCO_2}} \tag{9.45}$$

where $\hat{P}_{ECO_2}$ is the mean CO_2 partial pressure in the expirate, and $\hat{P}_{aCO_2}$ is the mean partial pressure of systemic arterial CO_2, each determined over several breaths.

The V_D/V_T is important in inferring the etiology of diseases involving poor gas exchange.

DIFFUSION PROCESSES

To complete the description of gas exchange in the lungs, we need to characterize the transport processes that occur between the alveolar gas space and the pulmonary capillary blood. Gas moves between these regions by diffusion. A parameter used to describe the diffusion processes in the vicinity of the alveolar membrane is the *diffusing capacity of the lung* D_L. This is the constant of proportionality between the rate of uptake of tracer gas by the blood (obtained from measurements at the airway opening) and the difference of partial pressure of the tracer gas between the alveolar gas space and the capillary blood. Because it is impossible to measure alveolar partial pressures directly, and because they are affected by convective transport between the airway opening and the alveoli, diffusing capacity is not independent of the pattern of ventilation distribution in the lungs.

Carbon Monoxide Diffusing Capacity CO_2 diffuses across the alveolar membrane much more easily than O_2, so a diffusion defect would affect O_2 transfer first. Because of this and the prime role O_2 plays in sustaining life, it is important to evaluate a diffusing capacity for O_2. However, obtaining the required $P_{a}O_2$ requires a sample of arterial blood. To avoid this, clinicians use CO as the tracer gas because its properties are sufficiently close to those of O_2 for its diffusing capacity, $D_{L}CO$, to provide a meaningful estimate of $D_{L}O_2$. In addition, because of its affinity for hemoglobin (at low concentrations essentially all CO that enters the blood chemically combines with the hemoglobin in the red blood cells), PCO exhibited by the blood is negligibly small and need not be measured.

One of the various methods that have been used to estimate $D_{L}CO$ involves a single-breath maneuver. The subject inspires a mixture of air, 0.3% CO (or less), and He (approximately 10%) from RV to TLC. The subject holds his or her breath at TLC for about 10 s and then forcefully exhales down to RV. Even though it necessitates subject cooperation, this procedure can be performed quickly and can be repeated easily. In addition, it does not require samples of arterial blood or an estimate of dead space before the clinician can obtain a value of alveolar $P_{A}CO$. The computation of $D_{L}CO$ from measurements made during this single-breath maneuver are based on a one-compartment model of the lung. If a well-mixed alveolar compartment is filled with a mixture of gases containing some initial $F_{A}CO$, then during breath-holding with the airway open, the CO diffuses into the blood in the pulmonary capillaries, and the alveolar $F_{A}CO$ decreases exponentially with time.

$$F_{A}CO(t_2) = F_{A}CO(t_1) \times \exp\left[-\frac{D_{L}CO(P_{atm} - P_{A}H_2O)(t_2 - t_1)}{V_A}\right] \tag{9.46}$$

where V_A is the equivalent volume of the alveolar gas space throughout which the inspired CO is assumed to be distributed; and t_1 and t_2 are the times corresponding to the end of the inspiration to TLC and the beginning of expiration to RV, respectively. That is, $t_2 - t_1$ is the duration of the breath-holding.

The fraction of alveolar CO at the end of the breath-holding, $F_{A}CO(t_2)$, is taken as that of the gas expired at the end of the expiration to RV. The FHe of this end-expiratory gas is also measured. Because the inspired He is insoluble in the lung tissues and blood, none of it should have left the lung during the breath-holding. It can be assumed that, during inspiration, both the He and the CO were similarly distributed throughout the lungs and that the dilution of the He in the alveoli at the end of inspiration is the same as that for the CO. The end-inspiratory $F\text{He}$ should not change during breath-holding, so it can be estimated from the measured end-expiratory value, $F_{\text{EE}}\text{He}$. The end-inspiratory alveolar $F_{\text{A}}\text{CO}$ can be estimated from

$$\frac{F_{\text{A}}\text{CO}(t_1)}{F_{\text{I}}\text{CO}(t_0)} = \frac{F_{\text{A}}\text{He}(t_1)}{F_{\text{I}}\text{He}(t_0)} = \frac{F_{\text{EE}}\text{He}}{F_{\text{I}}\text{He}(t_0)} \tag{9.47}$$

where $F_{\text{I}}\text{CO}(t_0)$ and $F_{\text{I}}\text{He}(t_0)$ are the fractions of CO and He, respectively, in the inspired gas. In addition, the equivalent alveolar volume to which the gas is distributed can be estimated using a mass balance on the He:

$$V_{\text{A}} \cdot F_{\text{EE}}\text{He} = \text{VC} \cdot F_{\text{I}}\text{He}(t_0) \tag{9.48}$$

Diffusing capacity depends on many factors besides the distribution of ventilation to the alveoli and the properties of the alveolar membrane. The distribution of pulmonary perfusion, cardiac output (flow rate), and volume of blood in the pulmonary capillaries also affect it. These, in turn, are related not only to disease processes but also to exertion during exercise or excitement and to the patient's body position when the measurements are made. $D_{\text{L}}\text{CO}$ also varies with hematocrit and type of hemoglobin in the blood. The American Thoracic Society has recommended a standard technique for the single breath $D_{\text{L}}\text{CO}$ and has extensively discussed practical considerations for obtaining reproducible, consistent values (Anonymous, 1995b).

PROBLEMS

9.1 State three reasons why abstract models are important in respiratory physiology, pulmonary function testing, and patient monitoring.

9.2 For a single mechanical unit lung, assume that the relationship among pressure, volume, and number of moles of ideal gas in the lung is given by

$$P_{\text{A}}\left(\frac{V_{\text{L}}}{N_{\text{L}}}\right)^{\alpha} = K$$

where $\alpha = 1$ and K is a constant. Derive the lowest-order (linear) approximation to the relationship among changes in pressure, changes in volume, and changes in moles of gas within the lung.

9.3 Define the acoustic compliance of a gas mixture as

$$C_{\text{g}} = -\frac{\partial V}{\partial P}$$

Using the relationship given in Problem 9.2, evaluate the acoustic compliance, in liter/cm H_2O, for a lung the volume of which is 2 liters with an ideal-dry-gas alveolar pressure of 713 mm Hg (95 kPa).

9.4 Assuming zero net gas exchange with the blood and using (9.2), (9.3a, b, and c), and the results of Problems 9.2 and 9.3, draw an analogous equivalent circuit model for the lung represented by Figure 9.2(a) with gas compression included.

9.5 Gas concentration is interchangeably expressed in units of mass density γ (such as mg/m^3), and molar fraction F (such as parts per million, ppm). For a given gas at 1 atm and 25 °C, mass density in mg/m^3 can be converted to molar fraction in ppm as follows:

$$\gamma\,(\text{mg/m}^3) \times \frac{24.44}{\text{MW}} = F\,(\text{ppm})$$

where MW is the gram molecular weight of the gas in question. Derive this equation from the ideal gas law, using the fact that 1 mole of gas occupies 24.44 dm^3 at 1 atm and 25 °C.

9.6 Show that the output of an instrument that is related to test-gas molar density is related in the same way to test-gas molar fraction (on a wet-gas basis) only if the ratio of the pressure of the total gas mixture to its absolute temperature remains constant.

9.7 Assume that the chest-wall static compliance is 0.31 liter/cm H_2O and that the pulmonary static compliance is 0.19 liter/cm H_2O when the ventilatory system described by Figure 9.2 is at its resting volume (FRC). Evaluate the static compliance for the total respiratory system.

9.8 A Fleisch pneumotachometer has 100 capillary tubes, each with a diameter of 1 mm and a length of 5 cm. What pressure drop occurs for a flow of 1 liter/s?

9.9 Discuss the effects (qualitative) of the placement of a container filled with 6-mm-diameter beads of soda lime in (a) the inlet side of a spirometer and (b) the outlet side of a spirometer, on the dynamics of the spirometer response and the accuracy of continuous measurements of changes in lung volume (1) during very slow maneuvers (in the limit) from one static volume to another; and (2) during very fast maneuvers, such as a forced-vital-capacity expiration—that is, a maximal-effort exhalation from TLC down to RV.

9.10 Evaluate the effect on an estimate of FRC from (9.23) when the experiment is terminated at a lung volume other than FRC. Evaluate the effect on an estimate of FRC from (9.24) when the final volume of the spirometer does not equal its original volume.

9.11 Discuss the subject of accuracy of measurement, especially of flow rate, by a single instrument over the wide range required for the FVC and N_2-washout experiments.

9.12 In an N_2-washout experiment, the subject's cumulative expired volume into a spirometer is 5 liters. At the beginning of the experiment, the spirometer has a volume of 7 liters but contains no N_2. At the end of the experiment, the molar fraction of N_2 in the spirometer is 0.026 and the F_{AN_2} of the subject has decreased by 0.1. The final temperature of the spirometer is 303 K. What was the lung volume at which the subject was breathing?

9.13 Derive an expression for TLC, using static mass balances on the lungs between the beginning and end of a slow inspiration of O_2 from RV to TLC, and between the subsequent TLC and the end of the slow expiration to RV at the finish of the N_2-washout test. Discuss the accuracy of this estimate.

(Answer):

$$\mathrm{TLC} = \frac{\mathrm{VC_I}\, F_{\mathrm{A}}\mathrm{N_2}(t_0) - V'_{\mathrm{D}}\, F_{\mathrm{A}}\mathrm{N_2}(t_1)}{F_{\mathrm{A}}\mathrm{N_2}(t_0) - F_{\mathrm{A}}\mathrm{N_2}(t_1)}$$

in which

$$F_{\mathrm{A}}\mathrm{N_2}(t_1) = \frac{\int_{t_1}^{t_2} F_{\mathrm{E}}\mathrm{N_2} Q_{\mathrm{AWO}}\, dt}{\mathrm{VC_E} - V'_{\mathrm{D}}}$$

where $\mathrm{VC_I}$ is the volume inspired from RV at t_0 to TLC at t_1, and $\mathrm{VC_E}$ is the volume expired from TLC at t_1 to RV at t_2. $F_{\mathrm{E}}\mathrm{N_2}$ is the instantaneous expired N_2 molar fraction measured at the AWO between t_1 and t_2. $F_{\mathrm{A}}\mathrm{N_2}(t_0)$ is the N_2 molar fraction in the alveolar gas before the test began (t_0), and V'_{D} is the anatomical dead space. Which terms must be estimated and which are accessible for measurement?

9.14 In an He-dilution experiment, a spirometer is preloaded with 10 liters of 5% He at room temperature, 25 °C. After the patient has rebreathed, the He concentration in the spirometer is 4%. What is the FRC? [Assume $T_{\mathrm{S}}(t_2) = 305$ K.]

9.15 In a 1000-liter body plethysmograph, a 100-liter person blows into a pressure sensor, raising mouth pressure 30 cm H_2O (3 kPa). The pressure in the box drops 0.1 cm H_2O (10 Pa). Calculate the person's lung volume, assuming that $\alpha_{\mathrm{B}} = 1.4$ and that atmospheric pressure is 760 mm Hg (101 kPa).

9.16 Derive the governing equation for the TBP for breathing within the box, including nonnegligible exchange of gases between the alveoli and pulmonary capillary blood.

9.17 Derive the governing equation for the TBP for breathing within the box for a lung characterized by two alveolar compartments (Figure P9.1). Assume that the net exchange of gases through the alveolar–capillary membrane is zero.

9.18 We want to measure the tissue density of a human being, from which we can estimate the volume of body fat as required in metabolic studies. Describe a dry method, using three steps. (a) Measure the total volume of the body. (b) Measure the volume of the lungs. (c) Make an additional measurement, then calculate tissue density.

9.19 Name three instruments for monitoring changes in thoracic volume, and cite the physical dimension from which each infers volume change.

9.20 Describe the experimental tests that yield a measure of the obstruction of the large airways.

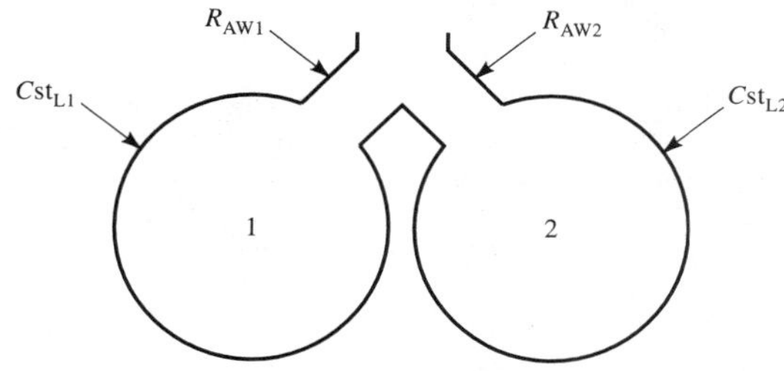

Figure P9.1

9.21 Describe three tests that yield measures of obstruction in small airways.

9.22 Design an experiment that requires no surgery and measures *in vivo* the mechanical time constant ($\tau = RC$) of the lungs of an anesthetized, paralyzed animal (assume normal lungs).

9.23 Describe an experiment that could be used to estimate R_L from (9.38). Why would the estimate of R_L be expected to differ from that obtained from (9.37)?

9.24 Assume that an abnormal pulmonary system can be represented as two airway-alveolar-compartment units (negligible gas compression) connected in parallel (Figure P9.1). Derive expressions for the effective compliance $Ceff_L$ and effective resistance $Reff_L$ for voluntary breathing frequencies.

9.25 Which gases can be measured by IR absorption spectroscopy?

9.26 Using Beer's Law (9.43), show that the power absorbed by a component of a gas mixture is approximately proportional to the concentration of that component in the mixture.

9.27 An instrument requires 10 ml/min of flow to analyze a gas properly. Assuming a blunt flow profile, and given that the transit delay time of a sample is to be 200 ms, what is the relationship between length and diameter for an inlet tube?

9.28 Name three instruments that can be used to measure oxygen in gas mixtures in respiratory experiments and that do not involve chemical reactions. On which physical phenomenon or property is each based?

9.29 Sketch a normal single-breath N_2-washout curve and explain why it has this shape.

9.30 Derive and discuss (9.44) and (9.45).

9.31 Derive (9.46) and describe how you could evaluate diffusing capacity of the lung.

REFERENCES

Anonymous, "Standardization of spirometry—1994 update." *Amer. J. Respir. Crit. Care Med.*, 1995a, 152, 1107–1136.

Anonymous, "Single-breath carbon monoxide diffusing capacity (transfer factor), recommendations for a standard technique—1995 update." *Amer. J. Respir. Crit. Care Med. Dis.*, 1995b, 152, 2185–2198.

Anonymous, "Adult Star ventilator. Operating instructions." Form 9910142, Rev. C(2/91), 1991.

Broughton, R. J., "Polysomnography: Principles and applications in sleep and arousal disorders," in E. Niedermeyer and F. L. da Silva (eds.), *Electroencephalography: Basic Principles, Clinical Applications, and Related Fields,* 3rd ed., Baltimore: Williams & Wilkins, 1993.

Buess, C., U. Boutellier, and E. A. Koller, "Pneumotachometers," in J. G. Webster (ed.), *Encyclopedia of Medical Devices and Instrumentation.* New York: Wiley, 1988, pp. 2319–2324.

Chatburn, R. L., and F. P. Primiano, Jr., "Mathematical models of respiratory mechanics," in R. L. Chatburn and K. C. Craig (eds.), *Fundamentals of Respiratory Care Research.* Norwalk, CT: Appleton & Lange, 1988, pp. 59–100.

Christensen, J., "The Brüel & Kjaer photoacoustic transducer system and its physical properties." *Brüel & Kjaer Technical Review,* 1990, 1, 1–39.

Coombes, R. G., and D. Halsall, "Carbon dioxide analyzers," in J. G. Webster (ed.), *Encyclopedia of Medical Devices and Instrumentation.* New York: Wiley, 1988, pp. 556–564.

DuBois, A. B., S. Y. Botelho, G. N. Bedell, R. Marshall, and J. H. Comroe, Jr., "A rapid plethysmographic method for measuring thoracic gas volume. A comparison with a nitrogen-washout method for measuring functional residual capacity in normal subjects." *J. Clin. Inv.,* 1956a, 35, 322–326.

DuBois, A. B., S. Y. Botelho, and J. H. Comroe, Jr., "A new method for measuring airway resistance in man using a body plethysmograph: Values in normal subjects and in patients with respiratory disease." *J. Clin. Inv.,* 1956b, 35, 327–336.

East, T. D., "What makes noninvasive monitoring tick? A review of basic engineering principles." *Respiratory Care,* 1990, 35, 500–519.

East, T. D., and K. A. East, "Nitrogen analyzers," in J. G. Webster (ed.), *Encyclopedia of Medical Devices and Instrumentation.* New York: Wiley, 1988, pp. 2052–2058.

Finucane, K. E., B. A. Egan, and S. V. Dawson, "Linearity and frequency response of pneumotachographs." *J. Appl. Physiol.,* 1972, 10(2), 210–214.

Fleisch, A., "Der Pneumotachograph: ein Apparat zur Beischwindigkeitregstrierung der Atemluft." *Arch. Ges. Physiol.,* 1925, 209, 713–722.

Hänninen, H., "Continuous patient spirometry during anesthesia," Presented at the 1st European Conference on Biomedical Engineering, Nice, France. February 17–20, 1991.

Hyatt, R. E., "Forced expiration," Chapter 19 in P. T. Macklem and J. Mead (eds.), *Handbook of Physiology,* Section 3: "The Respiratory System," Vol. III: *Mechanics of Breathing,* Part 1. Bethesda, MD: American Physiological Society, 1986, pp. 295–314.

Jones, W., "Characteristics of the disposable transducer/mouthpiece for the new Jones Satellite spirometer." Product Literature, 1990, Jones Medical Instrument Co., Oakbrook, IL.

Kocache, R., "Oxygen analyzers," in J. G. Webster (ed.), *Encyclopedia of Medical Devices and Instrumentation.* New York: Wiley, 1988, pp. 2154–2161.

Kreit, J. W., and F. C. Sciurba, "The accuracy of pneumotachograph measurements during mechanical ventilation." *Am. J. Respir. Crit. Care Med.,* 1996, 154, 913–917.

Kryger M. H., "Monitoring respiratory and cardiac function," in M. H. Kryger, T. Roth, and W. C. Dement (eds.), *Principles and Practice of Sleep Medicine,* 2nd ed., Philadelphia: W. B. Saunders, 1994.

Ligas, J. R., and F. P. Primiano, Jr., "Respiratory mechanics," in J. G. Webster (ed.), *Encyclopedia of Medical Devices and Instrumentation.* New York: Wiley, 1988, pp. 2550–2573.

Lord, R. C., "Infrared spectroscopy," in *McGraw-Hill Encyclopedia of Science and Technology,* 6th ed. New York: McGraw-Hill, 1987, Vol. 9, pp. 162–167.

Macklem, P. T., *Procedures for Standardized Measurements of Lung Mechanics.* Bethesda, MD: NHLI, Division of Lung Diseases, 1974.

Macklem, P. T., "Symbols and abbreviations," in P. T. Macklem and J. Mead (eds.), *Handbook of Physiology,* Section 3: "The Respiratory System," Vol. III: *Mechanics of Breathing,* Part 1. Bethesda, MD: American Physiological Society, 1986, p. ix.

Miller, W. F., R. Scacci, and L. R. Gast, *Laboratory Evaluation of Pulmonary Function.* Philadelphia: J. B. Lippincott, 1987.

Møllgaard, K., "Acoustic gas measurement." *Biomed. Instrum. Technol.,* 1989, 23, 495–497.

Peslin, R., J. Morinet-Lambert, and C. Duvivier, "Frequency response of pneumotachographs." *Bull. Physio-path. Resp.,* 1972, 8, 1363–1376.

Petrini, M. F., "Pulmonary function testing," in J. G. Webster (ed.), *Encyclopedia of Medical Devices and Instrumentation.* New York: Wiley, 1988, pp. 2379–2395.

Primiano, F. P., Jr., "A conceptual framework for pulmonary function testing." *Ann. Biomed. Eng.,* 1981, 9, 621–632.

Primiano, F. P., Jr., and R. L. Chatburn, "The use of models in pulmonary physiology," in R. L. Chatburn and K. C. Craig (eds.), *Fundamentals of Respiratory Care Research,* Norwalk, CT: Appleton and Lange, 1988, pp. 39–57.

Primiano, F. P., Jr., and I. Greber, "Analysis of plethysmographic estimation of alveolar pressure." *IEEE Trans. Biomed. Eng.,* 1975, 22(5), 393–399.

Sackner, M. A., "Monitoring of ventilation without a physical connection to the airway," in M. A. Sackner (ed.), *Diagnostic Techniques in Pulmonary Disease,* Part I. New York: Marcel Dekker, 1980, pp. 503–537.

Silverman, L., and J. L. Whittenberger, "Clinical pneumotachograph," in J. H. Comroe, Jr. (ed.), in *Methods in Medical Research,* Vol. 2. Chicago: Year Book, 1950, pp. 104–112.

Sodal, I. E., J. S. Clark, and G. D. Swanson, "Mass spectrometers in medical monitoring," in J. G. Webster (ed.), *Encyclopedia of Medical Devices and Instrumentation.* New York: Wiley, 1988, pp. 1848–1859.

Sullivan, W. J., G. M. Peters, and P. L. Enright, "Pneumotachographs: Theory and clinical application," *Respiratory Care,* 1984, 29, 736–749.

Turney, S. Z., and W. Blumenfeld, "Heated Fleisch pneumotachometer: A calibration procedure." *J. Appl. Physiol.,* 1973, 34, 117–121.

Verschakelen, J. A., K. Deschepper, I. Clarysse, and M. Demedts, "The effect of breath size and posture on calibration of the respiratory inductive plethysmograph by multiple linear regression." *Eur. Respir. J.,* 1989, 2, 71–77.

West, G. A., J. J. Barrett, D. R. Siebert, and K. V. Reddy, "Photoacoustic spectroscopy." *Rev. Sci. Instrum.,* 1983, 54, 797–817.

10

CHEMICAL BIOSENSORS

Robert A. Peura

A chemical biosensor is a sensor that produces an electric signal proportional to the concentration of biochemical analytes. These biosensors use chemical as well as physical principles in their operation.

The body is composed of living cells. These cells, which are essentially chemical factories the input to which is metabolic food and the output waste products, are the building blocks for the organ systems in the body. The functional status of an organ system is determined by measuring the chemical input and output analytes of the cells. As a consequence, the majority of tests made in the hospital or the physician's office deal with analyzing the chemistry of the body.

The important critical-care analytes are the blood levels of pH; $P\text{O}_2$; $P\text{CO}_2$; hematocrit; total hemoglobin; O_2 saturation; electrolytes including sodium, potassium, calcium, and chloride; and various metabolites including glucose, lactate, creatinine, and urea. Table 10.1 gives the normal ranges in blood for these critical-care analytes.

These variables are normally analyzed in a central clinical-chemistry laboratory remote from the patient's bedside. This conventional approach provides only historical values of the patient's blood chemistry, because there is a delay between when the sample is obtained and when the result is reported. (The sample must be transported to the main clinical-chemistry laboratory, and the appropriate analyses must be performed.) This inherent delay is approximately 30 min or more. Other significant drawbacks plague central-laboratory analyses of patient chemistry, including potential errors in the origin of the sample and in sample-handling techniques, and (because of the delay) the timeliness of the therapeutic intervention.

For these reasons, there has been a movement to decentralize clinical testing of the patient's chemistry (Collison and Meyerhoff, 1990). This is particularly important in the critical-care and surgical settings. The decentralized approach has resulted from a number of improvements in biosensor technology, including the development of blood-gas and electrolyte monitoring systems equipped with self-calibration for measuring the patient's blood chemistry at the bedside.

Economic pressures have also encouraged movement of sophisticated

Table 10.1 Critical-Care Analytes and Their Normal Ranges in Blood

Blood Gases and Related Parameters		Electrolytes		Metabolites	
PO_2	80–104 mm Hg	Na^+	135–155 mmol/l	Glucose	70–110 mg/100 ml
PCO_2	33–48 mm Hg	K^+	3.6–5.5 mmol/l	Lactate	3–7 mg/100 ml
pH	7.31–7.45	Ca^{2+}	1.14–1.31 mmol/l	Creatinine	0.9–1.4 mg/100 ml
Hematocrit	40–54%	Cl^-	98–109 mmol/l	Urea	8–26 mg/100 ml
Total hemoglobin	13–18 g/100 ml				
O_2-saturation	95–100%				

SOURCE: M. E. Collison and M. E. Meyerhoff, "Chemical sensors for bedside monitoring of critically ill patients." *Anal. Chem.*, 1990, 62, 425A–437A.

chemical-analysis and diagnostic equipment from the central laboratory to specific clinical areas. Such sites include the operating room, where patient blood gases and electrolytes must be monitored continuously, and dialysis centers, where patients are treated on an out-patient basis and measurements of uric acid and other blood analytes must be made in a timely manner. In addition, self-contained, small, economical blood-chemistry units have been developed for use in the physician's office and the patient's home.

In the future, integrated-circuit and optoelectronic technology will be used to develop miniaturized biosensors, which are sensitive to body analytes for real-time, *in vivo* measurements of body chemistry (Turner *et al.*, 1987). Self-contained biosensor units for closed-loop drug-delivery systems will also become available. Examples of future applications of closed-loop systems with chemical biosensors include (1) control of implantable pacemakers and defibrillators, (2) regulation of anesthesia during operations, and (3) control of insulin secretion from an artificial pancreas. Note that moving laboratory devices from a central location to a decentralized location in the hospital, physician's office, or patient's home poses significant challenges. These involve stability, calibration, quality control of the measurements and ease of instrument use.

Noninvasive measurement of the biochemistry of the body will increase tremendously in the future. The advances in and burgeoning applications of pulse oximetry offer just one example of the impact that noninvasive measurement can have on patient monitoring. Pulse oximetry has become the standard of care in a number of clinical situations, which include monitoring during administration of anesthesia (to assess functioning of the cardiopulmonary system) and during the administration of oxygen to neonates (to avoid high arterial oxygen levels, which can lead to serious damage to retinal and pulmonary tissue). The future will see applications for noninvasive monitoring of the blood biochemistry in the standard blood-chemistry tests for glucose, cholesterol, urea, electrolytes, and so on.

10.1 BLOOD-GAS AND ACID–BASE PHYSIOLOGY

The fast and accurate measurements of the blood levels of the partial pressure of oxygen (P_{O_2}), the partial pressure of CO_2 (P_{CO_2}), and the concentration of hydrogen ions (pH) are vital in the diagnosis and treatment of many pathological conditions. Significant abnormalities of these quantities can rapidly be fatal if not treated appropriately. These measurements are usually made on specimens of arterial blood, though "arterialized" venous samples are often obtained from infants.

Oxygen is carried in the blood in two separate states. Normally, approximately 98% of the O_2 in the blood is combined with hemoglobin (Hb) in the red blood cells. The remaining 2% is physically dissolved in the plasma. The amount (saturation, S) of O_2 bound to Hb in arterial blood is defined as the ratio of the concentration of oxyhemoglobin (HbO_2) to the total concentration of Hb. That is,

$$S_{O_2}\,(\%) = \frac{[\mathrm{HbO_2}]}{[\text{total Hb}]} \times 100 \tag{10.1}$$

The sigmoid-shaped oxyhemoglobin dissociation curve (ODC), shown in Figure 10.1, graphically illustrates the relationship between the percent oxygen saturation of hemoglobin and the partial pressure of oxygen in the plasma. The total content of O_2 in blood is directly related to S_{O_2} for any given Hb concentration, because the amount of O_2 that is physically dissolved in the blood is relatively small.

Arterial P_{O_2} and S_{O_2} have different physiological meanings. Arterial P_{O_2} determines the efficiency of alveolar ventilation; S_{O_2} indicates the amount of O_2 per unit of blood. It is possible to derive S_{O_2} from P_{O_2} measurements by using an ODC, but significant errors result for abnormal physiological situations unless the temperature and pH of the blood, the type of Hb derivative, and 2,3-diphosphoglycerate (DPG) are known. Direct measurement of S_{O_2} is more accurate than an indirect calculation, because the affinity of Hb for O_2 is affected by these several variables.

For young adults, the normal range of P_{O_2} in arterial blood is from 90 to 100 mm Hg (12 to 13.3 kPa). As a result of the sigmoid nature of the O_2 disassociation curve, a P_{O_2} of 60 mm Hg (8 kPa) still provides an O_2 saturation of 85%. Decreases in P_{O_2} are seen in a variety of settings. These can be divided into two groups: (1) decreased delivery of O_2 to the site of O_2 exchange between the inspired air and the blood (the lung alveoli) and (2) decreased delivery of blood to the alveoli to which O_2 is being supplied. Examples of

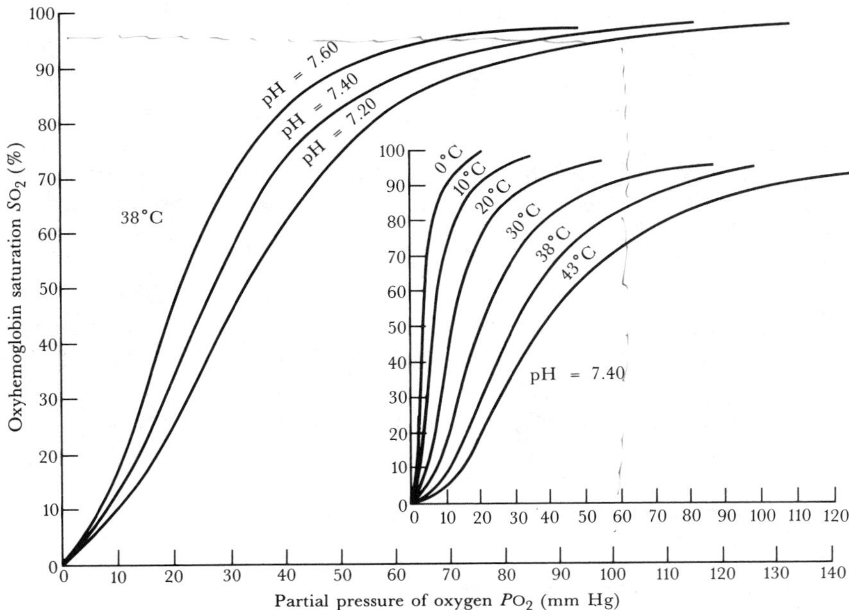

Figure 10.1 The oxyhemoglobin dissociation curve, showing the effect of pH and temperature on the relationship between SO_2 and PO_2.

the first group include decreased overall ventilation (such as caused by narcotic overdose or paralysis of the ventilatory muscles), obstruction of major airways (such as by aspirated foreign objects such as food; by spasm of the airway muscles, such as that which occurs in an acute attack of asthma); or by filling of the alveoli and small airways with fluid (such as in pneumonia or pulmonary edema). Examples of the second group include congenital cardiac abnormalities, in which blood is shunted past the lungs (the Tetralogy of Fallot, for example), and obstruction of flow through the pulmonary blood vessels (such as caused by pulmonary emboli). The important lung diseases of emphysema and chronic bronchitis usually display characteristics of both these types of abnormalities.

The PCO_2 level is an indicator of the adequacy of ventilation and is therefore increased in the first group of disorders discussed above, but it is generally normal in the second group unless the defect is massive in nature. In young adults, the normal range of PCO_2 in arterial blood is 35 to 40 mm Hg (4.7 to 5.3 kPa).

The acid–base status of the blood is assessed by measuring the hydrogen ion concentration $[H^+]$. It is conventional to use the negative logarithm to the base 10 (pH) to report this quantity; that is,

$$pH = -\log_{10} [H^+] \tag{10.2}$$

The normal range of pH in arterial blood is 7.38 to 7.44. Decreases in pH (increased quantity of hydrogen ions) occur with a decreased rate of excretion of CO_2 (respiratory acidosis) and/or with increased production of fixed acid (such as occurs in diabetic ketoacidosis) or abnormal losses of bicarbonate (the principal hydrogen ion buffer in the blood). Acidosis resulting from the last two processes is called metabolic acidosis. Increases in pH (decreased quantity of hydrogen ions) occur with an increased rate of excretion of CO_2 (respiratory alkalosis) and/or abnormal losses of acid (such as result from prolonged vomiting), which is called metabolic alkalosis. Table 10.2 gives examples of arterial-blood gases in different clinical situations. Note that a measurement of $P\text{CO}_2$, or the level of bicarbonate in the blood, along with a measurement of pH, must be done in order to classify the type of acid–base abnormality (Davenport, 1975).

The basic concepts of ions, electrochemical cells, and reference cells are discussed in Chapter 5. This section shows how these concepts are used to design electrodes for the measurement of pH, $P\text{CO}_2$ and $P\text{O}_2$.

10.2 ELECTROCHEMICAL SENSORS

MEASUREMENT OF pH

The measurement of pH is accomplished by utilizing a glass electrode that generates an electric potential when solutions of differing pH are placed on the two sides of its membrane (Cremer, 1906). Figure 10.2 is a schematic diagram of a pH electrode.

The glass electrode is a member of the class of ion-specific electrodes that react to any extent only with a specific ion.

The approach of a hydrogen ion to the outside of the membrane causes the silicate structure of the glass to conduct a positive charge (hole) into the ionic solution inside the electrode. The Nernst equation, (4.1), applies, so the voltage across the membrane changes by 60 mV/pH unit. Because the range of physiological pH is only 0.06 pH units, the pH meter must be capable of accurately measuring changes of 0.1 mV.

The basic approach is to place a solution of known pH on the inside of the membrane and the unknown solution on the outside. Hydrochloric acid is generally used as the solution of known pH. A reference electrode, usually an Ag/AgCl or a saturated calomel electrode, is placed in this solution. A second reference electrode is placed in the specimen chamber. A salt bridge is included within the reference to prevent the chemical constituents of the specimen from affecting the voltage of the reference electrode. The potential developed across the membrane of the glass electrode is read by a pH meter. This pH meter must have an extremely high input impedance, because the internal impedance of the pH electrode is in the 10- to 100-MΩ range.

The Nernst equation shows that the voltage produced by a pH electrode varies with the temperature of the specimen and the reference solution. Some

Table 10.2 Examples of Arterial Blood Gases in Different Clinical Situations

Example	P_{CO_2}, mm Hg	pH	P_{O_2}, mm Hg	Interpretation	Likely causes	Therapy
1	40 ± 3	7.40 ± 0.03	90 ± 5	Normal blood gas		None
2	44 ± 3	7.37 ± 0.03	88 ± 5	Normal blood gas while asleep		
3	22	7.57	106	Hyperventilation	Anxiety	None
4	68	7.10	58	Hypoventilation	Central nervous system depression; blockage of upper airway	Mechanical ventilation; relieve the cause
5	58	7.21	39	Hypoventilation and hypoxemia	Pneumonia; small-airway obstruction; severe asthma	Oxygen; bronchodilators; mechanical ventilation
6	61	6.99	29	Combined respiratory and metabolic acidosis and hypoxemia	Birth asphyxia; near-drowning	Oxygen; mechanical ventilation; buffers?
7	60	7.37	106	Chronic respiratory acidosis with metabolic compensation; patient is receiving supplemental oxygen	Patient has chronic lung disease and is on oxygen	Treat chronic disease; no additional therapy may be necessary
8	29	7.31	106	Metabolic acidosis with respiratory compensation	Diabetic; ketoacidosis; dehydration	Treat the cause; buffers?

SOURCE: B. G. Nickerson and F. Monaco, "Carbon dioxide electrodes, arterial and transcutaneous," in J. G. Webster (ed.), *Encyclopedia of Medical Devices and Instrumentation.* New York: Wiley, 1988, pp. 564–569.

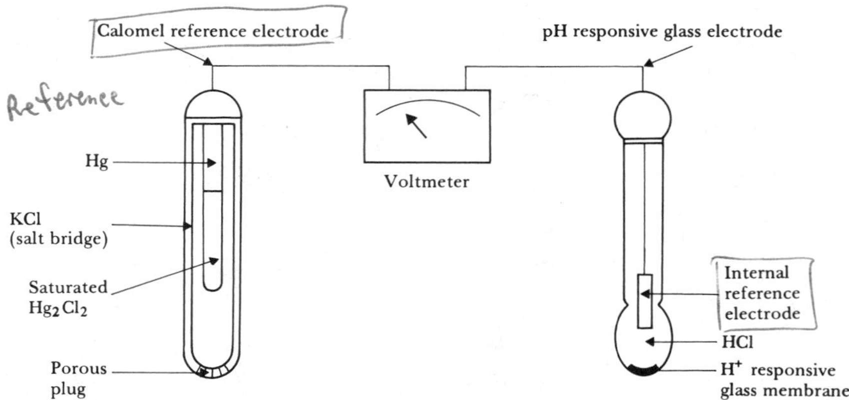

Figure 10.2 pH electrode (From R. Hicks, J. R. Schenken, and M. A. Steinrauf, *Laboratory Instrumentation.* Hagerstown, MD: Harper & Row, 1974. Used with permission of C. A. McWhorter.)

pH electrodes include a water bath that allows the pH determination to be made at 37 °C; others require a temperature correction. This temperature correction can be made by changing the constant used to convert from the electrode voltage to the meter scale reading in pH units by setting a temperature-control knob to the temperature at which the pH measurement is being made. A more complex correction includes the effect on instrument output of temperature and the CO_2 content of the specimen (Adamsons *et al.*, 1964; Burton, 1965). This type of correction can be made in modern devices that measure pH and $P\text{CO}_2$ and also have memory and computational capabilities.

Calibration with solutions of known pH is performed before measurements of patient specimens are made. Two solutions, one with a pH near 6.8 and one with the pH near 7.9, are normally used.

MEASUREMENT OF $P\text{CO}_2$

The measurement of $P\text{CO}_2$ is based on the fact that the relationship between log $P\text{CO}_2$ and pH is linear over the range of 10 to 90 mm Hg (1.3 to 12 kPa), which includes essentially all the values of clinical interest. This result can be established by examining some fundamental chemical relationships among H^+, H_2CO_3, HCO_3^-, and $P\text{CO}_2$. The first three quantities are related by the equilibrium equation

$$H_2O + CO_2 \rightleftharpoons H_2CO_3 \rightleftharpoons H^+ + HCO_3^- \tag{10.3}$$

In addition, the relationship between $P\text{CO}_2$ and the concentration of CO_2 dissolved in the blood, $[CO_2]$, is given by

$$[CO_2] = a(P\text{CO}_2) \tag{10.4}$$

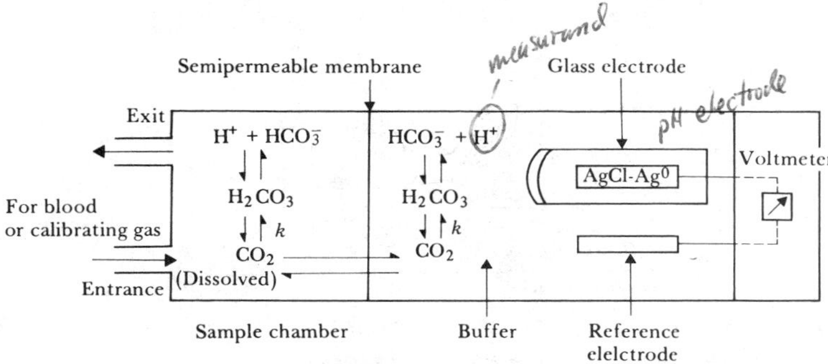

Figure 10.3 **P_{CO_2} electrode** (From R. Hicks, J. R. Schenken, and M. A. Steinrauf, *Laboratory Instrumentation.* Hagerstown, MD: Harper & Row, 1974. Used with permission of C. A. McWhorter.)

where $a = 0.0301$ mmol/liter per mm Hg P_{CO_2}. The mass relationship corresponding to (10.3) can then be written as

$$k' = \frac{[H^+][HCO_3^-]}{[H_2CO_3]} \tag{10.5}$$

Next we use the fact that $[H_2CO_3]$ is proportional to $[CO_2]$ to obtain the result

$$k = \frac{[H^+][HCO_3^-]}{[CO_2]} \tag{10.6}$$

where k represents the combined values of k' and the proportionality constant between $[H_2CO_3]$ and $[CO_2]$. Now, using (10.4), we obtain the following result:

$$k = \frac{[H^+][HCO_3^-]}{aP_{CO_2}} \tag{10.7}$$

Next, taking the base-10 logarithm of (10.7) and rearranging, we obtain

$$\log[H^+] + \log[HCO_3^-] - \log k - \log a - \log P_{CO_2} = 0 \tag{10.8}$$

Using the definition of pH yields

$$\text{pH} = \log[HCO_3^-] - \log k - \log a - \log P_{CO_2} \tag{10.9}$$

This shows that pH has a linear dependence on the negative of log P_{CO_2}.

This result is used in the construction of the P_{CO_2} electrode shown in Figure 10.3 (Severinghaus, 1965). The assembly includes two chambers, one for the specimen and a second containing a pH electrode of the type discussed. In contrast to the basic pH-measurement device in which the pH electrode

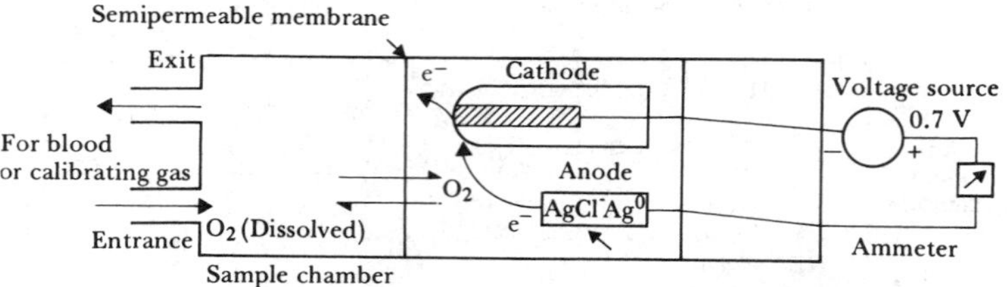

Figure 10.4 P_{O_2} electrode (From R. Hicks, J. R., Schenken, and M. A. Steinrauf, *Laboratory Instrumentation.* Hagerstown, MD: Harper & Row, 1974. Used with permission of C. A. McWhorter.)

is placed in the specimen, in this case the pH electrode is bathed by a buffer solution of bicarbonate and NaCl.

The two chambers are separated by a semipermeable membrane, usually made of Teflon or silicone rubber. This membrane allows dissolved CO_2 to pass through but blocks the passage of charged particles, in particular H^+ and HCO_3^-. When the specimen is placed in its chamber, CO_2 diffuses across the membrane to establish the same concentration in both chambers. If there is a net movement of CO_2 into (or out of) the chamber containing the buffer, $[H^+]$ increases (or decreases), and the pH meter detects this change. Because the relationship between pH and the negative log P_{CO_2} is only a proportional one, it is necessary to calibrate the instrument before each use with two gases of known P_{CO_2}.

Using the values of pH obtained by processing these two standards, we obtain a calibration curve of P_{CO_2} versus pH. We then use the measured pH value to obtain the specimen's P_{CO_2} from this curve. With some instruments, the capability of calibrating the P_{CO_2} electrode is built into the instrument so that the calibration curve is set up in the electronics of the instrument by setting the values of two potentiometers.

THE P_{O_2} ELECTRODE

Figure 10.4 shows the basic components of the Clark-type polarographic electrode. The measurement of P_{O_2} is based on the following reactions. At the cathode, reduction occurs:

$$O_2 + 2H_2O + 4e^- \rightarrow 2H_2O_2 + 4e^- \rightarrow 4OH^-$$

$$4OH^- + 4KCl \rightarrow 4KOH + 4Cl^- \qquad (10.10)$$

The hydroxyl ions created in this reaction are buffered by the electrolyte. At the anode, which in this P_{O_2} electrode is the reference electrode, oxidation occurs.

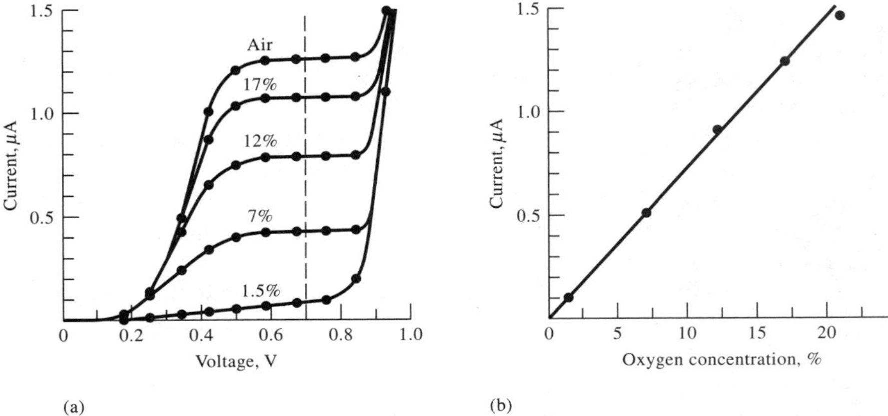

Figure 10.5 (a) Current plotted against polarizing voltage for a typical $P\text{O}_2$ electrode for the percents O_2 shown. (b) Electrode operation with a polarizing voltage of 0.68 V gives a linear relationship between current output and percent O_2.

$$4\text{Ag} + 4\text{Cl}^- \rightarrow 4\text{AgCl} + 4\text{e}^- \tag{10.11}$$

This produces the four electrons required for the reaction in (10.10).

The cathode is constructed of glass-coated Pt, and the reference electrode is made of Ag/AgCl.

The plot of current versus polarizing voltage of a typical $P\text{O}_2$ electrode (polarogram) is shown in Figure 10.5(a). The polarizing voltage is selected in the "plateau" region to provide a sufficient potential to drive the reaction, without permitting other electrochemical reactions that would be driven by greater voltages to take place. Thus the resulting current is linearly proportional to the number of O_2 molecules in solution; see Figure 10.5(b). The O_2 membrane is permeable to O_2 and other gases and separates the electrode from its surroundings.

A polarizing voltage of 600 to 800 mV is required for these reactions to occur. This voltage is usually supplied by a mercury cell.

We determine the value of $P\text{O}_2$ by using the fact that the flow of current through the external circuit connecting the electrodes is proportional to $P\text{O}_2$. The presence of O_2 and the resulting chemical reaction can be thought of as producing in the circuit a variable source of current the value of which is directly proportional to the $P\text{O}_2$ level. When the $P\text{O}_2$ level is zero, the current flowing through the circuit is called the background current. Part of the calibration sequence involves setting the $P\text{O}_2$ meter to zero when a CO_2/N_2 gas is bubbled through the specimen chamber. Slow bubbling is used to ensure proper temperature equilibration.

Equation (10.10) shows that the reaction consumes O_2. This loss is a direct function of the area of the Pt electrode that is exposed to the reaction solution

Platinum

and the permeability of the semipermeable membrane to O_2. The exposed area of the Pt electrode usually has a diameter of 20 μm.

The choice of the semipermeable membrane is based on a tradeoff between consumption of O_2 and the time required for the Po_2 values in the specimen and measurement chambers to equilibrate. The more permeable the membrane to O_2, the higher the consumption of O_2 and the faster the response. Polypropylene is less permeable than Teflon and is preferable in most applications. Polypropylene is also quite durable, and it maintains its position over the electrode more reliably than other membrane materials.

The membrane thickness and composition determine the O_2 diffusion rate; thicker membranes extend the sensor time response by significantly increasing the diffusion time and produce smaller currents.

Because the electrode consumes O_2, it partially depletes the oxygen in the immediate vicinity of the membrane. If movement of the sample takes place, undepleted solution brought to the membrane causes a higher instrument reading—the "stirring" artifact. This is avoided by waiting for a stagnant equilibrium to occur.

The reaction is very sensitive to temperature. To maintain a linear relationship between Po_2 and current, the temperature of the electrode must be controlled to ±0.1 °C. This has been traditionally accomplished by using a water jacket. However, new blood-gas analyzers are now available that use precision electronic heat sources. The current through the meter is approximately 10 nA/mm Hg (75 nA/kPa) O_2 at 37 °C, so the instruments must be designed to be accurate at very low current levels.

The system is calibrated by using two gases of known O_2 concentration. One gas with no O_2 (typically a CO_2–N_2 mixture) and a second with a known O_2 content (usually an O_2–CO_2–N_2 mixture) are used. The specimen chamber is filled with water, and the calibrating gas containing no O_2 is bubbled through it. The Po_2 meter output is set to zero after equilibrium of O_2 content is achieved—usually in about 90 s. Next the second calibrating gas is used to determine the second point on the Po_2-versus-electrode-current calibration scale, which is electrically set in the machine. Then the value of the specimen Po_2 can be measured. Note that the time required to reach equilibrium is a function of the Po_2 of the specimen. It may take as long as 360 s for a specimen with a Po_2 of 430 mm Hg (57 kPa) to reach equilibrium (Moran *et al.*, 1966). Drägerwerk Aktiengesellschaft, Lübeck, Germany, manufactures a gas O_2 sensor with 2-s response in which gas diffuses into a Po_2 electrode.

10.3 CHEMICAL FIBROSENSORS

Rapid advances in the communications industry have provided appropriate small optical fibers, high-energy sources such as lasers, and wavelength detectors. The fiber-optic sensors that were developed were called optodes, a term coined by Lübbers and Opitz (1975), which implies that optical sensors are very similar to electrodes. As we shall see, however, the properties and operating principles for optical fibrosensors are quite different from those for electrodes. The term *optrode,* with an r, is currently used.

Chemical fibrosensors offer several desirable features.

1. They can be made small in size.
2. Multiple sensors can be introduced together, through a catheter, for intracranial or intravascular measurements.
3. Because optical measurements are being made, there are no electric hazards to the patient.
4. The measurements are immune to external electric interference, provided that the electronic instrumentation is properly shielded.
5. No reference electrode is necessary.

In addition, fibrosensors have a high degree of flexibility and good thermal stability, and low-cost manufacturing and disposable usage are possible. In reversible sensors, the reagent phase is not consumed by its reaction with the analyte. In nonreversible sensors, the reagent phase is consumed. The consumption of the reagent phase for nonreversible sensors must be small, or there must be a way to replenish the reagent.

Optical-fiber sensors have several limitations when compared with electrode sensors. Optical sensors are sensitive to ambient light, so they must be used in a dark environment or must be optically shielded via opaque materials. The optical signal may also have to be modulated in order to code it and make it distinguishable from the ambient light. The dynamic response of optical sensors is normally limited compared with that of electrodes. Reversible indicator sensors are based on an equilibrium measurement rather than a diffusion-dependent one, so they are less susceptible to changes in flow concentration at the sensor (Seitz, 1988).

Long-term stability for optical sensors may be a problem for reagent-based systems. However, this can be compensated for by the use of multiple-wavelength detection and by the ease of changing reagent phases. In addition, because the reagent and the analyte are in different phases, a mass-transfer step is necessary before constant response is achieved (Seitz, 1988). This limits the temporal response of an optical sensor. Another consideration with optical sensors is that for several types of optical sensors, the response is proportional to the amount of reagent phase. For small amounts of reagent, an increased response can be achieved by increasing the intensity of the source. An increased response, however, results in an increase in the photodegradation process of the reagent. Designers of optical sensors, then, must consider amount of the reagent phase, intensity of the light source, and system stability (Seitz, 1984).

These limitations can be alleviated by an appropriate design of the optical sensor and instrumentation system (Wise, 1990). The systems described in the following paragraphs incorporate many features specifically for this purpose.

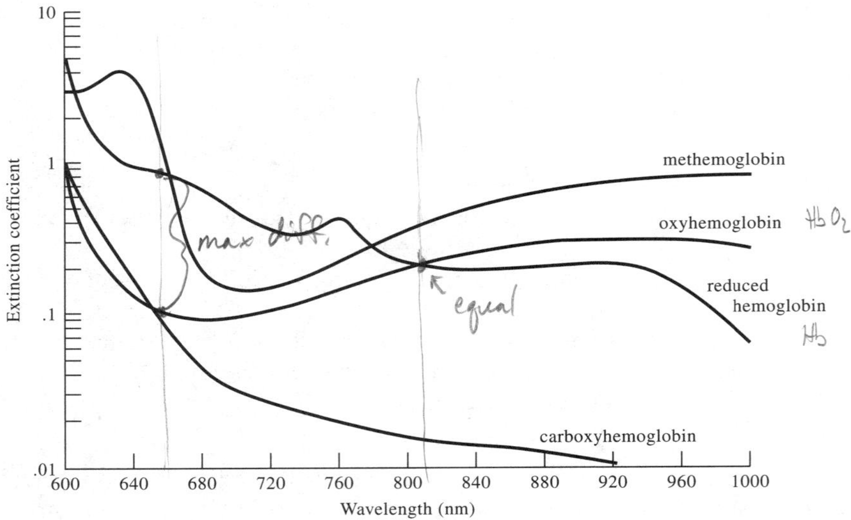

Figure 10.6 Absorptivities (extinction coefficients) in L/(mmol·cm) of the four most common hemoglobin species at the wavelengths of interest in pulse oximetry. (Courtesy of Susan Manson, Biox/Ohmeda, Boulder, CO.)

INTRAVASCULAR MEASUREMENTS OF OXYGEN SATURATION

Blood oxygen can be monitored by means of an intravascular fiber-optic catheter. These catheters are used to monitor mixed venous oxygen saturation during cardiac surgery and in the intensive-care unit. A Swan-Ganz catheter is used (see Section 7.11), in which a flow-directed fiber-optic catheter is placed into the right jugular vein. The catheter is advanced until its distal tip is in the right atrium, at which time the balloon is inflated. The rapid flow of blood carries the catheter into the pulmonary artery.

Measurements of mixed venous oxygen saturation give an indication of the effectiveness of a cardiopulmonary system. Measurements of high oxygen saturation in the right side of the heart may indicate congenital abnormalities of the heart and major vessels or the inability of tissue to metabolize oxygen. Low saturation readings on the left side of the heart may indicate a reduced ability of the lungs to oxygenate the blood or of the cardiopulmonary system to deliver oxygen from the lungs. Low saturation readings in the arterial system indicate a compromised cardiac output or reduced oxygen-carrying capacity of the blood.

Figure 10.6 shows the optical-absorption spectra for oxyhemoglobin, carboxyhemoglobin, hemoglobin, and methemoglobin. Measurements in the red region are possible because the absorption coefficient of blood at these wavelengths is sufficiently low that light can be transmitted through whole blood over distances such that feasible measurements can be made with fiber-optic catheters. Note that the 805-nm wavelength provides a measurement independent of the degree of oxygenation. This *isosbestic wavelength* is used to compensate for the scattering properties of the whole blood and to normalize the measurement signal with any changes in hemoglobin from patient to patient.

Oxygen saturation is measured by taking the ratio of the diffusely backscattered light intensities at two wavelengths. The first wavelength is in the red region (660 nm); the second is in the infrared region (805 nm), which is known as the isosbestic point for Hb and HbO_2. Oxygen saturation is given by (10.1), which considers the optical density of the blood—the light transmitted through the blood—according to Beer's law. For *hemolyzed blood* (blood with red cells ruptured), Beer's law (Section 11.1) holds, and the *absorbance* (optical density) at any wavelength is (Allan, 1973)

$$A(\lambda) = WL[a_o(\lambda)C_o + a_r(\lambda)C_r] \tag{10.12}$$

where

W = weight of hemoglobin per unit volume
L = optical path length
a_o and a_r = absorptivities of HbO_2 and Hb
C_o and C_r = relative concentrations of HbO_2 and Hb ($C_o + C_r = 1.0$)

Figure 10.6 shows that a_o and a_r are equal at 805 nm, called the isosbestic wavelength. If this wavelength is λ_2, then

$$WL = \frac{A(\lambda_2)}{a(\lambda_2)} \tag{10.13}$$

where

$$a(\lambda_2) = a_o(\lambda_2) = a_r(\lambda_2) \tag{10.14}$$

Therefore

$$A(\lambda) = \frac{A(\lambda_2)}{a(\lambda_2)}[a_o(\lambda)C_o + a_r(\lambda)C_r] \tag{10.15}$$

When absorbance is measured at a second wavelength λ_1, the oxygen saturation is given by

$$C_o = x + \frac{yA(\lambda_1)}{A(\lambda_2)} \tag{10.16}$$

where x and y are constants that depend only on the optical characteristics of blood. In practice, λ_1 is chosen to be that wavelength at which the difference between a_o and a_r is a maximum, which occurs at 660 nm [see Figure 10.6].

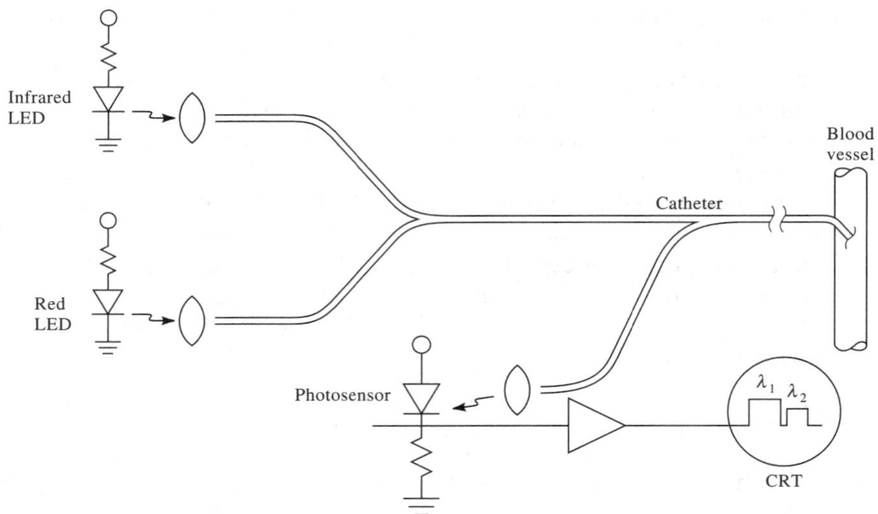

Figure 10.7 The oximeter catheter system measures oxygen saturation *in vivo,* using red and infrared light emitting diodes (LEDs) and a photosensor. The red and infrared LEDs are alternately pulsed in order to use a single photosensor.

Figure 10.7 shows a fiber-optic instrument devised to measure oxygen saturation in the blood. This device, which could also be used for measuring cardiac output with a dye injected, is described here. The instrument consists of red and infrared light-emitting diodes (LEDs) and a photosensor. Plastic optical fibers are well adapted to these wavelengths. Figure 10.8 shows a fiber-optic oximeter catheter that is flow-directed. After insertion, the balloon is inflated, and blood flow drags the tip through the chambers of the heart.

In addition to measuring blood-oxygen saturation through reflectance, the same dual-wavelength optics can be used to measure blood flow by dye dilution. Indo/cyanine/green, which absorbs light at 805 nm (the isosbestic wavelength of oxyhemoglobin), is used as the indicator. This is a dual-fiber system. Light at 805 nm is emitted from one fiber, scattered by the blood cells, attenuated by the dye in the blood, and partially collected by the other fiber for measurement. The second wavelength, above 900 nm, is used as a reference; this is the region where the light is absorbed by the dye. It is used to compare the effect of flow-rate light scattering. In effect, a dual-beam ratiometric system is developed for dye-dilution measurements of blood flow. Cardiac output is determined via the dye-dilution method described in Section 8.2.

A significant difference exists between two-wavelength oximetry systems and the Abbott three-wavelength Oximetry Opticath® System. In two-wavelength systems an important limitation, in the *in vivo* measurement of oxygen saturation below 80%, is the dependence of the reflected light's intensity on the patient's hematocrit. Hematocrit varies from subject to subject, and within one subject it varies for different physiological conditions. Catheter tip oxime-

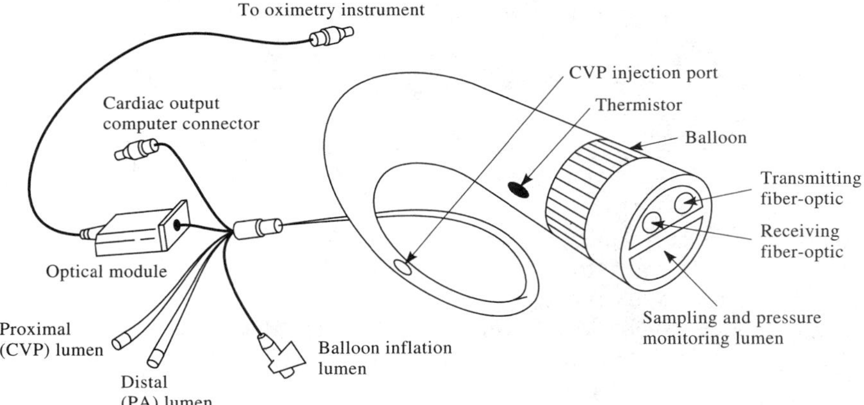

Figure 10.8 The catheter used with the Abbott Opticath Oximetry System transmits light to the blood through a transmitting optical fiber and returns the reflected light through a receiving optical fiber. The catheter is optically connected to the oximetry processor through the optical module. (From Abbott Critical Care Systems. Used by permission.)

ters require frequent updates of a patient's hematocrit. Various correction techniques have been devised to correct the oxygen-saturation measurements for errors due to hematocrit variations. (This limitation is eliminated in the three-wavelength Abbott Opticath Oximetry System.) False readings occur in situations in which hemoglobin combines with another substance besides oxygen, such as carbon monoxide. Hemoglobin has a strong affinity for carbon monoxide, so oxygen is displaced. The optical spectra for HbO_2 and HbCO overlap at 660 nm (Figure 10.6), causing an error in So_2 if CO is present in the blood.

A three-fiber intravascular fiber-optic catheter that measures mixed venous oxygen saturation and hematocrit simultaneously has been developed and tested (Mendelson *et al.*, 1990). The system consists of a catheter with a single light source in two equally spaced, near and far detecting fibers. The ratio of backscattered-light intensities measured at the isosbestic wavelength (805 nm) by the two detecting fibers (IR near/IR far) serves a correction factor that reduces the dependence of oxygen-saturation measurements on hematocrit.

This approach also provides a means for determining hematocrit independently. The principle of the measurement is based on the fact that variations in blood pH and osmolarity affect the shape and volume of the red blood cells. The IR near/IR far ratio is affected by variations in red blood cell volume and thus in hematocrit. The reflected-light intensities, measures by the two detecting fibers, are due to the higher-order multiple scattering. The intensity of the reflected light becomes more pronounced as source-to-detector separation distance increases. Details concerning the transcutaneous measurement of arterial oxygen saturation via pulse oximetry are given in Section 10.6.

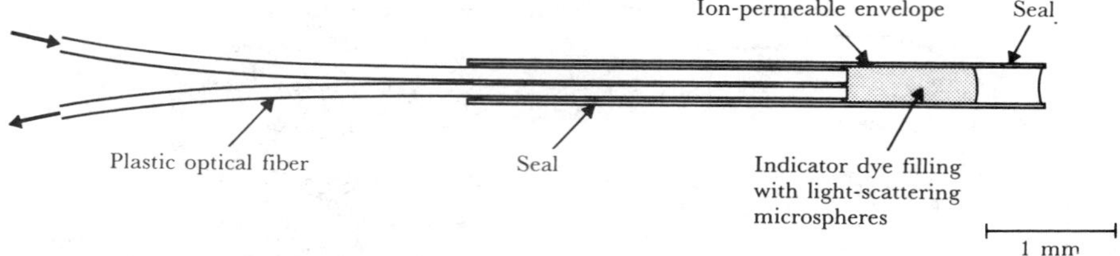

Figure 10.9 A reversible fiber-optic chemical sensor measures light scattered from phenol red indicator dye to yield pH. [From J. I. Peterson, "Optical sensors," in J. G. Webster (ed.), *Encyclopedia of Medical Devices and Instrumentation.* New York: Wiley, 1988, pp. 2121–2133. Used by permission.]

REVERSIBLE-DYE OPTICAL MEASUREMENT OF pH

The continuous monitoring of blood pH is essential for the proper treatment of patients who have metabolic and respiratory problems. Small pH probes have been developed for intravascular measurement of the pH of the blood (Peterson *et al.*, 1980). These instruments require a range of 7.0 to 7.6 pH units and a resolution of 0.01 pH unit.

Figure 10.9 shows an early version of a pH sensor, in which a reversible colorimetric indicator system is fixed inside an ion-permeable envelope at the distal tip of the two plastic optical fibers. Light-scattering microspheres are mixed with the indicator dye inside the ion-permeable envelope in order to optimize the backscattering of light to the collection fiber that leads to the detector.

The reversible indicator dye, phenol red, is a typical pH-sensitive dye. The dye exists in two tautomeric (having different isomers) forms, depending on whether it is in an acidic or a basic solution. The two forms have different optical spectra. In Figure 10.10, the absorbance is plotted against wavelength for phenol red for the base form of the dye, indicating that the optical-absorbance peak increases with increasing pH. The ratio of green to red light transmitted through the dye is (Peterson, 1988)

$$R = k \times 10^{[-C/(10^{-\Delta}+1)]} \tag{10.17}$$

where

Δ = difference between pH and pK of the dye
$R = I(\text{green})/I(\text{red})$ = measured ratio of light intensities
$k = I_0(\text{green})/I_0(\text{red})$ = a constant (I_0 = initial light intensity)
C = a constant determined by (1) the probe geometry, (2) the total dye concentration, and (3) the absorption coefficient of the dye's basic tautomer

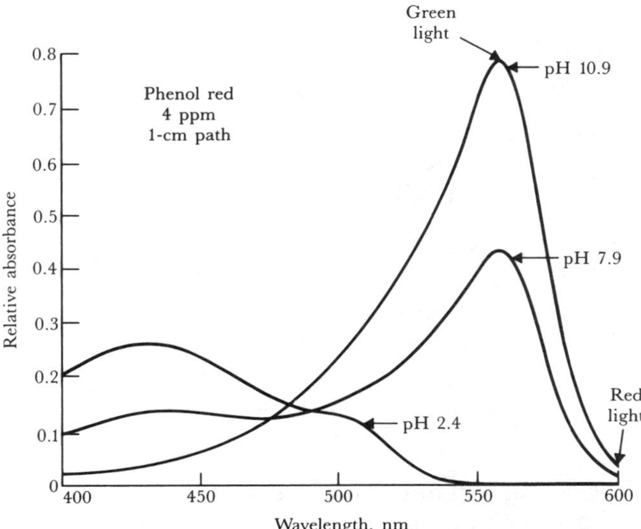

Figure 10.10 The plot of absorbance against wavelength of phenol red (base form) increases with pH for green light but is constant for red light.

Equation (10.17) shows that the ratio of green to red light transmitted through the dye can be expressed as a function of (1) the ionization constant of the dye—that is, the pK_a where "a" indicates the dye is a weak acid; (2) Beer's law for optical absorption; and (3) the use of the definition of pH. The constants are k, the optical constant; A, the absorbance of the probe when the dye is completely in the base form; and pK, the inverse log of the ionization constant of the dye. The ratio of green to red light is used because the green light transmitted varies with pH, whereas the red light is an isosbestic wavelength and does not vary with pH. In effect, this system is a dual-beam spectrometer.

Figure 10.11 shows a plot of R, the ratio of green to red light, against Δ, the deviation of the pH from the pK of the dye. The curve shows that over a range of about 1 pH unit, a nearly linear region for the S-shaped curve results. The instrument for pH measurement via the fiber-optic sensor uses a 100-W quartz halogen light as the source, and a rotating filter wheel selects between green and red light to illuminate the sample under study. Light passes down the fiber-optic input fiber and is scattered from the polystyrene light-scattering microspheres so that adequate light is collected and sent back to the receiving fiber (Peterson and Vurek, 1984).

The green light returning to the sensor varies as a function of the pH, whereas the red light does not vary with pH. Because the red light is generated by the same source as the green light and travels the same optical path to the detector, any changes in the optical system are reflected in changes in the red light received by the detector. Thus, when the intensity of the green light received by the detector is divided by the intensity of the red light received, any changes in the optical system are compensated for by this ratiometric method.

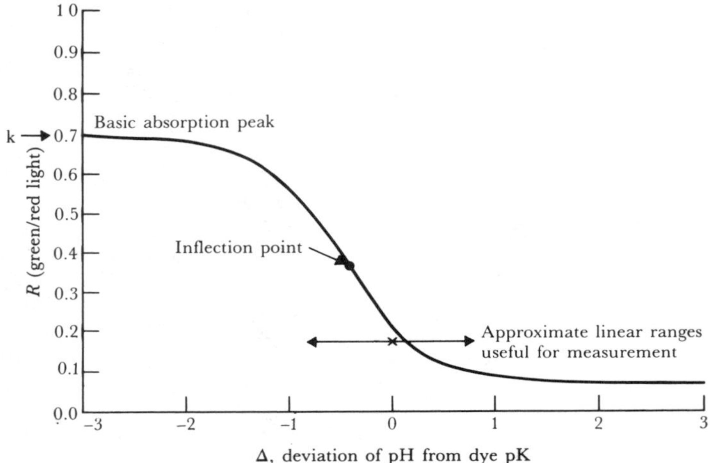

Figure 10.11 The ratio (R) of green to red light transmitted through phenol red for basic and acidic forms of the dye. Δ = deviation of pH from dye pK. (From J. I. Peterson, S. R. Goldstein, and R. V. Fitzgerald, "Fiber-optic pH probe for physiological use," *Anal. Chem.,* 1980, 52, 864–869. Used by permission.)

FLUORESCENCE OPTICAL pH SENSOR (IRREVERSIBLE)

Many colorimetric or fluorometric approaches are irreversible because of the tight binding between reagent and analyte or the formation of an irreversible product of the reaction. The pH sensor described below is based on irreversible chemistry, so either a long-lasting reagent or a continuous reagent-delivery system is necessary for long periods of operation. A fluorescence pH sensor based on the pH-sensitive dye hydroxypyrene trisulfonic acid (HPTS), which is a water-soluble fluorescent dye with a pK_a of 7.0, has been used as an intravascular blood-gas probe for pH (Gehrich *et al.*, 1986). The pH-sensitivity range is approximately equal to $pK_a \pm 1$.

Figure 10.12 is a diagram of the intravascular blood-gas sensor, in which chemistries are covalently bonded through a cellulose matrix attached to the fiber tip. An opaque cellulose overcoat formed over the matrix provides mechanical integrity and optical isolation from the environment.

The underlying principle of fluorescent measurement is that fluorescent dyes emit light energy at a wavelength different from that of the excitation wavelength, which they absorb. This can be seen in Figure 10.13, which gives the fluorescence spectra of a pH-sensitive dye. The excitation peak wavelength for the acidic form of the dye is 410 nm, whereas the excitation peak wavelength for the basic form of the dye is 460 nm. It is also apparent that the emission spectra for both the acidic and the basic forms of the dye have a peak at 520 nm. Because of the separation between the excitation and emission wavelengths, it

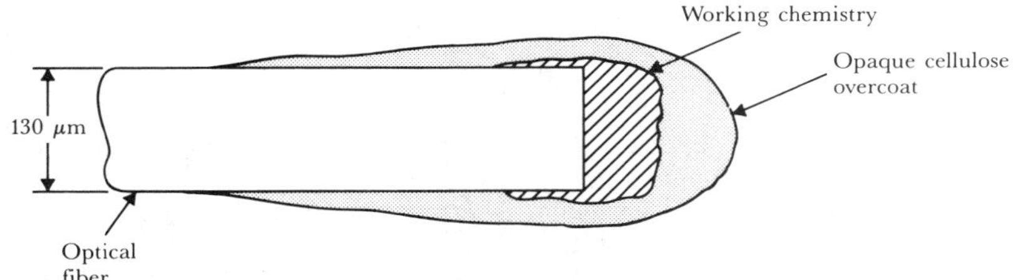

Figure 10.12 A single-fiber intravascular blood-gas sensor excites fluorescent dye at one wavelength and detects emission at a different wavelength. The following modifications are made to the sensor tip:
pH: Chemistry—pH-sensitive dye bound to hydrophilic matrix.
$P\text{CO}_2$: Chemistry—Bicarbonate buffer containing pH-sensitive dye with silicone.
$P\text{O}_2$: Chemistry—Oxygen-sensitive dye in silicone.
(From J. L. Gehrich, D. W. Lübbers, N. Optiz, D. R. Hansmann, W. W. Miller, J. K. Tusa, and M. Yafuso, "Optical fluorescence and its application to an intravascular blood gas monitoring system," *IEEE Trans. Biomed. Eng.*, 1986, BME-33, 117–132. Used by permission.)

is possible to use a single optical fiber both for the delivery of light energy to the sensor and for its reception from that sensor.

Intravascular dye fluorescence sensors must be stable enough to maintain accuracy for up to three days of use within the patient. Cost and shelf life of this disposable product must also be considered. In addition, the dye must be able to follow physiological changes in the blood-gas parameters and thus must have sufficient dynamic range and time response (Gehrich *et al.*, 1986).

The ratiometric principle, or two-wavelength approach, is used to design an optical measurement system that is independent of system and other parameters, which include (1) loss of the optical signal as a result of fiber bending, (2) optical misalignment, and (3) other changes in the optical path that could be incorrectly interpreted as changes in the concentration of the analyte being measured. The ratiometric approach is undertaken by selecting fluorescent dyes with two absorption or emission peaks or by providing a mixture of dyes at the sensor tip, one that is sensitive to the measured parameter and one that is not (reference wavelength, which is affected only by the optical system parameters). In the foregoing example, the emission due to excitation at 410 nm represents the relative amount of the basic phase, and the emission due to excitation at 460 nm represents the relative amount of the acidic phase. The ratio of these phases represents the pH.

FLUORESCENCE OPTICAL $P\text{CO}_2$ SENSOR

The $P\text{CO}_2$ sensor uses the same pH-sensitive fluorescent dye as the pH sensor described before. The operation of this sensor is similar to that of the electro-

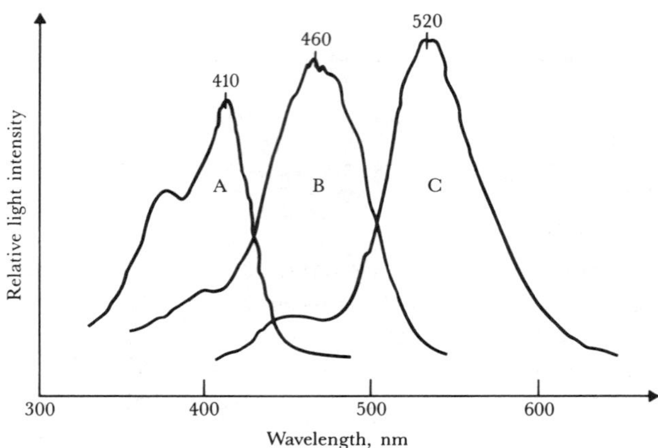

Figure 10.13 This pH-sensitive dye is excited at 410 and 460 nm and fluoresces at 520 nm. (A) The excitation spectrum of the acidic form of the dye; (B) the excitation spectrum of the basic form of the dye; (C) the emission spectrum of the acidic and basic forms of the dye. (From J. L. Gehrich, D. W. Lübbers, N. Opitz, D. R. Hansmann, W. W. Miller, J. K. Tusa, and M. Yafuso, "Optical fluorescence and its application to an intravascular blood gas monitoring system," *IEEE Trans. Biomed. Eng.,* 1986, BME-33, 117–132. Used by permission.)

chemical Severinghaus $P\text{CO}_2$ electrode described in the previous section in that a pH-type sensor is used as the basic sensing element to detect $P\text{CO}_2$. Carbon dioxide comes to an equilibrium with a mixture of a pH indicator in bicarbonate buffer. There is a direct relationship, based on the Henderson–Hasselbach equation, between the pH change in a bicarbonate solution and the CO_2 concentration in that solution. Thus a change in pH in an isolated bicarbonate buffer with a changing $P\text{CO}_2$ is measured. This buffer is encapsulated by a hydrophobic gas-permeable silicone matrix that provides ionic isolation and mechanical stability for the measurement system.

As before, an optical cellulose overcoat ensures optical isolation of the sensor chemistry from the environment. CO_2 equilibrates rapidly across the silicone membrane and causes a change in the pH. The concentration of the bicarbonate buffer is selected such that a sufficient pH change is detectable with appropriate accuracy and sensitivity over the physiological range for CO_2, which is 10–100 mm Hg. Dye strength must be optimized to increase the signal-to-noise ratio, and there is a tradeoff between ionic strength and pK.

FLUORESCENCE OPTICAL $P\text{O}_2$ SENSOR

One approach for a fiber-optic $P\text{O}_2$, or oxygen partial pressure, sensor makes use of the principle of fluorescence or luminescence quenching of oxygen. In

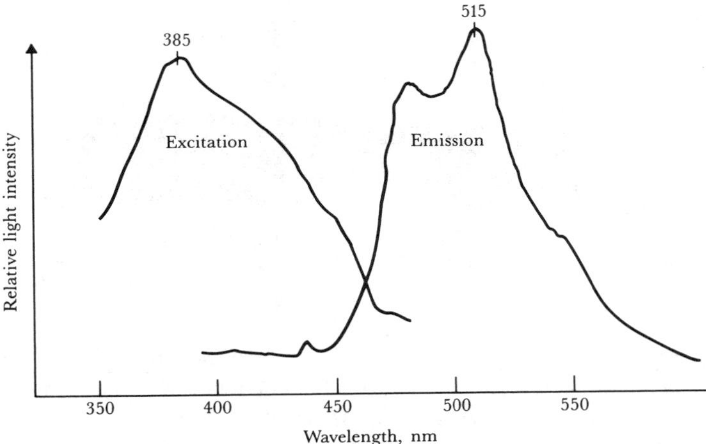

Figure 10.14 The emission spectrum of oxygen-sensitive dye can be separated from the excitation spectrum by a filter. (From J. L. Gehrich, D. W. Lübbers, N. Opitz, D. R. Hansmann, W. W. Miller, J. K. Tusa, and M. Yafuso, "Optical fluorescence and its application to an intravascular blood gas monitoring system," *IEEE Trans. Biomed. Eng.,* 1986, BME-33, 117–132. Used by permission.)

this quenching process, energy is absorbed and lost by various processes, such as vibration of the molecule (heat) and emission of the light as fluorescence or phosphofluorescence. With oxygen present, these molecules provide collision paths and transfer of energy to the oxygen molecule, which competes with the energy decay modes, and luminescence is decreased by the increasing loss of energy to oxygen.

Figure 10.14 shows the fluorescent spectra of oxygen-sensitive dye for both the excitation and the emission.

The Po_2 probe is similar in design to the pH sensor. The principle of its operation is that when these fluorescent quenching dyes are irradiated by light at an appropriate wavelength, they fluoresce in a nonoxygen atmosphere for a given period of time. However, when oxygen is present the fluorescence is quenched—that is, the dye fluoresces for a shorter period of time. The period of dye fluorescence is inversely proportional to the partial pressure of oxygen in the environment. This leads to a poor signal-to-noise ratio at high Po_2 values, because the high O_2 levels quench the luminescence, which results in a small signal at the detector. In Figure 10.15, fibers and inert beads are enclosed in an oxygen-permeable hydrophobic sheet such as porous polypropylene.

The Po_2-measurement instrument, includes both optical and electronic systems. The instrumentation system as designed uses plastic optical fibers because of their mechanical strength and flexibility; they allow for a sharp bending radius. The light returning from the sensor passes through a dichroic

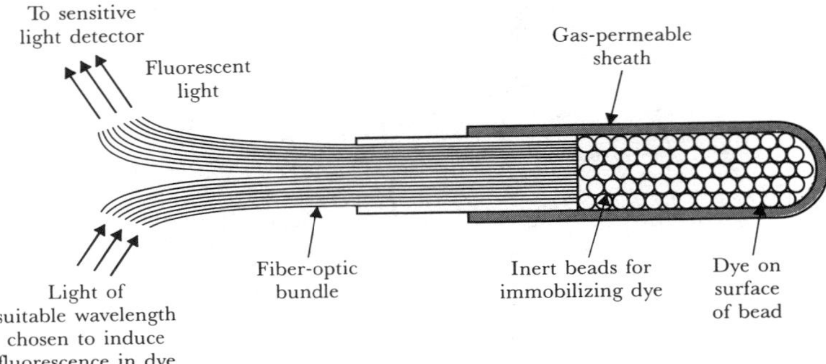

Figure 10.15 In a fiber-optic oxygen sensor, irradiation of dyes causes fluorescence that decreases with $P\mathrm{O}_2$. [From R. Kocache, "Oxygen analyzers," in J. G. Webster (ed.), *Encyclopedia of Medical Devices and Instrumentation.* New York: Wiley, 1988, pp. 2154–2161. Used by permission.]

filter, which separates the green fluorescent light from the blue excitation light, and the latter is scattered by the probe back into the return fiber. Photomultiplier tubes are used in this application to convert the light signal into a current, and then a current-to-voltage converter is used to provide the voltage proportional to the blue and the green light. The blue/green ratio is taken, and the $P\mathrm{O}_2$ output is calculated according to the Stern–Volmer equation.

There is a range of quenching-base sensors. They include sensors based on transition metal quenching of ligand fluorescence and on iodine quenching of rubrene fluorescence (Seitz, 1984).

DESIGN OF AN INTRAVASCULAR BLOOD-GAS MONITORING SYSTEM

Significant challenges confront the designer of a blood-gas probe and supporting instrumentation for clinical measurements. An optical fluorescence intravascular blood-gas monitoring system for critical care in surgical settings has been designed that uses a sensor probe introduced into the patient via a radial-artery catheter (Gehrich *et al.*, 1986). The same group has developed an extracorporeal circuit to monitor oxygenator performance and the patient's status during cardiopulmonary bypass surgery by means of an optical fluorescence-based blood-gas monitoring system. The following discussion deals with the development of an intravascular blood-gas monitoring system intended for continuous monitoring of arterial pH, $P\mathrm{CO}_2$, and $P\mathrm{O}_2$ in critical-care and surgical settings. The fluorescence-based blood-gas probe is introduced into the patient's vasculature by means of the radial-artery catheter. This approach is normally used for drawing blood-gas samples and for arterial pressure measurements (see Section 7.1).

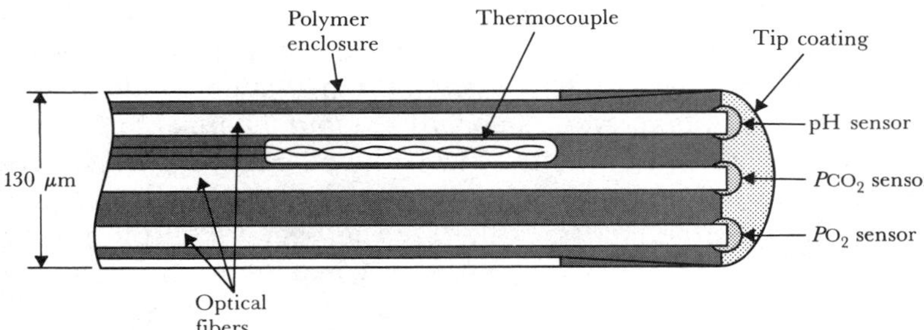

Figure 10.16 An intravascular blood-gas probe measures pH, P_{CO_2}, and P_{O_2} by means of single fiber-optic fluorescent sensors. (From J. L. Gehrich, D. W. Lübbers, N. Opitz, D. R. Hansmann, W. W. Miller, J. K. Tusa, and M. Yafuso, "Optical fluorescence and its application to an intravascular blood gas monitoring system," *IEEE Trans. Biomed. Eng.,* 1986, BME-33, 117–132. Used by permission.)

SYSTEM DESIGN CONSIDERATIONS

The system design considerations are given for the intravascular blood-gas monitoring system, which comprises a blood-gas probe, an optoelectronic instrument, and a probe calibration (Gehrich *et al.*, 1986).

Blood-Gas Probe Design The design requirements for an ideal blood-gas probe include the following: (1) operating temperature range of 15 to 42 °C, (2) pH from 6.8 to 7.8, (3) P_{CO_2} from 10 to 100 mm Hg, and (4) P_{O_2} from 20 to 300 mm Hg. The P_{O_2} value may reach 500 mm Hg for procedures that require high levels of supplemental oxygen, such as open-heart surgery. The probe must be fabricated from materials that are sterilizable and biocompatible. Carcinogenicity and toxicity must be avoided, and the blood-contact surfaces must exhibit nonthrombogenic and nonhemolytic properties.

One of the most significant requirements in designing an intravascular probe is that it not be affected by such naturally occurring substances as proteins in the blood and those introduced during the surgical or therapeutic procedures (Regnault and Picciolo, 1987). In addition, the probe must be immune to absorption of the components in the blood and to their deposition on the sensor surfaces. The probe must have a small diameter so that it can be introduced into the radial artery. At the same time, blood pressure must be measured through the lumen of the blood-gas probe.

Mechanical Design Considerations Figure 10.16 shows the design of the intravascular blood-gas probe. It consists of three single fiber-optic sensors and a thermocouple integral to a polymer structure that achieves the required strength. Fused silicon fibers are used for the three fiber-optic sensors, which measure pH, P_{CO_2}, and P_{O_2}, respectively. The thermocouple gives a direct readout of the probe and of blood temperature at the probe tip. Temperature measurements are important in that the blood solubilities of O_2 and CO_2 are temperature dependent. In addition, the fluorescence chemistry varies slightly with temperature and requires temperature compensation. *In vitro* blood-gas measurements in a laboratory are standardized to the normal body temperature of 37 °C. In the case of the intravascular blood-gas measurement system, the patient's core temperature during surgery may vary from hypothermia (say, 15 °C) to hyperthermia (say, 42 °C). The operator of the instrument must know the patient's temperature in order to make adjustments and report blood-gas values at the standardized temperature of 37 °C.

It is essential that the catheter be small, because the intravascular blood-gas probe will be inserted into a radial-artery catheter of a size consistent with clinical practice. That is, it must be possible to determine blood pressure, as well as to withdraw blood-gas samples, with the catheter in place. A viable blood pressure signal can be maintained by using a 20-gage radial-artery catheter if the diameter of the blood-gas probe is limited to 600 μm. With the probe described in Figure 10.16, this restricts the fiber diameter to 130 μm when three optical fibers and a thermocouple are included.

A major challenge is the selection of nontoxic materials the physical configuration and composition of which minimize the formation of blood clots on the blood-contact surfaces. This is important in order to prevent blockage of the residual lumen between the probe and catheter wall, which would compromise blood pressure measurements, and to reduce the risk that an embolus might form, slough off the probe/catheter, and cause trauma "downstream" in a cerebral or pulmonary capillary bed. In addition, formation of a thrombus at the site of the fluorescence sensors would affect the blood-gas measurement itself (Gehrich *et al.*, 1986).

This last issue is of least concern, because the fluorescence sensors are characterized as equilibrium sensors; that is, the parameter being measured is in equilibrium with the dye but is not being consumed. Thus thrombosis buildup on the probe would increase the time response of the sensors but would not affect the equilibrium accuracy. It has been proposed that the local metabolism of the cells coating the sensor, rather than the vascular blood gases, may affect the sensor output. This is in contrast to the behavior of electrochemical sensors (Section 10.1), which consume the analyte being measured. An example is oxygen measurement via a Clark electrode. With oxygen consumption, the buildup of fibrin causes a change in the diffusion gradient—and thus in the output current of the Clark oxygen electrode.

The issue of thrombogenicity is addressed by designing an assembly of blood surfaces that are smooth and present little opportunity for fibrin buildup. In addition, a heparin-bonding process is incorporated whereby heparin (an anticoagulant) is covalently bonded to the entire exposed surface of the probe (Gehrich *et al.*, 1986).

Fluorescence Sensor Design The system design requires a single optical fiber for both the delivery of light energy to the sensor dye and its reception from that dye. Two fibers are not necessary, because the sensor input and output signals are of different wavelengths. The design challenge is to select dyes that offer the appropriate absorption and emission wavelength characteristics, are nontoxic, can be attached to an optical fiber, have sufficient sensitivity to the physiological parameters being measured, and exhibit high fluorescent intensity for signal strength over the physiological measurement range of interest. In addition, fluorescent dyes must not be affected by drugs or other blood constituents and must be stable enough to maintain accuracy for up to three days. Because this dye is a disposable product, consideration must also be given to its cost and shelf life. Finally, the dye must have a dynamic time response such that physiological changes in the blood-gas parameters can be followed (Gehrich *et al.*, 1986).

Instrument Design The intravascular blood-gas system instrument design has three sections (Gehrich *et al.*, 1986). The first section is an analyzer module; the second is a patient interface module (PIM); and the third is the display. The illuminator consists of a broadband xenon-arc source lamp (350–750 nm), a collimating lens system, a filter wheel, and a condensing lens to direct the xenon emission onto the interface fibers. The xenon arc and filter wheel are synchronized at a flash rate of 20 Hz. The pulsating light source provides a more stable energy source than can be achieved with a constant, steady-state input signal. Light energy at specific wavelengths travels along the fiber optics to the PIM and is coupled by the graded index (GRIN) lens to the interface optics.

In order to maximize the energy delivered to and from each fiber-optic sensor, the following design approach was taken: (1) The number of optical connections was kept to a minimum. (2) The length of the fibers, especially those returning the fluorescent energy from the sensors, was kept to a minimum. (3) Transduction of the optical signal to an electric signal was made to occur at the distal end of the subsystem as near as possible to the patient. (4) The analog front-end circuitry in the patient interface module was located such that the analog signal is converted into a digital signal and multiplexed to be sent along approximately 4 m of cable to the analyzer section. All signals are normalized against the intensity, and ratiometric techniques are used to compare the active fluorescence wavelength to the reference wavelength before the blood-gas concentration is calculated.

Calibration Device For all blood-gas detection systems, it is essential that an independent calibration of the probe be made prior to its use in the patient. This is done by utilizing tonometric techniques and a fluid-filled calibration cuvette that is an integral part of the packaging of the probe, in that the sensors must remain hydrated. The calibration device uses two gas cylinders, each with appropriate, precisely controlled values of oxygen and carbon dioxide (Gehrich *et al.*, 1986).

10.4 ION-SENSITIVE FIELD-EFFECT TRANSISTOR (ISFET)

The potential for low-cost, reliable microminiature sensors that utilizes ion-sensitive field-effect transistors (ISFET) was first recognized over 20 years ago (Bergveld, 1970). ISFETs employ the same electrochemical principles in their measurement as ion-sensitive electrodes (ISE). The ISFET is produced by removal of the metal gate region that is normally present on a field-effect transistor (Rolfe, 1988).

A metal oxide–semiconductor field-effect transistor (MOSFET) is composed of two diodes separated by a gate region. The gate is a thin insulator—usually silicon dioxide—upon which a metallic material is deposited. This gate material can be any conducting material that is compatible with IC processing. Voltage applied to the gate controls the electric field in the dielectric and thus the charge on the silicon surface. This field effect is the basis of operation of the MOSFET and ISFET. The high-input impedance results from the gate insulator, which is essential for operation of the ISFET device (Janata, 1989).

Figure 10.17(a) is a schematic diagram of an ion-sensitive field-effect transistor with the sample under measurement in contact with an ion-selective membrane and a reference electrode. To improve the pH-sensitivity and stability of the silicon dioxide layer, a silicon nitride layer is placed over the silicon dioxide.

The potential developed across the insulator depends on the electrolyte concentration of the solution in contact with the ion-selective membrane. The ISFET measures the potential at the gate; this potential is derived through an ion-selective process, in which ions passing through the ion-selective membrane modulate the current between the source and the drain. The voltage across the gate region changes, and thus the field-effect transistor current flows (Arnold and Meyerhoff, 1988).

The ISFET is of considerable interest because it offers the potential for low-cost microminiature sensors. These devices can be produced by microfabrication of silicon integrated circuits (ICs). Figure 10.17(b) shows a plan view, with dimensions, for a microfabricated ISFET. The IC manufacturing technology makes use of photolithographic techniques for producing unique properties of IC silicon substrates. ISFETs are particularly attractive, because they can be made in very small sizes and because multiple analytes can be measured on a single chip. Note that ISFET sensors are in the development stage.

In one device for measuring CO_2, an Ag/AgCl reference is incorporated on the ISFET chip, and polyvinyl alcohol gel (which contains NaCl and $NaHCO_3$) is deposited over the ISFET and reference (Rolfe, 1990). These regions are then coated with a thin silicone resin. Measurements have been made for a 24-h period for intravascular experiments with animals and humans. However, encapsulation problems arose. Other ISFET sensors have been developed for potassium ion measurements; here the gate region is covered

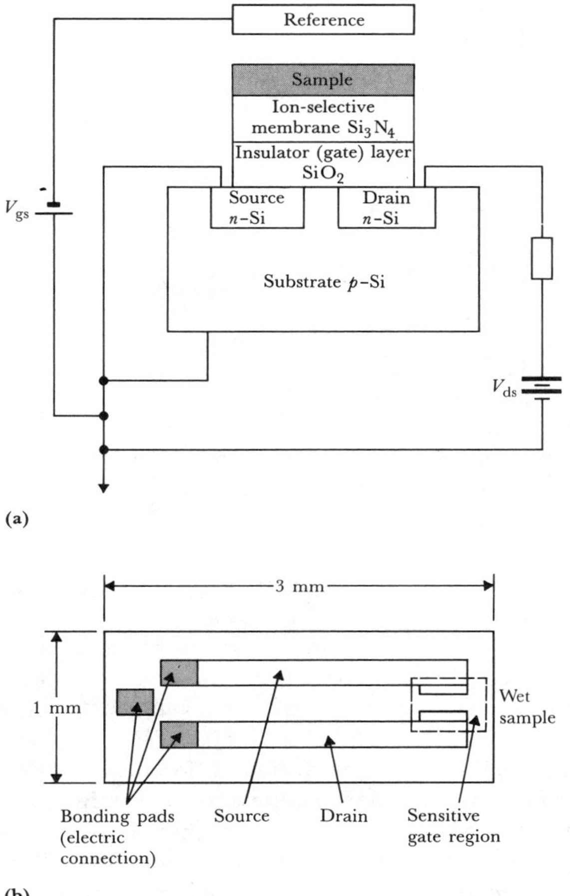

Figure 10.17 (a) In a chemically sensitive field-effect transistor, the ion-selective membrane modulates the current between the source and the drain. (b) A stretched ISFET maximizes the spacing between the "wet" sample region and the electric connections. (Part (b) from P. Rolfe, "*In vivo* chemical sensors for intensive-care monitoring," *Med. Biol. Eng. Comput.,* 1990, 28. Used by permission.)

with a glass potassium-selective membrane or with a balinomycin–PVC polymer membrane. Figure 10.18 is a plot of drain current versus potassium ion activity for an ISFET. Calcium ISFET sensors have been developed to monitor Ca^{2+} activity in venous blood of dogs.

The initial use of ISFETS will involve small volumes of analytes and measurement times of only a few seconds (Van der Spiegel and Zemel, 1988). This measurement speed is fast compared to the several minutes required in a typical laboratory analysis. ISFETs are suited for monitoring blood electrolytes

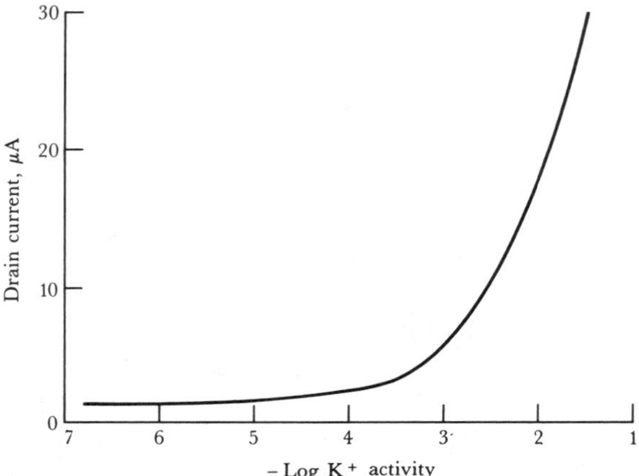

Figure 10.18 Dependence of current on potassium ion activity for a potassium ion-sensitive field-effect transistor.

and could perhaps be used for measurements inside a cell, provided that workable fabrication techniques are developed.

The main challenge of designing ISFET devices is satisfactory encapsulation of the ISFETs in order to protect the electric characteristics of the ISFET, which deteriorate as a result of water vapor entering from the environment.

Multiple-species ISFETs for up to eight different sensors have been fabricated on silicon chips a few square millimeters in size. In addition, probes 50 μm in diameter have been fabricated for on-chip circuitry that can measure pH, glucose, oxygen saturation, and pressure for biomedical applications. Another attractive feature of the ISFET is that on a single chip, in addition to the ISFET sensor, integrated circuits can also be deposited and used for signal processing (Van der Spiegel and Zemel, 1988).

10.5 IMMUNOLOGICALLY SENSITIVE FIELD-EFFECT TRANSISTOR (IMFET)

The immunologically sensitive field-effect transistor (IMFET) is an extension of the ISFET. As we noted, the ISFET takes advantage of the ion-sensitive or chemical sensitive properties of the field-effect transistor. As described above, the ISFET design makes use of the properties of the metal-insulator-semiconductor structure, in which the gate metal layer and the semiconductor layer form a capacitive sandwich by framing an insulating layer—normally SiO_2. Essentially, the system is a capacitor with a totally impermeable dielectric through which no charge passes.

The IMFET is similar in structure to the ISFET except that the solution-membrane interface is polarized rather than unpolarized; that is, charged species cannot cross the membrane (Liu, 1988). The ISFET interacts through an ion-exchange mechanism with the chemical analyte that is being measured, whereas the IMFET operation is based on an antigen–antibody reaction. An antibody is immobilized on the membrane that is attached to the insulator of a FET. In this way the device is used as an antigen sensor. An antibody could be detected in a similar way: by immobilizing an antigen on the membrane. The IMFET measures charge, so in order to be sensed, the absorbing species on the membrane must possess a net electric charge.

10.6 NONINVASIVE BLOOD-GAS MONITORING

Blood-gas determination can provide valuable information about the efficiency of pulmonary gas exchange, the adequacy of alveolar ventilation, blood-gas transport, and tissue oxygenation. Although invasive techniques to determine arterial blood gases are still widely practiced in many clinical situations, it is becoming apparent that simple, real-time, continuous, and noninvasive techniques offer many advantages. Most important, intermittent blood sampling provides historical data valid only at the time the sample was drawn. Delays between when the blood sample is drawn and when the blood-gas values are reported average about 30 min. Furthermore, invasive techniques are painful and have associated risks.

These limitations are particularly serious in critically ill patients for whom close monitoring of arterial blood gases is essential. Continuous noninvasive monitoring of blood gases, on the other hand, makes it possible to recognize changes in tissue oxygenation immediately and to take corrective action before irreversible cell damage occurs.

Various noninvasive techniques for monitoring arterial O_2 and CO_2 have been developed. This section describes the basic sensor principles, instrumentation, and clinical applications of the noninvasive monitoring of arterial oxygen saturation ($S\text{O}_2$), oxygen tension ($P\text{O}_2$), and carbon dioxide tension ($P\text{CO}_2$).

SKIN CHARACTERISTICS

In order to appreciate the challenges of noninvasive measurement of the blood chemistry, it is important to understand the structure of the human skin. The human skin has three principal layers: the stratum corneum, epidermis, and dermis (Mendelson and Peura, 1984). These layers form a cohesive structure that typically varies in thickness from 0.2 to 2 mm, depending on the position on the body. Figure 5.7 is a schematic diagram that represents a cross section of the human skin.

The stratum corneum is the nonliving, outer layer of the skin. It is composed of a supple, protective layer of dehydrated cells. The nonvascular epidermis layer is a living tissue underneath the stratum corneum. It consists of proteins, lipids, and the melanin-forming cells (melanocytes) that give skin its color. The average thickness of the epidermis is 0.1–0.2 mm.

Dense connective tissue, hair follicles, sweat glands, nerve endings, fat cells, and a profuse system of capillaries make up the dermis. Here vertical capillary loops approximately 200–400 μm in length provide nutrients for the upper layers of the skin. Blood is supplied to these capillaries by arterioles that form a flat network parallel to the surface of the skin below the dermis. Larger arteries located in the subcutaneous tissue supply these arterioles. Venous blood in the skin is drained by venules in the upper and middle dermis and by larger veins in the subcutaneous tissue.

Arteriovenous anastomoses are innervated by nerve fibers. These shunts are found largely in the dermis of the palms, ears, and face. They regulate blood flow through the skin in response to heat; blood flow through these channels can increase to nearly 30 times the basal rate. Normal gas diffusion through the skin is low, but with increased heat—at 40 °C and above—the skin becomes more permeable to gases.

TRANSCUTANEOUS ARTERIAL OXYGEN SATURATION MONITORING (PULSE OXIMETRY)

Attempts to apply the non-pulsed two-wavelengths approach that we have discussed, which was successful for intravascular oximetry applications, to the transilluminated ear or fingertip resulted in unacceptable errors due to light attenuation by tissue and blood absorption, refraction, and multiple scattering. And because of differences in the properties of skin and tissue, variation from individual to individual in attenuation of light caused large calibration problems. Oximeters can be used to measure $S\text{O}_2$ noninvasively by passing light through the pinna of the ear (Merrick and Hayes, 1976). Because of the complications caused by the light-absorbing characteristics of skin pigment and other absorbers, measurements are made at eight wavelengths and are computer-processed. The ear is warmed to 41 °C to stimulate arterial blood flow.

A two-wavelength transmission noninvasive pulse oximeter was introduced (Yoshiya *et al.*, 1980). This instrument determines $S\text{O}_2$ by analyzing the time-varying, or ac, component of the light transmitted through the skin during the systolic phase of the blood flow in the tissue (Figure 10.19). This approach achieves measurement of the arterial oxygen content with only two wavelengths (660 and 940 nm, for instance). The dc component of the transmitted light, which represents light absorption by the skin pigments and other tissues, is used to normalize the *ac* signals.

A transcutaneous reflectance oximeter based on a similar photoplethysmographic technique has been developed (Mendelson *et al.*, 1983). The advantage of the reflectance oximeter is that it can monitor $S\text{O}_2$ transcutaneously at various locations on the body surface, including more central locations

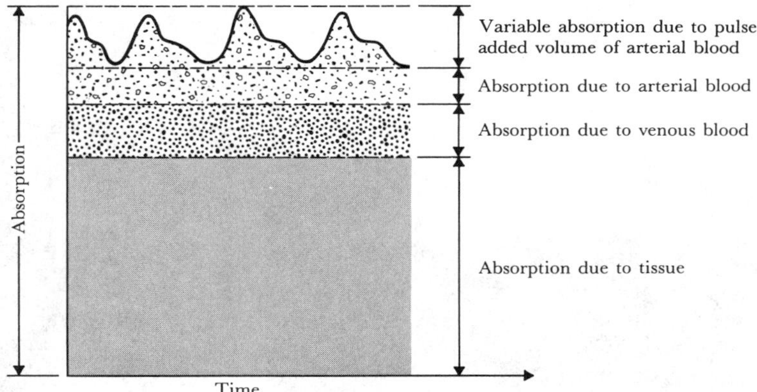

Figure 10.19 The pulse oximeter analyzes the light absorption at two wavelengths of only the pulse-added volume of oxygenated arterial blood. [From Y. M. Mendelson, "Blood gas measurement, transcutaneous," in J. G. Webster (ed.), *Encyclopedia of Medical Devices and Instrumentation.* New York: Wiley, 1988, pp. 448–459. Used by permission.]

(such as the chest, forehead, and limbs) that are not accessible via conventional transmission oximetry.

Because of these and other significant improvements in the instruments, measurements of ear, toe, and fingertip oximetry are widely used. Noninvasive measurements of S_{O_2} can be made with a 2.5% accuracy for saturation values from 50 to 100%.

Transcutaneous S_{O_2} Sensor The basic transcutaneous S_{O_2} sensor, for both the transmission and the reflective mode, makes use of a light source and a photodiode. In the transmission mode, the two face each other and a segment of the body is interposed; in the reflection mode, the light source and photodiode are mounted adjacent to each other on the surface of the body.

Figure 10.20 shows an example of a transcutaneous transmission S_{O_2} sensor and monitor. These transmission sensors are placed on the fingertips, toes, ear lobes, or nose. A pair of red and infrared light-emitting diodes are used for the light source, with peak emission wavelengths of 660 nm (red) and 940 nm (infrared). These detected signals are processed, in the form of transmission photoplethysmograms, by the oximeter, which determines the S_{O_2}.

Applications of S_{O_2} Monitoring As we have noted, the applications of noninvasive S_{O_2} monitoring have blossomed rapidly to the point where it has become the standard of clinical care in a number of areas. Direct assessment and trending of the adequacy of tissue oxygenation can be made by determining the S_{O_2} value. Oximetry is applied during the administration of anesthesia, pulmonary function tests, bronchoscopy, intensive care, and oral surgery, in neonatal monitoring, in sleep apnea studies, and in aviation medicine.

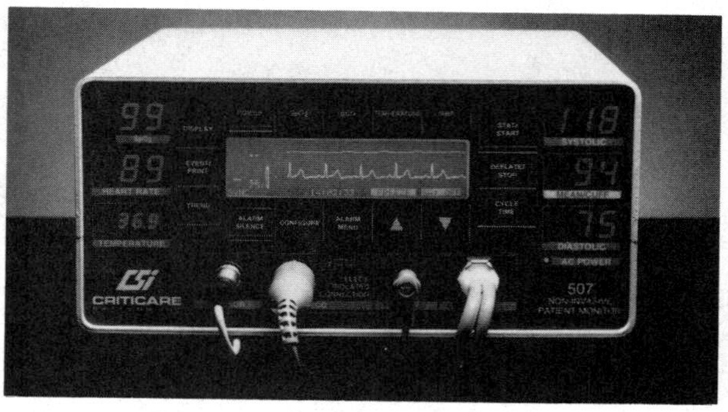

(a)

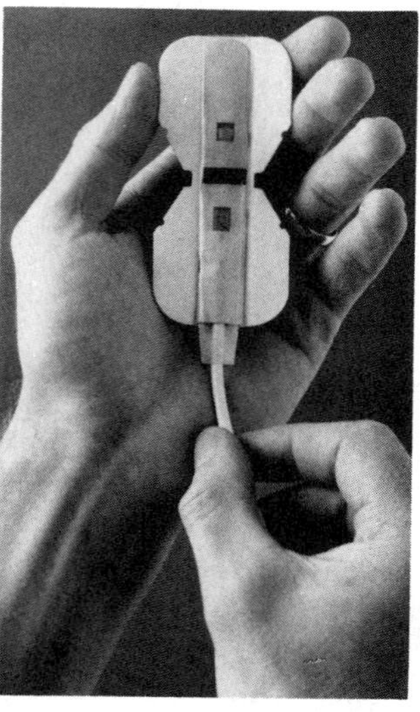

(b)

Figure 10.20 (a) Noninvasive patient monitor capable of measuring ECG, noninvasive blood pressure (using automatic oscillometry), respiration (using impedance pneumography), transmission pulse oximetry, and temperature. (From Criticare Systems, Inc. Used by permission.) (b) Disposable transmission SO_2 sensor in open position. Note the light sources and detector, which can be placed on each side of the finger. (From Datascope Corporation. Used by permission.)

Noninvasive oximetry is also used in the home for monitoring self-administered oxygen therapy. Noninvasive oximetry provides time-averaged blood oxygenation values and can be used to determine when immediate therapeutic intervention is necessary. A lightweight (less than 3 g) and small (20-mm diameter) optical sensor makes this transcutaneous reflectance sensor appropriate for monitoring newborns, ambulatory patients, and patients in whom a digit or ear lobe is not accessible. Problems with both transmission and reflectance oximetry include poor signal with shock, interference from lights in the environment and from the presence of carboxyhemoglobin, and poor trending of transients (Payne and Severinghaus, 1986, Moyle, 1994).

TRANSCUTANEOUS ARTERIAL OXYGEN TENSION (tcPO_2) MONITORING

Measurement of tcPO_2 is similar in principle to the conventional *in vitro* PO_2 determination we have described. A Clark electrode is used in a sensor unit that is placed in contact with the skin. The oxygen electrode principle of operation has already been discussed.

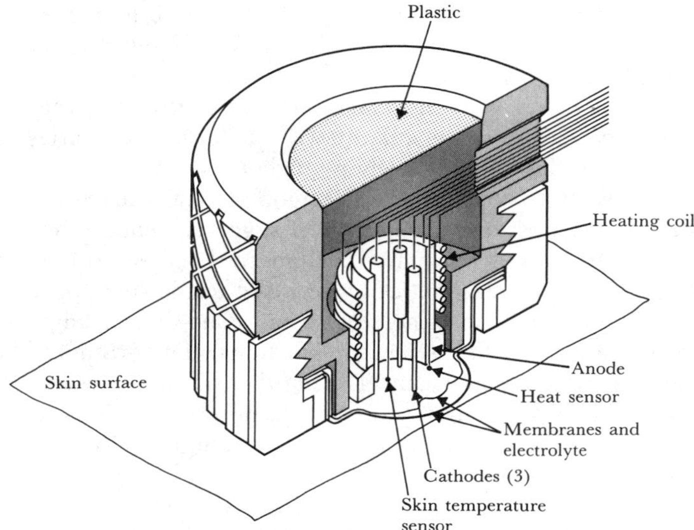

Figure 10.21 Cross-sectional view of a transcutaneous oxygen sensor. Heating promotes arterialization. (From A. Huch and R. Huch, "Transcutaneous, noninvasive monitoring of Po_2," *Hospital Practice,* 1976, 6, 43–52. Used by permission.)

Only two known gas mixtures are required to calibrate the sensor, because the relationship between O_2-dependent current and Po_2 is linear. Two calibration procedures are commonly used. One employs two precision medical gas mixtures, such as nitrogen and oxygen. The other employs sodium sulfite, which is a "zero-O_2 solution," and ambient air. Good stability of the sensor is usually maintained; a drift of 1–2 mm Hg/h for the tcPo_2 sensor is typical.

Transcutaneous Po_2 Sensor Figure 10.21 shows a cross-sectional view of a typical Clark-type tcPo_2 sensor in which three glass-sealed Pt cathodes are separately connected via current amplifiers to an Ag/AgCl anode ring (Huch and Huch, 1976). A buffered KCl electrolyte, which has a low water content to reduce drying of the sensor during storage, is used to provide a medium in which the chemical reactions can occur. Under normal physiological conditions, the Po_2 at the skin surface is essentially atmospheric regardless of the Po_2 in the underlying tissue.

Hyperemia of the skin causes the skin Po_2 to approach the arterial Po_2. Hyperemia can be induced by the administration of certain drugs, by the heating or abrasion of the skin, or by the application of nicotinic acid cream. Because heating gives the most readily controllable and consistent effect, a heating element and a thermistor sensor are used to control the skin temperature beneath the tcPo_2 sensor. Sufficient arterialization results when the skin is heated to temperatures between 43 and 44 °C. These temperatures cause minimal skin damage, but with neonates it is still necessary to reposition the sensor frequently to avoid burns.

Heating the skin has two beneficial effects: O_2 diffusion through the stratum corneum increases, and vasodilation of the dermal capillaries increases blood flow to skin at the sensor site where the heat is applied. Increased blood flow delivers more O_2 to the heated skin region, making the excess O_2 diffuse through the skin more easily. As Figure 10.1 suggests, heating the blood also causes the ODC to shift to the right, resulting in a decreased binding of Hb with O_2. Accordingly, the amount of O_2 released to the cells for a given $P\mathrm{O}_2$ is increased. Note that heat also increases local tissue O_2 consumption, which tends to decrease oxygen levels in the skin tissue. Opportunely, these two opposing factors approximately cancel each other. Duration of monitoring is a function of the skin's sensitivity to possible burns, as well as to electrode drift. Typically, continuous monitoring is recommended for 2–6 h before moving to a different skin site.

Applications of tc$P\mathrm{O}_2$ Monitoring Monitoring the tc$P\mathrm{O}_2$ has found many applications in both clinical medicine and physiological research in situations where tissue oxygenation values are important. Although tc$P\mathrm{O}_2$ measurements are used routinely for neonates because of their thin skin, clinical results with adults have proved less valuable. The prime application of tc$P\mathrm{O}_2$ is for newborn infants, especially those in respiratory distress (Cassady, 1983). The main reason for this application is that the need often arises to administer O_2 to sick infants, while at the same time avoiding high arterial $P\mathrm{O}_2$, which, in preterm infants, can lead to serious damage to retinal and pulmonary tissues. Under the opposite condition of low $P\mathrm{O}_2$, fetal circulation paths may be re-established in the neonate (Huch *et al.*, 1981). Even so, because of its simpler operation, lower cost, absence of calibration, and increased reliability, noninvasive pulse oximetry has supplanted the use of tc$P\mathrm{O}_2$ measurements in the neonate.

Good correlations between tc$P\mathrm{O}_2$ and arterial $P\mathrm{O}_2$ are possible when the patient is not in shock or in hypothermia. With patients who are hemodynamically compromised, tc$P\mathrm{O}_2$ does not always equal arterial $P\mathrm{O}_2$. Skin heating in situations where there are significant decreases in skin blood perfusion cannot compensate for the low blood flow and the attendant low delivery of oxygen to the tissue. Low transcutaneous $P\mathrm{O}_2$ readings result. Examples of conditions in which skin perfusion is compromised—and tc$P\mathrm{O}_2$ readings therefore do not represent tissue $P\mathrm{O}_2$ values—include severe hypothermia, acidemia, anemia, and shock. Adult tc$P\mathrm{O}_2$ values have not been found to equal arterial $P\mathrm{O}_2$, even when the skin is heated to 45 °C. This is due to the greater skin thickness of the adult; heating of the skin to intolerably high temperatures would be necessary to compensate for the increased metabolism. Studies have, however, demonstrated the clinical usefulness of this technique for evaluating the adequacy of cutaneous circulation in patients with peripheral resuscitation (Huch *et al.*, 1981).

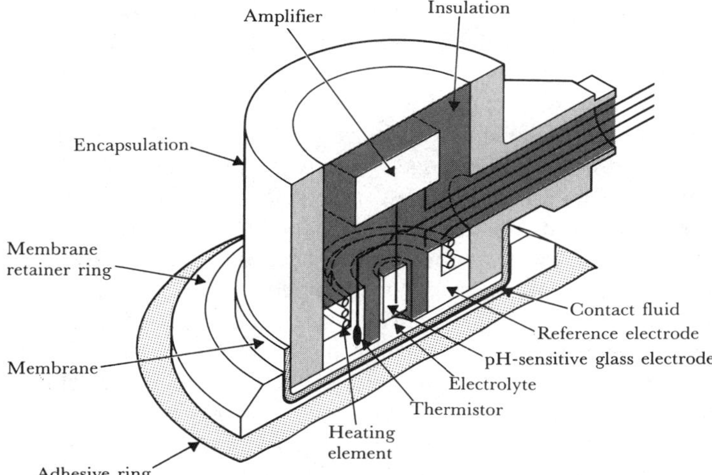

Figure 10.22 Cross-sectional view of a transcutaneous carbon dioxide sensor. Heating the skin promotes arterialization. (From A. Huch, D. W. Lübbers, and R. Huch, "Patientenuberwachung durch transcutane $P\text{CO}_2$ Messung bei gleiechzeiliger koutrolle der relatiuen Iokalen perfusion," *Anaesthetist,* 1973, 22, 379. Used by permission.)

Maintaining the seal between the tc$P\text{O}_2$ probe and the skin surface can be a problem with long-term monitoring. If the seal is compromised, the sensor is exposed to the atmosphere and will yield a $P\text{O}_2$ of approximately 155 mm Hg, instead of lower physiological values.

TRANSCUTANEOUS CARBON DIOXIDE TENSION (tc$P\text{CO}_2$) MONITORING

Monitoring tc$P\text{CO}_2$ gives more accurate results than tc$P\text{O}_2$ measurements in adult patients, because tc$P\text{CO}_2$ measurements are much less dependent on skin blood flow.

Transcutaneous $P\text{CO}_2$ Sensor Figure 10.22 shows a typical tc$P\text{CO}_2$ sensor, which is similar to a tc$P\text{O}_2$ sensor except for the sensing element. Its operation is similar to that of the electrochemical $P\text{CO}_2$ sensor described earlier. The CO_2 sensor is a glass pH electrode with a concentric Ag/AgCl reference electrode that is used as a heating element. The electrolyte, a bicarbonate buffer, is placed on the electrode surface. A CO_2-permeable Teflon membrane separates the sensor from its environment.

As we noted before, the tc$P\text{CO}_2$ sensor operates according to the Stow–Severinghaus principle; that is, a pH electrode senses a change in the CO_2 concentration. The system is calibrated with a known CO_2 concentration solution. Because a CO_2 electrode has a negative temperature coefficient, calibration must be performed at the temperature at which the device will be used. The effects of heating the skin beneath the tcP_{CO_2} sensor must be determined before the measurements can be properly interpreted.

Heating the skin beneath the sensor causes an increase in (1) P_{CO_2}, because the solubility of CO_2 decreases with an increase in temperature; (2) local tissue metabolism, because cell metabolism is directly correlated with temperature; and (3) the rate of CO_2 diffusion through the stratum corneum, which increases with temperature. As a consequence of these three effects, which all work in the same direction to increase tcP_{CO_2} values, heating the skin yields tcP_{CO_2} values larger than the corresponding arterial P_{CO_2}. Nevertheless, the correlation between tcP_{CO_2} and arterial P_{CO_2} is usually satisfactory. Because the slope of the CO_2 electrode calibration line is essentially that of the Nernst equation, a two-point calibration (as for the P_{O_2} electrode) is not needed.

Transcutaneous P_{CO_2} sensors have longer time constants than tcP_{O_2} sensors. The response time of a tcP_{CO_2} electrode varies inversely with temperature (Herrell *et al.*, 1980). *In vitro* tests have shown that the 90% response time is less than 60 s for a sensor at 44 °C. Measurements of the tcP_{CO_2} sensor response time, with step increases in the inspired CO_2, give longer time constants (Tremper *et al.*, 1981). Increasing CO_2 concentrations from 0 to 7% at different sensor and skin temperatures resulted in the time constants 15, 7.5, 5, and 3.5 min for electrode temperatures of 37, 39, 41, and 44 °C, respectively. Note, however, that the measured response time included the response times due to CO_2 diffusion in the alveoli, capillary blood, skin, and sensor. These pronounced temperature effects can be attributed to significant changes in the structure of the stratum corneum caused by temperatures greater than 40 °C. Heating the electrode has little effect in neonates, because the stratum corneum is not fully developed.

Applications of tcP$_{CO_2}$ Monitoring The tcP_{CO_2} is higher than blood P_{CO_2} because epidermal cell CO_2 diffuses to the dermal capillaries in response to a diffusion gradient. A countercurrent-exchange mechanism in the dermal capillaries causes CO_2 diffusion between the parallel arterial and venous sides of the capillary bed. Arterial blood entering the rising segment of the capillary loop picks up CO_2 from the exiting venous side. As a consequence, the venous P_{CO_2} is lowered, and a maximal P_{CO_2} gradient is established at the top of the countercurrent capillary loops. Because of this phenomenon, P_{CO_2} at the skin surface is higher than venous P_{CO_2}, even when the electrode is not heated (Tremper *et al.*, 1981).

Generally, it is accepted that tcP_{CO_2} is a valuable trend monitor in neonates and adults who are not in shock. Since arterial P_{CO_2} varies linearly with alveolar ventilation, tcP_{CO_2} provides information concerning the effectiveness of spontaneous or mechanical ventilation for individuals. The extent of impaired tissue perfusion, i.e. circulation to a limb, or response to therapy may be monitored by observing the change in tcP_{CO_2}.

10.7 BLOOD-GLUCOSE SENSORS

Accurate measurement of blood glucose is essential in the diagnosis and long-term management of diabetes. This section reviews the use of biosensors for continuous measurement of glucose levels in blood and other body fluids.

Glucose is the main circulating carbohydrate in the body. In normal, fasting individuals, the concentration of glucose in blood is very tightly regulated—usually between 80 and 90 mg/100 ml, during the first hour or so following a meal. The hormone insulin, which is normally produced by beta cells in the pancreas, promotes glucose transport into skeletal muscle and adipose tissue. In those suffering from diabetes mellitus, insulin-regulated uptake is compromised, and blood glucose can reach concentrations ranging from 300 to 700 mg/100 ml (hyperglycemia).

Accurate determination of glucose levels in body fluids, such as blood, urine, and cerebrospinal fluid, is a major aid in diagnosing diabetes and improving the treatment of this disease. Blood glucose levels rise and fall several times a day, so it is difficult to maintain normoglycemia by means of an "open-loop" insulin delivery approach. One solution to this problem would be to "close the loop" by using a self-adapting insulin infusion device with a glucose-controlled biosensor that could continuously sense the need for insulin and dispense it at the correct rate and time. Unfortunately, present-day glucose sensors cannot meet this stringent requirement (Peura and Mendelson, 1984).

Electroenzymatic Approach Electroenzymatic sensors based on polarographic principles utilize the phenomenon of glucose oxidation with a glucose oxidase enzyme (Clark and Lyons, 1962). The chemical reaction of glucose with oxygen is catalyzed in the presence of glucose oxidase. This causes a decrease in the partial pressure of oxygen (Po_2), an increase in pH, and the production of hydrogen peroxide by the oxidation of glucose to gluconic acid according to the following reaction:

$$\text{Glucose} + O_2 \xrightarrow[\text{oxidase}]{\text{Glucose}} \text{Gluconic acid} + H_2O_2 \tag{10.18}$$

Investigators measure changes in all of these chemical components in order to determine the concentration of glucose. The basic glucose enzyme electrode utilizes a glucose oxidase enzyme immobilized on a membrane or a gel matrix, and an oxygen-sensitive polarographic electrode. Changes in oxygen concentration at the electrode, which are due to the catalytic reaction of glucose and oxygen, can be measured either amperometrically or potentiometrically.

Because a single-electrode technique is sensitive both to glucose and to the amount of oxygen present in the solution, a modification to remove the oxygen response by using two polarographic oxygen electrodes has been sug-

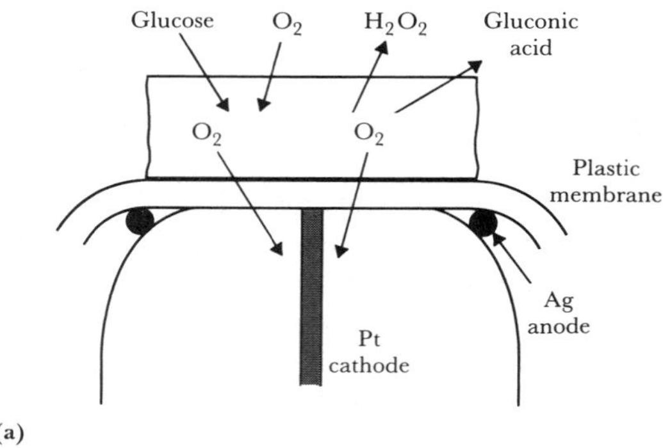

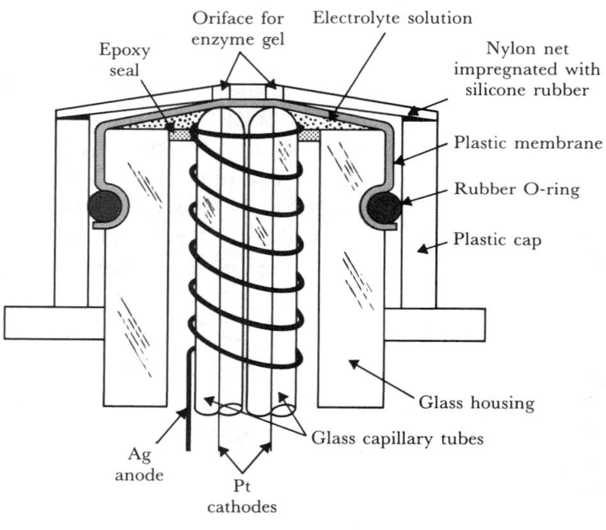

Figure 10.23 (a) In the enzyme electrode, when glucose is present it combines with O_2, so less O_2 arrives at the cathode. (b) In the dual-cathode enzyme electrode, one electrode senses only O_2, and the difference signal measures glucose independent of O_2 fluctuations. (From S. J. Updike and G. P. Hicks, "The enzyme electrode, a miniature chemical transducer using immobilized enzyme activity," *Nature,* 1967, 214, 986–988. Used by permission.)

gested (Updike and Hicks, 1967). Figure 10.23 illustrates both the principle of the enzyme electrode and the dual-cathode enzyme electrode. An active enzyme is placed over the glucose electrode, which senses glucose and oxygen. The other electrode senses only oxygen. The amount of glucose is determined as a function of the difference between the readings of these two electrodes. More recently, development of hydrophobic membranes that are more perme-

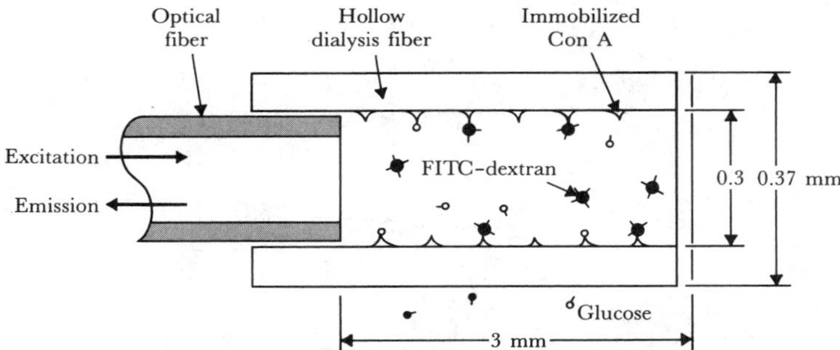

Figure 10.24 The affinity sensor measures glucose concentration by detecting changes in fluorescent light intensity caused by competitive binding of a fluorescein-labeled indicator. (From J. S. Schultz, S. Manouri, *et al.*, "Affinity sensor: A new technique for developing implantable sensors for glucose and other metabolites," *Diabetes Care,* 1982, 5, 245–253. Used by permission.)

able to oxygen than to glucose has been described (Updike *et al.*, 1982). Placing these membranes over a glucose enzyme electrode solves the problem associated with oxygen limitation and increases the linear response of the sensor to glucose.

The major problem with enzymatic glucose sensors is the instability of the immobilized enzyme and the fouling of the membrane surface under physiological conditions. Most glucose sensors operate effectively only for short periods of time. In order to improve the present sensor technologies, more highly selective membranes must be developed. The features that must be taken into account in designing and fabricating these membranes include the diffusion rate of both oxygen and glucose from the external medium to the surface of the membrane, diffusion and concentration gradients within the membrane, immobilization of the enzyme, and the stability of the enzymatic reaction (Jaffari and Turner, 1995).

Optical Approach A number of innovative glucose sensors, based on different optical techniques, has been developed in recent years. A new fluorescence-based affinity sensor has been designed for monitoring various metabolites, especially glucose in the blood plasma (Schultz *et al.*, 1982). The method is similar in principle to that used in radioimmunoassays. It is based on the immobilized competitive binding of a particular metabolite and fluorescein-labeled indicator with receptor sites specific for the measured metabolite and the labeled ligand (the molecule that binds).

Figure 10.24 shows an affinity sensor in which the immobilized reagent is coated on the inner wall of a glucose-permeable hollow fiber fastened to the end of an optical fiber. The fiber-optic catheter is used to detect changes in fluorescent light intensity, which is related to the concentration of glucose. These researches have demonstrated the simplicity of the sensor and the feasibility of its miniaturization, which could lead to an implantable glucose

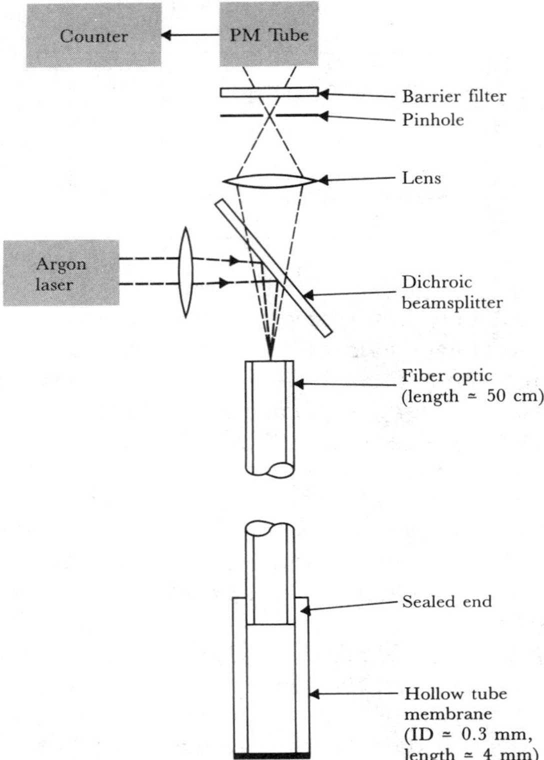

Figure 10.25 The optical system for a glucose affinity sensor uses an argon laser and a fiber-optic catheter. (From J. S. Schultz, S. Manouri, *et al.*, "Affinity sensor: A new technique for developing implantable sensors for glucose and other metabolites," *Diabetes Care,* 1982, 5, 245–253. Used by permission.)

sensor. Figure 10.25 is a schematic diagram of the optical system for the affinity sensor. The advantage of this approach is that it has the potential for miniaturization and for implantation through a needle. In addition, as with other fiber-optic approaches, no electric connections to the body are necessary.

The major problems with this approach are the lack of long-term stability of the reagent, the slow response time of the sensor, and the dependence of the measured light intensity on the amount of reagent, which is usually very small and may change over time.

Attenuated Total Reflection and Infrared Absorption Spectroscopy The application of multiple infrared ATR spectroscopy to biological media is another potentially attractive noninvasive technique. By this means, the infrared spectra of blood can be recorded from tissue independently of the sample thickness, whereas other optical-transmission techniques are strongly depen-

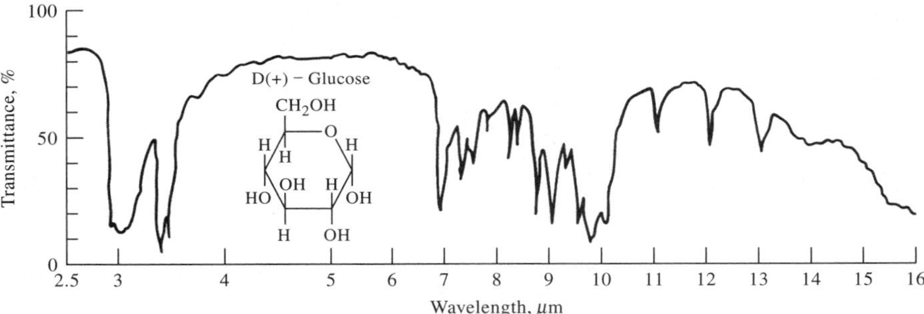

Figure 10.26 The infrared absorption spectrum of anhydrous D-glucose has a strong absorption peak at 9.7 μm. (From Y. M. Mendelson, A. C. Clermont, R. A. Peura, and B. C. Lin, "Blood glucose measurement by multiple attenuated total reflection and infrared absorption spectroscopy," *IEEE Trans. Biomed. Eng.,* 1990, 37, 458–465. Used by permission.)

dent on the optical-transmission properties of the medium. Furthermore, employing a laser light source makes possible considerable improvement of the measuring sensitivity. This is of particular interest when one is measuring the transmission of light in aqueous solutions, because it counteracts the intrinsic attenuation of water, which is high in most wavelength ranges.

Absorption spectroscopy in the infrared (IR) region is an important technique for the identification of unknown biological substances in aqueous solutions. Because of vibrational and rotational oscillations of the molecule, each molecule has specific resonance absorption peaks, which are known as fingerprints. These spectra are not uniquely identified; rather, the IR absorption peaks of biological molecules often overlap. An example of such a spectrum is shown in Figure 10.26, which is the characteristic IR spectrum of anhydrous D-glucose in the wavelength region 2.5–10 μm. The strongest absorption peak, around 9.7 μm, is due to the carbon–oxygen–carbon bond in the molecule's pyran ring.

The absorption-peak magnitude is directly related to the glucose concentration in the sample, and its spectral position is within the wavelength range emitted by a CO_2 laser. Thus a CO_2 laser can be used as a source of energy to excite this bond, and the IR absorption intensity at this peak provides, via Beer's law, a quantitative measure of the glucose concentration in a sample.

Two major practical challenges must be overcome in order to measure the concentration of glucose in an aqueous solution, such as blood, by means of conventional IR absorption spectroscopy. (1) Pure water has an intrinsic high background absorption in the IR region, and (2) the normal concentration of glucose and other analytes in human blood is relatively low (for glucose, it is typically 90–120 mg/dl, or mg%).

Significant improvements in measuring physiological concentrations of glucose and other blood analytes by conventional IR spectrometers have resulted from the use of high-power sources of light energy at specific active wavelengths. In the case of glucose, the CO_2 laser serves as an appropriate IR source.

10.8 SUMMARY

Many biosensors produce signals that are correlated with the concentration of glucose in body fluids. It may be possible to miniaturize some small sensors for implantation. Nevertheless, further progress must be made before these sensors can be used reliably for long-term monitoring of glucose in the body. The problems that have yet to be solved involve operating implanted sensors in the chemically harsh environment of the body, where they are subject to continuous degradation by blood and tissue components. The device must be biocompatible, properly encapsulated, and well protected against elevated temperatures and saline conditions. Furthermore, it should be possible to calibrate the sensor *in situ*.

PROBLEMS

10.1 Design an amplifier for use with the pH electrode. An output in the range of 1 to 2 mV is desired for the normal pH variation of blood.
10.2 Sketch the arrangement of a $P\text{CO}_2$ electrode. Explain briefly how it works.
10.3 What affects the response time of the CO_2 electrode?
10.4 Because the average venous-arterial oxygen tension is about 70 mm Hg and that of air is about 155 mm Hg, there exists an inward flux of oxygen from the air to all surfaces of the mammalian body. Normally insignificant compared to that of the lungs, this oxygen uptake *is*, however, significant for the cornea, which obtains its metabolic oxygen not from blood but rather from the inward flux of oxygen from the air. Design a system to measure the inward flux of oxygen across the cornea. Specify what parameters you would monitor, and indicate how you would determine the oxygen influx across the cornea in liters of O_2 per square centimeter of cornea surface per hour.
10.5 Design an amplifier and a power source for an O_2 electrode. The output of your device should range from 0 to 10 V for an oxygen range from 0 to 100%. At a 20% O_2 level, the electrode current is 50 nA.
10.6 What affects the response time of the O_2 electrode?
10.7 As described in the text, glucose concentration can be determined enzymatically by a glucose oxidase procedure. An oxygen electrode can be used if the plastic electrode membrane is coated with a layer of glucose oxidase immobilized in acrylamide gel. When the electrode is placed in a solution containing glucose and oxygen, the glucose and oxygen diffuse into the gel

layer of immobilized enzyme. The diffusion flow of oxygen through the plastic membrane to the oxygen electrode is decreased in the presence of the glucose. One difficulty with this electrode design is that it responds to changes in oxygen concentration as well as to changes in glucose concentration. Design an instrumentation system for *in vivo* measurement that responds only to the change in glucose concentration and not to changes in oxygen concentration. Your design should include circuit diagrams, the equations for all reactions occurring at the electrodes, and explanations of how your system would work.

10.8 Explain what a double-beam optical instrument is. Give an example of a medical instrument that operates on this principle, and explain how it improves the instrument's performance.

10.9 A blood specimen has a hydrogen ion concentration of 40 nmole/liter and a $P\text{CO}_2$ of 60 mm Hg. What is the pH? What type of acid–base abnormality does the patient exhibit?

REFERENCES

Adamsons, K., S. D. Salha, G. Gillian, and A. Games, "Influence of temperature on blood pH of the human adult and newborn." *J. Appl. Physiol.,* 1964, 19, 894.

Allan, W. B., *Fibre Optics Theory and Practice.* New York: Plenum, 1973.

Arnold, M. A., and M. E. Meyerhoff, "Recent advances in the development and analytical applications of biosensing probes." *CRC Crit. Rev. Anal. Chem.,* 1988, 20, 149–196.

Bergveld, P., "Development of an ion-sensitive solid-state device for neurophysiological measurement." *IEEE Trans. Biomed. Eng.,* 1970, BME-17, 70–71.

Burton, G. W., "Effects of the acid-base state upon the temperature coefficient of pH of blood." *Brit. J. Anesth.,* 1965, 37, 89.

Cassady, G., "Transcutaneous monitoring in the newborn infant." *J. Pediatr.,* 1983, 103, 837–848.

Clark, L. C., "Monitor and control of blood and tissue oxygen tensions." *Trans. Am. Soc. Artif. Intern. Organs,* 1956, 2, 41–47.

Clark, L. D., Jr., and C. Lyons, "Electrode systems for continuous monitoring in cardiovascular surgery." *Ann. N.Y. Acad. Sci.,* 1962, 102, 29–45.

Collison, M. E., and M. E. Meyerhoff, "Chemical sensors for bedside monitoring of critically ill patients." *Anal. Chem.,* 1990, 62, 7, 425A–437A.

Davenport, H. W., *The ABC of Acid-Base Chemistry,* 6th ed. Chicago: University of Chicago Press, 1975.

Fogt, E. J., "Electrochemical sensors," in J. G. Webster (ed.), *Encyclopedia of Medical Devices and Instrumentation.* New York: Wiley, 1988, pp. 1062–1072.

Gehrich, J. L., D. W. Lübbers, N. Opitz, D. R. Hansmann, W. W. Miller, J. K. Tusa, and M. Yafuso, "Optical fluorescence and its application to an intravascular blood-gas monitoring system." *IEEE Trans. Biomed. Eng.,* 1986, BME-33, 117–132.

Hahn, C. E. W., "Blood gas measurement." *Clin. Phys. Physiol. Meas.*, 1987, 8, 3–38.

Harrick, N. J., *Internal Reflection Spectroscopy.* Ossining, N.Y.: Harrick Scientific Co., 1979.

Herrell, N., R. J. Martin, M. Pultusker, M. Lough, and A. Fanaroff, "Optimal temperature for the measurement of transcutaneous carbon dioxide tension in the neonate." *J. Pediatr.,* 1980, 97, 114–117.

Hicks, R., J. R. Schenken, and M. A. Steinrauf, *Laboratory Instrumentation.* New York: Harper & Row, 1974.

Huch, A., and R. Huch, "Transcutaneous, noninvasive monitoring of $P\text{O}_2$." *Hospital Practice,* 1976, 6, 43–52.

Huch, R., A. Huch, and D. W. Lübbers, *Transcutaneous* Po_2. New York: Thieme-Stratton, 1981.

Jaffari, S. A., and A. P. F. Turner, "Recent advances in amperometric glucose biosensor for in vivo monitoring." *Physiol. Meas.,* 1995, 16, 1–15.

Janata, J., *Principles of Chemical Sensors.* New York: Plenum, 1989.

Kocache, R., "Oxygen analyzers," in J. G. Webster (ed)., *Encyclopedia of Medical Devices and Instrumentation.* New York: Wiley, 1988, pp. 2154–2161.

Liu, B. L., "Immunologically sensitive field-effect transistors," in J. G. Webster (ed.), *Encyclopedia of Medical Devices and Instrumentation.* New York: Wiley, 1988, pp. 1614–1618.

Lübbers, D. W., and N. Opitz, "Die pCO_2/pO_2 Optode: Eine Neue pCO_2—bzw. pO_2—Messonde zur Messung des pCO_2 oder pO_2 von Gasen und Flüssigkeiten." *Z. Naturfors,* 1975, 30c, 532–533.

Mendelson, Y. M., A. C. Clermont, R. A. Peura, and B. C. Lin, "Blood glucose measurement by multiple attenuated total reflection and infrared absorption spectroscopy." *IEEE Trans. Biomed Eng.,* 1990, 37, 458–465.

Mendelson, Y. M., "Blood gas measurement, transcutaneous," in J. G. Webster (ed.), *Encyclopedia of Medical Devices and Instrumentation.* New York: Wiley, 1988, pp. 448–459.

Mendelson, Y. M., J. J. Galvin, and Y. Wang, "*In-vitro* evaluation of a dual oxygen saturation/hematocrit intravascular fiberoptic catheter." *Biomed. Instrum. & Technol.,* 1990, 24, 199–206.

Mendelson, Y. M., and R. A. Peura, "Noninvasive transcutaneous monitoring of arterial blood gases." *IEEE Trans. Biomed. Eng.,* 1984, BME-31, 792–800.

Mendelson, Y. M., P. Cheung, M. R. Neuman, D. G. Fleming, S. D. Cahn, *et al.,* "Spectrophotometric investigation of pulsatile blood flow for transcutaneous reflectance oximetry." *Adv. Exp. Med. Biol.,* 1983, 159, 93–102.

Merrick, E. B., and T. J. Hayes, "Continuous, noninvasive measurements of arterial blood oxygen levels." *Hewlett-Packard J.,* 1976, 28(2), 2–9.

Moran, F., L. J. Kettel, and D. W. Dugell, "Measurement of blood Po_2 with the microcathode electrode." *J. Appl. Physiol.,* 1966, 21, 725–728.

Moyle, J. T. B., *Pulse Oximetry.* London: BMJ Publishing, 1994.

Murray, R. W., R. E. Dessy, W. E. Heineman, W. R. Seitz, J. Janata, and R. W. Seitz, *Chemical Sensors and Microinstrumentation.* Washington, D.C.: American Chemical Society, 1989.

Nickerson, B. G., and F. Monaco, "Carbon dioxide electrodes, arterial and transcutaneous," in J. G. Webster (ed.), *Encyclopedia of Medical Devices and Instrumentation.* New York: Wiley, 1988, pp. 564–569.

Payne, J. P., and J. W. Severinghaus, *Pulse Oximetry.* Berlin: Springer, 1986.

Peterson, J. I., "Optical sensors," in J. G. Webster (ed.), *Encyclopedia of Medical Devices and Instrumentation.* New York: Wiley, 1988, pp. 2121–2133.

Peterson, J. I., and Vurek, G. G., "Fiber-optic sensors for biomedical applications." *Science,* 1984, 224, 123–127.

Peterson, J. I., S. R. Goldstein, and R. V. Fitzgerald, "Fiber-optic pH probe for physiological use." *Anal. Chem.,* 1980, 52, 864–869.

Peura, R. A., and Y. Mendelson, "Blood glucose sensors: An overview," in *Proceedings of the Symposium on Biosensors.* Piscataway, N.J.: IEEE, 1984, pp. 63–68.

Regnault, W. R., and G. L. Picciolo, "Review of medical biosensors and associated materials problems." *J. Biomed. Mater. Res.: Applied Biomaterials,* 1987, 21, 163–180.

Rolfe, R., "Review of chemical sensors for physiological measurement." *J. Biomed. Eng.,* 1988, 10, 138–145.

Rolfe, P., "*In vivo* chemical sensors for intensive-care monitoring." *Med. Biol. Eng. Comput.,* 1990, 28, B34–B47.

Schultz, J. S., S. Manouri, *et al.,* "Affinity sensor: A new technique for developing implantable sensors for glucose and other metabolites." *Diabetes Care,* 1982, 5, 245–253.

Schultz, J. S., "Biosensors," *Sci Am.,* 1991, 265(2), 64–69.

Seitz, W. R., "Chemical sensors based on immobilized indicators and fiber optics." *CRC Crit. Rev. Anal. Chem.,* 1988, 19, 135–173.

Seitz, W. R., "Chemical sensors based on fiber optics." *Anal. Chem.,* 1984, 56, 16A–34A.

Severinghaus, J. W., "Blood gas concentrations," in W. O. Fenn and H. Rahn (eds.), *Handbook of Physiology,* Vol. II, Sec. 3. Washington, D.C.: American Physiological Society, 1965, pp. 1475–1482.

Tremper, K. K., R. A. Mentelos, and W. C. Shoemaker, "Clinical and experimental transcutaneous $P\text{CO}_2$ monitoring." *J. Clin. Eng.,* 1981, 6, 143–147.

Turner, A. P. F., I. Karube, and G. Wilson, "Biosensors: Fundamentals and Applications." New York: Oxford University Press, 1987.

Updike, S. J., M. Shults, and B. Ekman, "Implanting the glucose enzyme electrode: Problems, progress and alternative solutions." *Diabetes Care,* 1982, 5, 207–212.

Updike, S. J., and G. P. Hicks, "The enzyme electrode, a miniature chemical transducer using immobilized enzyme activity." *Nature,* 1967, 214, 986–988.

Van der Spiegel, J., and J. N. Zemel, "Ion-sensitive field-effect devices," in J. G. Webster (ed.), *Encyclopedia of Medical Devices and Instrumentation.* New York: Wiley, 1988, pp. 1671–1683.

Von Cremer, M., "Uber die Ursache der elektromotorischen Eigenschaften der Gewbe, zuglelch ein Beitrag zur Lehre von den polyphasischen Elektrolytketten." *Zeitschrift fuer Biologie,* 1906, 47, 564–608.

Walt, D. R., C. Munkholm, P. Yuan, S. Luo, and S. Barnard, "Design, preparation, and applications of fiber-optic chemical sensors for continuous monitoring," in R. W. Murray, R. E. Dessy, W. R. Heineman, J. Janata, and W. R. Seitz (eds.), *Chemical Sensors and Microinstrumentation.* Washington, D.C.: American Chemical Society, 1989, pp. 252–272.

Wise, D. L. (ed.), "*Bioinstrumentation: Research, Developments and Applications.*" Stoneham, Mass.: Butterworth, 1990.

Yoshiya, I., Y. Shimada, and K. Tanaka, "Spectrophotmetric monitoring of arterial oxygen saturation in the fingertip." *Med. Biol. Eng. Comput.,* 1980, 18, 27–32.

Webster, J. G. (ed.), Design of Pulse Oximeters. Bristol, UK: IOP Publishing, 1997.

11

CLINICAL LABORATORY INSTRUMENTATION

Lawrence A. Wheeler

The clinical laboratory is responsible for analyzing patient specimens in order to provide information to aid in the diagnosis of disease and evaluate the effectiveness of therapy. The hospital department that performs these functions may also be called the department of clinical pathology or the department of laboratory medicine. The major sections of the clinical laboratory are the chemistry, hematology, and microbiology sections and the blood bank.

The chemistry section performs analyses on blood, urine, cerebrospinal fluid (CSF), and other fluids to determine how much of various clinically important substances they contain. Most applications of electronic instrumentation in the clinical laboratory take place in the chemistry section. The hematology section performs determinations of the numbers and characteristics of the formed elements in the blood (red blood cells, white blood cells, and platelets) as well as tests of the function of physiological systems in the blood (clotting studies are an example). Many of the most frequently ordered of these tests have been automated on the Coulter Counter (see Section 11.5). The microbiology section performs studies on various body tissues and fluids to determine whether pathological microorganisms are present. Until quite recently, there were essentially no applications of electronic instrumentation in microbiology. However, devices that automatically monitor the status of blood cultures (tests for the presence of microorganisms) and tests that semiautomatically measure the sensitivity of microorganisms to antibiotics (susceptibility tests) are now being used in many microbiology laboratories. The application of electronic instrumentation for the blood bank is in its infancy. A few systems that automate the basic classification of the type of the blood product (ABO grouping) are currently being developed.

Because many critical patient-care decisions are based on test results supplied by the clinical laboratory, the accuracy and precision of these results are of great importance. Excellent equipment design and effective quality-control programs are essential. Everyone involved in the design or use of clinical laboratory instruments must be constantly aware that erroneous test results can lead to a tragic outcome.

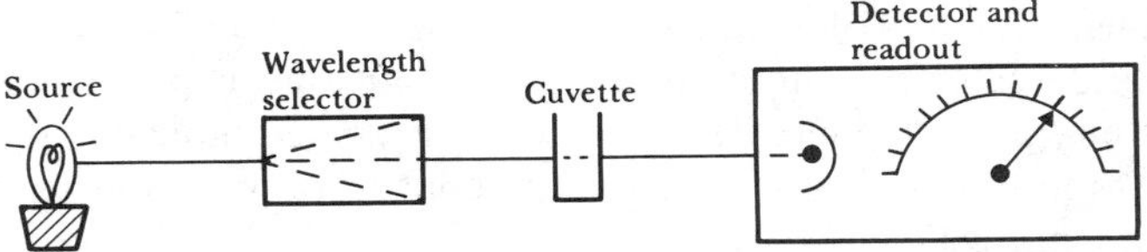

Figure 11.1 Block diagram of a spectrophotometer (Based on R. J. Henry, D. C. Cannon, and J. W. Winkelman, eds., *Clinical Chemistry,* 2nd ed. Hagerstown, MD: Harper & Row, 1974.)

A second important characteristic of many test procedures is fast response, because in many critical clinical situations, the therapy selected by the physician depends on the test results. The application of electronics in the clinical laboratory has greatly reduced the time required to perform a wide variety of crucial tests.

A major application of electronics in the clinical laboratory is the use of computer systems for information management. Mainframes, minicomputers, and microcomputers are used in commercial systems. Laboratory information systems keep track of patient specimens, organize the flow of work, automatically acquire test results from some types of instruments, maintain test-result databases, report results to on-line devices in patient-care areas, prepare printed reports, assist in quality control, and support a variety of management functions. We shall not discuss laboratory information management systems in detail. The current design trend in laboratory instrumentation, however, is to include data-processing capability in essentially every instrument. Therefore, our discussions of specific laboratory instruments will cover some aspects of laboratory information management.

11.1 SPECTROPHOTOMETRY

Spectrophotometry is the basis for many of the instruments used in clinical chemistry. The primary reasons for this are ease of measurement, satisfactory accuracy and precision, and the suitability of spectrophotometric techniques to use in automated instruments. In this section, *spectrophotometer* is used as a general term for a class of instruments. Photometers and colorimeters are members of this class.

Spectrophotometry is based on the fact that substances of clinical interest selectively absorb or emit electromagnetic energy at different wavelengths. For most laboratory applications, wavelengths in the range of the ultraviolet (200 to 400 nm), the visible (400 to 700 nm), or the near infrared (700 to 800 nm) are used; the majority of the instruments operate in the visible range.

Figure 11.1 is a general block diagram for a spectrophotometer-type instrument. The source supplies the radiant energy used to analyze the sample. The wavelength selector allows energy in a limited wavelength band to pass through. The cuvette holds the sample to be analyzed in the path of the energy. The detector produces an electric output that is proportional to the amount of energy it receives, and the readout device indicates the received energy or some function of it (such as the concentration, in the sample, of a substance of interest).

The basic principle of a spectrophotometer is that if we examine an appropriately chosen, sufficiently small portion of the electromagnetic spectrum, we can use the energy-absorption properties of a substance of interest to measure the concentration of that substance. In the vast majority of cases, these substances, as they are normally found in a patient's samples (of serum, urine, or CSF, for instance), do *not* exhibit the desired energy-absorption characteristics. In such cases, reagents are added to the sample, causing a reaction to occur. This reaction yields a product that *does* have the desired characteristics. The reaction products are then placed in the cuvette for analysis. The instrument-calibration procedures take into account the possible difference in concentration between the reaction product and the original quantity of interest.

Let us discuss in detail the characteristics of each of the subsystems shown in Figure 11.1.

Power Sources Hydrogen or deuterium discharge lamps are used to provide power in the 200-to-360-nm range, and tungsten filament lamps are used for the 360-to-800-nm range. Hydrogen and deuterium lamps both produce a continuous spectrum; but a problem with these power sources is that they produce about 90% of their power in the infrared range. The output in the ultraviolet and visible ranges can be increased by operating the lamp at voltages above the rated value, but this stratagem significantly reduces the expected life of the lamp. Another problem with tungsten lamps is that, during operation, the tungsten progressively vaporizes from the filaments and condenses on the glass envelope. This coating, which is generally uneven, alters the spectral characteristics of the lamp and can cause errors in determinations.

Wavelength Selectors A variety of devices are used to select those portions of the power spectrum produced by the power source that are to be used to analyze the sample. These devices can be divided into two classes: filters and monochromators. There are two basic types of filters: glass filters and interference filters.

Glass filters function by absorbing power. For example, a blue-colored filter absorbs in the higher-wavelength visible range (red region) and transmits in the lower-wavelength visible range (blue-green region). These filters (consisting of one or more layers of glass plates) are designed to be low-pass, high-pass, or bandpass (a combination of low- and high-pass) filters.

Interference filters are made by spacing reflecting surfaces such that the incident light is reflected back and forth a short distance. The distance is selected such that light in the wavelength band of interest tends to be in phase and to be reinforced; light outside this band is out of phase and is canceled (the interference effect). Harmonics of the frequencies in this band are also passed and must be eliminated by glass cutoff filters.

Glass filters are used in applications in which only modest accuracy is required. Interference filters are used in many spectrophotometers, including those used in the SMAC (Technicon Instrument Corporation) and the CentrifiChem (Union Carbide). Devices that use filters as their wavelength selectors are called colorimeters or photometers.

Monochromators are devices that utilize prisms and diffraction gratings. They provide very narrow bandwidths and have adjustable nominal wavelengths. The basic principle of operation of these devices is that they disperse the input beam spatially as a function of wavelength. A mechanical device is then used to allow wavelengths in the band of interest to pass through a slit.

Prisms are constructed from glass and quartz. Quartz is required for wavelengths below 350 nm. A convergent lens system is used to direct the light from the source through an entrance slit. The prism bends the light as a function of wavelength. The smaller wavelengths (ultraviolet) are bent the most. This produces an output beam in which the wavelength band of interest can be selectively passed by placing in the light path an opaque substance with a slit in it. The wavelength spectrum of the power passing through the slit is nominally triangle-shaped. In prisms, as in filters, the wavelength at which maximal transmittance occurs is the nominal central wavelength. Bandwidths of 0.5 nm can be obtained with this type of device. Prisms have been used over the wavelength range of 220 to 950 nm. The nonlinear spatial distribution of the power emerging from a prism requires relatively complex mechanical devices for control of the slit position to select different nominal wavelengths.

Diffraction gratings are constructed by inscribing a large number of closely spaced parallel lines on glass or metal. A grating exploits the fact that rays of light bend around sharp corners. The degree of bending is a function of wavelength. This results in separation of the light into a spectrum at each line. As these wavefronts move and interact, reinforcement and cancellation occur. The light emerging from a grating is resolved spatially in a linear fashion, unlike the light from a prism, in which the separation of wavelengths is less at longer wavelengths. As in the case of the prism, a slit is used to select the desired bandwidth. The mechanics of the slit-positioning mechanism of a grating are less complicated than those of a prism because of the linearity of the spatial separation of the wavelengths. Gratings can achieve bandwidths down to 0.5 nm and can operate over the range of 200 to 800 nm.

Cuvette The cuvette (Figure 11.1) holds the substance being analyzed. Its optical characteristics must be such that it does not significantly alter the spectral characteristics of the light as that light enters or leaves the cuvette. The degree of care and expense involved in cuvette design is a function of the overall accuracy required of the spectrophotometer.

Sample The sample (actually, in most cases, the substances resulting from the interaction of the patient specimen and appropriate reagents) absorbs light selectively according to the laws of Lambert, Bouguer, Bunsen, Roscoe, and Beer. The principles stated in these laws are usually grouped together and called Beer's law. The essence of the law was stated by Bouguer: "Equal thickness of an absorbing material will absorb a constant fraction of the energy incident upon it." This relationship can be stated formally as follows:

$$P = P_0 10^{-aLC} \tag{11.1}$$

where

P_0 = radiant power arriving at the cuvette
P = radiant power leaving the cuvette
a = absorptivity of the sample (extinction coefficient)
L = length of the path through the sample
C = concentration of the absorbing substance

Absorptivity is a function of the characteristics of the sample and the wavelength content of the incident light. This relationship is often rewritten in the form

$$\%T = 100P/P_0 = (100)10^{-aLC} \tag{11.2}$$

where $\%T$ is the percent transmittance. The value of a is constant for a particular unknown, and the cuvette and cuvette holder are designed to keep L as nearly constant as possible. Therefore, changes in P should reflect changes in the concentration of the absorbing substance in the sample.

Percent transmittance is often reported as the result of the determination. However, because the relationship between concentration and percent transmittance is logarithmic, it has been found convenient to report absorbance. Absorbance A is defined as log (P_0/P), so

$$A = \log\left(\frac{P_0}{P}\right) = \log\left(\frac{100}{\%T}\right) = 2 - \log(\%T) \tag{11.3}$$

Note that the relationship

$$A = aLC \tag{11.4}$$

follows from (11.1) and (11.3). As previously stated, the spectrophotometer is designed to keep a and L as nearly constant as possible so that a particular determination A ideally varies only with C. Therefore, the concentration of an unknown can be determined as follows. The absorbance A_s of a standard with known concentration of the substance of interest, C_s, is determined. Next the absorbance of the unknown, A_u, is determined. Finally, the concentration of the unknown, C_u, is computed via the relationship

$$C_u = C_s \left(\frac{A_u}{A_s}\right) \tag{11.5}$$

If this relationship holds over the possible range of concentration of the unknown substance in patient samples, then the determination is said to obey Beer's law. This relationship may not hold, however, because of absorption by the solvent or reflections at the cuvette. Then a relatively large number of standards with concentration values spanning the range of interest must be used to compute a calibration curve of concentration versus absorbance. This curve is then employed to obtain a concentration value for the absorbance value of the unknown.

The amount of light absorbed by a compound is generally a function of wavelength. The chemical reaction used in preparing the sample for spectrophotometry is designed to produce a compound (1) the concentration of which is proportional to that of the compound of interest and (2) the peak of the absorption spectrum of which is separated from the absorption peaks of the other compounds in the sample.

The wavelength band of light allowed to pass through the wavelength selector is generally chosen to cover the peak of the absorption curve symmetrically. There are a number of other factors to consider, however, including the absolute level of absorbance at the peak and its wavelength value. If the absorbance is too great ($A > 1.0$) or too small ($A < 0.11$), the errors of the photometric system become unacceptably large (Henry, 1984). In the case in which the absorbance is very large, the sample can be diluted, but this procedure is time-consuming and can result in errors. The wavelength of the peak must be within the range of the spectrophotometer's capabilities.

Photometric System A spectrophotometer's photometric system includes detectors to measure the amount of power leaving the cuvette (Radiation sensors; Section 2.16), circuits for amplification of the low currents developed by detectors (amplifiers; Chapter 3), and devices to present the results of the determination to the technologist operating the instrument (meters or recorders). Commonly used detectors include barrier layer cells, phototubes, and photoconductive cells.

The design of meters for this application has been somewhat of a problem in the past as a result of the need for precision and the nonlinear relationship between the detected quantity (power) and the quantity of interest (absorption). Now, thanks to the development of low-cost digital electronics, the problems of computation and data presentation have been largely eliminated.

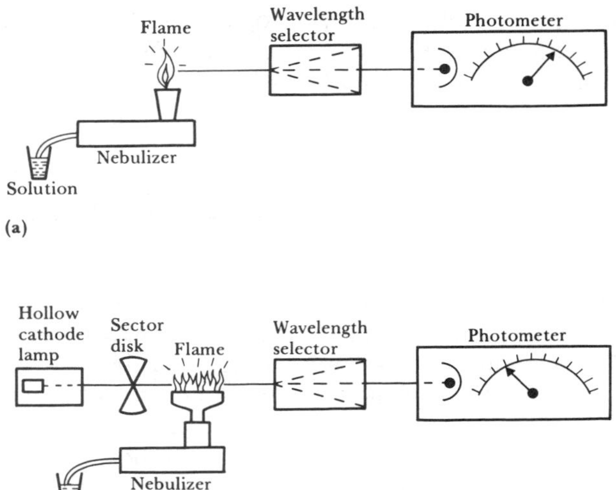

Figure 11.2 Block diagrams of instruments for (a) flame emission and (b) flame absorption. (Based on R. J. Henry, D. C. Cannon, and J. W. Winkelman, eds., *Clinical Chemistry,* 2nd ed. Hagerstown, MD: Harper & Row, 1974.)

FLAME PHOTOMETERS

Flame photometers differ in three important ways from the instruments we have already discussed. First, the power source and the sample-holder function are combined in the flame. Second, in most applications of flame photometry, the objective is measurement of the sample's emission of light rather than its absorption of light, although we shall also discuss atomic absorption-type flame photometers. (Schematic representations of these two types of instruments are shown in Figure 11.2.) Third, flame photometers can determine only the concentrations of pure metals.

ATOMIC EMISSION

At the normal levels of power used in flame photometers, only about 1% of the atoms are raised to an excited state. In addition, only a few elements produce enough power at a single wavelength as they move from higher-energy to lower-energy orbits. These two factors have limited the use of atomic-emission flame photometry largely to determinations of Na^+, K^+, and Li^+. Instruments have been developed that can make determinations of other elements, such as Ca^{2+}, but relatively complicated optical systems are required.

As shown in Figure 11.2(a), the sample, combined with a solvent, is drawn into a nebulizer that converts the liquid into a fine aerosol that is injected into the flame. Several types of fuels have been used in flame photometers. Currently, propane or natural gas mixed with compressed air is used. The solvent evaporates in the flame, leaving microscopic particles of the sample. These particles disintegrate to yield atoms. As we have noted, only a small proportion of these atoms are in the excited state. As the atoms fall to the ground state, they release power at their characteristic wavelength.

A simple optical system—including only a filter and a lens to focus the filtered light on the detector—is normally used for determinations of Na^+ and K^+. More sophisticated optical systems, including a monochromator, are required for other determinations.

Many modern atomic-emission flame photometers are designed to include an internal standard to compensate for variations in the rate of solution uptake, aerosol production, and flame characteristics. Lithium (Li^+) is used for this purpose. It is not normally found in biological samples, it has a high emission intensity, and its peak emission wavelength is well separated from those of sodium and potassium. A carefully controlled amount of Li^+ salt is added to the sample. An optical channel is provided to measure the power emitted by the Li^+, and this power, along with the known concentration of Li^+, is used to correct the determination of Na^+ or K^+ for variations in the instrument. Actually, in most applications, the determinations of Na^+, K^+, and Li^+ are done in parallel.

A few problems arise in the use of Li^+ as the internal standard. First, although correction for small variations in the characteristics of the instrument is possible, there is no way to correct for large variations. Second, Li^+ is being used increasingly for treatment of an important psychotic disorder, manic-depressive psychosis. If patients who are receiving Li^+ are not identified to the clinical laboratory, significant errors in determinations of Na^+ and K^+ can occur. It is an unfortunate fact that the clinical laboratory is rarely given any clinical information to use in assessing the accuracy of determinations.

ATOMIC ABSORPTION

This technique has shown great promise for the very accurate determination of the concentration of a variety of elements, including calcium, lead, copper, zinc, iron, and magnesium. It is based on the fact that the vast majority of atoms in a flame absorb energy at a characteristic wavelength. A special power source is used that emits power at the characteristic wavelength of the atom the concentration of which is being determined. This source is a hollow cathode lamp. Such lamps are constructed from the metal to be determined or are lined with a coating of it. In most cases, a separate lamp is needed for each metal determination, but the special characteristics of a few metals make it possible to use one lamp for combinations of two or three of them. The cathode is placed in an atmosphere of an inert gas. When the cathode is heated, the atoms of the cathode leave the surface of the cathode and fill the cathode cavity with an atomic vapor. These atoms become excited as a result of collisions with electrons and ions, and when they return to the ground state, each releases power at its characteristic wavelength, as previously discussed. This power is directed through the flame [see Figure 11.2(b)], and the amount of absorption is proportional to the amount of the atom present.

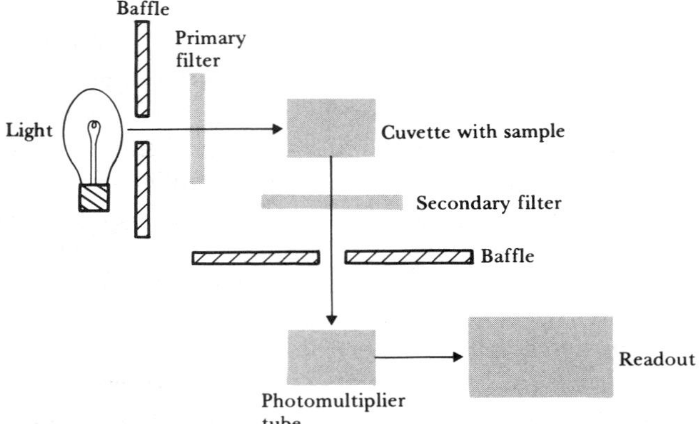

Figure 11.3 Block diagram of a fluorometer (From R. Hicks, J. R. Schenken, and M. A. Steinrauf, *Laboratory Instrumentation*. Hagerstown, MD: Harper & Row, 1974. Used with permission of C. A. McWhorter.)

Atomic-absorption flame photometers normally require a monochromator and use a photomultiplier as the detector. One additional feature of these devices is that, because the atoms in the flame emit as well as absorb power at the characteristic wavelength, it is necessary to be able to differentiate between the two sources of power reaching the detector. This is accomplished by designing the source to produce pulses of power rather than a steady output. A rotating-sector disk between the source and the flame is generally used for this purpose. The detection electronics incorporate the phase-sensitive demodulator described in Section 3.15 to eliminate the dc component and analyze only the ac signal.

FLUOROMETRY

Fluorometry is based on the fact that a number of molecules emit light in a characteristic spectrum—the emission spectrum—immediately after absorbing radiant energy and being raised to an excited state. The degree to which the molecules are excited depends on the amplitude and wavelength of the radiant power in the excitation spectrum. Small amounts of power are lost in this process, which results in the emission spectrum's being generally higher in wavelength content than the excitation spectrum.

Various power sources, wavelength selectors, and detection circuits are used in fluorometers of varying sensitivity. Mercury arc lamps are commonly used power sources. They produce major line spectra at 365, 405, 436, and 546 nm. Photomultipliers are normally used as detectors. A unique feature of these devices is the need to select operational bandwidths for two spectra—excitation and emission. Figure 11.3 shows a fluorometer block diagram. The detector is placed at a right angle to the power source to minimize the chance of direct transmission of light from source to detector. The wavelength characteristics of the wavelength selectors are also chosen such that there is little or no overlap in the wavelengths they pass.

One advantage fluorometry offers is its much greater sensitivity, which may exceed that of spectrophotometric methods by as much as four orders of magnitude. This is because in spectrophotometric methods, the difference between the absorption of a solution assumed to have a zero concentration ($\%T = 100$) of the unknown substance and the absorption of the sample is used as a measure of the concentration of the unknown substance. With highly dilute samples (such as one wherein $\%T = 98$), small errors in the process can cause large percent errors in the determination. In fluorometry, by contrast, a direct measurement of the fluorescence of the sample is used to determine the concentration of the unknown substance.

A special advantage of fluorometry is its great specificity. In spectrophotometric methods, the light absorbed in the wavelength band may be from substances other than the unknown. Only a relatively small number of substances, however, have the property of fluorescence. Thus many substances that might interfere with a spectrophotmetric measurement cannot interfere with a determination of fluorescence. Also, substances that have similar excitation spectra may have different emission spectra, and vice versa. Therefore, appropriate selection of the bandwidths of the two wavelengths selectors can provide additional rejection of noise.

The combination of these characteristics makes fluorometry capable of detecting picogram amounts of unknown substances. Highly dilute samples are used to prevent the light produced by fluorescence from being absorbed by other molecules as it passes through the sample solution.

The principal disadvantage of fluorometry is the sensitivity of its determinations to temperature and pH of the sample (fluorescence in general is pH-sensitive).

11.2 AUTOMATED CHEMICAL ANALYZERS

In this section we shall describe two important types of automated chemical analyzers. They are the Beckman Synchron CX4 and the Dupont Automatic Clinical Analyzer (ACA). Both of these devices greatly enhance productivity in the clinical laboratory and decrease the response time for emergency requests (called STAT requests). The Synchron CX4 is a relatively new instrument; the ACA is older but widely used. Both utilize spectrophotometric methods for making the actual measurements of interest. They differ in the way they perform the steps of specimen aspiration, dilution, combination of sample with reagents, physical movement of the sample, computation, and recording of the results.

SYNCHRON CX4

The Synchron CX4 system is a state-of-the-art, high-capacity specimen-processing chemistry analyzer (Anonymous, 1989). It is a discrete random-access clinical analyzer that can perform a wide variety of tests during a single run. It has a number of capabilities that are interesting from a biomedical engineering perspective. These include automated specimen handling, performance of a variety of analytical test techniques, extensive use of microcomputers, and bar code identification techniques. This instrument is a microcomputer-controlled random-access chemistry analyzer. It performs end-point and rate assays at 30 °C or 37 °C.

Operational Characteristics The operator controls the operation of the CX4 by means of a CRT and keyboard. A tree structure of screens is accessed by pressing one of eight function keys. (The action of each function key is displayed at the bottom of the CRT display.) The five main functions are sample (specimen) programming, reagent load, calibration, special functions, and system parameters.

1. *Sample programming* In this function the operator can assign a specimen or a control to a cup, enter information to identify the material, and specify which tests are to be performed on it. Special procedures are available to allow a STAT specimen to be analyzed as soon as possible. A listing of the type of material programmed to be placed in each cup in a sample sector (a load list) can be prepared on the system printer. The characteristics of a sample sector are discussed below. This list is used when the patient specimens and control materials are placed in the sector.
2. *Reagent load* This procedure allows the operator to insert, replace, or remove reagent cartridges from the 24 positions on the reagent carousel. Each reagent cartridge has a bar code label that provides complete information about it, including the type of test with which it is used and its expiration date. A bar code reader reads the label as the cartridge is transferred to or from the carousel, and this information is transferred to the main system CPU.
3. *Calibration* Materials with known concentrations of the test substance (standards) are used to allow adjustment of the instrument to provide accurate determinations. Calibration is performed both when new reagents are loaded into the system and at specified intervals.
4. *Special functions* These functions include system setup, system diagnostics, and maintenance procedures.
5. *System parameters* In this mode the operator can monitor instrument activity, the current system configuration (such as types of active tests), and the readings of various sensors within the instrument (such as temperatures and fluid levels).

Test Performance Overview The testing cycle consists of two types of processes, the service interval and the analytical spin cycle. Our discussion will follow the events occurring in one cuvette; each of the 80 cuvettes sequentially goes through these steps within the testing cycle. During the service interval (10 s) the test reaction is initiated. This process includes removing the products of the previous test from the cuvette and placing test reagents and test sample (patient specimen or control material) in the cuvette. During the analytical spin cycle, the sample carousel is spun at 90 rpm for 6 s. The cuvette passes through the optics station, where absorbance readings are made at five frequencies.

In an endpoint reaction, the absorbance of the reaction at the time of the last measurement is used to calculate the concentration of the test substance. The rate type of measurement exploits the fact that the rate of change of absorbance of the reaction mixture is related to the concentration of the test substance. The measurements taken during the most linear part of the reaction are used in this calculation.

The data analysis procedure involves calculating the concentration of the test substance from the absorbance readings. Each of these concepts is discussed in more detail below.

System Description The CX4 includes three major subsystems: specimen handling, reagent handling, and the cuvette reacton system.

1. *Specimen handling* The specimen-handling system consists of five modules: sample sectors (up to eight), autoloader assembly, sample turntable assembly, sample pipettor/mixer assembly, and probe/mixer wash cup. A sample sector holds up to ten patient specimens and control cups. The operator enters the identity of the material in each cup, using the instrument computer control system (see above). The autoloader assembly transfers sectors to the sample carousel under microcomputer control. This process is accomplished through the use of a stepper motor and a hydropneumatic system. The sectors are removed by a similar process after the testing cycle is completed. The sample turntable assembly includes the sample carousel and a reflective sensor reader. The sample carousel is a discrete-position stepper-motor-driven rotational device. It is rotated under microcomputer control to allow test material to be removed by the sample pipettor. The sample pipettor includes a liquid-level sensor that allows the pipettor probe to be positioned at the correct depth to make possible the withdrawal of the correct amount of test material. After the test material has been drawn into the pipettor, the crane rotates to position the probe over a cuvette on the reaction carousel. The material is then discharged into the cuvette.
2. *Reagent handling* The reagent-handling system includes the reagent cartridges, reagent carousel, and reagent pipettor. The reagent cartridges are disposable containers that hold the reagent needed for a particular test. They have bar code labels indicating the test with which they are used. The reagent carousel is moved under microcomputer control to allow the correct reagent to be pipetted. The reagent carousel holds 24 cartridges.
3. *Cuvette reaction system* The cuvette reaction system includes the reaction carousel and the photometer assembly. The reaction carousel holds

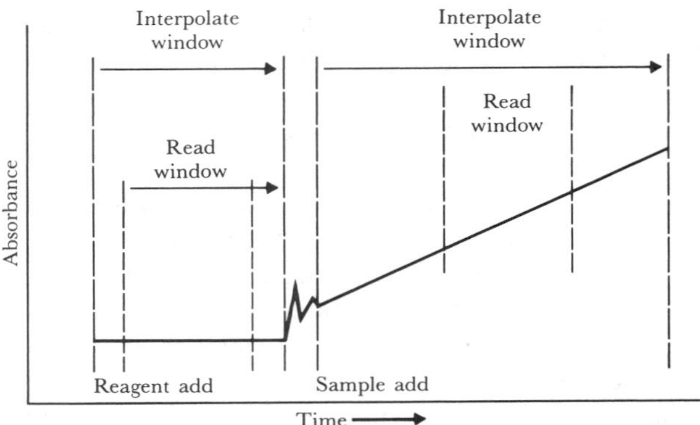

Figure 11.4 Synchron CX4 measurement read window for a rate-type measurement

the 80 cuvettes in which the chemical reactions occur. The temperature of the cuvettes is maintained at either 30 °C or 37 °C. The photometer assembly consists of a xenon light source, a light-collimating system, optical filters, and photodiode detectors. The xenon light source emits a flash of light as each cuvette passes through the optics station. Because of the variability of the intensity of flashes, a flash correction system is used. This system is based on making absorbance measurements at frequencies other than the main measurement frequency. The light that passes through the specimen is directed through a light-collimating system that presents light beams to each of ten filters (340, 380, 410, 470, 520, 560, 600, 650, 670, and 700 nm) and ten photodiode detectors. Each of the detectors produces an output proportional to the received light. The output is processed by a log amplifier signal-conditioning circuit, which supplies the input for a multiplexor. The multiplexor samples the signal at appropriate times under microcomputer control and delivers an analog signal to an analog-to-digital converter. The digital representation of the absorbance measurements is sent to the main central processing unit.

For most tests, only five of the ten possible absorbance frequency measurements are used in calculating the test result. During each spin cycle, each cuvette passes through the optics station eight times. The number of spin cycles varies with the type of test. For example, a glucose determination requires approximately 200 s from the time the sample is added to the reagent to completion of the reaction. Because a spin cycle occurs every 16 s, 13 spin cycles are required for this test. A mean absorbance is calculated for each spin cycle at its conclusion. After the reaction is completed, a cubic interpolation of the mean absorbance values is performed. This process also includes a digital filtering function to minimize noise in the data. As shown in Figure 11.4, this process is performed during the time when only the reagent is in the cuvette, to determine the background absorbance (blank value), and after the sample is added. The absorbance curve shown in Fig. 11.4 is for a test in which a rate type of measurement is occurring. In an endpoint reaction, by contrast, the measurement readings are taken after the reaction is completed and a flat absorbance curve is present. The final concentration of the test substance is calculated by means of Beer's law. The test result is printed on a report form that can be placed in the patient's chart and sent to his or her clinician. The instrument contains a serial computer port for transmitting test results to a laboratory computer.

AUTOMATIC CLINICAL ANALYZER (ACA)

The DuPont Automatic Clinical Analyzer (ACA) differs from high-capacity instruments such as the Synchron CX4 in that it is oriented toward flexibility rather than maximizing throughput. It performs determinations in serial rather than in parallel, but it can select any of 40 tests for each sample. A functional block of the ACA is shown in Figure 11.5 (Anonymous, 1975a).

The ACA uses the unique concept of combining the sample with the reagents in the analytical test pack (ATP). There is a different ATP for each determination. During operation of the ACA, the ATP moves from station to station on a conveyer; any sequence of ATPs can be selected. The time required to perform any of the ACA determinations is 7 min, and there is a 37-s spacing between completions of determinations. This means that the ACA can perform any of its determinations as STAT requests. This feature, which gives the clinical laboratory an important capability, results in a higher cost per test than that found in other automated methods.

We now turn to the general characteristics of the ACA. The patient sample is placed in a special holder called a sample kit, which has a patient-identification card attached to it. The ATPs corresponding to the determinations that are to be made on the sample are loaded behind the sample kit on the conveyer mechanism. Figure 11.6 shows the construction of the ATP. The last ATP is followed by a special end-of-run kit. The basic operations carried out by each subsystem are as follows:

Patient Identification When the sample kit first enters the ACA, it passes through a station where the patient-identification information that has been entered manually on the patient-identification card on the side of the sample kit is transferred to the printer paper.

Filling Station In the filling station, aliquots (measured liquid volumes) of the sample are withdrawn from the sample kit and mixed with a diluent (which may be different for different determinations). Then 5 ml of the combined solution is injected into each ATP. The ATP's binary code is read by electronic devices in the filling station to determine which diluent to use for that particular ATP. After the ATPs have been filled, they continue to the preheaters, and the sample kits are placed in the sample-kit exit tray.

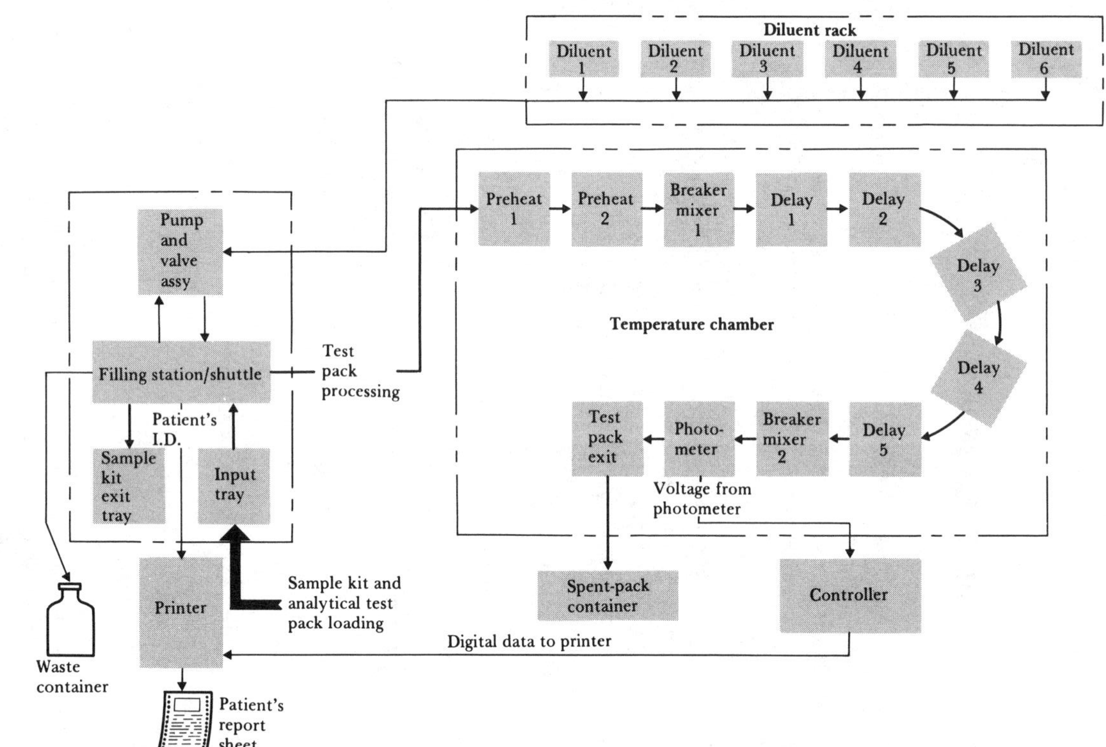

Figure 11.5 Block diagram of ACA (From *ACA Instrument Instruction Manual,* DuPont Company, Automatic Clinical Analysis Division, Wilmington, DE 19898.)

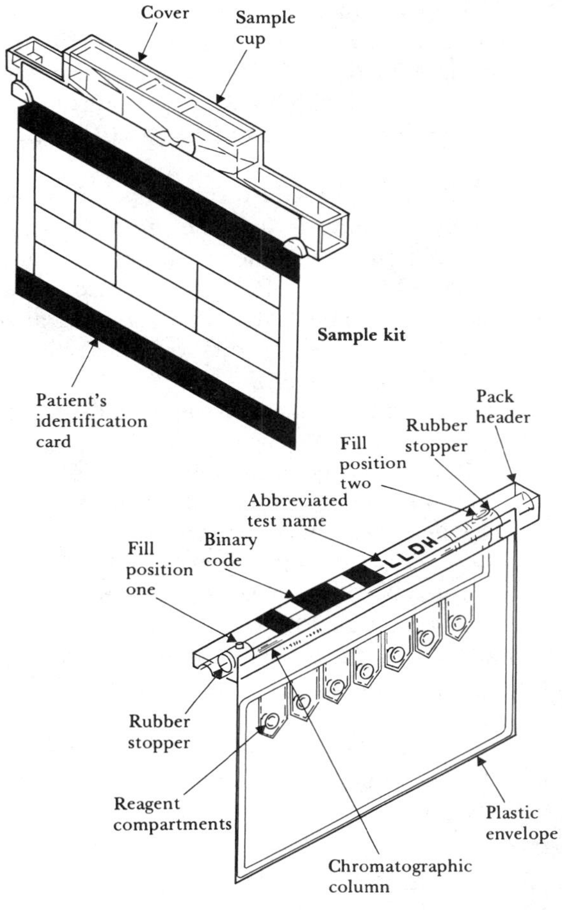

Figure 11.6 Sample kit and analytical test pack of ACA (From *ACA Instrument Instruction Manual,* DuPont Company, Automatic Clinical Analysis Division, Wilmington, DE 19898.)

Preheaters In the preheaters, the ATP is heated to 37 °C; it is maintained at this temperature for the remainder of the process.

Breaker-Mixer 1 At this station, the reagents in four of the plastic compartments (see Figure 11.6) are crushed and mixed with the diluted sample.

Delay Stations As the ATP successively passes through the five delay stations, the chemical reactions of the first four reagents, the diluent, and the sample occur.

Breaker-Mixer 2 Here the last three reagent compartments are crushed and mixed with the reaction solution. For some determinations, a delay is included here to allow sufficient time for the reactions to go to completion before the ATP enters the photometer station. This delay is controlled by the binary code that was read in the filling station and that identified the type of test.

Photometer In the photometer, the plastic envelope of the ATP is formed into a cuvette by a unique pressure device. The pressure in the plastic pack is measured and used to determine whether an adequate amount of sample and diluent has been inserted in the ATP. If the pressure is too low, the test result is flagged with the letter P. The ATP binary code is again decoded, and the photometer control board uses this information to select the measurement method, filter type, and ADC constants.

The ACA exhibits one of three measurement methods: rate, two-filter, and two-pack. In the rate method, the change in absorption of the cuvette during a fixed period is used as the measure of the concentration of the unknown. In the two-filter method, measurements of absorption are made with two different filters. The measurement of the concentration of the unknown is proportional to the difference between these absorption values. The two-pack method uses as the determination the difference between the absorptions of the combination of the unknown and the reagents in the two ATPs at the same wavelength band. The output of the photometer is converted to digital form (as specified by the photometer controller) and sent to the printer.

Printer The printer prepares the ACA report, which includes the patient-identification information obtained in the filling station and the photometer results for all the ATPs filled with the patient's sample.

11.3 CHROMATOLOGY

Chromatology is basically a group of methods for separating a mixture of substances into component parts. (Although the use of the term *chromatography* is firmly established, it is really a misnomer: In modern techniques, the colors of the mixture's components are not really used to identify substances.) One phase is fixed—liquid or solid—and the other is mobile—gas or liquid. When a liquid stationary phase is used, the process is called partition. When a solid stationary phase is used, the process is called adsorption.

In all chromatology, differences in the rate of movement of components of the mixture in the mobile phase, caused by interaction of these components with the stationary phase, are used to separate the components (Cawley, 1965). The four possible combinations of stationary and mobile phases have been used in chromatographic methods.

From the viewpoint of the clinical laboratory, these methods are used primarily for the detection of complex substances such as drugs and hormones.

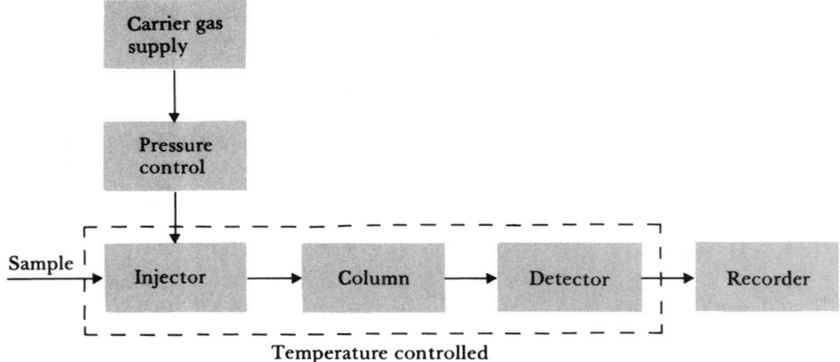

Figure 11.7 Block diagram of a gas-liquid chromatograph (GLC)

For example, gas–liquid chromatographs (GLC) and thin-layer chromatographs (TLC) have been useful in determining what drug or drugs have been taken in overdose cases. The availability of this information is vitally important to the clinician who must select appropriate therapy. The characteristics of the GLC are presented here as an important example of the use of chromatographic methods in the clinical laboratory.

GAS–LIQUID CHROMATOGRAPHS

The basic components of a GLC are shown in Figure 11.7. Prior to being injected into the GLC, the patient sample usually must undergo some initial purification, the extent of which depends on the determination that is being performed. The functions of the major subsystems are as follows:

Injector The injector is used to introduce into the GLC 1–5 ml of the patient sample including the solvent in which it is contained (usually a volatile organic solvent). The temperature of the injector is set to flash-evaporate the sample and solvent.

Carrier Gas The inert carrier gas (usually N_2 or He) is the mobile phase of the chromatograph. It sweeps the evaporated sample and solvent gas down the column.

Column The column typically is 1 m long and less than 7 mm in diameter. It is packed with the solid support material (such as diatomaceous earth). The solid support is coated with the liquid phase. The small size of the solid beads produces the separation of the components. The column is enclosed in an oven the temperature of which is carefully controlled. A temperature programmer gradually increases the temperature of the column in a sequence designed for maximal efficiency of separation for the type of substance being analyzed.

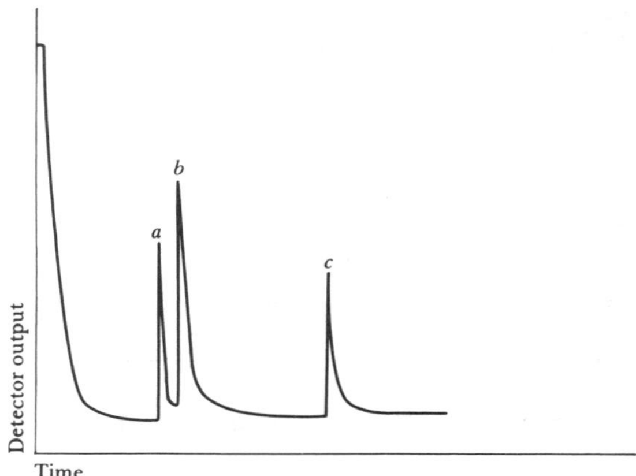

Figure 11.8 Example of a GLC recording for the analysis of blood levels of phenobarbital (peak *a*) and phenytoin (peak *c*). Peak *b* corresponds to the level of heptabarbital (the internal standard).

Detector The detector is located at the end of the column. Its function is to provide an electric output proportional to the quantity of the compound in the effluent gas. A number of types of detectors are available for use with different types of samples. They include ionization detectors, thermal-conductivity detectors, and electron-capture detectors. Ionization detectors are most commonly used in clinical laboratory applications (see Littlewood, 1970).

All the detectors are sensitive to classes of compounds, not only to some particular component of interest. Therefore, both the concentrations of the detected compounds and the times during the operation of the column when those concentrations occurred (that is, a plot of concentration versus time) are used in determining the types and quantities of components present in the sample. The output of the detector is connected to a recorder.

Recorder In the recorder, the x axis represents time, and the y axis the output of the detector. The recording thus provides a display of both the quantity of a component that was present (the area under the peak) and the time at which it was eluted off the column. From this information, the components present can be identified by the time they took to leave the column or, preferably, by comparison with recordings obtained by analyzing compounds of known composition with the GLC.

Figure 11.8 shows a recording obtained from the analysis of a blood specimen for the levels of the important anticonvulsant drugs phenobarbital and phenytoin. A measured amount of heptabarbital was added to the specimen to serve as an internal standard. The area under the phenobarbital and phenytoin peaks is compared with the area under the heptabarbital peak to compute the blood levels of these drugs.

Gas–liquid chromatography offers a number of important advantages in the analysis of complex compounds. They include speed, ability to operate with small amounts of sample, and great sensitivity. Most instruments can complete analyses of clinically important substances in less than 1 h, and often in 15 min or less. Only milliliter amounts of the sample are needed. The sensitivity of the device depends on the detector used, but high-quality instruments can detect 1-ng quantities of a compound.

11.4 ELECTROPHORESIS

Devices based on electrophoretic principles are used in the clinical laboratory to measure quantities of the various types of proteins in plasma, urine, and CSF; to separate enzymes into their component isoenzymes; to identify antibodies; and to serve in a variety of other applications.

BASIC PRINCIPLES

Electrophoresis may in general be defined as the movement of a solid phase with respect to a liquid (the buffer solution). The main functions of the buffer solution are to carry the current and to keep the pH of the solution constant during the migration. The buffer solution is supported by a solid substance called the medium.

Our discussion in this section is limited to zone electrophoresis. In this technique, the sample is applied to the medium; and under the effect of the electric field, groups of particles that are similar in charge, size, and shape migrate at similar rates. This results in separation of the particles into zones. The factors that affect the speed of migration of the particles in the field are discussed in the following paragraphs.

Magnitude of Charge The mobility of a given particle is directly related to the net magnitude of the particle's charge. Mobility is defined as "the distance in centimeters a particle moves in unit time per unit field strength, expressed as voltage drop per centimeter" [mobility = $cm^2/(V \cdot s)$] (Henry *et al.,* 1974).

Ionic Strength of Buffer The more concentrated the buffer, the slower the rate of migration of the particles. This is because the greater the proportion of buffer ions present, the greater the proportion of the current they carry. It is also due to interaction between the buffer ions and the particles.

Temperature Mobility is directly related to temperature. The flow of current through the resistance of the medium produces heat. This heat has two important effects on the electrophoresis. First, it causes the temperature of the medium to increase, which decreases its resistance and thereby causes the rate of migration to increase. Second, the heat causes water to evaporate from

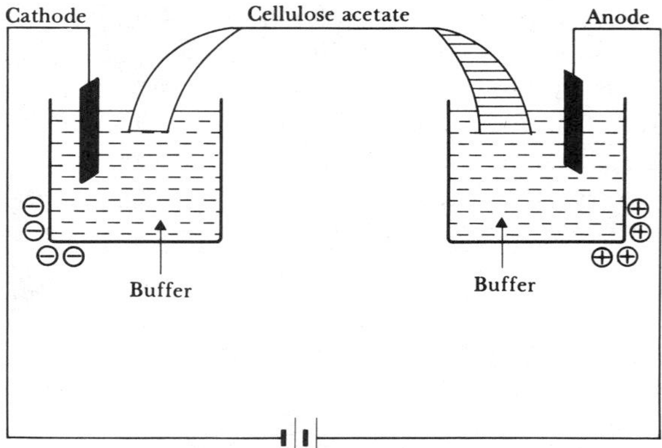

Figure 11.9 Cellulose acetate electrophoresis (From R. Hicks, J. R. Schenken, and M. A. Steinrauf, *Laboratory Instrumentation.* Hagerstown, MD: Harper & Row, 1974. Used with permission of C. A. McWhorter.)

the surface of the medium. This increases the concentration of the particles and further boosts the rate of migration. Because of these effects, either the applied voltage or the current must be held constant in order to maintain acceptable reproducibility of the procedures. For short runs at relatively low voltage levels, either can be held constant. But when a gel is used as the medium, heating is a significant problem. With this type of medium, constant-current sources are normally used to minimize the production of heat.

Time The distance of migration is directly related to the time the electrophoresis takes. Other factors that influence migration include electroendosmosis, chromatography, particle shape, "barrier" effect, "wick flow," and streaming potential (Henry *et al.*, 1974; Hicks *et al.,* 1974).

Types of Support Media A large variety of support media have been used in various electrophoretic applications. They include paper, cellulose acetate, starch gel, agar gel, acrylamide gel, and sucrose. We discuss cellulose acetate electrophoresis here, because it is used extensively in clinical laboratories and because the same general method is used with other media.

Cellulose acetate has a number of desirable properties compared with the paper that was the medium first used in electrophoresis.

Figure 11.9 illustrates the basic process of cellulose acetate electrophoresis. The cellulose acetate strip is saturated with the buffer solution and placed in the membrane holder (the "bridge"). The bridge is placed in the "cell" with both ends of the strip in the buffer wells.

A number of electrophoreses (typically eight) can be done on one strip. The sample for each test is placed on the strip at a marked location. Then the electric potential is applied across the strip. With this type of electrophoresis, constant-voltage-source power supplies are often used. A typical voltage is

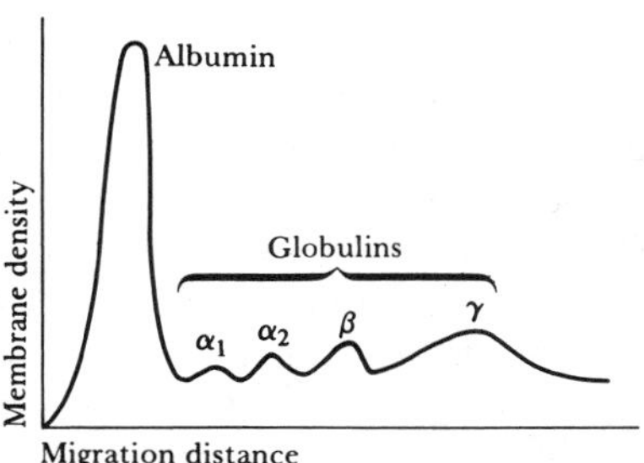

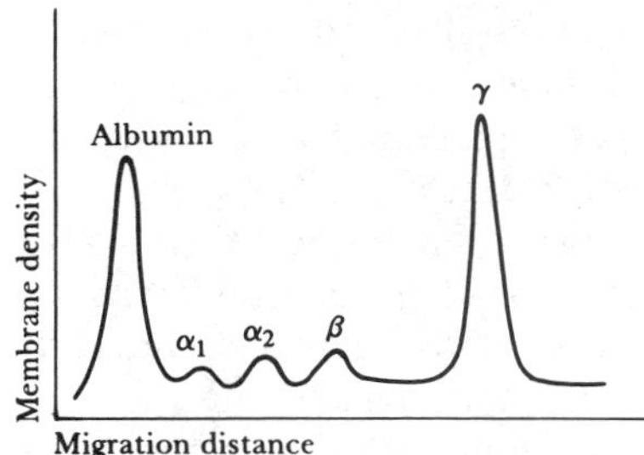

Figure 11.10 Examples of patterns of serum protein electrophoresis The left-hand pattern is normal; the right-hand pattern is seen when there is an overproduction of a single type of gamma globulin.

250 V, which results in an initial current of 4–6 mA. As we have noted, this current increases slightly during the procedure. After 15–20 min, depending on the device used, the electric voltage is removed. The next step is to fix the migrated protein bands to the buffer and to stain them so that they can be seen as well as subsequently quantified. This may be done in separate or combined exposures to a fixative and a dye. The membrane is now "cleared" to make it transparent. The densities due to the dyed-specimen fractions are not affected. The membrane is dried in preparation for densitometry.

The densitometer is a device that consists of a light source, filter, and detector (typically a photodiode). The design and operation of this type of device are discussed in Section 11.1.

The membrane is placed in a holder in the densitometer. The path of migration of one of the specimens is then scanned. The low-voltage output of the detector is amplified by a very stable analog preamplifier. The output of the preamplifier is sent to an analog x–y recorder and to an analog integrator circuit. The x–y recorder produces a plot whereon the x coordinate represents migration distance and the y coordinate represents membrane density (which is directly proportional to the amount of specimen component that has moved the corresponding migration distance). The integrator has circuitry that detects the beginning and end of each significant peak and computes the area under the peak. These numbers are printed on the analog recording next to the corresponding peak. This process is repeated for each of the specimens on the membrane.

Figure 11.10 shows examples of the types of plots that are obtained via electrophoresis. These plots are for serum protein electrophoresis.

11.5 HEMATOLOGY

BASIC CONCEPTS

The blood consists of formed elements, substances in solution, and water. This section covers only devices that measure characteristics of the formed elements: red blood cells (RBCs), white blood cells (WBCs), and platelets. The primary functions of the RBCs are to carry oxygen from the lungs to the various organs and to carry carbon dioxide back from these organs to the lungs for excretion. The primary function of the WBCs is to help defend the body against infections. Five types of WBCs are normally found in the peripheral blood. In order of decreasing numbers in the blood of adults, they are neutrophils, lymphocytes, monocytes, eosinophils, and basophils. In disease, the total number and the relative proportions of these types of WBCs can change; immature and malignant types of WBCs can also appear. Platelets plug small breaks in the walls of the blood vessels and also participate in the clotting mechanism.

The basic attribute of the formed elements in the blood that is measured is the number of elements of each type per microliter (μl). The normal range of the RBC count in an adult male is 4.6 to $6.2 \times 10^6/\mu$l and, in an adult female, 4.2 to $5.4 \times 10^6/\mu$l. The normal ranges of WBCs and platelet counts are the same for men and women. The normal range of the WBC count is 4,500 to 11,000/μl; that for the platelet count is 150,000 to 400,000/μl. The hematocrit (HCT) is the ratio of the volume of all the formed elements in a sample of blood to the total volume of the blood sample. It is reported as a percentage, the normal range in adult men being 40 to 54% and, in adult women, 35 to 47%. Hemoglobin (Hb) is a conjugated protein within the RBCs that transports most of the O_2 and a portion of the CO_2 that is carried in the blood. It is reported in grams per deciliter. The normal range in adult men is 13.5 to 18 g/dl, and that in adult women is 12 to 16 g/dl.

A second group of measurements is made to characterize the RBC volume and Hb concentration. These measurements include the mean corpuscular volume (MCV) in cubic micrometers, the mean corpuscular hemoglobin (MCH) content in picograms, and the mean corpuscular hemoglobin concentration (MCHC) in percent. These values are called the RBC indices. Normal ranges for these parameters are as follows:

MCV: 82–98 μm^3

MCH: 27–31 pg

MCHC: 32–36%

The RBC count (in millions per microliter), HCT (in percent), MCV (in cubic micrometers), Hb (in grams per deciliter), MCH (in picograms), and MCHC (in percent) are related as follows:

$$\text{MCV} = \frac{10\ \text{HCT}}{\text{RBC count}} \tag{11.6}$$

$$\text{MCH} = \frac{10\ \text{Hb}}{\text{RBC count}} \tag{11.7}$$

$$\text{MCHC} = \frac{100\ \text{Hb}}{\text{HCT}} \tag{11.8}$$

The units for RBC count, Hb, and HCT that are employed in these calculations are such that the units for MCV, MCH, and MCHC are those given above.

EXAMPLE 11.1 Calculate the RBC indices from the following data.

$$\text{RBC} = 5\ \text{million}/\mu\text{l}$$

$$\text{Hb} = 15\ \text{g/dl}$$

$$\text{HCT} = 45\%$$

ANSWER

$$\text{MCV} = \frac{10\ \text{HCT}}{\text{RBC count}} = \frac{450}{5} = 90\ \mu\text{m}^3$$

$$\text{MCH} = \frac{10\ \text{Hb}}{\text{RBC count}} = \frac{150}{5} = 30\ \text{pg}$$

$$\text{MCHC} = \frac{100\ \text{Hb}}{\text{HCT}} = \frac{1500}{45} = 33.3\%$$

A new RBC characteristic that is assuming increasing importance in hematology is the volume distribution width (called the RDW). Somewhat similar to the standard deviation of a Gaussian distribution, it is a measure of the spread of the RBC volume distribution. In many RBC disorders, RBC production is disordered and a wider range than usual of RBC sizes is produced, leading to an increased RDW value.

Using the RDW with other RBC parameters can aid in the diagnosis of RBC disorders. For example, in iron deficiency anemia (an acquired disorder in which hemoglobin production is reduced due to a lack of iron) the RDW is high and the MCV is low or normal, while in heterozygous thalassemia (an inherited disorder in hemoglobin synthesis is abnormal) the RDW is normal and MCV is low.

ELECTRONIC DEVICES FOR MEASURING BLOOD CHARACTERISTICS

There are two major classes of electronic devices for measuring blood characteristics. One type is based on changes in the electric resistance of a solution when a formed blood element passes through an aperture. The Coulter Corporation, Clay Adams, Lors & Lundberg, and Baker Diagnostics manufacture hematology instruments based on this technique. The other type utilizes deflections of a light beam caused by the passage of formed blood elements to make its measurements. Technicon Corporation is a leading manufacturer of hematology instruments that uses this approach. Coulter Corporation has been a leader in blood analyzers for many years and it has developed a large series of instruments. Let us review the most recent and most widely used instrument in this series, the Coulter STKS. The analyzed sample is blood that has been anticoagulated, with ethylenediaminetetraacetic (EDTA). Anticoagulants are substances that interfere with the normal clot-forming mechanism of the blood. They keep the formed elements from clumping together, which would prevent them from being counted accurately. EDTA does this by removing calcium from the blood. The initial step in the analysis procedure is the automatic aspiration of a carefully measured portion of the specimen. Next the specimen is diluted to 1:224 with a solution of approximately the same osmolality as the plasma in Diluter I, Figure 11.11. The diluted specimen is then split, part going to the mixing and lyzing chamber and part to Diluter II.

The function of the diluting and lyzing chamber is to prepare the specimen for the measurement of its hemoglobin content and WBC count. The lyzing agent causes the cell membranes of the RBCs to rupture and release their hemoglobin into the solution. The WBCs are not lyzed by this agent. Adding the volume of lyzing agent increases the dilution to 1:250. A second substance, Drabkin's solution, is present; it converts hemoglobin to cyanmethemoglobin. This is done to conform with the accepted standard method for determining hemoglobin concentration. The advantage of this method is that it includes essentially all forms of hemoglobin found in the blood. The specimen is next passed through the WBC bath, which functions as a cuvette for the spectrophotometric determination (see Section 11.1) of the hemoglobin content. The final step in this process is measurement of the WBC count.

Figure 11.12 outlines the method that is used in making this determination. The same method is used for counting RBCs. A vacuum pump draws a carefully controlled volume of fluid from the WBC-counting bath through the aperture. A constant current passes from the electrode in the WBC-counting bath through the aperture to the second electrode in the aperture tube. As each WBC passes through the aperture, it displaces a volume of the solution equal to its own volume. The resistance of the WBC is much greater than that of the fluid, so a voltage pulse is created in the circuit connecting the two electrodes. The magnitude of that voltage pulse is related to the volume of the WBC.

To increase the accuracy of the measurement, the system uses three parallel counting units. They share the common WBC-counting-bath electrode and have individual aperture-tube electrodes. The output of each of these circuits is connected to a preamplifier. The amplified voltage pulses pass through a threshold circuit. The threshold voltage is selected as part of the calibration procedure. Specimens whose WBC count values have been determined by reference methods are processed, and the threshold is set to give counts that agree with the reference values.

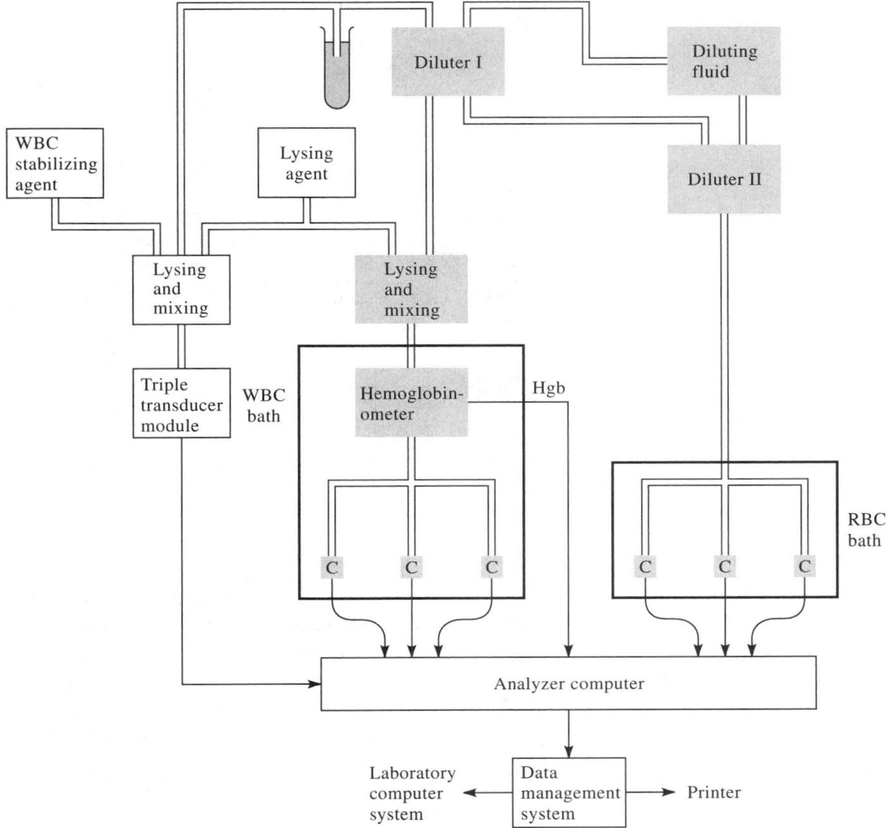

Figure 11.11 A block diagram of a Coulter Model STKS. (Modified from J. Davidsohn and J. B. Henry, Todd Sanford Clinical Diagnosis by Laboratory Methods, 15 ed. Philadelphia: W. B. Saunders Co.)

Pulses that exceed the threshold enter a pulse-integrator circuit, which produces a dc voltage proportional to the WBC count. The outputs of the three pulse-integrator circuits are sent to a voting circuit. If the three outputs agree within a specified range, they are averaged. If one output disagrees with the other two by more than the specified range, it is not used in computing the average. If all three outputs disagree by more than the specified range, an error indicator is set and a zero value is produced.

The next step in the signal processing is to correct the average-count signal for coincidence. Coincidence is the passage of two or more WBCs through the aperture at the same time. Statistical analysis is used to estimate the average level of coincidence for the aperture size and any uncorrected count level. An analog circuit makes this conversion. The digital WBC count value is displayed and also recorded on a printer.

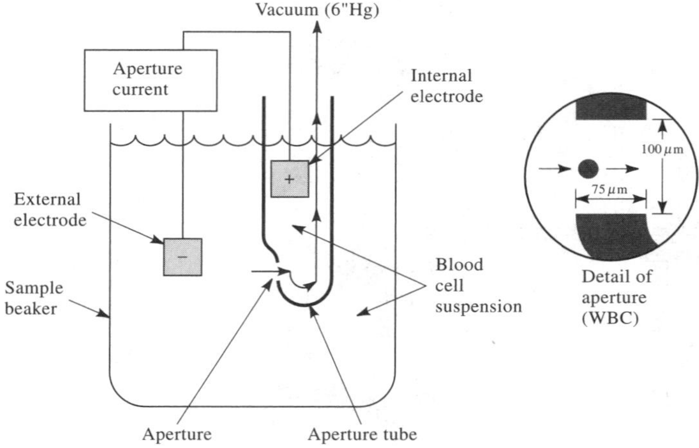

Figure 11.12 Coulter STKS aperture bath.

We will now examine the right side of Figure 11.11. The first step is the further dilution of the specimen to 1:224 in Diluter II. This second dilution is required because of the much greater concentration of RBC than of WBC in the blood. A system identical to the one described for the WBC count is used to obtain the RBC count.

Cells with volumes greater than 35.9 fl are classified as RBCs. A 256-channel RBC size histogram is prepared. The MCV and RDW are computed from this histogram. The RDW is the coefficient of variation of the RBC volume distribution.

Cells whose volumes are in the 2- to 20-fl range are classified as platelets. The volumes of these cells from each aperture is transformed into a 64-channel histogram. These histograms are statistically processed to yield a platelet count along with a mean platelet volume (MPV) and platelet distribution width (PDW) from each channel. A voting process similar to that described for the WBC count is used to determine the final values for these parameters. The MPV and PDW values are primarily used for quality control functions at this time.

The RBC count, Hb and MCV are input to a special-purpose computer circuit that calculates the values of HCT, MCH, and MCHC by using the relationships given in (11.6)–(11.8).

The Coulter STKS performs a WBC differential count using a flow cytometry approach. At the same time portions of the specimen are being delivered to the WBC and RBC baths, another portion of the specimen is sent to the WBC differential mixing and lysing chamber. Here the specimen is combined with (1) a lysing agent to remove the RBCs and (2) a WBC stabilizing agent. The WBC stabilizing agent preserves the characteristics of the WBCs as they are processed in the triple transducer flow cell. Flow cytometry consists of evaluating cells moving in a fluid stream. The triple transducer module includes

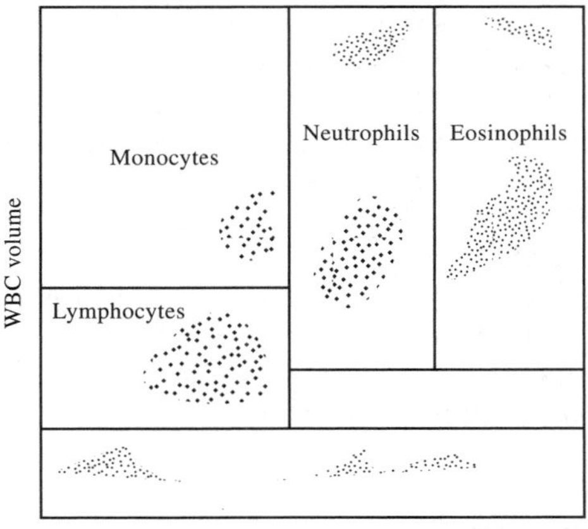

Figure 11.13 Two-dimensional scatterplot.

electronics to create a single file of cells which is passed through a measurement station. In the STKS measurements of low frequency impedance, high frequency conductivity and light scatter are made. The cell volume measurement is based on low frequency impedance (the same approach used to measure RBC and WBC volume), and the internal conductivity is derived from the high frequency conductivity. A laser illuminates the cells in the measurement station to create the light scatter. The light scatter measurement is made by a forward scatter detector. A measurement of cell internal structure and shape is based on the light scatter measurement. The cell volume, internal conductivity, and cell internal structure and shape measurements are sent to the Analyzer computer for processing. The percent lymphocytes, monocytes, neutrophils, basophils, and eosinophils is calculated based on the cell position in a three-dimensional scatter plot. Figure 11.13 is one of the two dimensional views of this three dimensional scatter plot. These two-dimensional views can be displayed at the instrument to allow the operator to monitor the functioning of the instrument. The values for basophils and eosinophils are less reliable than those for the other cell types. If the percentage of these cell types exceeds laboratory specified limits (e.g., 10% for eosinophils or 5% for basophils), a technologist will scan a peripheral blood smear slide to determine if the calculated value is correct. If it is not, a manual WBC differential will be performed. The STKS applies criteria to the scatter plot to determine if the computed WBC differential count appears to be accurate. In the majority of cases the count is judged to be accurate and the need to perform a labor intensive manual count is eliminated. Since the STKS cannot accurately identify immature WBCs (these cells are not normally found in the peripheral blood), the presence of these cells will trigger a message stating that a manual WBC count needs to be performed.

The STKS uses this same approach to measure the number of reticulocytes which are present. The RBCs are dyed before the specimen is placed in the STKS with New Methylene Blue to enhance the differences in characteristics measured in the flow cell between reticulocytes and mature RBCs. The reticulocyte count measurement is made as a separate run of the instrument. This is a very useful capability since a manual reticulocyte count is a time-consuming and relatively inaccurate process.

The blood parameters are printed on a result-report card. The printer includes a patient-identification number that is input to the STKS by the technologist. In computerized clinical laboratory systems, this identifying number, and the blood parameters are directly transmitted to the laboratory computer system.

AUTOMATED DIFFERENTIAL COUNTS

In the previous section, one approach to decrease the number of labor intensive manual WBC differential counts was presented. An alternative approach which has been used in clinical laboratories for more than ten years is based on a pattern recognition technique. The Hematrak (Geometric Data Corporation) is an example of this type of instrument and will be discussed below. The pattern recognition approach is currently also being used in the cytology laboratory to automatically screen pap smears (cytology preparations from the female cervix) for abnormal cells. Screening pap smears is another labor intensive procedure, and it is the most important method for detecting early cervical carcinoma.

The specimen that is evaluated is a blood smear stained with Wright's stain. Figure 11.14 shows the basic components of the system and the flow of information in the system (Levine, 1974). The smear is scanned by a color video scanner via microscope optics. The technologist selects the initial point for the scan. The device scans the smear until it finds a nucleated cell (mature RBCs of the type that are normally found in peripheral blood smears do not have nuclei). The digitized image of the cell is then transmitted into the image memory.

The morphological analyzer, a special-purpose digital device, next extracts important attributes of the cell, including nuclear morphology (shape), cytoplasmic morphology, nuclear/cytoplasm ratio, chromatin pattern, and cytoplasm characteristics. These attributes are transmitted to a recognition computer that compares them to the stored attributes of various types of cells (normal and abnormal) that can be found in the blood. The cell is classified as belonging to the group that its attributes most closely match. When a cell is classified as an abnormal form (for example, an immature form such as those found in the blood of people suffering from acute myelogenous leukemia), the scan is stopped and a displayed signal alerts the technologist that a "suspicious" cell has been found. The technologist then examines the cell and either accepts the Hematrak classification or enters a different classification on the keyboard.

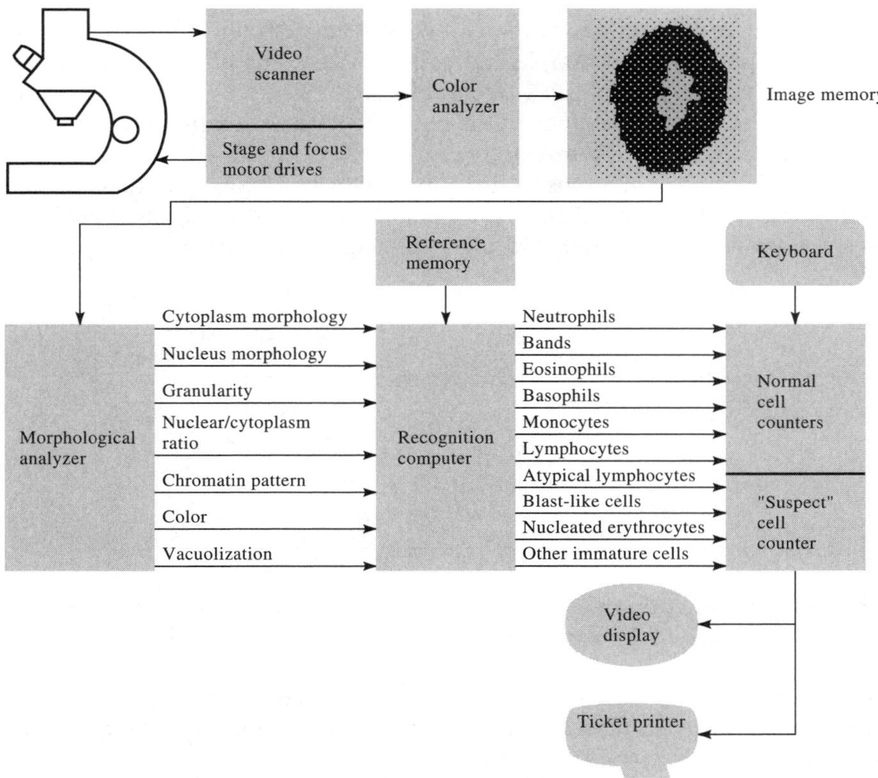

Figure 11.14 Block diagram of HEMATRAK (From M. Levine, "Automated differentials: Geometric Data's HEMATRAK," *Amer. J. Med. Tech.,* 40: 464, 1974.)

The differential count is displayed on a CRT, printed on a test-result card similar to that used by the Coulter Counter, and (in a computerized clinical laboratory) sent directly to the central computer. Further details on the system and the results of clinical trials of its effectiveness are available (Dutcher *et al.,* 1974).

PROBLEMS

11.1 Discuss the differences between photometers or colorimeters and monochromators. What are the factors to consider in selecting one or the other for a particular determination?

11.2 A filter photometer is being used to determine total concentration of serum protein (grams per deciliter). A technologist runs one standard with a known total protein concentration of 8 g/dl and obtains a $\%T$ reading of 20%, processes a patient sample and gets a $\%T$ reading of 30%, assumes the instrument's operation satisfies Beer's law, and calculates the patient value. What value should be obtained? Do you agree with the methodology? If not, what would you do differently and why?

11.3 A spectrophotometer is being calibrated before being used to determine concentration of serum calcium. Four standards (samples of known calcium concentration) are analyzed, and the following values emerge.

Standards	% Transmittance	Calcium Concentration, mg/dl
1	79.4	2
2	39.8	8
3	31.6	10
4	20.0	14

Does this determination follow Beer's law? If a patient sample were processed and a percentage of 35 were obtained, what would the calcium concentration be?

11.4 Sketch a double-beam spectrophotometer and explain its operation.

11.5 Explain why fluorometers can be used to detect much smaller quantities of substances than absorption spectrophotometers.

11.6 Assume that you are the biomedical engineer at a 300-bed hospital. The director of the clinical laboratory plans to buy an automated chemical analyzer and wants your advice on which type to buy. What factors would you consider in preparing your response? This is a broad question, but try to be as specific as possible.

11.7 The following values are obtained for a specimen of venous blood: MCV = 90 μm^3, HCT = 40%, and MCH = 30 pg. Compute the RBC count, the MCHC, and the Hb concentration.

11.8 Design a circuit to perform the RBC-counting function in a Coulter Counter. Include the voting logic.

REFERENCES

Anonymous, *Automatic Clinical Analyzer Instruction Manual.* Wilmington, DE: DuPont, Automatic Clinical Analysis Division, 1975a.

Anonymous, *Beckman Synchron CX4/CX5 Clinical Systems.* Brea, CA: Beckman Instruments, 1989.

Anonymous, *Microzone Electrophoresis Manual.* Fullerton, CA: Beckman Instruments, 1975b.

Brittin, G. M., and G. Brecher, "Instrumentation and automation in clinical hematology." *Prog. Hematol.,* 1971, 7, 299–341.

Dutcher, T. F., J. F. Benzel, J. J. Egan, D. F. Hart, and E. A. Christopher, "Evaluation of an automated differential leukocyte counting system." *Amer. J. Clin. Pathol.,* 1974, 62, 523–529.

Eggert, A. A., "Differential counts, automated," in J. G. Webster (ed.), *Encyclopedia of Medical Devices and Instrumentation.* New York: Wiley, 1988, pp. 944–956.

Ellis, K. J., and J. F. Morrison, "Some sources of error and artifacts in spectrophotometric measurements." *Clin. Chem.,* 1975, 21, 776–779.

Groner, W., "Cell counters, blood," in J. G. Webster (ed.), *Encyclopedia of Medical Devices and Instrumentation.* New York: Wiley, 1988, pp. 624–639.

Henry, J. B., *Todd Sanford Clinical Diagnosis by Laboratory Methods,* 17th ed. Philadelphia: Saunders, 1984.

Henry, R. J., D. C. Cannon, and J. W. Winkelman, *Clinical Chemistry.* New York: Harper & Row, 1974.

Hicks, R., J. R. Schenken, and M. A. Steinrauf, *Laboratory Instrumentation.* New York: Harper & Row, 1974.

Klebe, R. J., and P. M. Horowitz, "Fluorimetry," in J. G. Webster (ed.), *Encyclopedia of Medical Devices and Instrumentation.* New York: Wiley, 1988, pp. 1323–1331.

Levine, M., "Automated differentials: Geometric Data's HEMATRAK." *Amer. J. Med. Tech.,* 1974, 40, 462–468.

Littlewood, A. B., *Gas Chromatography.* New York: Academic, 1970.

Lutz, R. A., and M. Stojanov, "Flame photometry," in J. G. Webster (ed.), *Encyclopedia of Medical Devices and Instrumentation.* New York: Wiley, 1988, pp. 1314–1320.

Mandel, R., and W.-C. Shen, "Colorimetry," in J. G. Webster (ed.), *Encyclopedia of Medical Devices and Instrumentation.* New York: Wiley, 1988, pp. 771–779.

Miranda, H., and M. Hatziemmanuel, "Blood analyzer tests 30 samples simultaneously." *Electron.,* 1976, 49(8), 150–154.

Tilton, R. C., *Clinical Laboratory Medicine.* St. Louis: Mosby Yearbook, 1992.

12

MEDICAL IMAGING SYSTEMS

Melvin P. Siedband

For many years, photographic film was the principal means for storing medical images. Computers have provided a new means for storing, processing, transferring, and displaying images. First used for computerized tomography, computers and digital image processing have engendered a revolution in the way medical images are produced and manipulated. Now it is possible to acquire data, perform mathematical operations to produce images, emphasize details or differences of images, and store and retrieve images from remote sites, all without film. Telephone lines and other communication means such as the Internet with common standards of image communication make possible the near instant display of the information needed for diagnosis.

Photographic, x-ray, ultrasonic, radionuclide, television, and other imaging systems can be thought of as cameras. All camera images are limited by resolution, amplitude scale, and noise content. The photographic image can be enlarged until it becomes quite "grainy," and this graininess is the bound on both resolution and noise. The x-ray image is resolution-limited by the dimensions of the x-ray source and noise-limited by the beam intensity. The ultrasound image is limited by the angular resolution of the transducer and its ability to separate true signals from false signals and noise. The television image is limited by the electron storage capacity of the camera sensor.

An image can be studied without regard to the camera that produced it. A camera can be considered a device that transfers an image from one surface to another. It can be defined as an aperture through which the signals related to all elements of the original image must pass to appear in the final image. A camera—whether a television, x-ray, or other image-forming device—can be described in terms of its spatial-transfer function.

12.1 INFORMATION CONTENT OF AN IMAGE

In most cases, the information content of an image is the product of the number of discrete picture elements (pixels) and the number of amplitude levels of each pixel. Because pixels may not be quantized into neat boxes, but overlap each other, some method of defining their dimensions is needed.

Noise, whether originating from photographic grain, electric charge, or the statistical variations of the number of quanta (light, x ray, or gamma ray), limits the amplification permitted in a channel. For convenience, the number of amplitude steps within a channel is taken to be the same as the measured signal-to-noise ratio (SNR). The noise figure of a channel is the ratio of the theoretical or best possible SNR to the measured SNR.

RESOLUTION

An image can be considered a surface of given dimension that has a spatial resolution expressed in terms of line pairs per millimeter (lp/mm). Resolution is defined this way so that objects and the spaces between them are counted equally. If we examine copper mesh having 10 holes/cm or 100 holes/cm^2, the system must resolve 1 lp/mm and must have at least 2 pixels/mm to show each mesh hole and each mesh wire. A single countable object requires at least one line pair (2 pixels) on each axis so that the space between objects as well as the object itself may be resolved.

Figure 12.1 shows the image of a set of round objects in a television frame. The first problem is to determine the number of television scanning lines needed to resolve these objects. We assume that the television scanner comprises a sensor that can be swept one line at a time across the image, stepped down to the next line, and so on, until the entire image is raster scanned. We assume that the output of the sensor will eventually feed a scanning light projector which paints a beam of light onto a photographic film to reproduce the original image as the light beam is modulated by the output of the sensor.

The first requirement of the scanner is to direct a line through each of the round objects. If the object is white, the output signal of the sensor is positive; if dark, then negative. For a succession of white objects with dark spaces between them, each object will be represented by a positive half-cycle of the signal and each space by a negative half-cycle. In general, a countable object requires one whole cycle of spatial frequency passband: the positive half-cycle for the object and the negative half-cycle for the space between objects.

In the vertical axis of columns of objects, a scanning line is needed for each row of objects and another scanning line is needed to detect the spaces. In any real system, we may not know the location of the object relative to the scanning lines. For a vertical column of n objects, there must be at least $2n$ scanning lines. If the objects are disposed randomly, we must have more scanning lines to have a reasonable probability of having at least one scanning line through an object and one line through the space.

In general, the number of scanning lines is increased by $\sqrt{2}$ to allow for randomness. Thus, if a total of n^2 objects are randomly distributed in a square field of view, we assume that the vertical distribution is that of n objects. A system needs approximately $2n\sqrt{2}$ scanning lines to have a reasonable probability of detecting each of the objects and the spaces between them. For the

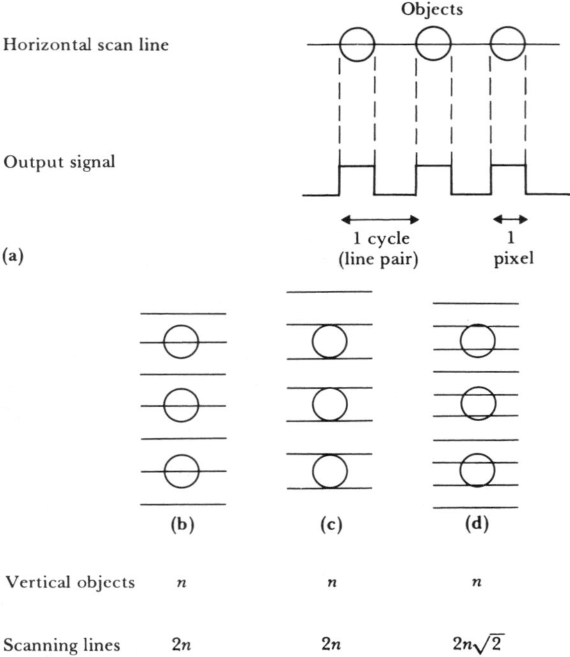

Figure 12.1 Scanning lines and round objects (a) Each object represents 1 pixel, but each cycle of output signal represents 2 pixels. (b) For $2n$ scanning lines, n vertical objects are required. (c) If objects are located between scanning lines, $2n$ lines are insufficient. (d) For adequate resolution, $2n\sqrt{2}$ lines are required.

n^2 objects within the field, a minimum of n cycles of passband is required to resolve these objects for each scanning line. The total image passband is thus $2n^2\sqrt{2}$ cycles per image.

Each object and space has equal weight. We define a pixel as a space on the image surface having a dimension of one-half cycle of bandwidth on the horizontal axis and the same dimension on the vertical axis (even though overscanned by factor of $\sqrt{2}$). We will usually choose the pixel dimension to be equal to the smallest dimension of objects we wish to resolve in the image. This does not mean that objects smaller than the pixel will be resolved. Rather, it defines the smallest size of an object for which amplitude information can be preserved. The amplitude information of small objects is averaged over the pixel dimension and amplitude information is compromised. For example, if a system had pixel dimensions corresponding to 0.5 mm and then 0.10 mm tungsten wires were examined, they would appear as 0.5 mm wires of lower contrast.

The bandwidth Δf required for a television system is given by

$$\Delta f = \frac{n_h n_v 2\sqrt{2}}{F_h F_v T} \tag{12.1}$$

where

n_h = maximal number of objects in a horizontal line
n_v = maximal number of objects in a vertical line
F_h = fraction of horizontal scan time spent on the picture (1.0 minus blanking fraction)
F_v = fraction of vertical time spent on the picture
T = total frame scanning time

EXAMPLE 12.1 A standard United States closed-circuit television system has 240 objects in the horizontal direction, 180 in the vertical, $F_h = 0.82$, $F_v = 0.92$, and $T = 1/30$ s. Calculate the bandwidth.

ANSWER

$$\Delta f = (240)(180)(2)(2)^{1/2}/(0.82)(0.92)(1/30) = 4.85 \text{ MHz}$$

IMAGE NOISE

All images are limited by both noise and spatial resolution. If we attempt to dissect an image into smaller and smaller areas, we soon find that the image is limited to some smallest element, or that the lens or scanning aperture imposes a bound on how small an element we can resolve within the image. Further, within the smallest element, we can say that the image is either on or off. The on–off criterion certainly applies when we are considering silver grains of film. It also applies when we are considering a beam of electrons hitting a cathode-ray-tube phosphor: The beam is not continuous but rather consists of discrete electrons. The phosphor particles vary in size and in probability of being illuminated. For elements of larger size, we can say that the image has a gray, or amplitude, level, which can be defined by a digital number. This is another way of saying that the image is characterized by the number of on or off states of smaller elements.

If for each of a large number of trial measurements q, there is a small probability p of a certain type of event, the average number of this type of event is simply $m = qp$. For example, if we wait q seconds for randomly distributed raindrops to fall within a certain area, and, for each second, the probability of observing a raindrop is p, then m is the average number observed in several observations of q seconds each. The relative probability of any *particular* number of drops K in a measurement having an average m, $p(K;m)$, is given by the Poisson probability density distribution.

$$p(K;m) = \frac{e^{-m}m^K}{K!} \tag{12.2}$$

We can check that this distribution really has the average m.

$$\sum_{K=0}^{\infty} Kp(K;m) = m \tag{12.3}$$

The sum of $p(K;m)$ for all outcomes K from zero to infinity is, of course, equal to 1. Finally, the variance of the distribution is also equal to m.

$$\sum_{K=0}^{\infty} (K - m)^2 p(K;m) = m \tag{12.4}$$

Thus the rms fluctuation of outcomes around the average value m is just $\sqrt{m}$. If we were to examine a succession of measurements that have an average outcome of 100 events, we would find very few with *exactly* 100 events, but they would be distributed about the average with a standard deviation of 10 events.

Independent raindrops falling on concrete squares and x-ray photons striking detector pixels have the same statistical properties. If the average number r of raindrops per square is 100, the probability of finding exactly 100 is 0.040, even though the *average* is 100/square. If we made a scanning voltmeter that read N volts for N events and used the scanning voltmeter to sample the output of the individual pixel, we would find a fluctuation of $\sqrt{N}$ rms volts.

Now we change from the steady signal N to a modulated signal $N(1 \pm \overline{M})$, where $0 \leqslant \overline{M} \leqslant 1$ is the modulation and $N\overline{M}$ is the incremental increase representing the information-containing signal S. The maximal signal exists when $\overline{M} = 1$, so the maximal SNR equals $N/\sqrt{N} = \sqrt{N}$. Thus, for a time-average value of 100 events per pixel in 1 s, the maximal SNR equals 10. If the linear dimension of the cell is doubled, the area is multiplied by 4, and, for the same time period, there are 400 events per larger pixel, for a maximal SNR of 20. Similarly, if only the integration time of the original cell is changed from 1 s to 4 s, there is also the possibility of detecting 400 events per pixel and a maximal SNR of 20. For purely random signals having the same time average, time integration and spatial integration have the same effect. In other words, decreasing the resolution of an image (increasing the area of the pixel) either increases the SNR of the image or, for the same noise level, reduces the requirement for number of events. For those cases in which $\overline{M}$ is not equal to 1, the signal is $\overline{M}N$ but the noise level is still $\sqrt{N}$, so the SNR is $\overline{M}\sqrt{N}$.

The detection of low-contrast signals in a noisy field requires a fairly high

SNR. See Figure 12.2, in which events correspond to gamma-ray photons, which register as counts on an image. We ask, "What is the probability that the number of photons per pixel randomly exceeds $N + J\sqrt{N}$ as a function of J units of standard deviations?"

By referring to a table of integral values of the standard deviation, we find that where $J = 1$, 16% of the pixels exceed the bound; for $J = 2$, 2.3% of the pixels exceed the bound; for $J = 3$, 0.14% of the pixels exceed the bound; and for $J = 4$, 0.003% of the pixels exceed the bound. However, if we are looking at, say, the field of view of a typical television camera having 180×240 countable objects, or $360 \times 480 = 1.7 \times 10^5$ pixels, then even for $J = 4$, each television frame randomly has 5 pixels exceeding the bound. In other words, the modulation of N must be greater than that required to produce the SNR $\overline{M}\sqrt{N}$ greater than 4 in order to simply detect the existence of a signal with fewer than five "false alarms," or random exceedances.

If we define C as the contrast of the $\overline{M}$ factor, the number of photons required must be at least equal to bJ^2/C^2, where b is the total number of pixels. J must be at least 4, for reasons cited earlier. If d is the linear dimension of the pixel and A is the area of the field, the total number of photons required equals $AJ^2/(d^2C^2)$. A good estimate for the number of photons of a grain-limited visual field would then be $N = 25\ A/(d^2C^2)$, under the assumption that $J = 5$. However, because of the presence of television scanning lines, phosphor granularity, and other factors, this number must actually be greater at low light levels and with low-contrast signals. The number approximately doubles to almost $50\ A/(d^2C^2)$. If we assume that, at the limit of resolution, an absolute-minimal contrast of approximately 5% is needed, then the number of photons required for detection is

$$N = A\left[\frac{7.2}{d(C - 0.05)}\right]^2 \tag{12.5}$$

Although (12.5) has been developed for the case in which d is the linear dimension of a single pixel, d can be extended to the dimension of any object. That is, as object size increases, the contrast required for the same detection probability decreases.

Experiments have shown that (12.5) is true in most cases where the contrast is greater than 5%. However, when the number of scintillations or grains/pixel is very high, the observer can see objects of contrast below 5%. It is as if radiation requirements are increased greatly for objects below 5% contrast. For example, if the exposure is calculated for a small object at 10% contrast, the exposure may have to be increased by 50 or more times to clearly see an object at 2% contrast. Remember that contrast as used for radiation-produced or other noisy images, is defined here as the number of scintillations or grains in the background reduced by a certain percentage (the contrast) by the presence of the object of interest. X-ray images are produced by subtracting quanta from the radiation field. Things would be far different if the objects produced quanta of their own against a dark field.

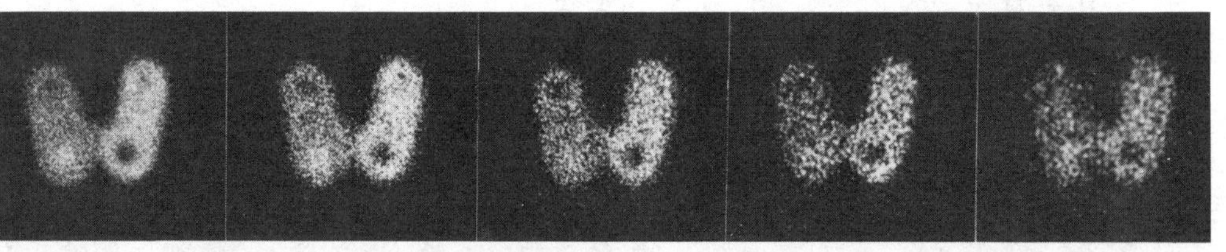

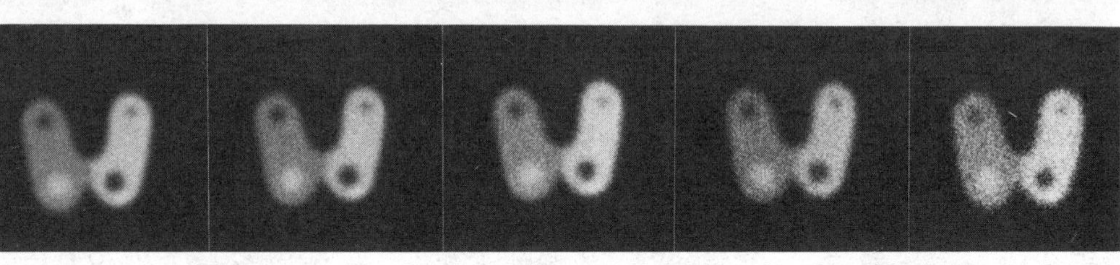

Figure 12.2 In each successive gamma-camera picture of a thyroid phantom, the number of counts is increased by a factor of 2. The number of counts ranges from 1563 to 800,000. The Polaroid camera aperture was reduced to avoid overexposure as the number of counts was increased.

One important exception to the 5% object contrast requirement of noisy images is seen when the eye is unburdened by the use of a computer. If several noisy images are integrated (accumulated and averaged), the contrast will remain the same but the ratio of contrast to rms noise will improve. If the SNR is increased, particularly for objects larger than *d,* then objects of contrast less than 5% can be seen. Note that contrast is the modulation of the background and SNR_{max} is the square root of the number of scintillations or grains in the background divided by the dimensions of the object of interest.

12.2 MODULATION TRANSFER FUNCTION

The *modulation transfer function* (MTF) is a modified form of the spatial-frequency response of an element or of the entire imaging system. The limiting resolution of a system, although it defines a system in terms of the smallest resolution element that can be seen, is not a sufficient indicator of system performance, because the system may resolve, say, a 2.0-mesh/mm copper screen at high contrast but perform poorly at showing the low-contrast image of a gall bladder. The MTF is plotted by measuring the amplitude response as a function of spatial frequency, assuming 100% response at zero frequency and ignoring phase shifts (see Figure 12.3). We assume that the amplitude is zero past the first phase rotation (crossover). The MTFs of each image transmission component may be combined as point-by-point products at each spatial frequency to obtain the MTF of the overall system.

To measure MTF, we use a sinusoidal test object to modulate the input signal to cover the band of frequencies from zero to the maximal frequency. Sinusoidal test objects are hard to make, so a square-bar pattern is generally used. In the case of x-ray systems, the bar pattern is usually made of lead or tungsten alloys and consists of a series of bars and spaces starting at a low frequency and increasing in frequency (that is, decreasing in spacing). Such test objects are placed in front of the x-ray image detector, and the system is irradiated. Similar photographic bar patterns can be imaged by the various lenses of the system and tested independently by optical means. The detector consists of a microscope with a small slit in the local plane of the microscope lens, behind which is mounted a photomultiplier tube. The output of the tube is fed to a recorder, which may record the square-wave amplitude as a function of frequency.

A square-wave function can be analyzed in terms of its sine-wave components by means of a Fourier-series expansion. A matrix inversion of the Fourier-series expansion of square-wave terms yields an equation that defines sine-wave amplitudes in terms of square-wave amplitudes. In the following formula, $S(f)$ represents the amplitude of the sine wave at frequency f, ob-

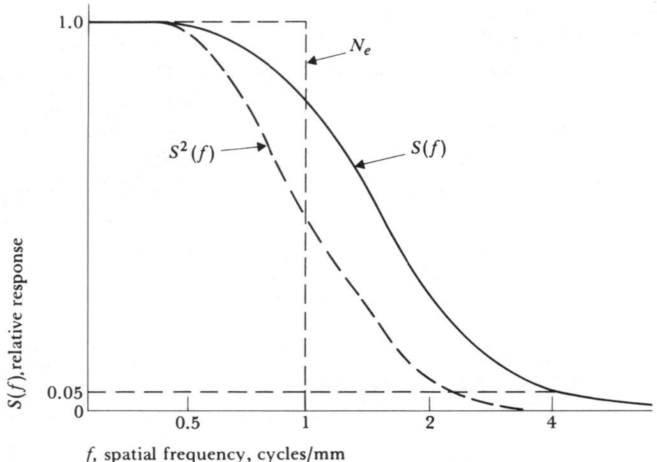

Figure 12.3 Modulation transfer function $S(f)$ for a typical x-ray system $S(f)$ is squared and integrated to yield N_e, the noise-equivalent bandwidth. The limiting resolution, 4 cycles/mm, is indicated at the 0.05 contrast level. The abscissa is plotted in cycles per millimeter, which is the same as line pairs per millimeter.

tained by substituting measured values $M(f)$ on the square-wave amplitudes at f, $3f$, $5f$, and so forth.

$$S(f) = \frac{\pi}{4}\left[M(f) + \frac{M(3f)}{3} - \frac{M(5f)}{5} + \frac{M(7f)}{7} - \ldots\right] \tag{12.6}$$

It is thus possible to convert the measured square-wave data to the sine-wave data required for obtaining the MTF. Other means exist for obtaining the MTF, such as observing the impulse or point-spread function and observing the response to a single fine line to determine the line-spread function.

Once we obtain the MTF of the system or elements of the system, we can derive from it other numbers of great value. The limiting resolution of the component or the system is often assumed to be the resolution measured at the 5% contrast level. In x-ray systems, the low-frequency contrast is assumed to be the amplitude value of the MTF at 0.1 lp/mm. The amplitude of a signal is related to the number of events per pixel. By taking equivalent bandwidths or by integrating over unit bandwidths, we can obtain a measure of the information content of the cell. However, we know that SNR is related to $\sqrt{N}$, where N is the number of events per pixel. In order to obtain a measure of the visual equivalence of various systems as a function of amplitude, we must obtain an rms equivalence. This can be done by integrating over the square of the amplitudes for all frequencies.

12.3 NOISE-EQUIVALENT BANDWIDTH

Another way of looking at the information content of an image is to note that N is related to area and that the square of the amplitude response is proportional to the number of events contained within an area defined by a given linear dimension. A noise-equivalent bandwidth N_e is that of an equivalent system that has 100% amplitude response from zero frequency to N_e and zero response above N_e when compared with a system having an amplitude response as a function of spatial frequency, $S(f)$ = MTF. The N_e is obtained by integrating the square of the MTF amplitudes.

$$N_e = \int_0^\infty S^2(f)df \tag{12.7}$$

It is as though N_e defines a mosaic of detectors of resolution N_e—say N_e pixels/cm—that has the same SNR properties as a continuum of detectors having the MTF from which N_e is derived. The value of N_e is an excellent measure of the equivalent spatial resolution from the point of view of the noise performance of any system.

$$\frac{1}{N_e} = \left[\left(\frac{1}{N_{e1}}\right)^2 + \left(\frac{1}{N_{e2}}\right)^2 + \ldots\right]^{1/2} \tag{12.8}$$

The system N_e can be estimated from elemental N_e's. The N_e concept becomes an extremely handy way to express the spatial-frequency response as a single number. As a useful and practical means for describing systems, N_e ought to be preferred to limiting resolution.

12.4 PHOTOGRAPHY

Common photography is based on the lattice properties of silver bromide crystals in the film emulsion. Light photons eject electrons from the bromine atoms. Some of these electrons are caught by the silver atoms and neutralize them. This enables the lattice to form metallic silver atoms. For a single silver atom, the neutralizing electron may be lost thermally and the silver atom bound again to the lattice. If say, five or more contiguous silver atoms are freed at almost the same time, the probability is high that they will remain in the free, metallic state. Chemical processing (developer) causes adjacent silver ions in the grain to accumulate around such a metallic silver atom to form a grain of about 10^9 metallic silver atoms, while the remaining silver bromide is carried away by the solution (fixer).

Let us assume that about 25 light photons are required to produce the

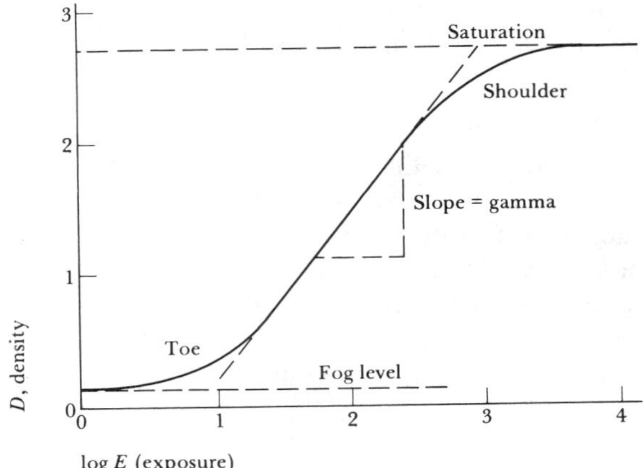

Figure 12.4 The characteristic curve of film is a plot of density versus log exposure. Higher gamma films have a higher contrast but a poorer ability to record a wide range of exposure. The slope of the straight-line portion of the D–log E curve is called the *gamma,* which is a measure of the relative contrast of the film.

first five silver atoms at a sensitivity site and to render a single grain capable of being developed. What this means is that, for a given type of emulsion and development, the number of grains per unit area and the size of the grains can be varied, but the amount of light required per grain is constant. Thus fine-grained films require more light per unit area than coarse-grained films. Actually, the number of photons required to render a grain developable is not constant but is distributed statistically about some mean value. Simple statistical averaging procedures can be used to show that the conclusion just drawn remains essentially true.

The image formed in the film prior to chemical development is called the *latent image.* It is still possible for some of the metallic silver atoms to return to their lattice coupling because of slow thermal effects. This is called *latent image fading.* It is important that the first five or so neighboring silver atoms be freed in a time that is not too long. All films have some range of exposure time such that a constant light-intensity–time product will bring the emulsion to a particular value of density, where density is defined as $D = \log 1/T$, and T is the light transmission of the film. Note that density D here is the same as absorbance A in (11.3). When this constant relationship of time and intensity fails to hold constant for longer or shorter exposures, reciprocity law failure is said to occur. Some films lose almost half their sensitivity for exposures of 10 min, compared with exposures of 100 ms.

Figure 12.4 shows the characteristic curve of film as a plot of density versus log exposure (D–log E). This curve is also often referred to as the *H and D curve,* after Hurter and Driffield, who used this description almost 100 years ago. The curve shows that a minimal exposure is needed before the film begins to respond, evidenced by the curvature at the toe of the curve. The saturation level above the shoulder is the level of maximal density for the processing conditions used. The minimal density is also called the *fog level.*

The N_e value for most photographic negative films is about 40 lp/mm. Because the resolving power of the eye is about 1 minute of arc, at a view distance of 36 cm the resolution of the eye is about 5 lp/mm. Under those conditions, enlargement of the negative image by a factor of 8 (40/5) results in no significant loss of information. Slower, detailed films permit greater enlargement. If some loss of information or increase of grain or noise can be tolerated, then greater enlargement is possible.

When films are exposed to high-energy electrons ($>$ 1 keV), excitation of more than five silver atoms takes place. In fact, the use of electrons to expose film also reduces the effects of light scatter (no light!) and permits the film to be developed to a value of N_e several times greater than that possible through exposure to visible light. *Electron-beam recording* (EBR) on film permits the use of 8-mm film to record about 30 times as much information as light exposures.

Equation (12.5) gives an estimate of the number N of events necessary to produce a just-detectable image. This estimate also holds when $1/C$ represents the number of *gray levels,* or discrete levels of intensity. We assume a rather grainy, low-resolution image of 100 $\times$ 100 mm; thus there are only three gray levels and a cell size of 3 mm. The number of events necessary (we assume that each event is a grain) is then

$$\begin{aligned} N &= A\left[\frac{7.2}{d(C-0.05)}\right]^2 \\ &= 100^2\left[\frac{7.2}{3(1/3-0.05)}\right]^2 \\ &= 7.3\times 10^5 \text{ grains} \quad \text{(poor picture)} \end{aligned} \tag{12.9}$$

When the number of gray levels is increased to 10 and the cell size is 0.5 mm, the number of events must be greater than

$$\begin{aligned} N &= 100^2\left[\frac{7.2}{0.5(0.1-0.05)}\right]^2 \\ &= 8.3\times 10^8 \text{ grains (good picture)} \end{aligned} \tag{12.10}$$

Because each grain requires about 25 photons captured, about 2 $\times$ 10^{10} photons are required to take a picture. A lesser exposure necessarily increases the noise or decreases the gray scale. Optical defocusing reduces the noise due to graininess, but it has the effect of increasing the cell size—that is, decreasing the resolution.

12.5 TELEVISION SYSTEMS

A television system consists of a camera, optional image storage and processing means, and a display monitor. In fluoroscopic systems, the television camera is coupled through a lens or fiber optic to the output of the x-ray image intensifier. The camera detects the light signal, blanks it at the edges of the image frame, and adds synchronizing pulses so that storage and display devices will be synchronized to the camera scanning. Magnetic tape and disks and optical disks are used to store images, and small computers are used to process them by emphasizing the edges of objects, by accumulating images to reduce noise, by showing differences within a group of images to enhance the outlines of blood vessels, or by showing the last image after x rays are turned off to reduce exposure. Computer processing makes possible rapid communication and diagnosis at a distance by way of the Internet or other data links. Data and image management systems combine the diagnostic images and their interpretation as well as other patient information in a complete hospital information system (HIS). During the diagnostic procedure, the display may show other vital information as well as the medical image. It is common to show the patient ECG, blood pressure, or other vital signs on the television screen during special or interventional fluoroscopic procedures.

TELEVISION CAMERAS

The sensor of a television camera comprises a matrix or surface of photodetector and capacitors. The sensor accumulates information all of the time even though each element is interrogated to yield its information only once during each scanning frame. The characteristics and thickness of the photodetector determine the light sensitivity and the capacitance determines the dynamic range and the noise properties of the sensor. The two most common geometries are of the form of an electron-beam-scanned camera tube or charge-coupled device (CCD). The sensor or target of a camera tube is a thin layer of a photoresistor such as antimony trisulphide where the sensitivity can be varied by adjustment of the target voltage V_t or a photoconductor such as lead oxide where the sensitivity is constant. An analogy can be made with film where variable gain film would have an ASA rating from 10 to 10,000 and fixed gain film would be fixed at, say, 1000. Electron charges are stored in the capacity of the target determined by the thickness of the layer. The material is deposited as a porous layer to increase its average resistance so that its dielectric constant is a minor factor in determining capacitance.

Figure 12.5 shows the construction of the vidicon and similar camera tubes. The tube has a hot cathode electron source which operates at near ground potential. The control grid G_1 surrounds the cathode and modulates the electron beam. An accelerating electrode G_2 provides an electric field, which attracts the cathode electrons. The electrode has a very small hole: the beam forming aperture. A metal cylinder G_3 may have a separate mesh at

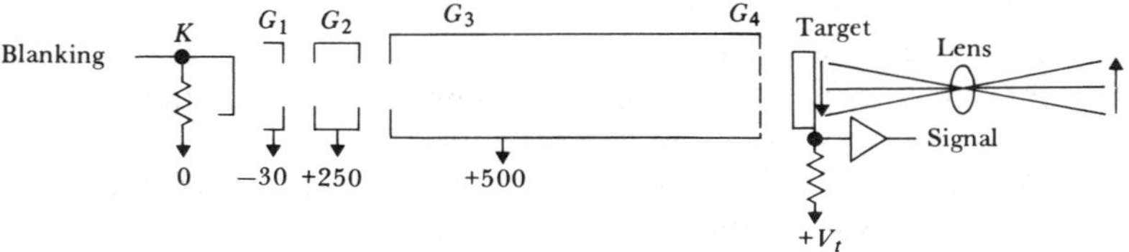

Figure 12.5 The vidicon has a cathode and a series of grids to form, shape, and control an electron beam. Magnetic deflection (not shown) scans the beam over the target, which is mounted on the interior of the glass face plate. From the target, light-modulated signal current flows through the load resistor and is amplified.

the end of the cylinder G_4 to provide a uniform electric field with room to magnetically deflect the beam by two orthogonal deflection coils and an axial magnetic field coil for focusing. The principal difference between a regular vidicon and a lead oxide vidicon is the material used for the target. In the vidicon, the material is often antimony trisulfide, which functions as a light-modulated resistor. The sensitivity can be varied by adjustment of the supply voltage to the target. Lead oxide camera tubes work as depleted junction photoconductors, and their sensitivity cannot be varied by adjustment of the supply voltage to the target.

Integrated circuit detectors comprising an array of silicon photodiodes, amplifiers, capacitors, and charge-transfer devices are in common use in video-cassette cameras. Each elemental circuit has a photodiode, a charging circuit, a capacitor, and a charge transfer circuit. The complete mosaic of detectors is called a *charge-coupled detector* (CCD). While such mosaics of up to 1024 × 2048 elements have been fabricated, the more common CCDs have about 250 × 350 elements. Because of their sensitivity to direct exposure to x rays as well as light, CCDs have found an unusual application in dental radiography. The device is small enough to be placed within a sanitary cover and then positioned within the mouth of the patient. Image information is obtained and stored in a computer for processing and retrieval. Another advantage of CCDs is that they can be cemented directly to fiber-optic plates which can be directly coupled to other fiber-optic devices such as one type of x-ray image intensifier.

IMAGE PROCESSING

CCD images are quantized as pixel-by-pixel information and analog camera tube information can be quantized by sampling and digitized for input to computers. A single frame can be stored in the computer and viewed continuously. Patient exposure can be reduced if the x-ray beam is pulsed on just long enough to acquire a picture and the images presented as a "slide show" rather than as a continuous image stream. During fluoroscopy, when the x rays are turned off, the last image can be retained as a "last image hold" for close examination without irradiating the patient.

Pairs of images can be stored and compared. If one image is stored followed by an injection of a contrast medium (usually an iodine compound tolerated by the body) in a blood vessel, the differences of that first and subsequent images will outline the blood vessels. This technique is called *digital subtraction angiography* (DSA). If one image is stored and defocused by "blending" neighboring stored pixels and compared to subsequent images, the resulting "harmonized" images will enhance the appearance of edges and small objects. Paired images can be used to show differences of energy absorption (images taken at different x-ray beam energies), useful for identifying tissue types. Images can be accumulated and averaged to reduce image noise.

12.6 RADIOGRAPHY

Radiography produced the first medical images of the inside of the body. X-ray photons are electromagnetic radiation, as are light photons. However, light photons have an energy of 2 to 4 eV and x-ray photons have an energy of 20 to 150 keV, about 10^4 times more energy than light photons. This higher energy makes x-ray photons more penetrating than light photons. Streams of x-ray photons can dissociate molecules by ionization and are thus called ionizing radiation. X rays damage the body in proportion to both the amount and rate of radiation as the body is normally in a constant state of damage and repair. X rays were first observed in 1895 by Roentgen as he experimented with a device which had a beam of electrons striking a metal target and he observed the fluorescence of some crystals several meters away.

MEASUREMENT OF X RAYS

The Roentgen is defined as 2.58×10^{-4} C/kg in air, a definition which accounts for changes of temperature and pressure. The original unit of absorbed dose is the *rad*, defined as an absorbed energy of 10^{-2} J/kg. Because about 33.7 eV is needed to form one ion pair in dry air, one R is the equivalent of 0.87 r. Not all photon energies of ionizing radiation incident to a patient have the same effect: Beams of more energetic photons can produce greater tissue damage for the same measured radiation. To account for different biological effects, the absorbed dose in rad is multiplied by a "biological effect" factor of about 1.0 in the low energy range (up to 150 keV/photon) and about 4.0 (in the 4-MeV range) and the resulting absorbed dose is expressed in rem (radiation effect in man).

The modern unit of radiation replacing the rad is the *gray* (Gy), defined as 1 J/kg in dry air and the modern unit of exposure is the *sievert* (Sv), the number of grays times the biological effect factor. Both units are 100 times greater than their older equivalents, the rad and the rem.

BACKGROUND RADIATION

We live in a world of background radiation, cosmic radiation, including that of the sun, and the natural radioactivity of the earth caused by disintegration of the heavier elements. A product of these disintegrations, *radon,* may seep through rock formations and invade the lower levels of our homes. Radon disintegration products may attach themselves to dust or smoke particles and find their way inside our bodies. The radiation of medical and dental x-ray machines, smoke detectors, package inspection machines, and other devices also affect the amount of background radiation. Like chlorine in water, very small amounts of radiation are not harmful, and may be beneficial. Large amounts can harm tissue, and this fact has been used to devise schemes to destroy cancerous tissue.

The natural background radiation ranges from 5×10^{-3} to 2×10^{-2} Sv/year. A few areas of the world have levels more than ten times higher. There is no statistically significant increase in deaths due to cancer for those living in areas of higher background. For people working with radiation, safe levels of occupational exposure are limited to 5×10^{-2} Sv/year. Nonoccupational exposure is limited to 5×10^{-3} Sv/year, a little above the background level. While ionizing radiation damages tissue, the slow rate of background exposure permits the body to repair the damage. A single whole body exposure of 6 Sv is the *mean lethal dose* (MLD), half of the people so exposed will die within a month. Yet, a lifetime exposure of 10 Sv appears harmless in that population exposures at this level will increase the incidence of cancer less than 1%. Not all tissue types are equally sensitive to the effects of exposure to ionizing radiation. A fetus is particularly sensitive as are the lens of the eye, bone marrow, the breast, and lung tissue. As a result, techniques have been developed to obtain radiographs with exposures *a*s *l*ow *a*s *r*easonably *a*chievable, the ALARA principle.

GENERATION OF X RAYS

A simple x-ray system consists of a high voltage generator, an x-ray tube, a collimator, the object or patient, an intensifying screen, and the film (see Figure 12.6). A simple x-ray generator has a line circuit breaker, a variable autotransformer, an exposure timer and contactor, a step-up transformer and rectifier, and a filament control for the tube. Medical exposures are of the order of 80 kVp (peak kilovolts), 300 mA, 0.1 s. Power levels range up to more than 100 kW (Balter, 1988).

The x-ray tube is a temperature-limited diode. Emission current is the smaller of the value defined by the Richardson-Dushman equation,

$$J_1 = AT^2 e^{-u/kT} \tag{12.11}$$

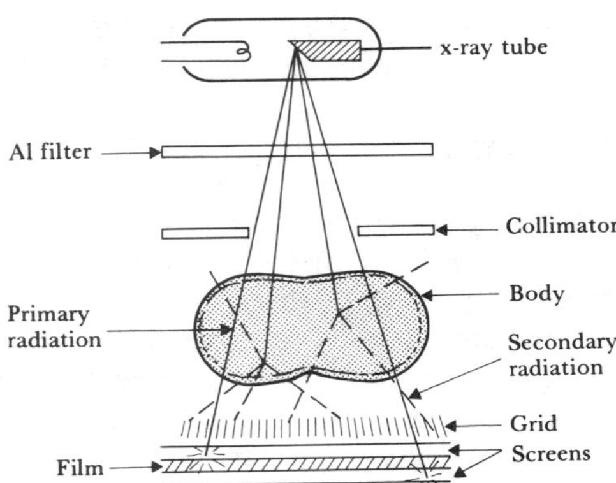

Figure 12.6 The x-ray tube generates x rays that are restricted by the aperture in the collimator. The Al filter removes low-energy x rays that would not penetrate the body. Scattered secondary radiation is trapped by the grid, whereas primary radiation strikes the screen phosphor. The resulting light exposes the film.

and that defined by the Langmuir equation,

$$J_2 = BV^{3/2} \tag{12.12}$$

where J is the current density, T is the filament temperature, u is the work function of the filament, k is Boltzmann's constant, A and B are constants usually determined by experiment, and V is the anode–cathode voltage. Because the filament cools primarily by radiation, filament power equals radiative heat dissipation,

$$\sigma T^4 = I^2 R \tag{12.13}$$

where σ is the Stefan-Boltzmann constant, and I and R are the current and resistance of the filament. Thus, electron beam current is controlled by adjusting the filament current, with compensation of filament current for variations of anode current. The fractional change of anode current is an order of magnitude greater than the change in filament current, so that circuits for filament control must be precisely regulated.

The beam electrons strike the anode and produce x rays through two mechanisms: *bremsstrahlung,* produced by the deceleration of the arriving electrons by the positively charged nuclei of the anode atoms, and *characteristic radiation,* produced when the anode's innermost electrons, knocked out of orbit by the arriving electrons, are replaced by outer shell electrons. Because the deceleration is proportional to the density of the protons of the nuclei, in turn proportional to the atomic number of the anode material Z, the efficiency of x-ray production is proportional to ZV. Because the energy of the photons produced will be proportional to V, the total energy produced will be proportional to ZV^2.

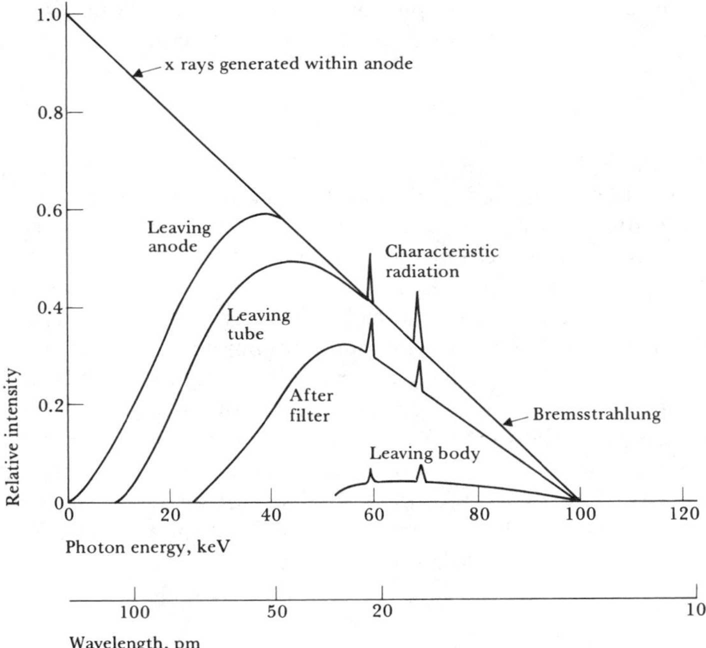

Figure 12.7 The lowest-energy x rays are absorbed in the anode metal and the tube glass envelope. An Al filter further reduces the low-energy x rays that do not pass through the body and would just increase the patient dose. Only the highest-energy x rays are capable of penetrating the body and contributing to the film darkening required for a picture. Note that the average energy increases with the amount of filtration.

Figure 12.7 shows that operating an x-ray tube at a fixed anode voltage V produces x-ray photons that have a distribution of energies. Transmission of x rays through thick layers is roughly proportional to E^3, where E is the energy of the x-ray photon. Thus, lower energy x-ray photons are less readily able than others to penetrate the anode or the glass envelope of the tube. For electrons of 100 keV energy striking the tungsten anode, the energy distribution of exiting x-ray photons shows two characteristic radiation peaks at 58 and 68 keV (68 keV radiation is the energy given up by a distant electron from well outside the l-shell as it fills the hole left by an electron emitted from the k-shell and the 58 keV is the energy given up by an l-shell electron as it fills the hole in k-shell), the *bremsstrahlung,* and the absorption of the lower energies.

ATTENUATION OF X RAYS

A beam of x rays covers a broad spectrum of energies similar to a beam of white light, which is the sum of beams of many colors. The attenuation of a beam of monochromatic x-ray photons, photons of the same energy, follows the same rules for, say, attenuation of light of one color in sea water:

$$I = I_0 e^{-\mu x} \tag{12.14}$$

where I is the final beam intensity, I_0 is the initial beam intensity, μ is the attenuation coefficient, and x is the thickness of the attenuator or layer of tissue. The attenuation coefficient depends on the photon energy, and the elemental composition and density of the layer of tissue. Differences in composition and density of body tissues, bones, muscles, and fluids are seen in the radiographs. One estimate of the average energy of the x-ray beam, is the half-value layer HVL, the thickness of aluminum needed to just attenuate the beam intensity by half. An ion chamber is used to measure the beam intensity, and thin aluminum sheets are placed close to the x-ray source to attenuate the beam. Because the thickness of the aluminum cannot be varied continuously, the thickness values closest to the HVL are recorded, just above and below the HVL, the average of the two values of the attenuation coefficient, μ, is calculated and the HVL calculated. The HVL of a medical x-ray system operated at 80 kV is close to 3.0 mm of Al.

DETECTION OF X RAYS

As the radiation passes through the patient, a portion is scattered as secondary radiation, most is absorbed, and 1 to 4% of the primary radiation is transmitted to the detector. A *grid* of fine lead strips, analogous to a miniature Venetian blind, is often placed in front of the detector. The short axis of these lead strips is aimed toward the focal spot of the x-ray tube, so that most of the primary radiation passes between the strips and most of the secondary radiation is intercepted.

After passing through the grid, the radiation is detected. Film may be used as the detector, but because of the low Z of the film and the thin emulsion, the film is relatively radiolucent. To improve the probability of detecting the x-ray photons, *intensifying screens* consisting of plastic sheets loaded or coated with high-Z scintillation powders (for example, $CaWO_4$) are placed against each surface of a *double-emulsion film*. The use of screen–film techniques increases the sensitivity and reduces the exposure by a factor of 20 to 100, depending on the screens used. Two half-thick screens will have higher resolution than one thick screen.

From consideration of the radiation-noise limit, we know that there is a minimal value for the number of x-ray quanta required to produce an image at a given resolution. The object is to choose a detector that has the required resolution, determine an acceptable noise limit, and operate the detector in such a way as to meet those objectives.

The number of photons detected per square millimeter required to just see a fine object of dimension d and contrast C is found from (12.5). We assume that 1 R of typical radiation exposure is equivalent to about 3×10^8

ϕ_x/mm^2, where ϕ_x/mm^2 is the number of x-ray photons per square millimeter when the x-ray beam is not filtered by much material. Then the radiation exposure required to produce an image can be estimated by using

$$\text{cGy/image} = \frac{2 \times 10^{-7}}{(\text{QDE})(\text{RL})d^2(C - 0.05)^2} \tag{12.15}$$

where QDE is the *quantum detection efficiency* (the fraction of x-ray photons detected) and RL is the *radiolucency* (the average fraction of incident x-ray photons that exit the object or patient and contribute to the image). Operation at values of x-ray exposure below that estimated by means of this formula results in noisy images. Operation at above that value produces quieter images, but at the cost of unnecessary exposure of the patient to radiation.

In choosing an x-ray film–screen combination, we should choose a screen that will give the necessary resolution and a film that will provide adequate film density when sufficient radiation has been received to meet statistical requirements. If the film chosen is too sensitive, the film reaches maximal density before enough photons have been detected to meet the statistical requirements, and when the film is correctly exposed, it appears noisy. A better procedure is to use film of lower sensitivity, which permits an *increase* in the x radiation reaching the film. Then the statistical requirements are exceeded, and the film noise is acceptably low at correct exposure.

The more sensitive screens are thicker than the less sensitive ones. Screen thickness varies from 300 μm or more for high-sensitivity/low-resolution screens to less than 70 μm for detail screens of higher resolution. The value for N_e of the screens is a function of screen thickness. The N_e has values ranging from about 2.5 lp/mm for the most sensitive screens to about 7 lp/mm for detail screens. The film has a value of N_e ranging from about 15 lp/mm for double-emulsion x-ray film to about 40 lp/mm for common photographic single-emulsion films. The screens obviously take precedence over the film when we are determining the MTF and N_e of a radiographic system.

The use of detail screens and high-resolution film to view low-resolution objects results in unnecessary exposure of the patient. Certainly, high-resolution film–screen combinations are necessary for certain forms of arteriography, in which fine blood vessels must be seen, but they are not needed for discovering broad, soft-edge lesions of the lung. Measured in the plane to the film, there are very few body parts that have spatial resolutions in excess of 1 lp/mm. High resolution may not be the proper objective in the designing of a system. Trying to achieve high-contrast performance and the proper relationship between contrast and SNR may be a more worthwhile goal. Of course, there are exceptions. Mammography (radiography of the breast), for instance, requires resolution greater than 3 lp/mm.

The thicker, high sensitivity screens have a QDE of about 10%, and the thinner, detail screens have a QDE as low as 2%. Screens used for mammogra-

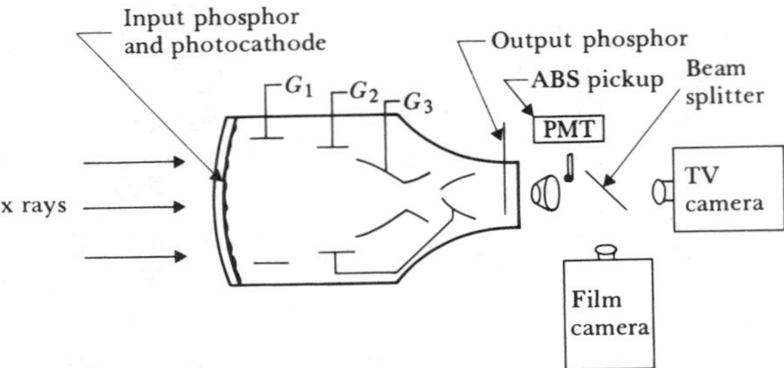

Figure 12.8 In the image intensifier, x rays strike the input phosphor screen, thus generating light. Light stimulates the photocathode to emit electrons, which are accelerated through 25 kV to strike the output phosphor screen. Brightness gain is due to both geometric gain and electronic gain.

phy are thin, but beam energies for that procedure are low and the screens have a higher QDE, about 30%, at those energies.

The choice of screens for any application is a compromise: thinner screens/higher resolution/lower QDE and thicker screens/lower resolution/higher QDE. However, cesium iodide (sodium doped) screens have the unusual property of forming as a fiber optic and can be deposited as a thick layer, sufficient for QDE of 50 to 80%, with little lateral light scatter and high resolution. Unfortunately they absorb water and will deteriorate in air and cannot be used with film but function well in the vacuum of image intensifiers.

IMAGE INTENSIFIERS

X-ray image intensifiers are used in fluoroscopic systems and have replaced the old-fashioned fluoroscopic screens. One disadvantage of the fluoroscopic screen was that the radiologist's eyes had to be dark adapted to see low-contrast objects. Other disadvantages included difficulties associated with photographing the image, particularly when simultaneous viewing and recording were required.

The x-ray image intensifier combines the functions of x-ray detection and light amplification in a single glass envelope. Figure 12.8 shows the construction of an x-ray image intensifier tube. X rays strike the input screen—usually a layer of cesium iodide—which fluoresces in proportion to the x-ray intensity. The input phosphor is in close proximity to a photocathode, so the light stimulates the emission of electrons. These electrons are accelerated through the 25-kV electric field and focused by shaping the electric field. They strike the output phosphor, which produces an image that is smaller but brighter than that produced at the input phosphor. The ratio of image brightness of the two phosphors is called the *brightness gain* of the intensifier tube. The brightness gain is the product of the geometric gain (the ratio of the areas of the input and output phosphors) and the electronic gain (the product of input quantum efficiency, photocathode efficiency, potential difference between input and output phosphor, and output phosphor efficiency).

A lens mounted on the image intensifier serves to collimate or focus the output image to infinity. A lens used as a collimator has maximal light-gathering power when compared with the same lens used as a single reimaging element. The objective lens for each camera collects the light of the collimating lens and refocuses it on the film plane. The advantage of the two-lens system, in addition to its optical speed, is that the distance between the two lenses does not influence the focusing of the systems and, furthermore, makes possible the use of a beam-splitting mirror so that light can be directed to more than one output port at a time and proportioned appropriately.

For example, in a two-port fluoroscopic system, all the light may be directed to the television camera during fluoroscopy. During cineradiography, the individual cine frames require a brighter image than that needed for the television camera, so the beam splitter is positioned to direct 90% of the light to the cine camera and 10% of the light to the television camera. Thus the radiologist may observe the images during the time when the motion pictures are being made.

IMAGE NOISE

If we use a *fluorospot* filming technique (a film system that photographs the output phosphor) with the objective-lens aperture wide open, the film becomes too dark. An operator might reduce the dose of x radiation to achieve proper darkening of the film. This would be a mistake, however, because the image would be statistically limited at the photocathode, and it would be noisy. The proper technique is to set the dose of radiation to achieve satisfactory image quality and then stop down the film objective lens to achieve proper darkening of the film.

A radiographic–fluoroscopic system consists of an x-ray table that contains an x-ray tube coupled mechanically to the spot-film device. The spot-film device holds a cassette that contains a film and intensifying screens in a "parked" position in a lead-shielded enclosure. During fluoroscopy, the radiologist observes the televised output image of the x-ray image intensifier. When the object or event of interest is discovered, a motor drives the cassette to a position in front of the image intensifier, and a radiographic exposure is made. The motor then returns the cassette to "park," and the system reverts to a fluoroscopic mode.

Tomographic systems are arranged such that the generator tube and cassette move about a pivot point or fulcrum during the exposure. The effect is to blur the image of objects outside the plane of the fulcrum. This procedure makes it possible to detect small lung or kidney tumors by blurring the images of overlying structures. There are tomographic machines that move the tube and cassette in circular, spiral, or hypocycloidal trajectories of such precision that they easily resolve even the bone of the inner ear.

PATIENT EXPOSURE TO X RAYS

As described earlier, background exposure is of the order of 10^{-2} Sv/yr. A typical chest x ray exposes the patient to 4×10^{-4} Sv, about the equivalent of 25 days of background. Other diagnostic x-ray procedures involve single exposures ranging from 10^{-4} Sv (fingers) to 10^{-1} Sv (some heads) and are considered relatively safe. An excessive number of exposures or very long fluoroscopic times can subject the patient to high exposures, and the risk versus benefit of the procedure must be considered. Cineradiography or electronic serial radiography (where a video recorder is used) records a succession of images, and each image is a separate radiograph. An extended study of the heart can subject the patient to very high incident radiation. We know that the MLD of a single exposure to the entire body is 6 Sv. Fractionated exposures or exposures to limited areas give the body time to recover but may increase the lifetime probability of cancer.

Interventional fluoroscopy permits the repair or modification of blood vessels or organs without conventional surgery. Guide wires and cannulae (small tubes) are threaded through blood vessels to reach the defect, and a small balloon is expanded to increase the effective diameter of a restricted blood vessel. Or a stent (an expanded metal retainer) can be threaded through a cannulus and positioned to keep a vessel from collapsing, all under the guidance of a fluoroscope. Unhappily, some interventional procedures take a long time and the exposure to the skin of the patient can exceed 1 Sv and cause skin reddening, *erythema,* and loss of hair, *epilation.* Higher exposures can cause ulceration which can be severe enough to require skin grafts.

12.7 COMPUTED RADIOGRAPHY

Some early phosphors used in x-ray intensifier screens produced residual or afterimages. These appeared as faint shadows of previous images on new exposures. During the exposure to x rays, most of the excited phosphor electrons returned directly to the valence band. Some returned via an *f*-center with an energy gap in the visible light region so that the screen fluoresced. A few electrons were trapped and released thermally or by light excitation at a later time. Heating the screen removed electrons from the traps to erase the potential afterimage. Most modern phosphors are deficient in traps to reduce the effect.

However, the afterimage effect can be increased by making the screen rich in traps and then used in a practical way. After the screen has been exposed it can be scanned with a laser of energy close to the trap level. Then the trapped electrons will be released and will return to the valence band causing the screen to fluoresce in proportion to the initial exposure. When screens are used in this way without film they are called *storage phosphors* and make possible computed radiography.

The storage phosphor screen can be mounted in a cassette of the same dimensions as a conventional x-ray cassette and exposed in the same way. The cassette is then placed in a laser scanner where the screen fluorescence is sensed by a phototube or photodiode, amplified, and converted to a digital signal for computer storage and processing. The screen can be scanned at a low level so that few electrons are removed from their traps in a single pass; but sufficient electrons are removed to estimate the level of the stored signal, i.e., the exposure level to set the gain of the computer circuits which will process the signal. By scanning several times and accumulating the signals in the computer, it is possible to use an analog-to-digital converter (ADC) of lower bit capacity but faster response to generate clean signals of wide dynamic range which approach the quality of x-ray film. For example, a fast 10-bit ADC can be used over four scans to accumulate a 12-bit image in the computer.

Once high-quality x-ray images are stored within the computer, image processing is used to emphasize certain features of single images by unsharp masking (subtracting an out-of-focus image and amplifying the difference), line averaging, or interpolation to convert images from one scanning standard to another, or image fusion standardization where the dimensions and scan parameters are converted to standard values so that images from different sources can be superimposed. Other imaging systems, e.g., CT, ultrasound, MRI, nuclear medical cameras, digital fluoroscopy, store images in computer memory. However, the dynamic range and resolution requirements of these images vary with their source. It is simple to convert one image type to another. For example, a 512-line CT image can be converted to a 1024-line image by interpolation to produce the additional lines. While the new image will not contain any more information, it is compatible with high-resolution television images or those obtained from storage phosphor screens and then images from different sources may be compared, differenced, and displayed on the same bank of computer monitors.

The computer can be used for writing the diagnostic report and images and reports sorted and controlled in a database management system. The same methods used for tranferring business information, technical drawings, or news photographs can be adapted for use in medical imaging for teleradiology or telemedicine. Using these methods, expert consultation is possible in the most remote areas of the world. However, image standards for communication and archiving engineering drawings or credit card information may not be adequate for the variety of medical images, or the range of image compression schemes now in use.

Medical image diagnosis often requires examination of an entire high-resolution image followed by scrutiny of a portion of the image, the region of interest (ROI). A 1024-line display monitor may be used to display the entire image and the ROI magnified to fill a portion of the display. One system displays the ROI as if under a magnifying glass in its area of the original image. This means that the original image must have much higher resolution than the display monitor so that magnified portions will not be compromised.

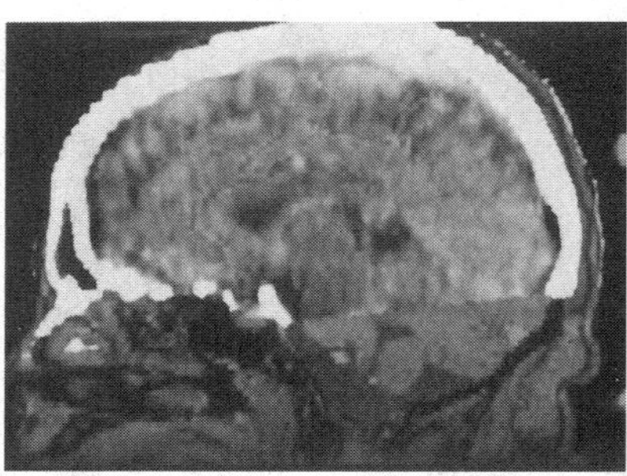
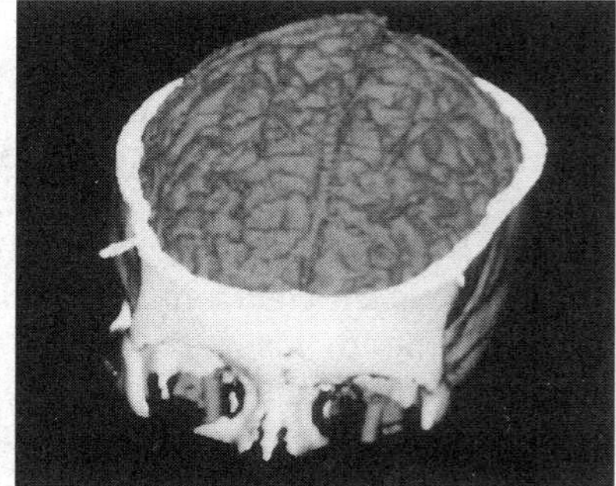

Figure 12.9 Images of the skull taken using CT and images of the brain taken with MRI, fused into composite images. (Courtesy of Rock Mackie, University of Wisconsin.)

One system averages lines of a 4096-line image to 1024 lines for display and magnifies 1/4 field sections by a factor of 4 for ROI display.

Images can be transformed from real space to frequency space by means of the fast Fourier transformation, FFT. The coefficients of the FFT image are smaller than those of the real image and require less computer memory. Such "compressed" images can be converted back to the original form by an inverse transformation. Another compression scheme considers the original image in blocks of, say, 4×4 pixels of one byte/pixel. The mean and standard deviation of the 16-pixel block are calculated, and each pixel equal or greater than the mean is given a value of "1" or is otherwise valued as "0." The matrix of 1's and 0's requires two bytes to define the 16 pixels (the bit map) and the mean and standard deviation require 1.5 bytes (the standard deviation requires only 4 bits) to transmit. When reconstructed, the 1's are given the value of the mean plus the standard deviation and the 0's are given the value of the mean minus the standard deviation. Simple image filters are used to give the reconstructed image a more pleasing appearance. Thus, to store or transmit 16 pixels requires only 3.5 bytes where the original image requires 16 bytes, a compression ratio of about 4.7. This technique is called *filtered block compression.*

Computer techniques are also used to convert the dimensions of one image to conform to those of another. By mixing or comparing images from different medical imaging systems, a coarse image of the absorption of a metabolite taken by a nuclear camera can be seen against the high-resolution image of the brain taken by a CT machine. The technique of generating a composite image from more than one source is called *image fusion.* Figure 12.9 shows images from MRI and CT systems fused to show the details of the skull (CT) with enclosed brain tissue (MRI). Details, such as the distortion of the brain tissue flowing into an artifact of the skull are seen with clarity and certainty not possible in the separate images.

Computed radiography has been incorporated into *local area networks*

(LANs) as part of a filmless medical imaging system. Systems that use a network of computers, optical disk archiving devices, and long line transmission have been assembled as *digital imaging network/picture archiving and communication systems* (DIN/PACS). In a DIN/PACS, an operator can call up any image and text file in the system for interactive display. *Teleradiology* enables a radiologist or other specialist at a distant site to view an image, make a diagnosis, write a report, or obtain expert consultation.

Standardization of DIN/PACS parameters is frequently done using the Digital Communications Standard, DICOM 3.0, to define the data header. This describes the characteristics of the image and the compression method, and identifies the image. Manufacturers of imaging systems incorporate DICOM interfaces so that standard display consoles can show images from several sources for comparison. Optical scanners are used to scan and digitize film radiographs for entry into a DIN/PACS.

12.8 COMPUTED TOMOGRAPHY

A conventional x-ray image is limited because it is generated by projection from the x-ray source through the object onto the film. If there are regions of small and large variations in electron density along the same beam path, then small variations cannot be detected. An example of this is the conventional chest radiograph where dense bony structures make it difficult to derive information about the less-dense information in the lung fields.

One way to minimize this obstruction of one structure by another is to expose radiographs from several directions. This may not be practical because of the higher exposure received by the patient. In recent years, however, technology has yielded new ways of extracting more information from each transmitted photon so that better information on electron density of objects may be determined to reveal otherwise-hidden structures through multidirectional exposures.

Computed tomography (CT) is the name given to the diagnostic imaging procedure in which anatomical information is *digitally reconstructed* from x-ray transmission data obtained by scanning an area from many directions in the same plane to visualize information in that plane. The ideas involved were originally developed for imaging the brain. The dynamic range of densities in the brain is only a few percent, but the brain is encased in a bony structure so dense that most of the x rays are absorbed by the bony structure. Imaging of the brain by conventional radiography is difficult even when contrast is enhanced by injection of contrast materials or air. In concept, CT solves the set of simultaneous equations involving thousands of attenuation coefficients, u_{ij}, for each ij element over the dozens of directions ("projections") used. Along a line of a given direction, the total attenuation is related to the sum of the individual attenuation coefficients:

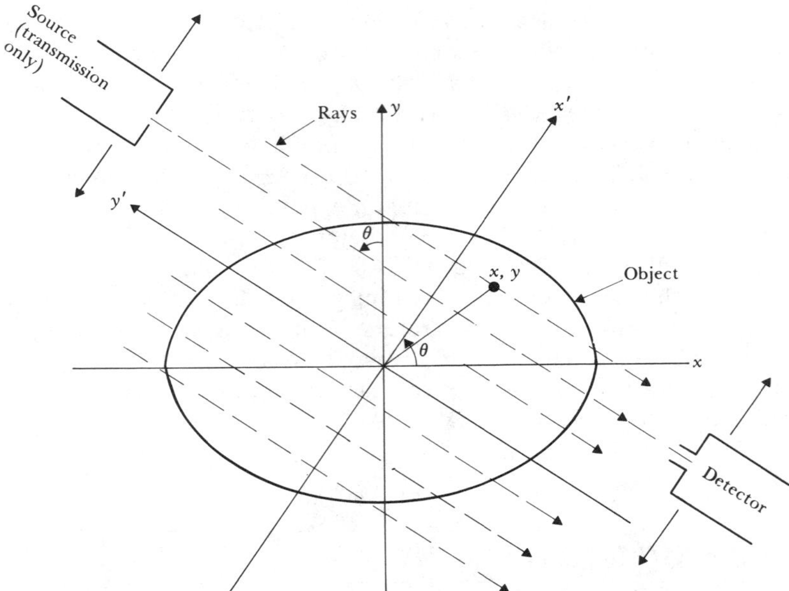

Figure 12.10 Basic coordinates and geometry for computed tomography The projection rays shown represent those measured at some angle θ. The source and detector pair are rotated together through a small angle, and a new set of rays measured. The process is repeated through a total angle of 180°. (From R. A. Brooks and G. Di Chiro, "Theory of image reconstruction in computed tomography." *Radiology,* 1975, 117, 561–572.)

For a single element: $l = l_0 e^{-\mu x}$ or $\ln I/I_0 = -\mu x$. For a series of elements of equal thickness,

$$\ln l/l_0 = -\Delta xe - Dx(\mu_1 + \mu_2 + \mu_3 + \mu_4 \ldots)$$

where l is the exit beam intensity, l_0 is the initial beam intensity, x is the layer thickness, Δx is the thickness of an element of constant size, μ is the attenuation coefficient, and μ_i is the absorption coefficient of a particular series element.

Figure 12.10 is a schematic diagram of the scanning operation of the "first-generation" CT machines. A collimated beam of x rays is passed through the patient's head in a direction transverse to the longitudinal axis. The emerging beam flux on the opposite side of the patient is constantly monitored in a scintillation detector. The x-ray source and detector move together, perpendicular to the beam direction, and roughly 160 distinct measurements of total attenuation of the x-ray beam are made at evenly spaced points along the scan path. The configuration of x-ray beam source and detector is then rotated through a small angle, usually 1°, and the procedure is repeated. The acquisition of absorption information by scanning is continued until an angle of 180° has been swept.

The procedure is called tomography (the Greek root *tomos* means "cut" or "section") because only those structures lying in the narrow anatomical slice traversed by the beam are imaged. The tightly collimated beam results in imaging that is essentially scatter-free and is efficient enough to make applying the procedure practical.

Fundamental to the procedure is the mathematical discovery that a two-dimensional function is determined by its *projections* in all directions. A sampling of projections at angles uniformly distributed about the origin can provide an approximate reconstruction of the function. How much detail can be reconstructed is straightforwardly dependent on the number of angles sampled and the sampling coarseness at each angle. If beam absorption is measured at 160 distinct points along each scanning path and a 1° increment in angle is used, nearly 29,000 distinct pieces of x-ray absorption data are acquired. These are employed to reconstruct a two-dimensional map of x-ray absorption as a function of position, presented as a 160×160 matrix of uniform square-picture elements.

The reconstruction of images from the scanning data is performed by means of a small digital computer. The time required for reconstructing the picture is of the same order of magnitude as that for acquiring the data. Some of the mathematical reconstruction algorithms permit reconstruction to begin as soon as the first projection data come in. These algorithms clearly provide a considerable saving in time by allowing the mathematical reconstruction to take place during the scan operation.

The mathematical algorithms fall into two general classes, the iterative and the analytic. In the *iterative* methods, an initial guess about the two-dimensional pattern of x-ray absorption is made. The projection data predicted by this guess are then calculated and these predictions compared with the measured results. Discrepancies between the measured values and the model predictions are employed in a continuous iterative improvement of the model array.

Figure 12.11 shows the scheme by which the model projections are generated and by which the discrepancies between model and measurement are used to best improve the model at each iteration. Each reconstructed picture element is represented by an average attenuation coefficient μ_{ij}, where the subscripts i and j specify the position of the picture element in the image. The relative degree to which each element can remove x-ray flux from the ray at the kth beam position at scan angle θ is expressed by the four-label quantity $W_{ij}^{\theta k}$. These quantities are essentially determined by the geometrical overlap between the finite-width x-ray beam at scan position θk and the square picture element at position ij. Clearly, the overwhelming majority of the more than 8×10^8 quantities $W_{ij}^{\theta k}$ are zero, because in most cases the ray θk does not pass through the element ij at all. The model array μ_{ij} determines the model projection data at each iteraction according to

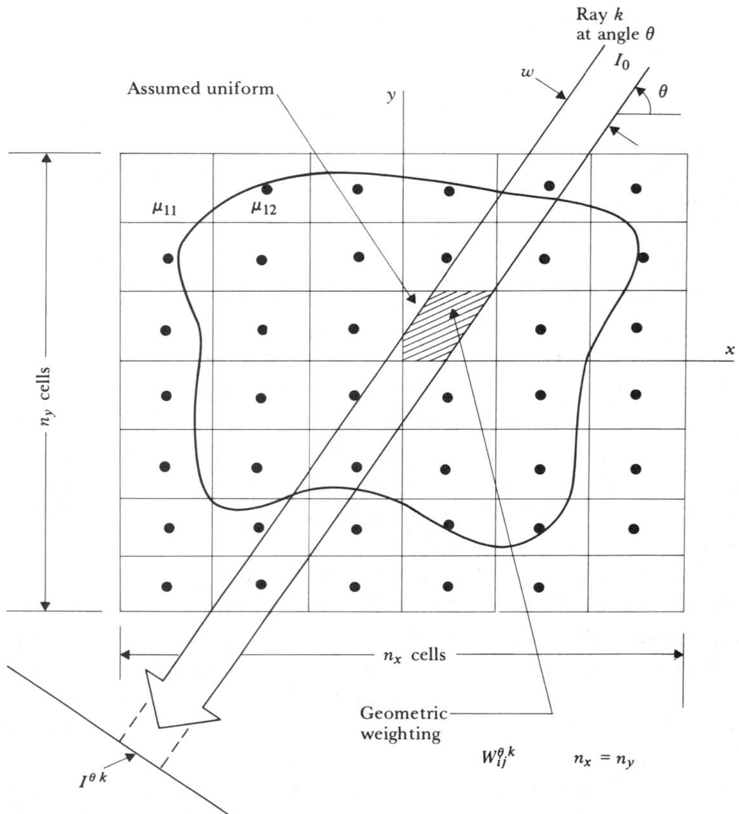

Figure 12.11 The basic parameters of computerized image reconstruction from projections. Shown are the picture element cells μ_{ij}, a typical projection ray $I^{\theta k}$, and their geometrical overlap $W_{ij}^{\theta k}$. (From Ernest L. Hall, *Computer Picture Processing and Recognition.* New York: Academic, 1978.)

$$I^{\theta k} = I_0 \exp\left(-\sum_{ij} W_{ij}^{\theta k} \mu_{ij}\right) \tag{12.16}$$

$$P^{\theta k} = \ln\left(\frac{I_0}{I^{\theta k}}\right) = \sum_{ij} W_{ij}^{\theta k} \mu_{ij}$$

where I_0 is the constant intensity of the input beam and $I^{\theta k}$ is the intensity transmitted of position k at angle θ. The quantities $P^{\theta k}$, conventionally called the projection data for the position k at angle θ, are calculated in the manner shown in order that the observed measurements be converted into quantities that are simple linear combinations of the unknown quantities μ_{ij}.

Just as the weights $W_{ij}^{\theta k}$ determine which picture elements are involved in the generation of the model projections, they also determine the manner

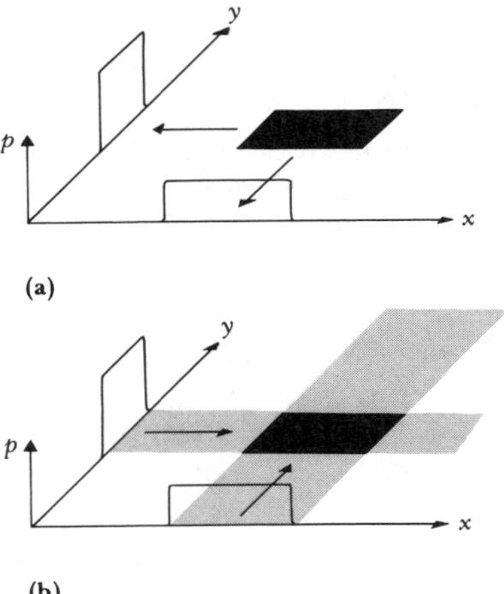

Figure 12.12 Back projection (a) Projections of this object in the two directions normal to the x and y axes are measured. (b) These projection data are projected back into the image plane. The area of intersection receives their summed intensities. It is apparent that the back-projected distribution is already a crude representation of the imaged object. (From R. A. Brooks and G. Di Chiro, "Theory of image reconstruction in computed tomography," *Radiology,* 117, 1975, 561–572.)

in which the model array is changed at each iteration. In the simplest iterative reconstruction techniques, the discrepancy between the measured and model values of $P^{\theta k}$ is attributed equally to all the elements ij traversed by the ray θk, and each model-picture element is changed, according to the geometrical weights $W_{ij}^{\theta k}$, to make the model value and the measured value for the scan line under consideration come into agreement. In actual practice, many time-saving variations of this fundamental idea have been successfully tried. Regardless of the exact way in which the model array is modified, the rays θk are cyclically iterated until the model values and the measurements of all ray projections are in adequate agreement.

Analytic methods differ from iterative methods in a very important way. In analytic methods, the image is reconstructed directly from the projection data without any recourse to comparison between the measured data and the reconstructed model. Fundamental to analytic methods is the concept of *back projection.*

Figure 12.12 illustrates back projection for the case of projections at two angles separated by 90°. As shown, a back-projected image is made by projecting the scan data $P^{\theta k}$ back onto the image plane such that the projection value measured for a given ray is applied to all the points in the image plane that lie on that ray.

The total back-projected image is made by summing the contributions from all the scan angles θ. The summing is carried out by using the same geometrical weights defined in the preceding paragraph. The back-projected image is already a crude reconstruction of the imaged object. More important, the exact relationship between the back-projected image and the desired actual array of attenuation is well understood; the latter can be calculated from the former via Fourier analysis. The back-projected image is Fourier-transformed into the frequency domain and filtered with a filter proportional to spatial frequency up to some frequency cutoff. The result is then transformed back into Cartesian coordinates.

It can be shown by Fourier analysis that the same result is obtained when the projection data are filtered first. These filtered projections are used to construct the final back-projected image. The filtering operations can also be done in Cartesian space by means of analytic algorithms called convolution techniques (Bracewell, 1965).

The first-generation machines used a single pencil beam and a single detector in a translate–rotate scan. Each translation took about 5 s followed by a rotation of 1°, then 1 s delay for the machine to stop vibrating, followed by another translation of the x-ray tube and detector. To rotate over the full 180° took about 20 min. Because x rays cannot be focused, the x-ray beam was limited to the dimensions of the detector by "coning," using a leaded box in front of the x-ray tube with a hole scaled to the dimension of the detector.

The second generation used an array of 100 or more detectors spaced every 5 mm and the x-ray tube output shaped as a fan beam. The detectors and the x-ray tube rotated together, and there was no need to translate the assembly to cover the field. For the same number of photons detected, the multiple detectors reduced the scan time by a factor of 100 or more. The time to collect information for one slice was less than 10 s. Because a patient could hold his breath for 10 s, it was now possible to take CT images of the chest, but heart motion was a limiting factor in many applications.

The next improvement lined the third-generation CT machine gantry with several hundred stationary detectors and rotated the x-ray tube. Improvements of the computers, higher output x-ray tubes reduced the scan time to 2 s or less and the slice thickness to around 2 mm. Cables connected the rotating tube to the power supply so that a scan series consisted of a number of winds and unwinds of the cables to make the exposures. For example, the tube would rotate once to make an exposure, the patient table would translate 1 cm, the tube would rotate in the opposite direction while making the next exposure, the table would translate 1 cm, etc. A series of 10 or 15 exposures would permit a study of, say, the liver or abdomen. By using power supplies which rotated with the x-ray tube and slip rings to carry power to the assembly, it was not necessary to rotate/reverse/rotate, etc. in order to wind/unwind the x-ray tube cables, and the scan process could proceed with a series of rotate the tube assembly, translate the table, rotate, etc., to complete the exposures for the study.

The most recent improvement is to move the table in a smooth, stepless motion while the tube assembly rotates continuously. This helical or spiral scanning method obtains image data faster than five images/s. Another im-

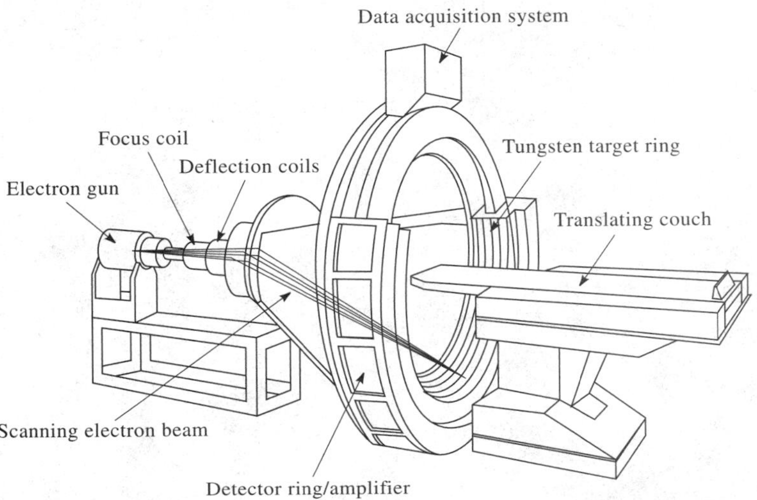

Figure 12.13 IMATRON electron beam CT system. (Courtesy of Doug Boyd, IMATRON Corp.)

provement uses paired detectors so that two cuts 1 cm apart are made simultaneously. These fast image acquisition systems make possible CT angiography of the heart, i.e., dynamic studies of the blood vessels of a beating heart. A practical limitation of any of these systems is the difficulty of rotating a 200 kg mass at several rpm while maintaining precision of the position of the source to within a fraction of a millimeter.

Figure 12.13 shows one novel machine, which avoids the practical problem of accelerating the mass of the x-ray tube and power supplies by building the machine in the form of a "demountable" x-ray tube. Here, the electron beam strikes a circular anode almost 2 m diameter, the patient is within the circle, and the ring of detectors is the same as the fourth-generation machines. Such machines take CT angiographic images faster than 30 images/s.

Figure 12.14 shows that the resolution of the third- and fourth-generation machines can be as high as 512×512 pixels. In some applications, the operator can set the machine to lower values when high resolution is not required and so reduce patient exposure. The early machines used scintillation detectors based on the technology of images for nuclear medicine. In order to pack many detectors into the small space required for high resolution, new detectors were developed. One type uses an "egg crate" assembly of pressurized xenon gas ion chambers. Several hundred cylinder equivalents are arranged in an arc, each feeding an amplifier. The dimensions of each cylinder and the gas pressure are such that the x-ray absorption of the mass of xenon gas along the length of the cylinder is sufficient for detection at reasonable exposure rates (about 10% stopping power). Solid-state detector technology has improved so that small scintillators can be coupled to arrays of photodiodes and still have

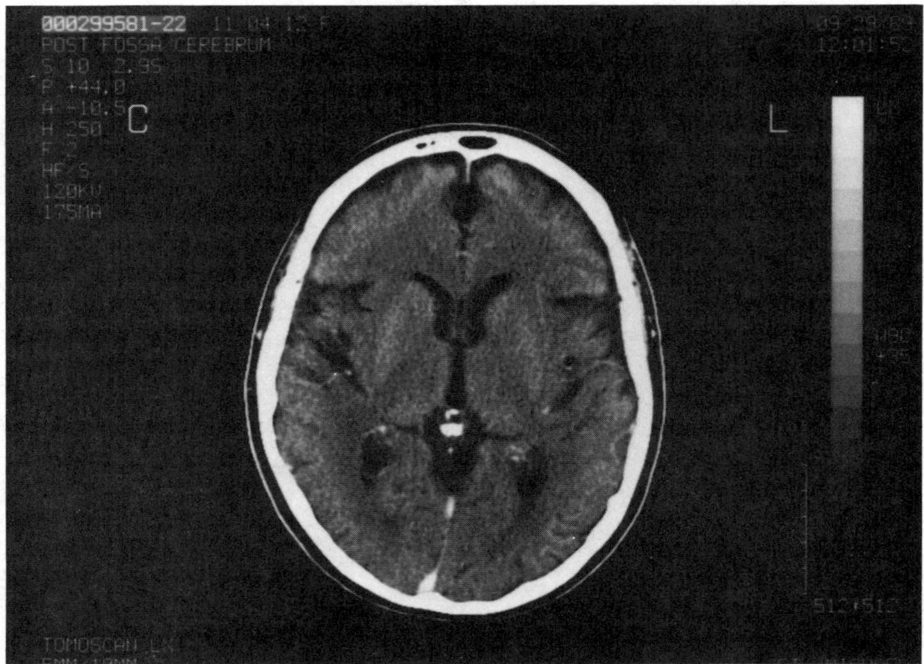

Figure 12.14 512 × 512 pixel CT image of the brain Note that the increased number of pixels yields improved images. (Photo Courtesy of Philips Medical Systems.)

excellent absorption. Newer detectors convert the x radiation directly into electric signals to multiple integrated-circuit amplifiers.

The linear attenuation coefficient of the x-ray beam of the patient tissue can be expressed in dB/cm. However, this term is dependent on beam energy and is often inconvenient to use. Of greater interest to the diagnostician is the relative attenuation corrected for beam energy and other effects. Tissue attenuation can be characterized by values from −1000 H (air) through 0 H (water) to +1000 H (bone). The units are expressed as H units, or Hounsfield units, and are compensated for beam and other effects. Abnormal tissue, bone density, and body fluids are often found by noting deviations of the attenuation in H units.

Figure 12.15 shows that images are presented at the computer console. The operator controls the window width and level (WWL, brightness and contrast with a wedge showing H units versus gray levels). Patient information and machine settings are also displayed. The patient table moves a preset distance between scans so that several planes or "cuts" can be imaged. The physician can observe the appearance of objects in several different cuts to get an idea of their shape. Computer programs have also been developed that take the information from several cuts and display that information as a three-

Figure 12.15 Control console and gantry assembly of a CT system (Photo courtesy of Philips Medical Systems.)

dimensional image object. For example, the images of cranial bones in several planes can be processed to generate a synthetic image of a complete skull. A duplicate monitor is used in a film camera assembly to photograph the CT images of several cuts and WWLs. Several images are recorded on a single film for the patient record. Physicians often examine CT images at the same time as other type of images, because objects that are obscure in one type of system may be obvious in others.

12.9 MAGNETIC RESONANCE IMAGING

Spinning charged particles have a magnetic moment and, when placed in an external magnetic field, tend to align with the field. The usual state would be for the field of the charged particles to align itself N to S, where N refers to the north pole of the particle's field and S refers to the south pole of the external field. However, it is possible for the particles to be oriented N to N and have the property that a slight perturbation causes the particle to flip back to the lower-energy state, N to S, and thereby return energy to the system. The N-to-N state is a high-energy state; it corresponds to an ionized state (or other excited state) of other particles (Fullerton and Cameron, 1988).

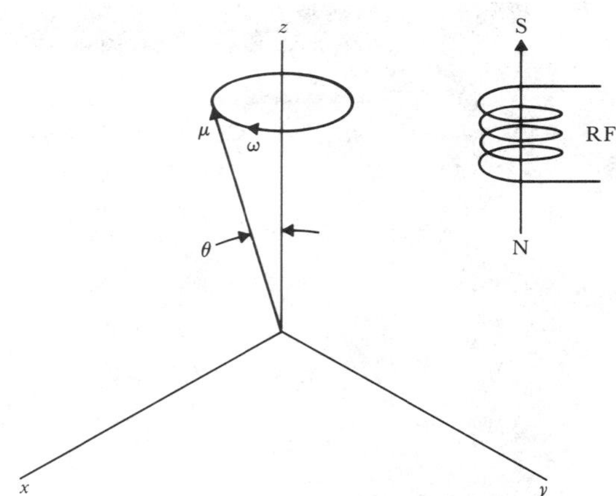

Figure 12.16 Precession of charged particles in a magnetic field

At any instant, there are particles in the normal state, or rest state, and particles in the excited state. These two states are also called the "spin up" or parallel state and the "spin down" or antiparallel state. The ratio of excited particles to particles at rest is a function of the energy difference between the states and of the temperature. The quantum energy difference is $\Delta E = hv$ where h is Planck's constant and v is the frequency. The ratio of the populations of normal to excited particles is $N_n/N_e = e^{hv/kT}$ where k is Boltzmann's constant and T is the temperature. Actually, the axes of the spinning particles do not remain fixed in the magnetic field but rather precess, or wobble just as a spinning top does in a gravitational field (see Figure 12.16). The precessional frequency can be found from the Larmor relationship, $\omega = \gamma B$, where ω is 2π times the precessional frequency, B is the magnetic field, and γ is a property of the particle called the gyromagnetic ratio. Resonance—the absorption of energy—occurs when radio-frequency energy is applied at the Larmor frequency and causes particles to change state and become excited.

The particles exist in three-dimensional space, and we can assume that the external magnetic field is along the z axis. Precession also occurs around the z axis. If a pulse of radio-frequency (RF) energy at the precessional or Larmor frequency is applied to the system, the particles absorb energy and the precessional axes rotate. They could rotate just 90° or to the point where they precisely reverse direction of alignment (a full 180°). Because the RF energy and the pulse width determine this angular shift, the pulses are called simply 90° or 180° RF pulses. After the pulse, the particles return to the equilibrium ratio at rates determined by thermal coupling to the lattice and by the exchange of magnetic energy between excited and nonexcited particles. The two types of energy decay are called "spin-lattice decay" (with time constant T_1) and "spin-spin decay" (with time constant T_2). These time constants are quite long, ranging from several milliseconds to seconds, and they depend on the type of particles and the surrounding material. By varying the time between the RF pulses, their type (90° or 180°), and the placement of receiver coils, it is possible to determine the values of T_1 and T_2.

The patient axis is the y axis. The patient is placed in the constant magnetic field (z axis) and then the field is perturbed to produce a small magnetic gradient along the y axis. There will be only one small section or slice at a particular magnetic field value. The surrounding RF coils are then pulsed at the frequency corresponding to the Larmor frequency of the particle being studied, usually the hydrogen nucleus. Because the resonant frequency is sharply defined, only those particles within this slice will be excited. The magnetic field is then quickly perturbed across the patient, along the x axis, and the relaxation decay frequency will vary along the x axis as a function of the magnetic field. The selectivity of the radio receiver will separate signals as if they were scan lines orthogonal to both gradient fields. A more elegant receiver, a spectrum analyzer, produces multichannel signals as functions of frequency to feed to the computer. The magnetic field can be either rotated slightly by introducing perturbations in the z axis or can simply produce a single gradient in the z axis to produce additional sets of scan lines. The similarity to CT systems is obvious, and the multiple line signals are processed in the same way.

Because of the reactance of the large magnetic coils, ingenious methods are used in the form of magnetic shunts to perturb the field. These produce mechanical noises as they drop in place. The cleverness of the system is impressive: produce an axial magnetic gradient so that only one slice is excited, produce an x-axis magnetic gradient and tune the receiver to scan lines across the slice, rotate the gradient field to scan additional sets of lines, receive the signals in a spectrum analyzer, and repeat several hundred times to reduce the noise and process them as done in CT systems to produce the images.

Because of the time relationships of T_1 and T_2 recovery signals to the initial RF exciting pulse, these may be enhanced by various pulse gate recovery schemes or pulse code systems. By observing the time differences of these signals, the receiver can be gated to accumulate signals of particular time relationships to enhance particular features of the images.

The spinning charged particles could be spinning electrons, either single or unpaired, or charged nuclei—in particular the simplest nucleus, the proton of ionized (in solution) hydrogen. The ratio of excited particles to particles at rest and other properties of particular nuclei determine the sensitivity to nuclear magnetic resonance sensing methods: the NMR sensitivity. This is a measure of the ease of obtaining useful signals. Table 12.1 characterizes some of the common biological elements.

The nuclear magnetic resonance effects of each of these elements can be measured when a sample is placed in the apparatus with a uniform magnetic field and the excitation frequency is varied. To image a cross section of tissue—in particular, living tissue of a patient—a gradient field is used. For example, if the field were varied about 1.0 T, the NMR frequency of hydrogen would vary about the 42.57-MHz value. In a large tissue section, hydrogen

Table 12.1 NMR Frequencies of Common Biological Elements

Element	% of Body Weight	Isotope	Relative Sensitivity	NMR Frequency, MHz/T
Hydrogen	10	^{1}H	1.0	42.57
Carbon	18	^{13}C	1.6×10^{-2}	10.70
Nitrogen	3.4	^{14}N	1.0×10^{-3}	3.08
Sodium	0.18	^{23}Na	9.3×10^{-2}	11.26
Phosphorous	1.2	^{31}P	6.6×10^{-2}	17.24

would be present in various densities throughout the sample, and a band of returned frequencies would be detected. Fourier analysis is used to determine the amplitude distribution of the returned frequencies, and one "pass" of the back projection (similar to that of a computerized tomographic image) is determined. Unlike CT, the entire scanner does not have to be rotated; the direction of the magnetic gradient is rotated slightly, and the process is repeated to get the next back projection. The back-projection signals are analyzed by the computer to generate an image of the density distribution of hydrogen in that plane or "cut." The cut thickness is determined by carefully restricting the field of the RF antenna of the transmitted signal and the return signal. When the NMR signals are used to produce an image in this way, the technique is called magnetic resonance imaging (MRI).

Unlike CT, MRI uses no ionizing radiation, and no measureable biological after-effects have been seen. MRI appears to be safe, so repeated images of delicate tissue can be made without harm or concern for exposure. By varying the sequence code of the 90° or 180° pulse train, the displayed contrast can be intensified for materials of slightly different T_1 and T_2 values.

The magnetic field can be quite strong—on the order of 2.0 T or above. Because most ferromagnetic materials saturate close to that level, superconducting coils are used when very strong fields are required. An alternative is to use either a conventional permanent magnet or a resistive-coil electromagnet when the field is 1.0 T or less. A lower magnetic field means that the Larmor frequency is also reduced and that the lower frequencies degrade the intensity of the received signal and the resolution of the final image. However, resistive-magnet MRI machines are suitable for many imaging applications.

The positioning of the patient in the gantry of the MRI machine is similar to the patient's positioning in a CT machine. The MRI gantry is deeper to accommodate the magnet over dimensions that will ensure uniformity of the field and to provide RF shielding for the receiving coils. MRI scans require up to several minutes, but, as Figure 12.17 shows, they are similar in general appearance to the tomographic slices or sections of CT images. Both machines can make a number of slices to determine the level of the anatomical object or anomaly. The diagnostician can examine films of several slices from both CT and MRI machines to obtain the necessary information.

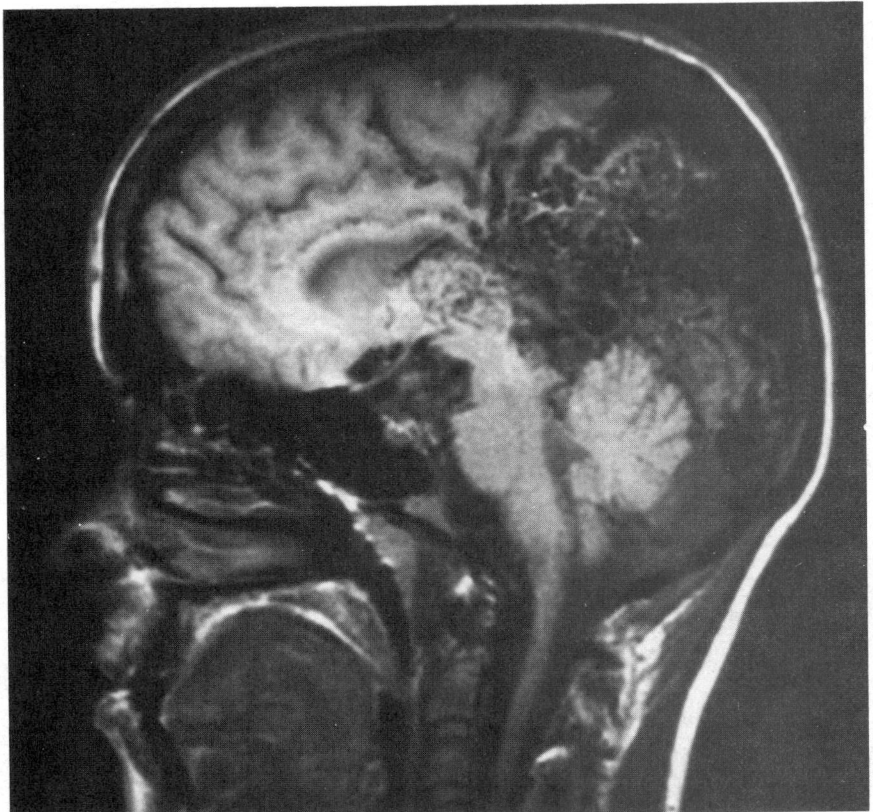

Figure 12.17 MRI image of the head. (Photo courtesy of Philips Medical Systems.)

12.10 NUCLEAR MEDICINE

Nuclear medicine enlists radioactive material for the diagnosis of disease and for assessment of the patient. Thus it differs from radiography in that the source of gamma rays is not external but rather *within* the patient. It also differs in a second very important way: The radioactivity can be attached to materials that are biochemically active in the patient. Therefore, nuclear medicine is said to image *organ function* as opposed to simple organ morphology. The basic imaging situation in nuclear medicine, then, is the measurement of a distribution of radioactivity inside the body of the patient. These distributions can be either static or changing in time (Benedetto, 1988).

Common to nearly all instruments employed in nuclear-medicine imaging is the *sodium iodide detector,* shown in Figure 12.18. The detector consists of three main components: (1) the crystal itself, which scintillates with blue light

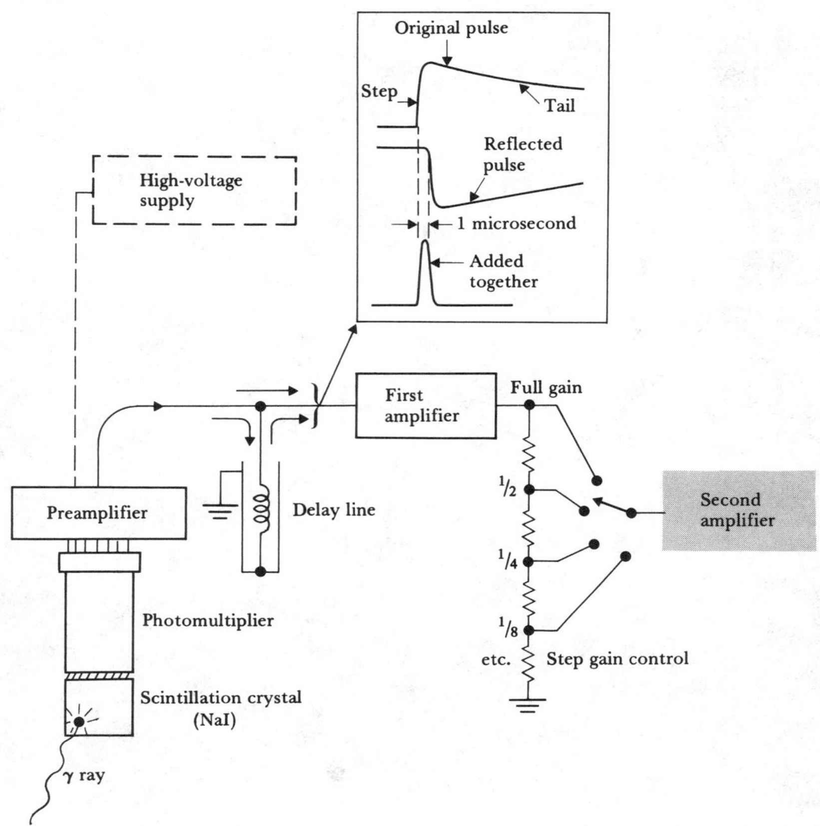

Figure 12.18 Basic implementation of a NaI scintillation detector, showing the scintillator, light-sensitive photomultiplier tube, and support electronics. (From H. N. Wagner, Jr., ed., *Principles of Nuclear Medicine.* Philadelphia: Saunders, 1968. Used with permission of W. B. Saunders Co.)

in linear proportion to the energy a gamma ray loses in it; (2) a photomultiplier tube, which converts this light into a proportional electric signal; and (3) the support electronics, which amplify and shape this electric signal into a usable form. The simplest nuclear-medicine procedures do not involve images at all but consist simply of placing such a detector near the surface of the patient's skin and counting the gamma-ray flux.

The first nuclear-medicine imaging device involved the operator taking such a simple detector and moving it in rectilinear paths relative to the patient, in much the same way as we put together an aerial map of the earth. Any process that involves this rectilinear motion of the detector is called *scanning*.

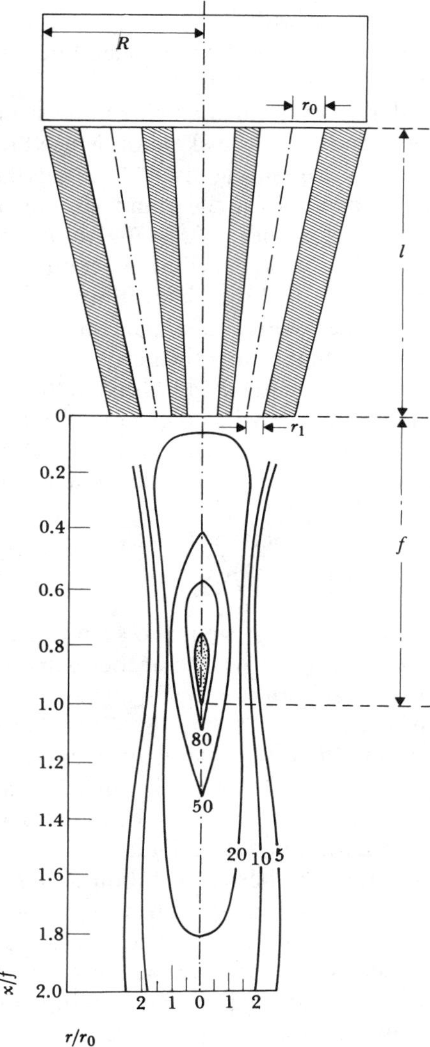

Figure 12.19 Cross section of a focusing collimator used in nuclear-medicine rectilinear scanning. The contour lines below correspond to contours of similar sensitivity to a point source of radiation, expressed as a percentage of the radiation at the focal point. (From G. J. Hine, ed., *Instrumentation in Nuclear Medicine.* New York: Academic, 1967.)

Figure 12.19 shows how the sodium iodide detector must be *collimated* in a scanning procedure in order to restrict its field of view, both along its longitudinal axis and transverse to it. The extension of the transverse field of view is the primary determinant of image resolution in a nuclear-medicine scan. If we know the approximate strength of the source, we can combine this information about strength with the basic resolution-element size determined by the collimator to calculate the scan speed necessary to provide the statistical accuracy desired in the image.

For example, consider the nuclear-medicine bone scan, in which an agent that has high metabolic turnover in the skeleton is used to locate sites of the bones involved with cancers originating elsewhere. A typical dose of such an agent is 15 mCi (Ci is the symbol for the standard unit of radioactivity, the curie; 1 Ci $= 3.7 \times 10^{10}$ nuclear decays/s). The total activity is distributed broadly over the skeleton. This wide distribution, combined with the isotropic distribution of the gamma rays and the absorption of many of those gamma rays in the patient's own tissue, yields a maximal count rate at the body surface of about 1000 cts/s. Typical dimensions of the pixel are 0.5 cm $\times$ 0.5 cm. If we demand a statistical fluctuation of 10% (100 cts) in this pixel for the position of greatest count rate, the speed is easily determined; counts per pixel = count rate/(horizontal speed $\times$ vertical dimension of pixel). Horizontal speed (cm/min) = count rate/(counts per pixel $\times$ vertical dimension of pixel):

$$\frac{1000\ \text{ct}}{\text{s}}\,\frac{60\ \text{s}}{\text{min}}\,\frac{0.25\ \text{cm}^2}{100\ \text{cts}}\,\frac{1}{0.5\ \text{cm}} = \frac{300\ \text{cm}}{\text{min}}$$

In the basic stand-alone nuclear-medicine scanning instrument, a *pulse-height analyzer,* selects events that have the proper gamma-ray energy. These events are used to gate a light source that scans across a film in the same rectilinear fashion in which the detector scans across the patient. The images are smoothed by integrating the rate of detector count in a simple rate-meter circuit. Alternatively, events selected by the pulse-height analyzer may be scaled digitally as a function of position of the detector. The spatial frequency of storage of scaled information is determined by the desire to have two to three picture elements within the basic resolution dimension defined by the detector collimator. The resulting image can be seen on a computer image display shown on a storage oscilloscope or collected on film. Figure 12.20 gives examples of rectilinear scans that were acquired by these two methods.

A second type of nuclear-medicine imaging instrument, introduced about ten years after the rectilinear scanner, has since become the workhorse of the typical nuclear-medicine laboratory. This is the so-called *gamma camera,* sometimes called the Anger camera after its original developer (Anger, 1958). The gamma camera is a stationary imaging system that is simultaneously sensitive to all the radioactivity in a large field of view. It does not depend on motion of the detector to piece together an image.

Figure 12.21 shows a simplified cross section of such an imaging system. The radiation detector is a single sodium iodide crystal 30 to 40 cm in diameter and 1.2 cm thick. This detector is viewed simultaneously by an array of photomultiplier tubes arranged in a hexagonal pattern at the rear of the detector. When a gamma ray enters the sodium iodide crystal, the resulting scintillation light spreads through the crystal, and each photomultiplier tube receives some portion of the total light. The fraction of the total light seen by each tube

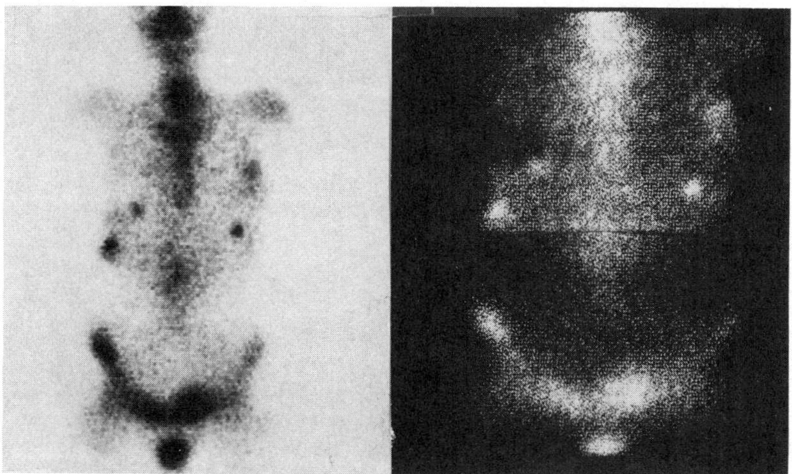

Figure 12.20 Images of a patient's skeleton obtained by a rectilinear scanner, in which a technetium-labeled phosphate compound reveals regions of abnormally high metabolism. The conventional analog image is on the left, the digitized version on the right.

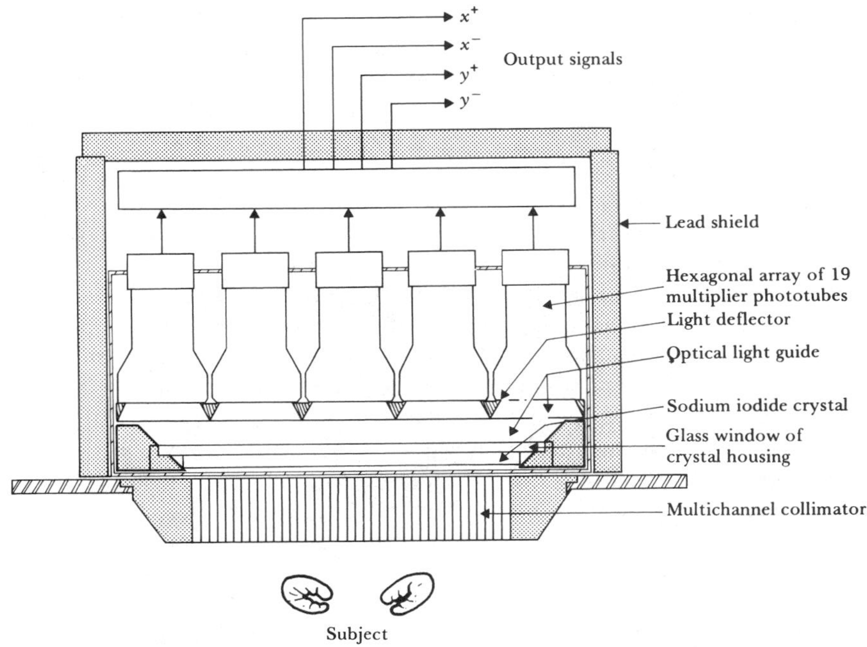

Figure 12.21 **Cross-sectional view of a gamma camera** (From G. J. Hine, ed., *Instrumentation in Nuclear Medicine.* New York: Academic, 1967.)

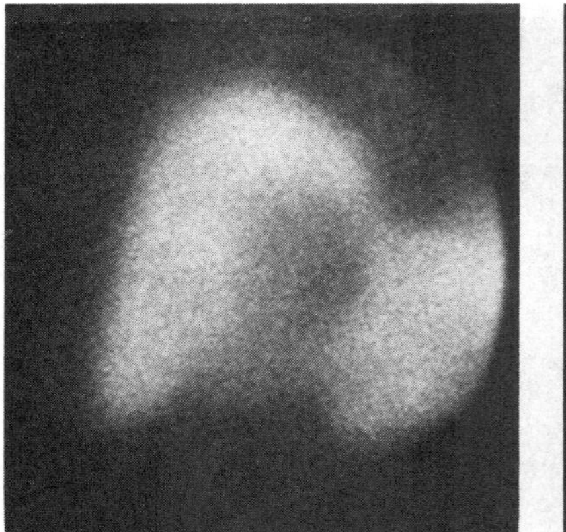
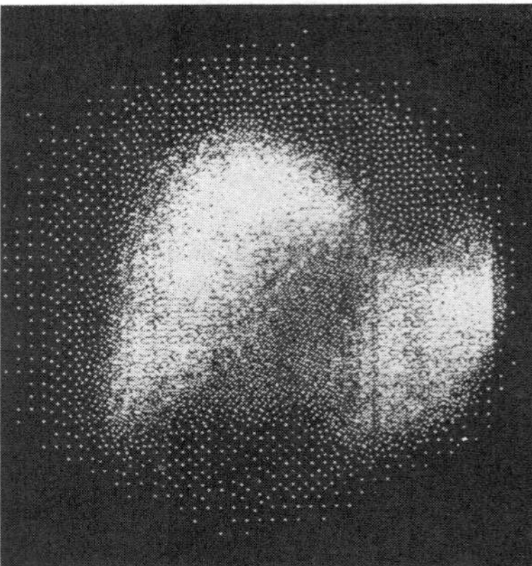

Figure 12.22 Gamma-camera images of an anterior view of the right lobe of a patient's liver. A colloid labeled with radioactive technetium was swept from the blood stream by normal liver tissue. Left: conventional analog image. Right: digitized version of the same data.

depends on the proximity of that tube to the original point of entry of the gamma ray.

The fundamental principle of operation of the gamma camera is that the relative fraction of the total light seen by each tube uniquely determines the position of the original point of entry of the gamma ray. Voltages corresponding to the x and y coordinates of the gamma-ray event are reconstructed from the signals of several photomultiplier tubes in an analog electronic circuit. In a modern instrument, this circuit employs operational amplifiers the gain of which reflects the position of the given photomultiplier tube in the array. The center of the sodium iodide detector is conventionally assigned the position $x = 0$ and $y = 0$.

For example, a photomultiplier tube positioned far from the center in the conventional $+x$ direction but centered in the y direction would have amplifier gains such that the signal from that tube would provide a $+x$ signal disproportionately large relative to the $-x$ signal but would provide equal contribution to the total $+y$ and $-y$ signals. The image can be recorded in both analog and digital form. In the analog form, the $+x$, $-x$, $+y$, and $-y$ signals are used as deflection voltages on the plates of an oscilloscope.

Events that satisfy an energy-discrimination condition briefly unblank the oscilloscope and expose a film on which the composite image is formed. Alternatively, the signals can be digitized, and the resulting digitized x and y coordinates used to determine a computer address corresponding to the position of the event. A digital image is built up by incrementing the appropriate computer address at each event.

Figure 12.22 shows gamma-camera images acquired via these two methods. The most important advantages of the gamma camera, compared to the rectilinear scanner, are its capability of measuring changes in the distribution

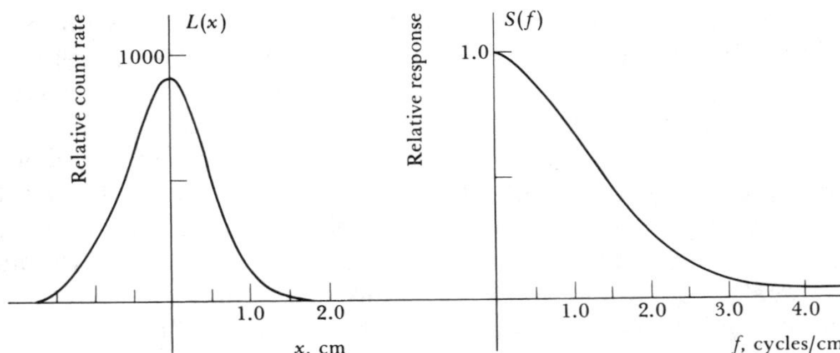

Figure 12.23 Line-spread response function obtained in a gamma camera under computer control, together with the corresponding modulation transfer function.

of radioactivity as a function of time and the increased use of the emitted photons. Gamma cameras interfaced to computer systems can acquire data at frame rates in excess of 30 frames/s. In fact, even higher frame rates would be conceivable if it were not for safety rules limiting exposure of the patient to radiation.

Because nuclear-medicine images typically have a fundamental resolution distance that is roughly 1% of the image dimension and are composed of only 10^5 to 10^6 photons, they are simple to investigate quantitatively and provide a convenient way of studying the basic imaging concepts listed earlier in this chapter. Figure 12.23 shows the measurement of the response of a gamma camera to a line source of radioactivity. The line-spread function of the camera was easily generated by summing over one dimension of the digital image acquired in a nuclear-medicine computer. This line-response function was then Fourier-transformed to generate the modulation transfer function.

Figure 12.2 dramatizes the relationship between quality of image and number of photons. It shows images of a so-called *thyroid phantom,* a small lucite tank that—when filled with radioactive material—provides a distribution of radionuclides approximating that of a similarly labeled human thyroid. Resolution is demonstrated by the defects of holes intentionally placed in the radioactivity distribution pattern. Each phantom image shown represents an increase in number of photons by a factor of 2 from the one before it.

12.11 SINGLE-PHOTON EMISSION COMPUTED TOMOGRAPHY

Single-photon emission computed tomography (SPECT) uses a large-area scintillation assembly similar to that of an Anger camera and rotates it around the patient. Many of the isotopes used in nuclear medicine produce a single photon or gamma ray in the useful range of energies. The collimator is often designed to collect radiation from parallel rays (the collimator is usually focused at infinity), and the collection of signals is used in an image-reconstruction process analogous to methods used in computed tomography (see Figure 12.15). Several planes or slices of activity are reconstructed at the same time. These multiple slices of isotope activity allow the system to show depth information in the volume of interest by constructing several cross-sectional views in addition to the head-on, or planar, view offered by a conventional nuclear camera. The method can also resolve the activity to a smaller volume than conventional Anger camera images. In order to achieve the greater image resolution, the acceptance angle of each hole of the collimator must be restricted, and the collection time must be considerably longer than that of the Anger camera. SPECT systems must compensate for variations in attenuation of the patient. As a result, these systems are not used for conducting dynamic studies but only for imaging near-static structures, such as tumors and subtle bone disease. By obtaining three-dimensional information and increased resolution, it is possible to see certain anomalies that are not so clear in conventional x-ray and other nuclear images.

12.12 POSITRON EMISSION TOMOGRAPHY

Certain isotopes produce positrons that react with electrons to emit two photons at 511 keV in opposite directions. Positron emission tomography (PET) takes advantage of this property to determine the source of the radiation. If one path is shorter, then the opposite path is longer, and the average signal level is the same without regard to patient attenuation or point of origin. These isotopes have two means of decay, which result in the annihilation of an electron. In one case, the nucleus can capture an orbital electron that combines with one positive charge; alternatively, the nucleus can emit the positive charge as a positron that travels a short distance to combine with an external electron. The combination of the negative and positive particles annihilates the charges and masses of each, energy and momentum are conserved, and two 511-keV gamma rays are emitted in opposite directions.

Positrons emitted by the nucleus have kinetic energy, so they travel a few millimeters before the annihilation emission event. The travel distance and interaction effects blur the dimensions of the region of origin when it is detected. The broadening effect shown in Table 12.2 is the width of the pulse measured at the 10% level. The dimensions of a useful picture element would be about twice these values because of other spreading effects, such as system bandwidth optical effects.

The property of simultaneous emission of two gamma rays in opposite directions gives PET the ability to locate the region of origin. Instead of the multihole collimator found in most gamma cameras, two imaging detectors capable of determining x–y position are used. Each x–y pair is accepted if

Table 12.2 Characteristics of Five Isotopes for PET

Isotope	Maximal Kinetic Energy	Half-life	Broadening
^{10}F	640 keV	110 min	1.1 mm
^{11}C	960 keV	20.4 min	1.9 mm
^{13}N	1.2 MeV	10.0 min	3.0 mm
^{60}Ga	1.9 MeV	62.3 min	5.9 mm
^{82}Rb	3.4 MeV	1.3 min	13.2 mm

the two scintillation effects are coincident and have energy levels (pulse heights) close to the expected value of 511 keV.

In the simplest PET camera, two modified Anger cameras are placed on opposite sides of the patient [Figure 12.24(a)]. The modification removes the multihole collimator and adds the coincidence and computing circuits. Removing the collimator increases the collection angle and reduces the collection time, which are limitations of SPECT. The camera is rotated slowly around the patient to obtain the additional views needed for reconstruction and to obtain better images. Images can be built up faster when additional pairs of detectors are used. Figure 12.24(b) shows the three pairs of cameras used in the hexagonal-ring camera. Incremental lateral translation and rotation improve the images by compensating for inhomogeneities and gaps of detection. Very elegant cameras can be constructed by using a circular ring of many detectors that surround the patient [Figure 12.24(c)]. The ring detector does not have to be rotated; image positions are resolved by computer analysis of the signals. As in all radionuclide imagers, the level of the radioactivity places a noise bound on the images. Obviously, compromises must be made in the amount of radionuclide administered: It must be large enough to obtain a good

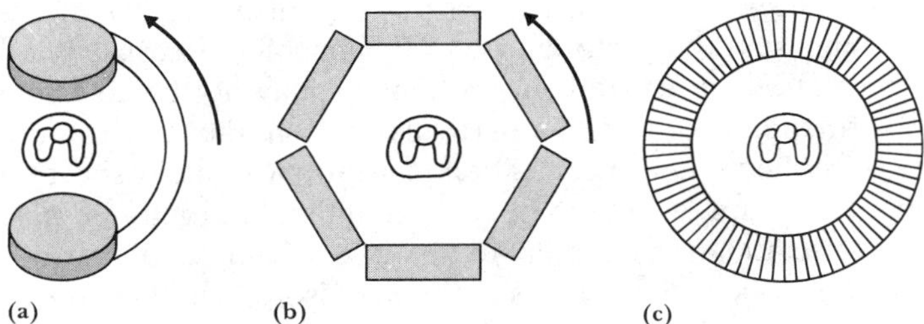

Figure 12.24 Evolution of the circular-ring PET camera (a) The paired and (b) the hexagonal ring cameras rotate around the patient. (c) The circular ring assembly does not rotate but may move slightly—just enough to fill in the gaps between the detectors. The solid-state detectors of the ring camera are integrated with the collimator and are similar in construction to detectors used in CT machines.

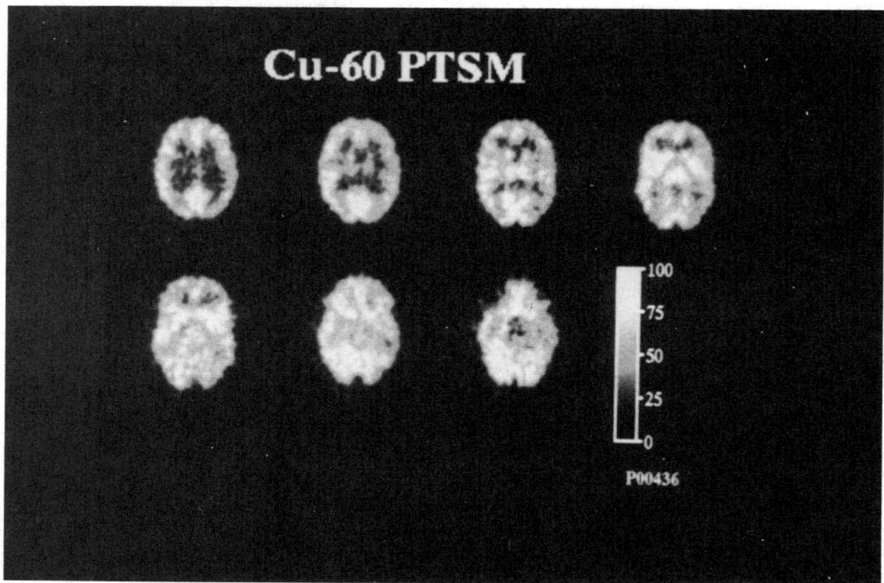

Figure 12.25 PET image The trapping of ^{60}Cu-PTSM (a thiosemicarbazone) reflects regional blood flow, modulated by a nonunity extraction into the tissue. (Photo courtesy of Dr. R. J. Nickles, University of Wisconsin.)

image in the time required and small enough to minimize patient exposure to radiation.

The advantages of PET over conventional nuclear imaging include the clarity of the cross-sectional views and the availability of positron emitters that can be compounded as metabolites. It is possible to map metabolic activity in the brain by using tagged compounds to observe uptake and clearance.

The measured quantity in PET imaging is the concentration, in tissue, of the positron emitter. To obtain the actual concentration, it is necessary to calibrate and measure the performance of the machine. Because knowing the actual concentration (in μCi/ml) in the patient may not be so important as knowing the fraction taken up in a particular region, the PET camera can be used to measure the tissue concentration in arbitrary units. A short time after the imaging procedure, a sample of the patient's blood may be placed in a well counter (a scintillation counter) to obtain a reference value. Comparison of the tissue and blood activity yields the ratio of isotope uptake. For example, the local cerebral blood volume and the distribution of activity are measured this way. Because the brain adjusts uptake as a function of the use of various metabolites, brain activity can be measured. A rapid sequence of brain images shows the response of the brain to various stimuli and pinpoints areas of abnormal activity.

As different parts of the brain respond to different stimuli, the PET image shows this activity (Figure 12.25). Normal brains generate one image of brain activity, but abnormal functioning, tumors, seizure, and other anomalies may also be clearly visible in the map of activity. The PET image of the brain shows the patient's responses to noise, illumination, changes in mental concentration, and other activities. One method of introducing a suitable isotope for brain imaging is for the patient to breathe air containing CO made with ^{11}C. Such short-lived positron-emitting isotopes do not occur in nature but can be created in a small cyclotron. This is done by introducing into the nucleus a proton that, in turn, emits an alpha particle or neutron. For certain elements, the nucleus becomes unstable and emits a positron in a short time. For example, ^{11}C is prepared in a small cyclotron and has a half-life of 20.4 min. The short half-life means that the isotope must be prepared near the point of use. The radionuclide decays rapidly, exhibits high activity during the time necessary to obtain images, and clears the patient in a short time. Clearance is a function of both the radioactive decay and the biological excretion of the material.

12.13 ULTRASONOGRAPHY

We know from old war movies that pulses of sound waves are used to detect submarines. Because wavelength λ, frequency f, and velocity u are related ($u = f\lambda$), it is easy to show that wavelengths in the audible spectrum are only a small fraction of the length of a submarine. A phase change of less than one cycle (360°) would result in a maximum error of position equal to the wavelength: $\lambda = u/f$. To find the error of position of a detected submarine, substitute the velocity of sound in water (1480 m/s) and let $f = 1$ kHz and $\lambda = 1.48$ m. This precision would be adequate for detecting submarines but not for, say, visualizing a human fetus. To obtain precision of 1.48 mm, the frequency of the pulse would have to be increased to 1.0 MHz in the ultrasonic range.

Sound and ultrasound follow rules of propagation and reflection similar to those that govern electric signals. A transmission line must be terminated in its characteristic impedance to avoid reflections. The acoustic impedance Z is a fundamental property of matter and is related to the density ρ and the velocity of sound u: $Z = \rho u$. The fraction of energy R reflected at the normal interface of two different tissue types is

$$R = \left[\frac{(Z_2 - Z_1)}{(Z_2 + Z_1)}\right]^2$$

and the impedances are those of the tissues on either side of the interface (Goldstein, 1988).

Acoustic signals diminish as a function of distance, geometry, and attenuation. In free space, the signals decrease as a result of the inverse square law

Table 12.3 Acoustic Properties of Some Tissues at 1.0 MHz

Tissue	*u*, m/s	*Z*, g/(cm²-s)	HVL, cm		*R* at Interface
Water	1496	1.49×10^5	4100	Air/water	0.999
Fat	1476	1.37×10^5	3.8	Water/fat	0.042
Muscle	1568	1.66×10^5	2.5	Water/muscle	0.054
Brain	1521	1.58×10^5	2.5	Water/brain	0.029
Bone	3360	6.20×10^5	0.23	Water/bone	0.614
Air	331	4.13	1.1	Tissue/air	0.999

because the energy per unit area is a function of the total area of the imaginary sphere at distance r. The signals also decrease as a result of attenuation by the medium. Where α is the coefficient of attenuation and I_0 is the incident signal intensity, the signal intensity is

$$I = \frac{I_0 e^{-\alpha r}}{r^2}$$

When α is large compared to r, the exponential term dominates, and it is convenient to define the thickness of material where the attenuation of the medium decreases the signal by half (the half-value layer or HVL) independently of the geometrical effects. Table 12.3 lists the HVL for water and some tissues. Note that water is not very "lossy" and that the signal decreases 50% for 41 m of water. However, a 50% decrease occurs through only 2.5 cm of muscle. Most biological tissues have high coefficients of attenuation and low HVLs. Attenuation also increases with frequency.

Ultrasound transducers use the piezoelectric properties of ceramics such as barium titanate or similar materials. When stressed, these materials produce a voltage across their electrodes. Similarly, when a voltage pulse is applied, the ceramic deforms. If the applied pulse is short, the ceramic element "rings" at its mechanical resonant frequency. With appropriate electronic circuits, the

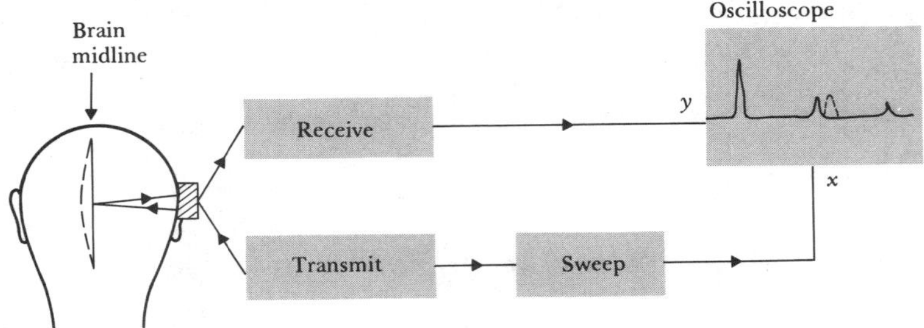

Figure 12.26 A-mode scan of the brain midline

ceramic can be pulsed to transmit a short burst of ultrasonic energy as a miniature loudspeaker and then switched to act as a microphone to receive signals reflected from the interfaces of various tissue types. The gain of the receiver can be varied as a function of time between pulses to compensate for the high attenuation of the tissues. Ultrasonic energy at the levels used for medical imaging appears to cause no harm to tissue, unlike the ionizing radiation of x rays.

The time delay between the transmitted pulse and its echo is a measure of the depth of the tissue interface. Fine structures of tissues (blood vessels, muscle sheaths, and connective tissue) produce extra echoes within "uniform" tissue structures. At each change of tissue type, a reflection results. Figure 12.26 on p. 566 shows how the interfaces of bodily structures produce the echoes that reveal their locations. This type of simple ultrasonic scanner, the A-mode device, was an early device used to measure the displacement of the brain midline. An A-mode device shows echo intensity as an x–y plot. The transducer is placed against the skull and the display gives the echo time of the brain midline (proportional to depth). The transducer is then moved to the other side of the skull and the procedure repeated. The images of normal patients are symmetric so that the brain midline should appear in the same

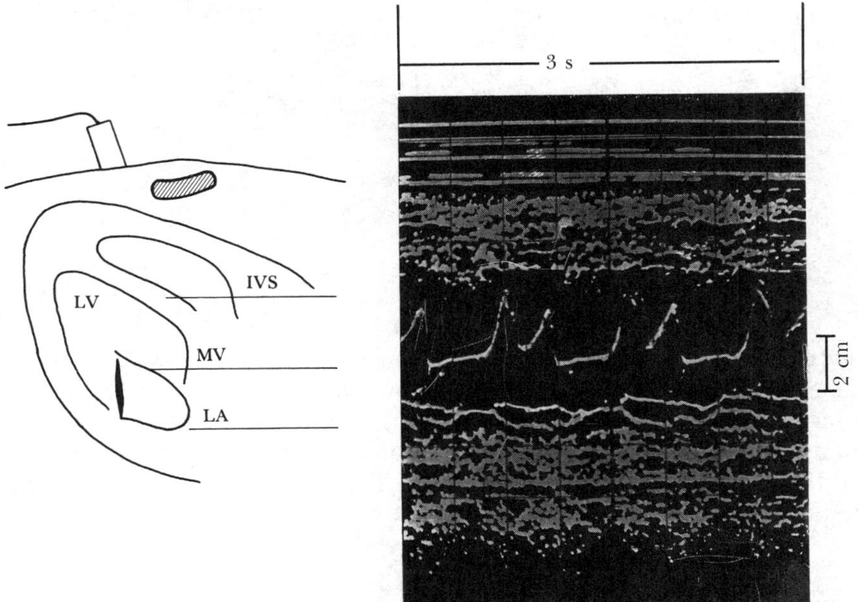

Figure 12.27 Time-motion ultrasound scan of the mitral valve of the heart The central trace follows the motions of the mitral valve (MV) over a 3-s period, encompassing three cardiac cycles. The other traces correspond to other relatively static structures, such as the interventricular septum (IVS) and the walls of the left atrium (LA).

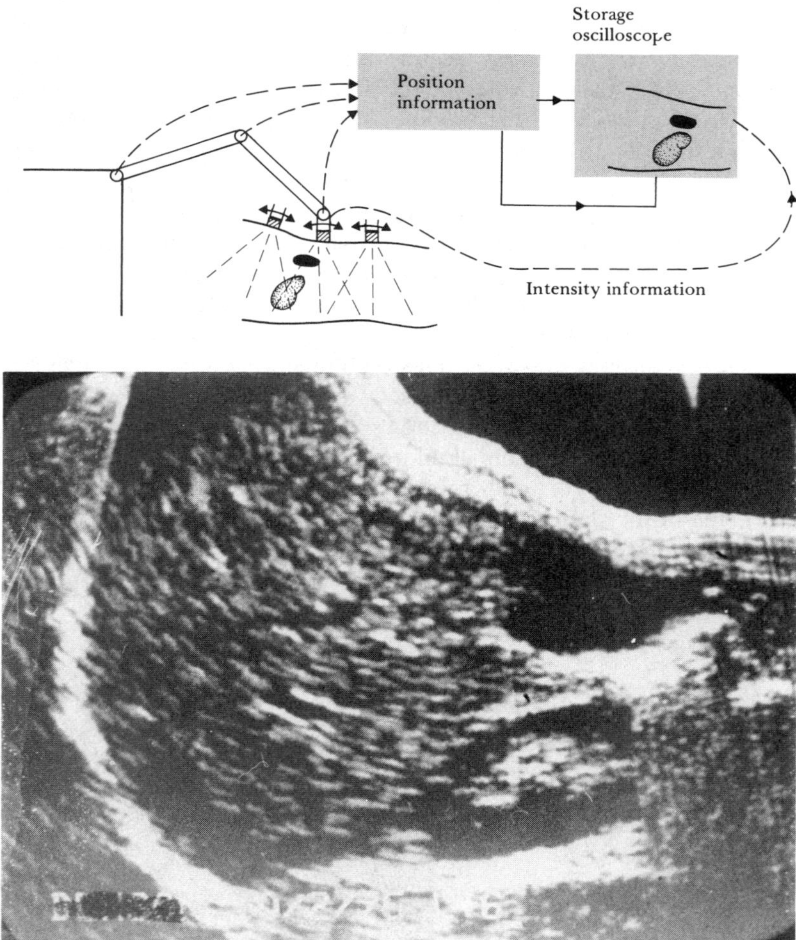

Figure 12.28 (a) B-mode ultrasonic imaging shows the two-dimensional shape and reflectivity of objects by using multiple-scan paths. (b) This B-mode ultrasonic image, which corresponds to (a), shows the skin of the belly at the top right, the liver at the left center, the gall bladder at the right above center, and the kidney at the right below center. The bright areas within the kidney are the collecting ducts.

position in the two images. A tumor or large blood clot could move the cerebral hemispheres to shift the midline. This type of simple device is now seldom used and has been replaced by more elegant systems which show far more detailed structures as well as the brain midline.

If the strength of the echo signal is used to modulate the intensity of the display against the echo-return time, with zero time at the top of the display and with the appropriate image storage circuits, the image can be shown as moving to the right. This technique presents the position of tissues as a function of time as a time-motion or TM scan. Figure 12.27 on p. 567 shows the motion of the mitral valve of the heart over three cardiac cycles.

Computer image storage displays or long persistence phosphor display screens can be intensity modulated as the position of the transducer is varied. The display will show the two-dimensional shape of objects. For older systems, the position and direction of the transducer are coupled to the display circuits by a system of pulleys and potentiometers. Newer systems use mechanical scanning or phased arrays within the transducer assembly and display the two-dimensional image relative to the fixed position of the transducer assembly. A computer stores the echo signals for display as a sector. Some sophisticated systems correlate sectors taken from a number of directions and display them as a single image, much improved in quality over an image taken from only one direction. Sector images and most two-dimensional images are called B-mode images (see Figure 12.28 on p. 568).

Frequencies from 1.0 MHz to 15 MHz are used for most medical ultrasonography. The operating frequency is chosen to meet the imaging task. Higher frequencies will improve resolution, but increased HVL limits the depth of

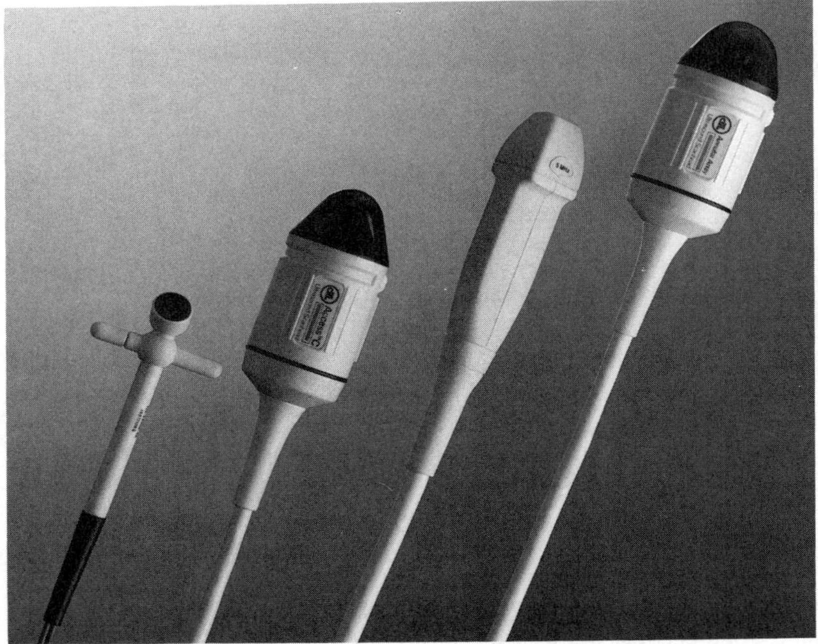

Figure 12.29 Different types of ultrasonic transducers range in frequency from 12 MHz for ophthalmic devices to 4 MHz for transducers equipped with a spinning head. (Photo courtesy of ATL.)

penetration. However, for special purposes (e.g., ophthalmic and neonatal imaging) where the objects are small, the operating frequency may be increased to 15 MHz or higher and the resolution will permit the observation of very small structures or anomalies.

Figure 12.29 on page 569 shows four ultrasonic transducers. The two larger devices use three transducers spinning in fluid-filled enclosures [Figure 12.30(a)]. The midsized device and the smaller, ophthalmic transducer, use phased arrays which can be steered by adjusting the timing of the pulses applied to sets of several ceramic elements. If all elements of the transducer are pulsed at the same instant, the ultrasonic energy will be projected in the forward direction [Figure 12.30(b)]. If the left side elements are pulsed in a delayed sequence, the energy will be projected toward the right with the angle proportional to the timing delay [Figure 12.30(d)]. By adjustment of the timing

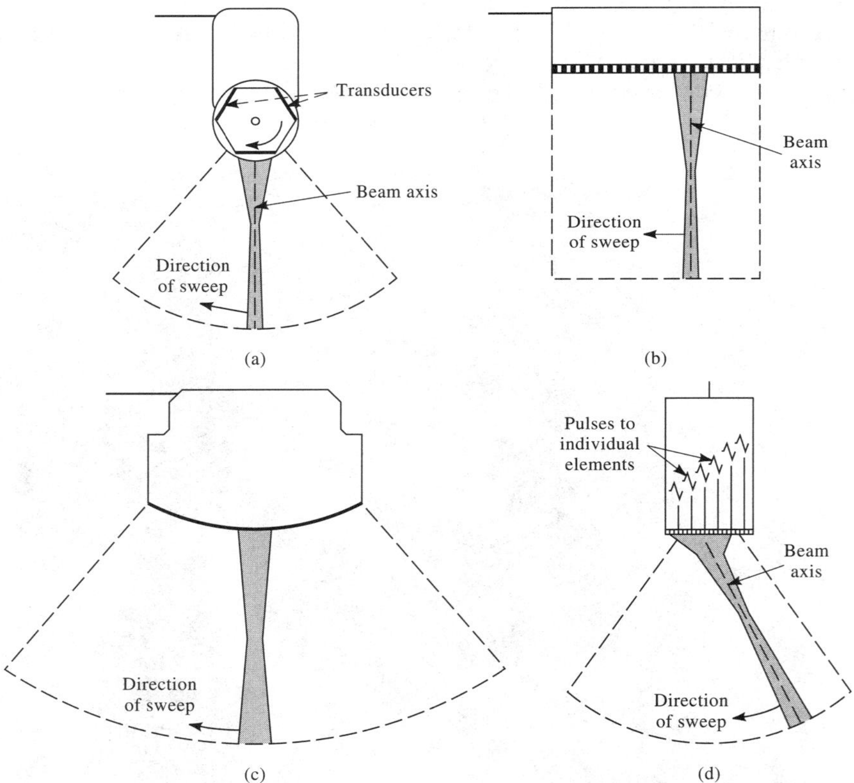

Figure 12.30 Ultrasound scan heads. (a) Rotating mechanical device. (b) Linear phased array which scans an area of the same width as the scan head. (c) Curved linear array can sweep a sector. (d) Phasing the excitation of the crystals can steer the beam so that a small transducer can sweep a large area.

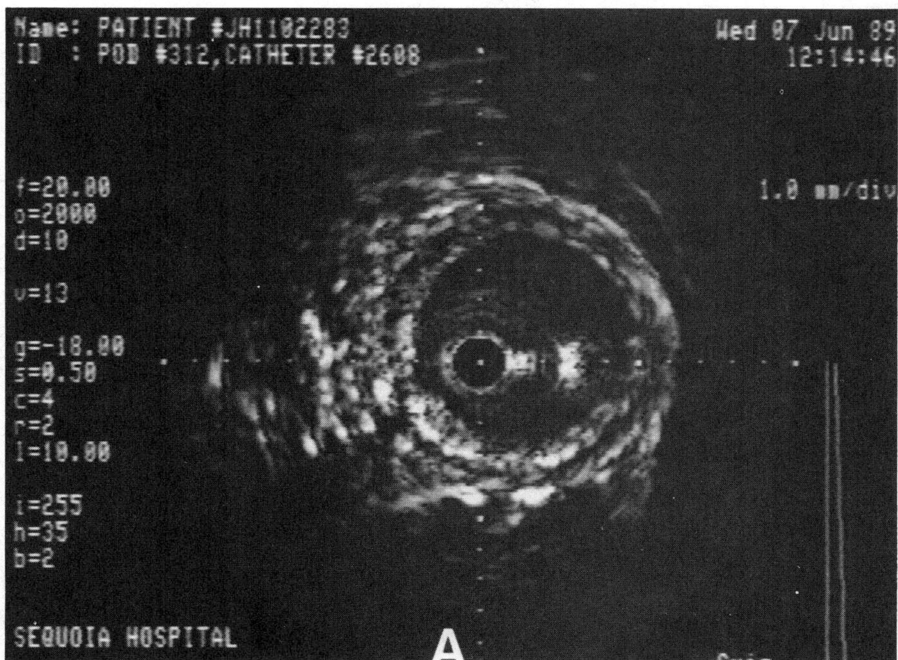

Figure 12.31 Intravascular ultrasonic image showing the characteristic three-layer appearance of a normal artery. Mild plaque and calcification can be observed at 7 o'clock. (Photo courtesy of Cardiovascular Imaging Systems, Inc.)

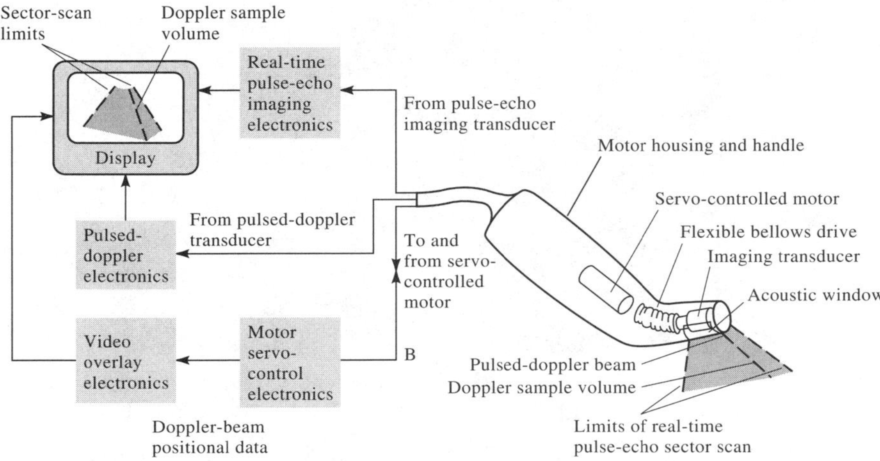

Figure 12.32 The duplex scanner contains a mechanical real-time sector scanner that generates a fan-shaped two-dimensional pulse-echo image. Signals from a selected range along a selected path are processed by pulsed Doppler electronics to yield blood velocity (from Wells, 1984).

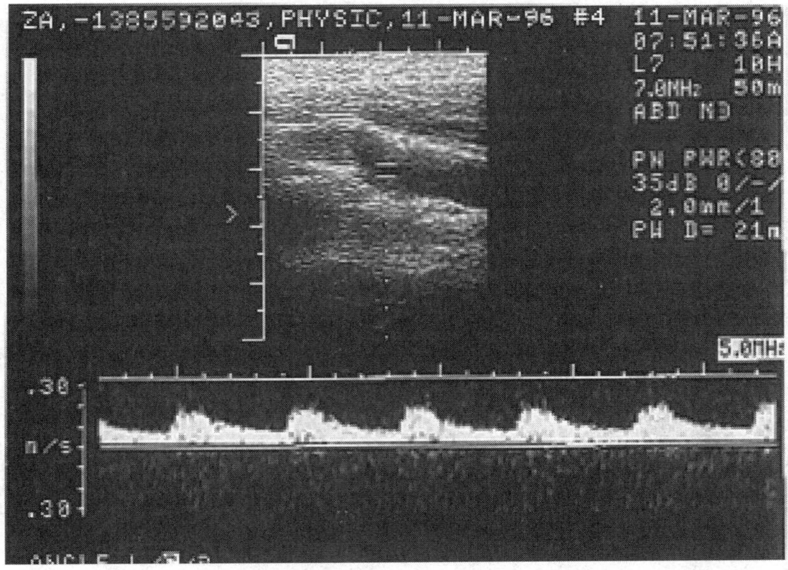

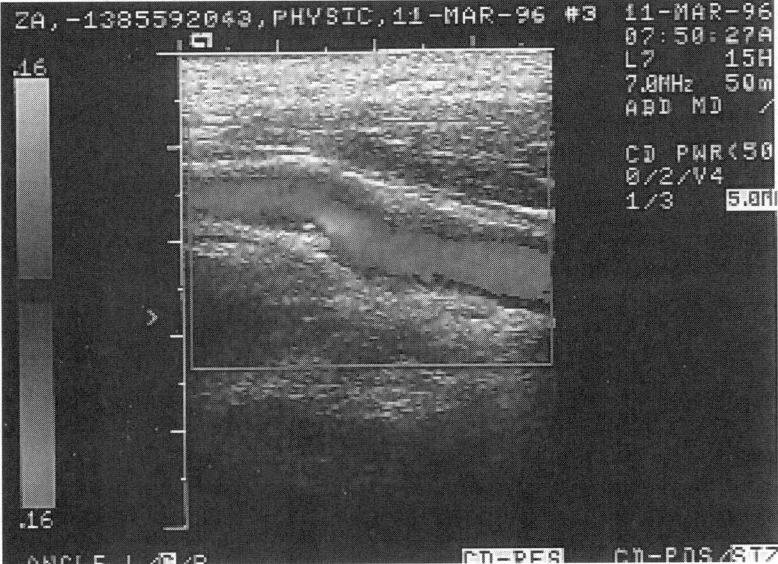

Figure 12.33 (a) Duplex scanner B-mode image and Doppler spectral analysis record for a normal carotid artery, near the bifurcation. The Doppler signals were recorded from the sample volume defined by the Doppler cursor, the two parallel lines located inside the carotid artery. (b) Color flow image of the vessel in (a). Higher velocity components (lighter color, reproduced here in black and white) are seen where the vessel direction courses more directly toward the transducer.

delay, the beam may be scanned from side to side. With either the spinning or phased array transducer placed against the skin, the image of a sector is displayed. Phased array transducers have been made small enough to be mounted at the ends of probes for insertion into body cavities such as the rectum for imaging the prostate or the vagina for showing the fetus or the condition of the reproductive organs.

Even smaller transducers have been made for high-frequency operation. These have been fitted at the tips of catheters and used for examining the characteristics of blood vessels prior to angioplasty. Figure 12.31 on p. 571 shows the appearance of a normal artery taken with a catheter tip transducer. In angioplasty, a balloon is introduced into ischemic or partially closed vessels and then inflated to stretch the walls of the vessel to increase the lumen or diameter and increase blood flow. If the vessel walls are weak, the probe images may show that angioplasty could jeopardize the life of the patient. Following balloon inflation, the probe can be pulled back to determine the dimensions of the stretched walls and verify the integrity of the vessel.

DUPLEX SCANNERS

Halberg and Thiele (1986) describe the design of a phased array ultrasonic duplex scanner that combines real-time two-dimensional imaging with the pulsed Doppler method to measure directional blood velocity noninvasively. Figure 12.30(d) shows how a mechanical real-time sector scanner can generate a fan-shaped beam. Figure 12.32 on p. 571 shows the system block diagram. A colored display from a duplex scanner shows flow into or out of the screen as red or blue against a monochrome background, with the intensity of the color approximating the velocity. This technique is called *color flow imaging* and yields images shown in Figure 12.33 on p. 572.

Because the duplex scanner can distinguish between moving blood and stationary soft plaque, it is useful for diagnosing obstruction in diseased carotid arteries. Pulsed Doppler techniques are useful in locating and determining in the heart the direction and extent of abnormal flow, valvular abnormalities, shunt lesions such as patent ductus arteriosis, and ventricular and septal defects.

PROBLEMS

12.1 How much time does it take to transmit a single television frame over telephone lines that have a bandwidth of 3 kHz?

12.2 A computer monitor uses a noninterlaced scheme with a 4:3 aspect ratio (the height is 3/4 of the width) frame rate of 67 frames/s, 480 visible scan lines, 640 pixels visible/line, 10% retrace time for each vertical and horizontal scan. Find the minimum bandwidth required. Find the bandwidth if the system were redesigned for high resolution and changed to 1024 total scan lines with 920 lines visible and with 1280 pixels visible per line (10%

retrace time). How would the bandwidth be affected if the systems were designed for interlaced scanning and the frame rate was 1/2 the field rate (equal to the original frame rate)?

12.3 A square 8 × 8 checkerboard is to be imaged using a raster scan system. The board is centered in this square image. When it is rotated 45° with respect to the image axes, its corners are just touching the centers of the image boundaries. Describe the horizontal and vertical characteristics of this raster scan and the total cycles/image required to detect all the checkerboard squares.

12.4 For the Poisson probability density distribution, calculate and plot $p(K;m)$ for $K = 0, 1, 2, 3, 4, 5$ and $m = 3$.

12.5 A 100 × 100 pixel array has an average of 25 photons per pixel. How many picture elements will randomly exceed this average by more than 17 photons?

12.6 How many photons are required to produce a 200 × 200 cell picture having 6 gray levels?

12.7 Calculate N_e for $S(f) = 1/(1 + 2f)$, where f is in cycles per millimeter.

12.8 Three elements of an optical system have $N_e = 1, 2,$ and 4 cycles/mm. Calculate the system N_e.

12.9 A new imaging system was tested for spatial frequency response, and it was observed that the amplitude response was constant from zero to 10 lp/mm and then fell linearly to zero response at 20 lp/mm. How would this system compare to one having a noise-equivalent bandwidth of 15 lp/mm?

12.10 A film having a gamma of 2.0 is exposed to light and shows a density of $D = 1.0$. What increase of light is required to expose the film to a density of 1.30?

12.11 If we make our measurements in the plane of the patient and must see the smallest pixel of about 0.25 mm (about 2 line pairs/mm), the contrast of the object is about 10%, the QDE of the image detector is 50%, the RL is about 3%, the image rate is 10/s, and the total exposure time is 10 min, what is the approximate total incident exposure? This level of exposure could be seen in an interventional radiographic procedure.

12.12 For all other variables fixed, including probability of detection, plot the dimension d of an object versus contrast C for an x-ray image.

12.13 When an x-ray machine is set for a normal film of the abdomen, around 100 kV, the energy distribution of the beam is such that 4% of the beam can penetrate and exit the patient. If we want to see objects 0.5 mm in diameter that can modulate the beam 50%, what level of incident exposure in R is required? What if the size of the object was 0.2 mm and the modulation (contrast) was 10%?

12.14 Radiation requirements are based on the statistical independence of the x-ray photons producing an image. If the SNR required is, say, 100, then 10,000 photons are required, on average, per pixel. If the image is to be digitized to 10 bits, 1024 levels, why is it necessary to have so many photons per pixel?

12.15 The Hounsfield units are a measure of amplitude resolution in computed tomography. If we need to resolve to 1.0% of amplitude, what does this mean in terms of radiation requirements when compared to conventional radiography (3–5% amplitude resolution)? What is the effect of taking thinner slices in CT in terms of surface (of the patient) incident exposure?

12.16 For the ray θk shown in Figure 12.11, estimate and list the value for each nonzero $W_{ij}^{\theta k}$. For ease of calculating, assume that a complete overlap of beam and pixel corresponds to a $W_{ij}^{\theta k} = 1.0$.

12.17 Our measurement for the ray shown in Figure 12.11 yields $I_0/I^{\theta k} = 2.0$. Calculate our best guess for μ_{ij} using the $W_{ij}^{\theta k}$ values from Problem 12.16.

12.18 Assume that the object in Figure 12.12(a) occupies the center square of a 3×3 square array. Assume that it has a density of 1.0 and that all other squares have a density of zero. Sketch the resulting curves for $p(x)$ and $p(y)$, the projection data for the directions normal to the x and y axes. Sketch the square array shown in Figure 12.12(b), and assign a density for each square in the resulting back projection.

12.19 A patient is placed in the strong magnetic field of an MRI imager. The field is 2.0 T, and the blood velocity (blood is an electric conductor) is 10 cm/s. What is the induced voltage gradient across a blood vessel? Could this be harmful?

12.20 In block-diagram form, show the design of a nuclear-medicine pulse-height analyzer. For each random pulse entering it that has an energy between two limits, it should give only one count. Note that for energies greater than both limits, the output pulse of the detector amplifier passes through both limits twice (rising and falling wave).

12.21 For the gamma camera, describe the x- and y-signal contribution from a photomultiplier that is located in the lower left of the detector array.

12.22 A gamma camera has a line-source response function of $k \exp(-2|x|)$, where k is a constant and x is in centimeters. Calculate the transfer function $S(f)$ of the system.

12.23 Draw a block diagram for an A-scan ultrasonic signal amplifier that corrects for ultrasonic attenuation with distance.

12.24 For Figure 12.26, estimate the sweep speed required for a 10×10 cm display. Estimate the maximal rate of repetition.

12.25 Why are ultrasound images of bone structures distorted? Why don't we observe a similar effect with air cavities?

12.26 In problem 12.11, if the geometry of the imaging system were such that the patient's skin was 70 cm from the x-ray source and the input of the detector, an image intensifier tube, was 100 cm from the source, how would this affect the exposure to the skin of the patient (assuming that the image features were measured at the image intensifier)?

12.27 Three measurements are made of the output of a medical x-ray machine using an ion chamber and electrometer (calibrated in mR) located 100 cm away from the focal spot (source) of the x-ray tube and several thin sheets of aluminum placed at the collimator. The machine was set to 80 kVp, 600 mA, and 0.1 s. The output of the machine was 380 mR (0.0 mm Al), 200 mR (3.0 mm Al), and 163 mR (4.0 mm Al). Find the HVL and output in mR/mAs.

REFERENCES

Balter, S., "X-ray equipment design," in J. G. Webster (ed.), *Encyclopedia of Medical Devices and Instrumentation.* New York: Wiley, 1988, pp. 2905–2912.

Benedetto, A. R., "Nuclear medicine instrumentation," in J. G. Webster (ed.), *Encyclopedia of Medical Devices and Instrumentation.* New York: Wiley, 1988, pp. 2072–2081.

Bracewell, R., *The Fourier Transform and Its Applications.* New York: McGraw-Hill, 1965.

Brooks, R. A., and G. Di Chiro, "Principles of computer-assisted tomography (CAT) in radiographic and radioisotopic imaging." *Phys. Med. Biol.,* 1976, 21, 689–732.

Christensen, D. A., *Ultrasound Bioinstrumentation.* New York: Wiley, 1988.

Foster, M. A., *Magnetic Resonance in Medicine and Biology.* Elmsford, N.Y.: Pergamon, 1984.

Fullerton, G. D., and I. L. Cameron, "Magnetic resonance imaging," in J. G. Webster (ed.). *Encyclopedia of Medical Devices and Instrumentation.* New York: Wiley, 1988, pp. 1833–1840.

Goldstein, A., "Ultrasonic imaging," in J. G. Webster (ed.), *Encyclopedia of Medical Devices and Instrumentation.* New York: Wiley, 1988, pp. 2803–2823.

Halberg, L. I., and K. E. Thiele, "Extraction of blood flow information using Doppler-shifted ultrasound." *Hewlett-Packard J.,* 1986, 37 (6), 35–40.

Hine, G. S., *Instrumentation in Nuclear Medicine.* New York: Academic, 1967.

Johns, H. E., and J. R. Cunningham, *The Physics of Radiology,* 5th ed. Springfield IL: Charles C. Thomas, 1990.

Phelps, M. E., J. C. Mazziota, H. R. Schelbert, *Positron Emission Tomography and Autoradiography.* New York: Raven, 1986.

Smith, H., and F. N. Ranallo, *A Non-mathematical Approach to Basic MRI.* Madison, Wis.: Medical Physics Publishing Co., 1989.

Sorenson, J. A., and M. E. Phelps, *Physics in Nuclear Medicine.* New York: Grune and Stratton, 1980.

Sprawls, P., *Physical Principles of Medical Imaging,* Madison, WI: Medical Physics Publishing Co., 1993.

Ter-Pogossian, M., *The Physics of Diagnostic Radiology.* New York: Harper & Row, 1971.

Wagner, H. N., Jr. (ed.), *Principles of Nuclear Medicine.* Philadelphia: Saunders, 1968.

Wells, P. N. T. (ed.), *Ultrasonics in Clinical Diagnosis,* 2nd ed. Edinburgh, Scotland: Churchill Livingstone, 1977.

Wells, P. N. T., "Medical ultrasonics." *IEEE Spectrum,* 1984, 21 (12), 44–51.

13

THERAPEUTIC AND PROSTHETIC DEVICES

Michael R. Neuman

As noted in earlier chapters of this book, a major use of medical electronic instrumentation is in diagnostic medicine. Most instruments sense various physiological signals, carry out some processing of these signals, and display or record them. There is, however, a class of medical electronic devices that are useful therapeutically, or as prostheses. Electric stimulators of one form or another represent an important subgroup in this area. Also available are other devices, such as incubators, ventilators, heart-lung machines, artificial kidneys, diathermy devices, and electrosurgical instruments. In this chapter we examine some of these devices and look briefly at their principles of operation.

13.1 CARDIAC PACEMAKERS AND OTHER ELECTRIC STIMULATORS

A wide variety of electric stimulators is used in patient care and research. They range from very low-current, low-duty-cycle stimulators, such as the cardiac pacemaker, to high-current single-pulse stimulators, such as defibrillators. In this section, we examine the pacemaker in detail and look at other applications of electric stimulators.

CARDIAC PACEMAKERS

The cardiac pacemaker is an electric stimulator that produces periodic electric pulses that are conducted to electrodes located on the surface of the heart (the epicardium), within the heart muscle (the myocardium), or within the cavity of the heart or the lining of the heart (the endocardium). The stimulus thus conducted to the heart causes it to contract; this effect can be used prosthetically in disease states in which the heart is not stimulated at a proper rate on its own. The principal pathologic conditions in which cardiac pacemakers are applied are known collectively as *heart block.* These are reviewed by Furman and Escher (1970). Schaldach (1992) and Webster (1995) review pacemaker design details.

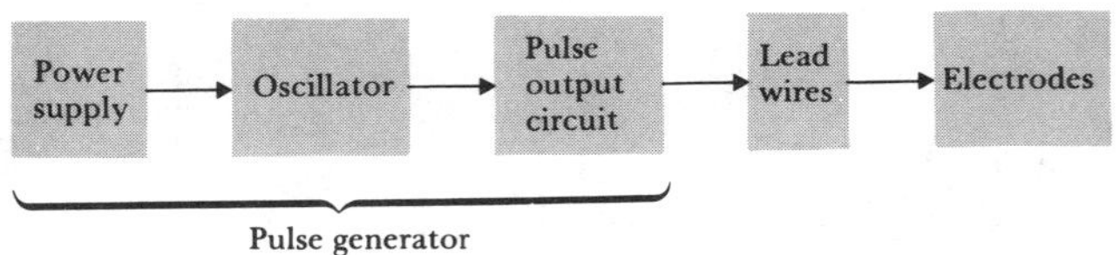

Figure 13.1 Block diagram of an asynchronous cardiac pacemaker

An *asynchronous* pacemaker is one that is free-running. Its electric stimulus appears at a uniform rate regardless of what is going on in the heart or the rest of the body. It therefore gives a fixed heart rate.

Figure 13.1 is a block diagram of an asynchronous pacemaker. The power supply is necessary to supply energy to the pacemaker circuit. Primary or secondary battery sources are used.

The oscillator establishes the pulse rate for the pacemaker; this, in turn, controls the pulse output circuit that provides the stimulating pulse to the heart. This pulse is conducted along lead wires to the cardiac electrodes.

Each of these blocks is important in the construction of the pacemaker, and each must be made highly reliable, because faulty operation of this device can cost a patient's life.

Another component of the overall construction of the pacemaker that is not included in Figure 13.1 is the package itself. Not only must the package of an implanted pacemaker be compatible and well tolerated by the body, but it must also provide the necessary protection to the circuit components in order to ensure their reliable operation. The body is a corrosive environment, so the package must be designed to operate well in this environment, while occupying minimal volume and mass.

Today, cardiac pacemakers are packaged in hermetically sealed metal packages. Titanium and stainless steel are frequently used for the package. Special electron beam or laser welding techniques have been developed to seal these packages without damaging the electronic circuit or the power source. These metal packages take up less volume and are more reliable than the earlier, polymer-based packages.

Although simple, asynchronous pacemakers such as that shown in Figure 13.1 are seldomly used anymore, we can learn about pacemakers in general by examining each of the blocks in more detail.

Power Supply The usual power supply for implantable pacemakers is a battery made up of primary cells. Customary practice in the early 1970s was to change the pacemaker generator every two years. It was not until the lithium iodide battery was introduced into use in pacemakers that the lifetime of the cardiac pacemaker was significantly increased (Greatbatch *et al.,* 1971). The fundamental lithium iodide cell involves the reactions

$$\text{Li} \rightarrow \text{Li}^+ + \text{e}^- \tag{13.1}$$

at the cathode and

$$I_2 + 2e^- \rightarrow 2I^- \tag{13.2}$$

at the anode, for the combined reaction

$$2Li + I_2 \rightarrow 2LiI \tag{13.3}$$

This cell has an open-circuit voltage of 2.8 V and is much more reliable than the batteries that had been used before. Its major limitation is its relatively high source resistance. Essentially all of presently applied pacemakers utilize various forms of lithium batteries as their power source.

Timing Circuit The asynchronous pacemaker represents the simplest kind of pacemaker, because it provides a train of stimulus pulses at a constant rate regardless of the functioning of the heart. A free-running oscillator is all that is required for the timing pulse in such a system. More advanced pacemakers, such as are used today, still have timing circuits to determine when a stimulus should be applied to the heart, but complex logic circuits, quartz crystal control, and even a microprocessor replace the simple, free-running oscillator. An overview of the function of some of these systems will be presented later in this chapter.

Output Circuit The pulse output circuit of the pacemaker generator produces the actual electric stimulus that is applied to the heart. At each trigger from the timing circuit, the output circuit generates an electric stimulus pulse that has been optimized for stimulating the myocardium through the electrode system that is being applied with the generator. Constant-voltage or constant-current amplitude pulses are the two usual types of stimuli produced by the output circuit. Constant-voltage amplitude pulses are typically in the range of 5.0 to 5.5 V with a duration of 500 to 600 μs. Constant-current amplitude pulses are typically in the range of 8 to 10 mA with pulse durations ranging from 1.0 to 1.2 ms. Rates for asynchronous pacemakers range from 70 to 90 beats per minute, whereas pacemakers that are not fixed-rate typically achieve rates ranging from 60 to 150 beats per minute.

LEAD WIRES AND ELECTRODES

Because, in most pacemaker designs, the generator is located at some position remote from the heart itself, there must be an appropriate conduit to carry the electric stimuli to the heart and to apply them in the appropriate place. The lead wires, in addition to being good electrical conductors, must be mechanically strong. Their distal ends must not only withstand the constant motion of the beating heart, but as the individual in whom the pacemaker is implanted moves about, these lead wires have to be able to withstand the stress of being flexed in various positions. A second requirement of the lead-wire system is that it must maintain good electrical insulation. If this is not the case, wherever faults in the insulation occur, there is effectively another stimulating electrode that, in addition to possibly stimulating the tissue in its vicinity, shunts important stimulating current away from its intended point of application on the heart.

To meet these requirements, the lead wires presently used consist of interwound helical coils of spring-wire alloy molded in a silicone-rubber or polyurethane cylinder. The helical coiling of the wire minimizes stresses applied to it, and the multiple strands serve as insurance against failure of the pacemaker following rupture of a single wire. The soft compliant silicone-rubber encapsulation both maintains flexibility of the lead-wire assembly and provides electrical insulation and biological compatibility.

Cardiac pacemakers are either of the *unipolar* or the *bipolar* type. In a unipolar one, a single electrode is in contact with the heart, and negative-going pulses are connected to it from the generator. A large indifferent electrode is located somewhere else in the body, usually mounted on the generator, to complete the circuit. In the bipolar system, two electrodes are placed within or on the heart, and the stimulus is applied across these electrodes. Both systems of electrodes require approximately the same stimulus for efficient cardiac pacing, as long as negative-going pulses are applied in the unipolar system.

There are clinically applied pacemakers utilizing each system. The electrodes themselves can be placed on the external surface of the heart (epicardial electrodes), buried within the heart wall (intramyocardial electrodes), or pressed against the inside surface of the heart (endocardial or intraluminal electrodes). In the latter case, it is possible to introduce the electrodes into the heart through a shoulder or neck vein so that it is not necessary to expose the heart surgically during the implantation process.

As with the lead wires, the materials of which electrodes are made are important. The electrodes must be able to stand up to the repeated stress they may encounter as a result of the mechanical activity of the heart, and they must remain in place to provide effective pacing. They must also be made of materials that do not dissolve during long-term implantation, cause undue irritation to the heart tissue adjacent to them, or undergo electrolytic reactions when the stimulus is applied. To avoid any junctional problems, these electrodes are often made of the same materials as the lead wires. Electrodes should also be made of materials that minimize biological interaction such as dense fibrous capsule formation around the electrode. Significant capsule formation can increase the threshold required for stimulation.

Several materials are used for pacemaker electrodes and lead wires. These include platinum and alloys of platinum with other materials; various formulations of stainless steel, carbon, and titanium; and specialized alloys such as Elgiloy* (40% cobalt, 20% chromium, 15% iron, 15% nickel, 7% molybdenum, 2% manganese, and traces of carbon and beryllium) and MP35N (35% nickel, 35% cobalt, 20% chromium, 10% molybdenum, and a trace of iron). In early

* Elgiloy was originally developed for wrist watch main springs that could endure repeated winding and unwinding without fatigue.

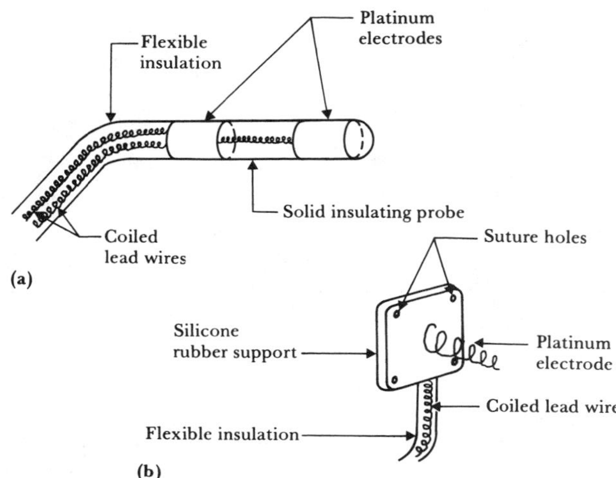

Figure 13.2 Two of the more commonly applied cardiac pacemaker electrodes (a) Bipolar intraluminal electrode. (b) Intramyocardial electrode.

pacemakers, a common type of failure was associated with breakage of the lead wire. Today, thanks to technological advances such as those just described, this problem has been greatly reduced, and lead wires and electrodes usually remain in place when the generator circuit and batteries are replaced.

Figure 13.2 shows the basic structure of a typical bipolar intraluminal electrode and single intramyocardial electrode. The conducting bands around the circumference of the solid intraluminal probe contact the endocardium (internal surface of the heart wall) and electrically stimulate it. The intramyocardial electrode is placed on the exterior surface of the heart. A puncture wound is made into the wall of the heart, and the helical spiral-shaped electrode is placed in this hole. To hold the electrode in place, the silicone-rubber supporting piece is then sutured to the epicardial (external) surface of the heart. This flexible back support provides a good mechanical match between the electrode and the heart wall. For bipolar intramyocardial stimulation, a pair of these electrodes is attached to the myocardium.

SYNCHRONOUS PACEMAKERS

Often patients require cardiac pacing only intermittently, because they can establish a normal cardiac rhythm between periods of block. For these patients, it is not necessary to stimulate the ventricles continuously; in some cases, continuous stimulation can even result in serious complications. For example, if an artificial stimulus falls in the repolarization period following a spontaneous ventricular contraction, ventricular tachycardia or fibrillation can result. Thus it is important in these cases that the artificial pacemaker not compete

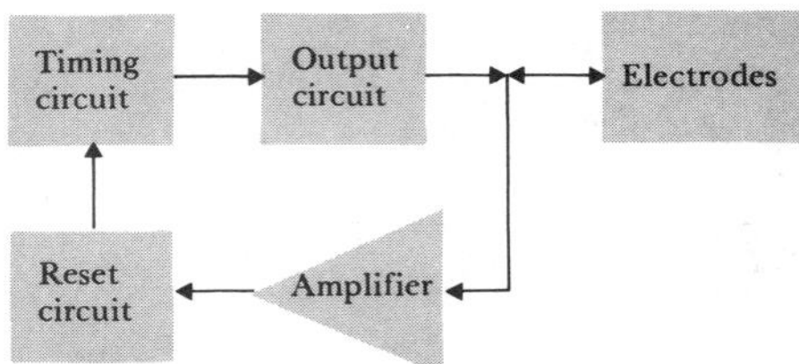

Figure 13.3 A demand-type synchronous pacemaker Electrodes serve as a means of both applying the stimulus pulse and detecting the electric signal from spontaneously occurring ventricular contractions that are used to inhibit the pacemaker's timing circuit.

with the heart's normal pacing action. Such a situation can be achieved with an asynchronous pacemaker by making the rate sufficiently high that the heart does not have a chance to beat on its own between pacemaker stimuli. A better solution, however, involves the use of synchronous pacemakers.

There are two general forms of synchronous pacemakers: the demand pacemaker and the atrial-synchronous pacemaker. A diagram of the *demand* pacemaker is shown in Figure 13.3. It consists of a timing circuit, an output circuit, and electrodes, just like those of the asynchronous pacemaker, but it has a feedback loop as well. The timing circuit is set to run at a fixed rate, usually 60 to 80 beats/min. After each stimulus, the timing circuit resets itself, waits the appropriate interval to provide the next stimulus, and then generates the next pulse. However, if during this interval a natural beat occurs in the ventricle, the feedback circuit detects the QRS complex of the ECG signal from the electrodes and amplifies it. This signal is then used to reset the timing circuit. It awaits its assigned interval before producing the next stimulus. If the heart beats again before this stimulus is produced, the timing circuit is again reset and the process repeats itself. Thus we see that, when the heart's conduction system is operating normally and the heart has a natural rate that is greater than the rate set for the timing circuit, the pacemaker remains in a standby mode, and the heart operates under its own pacing control. In this way the heart can respond to changing demands of the organism by changing its rate in the usual manner. If, on the other hand, temporary heart block occurs, the pacemaker takes over and stimulates the heart at the fixed rate of the timing circuit.

The *atrial-synchronous* pacemaker is a more complicated circuit, as shown in Figure 13.4. In this case, the pacemaker is designed to replace the blocked conduction system of the heart. The heart's physiological pacemaker, located at the SA node, initiates the cardiac cycle by stimulating the atria to contract and then providing a stimulus to the AV node which, after appropriate delay, stimulates the ventricles. If the SA node is able to stimulate the atria, the electric signal corresponding to atrial contraction (the P wave of the ECG) can be detected by an electrode implanted in the atrium and used to trigger

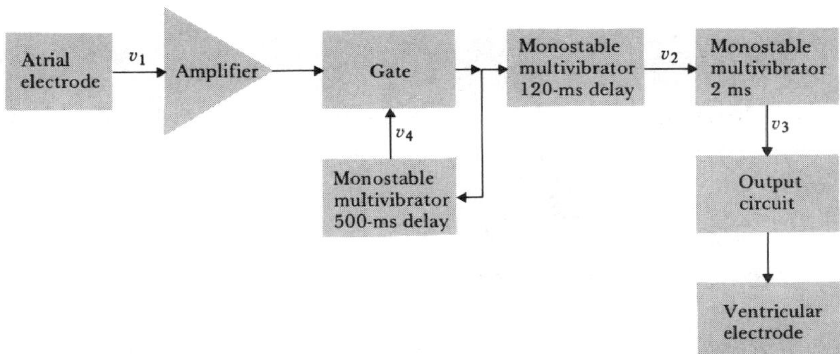

Figure 13.4 An atrial-synchronous cardiac pacemaker, which detects electric signals corresponding to the contraction of the atria and uses appropriate delays to activate a stimulus pulse to the ventricles. Figure 13.5 shows the waveforms corresponding to the voltages noted.

the pacemaker in the same way that it triggers the AV node. Figure 13.4 shows the voltage v_1 that is detected by the atrial electrodes.

This voltage is a pulse that corresponds to each beat. The atrial signal is then amplified and passed through a gate to a monostable multivibrator giving a pulse v_2 of 120-ms duration, the approximate delay of the AV node. Another monostable multivibrator giving a pulse duration of 500 ms is also triggered by the atrial pulse. It produces v_4, which causes the gate to block any signals from the atrial electrodes for a period of 500 ms following contraction. This eliminates any artifact caused by the ventricular contraction from stimulating additional ventricular contractions. Thus the pacemaker is refractory to any additional stimulation for 500 ms following atrial contractions.

The falling edge of the 120-ms-duration pulse, v_2, is used to trigger a monostable multivibrator of 2-ms duration. Thus the pulse v_2 acts as a delay, allowing the ventricular stimulus pulse v_3 to be produced 120 ms following atrial contraction. Then v_3 controls an output circuit that applies the stimulus to appropriate ventricular electrodes.

Often atrial-synchronous pacemakers have provisions to run at a fixed rate in case atrial stimulus is lost. This is achieved by combining the demand-pacemaker system with the atrial-stimulus pacemaker system so that an atrial stimulus disables a fixed-rate timing circuit. If the stimulus is absent, the fixed-rate timing circuit takes over and controls the output circuit in the same way as in the asynchronous pacemaker.

The pacing systems shown in Figures 13.3 and 13.4 are represented as having individual circuit blocks. As with the fixed-rate pacemaker, the blocks in these diagrams illustrate the functions carried out by the pacemaker system. These functions are now carried out by microprocessor systems within the pacemaker. Thus it is not possible to find individual components that make up a specific block in an actual device.

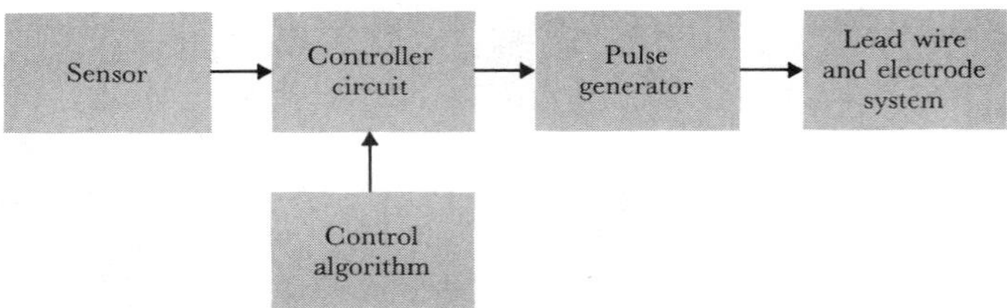

Figure 13.5 **Block diagram of a rate-responsive pacemaker**

RATE-RESPONSIVE PACING

Although synchronous pacemakers can meet some of the physiological demand for variation in heart rate and cardiac output, these devices still do not replicate the function of the heart in a physiologically intact individual. The demands of the body during stressful activities such as exercise cannot be fully met by these pacemakers. A new type of pacemaker system that can overcome these limitations is evolving. This pacemaker includes a control system (Figure 13.5). A sensor is used to convert a physiological variable in the patient to an electric signal that serves as an input to the controller circuit. This block of the pacemaker is programmed to control the heart rate on the basis of the physiological variable that is sensed. As with the demand pacemaker, this controller can determine whether any artificial pacing is required and can keep the pacemaker in a dormant state when the patient's natural pacing system is functional. The remainder of the pacing system is the same as described in this chapter for other generators.

The sensor can be located within the pacemaker itself, or it can be located at some other point within the body. In the latter case, it is necessary to connect the sensor to the pacemaker by using a lead-wire system.

Many different physiological variables have been used to control rate-responsive pacemakers (Smith and Fearnot, 1990). Table 13.1 lists some of these variables and, for each, a sensor that can be used to measure that variable in an implanted system. Each of these variables requires a different control algorithm for the control circuit. In some cases, simple proportional control can be used; in other cases, more complex control algorithms are necessary. For example, when the temperature of the venous blood is used as the control variable, derivative control has been found to be important. As a patient begins strenuous exercise, the venous blood temperature begins to decrease because the increased blood flow to the periphery returns cooler blood to the circulation. This decrease lasts for only about a minute, however, and then the venous blood temperature increases, as a result of the increased metabolic activity in the peripheral skeletal muscles, to a temperature that is above the patient's resting body temperature. If it is to perform similarly to the physiologically intact cardiovascular control system, the pacemaker must have a controller that can recognize these changes and respond to them with an increase in heart rate.

Table 13.1 Physiological Variables That Have Been Sensed by Rate-Responsive Pacemakers (*Not Commercially Available)

Physiological Variable	Sensor
Right-ventricle blood temperature	Thermistor
ECG stimulus-to-T-wave interval	ECG electrodes
ECG R-wave area	ECG electrodes
*Blood pH	Electrochemical pH electrode
*Rate of change of right ventricular pressure $\left(\frac{dp}{dt}\right)$	Semiconductor strain-gage pressure sensor
*Venous blood oxygen saturation	Optical oximeter
Intracardiac volume changes	Electric-impedance plethysmography (intracardiac)
Respiratory rate and/or volume	Thoracic electric-impedance plethysmography
Body vibration	Accelerometer

Although we generally think of cardiac pacemakers as implantable devices, there also are external versions of this electric stimulator. Fixed-rate, asynchronous pacemakers are appropriate for external devices, because controls for various pacing functions (such as rate) are located on the circuit and can be adjusted by the clinical staff. Intracardiac electrodes are used, and they are introduced percutaneously through a peripheral vein. The external pacemaker is used for patients who are expected to require pacing for only a few days while they are in the intensive-care unit or who are awaiting implantation of a permanent pacemaker. Frequently, external pacemakers are used for patients recovering from cardiac surgery to correct temporary conduction disturbances resulting from the surgery. As the patient recovers, normal conduction returns and use of the pacemaker is discontinued.

An external transcutaneous cardiac pacemaker can apply 80-mA pulses through 50-cm^2 electrodes on the chest. But this procedure is painful, so it is used only for emergency or temporary situations (Bocka, 1989).

BLADDER STIMULATORS

Urinary incontinence and other neurological bladder dysfunctions can, in some cases, be treated by electric stimulation. In the case of incontinence, the sphincter muscles surrounding the urethra are unable to contract sufficiently to occlude the urethra, and increased pressure within the bladder due to coughing, laughing, or neurologically excited excessive contraction of the detrusor muscle of the bladder wall can result in the uncontrollable passage of urine. Several investigators and manufacturers are looking for practical ways to control this problem through electric stimulation (Hill, 1973; Susset, 1973). These involve

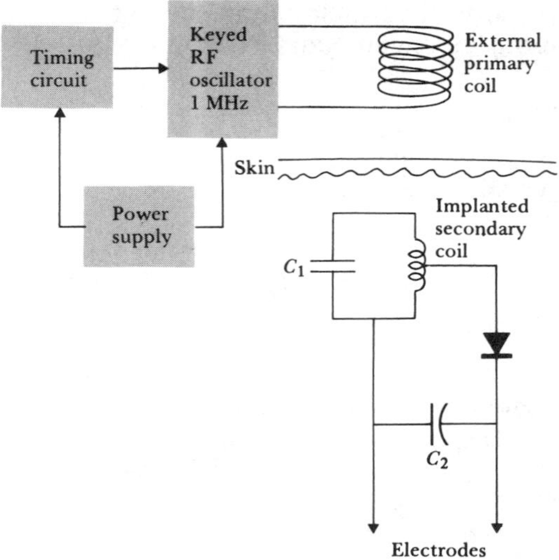

Figure 13.6 A transcutaneous RF-powered electric stimulator Note that the implanted circuit of this stimulator is entirely passive and that the amplitude of the pulse supplied to the electrodes is dependent on the coupling coefficient between the internal and external coils.

placing stimulating electrodes in or near the muscles involved in sphincteric control of the urethra or on the nerves supplying these muscles. The electrodes stimulate electrically, with pulses of durations of from 0.5 to 5 ms at a repetition rate from 20 to 100 pulses/s, depending on the individual investigator.

When neural electrodes are used, pulse duration can be shortened to be in the range of 100 to 400 μs. Average stimulating currents (during the pulse) are of the order of 1 mA. Electrodes are frequently placed directly on sphincter muscles, and optimal locations are determined during the implantation surgery by placing a balloon attached to a pressure sensor within the urethra in the region of the sphincters and locating the electrodes such as to give a maximal increase in pressure within the balloon during stimulation.

Noninvasive stimulating electrodes have also been described. In the case of women, such electrodes can be placed on a vaginal pessary that is positioned so as to place the electrodes against the anterior vaginal wall, posterior to the urethra. In men, an anal plug containing electrodes can be used to stimulate the sphincter muscles of the urethra. Although limited studies have shown these devices to be efficacious, they have not been popular with patients.

Some patients need continuous stimulation to avoid incontinence, and the high rate of stimulation calls for a greater power supply over a period of time than is required for the cardiac pacemaker. For this reason, techniques of transcutaneous stimulation are often used.

One kind of transcutaneous stimulator used in this application is the RF unit shown in Figure 13.6. The implanted circuit is entirely passive, with the internal secondary coil located just beneath the skin and coupled to an external primary coil placed over it. The primary coil is driven by a 1-MHz RF oscillator that is keyed by the timing circuit to produce the desired pulses. The power supply for this external circuitry is made up of replaceable or rechargeable batteries. The internal circuit consists of a capacitor C_1 to resonate the secondary coil to the oscillator frequency, a diode detector, and a filter capacitor C_2 to remove the RF component from the detected pulse waveform. The stimulus is then applied directly to the electrodes. Although the percutaneous transmission of energy is not very efficient, signal amplitudes of several volts can be obtained at the electrodes with primary-to-secondary coil spacings of approximately 1 cm.

Bladder stimulators are also used to help patients who otherwise cannot do so to void. Patients with certain neurological injuries find themselves unable to pass their urine because they cannot contract their detrusor muscle. Stimulators have been developed that can be controlled externally to cause the detrusor muscle to contract. This enables the patient to expel the contents of the bladder at will. To effect complete emptying, such stimulators require multiple electrodes on or within the bladder wall or on the spinal nerves that innervate the bladder. Again the RF percutaneous technique can be used; this time it is necessary only to bring the transmitting unit over the secondary coil when the patient desires to void. This type of stimulator has also been shown to be useful in helping to regain sexual function in men with spinal cord injuries. Advancement in microelectronics and microfabrication has made it possible to develop very small stimulators that receive their power from external sources. Unlike the passive circuit of Figure 13.6, these stimulators contain active circuit elements as well and can produce more carefully controlled electric pulses. Loeb *et al.* (1991) and Ziaie *et al.* (1993) have described stimulators based upon silicon intergrated circuit technology hermetically packaged in a glass cylinder that has the secondary coil and stimulating electrodes integrated into the device. These stimulators are so small (approximately 2 mm in diameter) that they can be injected into skeletal muscle rather than using open incision surgical techniques for implantation. The electronics on the implanted unit are used to detect a code on the transmitted signal from the primary coil to indicate when a particular implanted unit should produce a stimulus. In this way a single external unit can control several implanted units. The power to operate the implanted electronic circuit and to provide the stimulus pulse comes from the signal produced by the external unit.

MUSCLE STIMULATORS

There have been various applications of electric stimulation to muscle. Stimulators can be used in physical therapy to determine whether muscle groups are able to contract by applying external stimuli to these muscles and observing the results. Stimulators are especially useful in cases in which temporary paralysis can result from atrophy of the muscle caused by disuse, which significantly reduces the mass of the muscle. By periodic direct stimulation of the muscle, the clinician can exercise the muscle, even though the normal neurostimulation is not available. Electric stimulation of muscle can also be used to regain function of paralyzed muscles when the paralysis is a result of neurological injury. It has been demonstrated that patients with spinal-cord injury can regain some crude function of specific skeletal muscles by means of programmed electric stimulation. One such example is an individual whose muscles controlling a hand are completely paralyzed; electric stimulation can enable the person to gain or enhance some ability to grasp. This technique has been used in patients with spinal cord injuries secondary to traumatic accidents (Peckham, 1987; 1988). Fine-wire electrodes are placed in the flexor and extensor muscles controlling finger movement, and these electrodes are stimulated in different sequences to enable the patient to go through various grasping and releasing motions. The patient is able to control the stimulators by moving his or her contralateral shoulder.

Muscle stimulators have also been used to help patients gain function of the lower extremities. Functional electric-stimulation (FES) systems have been developed that enable paraplegic patients to stand, walk, and even climb stairs (Marsolais and Kobetic, 1987). Several electric stimulators are used, and a computer control system determines the timing of the stimulus bursts to the various muscles of the lower extremities. As with stimulation of the upper extremities, electrodes are placed in the muscle to be stimulated and actually stimulate the nerves that innervate the muscle. Stimulation of these nerves causes the muscle to contract.

Another example of a stimulator that is commercially available is one concerned with problems of stroke victims. Often these patients encounter gait problems that are evidenced in a condition known as dropfoot. In this case, an individual picks up the paralyzed foot to walk but is unable to lift the ball of the foot, so it drags along the ground. The person is thus particularly susceptible to tripping.

The *dropfoot* prosthesis shown in Figure 13.7 can help to minimize this problem. It consists of a switch in the heel of the patient's shoe. The contacts of the switch close when the patient takes weight off his foot. This switch controls a stimulator that continuously stimulates the muscles responsible for lifting the foot. When the individual again places weight on the foot, the switch contacts are opened and the stimulus is stopped. These devices are commercially available but have had only limited acceptance in the United States. In Europe they have been in use for several years.

Stimulus parameters for skeletal-muscle stimulators vary widely according to the type of stimulation, the number of channels, the type of electrode used, and the factor of safety chosen by the designer. Constant-current stimulators are popular in this application, because the charge transferred per stimulus pulse is constant regardless of the electrode load impedance. The pulse currents used range from 2 to 20 mA at pulse durations of 1 ms. These constant-current pulses can produce voltage peaks (Figure 5.23) ranging from 3 to 30 V at the electrodes. Thus if constant current is to be maintained, the power-supply voltage for the stimulator must exceed this.

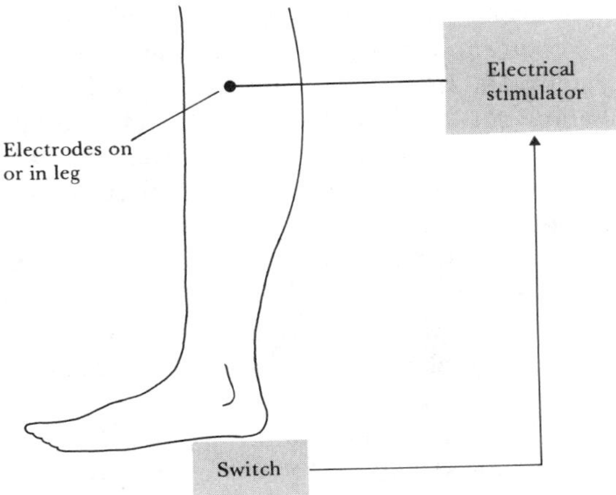

Figure 13.7 A stimulator system for use on stroke patients suffering from gait problems associated with dropfoot.

COCHLEAR PROSTHESIS

Profoundly deaf individuals whose hearing impairment results from dysfunction of the sound-transducing apparatus of the middle and inner ear can benefit from the use of a cochlear prosthesis (Spelman, 1988). This device consists of a set of electrodes that are passed into the scala tympani of the inner ear such that current from these electrodes can stimulate the nerves of the modiolus along the center of the cochlea. Prostheses that employ single-channel or multichannel systems give the patient the sensation of sound when the electrodes are stimulated. The prosthesis consists of two sections: the implanted stimulator and an external microphone and stimulus control unit. The block diagram for a simplified version of a multichannel cochlear prosthesis is given in Figure 13.8. The external unit includes a microphone to pick up the sound and a speech processor circuit. This block determines what specific characteristics of the sound picked up by the microphone will be used to control the stimulating electrodes. A simple form of such a processor is a set of bandpass filters arranged such that the output signal from these filters represents the sound intensity in a particular passband as a function of time. The signal from each bandpass filter is then used to control the stimulator such that it stimulates a particular set of electrodes in the cochlea.

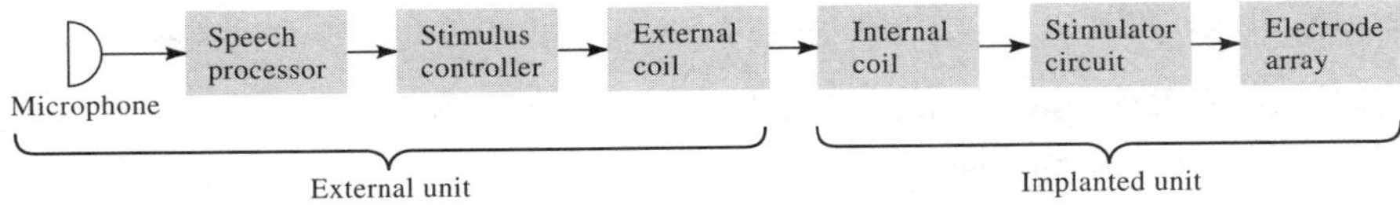

Figure 13.8 Block diagram of a cochlear prosthesis

The stimulus controller block takes the information from the speech processor and uses it to determine when a stimulus pulse should be applied to a particular set of electrodes. In multichannel systems, this block can carry out a multiplexing function (see Chapter 6) as well as a stimulator control function. The output from the stimulus controller circuit drives one or several external coils that are applied to the skin behind the ear directly over a similar coil that is implanted under the skin. The signal and enough power to drive the stimulator are thus transmitted by electric induction via a technique similar to that shown in Figure 13.6.

The input and power supply to the implanted unit come from the internal coil. The stimulator circuit applies an appropriate electric pulse to the electrode pair selected by the stimulus controller.

The electrode array consists of one or more pairs of very small electrodes that are placed on a flexible structure that can assume the spiral shape of the cochlea. Electrodes are positioned such that they face the basilar membrane when the electrode is inserted into the scala tympani. Electrodes have been made from fine wires cast in an elastomer such as silicone or as an array of thin-film deposited electrodes on a flexible polymeric substrate. Needless to say, given the size of the cochlea, these electrodes must be very small.

Although the cochlear prosthesis seems very similar to other types of electric stimulators used for therapeutic purposes, it is unique in one respect that is related to the speech processor and stimulus controller. The sound the patient "hears" is not the same as that heard by the intact ear. Methods of encoding speech into stimulus patterns are evolving, and patients currently have limited recognition of words. Nevertheless, many profoundly deaf patients have benefited from cochlear prostheses, and this technology has moved from the research laboratory to clinical care.

VISUAL PROSTHESIS

Electrical stimulation of regions of the occipital cortex of the brain and the optic nerve can give the sensation of light to individuals with certain types of blindness. In recent years investigators have begun to explore the potential of developing a visual prosthesis that would take advantage of this fact in creating rudimentary images for blind patients. Early work by Brindley and Lewin (1968) has shown that this is feasible. Research efforts employing stimulation of the occipital cortex of the brain and miniature arrays of stimulating electrodes on the retina have begun, and the neural prosthesis program of the National Institutes of Health in the United States has targeted the development of a visual prosthesis as an important area of research. Although we are far from achieving the potential of this new technology, it represents an exciting challenge for the future of electrical stimulation of tissue.

PAIN SUPPRESSION AND TRANSCUTANEOUS NERVE STIMULATION

Electric stimulation of tissues is sometimes used as a means of suppressing pain. Both battery-powered and transcutaneous RF-powered stimulators have been developed to alleviate severe intractable pain. Such devices have met with varying degrees of success. The amount of paraesthesia or anesthesia often depends on the selection of the patient. And in some cases the effectiveness of the stimulation decreases after periods of continuous stimulation. The gate-control theory of pain states that stimulation of certain neurons can have an inhibitory effect on the transmission of pain information from peripheral nerves to the spinal cord. This theory has led to the development of electronic circuits that can stimulate this inhibitory action and can, therefore, be used in the control of pain. Several investigators have studied these stimulators, which are applied to skin surface electrodes, and have found that in many cases (but not all), transcutaneous electric nerve stimulation does reduce perceived pain (Szeto and Nyquist, 1983). (Placebo effects were also seen in some cases where stimulators that provided no stimulus were used.)

A wide variety of stimulus waveforms are used with various TENS devices. They range from monophasic rectangular to biphasic spike pulses that are modulated in terms of amplitude, width, or rate. Burst modulation is another popular scheme. Outputs cover a wide range of voltages and currents up to 60 V and 50 mA, and pulse rates range from 2 to 200 pulses per second. Pulse widths also show a wide variation—from 20 to 400 μs. Burst rates are generally around 2 per second. Needless to say, this wide variety of stimulus variables demonstrates the current lack of information on just what the mechanism of TENS is and what the optimal stimulus should be. Nevertheless, many investigators have shown this phenomenon to be helpful to patients suffering from postoperative pain or pain associated with terminal cancer, and the use of TENS devices has been shown to reduce the amount of pain medication required by many patients.

TENS stimulators are similar to the asynchronous pacemaker illustrated in Figure 13.1. Electrodes come in various sizes and shapes, but they are in general similar to skin surface electrodes used for monitoring biopotential (see Chapter 5). A popular form of electrode consists of strips of silicone elastomer made conductive by loading with carbon particles. One electrode is placed on either side of an incision site by means of a conductive adhesive layer, and these electrodes are stimulated in the bipolar mode.

Although this technique is yet not well understood from the standpoint of physiological mechanisms, it offers potential relief to patients suffering from postoperative or intractable pain.

13.2 DEFIBRILLATORS AND CARDIOVERTERS

As we learned in Section 4.6, cardiac fibrillation is a condition wherein the individual myocardial cells contract asynchronously with only very local patterns relating the contraction of one cell and that of the next. This serious condition reduces the cardiac output to near zero, and it must be corrected as soon as possible to avoid irreversible brain damage to the patient and death. It is one of the most serious medical emergencies of the cardiac patient. Hence resuscitative measures must be instituted very quickly and definitely within 5 min after the attack.

Electric shock to the heart can be used to reestablish a more normal cardiac rhythm. Electric machines that produce the energy to carry out this function are known as *defribillators.* There are four basic types: the ac defibrillator, the capacitive-discharge defibrillator, the capacitive-discharge delay-line defibrillator, and the rectangular-wave defibrillator. We shall examine two of the most common types.

Defibrillation by electric shock is carried out either by passing current through electrodes placed directly on the heart or transthoracically, by using large-area electrodes placed against the anterior thorax (Tacker and Geddes, 1980; Tacker, 1988). The physician can achieve defibrillation of the heart with lower levels of current in the former case than in the latter, but electrodes can be placed directly on the heart only when the heart is exposed in a surgical procedure. Many defibrillators, however, have provisions for both types of defibrillation. They also incorporate appropriate safety features so that the high voltage used with surface electrodes cannot be applied accidentally when internal electrodes are being used, and so that the lower energy used with internal electrodes cannot be inadvertently connected to the surface electrodes, causing them not to produce effective defibrillation.

CAPACITIVE-DISCHARGE dc DEFIBRILLATORS

A short high-amplitude defibrillation pulse can be obtained by using the capacitive-discharge circuit shown in Figure 13.9. In this case, a half-wave rectifier driven by a step-up transformer is used to charge the capacitor C. The voltage to which C is charged is determined by a variable autotransformer in the primary circuit. A series resistance R limits the charging current to protect the circuit components, and an ac voltmeter across the primary is calibrated to indicate the energy stored in the capacitor. The resistor also helps to determine the time necessary to achieve a full charge on the capacitor. Five times the RC time constant for the circuit is required to reach 99% of a full charge. A good rule of thumb is to keep this time under 10 s, which means that the time constant must be less than 2 s.

The clinician discharges the capacitor when the electrodes are firmly in place on the body by momentarily changing the switch S from position 1 to position 2. The capacitor is discharged through the electrodes and the patient's torso, which represent a primarily resistive load, and the inductor L. The inductor tends to lengthen the pulse, producing a waveshape of the type shown in Figure 13.9. The slightly underdamped case is illustrated here, but overdamping and critical damping can also occur, the latter being most desirable. The situation is determined entirely by the resistance between the electrodes, which can vary from patient to patient. Once the discharge is completed, the switch automatically returns to position 1, and the process can be repeated if necessary.

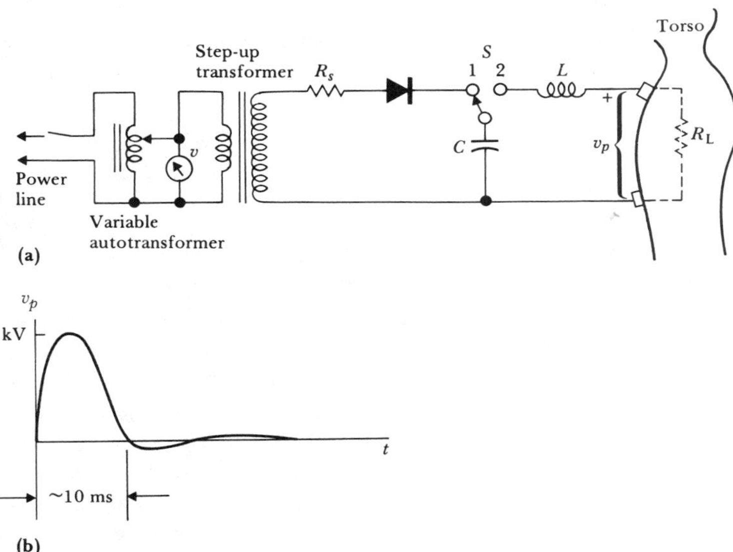

Figure 13.9 (a) Basic circuit diagram for a capacitive-discharge type of cardiac defibrillator. (b) A typical waveform of the discharge pulse. The actual waveshape is strongly dependent on the values of L, C, and the torso resistance R_L.

With a circuit such as this, 50 to 100 J (W · s) is required for defibrillation, using electrodes applied directly to the heart. When external electrodes are used, energies as high as 400 J may be required.

The energy stored in the capacitor is given by the well-known equation

$$E = \frac{Cv^2}{2} \tag{13.4}$$

where C is the capacitance and v is the voltage to which the capacitor is charged. Capacitors used in defibrillators range from 10 to 50 μF in capacitance. Thus we see that the voltage for a maximal energy of 400 J ranges from 2 to 9 kV, depending on the size of the capacitor. This stored energy is not necessarily the same energy that is delivered to the patient. Losses in the discharge circuit and at the electrodes result in an actual delivered energy that is lower.

RECTANGULAR-WAVE DEFIBRILLATORS

Geddes (1976) presents a comprehensive review of defibrillators and includes a description of rectangular-wave defibrillators. The capacitor is discharged through the subject by turning on a series *silicon-controlled rectifier* (SCR). When sufficient energy has been delivered to the subject, a shunt SCR short-circuits the capacitor and terminates the pulse. This eliminates the long discharge tail of the waveform. The output may be controlled by varying either the voltage on the capacitor or the duration of discharge. This design offers several advantages. (1) It requires less peak current. (2) It requires no inductor. (3) It makes it possible to use physically smaller electrolytic capacitors. And (4) it requires no relay.

DEFIBRILLATOR ELECTRODES

An important aspect of any defibrillator system is the electrodes. It is essential that they maintain excellent contact with the body so that the energy from the defibrillator reaches the heart and is not dissipated at the electrode–skin interface. If energy is dissipated at this interface, it can cause serious burns to the patient, further complicating a critical condition. To maintain good contact, the electrodes must be firmly placed against the patient. Often force-activated switches are contained within the electrode assembly, so that if firm-enough pressure is not applied to the electrodes, the circuit is interrupted and it is not possible to apply the defibrillation pulse.

A second key consideration is that the defibrillator electrodes must be safe to use. They must be sufficiently well insulated so that they do not allow any of the defibrillator output to pass through the hands of the operator. It would be tragic indeed if, in the process of the clinician's defibrillating a patient, her or his own heart were set into fibrillation. It is therefore important to consider the electrical safety of the defibrillator and electrodes.

Three types of electrodes are used for defibrillation. Figure 13.10(a) shows an internal type of electrode that can be used when the heart is surgically exposed. It consists of the metal electrode itself, which is spoon-shaped. The

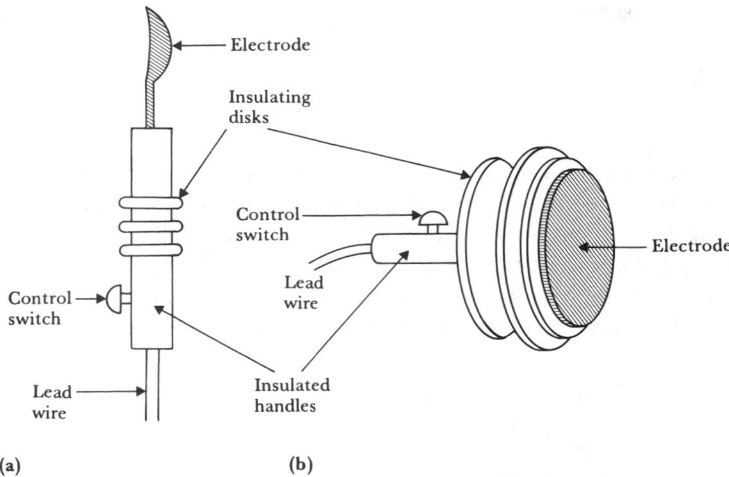

Figure 13.10 Electrodes used in cardiac defibrillation (a) A spoon-shaped internal electrode that is applied directly to the heart. (b) A paddle-type electrode that is applied against the anterior chest wall.

electrode is placed in a well-insulated handle with protecting corrugation between the electrode and hand positions so that body fluids cannot accidentally complete the circuit between the operator's hand and the electrode. A control switch is also often located on the handle, so that once the electrodes are in place, the operator can push the switch to initiate the pulse.

Figure 13.10(b) shows the type of electrode used for external defibrillation. It consists of a large metal disk in an insulated housing that is approximately 100 mm in diameter. The rest of the handle is similar to that for the internal electrodes, except that the insulating corrugation between the electrodes and the operator's hands are of greater diameter than the electrodes. The electrodes are sometimes called *paddles* because of their appearance. In operation, the electrodes (a pair must be used) are liberally coated with electrolyte gel of the type used with ECG-recording electrodes and are pressed firmly against the patient's chest. The operator then initiates the pulse with the control switch on the handle.

Disposable electrodes are also available. One type is like a large pregelled ECG electrode with a 50-cm^2 pregelled sponge backed by foil and surrounded by foam and pressure-sensitive adhesive to keep it in place. Another type is made from metal foil faced with conductive adhesive polymer so that the entire face is pressure-sensitive.

CARDIOVERTERS

When an operator applies an electric shock of the magnitude of that from a dc defibrillator to the patient's chest during the T wave of the ECG, there is a strong risk of producing ventricular fibrillation in the patient. Because the most frequent use of defibrillation is to terminate ventricular fibrillation, this problem does not occur; there is no T wave. If, on the other hand, the patient suffers from an atrial arrhythmia, such as atrial tachycardia or flutter, which in turn causes the ventricles to contract at an elevated rate, dc defibrillation can be used to help the patient revert to a normal sinus rhythm. In such a case, it is indeed possible accidentally to apply the defibrillator output during a T wave (ventricular repolarization) and cause ventricular fibrillation. To avoid this problem, special defibrillators are constructed that have synchronizing circuitry so that the output occurs immediately following an R wave, well before the T wave occurs.

Figure 13.11 is a block diagram of such a defibrillator, which is known as a *cardioverter*. Basically, the device is a combination of the cardiac monitor (Section 6.9) and the defibrillator. ECG electrodes are placed on the patient in the location that provides the highest R wave with respect to the T wave. The signal from these electrodes passes through a switch that is normally closed, connecting the electrodes to an appropriate amplifier. The output of the amplifier is displayed on a cardioscope so that the operator can observe the patient's ECG to see, among other things, whether the cardioversion was successful—or, in extreme cases, whether it produced more serious arrhythmias.

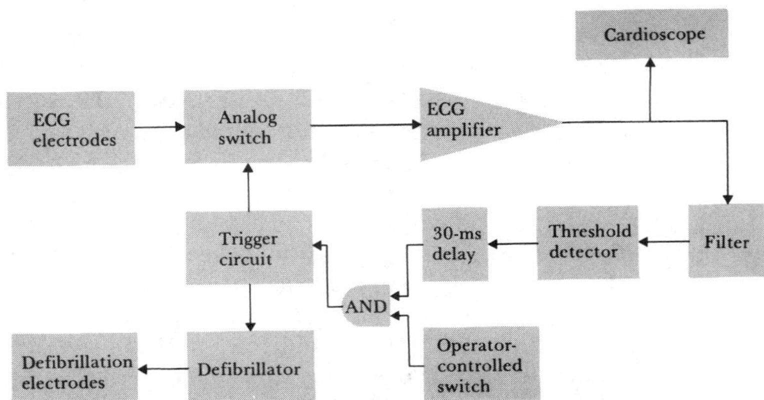

Figure 13.11 A cardioverter The defibrillation pulse in this case must be synchronized with the R wave of the ECG so that it is applied to a patient shortly after the occurrence of the R wave.

The output from the amplifier is also filtered and passed through a threshold detector that detects the R wave. This activates a delay circuit that delays the signal by 30 ms and then activates a trigger circuit that opens the switch connecting the ECG electrodes to the amplifier to protect the amplifier from the ensuing defibrillation pulse. At the same time, it closes a switch that discharges the defibrillator capacitor through the defibrillator electrodes to the patient. This R-wave-controlled switch discharges the defibrillator only once after the operator activates the defibrillator switch. Thus when the operator closes the defibrillator switch, it is discharged immediately after the next QRS complex. After the discharge of the defibrillator, the switch connecting the ECG electrodes to the amplifier is again closed, so that the operator can observe the cardiac rhythm on the cardioscope to determine the effectiveness of the therapy.

IMPLANTABLE AUTOMATIC DEFIBRILLATORS

It is well known that electrodes placed directly on the myocardium require less energy to defibrillate the heart than do transthoracic electrodes. It is also widely accepted that the earlier defibrillation can be achieved, the better it is for the patient. Thus it was inevitable that implantable automatic defibrillators would be developed (Singer, 1994). These devices, similar in appearance to the implantable pacemaker although somewhat larger, consist of a means of sensing cardiac fibrillation or ventricular tachycardia, a power-supply and energy-storage component, and electrodes for delivering a stimulus pulse should it be needed. The need for defibrillation is determined by processing the electric signals picked up by the implanted defibrillation electrodes. In some experimental devices, mechanical signals related to ventricular tachycardia and myocardial fibrillation were used to determine the need for a fibrillation pulse.

Energy storage adequate to provide pulses of from 5 to 30 J is necessary in the implantable defibrillator. This is considerably less than the amount of energy required for both the transthoracic defibrillator and those used on the exposed heart. The electrodes used with the implantable defibrillator are similar to those used with a cardiac pacemaker. They can be located on the epicardium, within the myocardium, or in the ventricular lumen.

Although the first human implant of an automatic defibrillator occurred in 1980, it was not until eight years later that commercial devices appeared on the market. Today, the implantable defibrillator is part of the cardiologist's armamentarium for treating cardiac fibrillation and tachyrhythmias, just as the pacemaker has become a major therapeutic method for treating bradyarrhythmias.

13.3 MECHANICAL CARDIOVASCULAR ORTHOTIC AND PROSTHETIC DEVICES

Cardiovascular orthotic prosthetic devices are primarily mechanical in nature, but they are always associated with various pieces of electronic instrumentation that are necessary to control the devices and to monitor their operation.

CARDIAC-ASSIST DEVICES

It has been the goal of cardiac surgeons and cardiologists to develop mechanical pumps that can be used to aid the failing heart after acute traumatic insults such as myocardial infarction or cardiac surgery. Devices ranging from pumps that can completely replace the heart to a device to reduce the load driven by the heart are being considered. In the latter case, physicians have developed an aortic-balloon system that is used clinically. It consists of a long sausage-shaped balloon that can be introduced into the aorta through a femoral artery and connected to an external drive apparatus (Jaron, 1988).

The operation of this intraaortic balloon pump is quite simple. Let us consider the balloon as being in the aorta in its inflated state, occupying a major portion of the aortic lumen but still allowing blood to flow past it. At the initiation of the next ventricular contraction, suction is applied to the balloon, causing it to collapse. The blood pumped by the left ventricle enters the aorta and replaces the volume previously occupied by the balloon. This requires only low pressure and demands less effort from the left ventricle. After the contraction, the aortic valve closes, and pressurized CO_2 is applied to the balloon, causing it to expand. CO_2 is used because it is more soluble in blood than air is. Thus if the balloon or its supply tubing should leak or rupture, there is less risk of fatal gas embolism.

As the balloon expands, it forces the blood surrounding it out of the aorta and into the rest of the body. Hence the balloon does much of the work normally done by the left ventricle and causes the blood to circulate to the periphery. The process is repeated after the next ventricular contraction.

This device must rely on a sophisticated system of electronic controls to detect ventricular contractions either from a pressure sensor at the arch of the aorta or, more commonly, from the ECG. The signal must then go through appropriate delay circuits to control the suction and pressurized CO_2 supplied to the balloon. Appropriate sensors must also be included in the system to ensure that alarms are sounded if any leaks occur.

PUMP OXYGENATORS

In cardiac surgery it is often necessary to stop the heart from pumping during certain procedures of the operation. In this case, to keep the patient alive it is necessary to replace the heart's pumping action and also the oxygenation normally provided by the lungs, because they are usually not functioning either. Machines known as *pump oxygenators* have been developed that can carry out these functions. They consist of pumps for maintaining arterial blood pressure connected in series with oxygenators that increase the blood O_2 content and remove CO_2. In surgery, the pump oxygenator is usually connected between the superior and inferior venae cavae or between the right atrium and a femoral artery, as shown in Figure 13.12. In some cases a femoral-artery-to-femoral-vein-bypass technique is used to keep all cannulae away from the heart.

Various types of pumps can be used. Roller pumps and multiple-finger pumps are often employed, because the pump itself does not come in contact with the blood. Disposable tubing can be used to contain the blood that is pinched between the propagating rollers or fingers. Pulsatile pumps—consisting of a chamber subjected to the reciprocating motion of a piston, membrane, or bladder—are also used, with appropriate check valves to direct the flow. Such pumps more closely follow the normal action of the heart and produce a pulsatile blood pressure.

Two general types of oxygenators are used with these prosthetic devices. One is the *film* type, in which a large-surface-area film of blood is drawn into contact with a nearly 100% O_2 atmosphere by rotating disks (Figure 13.12). The second type is the *membrane* oxygenator, in which blood flows through fine tubes of a membrane permeable to gas. This device has a large exchange-surface area to allow the gas transfer to take place.

Studies have shown the membrane oxygenator to have less deleterious effects on the blood than film and other direct types of oxygenators. The membrane separates the liquid blood phase from the gaseous oxygen phase, whereas in direct oxygenators such as that shown in Figure 13.12, blood and oxygen are in direct contact. This can denature some of the protein components of the blood, which can lead to formation of emboli in the patient.

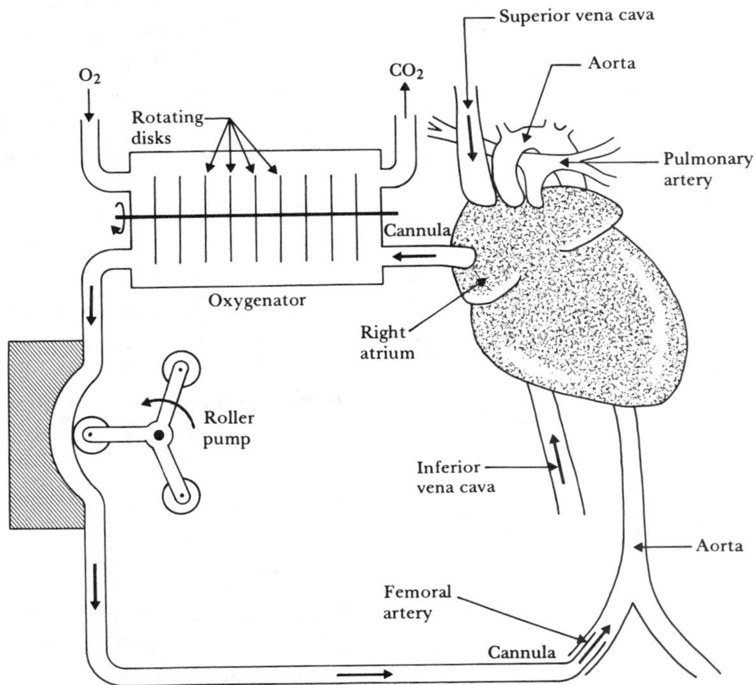

Figure 13.12 Connection of a pump oxygenator to bypass the heart A disk-type oxygenator is used with a roller pump. Venous blood is taken from a cannula in the right atrium, and oxygenated blood is returned through a cannula in the femoral artery.

Pump oxygenator systems are also applied in newborn intensive care. Infants who have severe lung disease that cannot be treated any other way have been put on pump oxygenators for several days to allow the diseased lung to "rest." The lungs recover in a significant number of these infants, who can then be removed from the pump oxygenator. This therapy, which is based on the extra corporeal membrane oxygenator (ECMO), has been demonstrated to be effective for seriously ill term infants, but the same approach has not shown any therapeutic efficacy in adults.

TOTAL ARTIFICIAL HEART

Blood pumps have been miniaturized and constructed of such materials that they can replace the natural hearts of patients. They are implanted in the thoracic cavity and operate via pneumatic and electric connections to an external drive apparatus (DeVries and Joyce, 1983; Yared and DeVries, 1988). Such devices were the subject of many popular press reports during the period in which they were studied in human subjects. The most notable of the devices was the Jarvic 7, which has been used as both a temporary and a permanent heart replacement (Jarvic, 1981). In the former case, the device was used to keep a patient alive until a suitable natural donor heart could be found for transplantation. Total heart replacements enabling patients to live for up to 620 days after surgery have been reported. Technical problems, however, continue to plague the device, and its use has been halted pending further technical improvement.

Electronic instrumentation is essential when pumps or pump oxygenators are used. It is necessary to monitor the hemodynamics of the patient during the procedure. In addition, the ECG, aortic, and central-venous pressure waveforms must be carefully watched. The pump oxygenator itself must also be monitored. The degree of oxygenation of the blood, as well as its pressure, must be recorded. And it is necessary to protect the patient from leaks in the system that could cause O_2 to enter the blood vessels or result in the serious loss of blood.

13.4 HEMODIALYSIS

One of the most important prosthetic devices in modern medicine is the artificial kidney, which is periodically connected to the circulatory systems of uremic patients to remove metabolic waste products from their blood. A general scheme for the operation of this device is shown in Figure 13.13.

There are two basic units in a hemodialysis system: the exchanger and the dialysate delivery system. The exchanger consists of the dialysis chamber itself, which is a compartment containing the patient's blood and a compartment containing the dialysate. These two compartments are separated by a semipermeable membrane that allows the waste components in the blood to diffuse through to the dialysate, which carries them away.

There are three basic types of exchangers in use. The *coil dialyzer* consists of a tube made of the semipermeable membrane material wound into a coil, in such a way that the dialysate can be circulated between individual turns of it. This is the most commonly used type of dialyzer. It has the limitation that the coil must be fairly long to provide a large effective surface area for mass transport, and it thus imposes a relatively high resistance to the flow of blood. For this reason, and to maintain effective blood flow, it is necessary to put a pump in series with the arterial blood supply to increase the pressure. However, the increased pressure improves the ultrafiltration rate of the membrane. It is important in this unit that the dialysate also be forcibly circulated to ensure rapid mixing. Because fresh dialysate must be available to the surface of the membrane throughout the coil, a pump is necessary for the dialysate as well.

The second type of exchanger used in artificial-kidney systems is the *parallel-plate dialyzer*. It is constructed similarly to a multilayer parallel-plate capacitor, with the plates made of semipermeable-membrane material. Blood

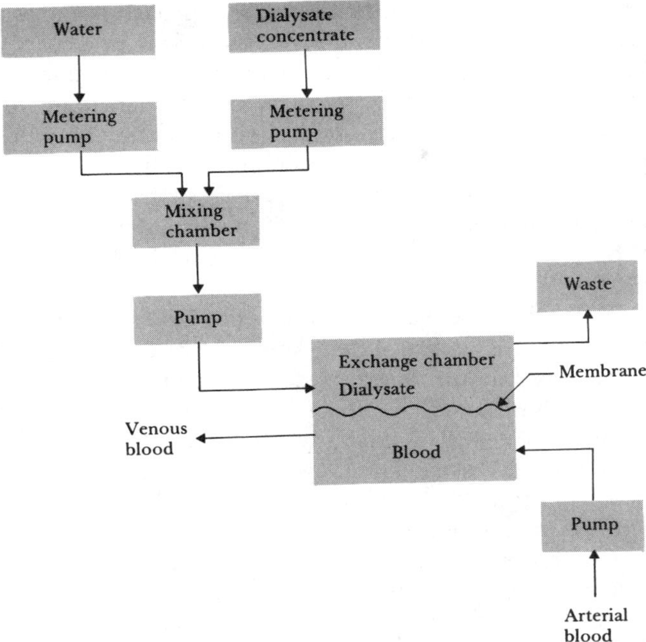

Figure 13.13 An artificial kidney The dialysate delivery system in this unit mixes dialysate from a concentrate before pumping it through the exchange chamber.

is circulated between alternate pairs of plates, and the dialysate is circulated between the other plates. Each plate thus serves as a membrane between dialysate and blood. The blood flows in thin sheets to maximize the surface-to-volume ratio for the dialyzer.

The third type of exchange apparatus is the *hollow-fiber kidney.* This consists of from 10,000 to 15,000 hollow fibers, with an internal diameter of approximately 0.2 mm and a length of approximately 150 mm, connected in parallel. The blood flows in the lumen of the fibers while the dialysate surrounds them. The walls of the fibers serve as the semipermeable membrane. The dialysate is pumped through the space surrounding the fibers to achieve the most efficient exchange.

The remainder of the artificial kidney consists of the dialysate-delivery system. The dialysate is made up of water with various solutes added. Either it is prepared in a batch and pumped through the dialysis chamber, or (in applications in which large amounts of dialysate are used) a concentrated solution of the solutes is made and automatically mixed with pure water to achieve the correct concentration (Figure 13.13). Metering pumps administer the correct amount of dialysate concentrate and water into a mixing chamber to produce the dialysate, which is then pumped through a dialysis chamber, where it picks up the metabolic waste products. It is then discarded.

As in the case of the cardiovascular prostheses, described in Section 13.3, the hemodialysis apparatus does not require any electronic instrumentation to function. However, in using this device with patients, the clinician finds that several pieces of electronic instrumentation greatly aid in its application and operation. Because only a membrane separates the patient's blood from the dialysate, it is important that any leaks in the membrane be detected immediately before serious losses of blood occur. In some cases, in fact, dialysate can leak into the patient's circulatory system. The blood is usually at a higher pressure than the dialysate, so loss of blood is the hazard of major concern. The dialysate is a clear liquid, and the presence of blood in it can be detected as a colorimetric or optical density change; thus optical systems are used to detect leaks. In addition, instruments monitor the pressure in the blood compartment to detect rapidly any abnormalities, such as major leaks or clotting phenomena, that might change the pressure.

The gross concentration of electrolytes of the dialysate is also monitored by electronic instrumentation. Because the solute is made up of electrolytes, the overall concentration of these in the dialysate is determined by impedance techniques. Thus, by measuring the conductivity of the dialysate in the mixing chamber, instruments can detect any major abnormalities in concentration before the dialysate enters the dialysis chamber.

Another problem common to both hemodialysis units and the pump oxygenator is that air bubbles cannot be tolerated in the blood that reenters the patient: This produces air emboli that may be life-threatening. Thus it is important that some type of bubble detector be included in the path of the blood before it reenters the body. If bubbles are detected, the blood pump is turned off until the technician operating the dialyzer solves the problem.

13.5 LITHOTRIPSY

Kidney stones can cause great discomfort to the patient as they are passed through the urinary tract, and their presence may eventually lead to loss of function of the affected kidney. An open incision surgical technique known as lithotomy can be used to remove the stone, and this procedure includes all the risks, complications, discomfort, and disability of major surgery. Lithotripsy refers to noninvasive or minimally invasive surgical techniques for removing kidney stones without these risks and complications. The technique involves disintegration of the stone *in vivo* so that it can pass through the urinary tract in the form of small particles the passage of which does not result in severe discomfort or disability (Bush and Brannen, 1988).

In percutaneous lithotripsy, a probe is guided under x-ray fluoroscopy through a small incision into the location of the kidney stone. Either mechanical shock waves are produced at the tip of the probe by a controlled electric

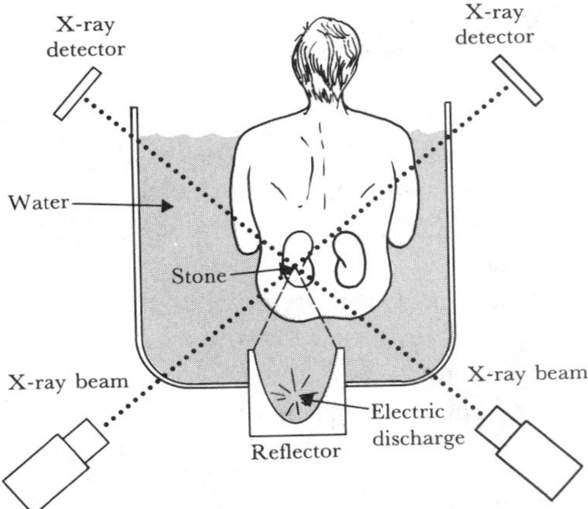

Figure 13.14 In extracorporeal shock-wave lithotripsy, a biplane x-ray apparatus is used to make sure the stone is at the focal point of spark-generated shock waves from the ellipsoidal reflector.

discharge (spark), or the probe contains an ultrasonic transducer that produces ultrasonic waves. Each of these forms of energy is used to break up the kidney stone so that it can be withdrawn in pieces through the probe guide element or can be allowed to pass through the urinary tract.

Extracorporeal shock-wave lithotripsy is an entirely noninvasive method that can be used to break up kidney stones. Figure 13.14 shows the basic structure of a shock-wave lithotripter. Many mechanical shock waves are produced at one focus of an ellipsoidal reflector such that these waves converge to another focal point several centimeters away from the reflector. The reflector and the patient are submerged in demineralized, degassed water in such a way that the patient can be moved until the stone is located at the focal point of the shock wave. This positioning of the patient is critical, and a biplane x-ray system is used to establish the position of the stone at the focal point as well as to monitor its disintegration. A high-voltage pulse (approximately 20 kV) is applied to the spark gap, and the discharge produces a shock wave that is propagated through the water to the focal point. The patient is placed on a gantry support that can be precisely positioned as the operator observes the stone on the biplane x-ray monitors. Once the patient is appropriately positioned, multiple-shock waves are generated by multiple discharges across the spark gap. Up to 2000 shock waves may be necessary to reduce a kidney stone to 1- to 2-mm fragments that can pass through the urinary tract.

With this treatment, most patients are able to resume full activity within two days. This is considerably less time than surgical treatment by lithotomy would entail. Although the apparatus is complex and expensive to purchase and operate, the overall savings for the patient and the health-delivery system are clear.

13.6 VENTILATORS

An important factor in respiratory therapy is being able to assist patients in ventilating their lungs. Various mechanical devices have been developed over the years to carry out this function. These devices, known as ventilators or respirators, can be separated into two general categories: the *controller* and the *assister.*

When a patient is connected to a controller type of ventilator, his or her respiratory ventilation is determined by the machine. The device sets the ventilatory cycle, and any tendency toward spontaneous ventilation on the patient's part does not affect the machine and can even oppose it. The assister, on the other hand, is controlled by the patient and used to augment his or her own ventilation activities. The assister detects the patient's attempt at ventilation and augments it mechanically. Thus it assists rather than controls the patient in ventilation.

We can further classify ventilators as negative- or positive-pressure devices. *Negative-pressure* ventilators are more physiological, in that the body of the patient is contained in a sealed chamber in which the pressure can be reduced. This negative pressure is transferred to the space within the thorax, producing a pressure gradient along the trachea that results in air entering the lungs. Pressure is then returned to atmospheric, allowing the lungs to recoil to their original shape and to expel some of their air. *Positive-pressure* ventilators, on the other hand, blow air into the lungs by increasing the pressure in the trachea. This causes the lungs to expand due to internal pressure and then to recoil naturally, expelling a portion of the air once the positive pressure is removed.

Ventilators can be time-cycled, volume-cycled, or pressure-cycled. Negative-pressure ventilators are usually time-cycled. This means that the negative pressure is applied to the body for a given period of time and then released for another given period of time before the process is repeated. Modern time-cycled ventilators are electronically controlled. Microprocessors are used to establish the cycling or the rate of ventilation, as well as the ratio between inspiration and expiration times or volumes. These electronic circuits activate solenoid valves that regulate the airflow.

In the *volume-controlled* ventilator, the progression of the cycles of the ventilator is controlled by the volume of air administered to the patient. Thus, if a machine is set to cycle on a given volume, it does not cycle until that volume of air has been administered to the patient. It also has a pressure-override valve, so that if, while the machine is in the process of administering the set volume, the pressure exceeds a predetermined maximal value, the ventilator will cycle whether or not the appropriate volume has been administered. This is an important safety consideration, because uncontrolled pressures could cause serious damage. In the *pressure-cycled* ventilator, air is administered to the patient until the pressure reaches a predetermined limit, at which time the ventilator switches to its expiratory portion of the cycle, and the process is repeated.

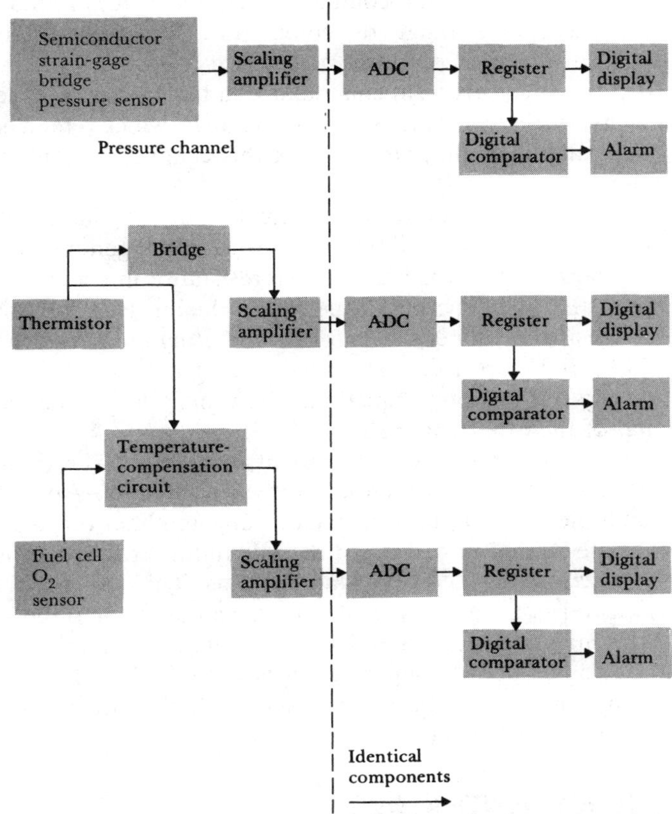

Figure 13.15 An instrumentation system for recording pressure, temperature, and percentage of O_2 in inspired air coming from a continuous-positive-airway-pressure apparatus.

In some instances, respiratory-assist devices are being developed with electronic instrumentation to evaluate their function as an integral part of the system. Figure 13.15 is a block diagram of an instrumentation system for a continuous-positive-airway-pressure device for use with newborns. The system consists of monitors for temperature, pressure, and O_2 fraction, including sensors located near the exit port of the device to sense the properties of the inspired air. The sensors are connected to electronic processing circuits that then make the physiological variables available for digital readouts. The signals are also compared with preset alarm levels, so that if they fall outside a predetermined normal range, alarms are sounded.

The device senses pressure by means of a semiconductor strain-gage pressure sensor. The output is amplified by a scaling amplifier to yield an appropriate level for digital conversion by the ADC that operates at a slow strobe rate. Digital signals are then stored in a register connected to the digital display. The contents of the register are also compared with present high-

and low-alarm limits in the digital-comparator circuit. If it is found that the signal is out of its acceptable range, the comparator sets off the alarm, which flashes the digital display and produces an audible beep.

The temperature-measurement channel uses a thermistor as its sensor. A bridge circuit converts the resistance signal to a voltage, which is then amplified through a scaling amplifier; the remainder of this channel is identical to the pressure channel.

The final channel on the monitor is used for sensing the fraction of oxygen in the inspired air. It uses a fuel-cell type of O_2 sensor that generates a current proportional to the P_{O_2}. It is loaded with the resistance in order to produce an output voltage that is a function of the P_{O_2} in the air. Because this sensor is temperature-sensitive, compensation for its operating temperature must be included, as shown in Figure 13.15. The circuit could also use an independent thermistor to carry out the same function. The remainder of the channel is identical to that of the other channels.

This system has several redundant circuit elements. Everything to the right of the scaling amplifier is identical for each channel. An alternative design could multiplex the output from each scaling amplifier through a single ADC and associated circuitry and could then demultiplex it at the output to operate the respective digital displays and alarms. This approach was not taken because it did not have any economic advantages for three channels and because the manufacturer wanted to have three independent modular channels that could be operated separately if desired. Although these factors are not necessarily important in the circuit design itself, they are very important with respect to the design of the entire product.

HIGH-FREQUENCY VENTILATORS

A technique for ventilating patients at frequencies much higher than the normal respiration rate has been investigated and has seen limited clinical use. Not only are the inspiration–expiration frequencies higher than normal, but the actual volume of air moved per "breath" is on the order of the anatomical dead space of the pulmonary system. Because the anatomical dead space constitutes the volume of the conducting airway, the upper airway, trachea, and bronchial tree, high-frequency ventilation is unusual from the standpoint of volume as well as frequency (Hamilton, 1988).

The physiological mechanisms of gas transport in high-frequency ventilation constitute various combinations of volume transport and molecular diffusion (Chang, 1984). Even though it is unlikely that a molecule of oxygen entering the airway would enter the alveoli at normal breathing frequencies and tidal volumes equivalent to the volumes used in high-frequency ventilation, when this process is carried out at high frequencies, increased mixing occurs in the airway. This can be the result of turbulence, asymmetric velocity profiles, and molecular diffusion that occurs at the elevated frequencies. The net effect is that the mixed gases in the airway do in fact reach the alveoli, making it possible for the alveoli to be effectively ventilated.

High-frequency ventilation principles can be put into practice in several different ways. The easiest method is to generate the high-frequency pressure waves by means of a large, low-frequency loud speaker that is driven by an electric signal of the desired frequency and waveshape. In high-frequency jet ventilators, a small-diameter tube is passed down a tracheal cannula and either is terminated at its distal end or extends into the trachea itself. Short pulses of higher-pressure oxygen are introduced into the airway through the cannula at frequencies well above the normal respiration rate. Another approach to achieving high-frequency ventilation is to use a piston pump that can alternately compress and decompress a column of gas that is coupled to the airway through an endotracheal cannula. A rotating valve that is driven by a motor can also be used to connect the airway alternately to a high-pressure and a low-pressure source of gas.

The frequencies used in high-frequency ventilation vary with the investigator and the approach used to obtain the high-frequency pressure oscillation. In general, these frequencies range from 60 to 3600 min^{-1} (1 to 60 Hz), though frequencies in the vicinity of 900 min^{-1} (15 Hz) are popular. Inspiration-to-expiration ratios can usually be varied from 1:1 to 1:4, and waveshapes ranging from sinusoidal to rectangular can be achieved. Like conventional ventilators, high-frequency ventilators are usually electronically controlled.

13.7 INFANT INCUBATORS

The care of premature newborns often requires that they be in an environment in which temperature is elevated and controlled, because they are unable to regulate their own temperatures. When infants are kept in a chamber maintained within a specific temperature range, O_2 requirements are minimized. This is especially important for premature newborns, who are more susceptible to respiratory problems than full-term infants, because their lungs may be unable to supply enough oxygen to meet elevated demands. Such controlled-temperature environments are maintained in infant incubators.

Temperature-controlled air is passed through the chamber in which the baby is located to maintain it at a set temperature. The temperature is controlled in modern units by means of the proportional control system shown in Figure 13.16. The temperature in the air-supply line varies a thermistor resistance that is compared with a fixed resistance that corresponds to the set temperature. If the temperature of the air entering the infant's chamber is lower than the set temperature, power is applied to the heater to correct for this difference. In the proportional-controller system, the amount of power applied to the heater is proportional to the difference between the actual air temperature and the set point. This means that the amount of power decreases as the temperature approaches the set point, an important feature in effecting more precise control and minimizing overshoot of the set point.

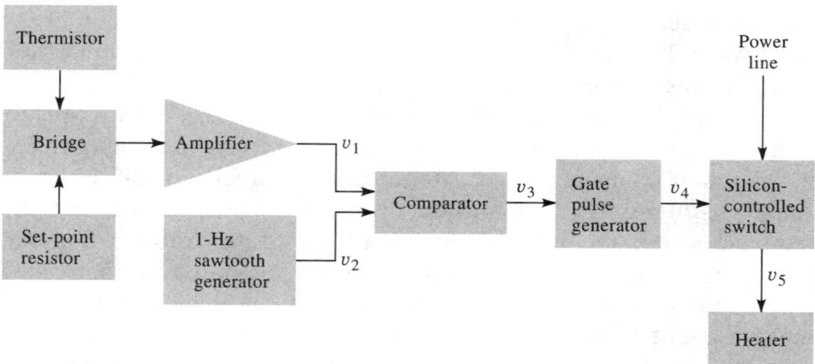

Figure 13.16 Block diagram of a proportional temperature controller used to maintain the temperature of air inside an infant incubator.

The control system shown in Figure 13.16 uses the thermistor in a bridge circuit, with the set-point resistance as another arm of the bridge. The bridge output is amplified, giving the voltage v_1 at the output, which is proportional to the difference in temperature between the thermistor and the set point.

Some incubators, instead of controlling the air temperature directly, use the skin temperature of the infant as a control parameter. The thermistor is place against the skin of the infant, and the controller is set to maintain the infant's skin at a given temperature. If the infant is cooler than the set point, the air entering the chamber of the incubator is heated an amount proportional to the difference between the set temperature and the baby's actual temperature.

Incubators also have a simple alarm system to alert the clinical staff if there is any dangerous overheating of the device. The system embodies a temperature-controlled switch that carries power to an audible alarm when, if ever, the temperature exceeds the safe limit. Often there is a buzzer connected in series with a switch that is activated by a bimetallic strip. This keeps the system simple and reliable. In some cases, this circuit also immediately reduces power to the heater to stop the overheating.

Infants lose a large percentage of their body heat by radiation. In some clinical situations, the intensive care of term and preterm infants requires that large portions of their skin surface be exposed to enable the clinical staff to observe the infant's condition, carry out special procedures, and attach various therapeutic and monitoring devices. It may be inconvenient to have such infants in an incubator where body temperature is maintained by means of heated air and convection. Accordingly, incubators consisting of radiant warmers have been developed. The infant is placed on a mattress under a radiant heating element. Low walls surround the mattress so that there is no risk of the infant falling off the device, but the remainder of the area surrounding the infant is open to allow access to the patient. This is in contrast to the warm-air convection incubator, where the infant must be accessed through arm ports.

The radiant element consists of an electric heating element such as a coil of the high-resistivity wire used in electric space heaters. A heat reflector above this heater helps focus the radiant energy onto the mattress at the infant's location. The current through the heating element must be carefully controlled so that sufficient heat to maintain the infant's temperature is provided but overheating is avoided. Proportional control is generally used to achieve this. A thermistor is placed on the exposed chest or abdomen of the infant to measure the skin surface temperature. Because this thermistor could be heated by the radiant source, thereby giving an erroneous temperature, it is attached to the infant via a special tape that consists of a foam pad with a shiny, metallized surface to reflect the incoming thermal radiation. Not only does this pad hold the thermistor in place, but the foam serves as a thermal insulator, and the reflecting surface minimizes the direct effect of radiation on the thermistor.

Premature and, in some cases, full-term infants often develop respiratory distress or other anomalies. These infants can stop breathing and such apnea is life-threatening. A tap on the incubator wall or a slap on the foot is usually all such infants need to remind them to take another breath. Thus it is important to detect when apnea has persisted for a given period of time and to alert clinical personnel so that they can immediately attend to the infant. Various types of apnea monitors have been developed to carry out this function. They detect ventilation by one of the following methods (Chapter 9): transthoracic impedance, movement of the baby on a displacement detecting pad, or movement of the baby's chest wall. An alarm is sounded when respiratory activity ceases for periods of greater than a preset time, in the range of 15 to 30 s.

13.8 DRUG DELIVERY DEVICES

One of the principal activities of clinical medicine is the pharmacologic treatment of diseases. Traditionally, this has been carried out by administering drugs and other materials in individual doses by the oral, subcutaneous, intramuscular, or intravenous route. In some cases, however, these methods of administration are not satisfactory, because drug levels vary from one administration to the next. For drugs the therapeutic levels of which are close to their toxic levels, this type of administration results in either toxic side effects or suboptimal therapy. Methods of using physical devices to administer drugs continuously within a narrow therapeutic range have been developed to overcome these problems.

DRUG INFUSION PUMPS

Controlled infusion of fluids and drugs is a well-established technique in the hospital. Intravenous (IV) therapy from a gravity-fed fluid reservoir is

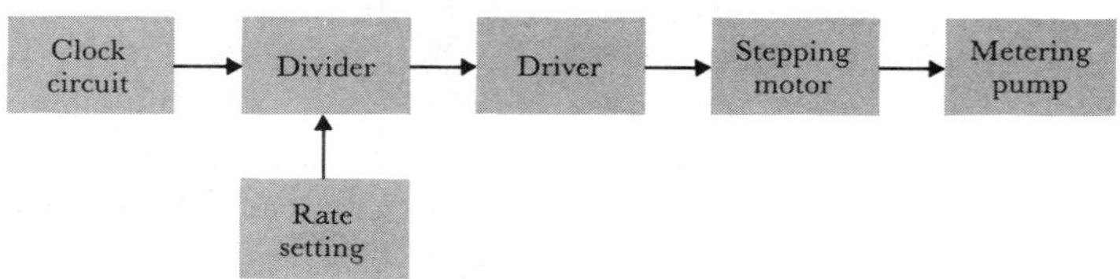

Figure 13.17 **Block diagram of the electronic control system for a fluid or drug delivery pump**

extensively used in cases ranging from the treatment of severe infections to intensive-care in the operating room and postoperatively. These fluids are frequently routed by gravity feed through some type of constriction that controls the total amount of fluid administered. Although this method is widely used in hospitals and is relatively inexpensive, it is limited in that careful control of the amount of fluid or drug administered is not possible. In those cases where precise control is necessary, some type of volumetric controller or a complete pumping system must be used.

In the case of a volumetric controller, the amount of fluid administered through a gravity-feed system is carefully regulated. Most clinical intravenous fluid administration sets include a drip chamber where the fluid passes through a fixed-diameter capillary tube and forms drops that drip from the tube into the chamber when they reach a certain volume. The clinician adjusts the flow rate to produce a certain number of drops per unit of time. A simple controller can achieve the same function by photoelectrically observing each drop as it interrupts a light beam passing from a light-emitting diode source to a photodetector such as a phototransistor. A device containing the light source and light detector can be clamped on the drip chamber of a conventional IV administration set and used to control a valve located between the fluid reservoir and the drip chamber. The clinician determines the number of drops required per unit time and sets the controller to deliver drops at the required rate.

Another approach is to use a peristalsis pump such as a roller pump (see Figure 13.12) to allow only a certain amount of fluid to pass from a gravity-fed reservoir to the patient. The inside diameter of the tubing and the propagation velocity of the peristalsis wave determine the amount of fluid administered.

There are other methods that can be used to meter the amount of fluid administered (Katona, 1982; 1988). All these systems require some type of electronic control to maintain steady infusion at a fixed, predetermined rate. Though earlier systems utilized a dc motor to drive a piston pump at a rate determined by a gear system, stepping motors are used today, and the angular velocity is controlled by digital electronics. Such a system is illustrated in Figure 13.17. The object of this system is to provide a series of pulses to the stepping motor at a precise frequency so that it drives the metering pump to supply the fluid at the desired rate. The pulse rate required for the desired infusion rate is set by the operator, who adjusts the rate-setting control. The advantage of using a stepping motor and this system is that the motor advances by a known amount for each applied pulse. Thus, by precisely controlling the number of pulses applied to the stepping motor, it is possible to control precisely its angular displacement. And by precisely controlling the rate at which pulses are applied to the motor, it is possible to control its angular velocity. This degree of control is not possible with open loop analog motors.

Infusion systems can be used in two general ways. For an open-loop device, the operator sets the desired infusion rate, and the fluid or drug is delivered at that rate until the setting is changed. The closed-loop system involves operating the pump in such a way as to keep a physiological variable as close as possible to a desired value. An example of such a system is the use of controlled infusion of the drug sodium nitroprusside, a vasodilator, to control the blood pressure of hypertensive patients during surgery and postoperatively. A pressure sensor measures the blood pressure, and this information is fed into a control algorithm that determines the number of pulses sent to the stepping motor—and hence the rate at which it infuses the drug into the patient. Investigators have shown that the automatic control of blood pressure in this way is more efficacious than the manual control of blood pressure via the same chemical agent.

AMBULATORY AND IMPLANTABLE INFUSION SYSTEMS

Miniature infusion pumps have been developed for patients with special therapeutic needs. Insulin pumps that are small enough to be worn by a patient are routinely used to help diabetic patients control their blood glucose. These insulin-delivery systems are currently run open-loop, because there is no reliable, continuously operating glucose sensor available for such a system. The patient controls the rate of infusion, after taking into account his or her activity level, meals, and occasional self-administered serum glucose tests.

Miniature pumps that are only slightly larger than an implantable cardiac pacemaker have been developed for implantable drug delivery. These pumps apply a known pressure to a reservoir of the drug, and there is a high-resistance connection between the pump and the site where the drug is to be infused, which is usually a vein. This high-resistance connection is generally a long, thin capillary tube that is wound around the periphery of the pump. The constant pressure in the reservoir and the fixed resistance of the tube maintain a steady but slow rate of infusion of the drug into the venous circulation. Implantable pumps, therefore, utilize a concentrated form of the agent to be infused. Nevertheless, it is periodically necessary to refill the reservoir. This is done percutaneously by means of a needle that can enter the reservoir and refill it without any of the drug leaking into the surrounding tissues. This technique is no more uncomfortable to the patient than receiving an injection.

Implantable pumps have also been used for delivering drugs to a specific tissue. This technique has been found useful in cancer chemotherapy, because systemic administration of the cytotoxic agents used would have far more

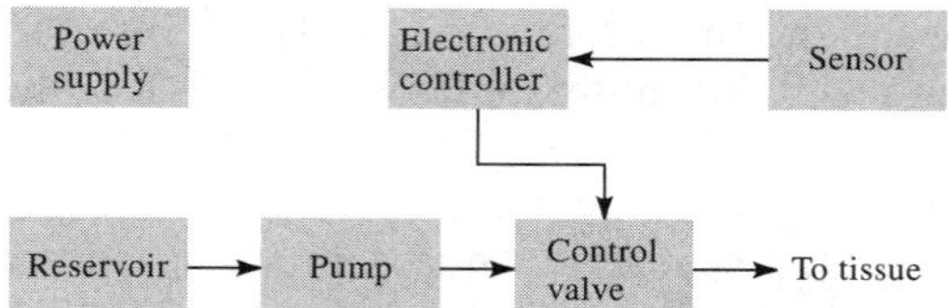

Figure 13.18 A block diagram of an implantable artificial pancreas showing the major components of the system.

serious side effects than the patient experiences when the agent is administered locally.

Implantable closed-loop drug delivery systems are of great interest for biomedical therapy because with the appropriate control algorithm they should be able to imitate the behavior of malfunctioning organs. The classic example of a closed-loop drug delivery system is the artificial pancreas. This device consists of an implantable pump of the type described in previous paragraphs. This pump, as shown in Figure 13.18, has a reservoir containing insulin and a control valve that determines the flow rate of insulin into the surrounding tissue from which it is picked up by the circulation. The valve in turn is controlled by an electronic control system which performs the control algorithm for the device. A glucose sensor determines the body's glucose level and serves as the input device for the control system. Elevated glucose levels cause the pump to dispense increased amounts of insulin, while reduced glucose levels have the opposite effect. Most of these components have been developed in implantable form, and miniature closed-loop insulin delivery systems should be quite feasible with the exception of one component: the sensor. Although various investigators have developed glucose sensors that function in vitro, reliable miniature in vivo sensors capable of measurements for many years are not available. Without this crucial component, the implantable artificial pancreas is constrained to being a dream for the future.

Long-term drug delivery from passive devices has been demonstrated and utilized in some products. The basic principle has been described and reduced to practice by Langer (1995) and consists of a polymer matrix in which the drug is dispersed that is implanted in tissue. The drug slowly leaches from this material and is taken up by the capillaries in the surrounding tissue whereupon it is dispersed throughout the body by the cardiovascular system. By choosing appropriate materials and forming them in a specific shape, the drug can be leached out to give a desired level in the body appropriate for the therapeutic situation. An example of such a device is a thin polymer cylinder that can be injected beneath the skin and slowly leaches the hormone progesterone to provide contraceptive protection lasting several months in women. Even though these devices do not require electronic control systems, and they must operate open loop, they have several clinical advantages due to their simplicity.

ANESTHESIA MACHINES

A special case of a controlled drug delivery system is the anesthesia machine. This device enables anesthesiologists and anesthetists to administer volatile anesthetic agents to patients in the operating room through their lungs. There are three sections to the typical anesthesia machine. The first is the gas supply and delivery system. Here oxygen and nitrous oxide from central hospital sources or small storage cylinders on the anesthesia machine are mixed in the desired proportions. Flow meters indicate the amount of each gas that is delivered, and the operator can adjust the flow rate to get the desired ratio and total volume.

The second section of the anesthesia machine is the vaporizer. In this section, pure oxygen or an oxygen–nitrous oxide mixture from the gas delivery system is bubbled through or passed over the volatile anesthetic agent in the liquid phase. The amount of anesthetic agent given is related to the flow rate of the gas through the vaporizer. The anesthesiologist or anesthetist controls this rate by adjusting the valves in a plumbing system and measuring, by means of flow meters, the flow through the vaporizer and the amount of gas that bypasses it.

The final section of the anesthesia machine is the patient breathing circuit. This section is responsible for delivery of the anesthesia-producing gases to the patient and removal of expired gases coming from the patient. This portion of the system is a closed circuit. That is, the gas administered to the patient is introduced via a one-way (check) valve through one section of tubing, and the expired gas passes through a different section of tubing, again via a one-way valve. Thus the expired gas is separated from the inspiratory line. The expired gas is passed through a carbon dioxide absorber to remove the carbon dioxide and is reintroduced into the inspiratory line. A reservoir bag is connected in the circuit to provide low-pressure gas storage and to enable the anesthesiologist or anesthetist to assist in ventilating the patient when necessary. Expiratory gas can also be removed from the patient breathing circuit and passed through a scavenging system to remove the anesthetic agent before the gas is vented to the atmosphere. The patient breathing circuit can be connected to a ventilator for those patients who need assistance in ventilation.

13.9 SURGICAL INSTRUMENTS

There are many devices that can be classified as surgical instruments; to consider all of them would require several volumes. There are, however, electric and electronic devices that are important in the surgical care of patients, in addition to those used for monitoring patients in the operating and recovery rooms. In the next two sections, we examine two of these: the electrosurgical unit and the laser.

ELECTROSURGICAL UNIT

Electric devices to assist in surgical procedures by providing cutting and hemostasis (stopping bleeding) are widely applied in the operating room. These devices are also known as electrocautery apparatuses. They can be used to incise tissue, to destroy tissue through desiccation, and to stop bleeding by causing coagulation of blood. The process involves the application of an RF spark between a probe and tissue to cause localized heating and damage to that tissue.

The basic electrosurgical unit is shown in Figure 13.19(a) (Gerhard, 1988). The high-frequency power needed to produce the spark comes from a high-power high-frequency generator. The power to operate the generator comes from a power supply, the output of which may in some cases be modulated to produce a waveform more appropriate for particular actions. In this case, a modulator circuit controls the output of the generator. The application of high-frequency power from the generator is ultimately controlled by the surgeon through a control circuit, which determines when power is applied to the electrodes to carry out a particular action. Often the output of energy from the high-frequency generator needs to be at various levels for various jobs. For this reason, a coupling circuit is inserted between the generator output and the electrodes to control this energy transfer.

The electric waveforms generated by the electrosurgical unit differ for its different modes of action. To bring about desiccation and coagulation, the device uses damped sinusoidal pulses, as shown in Figure 13.19(b). The RF sine waves have a nominal frequency of 250 to 2000 kHz and are usually pulsed at a rate of 120 per second. Open-circuit voltages range from 300 to 2000 V, and power into a 500-Ω load ranges from 80 to 200 W. The magnitude of both voltage and power depends on the particular application.

Cutting is achieved with a CW RF source, as shown in Figure 13.19(b). Often units cannot produce truly continuous waves, as shown in Figure 13.19(b), and some amplitude modulation is present. Cutting is done at higher frequency, voltage, and power, because the intense heat at the spark destroys tissue rather than just desiccating it, as is the case with coagulation. Frequencies range from 500 kHz to 2.5 MHz, with open-circuit voltages as high as 9 kV. Power levels range from 100 to 750 W, depending on the application.

The cutting current usually results in bleeding at the site of incision, and the surgeon frequently requires "bloodless" cutting. Electrosurgical units can achieve this by combining the two waveforms, as shown in Figure 13.19(b). The frequency of this *blended* waveform is generally the same as the frequency for the cutting current. For best results, surgeons prefer to operate at a higher voltage and power when they want bloodless cutting than when they want cutting alone.

Many different designs for electrosurgical units have evolved over the years. Modern units generate their RF waveforms by means of solid-state electronic circuits. Older units were based on vacuum tube circuits and even utilized a spark gap to generate the waveforms shown in Figure 13.19(b).

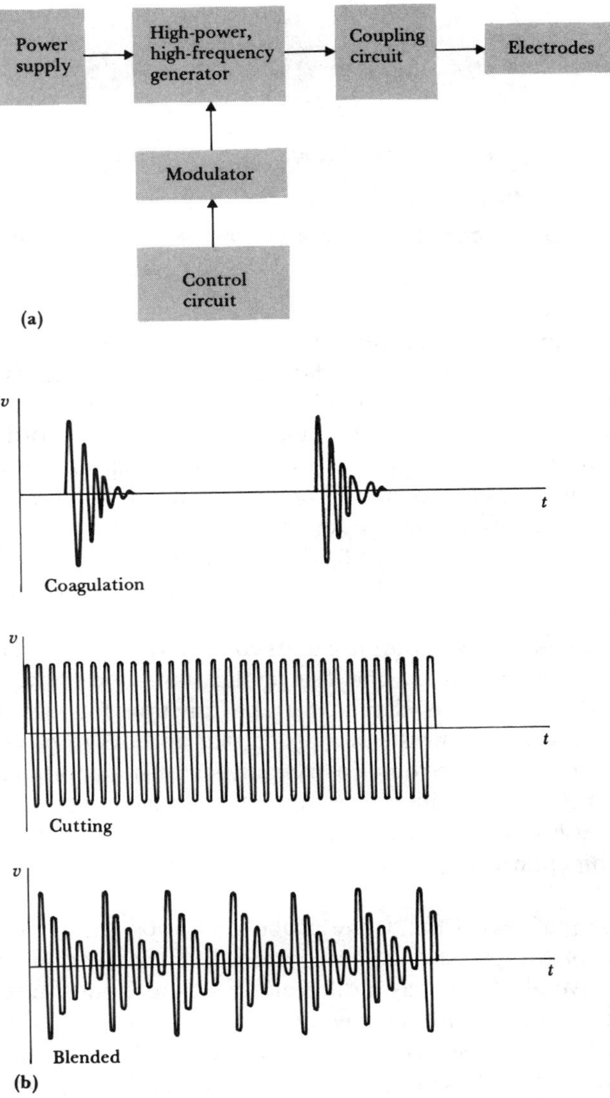

Figure 13.19 (a) Block diagram for an electrosurgical unit. High-power, high-frequency oscillating currents are generated and coupled to electrodes to incise and coagulate tissue. (b) Three different electric voltage waveforms available at the output of electrosurgical units for carrying out different functions.

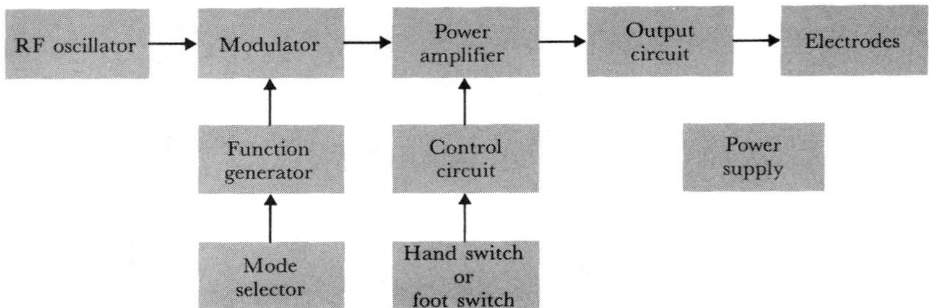

Figure 13.20 Block diagram of a typical electrosurgical unit

A block diagram of a typical electrosurgical unit is shown in Figure 13.20. The RF oscillator provides the basic high-frequency signal, which is amplified and modulated to produce the coagulation, cutting, and blended waveforms. A function generator produces the modulation waveforms according to the mode selected by the operator. The RF power output is turned on and off by means of a control circuit connected either to a hand switch on the active electrode or to a foot switch that can be operated by the surgeon. An output circuit couples the power generator to the active and dispersive electrodes. The entire unit derives its power from a power-supply circuit that is driven by the power lines.

Electrodes used with electrosurgical units come in various sizes and shapes, depending on the manufacturer and the application. The active electrode is a scalpel-like probe that is shaped for the function for which it is intended. The simplest form consists of a probe that appears to be similar to a test probe used with an electronic instrument such as a multimeter or an oscilloscope. A pointed metallic probe fits into an insulating handle and is held by the surgeon as one would hold a pencil. The hand switch located on the handle is momentarily depressed when the surgeon wants to apply power to the probe.

Whereas the purpose of the active probe is to apply energy to the local tissue at the tip of the probe and thereby to effect coagulation, cutting, or both, the dispersive electrode has a different function. It must complete the RF circuit to the patient without having current densities high enough to damage tissue. The simplest dispersive electrode is a large, reusable metal plate placed under the buttocks or back of the patient. Most procedures use a 70-cm^2 disposable dispersive electrode placed on the thigh. One type is like a disposable ECG electrode with a gel-soaked sponge backed by metal foil and surrounded by foam and pressure-sensitive adhesive. Another capacitive type has a thin Mylar insulator backed by foil and its entire face coated with pressure-sensitive adhesive. It is important that this electrode make good contact with the patient over its entire surface so that "hot spots" do not develop.

RADIOFREQUENCY (RF) CATHETER ABLATION

Extra-conductive pathways between the cardiac atria and ventricles may permit excitation to travel in circles and cause tachycardia. These extra pathway sites can be mapped by inserting a collapsed 64-electrode catheter into the heart chamber and expanding it into a basket to measure conduction times. Then a catheter is introduced to the sites and RF current at about 500 kHz heats the pathways to 50 °C to destroy (ablate) the conductive pathways. A temperature sensor at the electrode tip may be used to control the temperature (Huang, 1995).

13.10 THERAPEUTIC APPLICATIONS OF THE LASER

As we saw in Section 2.13, the laser makes available coherent light at high intensities that can be focused to a very fine point. When this focal point is projected on tissue and the tissue absorbs the radiation, the power may be great enough to vaporize immediately the tissue illuminated by the point, resulting in a coagulated incision. Thus the laser can be used as a surgical instrument in much the same way as the electrosurgical unit described in Section 13.9 (Auth, 1988). The laser beam requires powers of from 25 to 100 W to cut various tissues of the body ranging from skin to bone.

An important goal that must be achieved before lasers can be widely used in this application is quick and easy manipulation of the laser beam to the point where the surgeon wants it. Unlike the electrosurgical probe that can be connected to its generator by a flexible insulated lead wire, the laser must be coupled to its point of application through a system of mirrors and a lens. The powers required are high enough to make the laser too large to be manipulated itself, so this coupling scheme becomes important but cumbersome.

The laser has some important applications in other aspects of medicine. Because it can be used in the control of bleeding through photocoagulation and heating, the laser has been applied through a fiber-optic endoscope in the coagulation of bleeding gastric ulcers. In some cases, the power is sufficiently great to stop bleeding in experimentally induced lesions in dogs.

In ophthalmology, the laser has found an important therapeutic application. Laser photocoagulators are able to repair detached retinas much more quickly than conventional photocoagulators, thereby minimizing the risk of damage to the retina due to movement of the eyes. Such systems are now commercially available and are in use at many institutions. Lasers have also been investigated as diagnostic and therapeutic instruments in many centers. This work ranges from dentistry to oncology (the study and treatment of tumors). Laser radiation is even being investigated in the clinical laboratory for analysis of tissue and fluid specimens taken from patients.

PROBLEMS

13.1 An asynchronous cardiac pacemaker operates at a rate of 70 pulses/min. These pulses are of 2-ms duration and have an amplitude of 5 V when driving a 500-Ω load.

a. What is the total energy supplied to this load over a 10-year period?

b. Suppose that 35% of the energy from the power supply goes into the output pulses. What must be the capabilities of the power supply to operate the pacemaker for 10 years?

c. Assume that the power supply is two 2.8-V lithium cells. What must the milliampere-hour capacity of each cell be to operate this pacemaker for 10 years?

d. Even though a pacemaker such as this would have a 10-year capacity in its power supply, it is found that when it is implanted, the power supply becomes exhausted in a little over 5 years. Can you suggest some of the reasons why this theoretical calculation disagrees with actual practice?

13.2 An important problem in the clinical application of pacemakers is to determine when a given pacemaker is about to fail. Then surgeons may replace it before there is a failure that puts the patient at extreme risk. Describe a technique that could be used to determine the status of the battery of an implanted pacemaker that cannot be directly contacted with electric probes connected to test instruments. It is possible to place electrodes on the surface of the patient's body to observe signals from the pacemaker. Assume that you can use any of the test equipment available to the medical electronics engineer.

13.3 There are some limitations on the use of synchronous cardiac pacemakers, especially when the patient is in the vicinity of strong electromagnetic fields, such as those produced by nearby microwave ovens, powerful radio transmitters, and radar stations. Discuss these limitations and describe the mechanisms involved.

13.4 For the demand type of synchronous pacemaker shown in Figure 13.3, draw an appropriate waveform diagram that describes the operation of this pacemaker with and without spontaneous venticular beats.

13.5 How might the atrial-synchronous pacemaker shown in Figure 13.4 be modified to enable it to operate asynchronously in the absence of atrial contractions? Draw a block diagram for the modified circuit and explain its operation.

13.6 Compare the power requirements for a bladder stimulator that produces 5-V pulses—each having a duration of 2 ms, at a frequency of 100 pulses/s—with those of the cardiac pacemaker described in Problem 13.1. Do you think the power requirements of this stimulator make battery operation possible?

13.7 The circuit of the RF-powered bladder stimulator of Figure 13.6 produces pulses of RF energy that have a voltage of 10 V rms as seen across the primary coil. The source impedance of the generator may be considered to be very low. The secondary coil has 1.2 times the number of turns of the primary and is coupled to the primary with a coupling coefficient of 0.5. What is the pulse amplitude seen across the electrodes when the tissue appears as a 500-Ω resistive load? What is the load impedance that the RF oscillator sees?

13.8 The effective load resistance seen by the electrodes of an electric stimulator used for pain suppression can increase with time following the implantation of the stimulator and the healing of the site. Discuss the way this affects the efficacy of the stimulator. List the precautions that can be taken in the design of the circuit to minimize these problems.

13.9 It was stated in the text that there is often a discrepancy between the stored energy in a capacitive-discharge cardiac defibrillator and the energy that is actually given to the patient during discharge. Explain what you believe to be reasons for this discrepancy and what can be done to minimize it.

13.10 Capacitor-discharge cardiac defibrillators can be constructed with relatively small capacitances charged to high voltages or with comparatively large capacitances charged to lower voltages. Contrast the two types of defibrillators. List and discuss the advantages and disadvantages of each.

13.11 If a capacitive-discharge cardiac defibrillator is discharged on a patient when the electrodes are not firmly in contact with the patient's chest wall, serious complications can develop. Describe what these are from the standpoint of what happens to the patient and the waveform produced by the defibrillator.

13.12 What precautions should a clinician take when it is necessary to defibrillate a patient who has an implanted pacemaker of either the synchronous or the asynchronous type?

13.13 The 20-μF capacitor of the defibrillator circuit shown in Figure 13.9 is charged to an energy of 200 J. When the electrodes are attached to the patient, a 50-Ω resistive load is seen.

a. What value of inductance L is required for critical damping?

b. What is the peak current passing through the patient during the discharge under these conditions?

13.14 When it is necessary to defibrillate a patient, the clinician must be assured that the defibrillator will operate properly. Therefore, it is necessary for the clinical engineer to have an aggressive program of testing and preventive maintenance of all cardiac defibrillators in the hospital. The key to this program is a method of evaluating the performance of each defibrillator. Design an instrument that is capable of measuring what you believe to be the key parameters of a cardiac defibrillator to assure the clinician that it is operating properly.

13.15 Instrumentation is needed to control a balloon-type cardiac-assist device. It must detect the R wave of the ECG and use this to control the mechanism for deflation of the balloon, followed by reinflation. Design the necessary instrumentation and control system in block-diagram form.

13.16 Design an instrumentation system for use with a pump-oxygenator system such as that shown in Figure 13.12. This system should monitor the oxygen saturation and pressure of the blood being reinfused into the body, as well as the pressure of the venous-return blood. Specify the sensors to be used in the instrumentation system, and explain why your choice is the most desirable. Show the electronics in block-diagram form, and include alarm circuitry to detect elevated arterial and venous pressures as well as reduced oxygen saturation.

13.17 An artificial-kidney system must be protected from leaks across the dialyzing membrane. Design an instrumentation system that can provide this protection, and describe its operation.

13.18 Redesign the electronics for the continuous-positive-airway-pressure monitor shown in Figure 13.15 so that only a single ADC and associated circuitry are required.

13.19 A capacitive electrosurgical dispersive electrode has a Mylar insulator that is 0.002 in. thick and has an area of 70 cm^2. Find the relative dielectric constant for plastic, and calculate the impedance at 500 kHz.

13.20 A broken lead wire on the back-plate or thigh-plate electrode of an electrosurgical unit can produce serious burns. Can you describe a way to eliminate this problem using an ancillary electronic circuit?

REFERENCES

Auth, D. C., "Laser scalpel." in J. G. Webster (ed.), *Encyclopedia of Medical Devices and Instrumentation.* New York: Wiley, 1988, p. 1717–1725.

Bocka, J. J., "External transcutaneous pacemakers." *Ann. Emerg. Med.,* 1989, 18, 1280–1286.

Brindley, G. S., and W. S. Lewin, "The sensations produced by electrical stimulation of the visual cortex." *J. Physiol.,* 1968, 196, 479–493.

Brindley, G. S., "Sensations produced by electrical stimulation of the occipital poles of the cerebral hemispheres, and their use in constructing visual prostheses." *Ann. Royal Coll. Surg. England,* 1970, 47, 106–108.

Brindley, G. S., C. E. Polkey, and R. D. Rushton, "Sacral anterior root stimulators for bladder control in paraplegia." *Paraplegia,* 1982, 20, 365.

Burton, C. V., "Pain suppression through peripheral nerve stimulation," in W. S. Fields and L. A. Leavitt (eds.). *Neural Organization and Its Relevance to Prosthetics.* New York: Intercontinental, 1973, pp. 241–250.

Bush, W. H., and G. E. Brannen, "Lithotripsy," in J. G. Webster (ed.), *Encyclopedia of Medical Devices and Instrumentation.* New York: Wiley, 1988, pp. 1806–1820.

Cicman, J. H., V. F. Skibo, and J. M. Yoder, "Anesthesia systems. Part II: Operating principles of fundamental components." *J. Clin. Monitoring,* 1993, 9, 104–111.

Chang, H. K., "Mechanisms of gas transport during ventilation by high-frequency oscillation." *J. Appl. Physiol.,* 1984, 56, 553–563.

Cook, A. M., and J. G. Webster, *Therapeutic Medical Devices: Application and Design.* Englewood Cliffs, N.J.: Prentice-Hall, 1982.

Creasey, G. H., "Electrical stimulation of sacral roots for micturation after spinal cord injury." *Urol. Clin. N. Am.,* 1993, 20, 505–515.

DeVries, W. C., and L. D. Joyce, "The artificial heart." *Ciba Clinical Symposia,* 35, 1983.

Dorsch, J. A., and S. E. Dorsch, *Understanding Anesthesia Equipment: Construction, Care and Complications,* 2nd ed. Baltimore: Williams and Wilkins, 1984.

Dorson, W. J., and J. B. Loria, "Heart-lung machines," in J. G. Webster (ed.), *Encyclopedia of Medical Devices and Instrumentation.* New York: Wiley, 1988, pp. 1440–1457.

Ehrenwerth, J., and J. B. Eisenkraft, *Anesthesia Equipment: Principles and Applications.* St. Louis, Mosby, 1993.

Ellenbogen, K. (ed.), *Cardiac Pacing,* London: Blackwell Scientific, 1992.

Freehafer, A. A., P. H. Peckham, and M. W. Keith, "New concepts on treatment of the upper limb in the tetraplagic surgical restoration and functional neuromuscular stimulation." *Hand Clinics,* 1988, 4, 563–574.

Furman, S., and D. J. W. Escher, *Principles and Techniques of Cardiac Pacing.* New York: Harper & Row, 1970.

Geddes, L. A., "Electrical ventricular defibrillation," in D. W. Hill and B. W. Watson (eds.), *IEE Medical Electronics Monographs, 18–22.* Stevenage, Eng.: Peter Peregrinus, 1976, pp. 42–72.

Geddes, L. A., W. A. Tacker, J. Rosborough, P. Cabler, R. Chapman, and R. Rivers, "The increased efficacy of high-energy defibrillation." *Med. Biol. Eng.,* 1976, 14, 330–333.

Gerhard, G. C., "Electrosurgical technology: Quo vadis?" *IEEE Trans. Biomed. Eng.,* 1984, BME-31, 787–791.

Gerhard, G. C., "Electrosurgical unit," in J. G. Webster (ed.), *Encyclopedia of Medical Devices and Instrumentation.* New York: Wiley, 1988, pp. 1180–1203.

Girvin, J. P., "Current status of artificial vision by electrocortical stimulation." *Can. J. Neurol. Sci.,* 1988, 15, 58–62.

Greatbatch, W., J. Lee, W. Mathias, M. Eldridge, J. Moser, and A. Schneider, "The solid-state lithium battery." *IEEE Trans. Biomed. Eng.,* 1971, BME-18, 317.

Greatbatch, W., and L. J. Seligman, "Pacemakers," in J. G. Webster (ed.), *Encyclopedia of Medical Devices and Instrumentation.* New York: Wiley, 1988, pp. 2175–2203.

Hamilton, L. H., "Ventilators, high frequency," in J. G. Webster (ed.), *Encyclopedia of Medical Devices and Instrumentation.* New York: Wiley, 1988, pp. 2858–2864.

Hampers, C. L., E. Schubak, E. G. Lowrie, and J. M. Lazarus, "Clinical engineering in hemodialysis and anatomy of an artificial kidney unit," in *Long-term Hemodialysis.* New York: Grune & Stratton, 1973.

Hitchcock, E., "Development of a visual prosthesis." *Appl. Neurophysiol.,* 1982, 45, 21–31.

Huang, S. K. S. (ed.), *Radiofrequency Catheter Ablation of Cardiac Arrhythmias. Basic Concepts and Clinical Applications.* Armonk, NY: Futura, 1995.

Humayun, M. S., E. de Juan, G. Dagnelie, R. J. Greenberg, R. H. Probst, and D. H. Philips, "Visual perception elicited by electrical stimulation of retina in blind humans." *Arch. Ophthalmol.,* 1996, 114, 40–46.

Jaron, D., "Intraaortic balloon pump," in J. G. Webster (ed.), *Encyclopedia of Medical Devices and Instrumentation.* New York: Wiley, 1988, pp. 1656–1665.

Jarvik, R. K., "The total artificial heart," *Sci. Am.,* 1981, 244(1), 74–80.

Judy, M. M., "Biomedical Lasers," in Bronzino, J. D. (ed.), *The Biomedical Engineering Handbook.* Boca Raton, Fl.: CRC Press, 1995, pp. 1333–1345.

Katona, P. G., "Automated control of physiological variables and clinical therapy." *CRC Crit. Rev. Bioeng.,* 1982, 8, 281–310.

Katona, P. G., "Drug infusion systems," in J. G. Webster (ed.), *Encyclopedia of Medical Devices and Instrumentation.* New York: Wiley, 1988, pp. 971–980.

Keith, M. W., P. H., Peckham, G. B. Thrope, et al., "Implantable functional neuromuscular stimulation in the tetraplegic hand." *J. Hand Surg.,* 1989, 14A, 524.

Langer, R., "1994 Whitaker Lecture: polymers for drug delivery and tissue engineering." *Ann. Biomed. Eng.,* 1995, 23, 101–111.

Light, J. K., "Electrical stimulation to modify detrusor function." *Adv. Neurol.,* 1993, 63, 303–309.

Loeb, G. E., C. J. Zamin, J. H. Schulman, and P. R. Troyk, "Injectable microstimulator for functional electrical stimulation." *Med. Biol. Eng. Comput.,* 1991, 29, NS13-NS19.

Marsolais, E. B., and R. Kobetic, "Functional electrical stimulation for walking in paraplegics." *J. Bone Joint Surg.,* 1987, 69, 728–733.

McCarthy, P. M., "HeartMate implantable left ventricular assist device: bridge to transplantation and future applications." *Ann. Thoracic Surg.,* 1995, 59, S46–51.

McPherson, S. P. and C. B. Spearman, *Respiratory Therapy Equipment, 4th Edition,* St. Louis: C. V. Mosby, 1990.

Nardin, M., B. Ziaie, J. Von Arx, A. Coghlan, M. Dokmeci and K. Najafi, "An inductively powered microstimulator for functional neuromuscular stimulation," in Cristalli, C., C. J.

Amlaner, and M. R. Neuman (eds.), *Biotelemetry XIII: Proc. 13th Int. Symp. Biotelemetry,* Strasbourg: International Society on Biotelemetry, 1995, pp. 99–104.

Pearce, J. A., *Electrosurgery.* New York: John Wiley, 1986.

Peckham, P. H., "Functional electrical stimulation: Current status and future prospects of applications to neuromuscular system in spinal cord injury." *Paraplegia,* 1987, 25, 274–288.

Peckham, P. H., "Functional electrical stimulation," in J. G. Webster (ed.), *Encyclopedia of Medical Devices and Instrumentation.* New York: Wiley, 1988, pp. 1331–1352.

Peckham, P. H. and G. H. Creasey, "Neuroprostheses: clinical application of functional electrical stimulation in spinal cord injury." *Paraplegia,* 1992, 30, 96–101.

Plaisier, P. W., R. L. van der Hul, O. T. Terpstra, and H. A. Bruining, "Current role of extracorporeal shockwave therapy in surgery." *Br. J. Surg.,* 1994, 81, 174–181.

Powers, S. K., "Laser surgery," in J. G. Webster (ed.), *Encyclopedia of Medial Devices and instrumentation.* New York: Wiley, 1988, pp. 1725–1742.

Schaldach, M., *Electrotherapy of the Heart.* Berlin: Springer, 1992.

Scott, F. B., W. E. Bradley, and G. W. Timm, "Electromechanical restoration of micturition," in W. S. Fields and L. A. Leavitt (eds.), *Neural Organization and Its Relevance to Prosthetics.* New York: Intercontinental, 1973, pp. 311–318.

Seligman, L., "Stimulators, physiological," in J. G. Webster (ed.), *Encyclopedia of Medical Devices and Instrumentation.* New York: Wiley, 1988, pp. 2704–2715.

Shealy, C. N., "Pain suppression through posterior column stimulation," in W. S. Fields and L. A. Leavitt (eds.), *Neural Organization and Its Relevance to Prosthetics.* New York: Intercontinental, 1973, pp. 251–260.

Singer, I. (ed.). *Implantable Cardioverter-Defibrillator.* Mt. Kisco, New York: Futura Publishing, 1994.

Smith, H. J., and N. E. Fearnot, "Concepts of rate-responsive pacing." *IEEE Eng. Med. Biol. Mag.,* 1990, 9(2), 32–35.

Spelman, F. A., "Cochlear prosthesis," in J. G. Webster (ed.), *Encyclopedia of Medical Devices and Instrumentation.* New York: Wiley, 1988, pp. 720–727.

Susset, J. G., "The electrical drive of the urinary bladder and sphincter," in W. S. Fields and L. A. Leavitt (eds.), *Neural Organization and Its Relevance to Prosthetics.* New York: Intercontinental, 1973, pp. 319–342.

Sweet, W. H., and J. G. Wepsic, "Electrical stimulation for suppression of pain in man," in W. S. Fields and L. A. Leavitt (eds.), *Neural Organization and Its Relevance to Prosthetics.* New York: Intercontinental, 1973, pp. 219–240.

Szeto, A. Y. J., and J. K. Nyquist, "Transcutaneous electrical nerve stimulation for pain control." *IEEE Eng. Med. Biol. Mag.,* 1983, 2(4), 14.

Tacker, W. A. Jr. (ed.). *Defibrillation of the Heart: ICDs, AEDs, and Manual.* St. Louis: Mosby-Yearbook, 1994.

Tacker, W. A., Jr., "Defibrillators, electrical," in J. G. Webster (ed.). *Encyclopedia of Medical Devices and Instrumentation.* New York: Wiley, 1988, pp. 939–944.

Tacker, W. A., and L. A. Geddes, *Electrical Defibrillation.* Boca Raton, FL: CRC Press, 1980.

Troyk, P. R., M. A. K. Schwan, "Closed-loop class E transcutaneous power and data link for micro implants." *IEEE Trans. Biomed. Eng.* 1992, 39, 589–599.

Vodovnik, L., A. Kralj, U. Stanic, R. Acimovic, and N. Gros, "Recent applications of functional electrical stimulation to stroke patients in Ljubljna." *Clinical Orthopedics,* 1978, 131, 64–70.

Webster, J. G. (ed.). *Design of Cardiac Pacemakers.* Piscataway, NJ: IEEE Press, 1995.

Yared, S. F., and W. C. DeVries, "Heart, total artificial," in J. G. Webster (ed.), *Encyclopedia of Medical Devices and Instrumentation.* New York: Wiley, 1988, pp. 1429–1440.

Ziaie, B., M. Nardin, J. Von Arx, and K. Najafi, "A single channel implantable microstimulator for functional neuromuscular stimulation (FNS)," *Proc. 7th Int. Conf. Solid State Sensors Actuators,* 1993, Yokohama, Japan, pp. 450–453.

14

ELECTRICAL SAFETY

Walter H. Olson

Medical technology has substantially improved health care in all medical specialties and has reduced morbidity and mortality for critically ill patients. Nevertheless, the increased complexity of medical devices and their utilization in more procedures now result in about 10,000 device-related patient injuries in the United States each year. Most of these injuries are attributable to improper use of a device as a result of inadequate training and lack of experience. Medical personnel rarely read user manuals until a problem has occurred. Furthermore, all medical devices eventually fail, so engineers must develop fail-safe designs.

The safe design and the safe use of medical instrumentation are broad subjects that involve nearly all medical procedures, every conceivable form of energy, and the familiar concept that everything that *can* go wrong eventually *will* go wrong. Medical procedures usually expose the patient to more hazards than the typical home or workplace, because in medical environments the skin and mucous membranes are frequently penetrated or altered, and because there are many sources of potentially hazardous substances and energy forms that could injure either the patient or the medical staff. These sources include fire, air, earth, water, chemicals, drugs, microorganisms, vermin, waste, sound, electricity, natural and unnatural disasters, surroundings, gravity, mechanical stress, and people responsible for acts of omission and commission, not to mention radiation from x rays, ultrasound, magnets, ultraviolet light, microwaves, and lasers (Dyro, 1988). Although this chapter focuses on electrical safety, it is important to recognize that there are also many other aspects of medical instrumentation safety.

In the 1980s, many minimum performance standards were written for most medical devices. Issues for the 1990s include inappropriate use of electrical connectors, sterilization efficacy, medical waste, and laser safety.

In this final chapter we focus on electrical safety and discuss the physiological effects of electric current, shock hazards, methods of protection, electrical-safety standards, and electrical-safety testing procedures. Our objectives are to understand the possible hazards and to learn how safety features can be incorporated into the design of medical instruments we have studied in previous chapters.

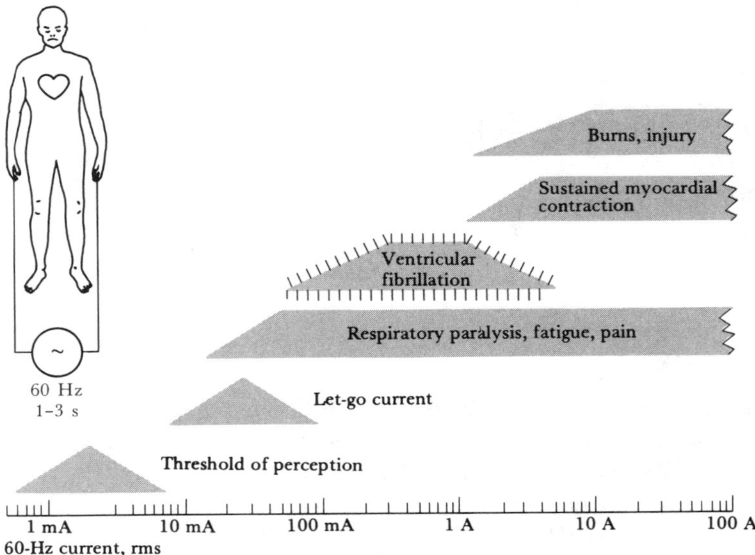

Figure 14.1 Physiological effects of electricity Threshold or estimated mean values are given for each effect in a 70-kg human for a 1- to 3-s exposure to 60-Hz current applied via copper wires grasped by the hands.

14.1 PHYSIOLOGICAL EFFECTS OF ELECTRICITY

For a physiological effect to occur, the body must become part of an electric circuit. Current must enter the body at one point and leave at some other point. The magnitude of the current is equal to the applied voltage divided by the sum of the series impedances of the body tissues and the two interfaces at the entry points. The largest impedance is often the skin resistance at the contact surface. Three phenomena can occur when electric current flows through biological tissue: (1) electric stimulation of excitable tissue (nerve and muscle), (2) resistive heating of tissue, and (3) electrochemical burns and tissue damage for direct current and very high voltages.

Let us now discuss the psychophysical and physiological effects that occur in humans as the magnitude of applied electric current progressively increases. The chart in Figure 14.1 shows the approximate range of currents needed to produce each effect when 60-Hz current is applied for 1 to 3 s via AGW No. 8 copper wires that a 70-kg human holds in each hand. Then, in the section that follows, we will examine the effect of each of these conditions (weight of the individual and so on).

THRESHOLD OF PERCEPTION

For the conditions just stated, when the local current density is large enough to excite nerve endings in the skin, the subject feels a tingling sensation.

Current at the *threshold of perception* is the minimal current that an individual can detect. This threshold varies considerably among individuals and with the measurement conditions. When someone with moistened hands grasps small copper wires, the lowest thresholds are about 0.5 mA at 60 Hz. Thresholds for dc current range from 2 to 10 mA, and slight warming of the skin is perceived.

LET-GO CURRENT

For higher levels of current, nerves and muscles are vigorously stimulated, and pain and fatigue eventually result. Involuntary contractions of muscles or reflex withdrawals by a subject experiencing any current above threshold may cause secondary physical injuries, such as might result from falling off a ladder. As the current increases further, the involuntary contractions of the muscles can prevent the subject from voluntarily withdrawing. The *let-go current* is defined as the maximal current at which the subject can withdraw voluntarily. The minimal threshold for the let-go current is 6 mA.

RESPIRATORY PARALYSIS, PAIN, AND FATIGUE

Still higher currents cause involuntary contraction of respiratory muscles severe enough to bring about asphyxiation if the current is not interrupted. During let-go experiments, respiratory arrest has been observed at 18 to 22 mA (Dalziel, 1973). Strong involuntary contractions of the muscles and stimulation of the nerves can be painful and cause fatigue if there is long exposure. (Today's human-subjects research committees probably would not approve these experiments.)

VENTRICULAR FIBRILLATION

The heart is susceptible to electric current in a special way that makes some currents particularly dangerous. Part of the current passing through the chest flows through the heart. If the magnitude of the current is sufficient to excite only part of the heart muscle, then the normal propagation of electric activity in the heart muscle is disrupted. If the cardiac electric activity is sufficiently disrupted, the heart rate can rise to 300 beats per minute as reentrant wavefronts of depolarization randomly sweep over the ventricles. The pumping action of the heart ceases and death occurs within minutes.

This rapid, disorganized cardiac rhythm is called *ventricular fibrillation,* and unfortunately, it does not stop when the current that triggered it is removed. Ventricular fibrillation is the major cause of death due to electric shock. The threshold for ventricular fibrillation for an average-sized human varies from about 75 to 400 mA. Normal rhythmic activity returns only if a brief high-current pulse from a defibrillator is applied to depolarize all the cells of the heart muscle simultaneously. After all the cells relax together, a normal rhythm usually returns. In the United States, approximately 1000 deaths per year occur in accidents that involve cord-connected appliances.

SUSTAINED MYOCARDIAL CONTRACTION

When the current is high enough, the entire heart muscle contracts. Although the heart stops beating while the current is applied, a normal rhythm ensues when the current is interrupted, just as in defibrillation. Data from ac-defibrillation experiments on animals show that minimal currents for complete myocardial contraction range from 1 to 6 A. No irreversible damage to the heart tissue is known to result from brief applications of these currents.

BURNS AND PHYSICAL INJURY

Very little is known about the effects of currents in excess of 10 A, particularly for currents of short duration. Resistive heating causes burns, usually on the skin at the entry points, because skin resistance is high. Voltages greater than 240 V can puncture the skin. The brain and other nervous tissue lose all functional excitability when high currents pass through them. And excessive currents may stimulate muscular contractions that are strong enough to pull the muscle attachment away from the bone.

14.2 IMPORTANT SUSCEPTIBILITY PARAMETERS

The physiological effects previously described are for an average 70-kg human and for 60-Hz current applied for 1 to 3 s to moistened hands grasping a No. 8 copper wire. The current needed to produce each effect depends on all these conditions, as explained below. Safety considerations dictate thinking in terms of minimal rather than average values for each condition.

THRESHOLD AND LET-GO VARIABILITY

Figure 14.2 shows the variability of the threshold of perception and the let-go current for men and women (Dalziel, 1973). On this plot of percentile rank versus rms current in milliamperes, the data are close to the straight lines shown, so a Gaussian distribution may be assumed. For men, the mean value for the threshold of perception is 1.1 mA; for women, the estimated mean is 0.7 mA. The minimal threshold of perception is 500 μA. When the current was applied to ECG gel electrodes, the threshold of perception averages only 83 μA with a range of 30 to 200 μA (Tan and Johnson, 1990).

Let-go currents also appear to follow Gaussian distributions, with mean let-go currents of 16 mA for men and 10.5 mA for women. The minimal threshold let-go current is 9.5 mA for men and 6 mA for women. Note that the range of variability for let-go current is much greater than the range for threshold-of-perception current.

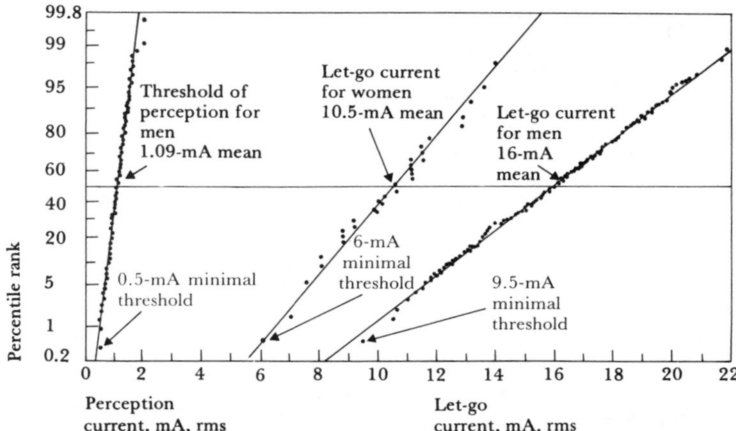

Figure 14.2 Distributions of perception thresholds and let-go currents These data depend on surface area of contact (moistened hand grasping AWG No. 8 copper wire). (Replotted from C. F. Dalziel, "Electric Shock," *Advances in Biomedical Engineering,* edited by J. H. U. Brown and J. F. Dickson III, 1973, 3, 223–248.)

FREQUENCY

Figure 14.3 shows a plot of let-go current versus frequency of the current. Unfortunately, the minimal let-go currents occur for commercial power-line frequencies of 50 to 60 Hz. For frequencies below 10 Hz, let-go currents rise, probably because the muscles can partially relax during part of each cycle. And at frequencies above several hundred hertz, the let-go currents rise again.

DURATION

A single electric stimulus pulse can induce ventricular fibrillation if it is delivered during the vulnerable period of cardiac repolarization that corresponds to the T wave on the ECG. For large-amplitude electric transients less than 100 μs in duration applied directly to the heart, the stimulation threshold approaches a constant charge transfer density of 3.5 $\mu C \cdot cm^{-2}$. The stimulation current threshold I_t is inversely related to the pulse duration t by the well-known strength–duration equation

$$I_t = I_r \left(1 + \frac{\tau}{t}\right) \tag{14.1}$$

where I_r is the constant rheobase current and τ is the chronaxie time constant. For normal hearts, the ratio of the fibrillation stimulation threshold to the single-beat stimulation threshold is 20:1 to 30:1 for electrodes on the heart

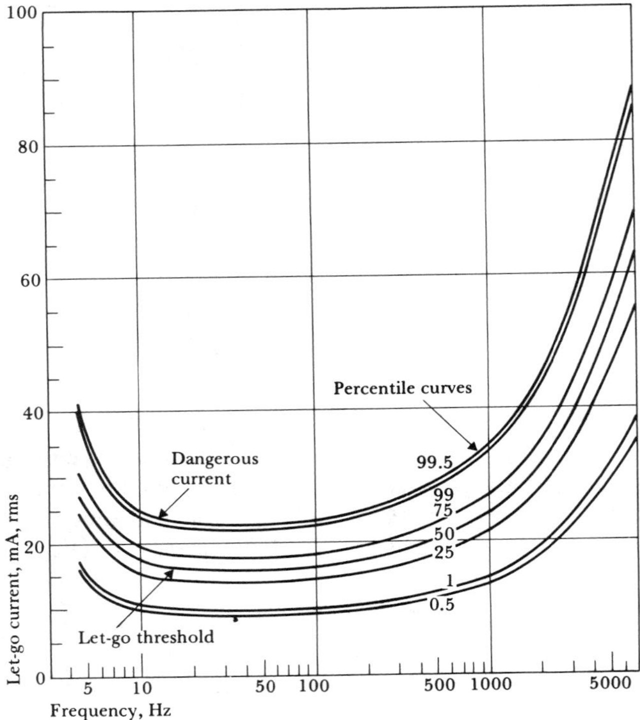

Figure 14.3 Let-go current versus frequency Percentile values indicate variability of let-go current among individuals. Let-go currents for women are about two-thirds the values for men. (Reproduced, with permission, from C. F. Dalziel, "Electric Shock," *Advances in Biomedical Engineering,* edited by J. H. U. Brown and J. F. Dickson III, 1973, 3, 223–248.)

and 10:1 to 15:1 for chest surface electrodes (Geddes, 1986). For 60-Hz current applied to the extremities, the fibrillation threshold increases sharply for shocks that last less than about 1 s, as shown in Figure 14.4. Shocks must last long enough to take place during the vulnerable period that occurs during the T wave in each cardiac cycle.

BODY WEIGHT

Several studies using animals of various sizes have shown that the fibrillation threshold increases with body weight. Fibrillating current increases from 50 mA rms for 6 kg dogs to 130 mA rms for 24 kg dogs. These findings deserve more study, because they are used to extrapolate fibrillating currents for humans.

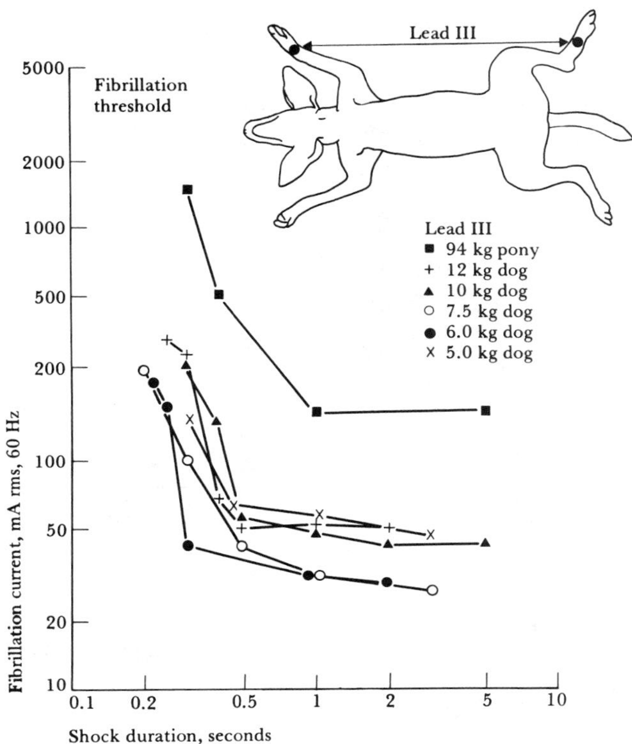

Figure 14.4 Fibrillation current versus shock duration. Thresholds for ventricular fibrillation in animals for 60-Hz ac current. Duration of current (0.2 to 5 s) and weight of animal body were varied. (From L. A. Geddes, *IEEE Trans. Biomed. Eng.,* 1973, 20, 465–468. Copyright 1973 by the Institute of Electrical and Electronics Engineers. Reproduced with permission.)

POINTS OF ENTRY

When current is applied at two points on the surface of the body, only a small fraction of the total current flows through the heart, as shown in Figure 14.5(a). These large, externally applied currents are called *macroshocks.* The magnitude of current needed to fibrillate the heart is far greater when the current is applied on the surface of the body than it would be if the current were applied directly to the heart. The importance of the location of the two macroshock entry points is often overlooked. If the two points are both on the same extremity, the risk of fibrillation is small, even for high currents. For dogs, the current needed for fibrillation is greater for ECG lead I (LA–RA) electrodes than for ECG leads II and III (LL–RA and LL–LA) (Geddes *et al.,* 1973). The protection afforded by the skin resistance (15 kΩ to 1 MΩ for 1 cm^2) is eliminated by many medical procedures that require insertion of

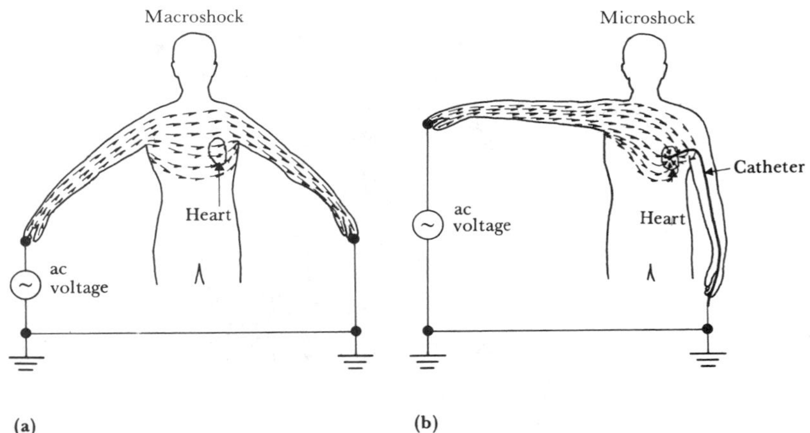

Figure 14.5 Effect of entry points on current distribution (a) *Macroshock,* externally applied current spreads throughout the body. (b) *Microshock,* all the current applied through an intracardiac catheter flows through the heart. (From F. J. Weibell, "Electrical Safety in the Hospital," *Annals of Biomedical Engineering,* 1974, 2, 126–148.)

conductive devices into natural openings or incisions in the skin. If the skin resistance is bypassed, less voltage is required to produce sufficient current for each physiological effect.

Patients are particularly vulnerable to electric shock when invasive devices are placed in direct contact with cardiac muscle. If a device provides a conductive path to the heart that is insulated except at the heart, then very small currents called *microshocks* can induce ventricular fibrillation. As Figure 14.5(b) shows, all the current flowing through such a conductive device flows through the heart. The current density at the point of contact can be quite high, and fibrillation in dogs can be induced by total currents as low as 20 μA. (See Roy, 1976.) The sparse data we have for human-heart fibrillation with an intracardiac catheter indicate that microshock currents ranging from 80 to 600 μA can cause fibrillation. The other connection can be at any point on the body. The widely accepted safety limit to prevent microschocks is 10 μA.

14.3 DISTRIBUTION OF ELECTRIC POWER

Electric power is needed in health-care facilities not only for the operation of medical instruments but also for lighting, maintenance appliances, patient conveniences (such as television, hair curlers, and electric toothbrushes), clocks, nurse call buttons, and an endless list of other electric devices. A first step in providing electrical safety is to control the availability of electric power and the grounds in the patients' environment. This section is concerned with methods

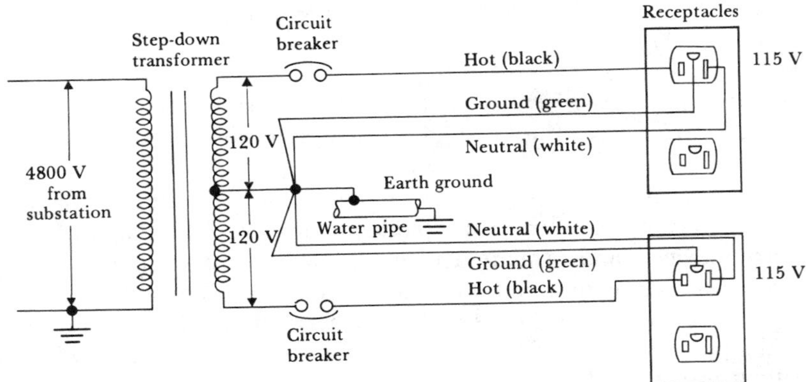

Figure 14.6 Simplified electric-power distribution for 115-V circuits. Power frequency is 60 Hz.

for safe distribution of power in health-care facilities. Then, in the sections that follow, we will discuss various macroshock and microshock hazards.

A simplified diagram of an electric-power-distribution system is shown in Figure 14.6. High voltage (4800 V) enters the building—usually via underground cables. The secondary of a stepdown transformer develops 240 V. This secondary has a grounded center tap to provide two 120-V circuits between ground and each side of the secondary winding. Some heavy-duty devices (such as air conditioners, electric dryers, and x-ray machines) that require 240 V are placed across the entire secondary winding; electricians do this by making connections to the two ungrounded terminals. Ordinary wall receptacles and lights operate on 120 V, obtained from either one of the ungrounded hot (black) transformer terminals and the neutral (white) grounded center tap. In addition, for health-care facilities, the National Electrical Code (NEC) for 1996 requires that all receptacles be "Hospital Grade" and be grounded by a separate insulated (green) copper conductor (Article 517-13). An additional redundant ground path through the metal raceway, conduit, or a separate cable is required for patient-care areas. Some older installations used metal conduit as the only ground conductor. Conduit grounds are generally unsatisfactory, because corrosion and loose conduit connections make them unreliable.

PATIENTS' ELECTRICAL ENVIRONMENT

Of course, a shock hazard exists between the two conductors supplying either a 240-V or a 120-V appliance. Because the neutral wire on a 120-V circuit is connected to ground, a connection between the hot conductor and *any* grounded object poses a shock hazard. Microshocks can occur if sufficient potentials exist between exposed conductive surfaces in the patients' environment. The following maximal potentials permitted between any two exposed conductive surfaces in the vicinity of the patient are specified by the 1996 NEC, Article 517-15:

1. General-care areas, 500 mV under normal operation
2. Critical-care areas, 40 mV under normal operation

In general-care areas, patients have only incidental contact with electric devices. For critical-care areas, hospital patients are intentionally exposed to electric devices, and insulation of externalized cardiac conductors from conductive surfaces is required. In critical-care areas, all exposed conductive surfaces in the vicinity of the patient must be grounded at a single patient-grounding point (Section 14.8). Also, periodic testing for continuity between the patient ground and all grounded surfaces is required.

Each patient-bed location in general-care areas must have at least four single or two duplex receptacles. Each receptacle must be grounded. At least two branch circuits with separate automatic overcurrent devices must supply the location of each patient bed. For critical-care areas, at least six single or three duplex receptacles are required for each location of a patient bed. Two branch circuits are also required, at least one being an individual branch circuit from a single panelboard. A patient-equipment grounding point (Section 14.8) is permitted for critical-care areas. For details, see NEC 70–1996, Article 517-19.

ISOLATED-POWER SYSTEMS

Even installing a good separate grounding system for each patient cannot prevent possibly hazardous voltages that can result from ground faults. A *ground fault* is a short circuit between the hot conductor and ground that injects large currents into the grounding system. These high-current ground faults are rare, and usually the circuit breakers open quickly. If the center tap of the stepdown transformer were not grounded, then very little current could flow, even if a short circuit to ground developed. So long as both power conductors are isolated from ground, a single ground fault will not allow the large currents that cause hazardous potentials between conductive surfaces.

Isolation of both conductors from ground is commonly achieved with an *isolation transformer.* A typical isolated-power system is shown in Figure 14.7. In an isolated system such as this, if a single ground fault from either conductor to ground occurs, the system simply reverts to a normal grounded system. A second fault from the other conductor to ground is then required to get large currents in the grounds.

A continually operating *line-isolation monitor,* LIM (also called a dynamic ground detector), must be used with isolation transformers to detect the occurrence of the first fault from either conductor to ground. This monitor alternately measures the total possible resistive and capacitive leakage current *(total hazard current)* that would flow through a low impedance *if it were*

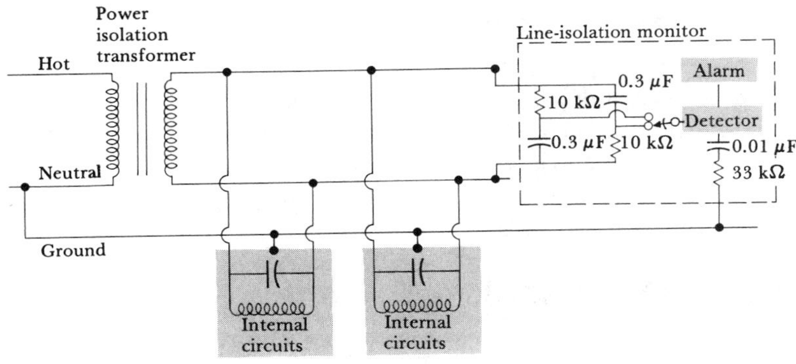

Figure 14.7 Power-isolation-transformer system with a line-isolation monitor to detect ground faults.

connected between either isolated conductor and ground. When the total hazard current exceeds 3.7 to 5.0 mA for normal line voltage, a red light and an audible alarm are activated. The LIM itself has a monitor hazard current of 1 mA. This makes the allowed fault total hazard current for all appliances served by the transformer somewhat less than 5 mA.

The kinds of corrective action that should be taken when the alarm goes off must be explained to medical personnel so that they do not overreact. The periodic switching in some line-isolation monitors produces transients that can interfere with monitoring of low-level physiological signals (ECG and EEG) and give erroneous heart rates. Or it can trigger synchronized defibrillators and aortic-balloon assist pumps during the wrong phase of the patient's heart cycle. Some LIMs avoid these problems by using continuous two-channel circuitry instead of measuring the total hazard current by switching between each line and ground.

Isolated-power systems were originally introduced to prevent sparks from coming into contact with flammable anesthetics such as ether. The NEC requires isolated-power systems only in those operating rooms and other locations where flammable anesthetics are used or stored.

EMERGENCY-POWER SYSTEMS

Article 517 of the National Electrical Code (1996) specifies the emergency electric system required for heath-care facilities. An emergency system is required that automatically restores power to specified areas within 10 s after interruption of the normal source. The emergency system may consist of two parts: (1) the life-safety branch (illumination, alarm, and alerting equipment), and (2) the critical branch (lighting and receptacles in critical patient-care areas). For additional details, see Article 517-25, 30–35.

14.4 MACROSHOCK HAZARDS

The high resistance of dry skin and the spatial distribution of current throughout the body when a person receives an electric shock are two factors that reduce the danger of ventricular fibrillation. Furthermore, electric equipment is designed to minimize the possibility of humans coming into contact with dangerous voltages.

SKIN AND BODY RESISTANCE

The resistance of the skin limits the current that can flow through a person's body when that person comes into contact with a source of voltage. The resistance of the skin varies widely with the amount of water and natural oil present. It is inversely proportional to the area of contact.

Most of the resistance of the skin is in the outer, horny layer of the epidermis. For one square centimeter of electric contact with dry, intact skin, resistance may range from 15 kΩ to almost 1 MΩ, depending on the part of the body and the moisture or sweat present. If skin is wet or broken, resistance drops to as low as 1% of the value for dry skin. By contrast, the internal resistance of the body is about 200 Ω for each limb and about 100 Ω for the trunk. Thus internal body resistance between any two limbs is about 500 Ω. These values are probably higher for obese patients, because the specific resistivity of fat is high. Actually, the distribution of current in various tissues in the body is poorly understood.

Any medical procedure that reduces or eliminates the resistance of the skin increases possible current flow and makes the patient more vulnerable to macroshocks. For example, biopotential electrode gel reduces skin resistance. Electronic thermometers placed in the mouth or rectum also bypass the skin resistance, as do intravenous catheters containing fluid that can act as a conductor. Thus patients in a medical-care facility are much more susceptible to macroshock than the general population.

ELECTRIC FAULTS IN EQUIPMENT

All electric devices are of course designed to minimize exposure of humans to hazardous voltages. However, many devices have a metal chassis and cabinet that medical personnel and patients may touch. If the chassis and cabinet are not grounded, as shown in Figure 14.8(a), then an insulation failure or shorted component between the black hot power lead and the chassis results in a 115-V potential between the chassis and any grounded object. If a person simultaneously touches the chassis and any grounded object, a macroshock results.

The chassis and cabinet can be grounded via a third green wire in the power cord and electric system, as shown in Figure 14.8(b). This ground wire is connected to the neutral wire and ground at the power-distribution panel. Then, when a fault occurs between the hot conductor and the chassis, the

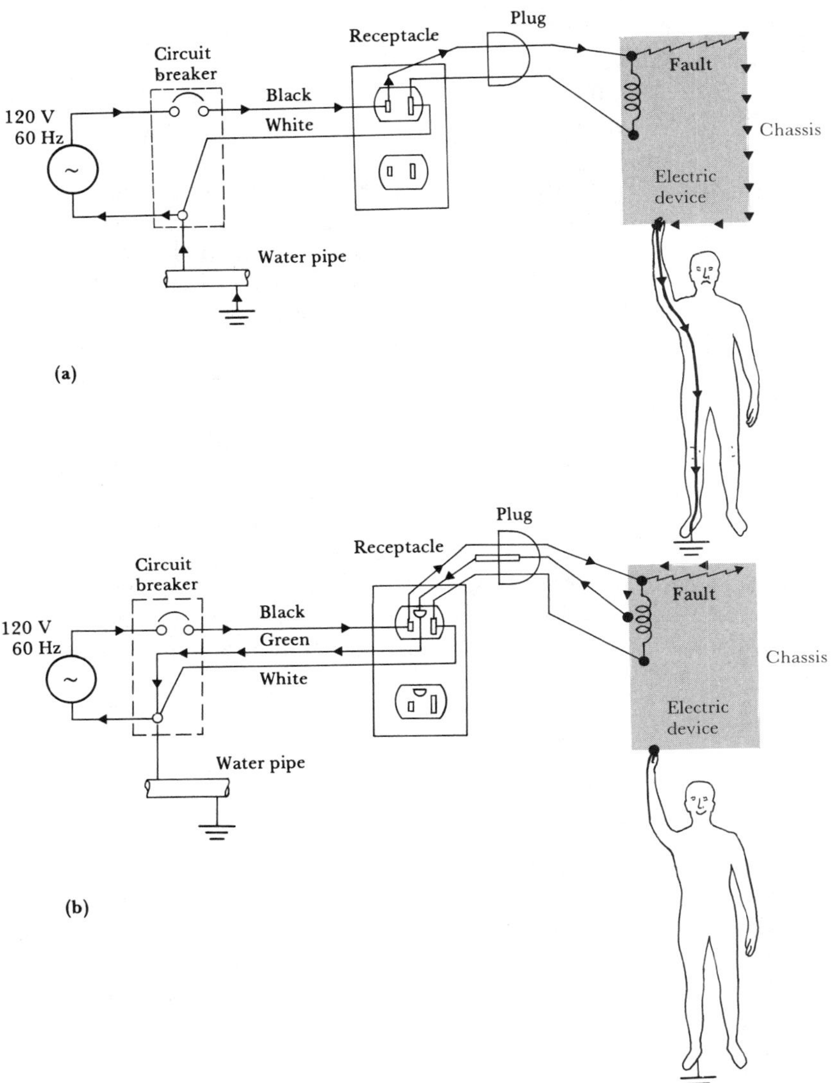

Figure 14.8 Macroshock due to a ground fault from hot line to equipment cases for (a) ungrounded cases and (b) grounded chassis.

current flows safely to ground on the green conductor. If the ground-wire resistance is very low, the voltage between the chassis and other grounded objects is negligible. If enough current flows through the ground wire to open the circuit breaker, this will call people's attention to the fault.

Note that direct faults between the hot conductor (or any high voltage in the device) and ground are not common. Little or no current flows through the ground conductor during normal operation of electric devices. The ground conductor is not needed for protection against macroshock until a hazardous fault develops. Thus a broken ground wire or a poor connection of a receptacle ground is not detected during normal operation of the device. For this reason, continuity of the ground wire in the device and the receptacle must be tested periodically.

Faults inside electric devices may result from failures of insulation, shorted components, or mechanical failures that cause shorts. Power cords are particularly susceptible to strain and physical abuse, as are plugs and receptacles. Ironically, it is possible for a device's chassis and cabinet to become hot because a ground wire is in the power cord. If the ground wire is open anywhere between the power cord and ground, then a frayed cord could permit contact between the hot conductor and the broken ground wire leading to the chassis. Often, macroshock accidents result from carelessness and failure to correct known deficiencies in the power-distribution system and in electric devices.

Fluids—such as blood, urine, intravenous solutions, and even baby formulas—can conduct enough electricity to cause temporary short circuits if they are accidentally spilled into normally safe equipment. This hazard is particularly acute in hospital areas that are subject to wet conditions, such as hemodialysis and physical therapy areas. The cabinets of many electric devices have holes and vents for cooling that provide access for spilled conductive fluids. The mechanical design of devices should protect patient electric connections from this hazard.

14.5 MICROSHOCK HAZARDS

Microshock accidents in patients who have direct electric connections to the heart are usually caused by circumstances unrelated to macroshock hazards. Microshocks generally result from *leakage currents* in line-operated equipment or from differences in voltage between grounded conductive surfaces due to large currents in the grounding system. The microshock current can flow either into or out of the electric connection to the heart.

LEAKAGE CURRENTS

Small currents (usually on the order of microamperes) that inevitably flow between any adjacent insulated conductors that are at different potentials are called *leakage currents.* Although most of the leakage current in line-operated equipment flows through the stray capacitance between the two conductors, some resistive leakage current flows through insulation, dust, and moisture.

The most important source of leakage currents is the currents that flow from all conductors in the electric device to lead wires connected either to the chassis or to the patient. Leakage current flowing to the chassis flows safely to ground if a low-resistance ground wire is available, as shown in

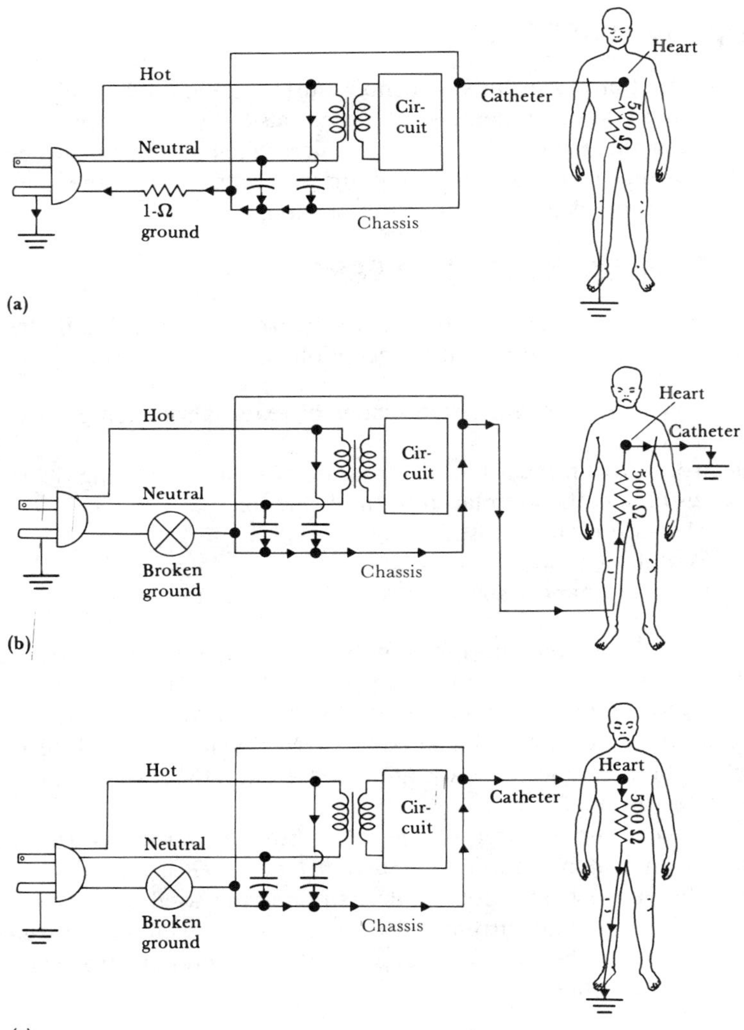

Figure 14.9 Leakage-current pathways Assume 100 μA of leakage current from the power line to the instrument chassis. (a) Intact ground, and 99.8 μA flows through the ground. (b) Broken ground, and 100 μA flows through the heart. (c) Broken ground, and 100 μA flows through the heart in the opposite direction.

Figure 14.9(a). If the ground wire is broken, then the chassis potential rises above ground, and a patient who touches the chassis *and* has a grounded electric connection to the heart may receive a microshock [Figure 14.9(b)]. If there is a connection from the chassis to the patient's heart *and* a connection to ground anywhere on the body, this could also cause a microshock [Figure 14.9(c)].

CONDUCTIVE SURFACES

The source that produces the microshock current need not be leakage current from line-operated equipment. Small potentials between any two conductive surfaces near the patient can cause a microshock if either surface makes contact with the heart and the other surface contacts any other part of the body. An example is given later in this section.

CONDUCTIVE PATHS TO THE HEART

Specific types of electric connections to the heart can be identified. The following clinical devices make patients susceptible to microshock.

1. Epicardial or endocardial electrodes of externalized temporary cardiac pacemakers
2. Electrodes for intracardiac electrogram (EGM) measuring devices
3. Liquid-filled catheters placed in the heart to:
   a. Measure blood pressure
   b. Withdraw blood samples
   c. Inject substances such as dye or drugs into the heart

It should be emphasized that a patient is in danger of microshock only when there is some electric connection to the heart. The internal resistance of liquid-filled catheters is much greater (50 kΩ to 1 MΩ) than the resistance of metallic conductors in pacemaker and EGM electrode leads. Internal resistance of the body to microshock is about 300 Ω, and the resistance of the skin can be quite variable.

In dogs, the surface area of the intracardiac electrode is an important determinant of minimal fibrillating current (Roy, 1976). Figure 14.10 shows that as catheter electrodes get smaller, so does the total current needed to fibrillate. This means that current density at the tip of the intracardiac electrode is the important microshock parameter. Smaller catheter electrodes may have larger internal resistance.

Microshock Via Ground Potential Differences An example of microshock illustrates the need for a single reference ground point of each patient in critical-care areas and the need for a 40-mV limit on the difference in potential between conductive surfaces in these areas.

Figure 14.11 shows a patient in the intensive-care unit (ICU) who is connected to an ECG monitor that grounds the right-leg electrode to reduce 60-Hz interference. Also, the patient's left-ventricular blood pressure is being monitored by an intracardiac saline-filled catheter connected to a metallic pressure sensor that is also grounded. Assume that these two monitors are connected to grounded three-wire wall receptacles on separate circuits that can come from a central power-distribution panel many meters away. A microshock can occur when any device with a ground fault that does not open the circuit breaker is operated on *either* circuit.

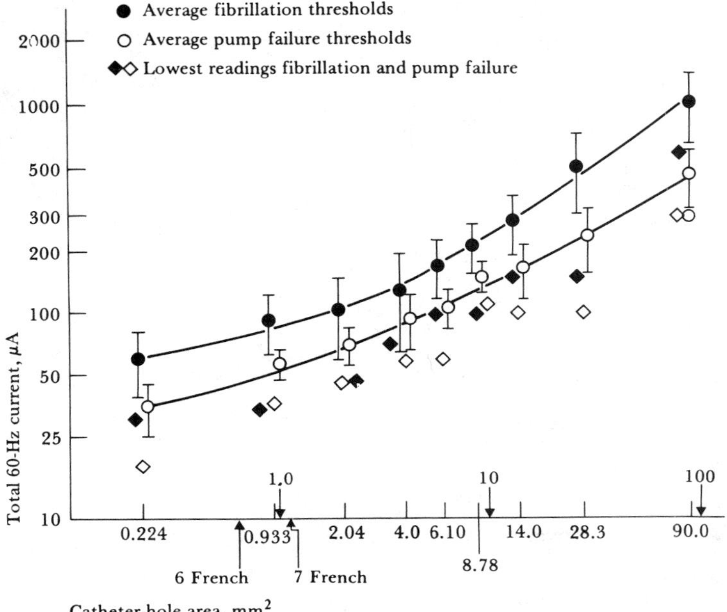

Figure 14.10 Thresholds of ventricular fibrillation and pump failure versus catheter area in dogs. (From O. Z. Roy, J. R. Scott, and G. C. Park, "Ventricular Fibrillation and Pump Failure Thresholds Versus Electrode Area," *IEEE Transactions on Biomedical Engineering,* 1976, 23, 45–48. Reprinted with permission.)

Figure 14.11(a) shows the scheme of this hazard; Figure 14.11(b) shows an equivalent circuit. Suppose that a faulty electric floor polisher, which is dusty and damp, allows 5 A to flow to the distribution panel on the ground wire. The floor polisher functions properly, so the fault is not noticed by the operator. The ground wire could easily have a 0.1-Ω resistance, so 500 mV could appear across the patient between the ECG-monitor ground and the pressure-monitor ground. If the resistance of the patient's body and of the liquid-filled catheter is less than 50 kΩ, a current in excess of the 10-μA safe limit could flow. Of course, more current would flow if the ground resistance or fault current were higher or if the catheter resistance were lower. If a grounded pacing catheter were to be used instead of the liquid-filled catheter in this example, then much smaller differences in ground potential would exceed the safe limit.

Most low-voltage hazards can be avoided if the grounds of all devices used in the vicinity of each patient are connected to a single patient-grounding point. This also prevents faults at one patient's bedside from affecting the safety of other patients. Modern pressure sensors and ECG monitors provide electrical isolation for all patient leads.

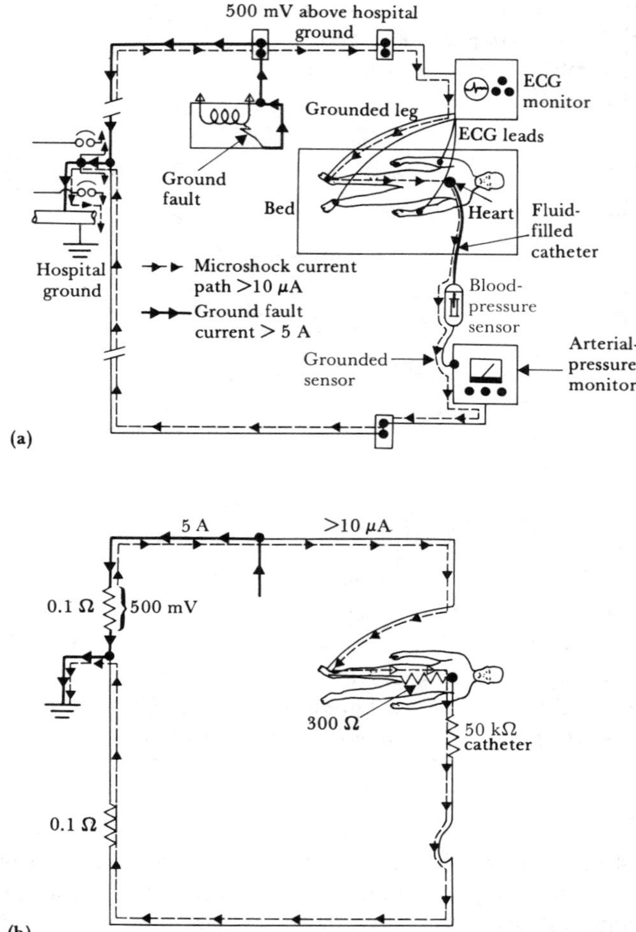

Figure 14.11 (a) Large ground-fault current raises the potential of *one* ground connection to the patient. The microshock current can then flow out through a catheter connected to a different ground. (b) Equivalent circuit. Only power-system grounds are shown.

14.6 ELECTRICAL-SAFETY CODES AND STANDARDS

A *code* is a document that contains only mandatory requirements. It uses the word *shall* and is cast in a form suitable for adoption into law by an authority that has jurisdiction. Explanations in a code must appear only in fine-print notes, footnotes, and appendices. A *standard* also contains only mandatory requirements, but compliance tends to be voluntary, and more detailed notes and explanations are given. A *manual* or *guide* is a document that is informative and tutorial but does not contain requirements.

The development, adoption, and use of standards and codes for electrical safety in health-care facilities has had a particularly arduous history that continues to the present day (Bruner and Leonard, 1989, Chapter 9). The process began following tragic explosions and fires resulting from electric ignition of flammable anesthetics such as ether. In the early 1970s, the microshock electrical-safety scare resulted in some proposals that were not practical. Implicit requirements for isolated-power systems and very low-leakage current requirements were hotly debated for many years. Finally, the National Fire Protection Association NFPA 99-1984 and ANSI/AAMI ES1-1985 standards were adopted.

The NFPA 99—Standard for Health Care Facilities—1996 has evolved from 12 NFPA documents that were combined in 1984 and revised every 3 years. In addition to electric equipment, this standard also describes gas, vacuum, and environmental systems and materials. It is the primary document that describes the requirements for patient-care-related electric appliances used for diagnostic, therapeutic, or monitoring purposes in a patient-care area. Chapter 7 "covers the performance, maintenance, and testing of electrical equipment" by personnel in health-care facilities. More detailed manufacturer requirements are given in Chapter 9 for "the performance, maintenance, and testing with regard to safety, required of manufacturers of equipment used within health-care facilities." Annex 2 concerns "the safe use of high-frequency (100 kHz to microwave frequencies) electricity in health-care facilities."

The National Electrical Code—1996, Article 517—Health Care Facilities is published by the NFPA and is widely adopted and enforced by state, county, and municipal authorities having jurisdiction. Requirements vary for general-care areas, critical-care areas, and wet locations. The major sections are A. General; B. Wiring Design and Protection; C. Essential Electrical System; D. Inhalation Anesthetizing Locations; E. X-Ray Installations; F. Communications, Signaling Systems, Data Systems, Fire-Protective Signaling Systems, and Systems Less than 120 Volts, Nominal; and G. Isolated Power Systems.

The Association for the Advancement of Medical Instrumentation (AAMI) developed an American National Standard on "Safe Current Limits for Electromedical Apparatus," ANSI/AAMI ES1—1993. This standard concerns limits on chassis and patient-lead leakage currents, which are fixed from dc to 1 kHz and increase from 1 kHz to 100 kHz. For single-fault conditions only the patient lead leakage current was relaxed from 10 μA to 50 μA and the chassis leakage current was relaxed from 100 μA to 300 μA. These changes have been hotly challenged by the American Heart Association committee on electrocardiography (Laks, 1994, 1996). Laks said this change "constitutes experimentation on humans without their consent to determine the safe range of such currents."

Underwriters Laboratories (UL) plans to adopt the International Electrotechnical Commission (IEC) 601-1 standard as far as practical, including the

Table 14.1 Limits on leakage current for electric appliances See Section 14.12 for specific test conditions and requirements

Electric Appliance	Chassis Leakage, μA	Patient-lead Leakage, μA
Appliances not intended to contact patients	100	NA
Appliances not intended to contact patients and single fault	500	NA
Appliances with *nonisolated* patient leads	100	10
Appliances with *nonisolated* leads and single fault	300	100
Appliances with *isolated* patient leads	100	10
Appliances with *isolated* leads and single fault	300	50

limit on leakage current for medical electric devices. This conformity to a widely supported international standard is endorsed by the Health Industry Manufacturers Association (HEMA), the National Electrical Manufacturers Association (NEMA), and the U.S. Food and Drug Administration (FDA). The IEC 601-1 standard allows a "patient auxiliary current" up to 100 μA at not less than 0.1 Hz to permit amplifier bias currents and impedance plethysmography if the current is not intended to produce a physiological effect.

The present limits on leakage current for the ANSI/AAMI ES1-1993 standards are shown in Table 14.1.

14.7 BASIC APPROACHES TO PROTECTION AGAINST SHOCK

There seem to be two fundamental methods of protecting patients against shock. First, the patient can be completely isolated and insulated from all grounded objects and all sources of electric current. Second, all conductive surfaces within reach of the patient can be maintained at the same potential, which is not necessarily ground potential. Neither of these approaches can be fully achieved in most practical environments, so some combination of the two methods must usually suffice.

Not only must all hospital patients be protected from macroshocks, but all visitors and staff must be protected as well. Patients with reduced skin resistance (perhaps coupled to electrodes), invasive connections (such as intravenous catheters), or exposure to wet conditions (as happens during dialysis) need extra protection. The small number of patients with accessible electric connections to the heart need additional protection from microshock currents. Many of the specific methods of protection described here can be used in combination to provide redundant safeguards. And it is also necessary to consider cost-benefit ratios with respect to both the purchase cost of safety equipment and the periodic maintenance costs of such equipment.

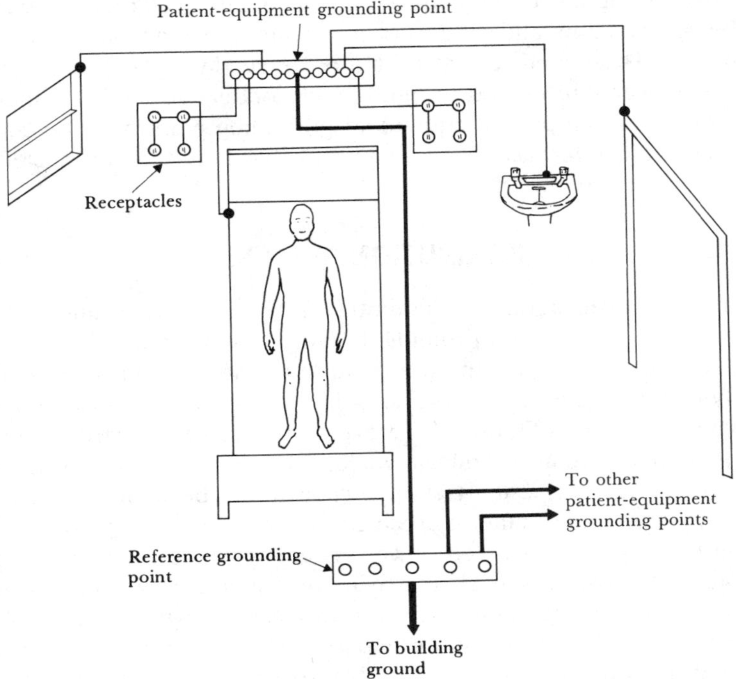

Figure 14.12 Grounding system All the receptacle grounds and conductive surfaces in the vicinity of the patient are connected to the patient-equipment grounding point. Each patient-equipment grounding point is connected to the reference grounding point that makes a single connection to the building ground.

14.8 PROTECTION: POWER DISTRIBUTION

GROUNDING SYSTEM

Low-resistance grounds that can carry currents up to circuit-breaker ratings are clearly essential for protecting patients against both macroshock and microschock, even when an isolated-power system is used. Figure 14.8 shows the importance of adequate grounds for protection against macroshock. Grounding is equally significant in preventing microshock (see Figure 14.9). A grounding system protects patients by keeping all conductive surfaces and receptacle grounds in the patient's environment at the same potential. It also protects the patient from ground faults at other locations.

The grounding system has a *patient-equipment grounding point,* a *reference grounding point,* and connections, as shown in Figure 14.12. The patient-equipment grounding point is connected individually to all receptacle grounds, metal beds, metal door and window frames, water pipes, and any other conductive surfaces. These connections should not exceed 0.15 Ω. The difference in potential between receptacle grounds and conductive surfaces should not exceed 40 mV. Each patient-equipment grounding point must be connected individually to a reference grounding point that is in turn connected to the building service ground.

ISOLATED POWER-DISTRIBUTION SYSTEM

Unfortunately, even a good equipotential grounding system cannot eliminate voltages produced between grounds by large ground faults that cause large ground currents. However, these ground faults are rare in high-quality and properly maintained equipment. The isolation transformers discussed in Section 14.3 and shown in Figure 14.7 prevent this unlikely hazard. The isolated power system also reduces leakage current somewhat, but not below the 10-μA safe limit. There is usually enough capacitance between the transformer secondary circuit and ground to preclude protection against microshocks with isolation transformers. Isolated power systems provide considerable protection against macroshocks, particularly in areas subject to wet conditions. Isolated power systems are only necessary in locations where flammable anesthetics are used. The additional protection against microshocks provided by isolation transformers does not generally justify the high cost of these systems.

GROUND-FAULT CIRCUIT INTERRUPTERS (GFCI)

Ground-fault circuit interrupters disconnect the source of electric power when a ground fault greater than about 6 mA occurs. In electric equipment that has negligible leakage current, the current in the hot conductor is equal to the current in the neutral conductor. The GFCI senses the difference between these two currents and interrupts power when this difference, which must be flowing to ground somewhere, exceeds the fixed rating. The devices make no distinction among paths the current takes to ground: That path may be via the ground wire or through a person to ground (Figure 14.8).

Most GFCIs use a differential transformer and solid-state circuitry, as shown in Figure 14.13(a). The trip time for the GFCI varies inversely with the magnitude of the ground-fault current, as shown in Figure 14.13(b). The GFCI is used with conventional three-wire grounded power-distribution systems. When power is interrupted by a GFCI, the manual reset button on the GFCI must be pushed to restore power. Most GFCIs have a momentary pushbutton that creates a safe ground fault to test the interrupter.

The National Electrical Code (1996) requires that there be GFCIs in circuits serving bathrooms, garages, outdoor receptacles, swimming pools, and construction sites (Articles 210-8, 680-5). NFPA 99 requires the use of GFCIs in wet locations, particularly hydrotherapy areas, where continuity of power is not essential.

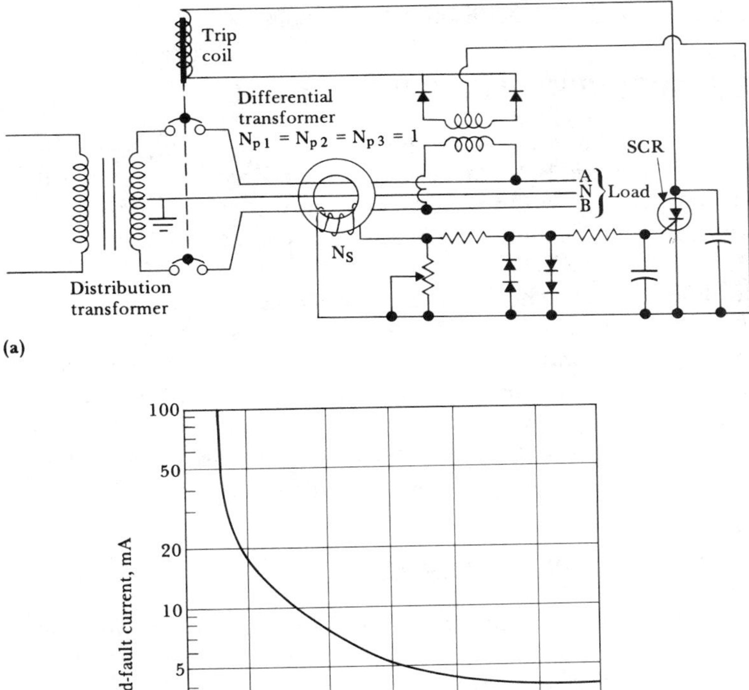

Figure 14.13 Ground-fault circuit interrupters (a) Schematic diagram of a solid-state GFCI (three wire, two pole, 6 mA). (b) Ground-fault current versus trip time for a GFCI. [Part (a) is from C. F. Dalziel, "Electric Shock," in *Advances in Biomedical Engineering,* edited by J. H. U. Brown and J. F. Dickson III, 1973, 3: 223–248.]

GFCIs are not sensitive enough to interrupt microshock levels of leakage current, so they are primarily macroshock-protection devices. They can, however, prevent some microshocks by interrupting the source of large ground-fault currents that cause differences in potential in grounding systems.

However, circuits in patient-care areas generally should not include GFCIs, because the loss of power to life-support equipment due to GFCIs is probably more hazardous to the patient than most small ground faults would be. Where brief power interruptions can be tolerated, the low cost of GFCIs ($10) make them an attractive alternative to isolated power-distribution systems ($2000).

14.9 PROTECTION: EQUIPMENT DESIGN

RELIABLE GROUNDING FOR EQUIPMENT

The importance of an effective grounding system for equipment has already been illustrated (Figure 14.8). Most failures of equipment grounds occur either at the ground contact of the receptacle or in the plug and cable leading to the line-powered equipment. Hospital-grade receptacles and plugs and "Hard Service" (SO, ST, or STO) or "Junior Hard Service" (SJO, SJT, or SJTO) power cords must be used in all patient areas. Molded plugs should be avoided, because surveys have shown that 40 to 85% of these plugs develop invisible breaks within 1 to 10 years of hospital service. Strain-relief devices are recommended both where the cord enters the equipment and at the connection between cord and plug. A convenient cord-storage compartment or device reduces cord damage. Equipment grounds are often deliberately interrupted by improper use of the common three-prong-to-two-prong adapter *(cheater adapter).*

REDUCTION OF LEAKAGE CURRENT

Reduction of leakage current in the chassis of equipment and in patient leads is an important goal for designers of all line-powered instruments. Special low-leakage power cords are available ($<1.0\ \mu A/m$). Leakage current inside the chassis can be reduced by using layouts and insulating materials that minimize the capacitance between all hot conductors and the chassis. Particular attention must be given to maximizing the impedance from patient leads to hot conductors and from patient leads to chassis ground. Most modern equipment meets the leakage-current limits given in Section 14.6. Old equipment with higher leakage should not be used with patients susceptible to microshocks unless proper grounding is ensured.

DOUBLE-INSULATED EQUIPMENT

The objective of grounding is to eliminate hazardous potentials by interconnecting all conductive surfaces. An equally effective approach is to use a separate layer of insulation to prevent contact of any person with the chassis or any exposed conductive surface. Primary insulation is the normal functional insulation between energized conductors and the chassis. A separate secondary layer of insulation between the chassis and the outer case protects personnel even if a ground fault to the chassis occurs. The outer case, if it is made of insulating material, may serve as the secondary insulation. All switch levers and control shafts must also be double-insulated (for example, plastic knobs may have recessed screws). Double insulation generally reduces leakage current. For medical instruments, both layers of insulation should remain effective, even when conductive fluid is spilled. Double insulation protects against both macroshock and microshock.

OPERATION AT LOW VOLTAGES

Most solid-state electronic diagnostic equipment can be powered by low-voltage batteries (<10 V) or low-voltage isolation transformers. Macroshock is avoided if the voltage is low enough to be safe even when the device is applied directly to wet skin. Low-voltage ac-powered equipment can still cause microshock if the current is applied directly to the heart. However, low-voltage ac equipment is generally safer than high-voltage ac equipment. See Section 164 in Article 517 of the National Electrical Code (1996) for requirements for low-voltage equipment used in inhalation-anesthetizing locations.

ELECTRICAL ISOLATION

Isolation amplifiers are devices that break the ohmic continuity of electric signals between the input and output of the amplifier. This isolation includes different supply-voltage sources and different grounds on each side of the isolation barrier. Isolation amplifiers usually consist of an instrumentation amplifier at the input followed by a unity-gain isolation stage. Figure 14.14(a) shows a general model for an isolation amplifier that has a triangular operational amplifier symbol split by a perfect isolation barrier (dashed line). The very high impedance across the barrier is modeled by the isolation capacitance and resistance. The isolation voltage v_{ISO} is the potential that can exist between the input common and the output common (note the different ground symbols) and is rated from 1 to 10 kV without breakdown. The rejection of this voltage by the amplifier is specified by the isolation-mode rejection ratio (IMRR). The desired input voltage v_{SIG}, the input common-mode voltage v_{CM}, and the common-mode rejection ratio (CMRR) are the same as for a nonisolated amplifier. Typical maximal ratings for v_{CM} are only ± 10 V. The input common may be connected to the source in applications that break ground loops or may be floated to make possible simpler, two-wire connections to the source and reference of the common-mode signal across the isolation barrier to the output common. The three main features of an isolation amplifier are high ohmic isolation between input and output (>10 MΩ), high isolation-mode voltage (>1000 V), and high common-mode rejection (>100 dB).

Three fundamental methods are used in the design of isolation amplifiers: transformer isolation, optical isolation, and capacitive isolation. The transformer approach illustrated in Figure 14.14(b) uses either a frequency-modulated or a pulse-width-modulated carrier signal with small signal bandwidths up to 30 kHz to carry the signal. It uses an internal dc-to-dc converter composed of a 25-kHz oscillator, transformer, rectifier, and filter to supply isolated power. The optical method uses an LED on the source side and a photodiode on the output side. No modulator/demodulator is needed, because the signal all the way to dc is transmitted optically. A matched photodiode on the source side is used with feedback to improve linearity. Increased light from the forward-biased LED CR_1 causes increased reverse leakage current through CR_2 and CR_3 (see Figure 2.22). The simplified circuit in Figure 14.14(c) operates only

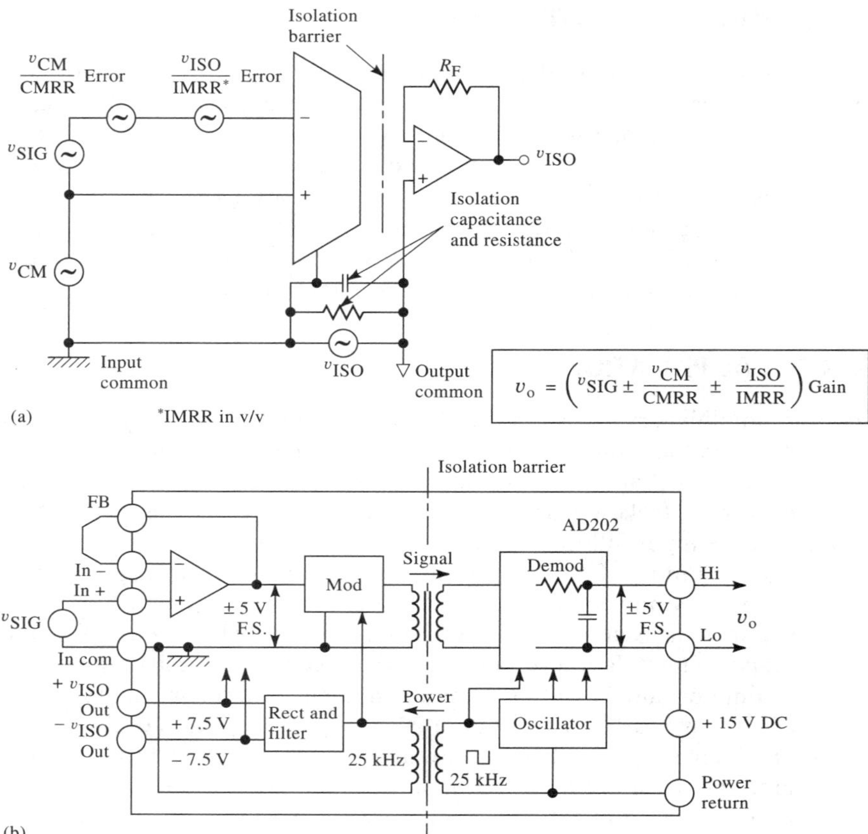

Figure 14.14 Electrical isolation of patient leads to biopotential amplifiers (a) General model for an isolation amplifier. (b) Transformer isolation amplifier (Courtesy of Analog Devices, Inc., AD202). (c) Simplified equivalent circuit for an optical isolator (Copyright © 1989 Burr-Brown Corporation. Reprinted in whole or in part, with the permission of Burr-Brown Corporation. Burr Brown ISO100). (d) Capacitively coupled isolation amplifier (Horowitz and Hill, Art of Electronics, Cambridge Univ. Press, Burr Brown ISO106).

for one polarity of input signal. The capacitive method, shown in Figure 14.14(d), uses digital encoding of the input voltage and frequency modulation to send the signal across a differential ceramic capacitive barrier. There is no feedback, though a power supply is needed on both sides of the barrier. The peak isolation voltage can be as high as 8 kV, and bandwidth up to 70 kHz is available.

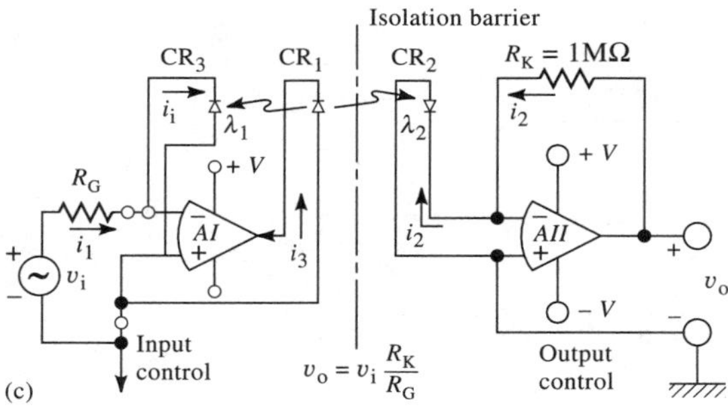

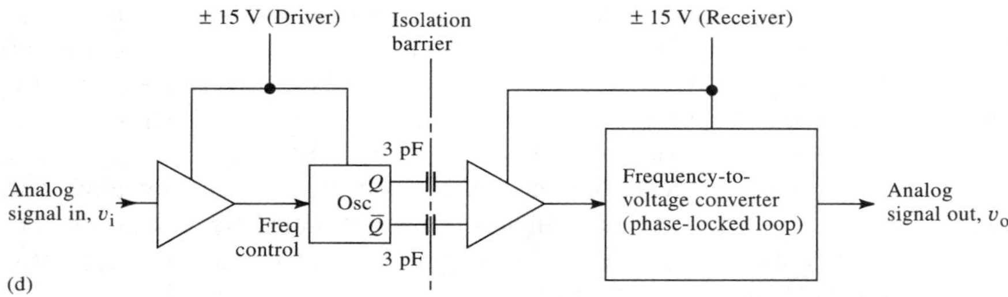

Figure 14.14 *(Continued)*

ISOLATED HEART CONNECTIONS

Undoubtedly the best way to minimize the hazards of microshock is to isolate or eliminate electric connections to the heart. Fully insulated connectors for external cardiac pacemakers powered by batteries have greatly reduced this hazard. Modern blood-pressure sensors are designed with triple insulation between the column of liquid, the sensor case, and the electric connections (Figure 14.15). Catheters with conductive walls have been developed that provide electric contact all along that part of the catheter that is inside the patient, so that microshock current is distributed throughout the body, not concentrated at the heart. Conductivity of the catheter wall does not affect measurements of pressure made with liquid-filled catheters. Catheters that contain sensors in the tip for measuring blood pressure and flow should have low leakage currents.

14.10 ELECTRICAL-SAFETY ANALYZERS

Commercially available instruments called electrical-safety analyzers are available for testing both medical-facility power systems and medical appliances (Anonymous, 1988 and 1989). These analyzers range in complexity from simple

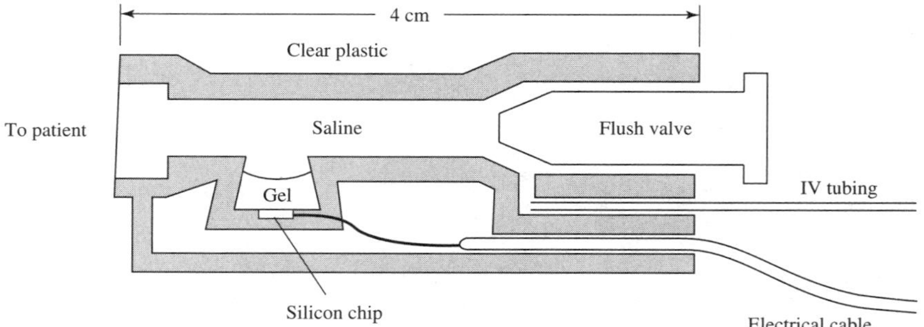

Figure 14.15 Isolation in a disposable blood-pressure sensor Disposable blood pressure sensors are made of clear plastic so air bubbles are easily seen. Saline flows from an intravenous (IV) bag through the clear IV tubing and the sensor to the patient. This flushes blood out of the tip of the indwelling catheter to prevent clotting. A lever can open or close the flush valve. The silicon chip has a silicon diaphragm with a four-resistor Wheatstone bridge diffused into it. Its electrical connections are protected from the saline by a compliant silicone elastomer gel, which also provides electrical isolation. This prevents electric shock from the sensor to the patient and prevents destructive currents during defibrillation from the patient to the silicon chip.

conversion boxes used with any volt–ohm-meter to computerized automatic measurement systems with bar code readers that generate written reports of test results. The features to consider are accuracy, ease of use, testing time, and cost. The analyzers also reduce errors caused by incorrect test setups and reduce the risk of shock to the person performing tests such as applying line voltage to patient leads to test isolation.

14.11 TESTING THE ELECTRIC SYSTEM

When we test systems of electric distribution and line-powered equipment, we must consider the safety of both the patients and the personnel conducting the tests. We shall briefly describe and comment on only the common tests.

TESTS OF RECEPTACLES

Receptacles should be tested for proper wiring, adequate line voltage, low ground resistance, and mechanical tension. The common three-light receptacle testers shown in Figure 14.16 are deficient in several respects. These devices were designed to check only the wiring, but even so, they can indicate only 8 (2^3) of 64 (4^3) possible states for an outlet. The three lights have only two states (2^3), whereas each of the three outlet contacts has four states (4^3)—hot, neutral, ground, and open.

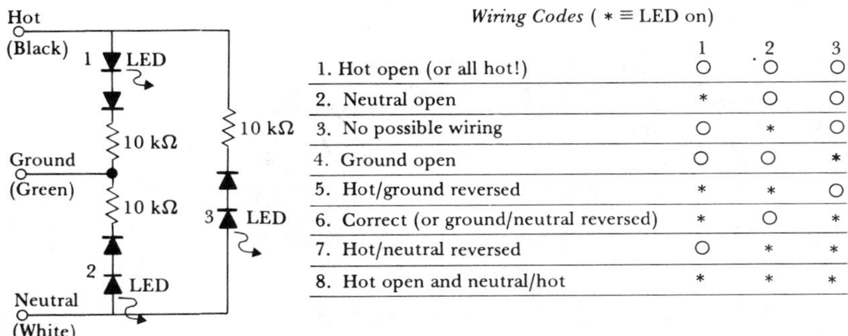

Wiring Codes (* ≡ LED on)

	1	2	3
1. Hot open (or all hot!)	O	O	O
2. Neutral open	*	O	O
3. No possible wiring	O	*	O
4. Ground open	O	O	*
5. Hot/ground reversed	*	*	O
6. Correct (or ground/neutral reversed)	*	O	*
7. Hot/neutral reversed	O	*	*
8. Hot open and neutral/hot	*	*	*

Figure 14.16 Three-LED receptacle tester Ordinary silicon diodes prevent damaging reverse-LED currents, and resistors limit current. The LEDs are ON for line voltages from about 20 V rms to greater than 240 V rms, so these devices should not be used to measure line voltage.

These testers give an OK reading when the ground and neutral wires are transposed and when the green and white wires are hot and the black wire is grounded. (Opening of the circuit breaker would probably call attention to the latter miswiring and to several others as well.)

Ground resistance can be measured by passing up to 1 A through the ground wire and measuring the voltage between ground and neutral. Anyone doing these ground-wire tests should take care not to incur the microshock hazards described in Section 14.5 (and shown in Figure 14.11). The resistance of neutral wiring can be tested similarly, by passing the current through the neutral conductor. Ground or neutral resistance should not exceed 0.2 Ω. The minimal mechanical retaining force for each of the three contacts is about 115 g (4 oz).

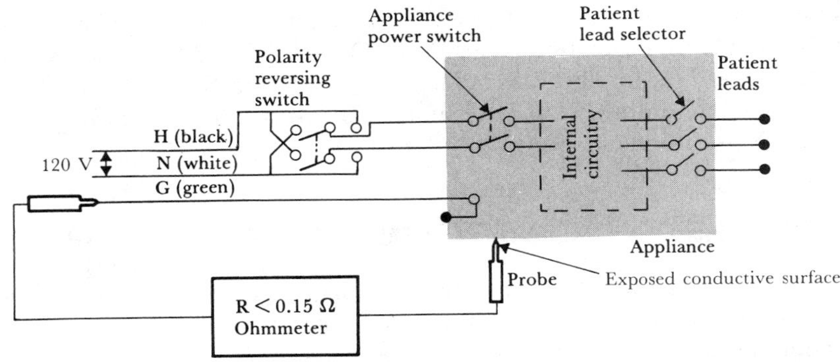

Figure 14.17 Ground-pin-to-chassis resistance test

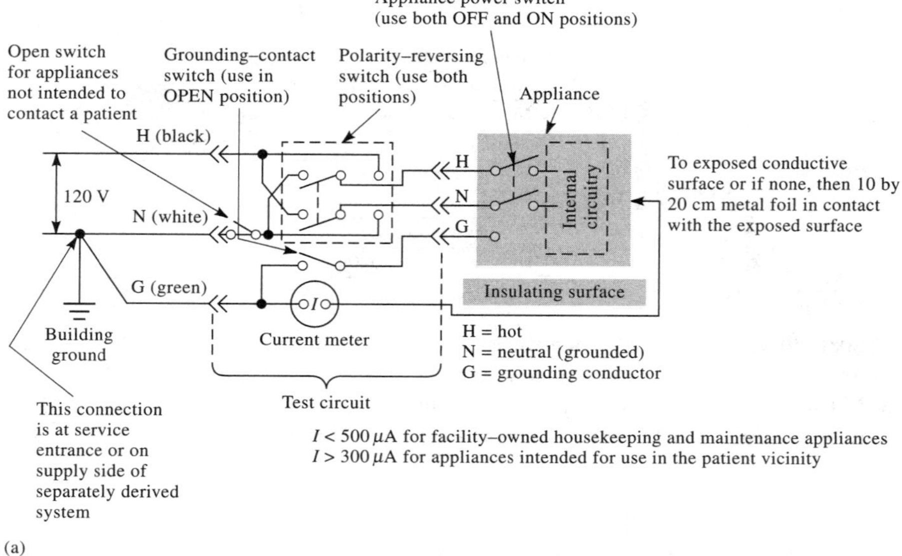

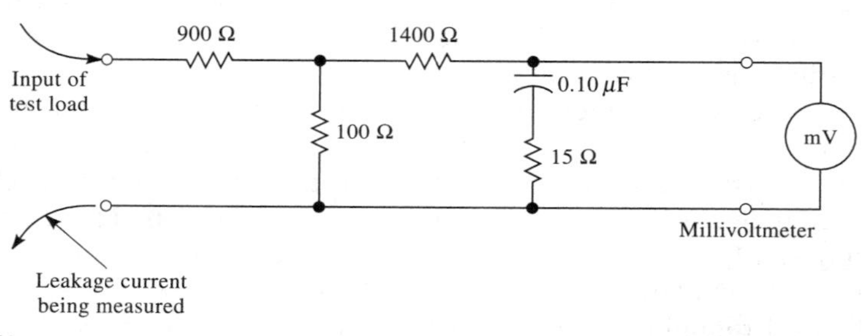

Figure 14.18 (a) Chassis leakage-current test. (b) Current-meter circuit to be used for measuring leakage current. It has an input impedance of 1 kΩ and a frequency characteristic that is flat to 1 kHz, drops at the rate of 20 dB/decade to 100 kHz, and then remains flat to 1 MHz or higher. (Reprinted with permission from NFPA 99-1996, "Health Care Facilities," Copyright © 1996, National Fire Protection Association, Quincy, MA 02269. This reprinted material is not the complete and official position of the National Fire Protection Association, on the referenced subject, which is represented only by the standard in its entirety.)

TESTS OF THE GROUNDING SYSTEM IN PATIENT-CARE AREAS

The NFPA 99 requires both voltage and impedance measurements with different limits for new and existing construction. The voltage between a reference grounding point (see Figure 14.12) and exposed conductive surfaces should not exceed 20 mV for new construction. For existing construction, the limit

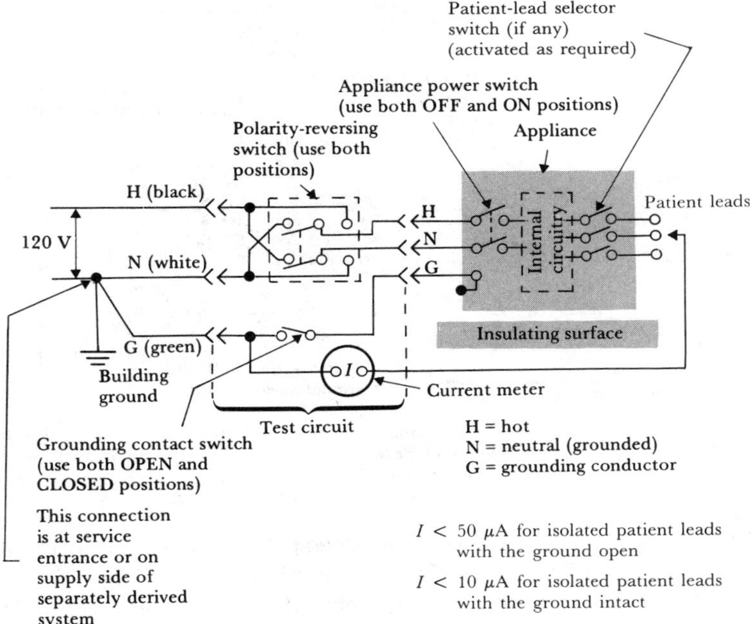

Figure 14.19 Test for leakage current from patient leads to ground (Reprinted with permission from NFPA 99-1996, "Health Care Facilities," Copyright © 1996, National Fire Protection Association, Quincy, MA 02269. This reprinted material is not the complete and official position of the National Fire Protection Association, on the referenced subject, which is represented only by the standard in its entirety.)

is 500 mV for general-care areas and 40 mV for critical-care areas. The impedance between the reference grounding point and receptacle grounding contacts must be less than 0.1 Ω for new construction and less than 0.2 Ω for existing construction.

TESTS OF ISOLATED-POWER SYSTEMS

Isolated-power systems should have equipotential grounding that is similar to that of unisolated systems (Figure 14.12). The line-isolation monitor (Figure 14.7) should trigger a visible (red) and an audible alarm when the total hazard current (resistive and capacitive leakage currents and LIM current) reaches a threshold of 5 mA under normal line-voltage conditions. The LIM should not trigger the alarm for a fault-hazard current of less than 3.7 mA. For complete specifications, see the latest NFPA 99 standard.

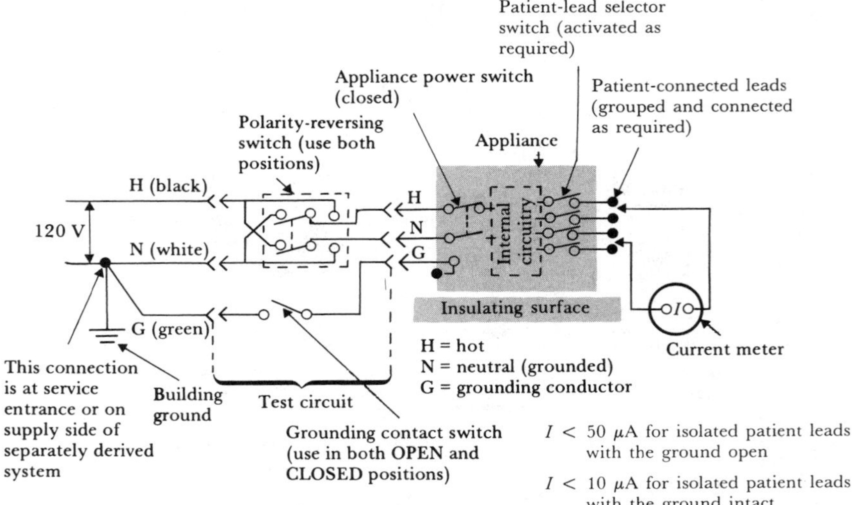

Figure 14.20 Test for leakage current between patient leads (Reprinted with permission from NFPA 99-1996, "Health Care Facilities," Copyright © 1996, National Fire Protection Association, Quincy, MA 02269. This reprinted material is not the complete and official position of the National Fire Protetion Association, on the referenced subject, which is represented only by the standard in its entirety.)

14.12 TESTS OF ELECTRIC APPLIANCES

GROUND-PIN-TO-CHASSIS RESISTANCE

The resistance between the ground pin of the plug and the equipment chassis and exposed metal objects should not exceed 0.15 Ω during the life of the appliance (Figure 14.17 on page 651).

During the measurement of resistance, the power cord must be flexed at its connection to the attachment plug and at its strain relief where it enters the appliance.

CHASSIS LEAKAGE CURRENT

Leakage current emanating from the chassis, as measured in Figure 14.18(a) on page 654, should not exceed 500 μA for appliances with single fault not intended to contact patients and should not exceed 300 μA for appliances that are intended for use in the patient care vicinity. These are limits on rms current for sinusoids from dc to 1 kHz, and they should be obtained with a current-measuring device of 1000 Ω or less. Figure 14.18(b) shows a suitable

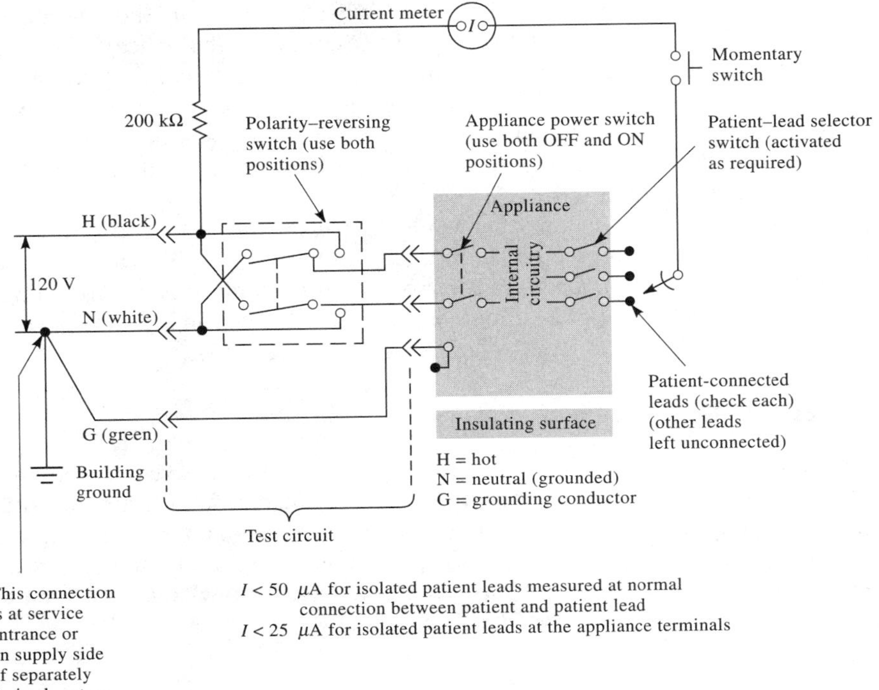

Figure 14.21 Test for ac isolation current (Reprinted with permission from NFPA 99-1996, "Health Care Facilities," Copyright © 1996, National Fire Protection Association, Quincy, MA 02269. This reprinted material is not the complete and official position of the National Fire Protection Association, on the referenced subject, which is represented only by the standard in its entirety.)

circuit. The limits on leakage current apply whether the polarity of the power line is correct or reversed, whether the power switch of the appliance is in the on or the off position, and whether or not all the control switches happen to be in the most disadvantageous position at the time of testing. The polarity-reversing switches in Figures 14.17–14.21 are required for equipment manufacturer testing but may be omitted for testing in health-care facilities. When several appliances are mounted together in one rack or cart, and all the appliances are supplied by one power cord, the complete rack or cart must be tested as one appliance.

LEAKAGE CURRENT IN PATIENT LEADS

Leakage current in patient leads is particularly important because these leads are the most common low-impedance patient contacts. Limits on leakage current in patient leads should be 50 μA. *Isolated* patient leads must have leakage current that is less than 10 μA. Only *isolated* patient leads should be connected to catheters or electrodes that make contact with the heart. Leakage current between individual or interconnected patient leads and ground should be measured with the patient leads active, as shown in Figure 14.19 on page 653.

In addition, leakage current between any pair of leads or between any single lead and all the other patient leads should be measured, as indicated in Figure 14.20 on page 654.

Finally, the leakage current that would flow through patient leads to ground if line voltage were to appear on the patient should be tested. This leakage current is called isolation current or risk current. Application of power-line voltage and frequency to the isolated patient leads should produce an isolation current to ground that is less than 50 μA (Figure 14.21 on page 655).

CONCLUSION

Adequate electrical safety in health-care facilities can be achieved at moderate cost by combining a good power-distribution system, careful selection of well-designed equipment, periodic testing of power systems and equipment, and a modest training program for medical personnel. Fortunately, the electrical-safety scare of the early 1970s has led to increased knowledge and greater safety for both patients and medical personnel.

PROBLEMS

14.1 From Figure 14.1, find the current required for arm-to-arm ventricular fibrillation. Assume that all this current passes through the area of the heart (about 10 $\times$ 10 cm). Calculate the current density through the heart. How does this compare with the lowest value calculated in Problem 14.8?

14.2 Assume that the cell membranes of a very large number of cells in parallel can be modeled by a 1-Ω resistor in parallel with a 100-μF capacitor. Determine the rms sinusoidal current versus frequency necessary to depolarize the cells. Assume that the peak potential of the cell membrane must be raised 20 mV above its resting potential to exceed threshold. Plot your results together with those shown in Figure 14.3, and compare.

14.3 From your knowledge of cardiac electrophysiology (Section 4.6), explain what rhythm would result from an intense 100-ms shock that occurred during (a) the P wave, (b) the R wave, (c) the T wave, (d) diastole. From these results, explain the shape of the curves shown in Figure 14.4.

14.4 Resketch Figure 14.5(b) for the case in which a catheter made of conductive plastic is used.

14.5 If the secondary earth ground in Figure 14.6 were not connected, would this prevent electrocution under no-fault conditions? What would be the result in case of a primary-to-secondary fault in the transformer?

14.6 The LIM in Figure 14.7 has a monitor hazard current that is too high.

Redesign it to achieve a lower monitor hazard current of 25 μA by changing the value of a *single* passive component shown and adjusting the detector threshold.

14.7 Design the simplest line-isolation monitor that would be capable of detecting a *single* fault from either line to ground.

14.8 Some authors hypothesize that it is current density flowing through the cell membrane that raises the resting potential of the cell to exceed threshold. Replot Figure 14.10 to show the average current density of the fibrillation threshold versus area of the catheter. Is the foregoing hypothesis correct? Explain any discrepancies.

14.9 Calculate the maximal safe capacitance between a liquid-filled catheter and dc-isolated pressure-sensor leads for a 120-V, 60-Hz fault in the sensor leads.

14.10 Compute the resistivity of the liquid necessary for safe operation of a liquid-filled catheter that is 1 m long and has a radius of 1.13 mm. Use the data given in Roy *et al.*, 1976 (shown in Figure 14.10). Assume that the patient is grounded and that a 120-V fault develops at the sensor.

14.11 Draw a complete equivalent circuit and compute the rms current through the patient's heart for the following situation. The patient's hand touches a faulty metal lamp that is 120 V rms above ground. A saline-filled catheter ($R = 50$ kΩ) for measuring blood pressure is connected to the patient's heart. Some of the pressure-sensor strain-gage wiring is grounded, and the sensor is somewhat isolated electrically. However, there is 20 MΩ of leakage resistance in the insulation between the ground and the saline in the sensor. There is also 100 pF of capacitance between the ground and the saline. Assume that the skin resistance of the patient is 1 MΩ. Is there a microshock hazard?

14.12 Show how a single electric instrument can, at the same time, be the path for microshock current flowing both to and from the patient. Use complete diagrams and do a sample calculation.

14.13 Devise your own hospital-patient microshock situation. Give complete details, including a diagram and equivalent circuit. Describe all tests, and give the standards for test results necessary to ensure the safety of the patient.

14.14 Most GFCIs have a momentary pushbutton that creates a safe ground fault to test the interrupter. On Figure 14.13, *design* the modifications to permit this test.

14.15 Figure 14.13 is designed for two-phase operation. Redesign it (draw a circuit diagram) for one-phase operation.

14.16 Design a tester for an electric receptacle that will indicate as many states as possible, including those not detected by the common three-LED receptacle testers (Figure 14.16).

14.17 A power engineer receives a lethal macroshock while standing in water and simultaneously touching the ungrounded metal casing on a high-voltage, 60-Hz power transformer. Assume that the resistance of the skin on the engineer's hand is 100 kΩ and that the resistance of the skin on the engineer's feet is negligible. A capacitance of 25 nF is measured between the transformer casing and the high-voltage conductors. Find the minimal value of the high voltage, assuming that 75 mA is the minimal fibrillating macroshock. Draw an equivalent circuit.

14.18 In Figure 14.14(c), the diodes are forward-biased for only *one* polarity of v_i. Redesign the circuit such that it works for *both* polarities of v_i. Consider the op-amp summer as a possibility.

REFERENCES

AAMI, American National Standard, Safe Current Limits for Electromedical Apparatus. (ANSI/AAMI ES1-1993) Arlington, VA: Association for the Advancement of Medical Instrumentation, 1993.

Anonymous, *AAMI Electrical Safety Manual,* Arlington, VA: Association for the Advancement of Medical Instrumentation, 1996.

Anonymous, "Electrical safety analyzers," *Health Devices,* 1988, 17, 283–309; "Update," *Health Devices,* 1989, 18, 411–413.

Anonymous, "Patient safety," *Application Note AN 718.* Waltham, MA: Hewlett-Packard Co., 1971.

Bruner, J. M. R., and P. F. Leonard, *Electrical Safety and the Patient.* Chicago: Year Book Medical Publishers, 1989.

Charney, W., J. Schirmer, *Essentials of Modern Hospital Safety.* Chelsea, MI: Lewis Publishing, 1990.

Dyro, J. F., "Safety program, hospital," in J. G. Webster (ed.), *Encyclopedia of Medical Devices and Instrumentation.* New York: Wiley, 1988, pp. 2575–2585.

Earley, M. W., R. H. Murray, J. M. Caloggero, and J. A. O'Connor, *The National Electrical Code 1996 Handbook,* 7th ed. Quincy, MA: National Fire Protection Association, 1996.

Fagerhaugh, S. Y., A. Strauss, B. Suczek, and C. L. Wiener, *Hazards in Hospital Care: Ensuring Patient Safety.* San Francisco, CA: Jossey–Bass, 1987.

Geddes, L. A., *Handbook of Electrical Hazards and Accidents.* Boca Raton, FL: CRC Press, Inc., 1995.

Geddes, L. A., J. D. Bourland, and G. Ford, "The mechanism underlying sudden death from electric shock." *Med Instrum.,* 1986, 20, 303–315.

Klein, B. R., *Health Care Facilities Handbook,* 5th ed. Quincy, MA: National Fire Protection Association, 1996.

Laks, M., R. Arzbaecher, J. Bailey, A. Berson, S. Briller, and D. Geselowitz, "Will relaxing safe current limits for electromedical equipment increase hazards to patients?" *Circulation,* 1994, 89, 909–910.

Laks, M., R. Arzbaecher, J. Bailey, D. Geselowitz, and A. Berson, "Recommendations for safe current limits for electrocardiographs," *Circulation,* 1996, 93, 837–839.

Lee, R. C., E. G. Cravalho, and J. F. Burke, *Electrical Trauma: The Pathophysiology, Manifestations and Clinical Management.* Cambridge, England: Cambridge University Press, 1992.

NFPA No. 99-1996, *Standard for Health Care Facilities.* Quincy, MA: National Fire Protection Association, 1996.

Roy, O. Z., "Summary of cardiac fibrillation thresholds for 60-Hz currents and voltages applied directly to the heart." *Med. Biol. Eng. Comput.,* 1980, 18, 657–659.

Roy, O. Z., A. J. Mortimer, B. J. Trollope, and E. J. Villeneuve, "Effects of short-duration transients on cardiac rhythm." *Med. Biol. Eng. Comput.,* 1984, 22, 225–228.

Roy, O. Z., J. R. Scott, and G. C. Park, "60-Hz ventricular fibrillation and pump failure thresholds versus electrode area." *IEEE Trans. Biomed. Eng.,* 1976, 23, 45–48.

Staewen, W. S., "Electrical safety reconsidered—The new AAMI electrical safety standard." *Biomed. Instrum. Tech.,* 1994, 28, 131–132.

Tan, K. S., and D. L. Johnson, "Threshold of sensation for 60-Hz leakage current: Results of a survey." *Biomed. Instrum. Tech.,* 1990, 24, 207–211.

APPENDIX A.1

Physical Constants

$g = 9.8\ m/s^2$	Acceleration due to gravity
$c = 3 \times 10^8\ m/s$	Velocity of light
$\sigma = 5.67 \times 10^{-12}\ W/(cm^2 \cdot K^4)$	Stefan-Boltzmann constant
$k = 1.38 \times 10^{-23}\ J/K$	Boltzmann's constant
$h = 6.63 \times 10^{-34}\ J \cdot s$	Planck's constant
$R = 8.31\ J/(mol \cdot K)$	Gas constant
$F = 96{,}500\ C/equivalent$	Faraday's constant (equivalent =mole/valence)
$q = -1.602 \times 10^{-19}\ C$	Charge on the electron
$\epsilon_0 = 8.8 \times 10^{-12}\ F/m$	Dielectric constant of free space
$N = 6.02 \times 10^{23}$ molecules/mol	Avogadro's number

APPENDIX A.2

SI Prefixes

Multiplication Factor	Prefix	Symbol
10^{24}	yotta	Y
10^{21}	zetta	Z
10^{18}	exa	E
10^{15}	peta	P
10^{12}	tera	T
10^{9}	giga	G
10^{6}	mega	M
10^{3}	kilo	k
10^{-1}	deci	d
10^{-2}	centi	c
10^{-3}	milli	m
10^{-6}	micro	μ
10^{-9}	nano	n
10^{-12}	pico	p
10^{-15}	femto	f
10^{-18}	atto	a
10^{-21}	zepto	z
10^{-24}	yocto	y

APPENDIX A.3

SI Units

To Convert From	To	Multiply By
degree	radian (rad)	0.0175
inch	meter (m)	0.0254
gallon	liter (l)	3.79
cycle/second	hertz (Hz)	1.0
minute	second (s)	60
hour	minute (min)	60
day	hour (h)	24
pound	kilogram (kg)	0.454
0.012 kg of carbon-12	mole (mol)	1.0
pound-force	newton (N)	4.45
degree Rankine	kelvin (K)	$t_K = t°_R/1.8$
calorie	joule (J)	4.186
British thermal unit	joule (J)	1055
horsepower	watt (W)	745
cm H_2O	pascal (Pa)	98.1
mm Hg (torr)	pascal (Pa)	133.3
psi	pascal (Pa)	6895
atmosphere	pascal (Pa)	101325
poise	pascal · second (Pa · s)	0.1
	volt (V)	
	ampere (A)	
	ohm (Ω)	
mho	siemens (S)	1.0
gauss	tesla (T)	0.0001
maxwell	weber (Wb)	10^{-8}
	farad (F)	
	decibel (dB)	
	candela (cd)	
roentgen (R)	coulomb per kilogram (C/kg)	0.000258
rad	gray (Gy)	0.01
curie (Ci)	becquerel (Bq)	3.7×10^{10}

REFERENCES

Taylor, B. N., Guide for the Use of the International System of Units (SI). NIST Special Publication 811. Gaithersburg, MD: National Institute of Standards and Technology, 1995.

APPENDIX A.4

Abbreviations

Abbreviation	Term
AAMI	Association for the Advancement of Medical Instrumentation
AAP	axon action potential
ac	alternating current
ACA	Automatic Clinical Analyzer
ADC	analog-to-digital converter
AF	audiofrequency
AIDS	acquired immune deficiency syndrome
AM	amplitude modulation
ANSI	American National Standards Institute
ATP	analytical test pack
ATR	attenuated total reflection
AV	atrioventricular
AWG	American wire gage
CAD	computer-aided design
CC	closing capacity
CCD	charge-coupled device
CMRR	common-mode rejection ratio
CNS	central nervous system
CPAP	continuous positive airway pressure
CPU	central processing unit
CSF	cerebrospinal fluid
CV	closing volume
cv	coefficient of variation
CVP	central venous pressure
CW	continuous wave
D	*d/dt*
DAC	digital-to-analog converter
dc	direct current
DNA	deoxyribonucleic acid
DPG	diphosphoglycerate
EBR	electron beam recording
ECG	electrocardiogram
ECMO	extracorporeal membrane oxygenator
ECO	engineering change order
ECoG	electrocorticogram
EEG	electroencephalogram
EGM	myocardial electrogram
EIA	Electronics Industries Association
ELISA	enzyme-linked immunosorbent assay
emf	electromotive force
EMG	electromyogram
ENG	electroneurogram
EOG	electro-oculogram
EPROM	erasable programmable read-only memory

Abbreviations *(Continued)*

Abbreviation	Term
ERG	electroretinogram
ERP	early-receptor potential
ERV	expiratory reserve volume
FDA	Food and Drug Administration
FEF	forced expiratory flow
FET	field-effect transistor
FEV	forced expiratory volume
FM	frequency modulation
FRC	functional residual capacity
FVC	forced vital capacity
GC	gas chromatograph
GFCI	ground-fault circuit interrupter
GLC	gas-liquid chromatograph
GM	geometric mean
GSR	galvanic skin response
GRIN	graded index
HCT	hematocrit
Hb	hemoglobin
HPTS	hydroxypyrene trisulfonic acid
IC	inspiratory capacity
IC	integrated circuit
ICU	intensive-care unit
ID	inside diameter
IMFET	immunologically sensitive field-effect transistor
IR	infrared
ISE	ion-sensitive electrode
ISFET	ion-sensitive field-effect transistor
IV	intravenous
j	$+\sqrt{-1}$
LED	light-emitting diode
LIM	line isolation monitor
lps	liters per second
LRP	late-receptor potential
MBC	maximal breathing capacity
MCH	mean corpuscular hemoglobin
MCHC	mean corpuscular hemoglobin concentration
MCV	mean corpuscular volume
MEFV	maximal expiratory flow volume
MEG	magnetoencephalogram
MMF	maximal mid-expiratory flow
MOSFET	metal-oxide–semiconductor field-effect transistor
MRI	magnetic resonance imaging
MTBF	mean time between failures
MTF	modulation transfer function
NDIR	nondispersive infrared analysis
NEC	National Electric Code

Abbreviations *(Continued)*

Abbreviation	Term
NEMA	National Electrical Manufacturers Association
NEP	noise-equivalent power
NFPA	National Fire Protection Association
NIST	National Institute of Standards and Technology
NREM	nonrapid eye movement
OD	outside diameter
ODC	oxyhemoglobin dissociation curve
PA	pulmonary artery
PCTA	percutaneous translumenal coronary angioplasty
PEEP	positive end expiratory pressure
PEF	peak expiratory flow
PEP	pre-ejection period
PFT	pulmonary function tests
PIM	patient interface module
PM	photomultiplier
p-p	peak-to-peak
PROM	programmable read-only memory
PSP	post-synaptic potential
PT	phototransistor
PVC	premature ventricular contraction
PVC	polyvinyl chloride
RAM	random-access memory
RAS	reticular activating system
RBC	red blood cell
RDW	erythrocyte volume distribution width
REM	rapid eye movement
RF	radiofrequency
RIA	radioimmunoassay
rms	root-mean-square
ROM	read-only memory
RV	residual volume
SA	sinoatrial
SCR	silicon-controlled rectifier
SEC	secondary-electron conduction
SEM	standard error of the mean
SHR	signal-to-hysteresis ratio
SMA	Sequential Multiple Analyzer
SMU	single motor unit
SNR	signal-to-noise ratio
SQUID	superconducting quantum interference device
SVP	surge-voltage protection
TBP	total-body plethysmograph
TCD	thermal-conductivity detector
TLC	total lung capacity
TLC	thin-layer chromatograph
TV	television

Abbreviations *(Continued)*

Abbreviation	Term
TVC	timed vital capacity
UL	Underwriters' Laboratory
VC	vital capacity
VCVS	voltage-controlled voltage source
VLSI	very large scale integration
WBC	white blood cell
WPW	Wolff-Parkinson-White
YAG	yttrium aluminum garnet

APPENDIX A.5

Chemical Elements

Element and Symbol	Atomic Number	Atomic Weight (C = 12)	Element and Symbol	Atomic Number	Atomic Weight (C = 12)
actinium (Ac)	89		mercury (Hg)	80	200.59
aluminum (Al)	13	26.9815	molybdenum (Mo)	42	95.94
americium (Am)	95		neodymium (Nd)	60	144.24
antimony (Sb)	51	121.75	neon (Ne)	10	20.179
argon (Ar)	18	39.948	neptunium (Np)	93	237.0482
arsenic (As)	33	74.9216	nickel (Ni)	28	58.71
astatine (At)	85		niobium (Nb)	41	92.9064
barium (Ba)	56	137.34	nitrogen (N)	7	14.0067
berkelium (Bk)	97		nobelium (No)	102	
beryllium (Be)	4	9.01218	osmium (Os)	76	190.2
bismuth (Bi)	83	208.9806	oxygen (O)	8	15.9994
boron (B)	5	10.81	palladium (Pd)	46	106.4
bromine (Br)	35	79.904	phosphorus (P)	15	30.9738
cadmium (Cd)	48	112.40	platinum (Pt)	78	195.09
calcium (Ca)	20	40.08	plutonium (Pu)	94	
californium (Cf)	98		polonium (Po)	84	
carbon (C)	6	12.011	potassium (K)	19	39.102
cerium (Ce)	58	140.12	praseodymium (Pr)	59	140.9077
cesium (Cs)	55	132.9055	promethium (Pm)	61	
chlorine (Cl)	17	35.453	protactinium (Pa)	91	231.0359
chromium (Cr)	24	51.996	radium (Ra)	88	226.0254
cobalt (Co)	27	58.9332	radon (Rn)	86	
columbium (Cb)	(see niobium)		rhenium (Re)	75	186.2
copper (Cu)	29	63.546	rhodium (Rh)	45	102.9055
curium (Cm)	96		rubidium (Rb)	37	85.4678
dysprosium (Dy)	66	162.50	ruthenium (Ru)	44	101.07
einsteinium (Es)	99		samarium (Sm)	62	150.4
erbium (Er)	68	167.26	scandium (Sc)	21	44.9559
europium (Eu)	63	151.96	selenium (Se)	34	78.96
fermium (Fm)	100		silicon (Si)	14	28.086
fluorine (F)	9	18.9984	silver (Ag)	47	107.868
francium (Fr)	87		sodium (Na)	11	22.9898
gadolinium (Gd)	64	157.25	strontium (Sr)	38	87.62

Chemical Elements (*Continued*)

Element and Symbol	Atomic Number	Atomic Weight (C = 12)	Element and Symbol	Atomic Number	Atomic Weight (C = 12)
gallium (Ga)	31	69.72	sulfur (S)	16	32.06
germanium (Ge)	32	72.59	tantalum (Ta)	73	180.9479
gold (Au)	79	196.9665	technetium (Tc)	43	98.9062
hafnium (Hf)	72	178.49	tellurium (Te)	52	127.60
helium (He)	2	4.00260	terbium (Tb)	65	158.9254
holmium (Ho)	67	164.9303	thallium (Tl)	81	204.37
hydrogen (H)	1	1.0080	thorium (Th)	90	232.0381
indium (In)	49	114.82	thulium (Tm)	69	168.9342
iodine (I)	53	126.9045	tin (Sn)	50	118.69
iridium (Ir)	77	192.22	titanium (Ti)	22	47.90
iron (Fe)	26	55.847	tungsten (W)	74	183.85
krypton (Kr)	36	83.80	uranium (U)	92	238.029
lanthanum (La)	57	138.9055	vanadium (V)	23	50.9414
lawrencium (Lr)	103		wolfram (W)	(see tungsten)	
lead (Pb)	82	207.2	xenon (Xe)	54	131.30
lithium (Li)	3	6.941	ytterbium (Yb)	70	173.04
lutetium (Lu)	71	174.97	yttrium (Y)	39	88.9059
magnesium (Mg)	12	24.305	zinc (Zn)	30	65.37
manganese (Mn)	25	54.9380	zirconium (Zr)	40	91.22
mendelevium (Md)	101				

INDEX

B

C